STUDENT'S SOLUTIONS MANUAL

JUDITH A. PENNA

INTERMEDIATE ALGEBRA
TWELFTH EDITION

Marvin L. Bittinger
Indiana University Purdue University Indianapolis

Judith A. Beecher

Barbara L. Johnson
Indiana University Purdue University Indianapolis

PEARSON

Boston Columbus Indianapolis New York San Francisco Hoboken
Amsterdam Cape Town Dubai London Madrid Milan Munich Paris Montreal Toronto
Delhi Mexico City São Paulo Sydney Hong Kong Seoul Singapore Taipei Tokyo

The author and publisher of this book have used their best efforts in preparing this book. These efforts include the development, research, and testing of the theories and programs to determine their effectiveness. The author and publisher make no warranty of any kind, expressed or implied, with regard to these programs or the documentation contained in this book. The author and publisher shall not be liable in any event for incidental or consequential damages in connection with, or arising out of, the furnishing, performance, or use of these programs.

Reproduced by Pearson from electronic files supplied by the author.

ISBN-13: 978-0-321-92474-2
ISBN-10: 0-321-92474-6

1 2 3 4 5 6 7 8 9 RRD 17 16 15 14

www.pearsonhighered.com

Contents

Chapter R

Review of Basic Algebra

RC1. The correct answer is (c). See page 2 in the text.

RC3. The correct answer is (a). See page 2 in the text.

RC5. The correct answer is (e). See page 4 in the text.

In Exercises 1-5, consider the following numbers: -6, 0, 1, $-\frac{1}{2}$, -4, $\frac{7}{9}$, 12, $-\frac{6}{5}$, 3.45, $5\frac{1}{2}$, $\sqrt{3}$, $\sqrt{25}$, $-\frac{12}{3}$, $0.131331333133331\ldots$.

1. The natural numbers are numbers used for counting. The natural numbers in the list above are 1, 12, and $\sqrt{25}$ ($\sqrt{25} = 5$).

3. The rational numbers can be named as quotients of integers with nonzero divisors. The rational numbers in the list above are -6, 0, 1, $-\frac{1}{2}$, -4, $\frac{7}{9}$, 12, $-\frac{6}{5}$, 3.45 ($3.45 = \frac{345}{100}$), $5\frac{1}{2}$ ($5\frac{1}{2} = \frac{11}{2}$), $\sqrt{25}$ ($\sqrt{25} = 5$), and $-\frac{12}{3}$.

5. The real numbers consist of the rational numbers and the irrational numbers. All of the numbers in the list are real numbers.

In Exercises 7-11, consider the following numbers: $-\sqrt{5}$, -3.43, -11, 12, 0, $\frac{11}{34}$, $-\frac{7}{13}$, π, $-3.565665666566665\ldots$

7. The whole numbers consist of the natural numbers and 0. The whole numbers in the list above are 12 and 0.

9. The integers consist of the whole numbers and their opposites. The integers in the list above are -11, 12, and 0.

11. The irrational numbers are the numbers that are not rational. The irrational numbers in the list above are $-\sqrt{5}$, π, and $-3.565665666566665\ldots$.

13. We list the members of the set.
    $\{m,a,t,h\}$

15. We list the members of the set.
    $\{1, 2, 3, 4, 5, 6, 7, 8, 9, 10, 11, 12\}$

17. We list the members of the set.
    $\{2, 4, 6, 8, \ldots\}$

19. We specify conditions by which we know whether a number is in the set.
    $\{x | x$ is a whole number less than or equal to 5$\}$, or
    $\{x | x$ is a whole number less than 6$\}$

21. We specify conditions by which we know whether a number is in the set.
    $\left\{\frac{a}{b} \middle| a \text{ and } b \text{ are integers and } b \neq 0\right\}$

23. We specify conditions by which we know whether a number is in the set.
    $\{x | x$ is a real number and $x > -3\}$, or $\{x | x > -3\}$

25. Since 13 is to the right of 0 on the number line, we have $13 > 0$.

27. Since -8 is to the left of 2 on the number line, we have $-8 < 2$.

29. Since -8 is to the left of 8, we have $-8 < 8$.

31. Since -8 is to the left of -3, we have $-8 < -3$.

33. Since -2 is to the right of -12, we have $-2 > -12$.

35. Since -9.9 is to the left of -2.2, we have $-9.9 < -2.2$.

37. Since $37\frac{1}{5}$ is to the right of $-1\frac{67}{100}$, we have $37\frac{1}{5} > -1\frac{67}{100}$.

39. We convert to decimal notation: $\frac{6}{13} = 0.461538\ldots$ and $\frac{13}{25} = 0.52$. Thus $\frac{6}{13} < \frac{13}{25}$.

41. $-8 > x$

    The inequality $x < -8$ has the same meaning.

43. $-12.7 \leq y$

    The inequality $y \geq -12.7$ has the same meaning.

45. $6 \leq -6$ False since neither $6 < -6$ nor $6 = -6$ is true.

47. $5 \geq -8.4$ True since $5 > -8.4$ is true.

49. $x < -2$

    We shade all numbers less than -2. We indicate that -2 is not a solution by using a parenthesis at -2.

51. $x \leq -2$

    We shade all the numbers to the left of -2 and use a bracket at -2 to indicate that it is also a solution.

53. $x > -3.3$

We shade all the numbers to the right of -3.3 and use a parenthesis at -3.3 to indicate that it is not a solution.

55. $x \geq 2$

We shade all the numbers to the right of 2 and use a bracket at 2 to indicate that it is also a solution.

57. The distance of -6 from 0 is 6, so $|-6| = 6$.

59. The distance of 28 from 0 is 28, so $|28| = 28$.

61. The distance of -35 from 0 is 35, so $|-35| = 35$.

63. The distance of $-\frac{2}{3}$ from 0 is $\frac{2}{3}$, so $\left|-\frac{2}{3}\right| = \frac{2}{3}$.

65. The distance of 42.8 from 0 is 42.8, so $|42.8| = 42.8$.

67. The distance of 986 from 0 is 986, so $|986| = 986$.

69. The distance of $\frac{0}{-7}$, or 0, from 0 is 0, so $\left|\frac{0}{-7}\right| = 0$.

71. $|-3| = 3$, so $|-3| \leq 5$.

73. $|-7| = 7$, so $|4| \leq |-7|$.

75. For comparison, we first write each number in decimal notation.

$\frac{1}{11} = 0.090909\ldots$

$1.1\% = 0.011$

$\frac{2}{7} = 0.285714285714\ldots$

$0.3\% = 0.003$

$0.11 = 0.11$

$\frac{1}{8}\% = 0.00125 \qquad \left(\frac{1}{8} = 0.125\right)$

$0.009 = 0.009$

$\frac{99}{1000} = 0.099$

$0.286 = 0.286$

$\frac{1}{8} = 0.125$

$1\% = 0.01$

$\frac{9}{100} = 0.09$

Then these rational numbers listed from least to greatest are $\frac{1}{8}\%$, 0.3%, 0.009, 1%, 1.1%, $\frac{9}{100}$, $\frac{1}{11}$, $\frac{99}{1000}$, 0.11, $\frac{1}{8}$, $\frac{2}{7}$, and 0.286.

Exercise Set R.2

RC1. A negative number has a <u>negative</u> sign.

RC3. The reciprocal of a negative number is <u>negative</u>.

RC5. When two negative numbers are multiplied, the result is <u>positive</u>.

RC7. The sum of two negative numbers is <u>negative</u>.

1. $-10 + (-18)$

The sum of two negative numbers is negative. We add their absolute values, $10 + 18 = 28$, and make the answer negative.

$-10 + (-18) = -28$

3. $7 + (-2)$

We find the difference of their absolute values, $7 - 2 = 5$. Since the positive number has the larger absolute value, the answer is positive.

$7 + (-2) = 5$

5. $-8 + (-8)$

The sum of two negative numbers is negative. We add the absolute values, $8 + 8 = 16$, and make the answer negative.

$-8 + (-8) = -16$

7. $7 + (-11)$

We find the difference of their absolute values, $11 - 7 = 4$. Since the negative number has the larger absolute value, the answer is negative.

$7 + (-11) = -4$

9. $-16 + 6$

We find the difference of their absolute values, $16 - 6 = 10$. Since the negative number has the larger absolute value, the answer is negative.

$-16 + 6 = -10$

11. $-26 + 0$

One number is 0. The sum is -26.

$-26 + 0 = -26$

13. $-8.4 + 9.6$

We find the difference of their absolute values, $9.6 - 8.4 = 1.2$. Since the positive number has the larger absolute value, the answer is positive.

$-8.4 + 9.6 = 1.2$

15. $-2.62 + (-6.24)$

The sum of two negative numbers is negative. We add the absolute values, $2.62 + 6.24 = 8.86$, and make the answer negative.

$-2.62 + (-6.24) = -8.86$

17. $-\dfrac{5}{9} + \dfrac{2}{9}$

We find the difference of their absolute values, $\dfrac{5}{9} - \dfrac{2}{9} = \dfrac{3}{9} = \dfrac{1}{3}$. Since the negative number has the larger absolute value, the answer is negative.

$$-\dfrac{5}{9} + \dfrac{2}{9} = -\dfrac{1}{3}$$

19. $-\dfrac{11}{12} + \left(-\dfrac{5}{12}\right)$

The sum of two negative numbers is negative. We add their absolute values, $\dfrac{11}{12} + \dfrac{5}{12} = \dfrac{16}{12}$, or $\dfrac{4}{3}$, and make the answer negative.

$$-\dfrac{11}{12} + \left(-\dfrac{5}{12}\right) = -\dfrac{4}{3}$$

21. $\dfrac{2}{5} + \left(-\dfrac{3}{10}\right)$

We find the difference of their absolute values.
$$\dfrac{2}{5} - \dfrac{3}{10} = \dfrac{2}{5} \cdot \dfrac{2}{2} - \dfrac{3}{10} = \dfrac{4}{10} - \dfrac{3}{10} = \dfrac{1}{10}$$
Since the positive number has the larger absolute value, the answer is positive.

$$\dfrac{2}{5} + \left(-\dfrac{3}{10}\right) = \dfrac{1}{10}$$

23. $-\dfrac{2}{5} + \dfrac{3}{4}$

We find the difference of their absolute values.
$$\dfrac{3}{4} - \dfrac{2}{5} = \dfrac{3}{4} \cdot \dfrac{5}{5} - \dfrac{2}{5} \cdot \dfrac{4}{4} = \dfrac{15}{20} - \dfrac{8}{20} = \dfrac{7}{20}$$
Since the positive number has the larger absolute value, the answer is positive.

$$-\dfrac{2}{5} + \dfrac{3}{4} = \dfrac{7}{20}$$

25. When $a = -4$, then $-a = -(-4) = 4$.

(The opposite, or additive inverse, of -4 is 4.)

27. When $a = 3.7$, then $-a = -3.7$.

(The opposite, or additive inverse, of 3.7 is -3.7.)

29. The opposite, or additive inverse, of 10 is -10, because $10 + (-10) = 0$.

31. The opposite, or additive inverse, of 0 is 0, because $0 + 0 = 0$.

33. $3 - 7 = 3 + (-7) = -4$

35. $-5 - 9 = -5 + (-9) = -14$

37. $23 - 23 = 23 + (-23) = 0$

39. $-23 - 23 = -23 + (-23) = -46$

41. $-6 - (-11) = -6 + 11 = 5$

43. $10 - (-5) = 10 + 5 = 15$

45. $15.8 - 27.4 = 15.8 + (-27.4) = -11.6$

47. $-18.01 - 11.24 = -18.01 + (-11.24) = -29.25$

49. $-\dfrac{21}{4} - \left(-\dfrac{7}{4}\right) = -\dfrac{21}{4} + \dfrac{7}{4} = -\dfrac{14}{4} = -\dfrac{7}{2}$

51. $-\dfrac{1}{3} - \left(-\dfrac{1}{12}\right) = -\dfrac{1}{3} + \dfrac{1}{12} = -\dfrac{4}{12} + \dfrac{1}{12} = -\dfrac{3}{12} = -\dfrac{1}{4}$

53. $-\dfrac{3}{4} - \dfrac{5}{6} = -\dfrac{3}{4} + \left(-\dfrac{5}{6}\right) = -\dfrac{9}{12} + \left(-\dfrac{10}{12}\right) = -\dfrac{19}{12}$

55. $\dfrac{1}{3} - \dfrac{4}{5} = \dfrac{1}{3} + \left(-\dfrac{4}{5}\right) = \dfrac{5}{15} + \left(-\dfrac{12}{15}\right) = -\dfrac{7}{15}$

57. $3(-7)$

The product of a positive number and a negative number is negative. We multiply their absolute values, $3 \cdot 7 = 21$, and make the answer negative.

$$3(-7) = -21$$

59. $-2 \cdot 4$

The product of a negative number and a positive number is negative. We multiply their absolute values, $2 \cdot 4 = 8$, and make the answer negative.

$$-2 \cdot 4 = -8$$

61. $-8(-3)$

The product of two negative numbers is positive. We multiply their absolute values, $8 \cdot 3 = 24$, and make the answer positive.

$$-8(-3) = 24$$

63. $-7 \cdot 16$

The product of a negative number and a positive number is negative. We multiply their absolute values, $7 \cdot 16 = 112$, and make the answer negative.

$$-7 \cdot 16 = -112$$

65. $-6(-5.7)$

The product of two negative numbers is positive. We multiply their absolute values, $6(5.7) = 34.2$, and make the answer positive.

$$-6(-5.7) = 34.2$$

67. $-\dfrac{3}{5} \cdot \dfrac{4}{7}$

The product of a negative number and a positive number is negative. We multiply their absolute values, $\dfrac{3}{5} \cdot \dfrac{4}{7} = \dfrac{12}{35}$, and make the answer negative.

$$-\dfrac{3}{5} \cdot \dfrac{4}{7} = -\dfrac{12}{35}$$

69. $-3\left(-\dfrac{2}{3}\right)$

The product of two negative numbers is positive. We multiply their absolute values, $3 \cdot \dfrac{2}{3} = \dfrac{6}{3} = 2$, and make the answer positive.

$$-3\left(-\dfrac{2}{3}\right) = 2$$

71. $-3(-4)(5)$

     $= 12(5)$ The product of two negative numbers is positive.

     $= 60$ The product of two positive numbers is positive.

73. $-4.2(-6.3)$

The product of two negative numbers is positive. We multiply the absolute values, $4.2(6.3) = 26.46$, and make the answer positive.

$-4.2(-6.3) = 26.46$

75. $-\dfrac{9}{11} \cdot \left(-\dfrac{11}{9}\right)$

The product of two negative numbers is positive. We multiply their absolute values, $\dfrac{9}{11} \cdot \dfrac{11}{9} = \dfrac{99}{99} = 1$, and make the answer positive.

$-\dfrac{9}{11} \cdot \left(-\dfrac{11}{9}\right) = 1$

77. $-\dfrac{2}{3} \cdot \left(-\dfrac{2}{3}\right) \cdot \left(-\dfrac{2}{3}\right)$

     $= \dfrac{4}{9} \cdot \left(-\dfrac{2}{3}\right)$ The product of two negative numbers is positive.

     $= -\dfrac{8}{27}$ The product of a positive number and a negative number is negative.

79. When a negative number is divided by a positive number, the answer is negative.

$\dfrac{-8}{4} = -2$

81. When a positive number is divided by a negative number, the answer is negative.

$\dfrac{56}{-8} = -7$

83. When a negative number is divided by a negative number, the answer is positive.

$-77 \div (-11) = \dfrac{-77}{-11} = 7$

85. When a negative number is divided by a negative number, the answer is positive.

$\dfrac{-5.4}{-18} = \dfrac{5.4}{18} = 0.3$

87. $\dfrac{5}{0}$ Not defined: Division by 0.

89. $\dfrac{0}{32} = 0$ because $0 \cdot 32 = 0$.

91. $\dfrac{9}{y - y}$ Not defined: $y - y = 0$ for any y.

93. The reciprocal of $\dfrac{3}{4}$ is $\dfrac{4}{3}$, because $\dfrac{3}{4} \cdot \dfrac{4}{3} = 1$.

95. The reciprocal of $-\dfrac{7}{8}$ is $-\dfrac{8}{7}$, because $-\dfrac{7}{8} \cdot \left(-\dfrac{8}{7}\right) = 1$.

97. The reciprocal of 25 is $\dfrac{1}{25}$, because $25 \cdot \dfrac{1}{25} = 1$.

99. The reciprocal of 0.2 is $\dfrac{1}{0.2}$.

This can also be expressed as follows:

$\dfrac{1}{0.2} = \dfrac{1}{0.2} \cdot \dfrac{10}{10} = \dfrac{10}{2} = 5.$

101. The reciprocal of $-\dfrac{a}{b}$ is $-\dfrac{b}{a}$, because $-\dfrac{a}{b} \cdot \left(-\dfrac{b}{a}\right) = 1$.

103. $\dfrac{2}{7} \div \left(-\dfrac{11}{3}\right) = \dfrac{2}{7} \cdot \left(-\dfrac{3}{11}\right) = -\dfrac{6}{77}$

105. $-\dfrac{10}{3} \div -\dfrac{2}{15} = -\dfrac{10}{3} \cdot \left(-\dfrac{15}{2}\right) = \dfrac{150}{6}$, or 25

107. $18.6 \div (-3.1) = \dfrac{18.6}{-3.1} = -\dfrac{18.6}{3.1} = -6$

109. $(-75.5) \div (-15.1) = \dfrac{-75.5}{-15.1} = \dfrac{75.5}{15.1} = 5$

111. $-48 \div 0.4 = \dfrac{-48}{0.4} = -\dfrac{48}{0.4} = -120$

113. $\dfrac{3}{4} \div \left(-\dfrac{2}{3}\right) = \dfrac{3}{4} \cdot \left(-\dfrac{3}{2}\right) = -\dfrac{9}{8}$

115. $-\dfrac{5}{4} \div \left(-\dfrac{3}{4}\right) = -\dfrac{5}{4} \cdot \left(-\dfrac{4}{3}\right) = \dfrac{20}{12} = \dfrac{5}{3}$

117. $-\dfrac{2}{3} \div \left(-\dfrac{4}{9}\right) = -\dfrac{2}{3} \cdot \left(-\dfrac{9}{4}\right) = \dfrac{18}{12} = \dfrac{3}{2}$

119. $-\dfrac{3}{8} \div \left(-\dfrac{8}{3}\right) = -\dfrac{3}{8} \cdot \left(-\dfrac{3}{8}\right) = \dfrac{9}{64}$

121. $-6.6 \div 3.3 = \dfrac{-6.6}{3.3} = -2$

123. $\dfrac{-12}{-13} = \dfrac{12}{13}$, or $0.\overline{923076}$

125. $\dfrac{48.6}{-30} = \dfrac{16.2}{-10} = -\dfrac{81}{50}$, or -1.62

127. $\dfrac{-9}{17 - 17}$ Not defined: $17 - 17 = 0$

129. $\dfrac{2}{3}$:   The opposite of $\dfrac{2}{3}$ is $-\dfrac{2}{3}$, because

$\dfrac{2}{3} + \left(-\dfrac{2}{3}\right) = 0.$

The reciprocal of $\dfrac{2}{3}$ is $\dfrac{3}{2}$, because

$\dfrac{2}{3} \cdot \dfrac{3}{2} = 1.$

$-\dfrac{5}{4}$:   The opposite of $-\dfrac{5}{4}$ is $\dfrac{5}{4}$, because $-\dfrac{5}{4} + \dfrac{5}{4} = 0.$

The reciprocal of $-\dfrac{5}{4}$ is $-\dfrac{4}{5}$, because

$-\dfrac{5}{4} \cdot \left(-\dfrac{4}{5}\right) = 1.$

$0:$ The opposite of 0 is 0, because $0 + 0 = 0$.

The reciprocal of 0 does not exist. (Only nonzero numbers have reciprocals.)

$1:$ The opposite of 1 is -1, because $1 + (-1) = 0$.

The reciprocal of 1 is 1, because $1 \cdot 1 = 1$.

$-4.5:$ The opposite of -4.5 is 4.5, because $-4.5 + 4.5 = 0$.

The reciprocal of -4.5 is $-\dfrac{1}{4.5}$, because $-4.5 \cdot \left(-\dfrac{1}{4.5}\right) = 1$. (Note that $-\dfrac{1}{4.5} = -\dfrac{1}{4.5} \cdot \dfrac{10}{10} = -\dfrac{10}{45}$, or $-\dfrac{2}{9}$.)

$x, x \neq 0:$ The opposite of x is $-x$, because $x + (-x) = 0$.

The reciprocal of x, $x \neq 0$, is $\dfrac{1}{x}$, because $x \cdot \dfrac{1}{x} = 1$.

131. The set of whole numbers is $\{0, 1, 2, 3, \ldots\}$. The whole numbers in the given list are 26 and 0.

133. The set of integers is $\{\ldots, -3, -2, -1, 0, 1, 2, 3, \ldots\}$. The integers in the given list are -13, 26 and 0.

135. The set of rational numbers is $\left\{\dfrac{p}{q} \,\middle|\, p \text{ is an integer, } q \text{ is an integer, and } q \neq 0\right\}$.

Decimal notation for rational numbers either terminates or has a repeating block of digits. The integers in the given list are -12.47, -13, 26, 0, $-\dfrac{23}{32}$, and $\dfrac{7}{11}$.

137. Since -7 is to the left of 8 on the number line, we have $-7 < 8$.

139. Since -45.6 is to the left of -23.8 on the number line, we have $-45.6 < -23.8$.

141.
$$\dfrac{1}{r_1} + \dfrac{1}{r_2}$$
$$= \dfrac{1}{12} + \dfrac{1}{6} \quad \text{Substituting 12 for } r_1 \text{ and 6 for } r_2$$
$$= \dfrac{1}{12} + \dfrac{2}{12}$$
$$= \dfrac{3}{12}, \text{ or } \dfrac{1}{4}$$

The conductance is $\dfrac{1}{4}$.

143. We want to find a number c such that $-625 = -0.02 \cdot c$. From the definition of division we know that $c = \dfrac{-625}{-0.02} = 31{,}250$

Exercise Set R.3

RC1. The statement is false. See Example 19 on page 24 in the text.

RC3. The statement is true. See page 26 in the text.

RC5. The statement is false. See page 22 in the text.

RC7. $3 \cdot 3^{-1} = 3^0 = 1$, so the given statement is true.

1. $\underbrace{4 \cdot 4 \cdot 4 \cdot 4 \cdot 4}_{5 \text{ factors}} = 4^5$

3. $\underbrace{5 \cdot 5 \cdot 5 \cdot 5 \cdot 5 \cdot 5}_{6 \text{ factors}} = 5^6$

5. $\underbrace{m \cdot m \cdot m}_{3 \text{ factors}} = m^3$

7. $\underbrace{\dfrac{7}{12} \cdot \dfrac{7}{12} \cdot \dfrac{7}{12} \cdot \dfrac{7}{12}}_{4 \text{ factors}} = \left(\dfrac{7}{12}\right)^4$

9. $\underbrace{(123.7)(123.7)}_{2 \text{ factors}} = (123.7)^2$

11. $2^7 = \underbrace{2 \cdot 2 \cdot 2 \cdot 2 \cdot 2 \cdot 2 \cdot 2}_{7 \text{ factors}} = 128$

13. $(-2)^5 = \underbrace{(-2) \cdot (-2) \cdot (-2) \cdot (-2) \cdot (-2)}_{5 \text{ factors}} = -32$

15. $\left(\dfrac{1}{3}\right)^4 = \underbrace{\dfrac{1}{3} \cdot \dfrac{1}{3} \cdot \dfrac{1}{3} \cdot \dfrac{1}{3}}_{4 \text{ factors}} = \dfrac{1}{81}$

17. $(-4)^3 = \underbrace{(-4) \cdot (-4) \cdot (-4)}_{3 \text{ factors}} = -64$

19. $(-5.6)^2 = \underbrace{(-5.6)(-5.6)}_{2 \text{ factors}} = 31.36$

21. $5^1 = 5$ (For any number a, $a^1 = a$.)

23. $34^0 = 1$ (For any nonzero number a, $a^0 = 1$.)

25. $(\sqrt{6})^0 = 1$ (For any nonzero number a, $a^0 = 1$.)

27. $\left(\dfrac{7}{8}\right)^1 = \dfrac{7}{8}$ (For any number a, $a^1 = a$.)

29. $\left(\dfrac{1}{4}\right)^{-2} = \dfrac{1}{\left(\dfrac{1}{4}\right)^2} = \dfrac{1}{\dfrac{1}{16}} = 1 \cdot \dfrac{16}{1} = 16$

31. $\left(\dfrac{2}{3}\right)^{-3} = \dfrac{1}{\left(\dfrac{2}{3}\right)^3} = \dfrac{1}{\dfrac{8}{27}} = 1 \cdot \dfrac{27}{8} = \dfrac{27}{8}$

33. $y^{-5} = \dfrac{1}{y^5}$

35. $\dfrac{1}{a^{-2}} = a^2$

37. $(-11)^{-1} = \dfrac{1}{(-11)^1} = \dfrac{1}{-11} = -\dfrac{1}{11}$

39. $\dfrac{1}{3^4} = 3^{-4}$

41. $\dfrac{1}{b^3} = b^{-3}$

43. $\dfrac{1}{(-16)^2} = (-16)^{-2}$

45. $\begin{aligned}[12 - 4(5-1)] &= [12 - 4(4)]\\ &= [12 - 16]\\ &= -4\end{aligned}$

47. $\begin{aligned}9[8 - 7(5-2)] &= 9[8 - 7 \cdot 3]\\ &= 9[8 - 21]\\ &= 9[-13]\\ &= -117\end{aligned}$

49. $\begin{aligned}[5(8-6) + 12] - [24 - (8-4)] &= [5 \cdot 2 + 12] - [24 - 4]\\ &= [10 + 12] - [24 - 4]\\ &= 22 - 20\\ &= 2\end{aligned}$

51. $\begin{aligned}[64 \div (-4)] \div (-2) &= -16 \div (-2)\\ &= 8\end{aligned}$

53. $\begin{aligned}19(-22) + 60 &= -418 + 60\\ &= -358\end{aligned}$

55. $\begin{aligned}(5+7)^2 &= 12^2 = 144\\ 5^2 + 7^2 &= 25 + 49 = 74\end{aligned}$

57. $\begin{aligned}2^3 + 2^4 - 20 \cdot 30 &= 8 + 16 - 600\\ &= 24 - 600\\ &= -576\end{aligned}$

59. $\begin{aligned}5^3 + 36 \cdot 72 &- (18 + 25 \cdot 4)\\ &= 5^3 + 36 \cdot 72 - (18 + 100)\\ &= 5^3 + 36 \cdot 72 - 118\\ &= 125 + 36 \cdot 72 - 118\\ &= 125 + 2592 - 118\\ &= 2717 - 118\\ &= 2599\end{aligned}$

61. $\begin{aligned}(13 \cdot 2 - 8 \cdot 4)^2 &= (26 - 32)^2\\ &= (-6)^2\\ &= 36\end{aligned}$

63. $\begin{aligned}4000 \cdot (1 + 0.12)^3 &= 4000(1.12)^3\\ &= 4000(1.404928)\\ &= 5619.712\end{aligned}$

65. $\begin{aligned}(20 \cdot 4 &+ 13 \cdot 8)^2 - (39 \cdot 15)^3\\ &= (80 + 104)^2 - (585)^3\\ &= 184^2 - 585^3\\ &= 33,856 - 200,201,625\\ &= -200,167,769\end{aligned}$

67. $\begin{aligned}18 - 2 \cdot 3 - 9 &= 18 - 6 - 9\\ &= 12 - 9\\ &= 3\end{aligned}$

69. $\begin{aligned}(18 - 2 \cdot 3) - 9 &= (18 - 6) - 9\\ &= 12 - 9\\ &= 3\end{aligned}$

71. $\begin{aligned}[24 \div (-3)] &\div \left(-\dfrac{1}{2}\right)\\ &= -8 \div \left(-\dfrac{1}{2}\right)\\ &= -8 \cdot (-2)\\ &= 16\end{aligned}$

73. $\begin{aligned}15 \cdot (-24) + 50 &= -360 + 50\\ &= -310\end{aligned}$

75. $\begin{aligned}4 \div (8 - 10)^2 + 1 &= 4 \div (-2)^2 + 1\\ &= 4 \div 4 + 1\\ &= 1 + 1\\ &= 2\end{aligned}$

77. $\begin{aligned}6^3 + 25 &\cdot 71 - (16 + 25 \cdot 4)\\ &= 6^3 + 25 \cdot 71 - (16 + 100)\\ &= 6^3 + 25 \cdot 71 - 116\\ &= 216 + 25 \cdot 71 - 116\\ &= 216 + 1775 - 116\\ &= 1991 - 116\\ &= 1875\end{aligned}$

79. $\begin{aligned}5000 \cdot (1 + 0.16)^3 &= 5000 \cdot (1.16)^3\\ &= 5000(1.560896)\\ &= 7804.48\end{aligned}$

81. $\begin{aligned}4 \cdot 5 - 2 \cdot 6 + 4 &= 20 - 12 + 4\\ &= 8 + 4\\ &= 12\end{aligned}$

83. $\begin{aligned}4 \cdot (6 + 8)/(4 + 3) &= 4 \cdot 14/7\\ &= 56/7\\ &= 8\end{aligned}$

85. $[2 \cdot (5 - 3)]^2 = [2 \cdot 2]^2$

$\qquad = 4^2$

$\qquad = 16$

87. $8(-7) + 6(-5) = -56 - 30$

$\qquad = -86$

89. $19 - 5(-3) + 3 = 19 + 15 + 3$

$\qquad = 34 + 3$

$\qquad = 37$

91. $9 \div (-3) + 16 \div 8 = -3 + 2$

$\qquad = -1$

93. $7 + 10 - (-10 \div 2) = 7 + 10 - (-5)$

$\qquad = 7 + 10 + 5$

$\qquad = 17 + 5$

$\qquad = 22$

95. $5^2 - 8^2 = 25 - 64$

$\qquad = -39$

97. $20 + 4^3 \div (-8) = 20 + 64 \div (-8)$

$\qquad = 20 + (-8)$

$\qquad = 12$

99. $-7(3^4) + 18 = -7 \cdot 81 + 18$

$\qquad = -567 + 18$

$\qquad = -549$

101. $9[(8 - 11) - 13] = 9[-3 - 13]$

$\qquad = 9[-16]$

$\qquad = -144$

103. $256 \div (-32) \div (-4) - -8 : (-4)$

$\qquad = 2$

105. $\dfrac{5^2 - |4^3 - 8|}{9^2 - 2^2 - 1^5} = \dfrac{5^2 - |64 - 8|}{81 - 4 - 1}$

$\qquad = \dfrac{5^2 - |56|}{77 - 1}$

$\qquad = \dfrac{5^2 - 56}{77 - 1}$

$\qquad = \dfrac{25 - 56}{76}$

$\qquad = \dfrac{-31}{76}$

$\qquad = -\dfrac{31}{76}$

107. $\dfrac{30(8 - 3) - 4(10 - 3)}{10|2 - 6| - 2(5 + 2)} = \dfrac{30 \cdot 5 - 4 \cdot 7}{10|-4| - 2 \cdot 7}$

$\qquad = \dfrac{150 - 28}{10 \cdot 4 - 2 \cdot 7}$

$\qquad = \dfrac{122}{40 - 14}$

$\qquad = \dfrac{122}{26}$

$\qquad = \dfrac{61}{13}$

109. The distance of $-\dfrac{9}{7}$ from 0 is $\dfrac{9}{7}$, so $\left|-\dfrac{9}{7}\right| = \dfrac{9}{7}$.

111. The distance of 0 from 0 is 0, so $|0| = 0$.

113. $23 - 56 = 23 + (-56) = -33$

115. $-23 - (-56) = -23 + 56 = 33$

117. $(-10)(2.3) = -23$

(The product of a negative number and a positive number is negative.)

119. $10(-2.3) = -23$

(The product of a positive number and a negative number is negative.)

121. $\qquad (-2)^0 - (-2)^3 - (-2)^{-1} + (-2)^4 - (-2)^{-2}$

$\qquad = 1 - (-8) - \left(-\dfrac{1}{2}\right) + 16 - \dfrac{1}{(-2)^2}$

$\qquad = 1 - (-8) - \left(-\dfrac{1}{2}\right) + 16 - \dfrac{1}{4}$

$\qquad = 1 + 8 + \dfrac{1}{2} + 16 - \dfrac{1}{4}$

$\qquad = 25\dfrac{1}{4}$

123. $9 \cdot 5 + 2 - (8 \cdot 3 + 1) = 22$

125. $(0.2)^{(-0.2)^{-1}} = (0.2)^{1/-0.2} = (0.2)^{-5} =$

$\dfrac{1}{(0.2)^5} = \dfrac{1}{0.00032} = 3125$

127. $(2 + 3)^{-1} = 5^{-1} = \dfrac{1}{5}$; $2^{-1} + 3^{-1} = \dfrac{1}{2} + \dfrac{1}{3} = \dfrac{3}{6} + \dfrac{2}{6} = \dfrac{5}{6}$;

thus $(2 + 3)^{-1} \neq 2^{-1} + 3^{-1}$.

Exercise Set R.4

RC1. (c)

RC3. (a)

RC5. (a)

RC7. (b)

1. 8 more than b

We have $b + 8$, or $8 + b$.

3. 13.4 less than c

We have $c - 13.4$.

5. 5 increased by q

We have $5 + q$, or $q + 5$.

7. b more than a

We have $a + b$, or $b + a$.

9. x divided by y

We have $x \div y$, or $\dfrac{x}{y}$.

11. x plus w

We have $x + w$, or $w + x$.

13. m subtracted from n

We have $n - m$.

15. The sum of p and q

We have $p + q$, or $q + p$.

17. Three times q

We have $3q$.

19. -18 multiplied by m

We have $-18m$.

21. The product of 17% and your salary

Let s represent your salary. We have $17\%s$, or $0.17s$.

23. We have $75t$.

25. We have $\$40 - x$.

27. Substitute -4 for z and carry out the multiplication.

$$23z = 23(-4) = -92$$

29. Substitute -24 for a and -8 for b and carry out the division.

$$\frac{a}{b} = \frac{-24}{-8} = 3$$

31. Substitute 36 for m and 4 for n and carry out the calculations.

$$\frac{m-n}{8} = \frac{36-4}{8} = \frac{32}{8} = 4$$

33. Substitute 9 for z and 2 for y and carry out the calculations.

$$\frac{5z}{y} = \frac{5 \cdot 9}{2} = \frac{45}{2}, \text{ or } 22\frac{1}{2}, \text{ or } 22.5$$

35. Substitute 4 for b and 6 for c and carry out the calculations.

$$\begin{aligned} 2c \div 3b &= 2 \cdot 6 \div 3 \cdot 4 \\ &= 12 \div 3 \cdot 4 \\ &= 4 \cdot 4 \\ &= 16 \end{aligned}$$

37. Substitute 3 for r and 27 for s and carry out the calculations.

$$\begin{aligned} 25 - r^2 + s \div r^2 &= 25 - 3^2 + 27 \div 3^2 \\ &= 25 - 9 + 27 \div 9 \\ &= 25 - 9 + 3 \\ &= 16 + 3 \\ &= 19 \end{aligned}$$

39. Substitute 15 for m and 3 for n and carry out the calculations.

$$\begin{aligned} m + n(5 + n^2) &= 15 + 3(5 + 3^2) \\ &= 15 + 3(5 + 9) \\ &= 15 + 3 \cdot 14 \\ &= 15 + 42 \\ &= 57 \end{aligned}$$

41. Substitute $\$7345$ for P, 6% or 0.06 for r, and 1 for t and carry out the calculations.

$$I = Prt$$
$$I = \$7345(0.06)(1)$$
$$I = \$440.70$$

43. Substitute 3.14 for π and 27 for r in each formula and carry out the calculations.

$$\begin{aligned} A &= \pi r^2 \\ &= 3.14(27)^2 \\ &= 3.14(729) \\ &= 2289.06 \text{ cm}^2 \end{aligned}$$

$$\begin{aligned} C &= 2\pi r \\ &= 2(3.14)(27) \\ &= 169.56 \text{ cm} \end{aligned}$$

45. $3^5 = 3 \cdot 3 \cdot 3 \cdot 3 \cdot 3 = 243$

47. $(-10)^4 = (-10)(-10)(-10)(-10) = 10,000$

49. $\left(\dfrac{3}{5}\right)^2 = \dfrac{3}{5} \cdot \dfrac{3}{5} = \dfrac{9}{25}$

51. $(4.5)^1 = 4.5$

53. $y < -1$

We shade all numbers to the left of -1 and use a parenthesis at -1 to indicate that -1 is not included.

55. Distance = speed × time

$$d = r \cdot t$$

57. Substitute 2 for x and 4 for y and carry out the calculations.

$$\frac{y+x}{2} + \frac{3y}{x} = \frac{4+2}{2} + \frac{3 \cdot 4}{2} = \frac{6}{2} + \frac{12}{2} = 3 + 6 = 9$$

Exercise Set R.5

RC1. The statement $3 + 7 = 7 + 3$ illustrates a <u>commutative</u> law. See page 39 in the text.

RC3. The statement $5 \cdot (6 \cdot 7) = (5 \cdot 6) \cdot 7$ illustrates an <u>associative</u> law. See page 40 in the text.

RC5. In the expression $7(5 + x)$, 7 and $5 + x$ are <u>factors</u>.

1. Substitute and find the value of each expression. For example, for $x = -2$, $2x + 3x = 2(-2) + 3(-2) = -4 - 6 = -10$.

	$2x + 3x$	$5x$	$2x - 3x$
$x = -2$	-10	-10	2
$x = 5$	25	25	-5
$x = 0$	0	0	0

The values of $2x + 3x$ and $5x$ are the same for the given values of x and, indeed, for any allowable replacement for x. Thus, they are equivalent.

The values of $2x + 3x$ and $2x - 3x$ do not agree for $x = -2$ and for $x = 5$. Since one disagreement is sufficient to show that two expressions are not equivalent, we know that these expressions are not equivalent. Similarly, $5x$ and $2x - 3x$ are not equivalent.

3. Substitute and find the value of each expression. For example, for $x = -1$, $4x + 8x = 4(-1) + 8(-1) = -4 - 8 = -12$.

	$4x + 8x$	$4(x + 3x)$	$4(x + 2x)$
$x = -1$	-12	-16	-12
$x = 3.2$	38.4	51.2	38.4
$x = 0$	0	0	0

The values of $4x + 8x$ and $4(x + 2x)$ are the same for the given values of x and, indeed, for any allowable replacement for x. Thus, they are equivalent.

The values of $4x + 8x$ and $4(x + 3x)$ do not agree for $x = -1$ and for $x = 3.2$. Since one disagreement is sufficient to show that two expressions are not equivalent, we know that these expressions are not equivalent. Similarly, $4(x + 3x)$ and $4(x + 2x)$ are not equivalent.

5. Since $8x = 8 \cdot x$, we multiply by 1 using x/x as a name for 1.
$$\frac{7}{8} = \frac{7}{8} \cdot 1 = \frac{7}{8} \cdot \frac{x}{x} = \frac{7x}{8x}$$

7. Since $8a = 4 \cdot 2a$, we multiply by 1 using $\dfrac{2a}{2a}$ as a name for 1.
$$\frac{3}{4} = \frac{3}{4} \cdot 1 = \frac{3}{4} \cdot \frac{2a}{2a} = \frac{6a}{8a}$$

9. $\dfrac{25x}{15x} = \dfrac{5 \cdot 5x}{3 \cdot 5x}$ We look for the largest common factor of the numerator and denominator and factor each.

$\quad = \dfrac{5}{3} \cdot \dfrac{5x}{5x}$ Factoring the expression

$\quad = \dfrac{5}{3} \cdot 1 \quad \left(\dfrac{5x}{5x} = 1\right)$

$\quad = \dfrac{5}{3}$

11. $\quad -\dfrac{100a}{25a}$

$\quad = -\dfrac{25a \cdot 4}{25a \cdot 1}$ Factoring numerator and denominator

$\quad = -\dfrac{25a}{25a} \cdot \dfrac{4}{1}$ Factoring the expression

$\quad = -1 \cdot \dfrac{4}{1} \quad \left(\dfrac{25a}{25a} = 1\right)$

$\quad = -\dfrac{4}{1}$

$\quad = -4$

13. $w + 3 = 3 + w$ Commutative law of addition

15. $rt = tr$ Commutative law of multiplication

17. $4 + cd = cd + 4$ Commutative law of addition

or

$4 + cd = cd + 4 = dc + 4$ Commutative laws of addition and multiplication

or

$4 + cd = 4 + dc$ Commutative law of multiplication

19. $yz + x = x + yz$ Commutative law of addition

or

$yz + x = x + yz = x + zy$ Commutative laws of addition and multiplication

or

$yz + x = zy + x$ Commutative law of multiplication

21. $m + (n + 2) = (m + n) + 2$ Associative law of addition

23. $(7 \cdot x) \cdot y = 7 \cdot (x \cdot y)$ Associative law of multiplication

25. $(a + b) + 8 = a + (b + 8)$ Associative law
$\qquad\qquad\quad = a + (8 + b)$ Commutative law

$\quad (a + b) + 8 = a + (b + 8)$ Associative law
$\qquad\qquad\quad = a + (8 + b)$ Commutative law
$\qquad\qquad\quad = (a + 8) + b$ Associative law

$\quad (a + b) + 8 = (b + a) + 8$ Commutative law
$\qquad\qquad\quad = b + (a + 8)$ Associative law

Other answers are possible.

27. $7 \cdot (a \cdot b) = 7 \cdot (b \cdot a)$ Commutative law

$\quad = (7 \cdot b) \cdot a$ Associative law

$7 \cdot (a \cdot b) = (7 \cdot a) \cdot b$ Associative law

$\quad = b \cdot (7 \cdot a)$ Commutative law

$\quad = b \cdot (a \cdot 7)$ Commutative law

$7 \cdot (a \cdot b) = 7 \cdot (b \cdot a)$ Commutative law

$\quad = (b \cdot a) \cdot 7$ Commutative law

Other answers are possible.

29. $4(a + 1) = 4 \cdot a + 4 \cdot 1$

$\quad = 4a + 4$

31. $8(x - y) = 8 \cdot x - 8 \cdot y$

$\quad = 8x - 8y$

33. $-5(2a + 3b) = -5 \cdot 2a + (-5) \cdot 3b$

$\quad = -10a - 15b$

35. $2a(b - c + d) = 2a \cdot b - 2a \cdot c + 2a \cdot d$

$\quad = 2ab - 2ac + 2ad$

37. $2\pi r(h + 1) = 2\pi r \cdot h + 2\pi r \cdot 1$

$\quad = 2\pi r h + 2\pi r$

39. $\frac{1}{2}h(a + b) = \frac{1}{2}h \cdot a + \frac{1}{2}h \cdot b$

$\quad = \frac{1}{2}ha + \frac{1}{2}hb$

41. $4a - 5b + 6 = 4a + (-5b) + 6$

The terms are $4a$, $-5b$, and 6.

43. $2x - 3y - 2z = 2x + (-3y) + (-2z)$

The terms are $2x$, $-3y$, and $-2z$.

45. $24x + 24y = 24 \cdot x + 24 \cdot y$

$\quad = 24(x + y)$

47. $7p - 7 = 7 \cdot p - 7 \cdot 1$

$\quad = 7(p - 1)$

49. $7x - 21 = 7 \cdot x - 7 \cdot 3$

$\quad = 7(x - 3)$

51. $xy + x = x \cdot y + x \cdot 1$

$\quad = x(y + 1)$

53. $2x - 2y + 2z = 2 \cdot x - 2 \cdot y + 2 \cdot z$

$\quad = 2(x - y + z)$

55. $3x + 6y - 3 = 3 \cdot x + 3 \cdot 2y - 3 \cdot 1$

$\quad = 3(x + 2y - 1)$

57. $4w - 12z + 8 = 4 \cdot w - 4 \cdot 3z + 4 \cdot 2$

$\quad = 4(w - 3z + 2)$

59. $20x - 36y - 12 = 4 \cdot 5x - 4 \cdot 9y - 4 \cdot 3$

$\quad = 4(5x - 9y - 3)$

61. $ab + ac - ad = a \cdot b + a \cdot c - a \cdot d$

$\quad = a(b + c - d)$

63. $\frac{1}{4}\pi rr + \frac{1}{4}\pi rs = \frac{1}{4}\pi r \cdot r + \frac{1}{4}\pi r \cdot s$

$\quad = \frac{1}{4}\pi r(r + s)$

65. Let x and y represent the numbers. Then we have

$(x + y)^2$.

67. $x^{-4} = \dfrac{1}{x^4}$

69. $4 \cdot 2 - 5^2 - 3^2 = 4 \cdot 2 - 25 - 9$

$\quad = 8 - 25 - 9$

$\quad = -17 - 9$

$\quad = -26$

71. Let $x = 1$ and $y = -1$.

$x^2 + y^2 = 1^2 + (-1)^2 = 1 + 1 = 2$

$(x + y)^2 = (1 + (-1))^2 = 0^2 = 0$

Since the expressions have different values, they are not equivalent.

73. For $x = -2$, $x^2 \cdot x^3 = (-2)^2 \cdot (-2)^3 = 4(-8) = -32$ and $x^5 = (-2)^5 = -32$.

For $x = 0$, $x^2 \cdot x^3 = 0^2 \cdot 0^3 = 0 \cdot 0 = 0$ and $x^5 = 0^5 = 0$.

For $x = 1$, $x^2 \cdot x^3 = 1^2 \cdot 1^3 = 1 \cdot 1 = 1$ and $x^5 = 1^5 = 1$.

Indeed, the expressions have the same value for all values of x, so they are equivalent.

Exercise Set R.6

RC1. $6x$ and $-7x$ have the same letter raised to the same power, so they are like terms. The given statement is true.

RC3. True; see page 46 in the text.

RC5. $6x + (-7x) = [6 + (-7)]x = -1 \cdot x = -x$. The given statement is true.

1. $7x + 5x = (7 + 5)x$

$\quad = 12x$

3. $8b - 11b = (8 - 11)b$

$\quad = -3b$

5. $14y + y = 14y + 1y$

$\quad = (14 + 1)y$

$\quad = 15y$

7. $12a - a = 12a - 1a$
$$= (12 - 1)a$$
$$= 11a$$

9. $t - 9t = 1t - 9t$
$$= (1 - 9)t$$
$$= -8t$$

11. $5x - 3x + 8x = (5 - 3 + 8)x$
$$= 10x$$

13. $3x - 5y + 8x = (3 + 8)x - 5y$
$$= 11x - 5y$$

15. $3c + 8d - 7c + 4d = (3 - 7)c + (8 + 4)d$
$$= -4c + 12d$$

17. $4x - 7 + 18x + 25 = (4 + 18)x + (-7 + 25)$
$$= 22x + 18$$

19. $1.3x + 1.4y - 0.11x - 0.47y$
$$= (1.3 - 0.11)x + (1.4 - 0.47)y$$
$$= 1.19x + 0.93y$$

21. $\dfrac{2}{3}a + \dfrac{5}{6}b - 27 - \dfrac{4}{5}a - \dfrac{7}{6}b$
$$= \left(\dfrac{2}{3} - \dfrac{4}{5}\right)a + \left(\dfrac{5}{6} - \dfrac{7}{6}\right)b - 27$$
$$= \left(\dfrac{10}{15} - \dfrac{12}{15}\right)a + \left(-\dfrac{2}{6}\right)b - 27$$
$$= \dfrac{2}{15}a - \dfrac{1}{3}b - 27$$

23. $P = 2(l + w)$
$$P = 2 \cdot l + 2 \cdot w$$
$$P = 2l + 2w$$

25. $-(-2c) = -1(-2c)$
$$= [-1(-2)]c$$
$$= 2c$$

27. $-(b + 4) = -1(b + 4)$
$$= -1 \cdot b + (-1)4$$
$$= -b - 4$$

29. $-(b - 3) = -1(b - 3)$
$$= -1 \cdot b - (-1) \cdot 3$$
$$= -b + [-(-1)3]$$
$$= -b + 3, \text{ or } 3 - b$$

31. $-(t - y) = -1(t - y)$
$$= -1 \cdot t - (-1) \cdot y$$
$$= -t + [-(-1)y]$$
$$= -t + y, \text{ or } y - t$$

33. $-(x + y + z) = -1(x + y + z)$
$$= -1 \cdot x + (-1) \cdot y + (-1) \cdot z$$
$$= -x - y - z$$

35. $-(8x - 6y + 13)$
$$= -8x + 6y - 13 \qquad \text{Changing the sign of every term inside parentheses}$$

37. $-(-2c + 5d - 3e + 4f)$
$$= 2c - 5d + 3e - 4f \qquad \text{Changing the sign of every term inside parentheses}$$

39. $-\left(-1.2x + 56.7y - 34z - \dfrac{1}{4}\right)$
$$= 1.2x - 56.7y + 34z + \dfrac{1}{4} \qquad \text{Changing the sign of every term inside parentheses}$$

41. $a + (2a + 5) = a + 2a + 5$
$$= 3a + 5$$

43. $4m - (3m - 1) = 4m - 3m + 1$
$$= m + 1$$

45. $5d - 9 - (7 - 4d) = 5d - 9 - 7 + 4d$
$$= 9d - 16$$

47. $-2(x + 3) - 5(x - 4)$
$$= -2(x + 3) + [-5(x - 4)]$$
$$= -2x - 6 + [-5x + 20]$$
$$= -2x - 6 - 5x + 20$$
$$= -7x + 14$$

49. $5x - 7(2x - 3) - 4$
$$= 5x + [-7(2x - 3)] - 4$$
$$= 5x + [-14x + 21] - 4$$
$$= 5x - 14x + 21 - 4$$
$$= -9x + 17$$

51. $8x - (-3y + 7) + (9x - 11)$
$$= 8x + 3y - 7 + 9x - 11$$
$$= 17x + 3y - 18$$

53. $\dfrac{1}{4}(24x - 8) - \dfrac{1}{2}(-8x + 6) - 14$
$$= \dfrac{1}{4}(24x - 8) + \left[-\dfrac{1}{2}(-8x + 6)\right] - 14$$
$$= 6x - 2 + [4x - 3] - 14$$
$$= 6x - 2 + 4x - 3 - 14$$
$$= 10x - 19$$

55. $7a - [9 - 3(5a - 2)] = 7a - [9 - 15a + 6]$
$$= 7a - [15 - 15a]$$
$$= 7a - 15 + 15a$$
$$= 22a - 15$$

57. $5\{-2 + 3[4 - 2(3 + 5)]\} = 5\{-2 + 3[4 - 2(8)]\}$
$$= 5\{-2 + 3[4 - 16]\}$$
$$= 5\{-2 + 3[-12]\}$$
$$= 5\{-2 - 36\}$$
$$= 5\{-38\}$$
$$= -190$$

59. $[10(x + 3) - 4] + [2(x - 1) + 6]$
$$= [10x + 30 - 4] + [2x - 2 + 6]$$
$$= 10x + 26 + 2x + 4$$
$$= 12x + 30$$

61. $[7(x + 5) - 19] - [4(x - 6) + 10]$
$$= [7x + 35 - 19] - [4x - 24 + 10]$$
$$= [7x + 16] - [4x - 14]$$
$$= 7x + 16 - 4x + 14$$
$$= 3x + 30$$

63. $3\{[7(x - 2) + 4] - [2(2x - 5) + 6]\}$
$$= 3\{[7x - 14 + 4] - [4x - 10 + 6]\}$$
$$= 3\{[7x - 10] - [4x - 4]\}$$
$$= 3\{7x - 10 - 4x + 4\}$$
$$= 3\{3x - 6\}$$
$$= 9x - 18$$

65. $4\{[5(x - 3) + 2^2] - 3[2(x + 5) - 9^2]\}$
$$= 4\{[5(x - 3) + 4] - 3[2(x + 5) - 81]\}$$
$$= 4\{[5x - 15 + 4] - 3[2x + 10 - 81]\}$$
$$= 4\{[5x - 11] - 3[2x - 71]\}$$
$$= 4\{5x - 11 - 6x + 213\}$$
$$= 4\{-x + 202\}$$
$$= -4x + 808$$

67. $2y + \{8[3(2y - 5) - (8y + 9)] + 6\}$
$$= 2y + \{8[6y - 15 - 8y - 9] + 6\}$$
$$= 2y + \{8[-2y - 24] + 6\}$$
$$= 2y + \{-16y - 192 + 6\}$$
$$= 2y + \{-16y - 186\}$$
$$= 2y - 16y - 186$$
$$= -14y - 186$$

69. $10 - x = 10 - (-3) = 10 + 3 = 13$

71. $a + n(n^2 - 1) = 100 + (-2)[(-2)^2 - 1]$
$$= 100 + (-2)(4 - 1)$$
$$= 100 + (-2) \cdot 3$$
$$= 100 - 6$$
$$= 94$$

73. When a negative number is divided by a positive number, the answer is negative.

$-256 \div 16 = -16$

75. When a positive number is divided by a negative number, the answer is negative.

$256 \div (-16) = -16$

77. $8(a - b) = 8 \cdot a - 8 \cdot b = 8a - 8b$

79. $6x(a - b + 2c) = 6x \cdot a - 6x \cdot b + 6x \cdot 2c =$
$6ax - 6bx + 12cx$

81. $24a - 24 = 24 \cdot a - 24 \cdot 1 = 24(a - 1)$

83. $ab - ac + a = a \cdot b - a \cdot c + a \cdot 1 = a(b - c + 1)$

85. $(3 - 8)^2 + 9 = 34$

87. $5 \cdot 2^3 \div (3 - 4)^4 = 40$

89. $[11(a - 3) + 12a] - \{6[4(3b - 7) - (9b + 10)] + 11\}$
$$= [11a - 33 + 12a] - \{6[12b - 28 - 9b - 10] + 11\}$$
$$= [23a - 33] - \{6[3b - 38] + 11\}$$
$$= 23a - 33 - \{18b - 228 + 11\}$$
$$= 23a - 33 - \{18b - 217\}$$
$$= 23a - 33 - 18b + 217$$
$$= 23a - 18b + 184$$

91. $z - \{2z + [3z - (4z + 5x) - 6z] + 7z\} - 8z$
$$= z - \{2z + [3z - 4z - 5x - 6z] + 7z\} - 8z$$
$$= z - \{2z - 7z - 5x + 7z\} - 8z$$
$$= z - \{2z - 5x\} - 8z$$
$$= z - 2z + 5x - 8z$$
$$= -9z + 5x$$

93. $x - \{x + 1 - [x + 2 - (x - 3 - \{x + 4 -$
$$[x - 5 + (x - 6)]\})]\}$$
$$= x - \{x + 1 - [x + 2 - (x - 3 - \{x + 4 - [2x - 11]\})]\}$$
$$= x - \{x + 1 - [x + 2 - (x - 3 - \{x + 4 - 2x + 11\})]\}$$
$$= x - \{x + 1 - [x + 2 - (x - 3 - \{-x + 15\})]\}$$
$$= x - \{x + 1 - [x + 2 - (x - 3 + x - 15)]\}$$
$$= x - \{x + 1 - [x + 2 - (2x - 18)]\}$$
$$= x - \{x + 1 - [x + 2 - 2x + 18]\}$$
$$= x - \{x + 1 - [-x + 20]\}$$
$$= x - \{x + 1 + x - 20\}$$
$$= x - \{2x - 19\}$$
$$= x - 2x + 19$$
$$= -x + 19$$

Exercise Set R.7

RC1. $4^5 \cdot 4^3 = 4^{5+3}$; the correct answer is (c).

RC3. $\dfrac{4^5}{4^3} = 4^{5-3}$; the correct answer is (a).

RC5. $\left(\dfrac{4}{5}\right)^{-3} = \left(\dfrac{5}{4}\right)^3 = \dfrac{5^3}{4^3}$; the correct answer is (f).

1. $3^6 \cdot 3^3 = 3^{6+3} = 3^9$

3. $6^{-6} \cdot 6^2 = 6^{-6+2} = 6^{-4} = \dfrac{1}{6^4}$

5. $8^{-2} \cdot 8^{-4} = 8^{-2+(-4)} = 8^{-6} = \dfrac{1}{8^6}$

7. $b^2 \cdot b^{-5} = b^{2+(-5)} = b^{-3} = \dfrac{1}{b^3}$

9. $a^{-3} \cdot a^4 \cdot a^2 = a^{-3+4+2} = a^3$

11. $\begin{aligned}(2x)^3(3x)^2 &= 8x^3 \cdot 9x^2 \\ &= 8 \cdot 9 \cdot x^3 \cdot x^2 \\ &= 72x^{3+2} \\ &= 72x^5\end{aligned}$

13. $\begin{aligned}(14m^2n^3)(-2m^3n^2) &= 14 \cdot (-2) \cdot m^2 \cdot m^3 \cdot n^3 \cdot n^2 \\ &= -28m^{2+3}n^{3+2} \\ &= -28m^5n^5\end{aligned}$

15. $\begin{aligned}(-2x^{-3})(7x^{-8}) &= -2 \cdot 7 \cdot x^{-3} \cdot x^{-8} \\ &= -14x^{-3+(-8)} \\ &= -14x^{-11} = -\dfrac{14}{x^{11}}\end{aligned}$

17. $\begin{aligned}(15x^{4t})(7x^{-6t}) &= 15 \cdot 7 \cdot x^{4t} \cdot x^{-6t} \\ &= 105x^{4t+(-6t)} \\ &= 105x^{-2t} \\ &= \dfrac{105}{x^{2t}}\end{aligned}$

19. $\begin{aligned}(2y^{3m})(-4y^{-9m}) &= 2 \cdot (-4) \cdot y^{3m} \cdot y^{-9m} \\ &= -8y^{3m+(-9m)} \\ &= -8y^{-6m} \\ &= -\dfrac{8}{y^{6m}}\end{aligned}$

21. $\dfrac{8^9}{8^2} = 8^{9-2} = 8^7$

23. $\dfrac{6^3}{6^{-2}} = 6^{3-(-2)} = 6^{3+2} = 6^5$

25. $\dfrac{10^{-3}}{10^6} = 10^{-3-6} = 10^{-3+(-6)} = 10^{-9} = \dfrac{1}{10^9}$

27. $\dfrac{9^{-4}}{9^{-6}} = 9^{-4-(-6)} = 9^{-4+6} = 9^2$

29. $\dfrac{x^{-4n}}{x^{6n}} = x^{-4n-6n} = x^{-10n} = \dfrac{1}{x^{10n}}$

31. $\dfrac{w^{-11q}}{w^{-6q}} = w^{-11q-(-6q)} = w^{-11q+6q} = w^{-5q} = \dfrac{1}{w^{5q}}$

33. $\dfrac{a^3}{a^{-2}} = a^{3-(-2)} = a^{3+2} = a^5$

35. $\dfrac{2yx^7z^5}{-9x^2z} = \dfrac{27}{-9}x^{7-2}z^{5-1} = -3x^5z^4$

37. $\begin{aligned}\dfrac{-24x^6y^7}{18x^{-3}y^9} &= \dfrac{-24}{18}x^{6-(-3)}y^{7-9} \\ &= -\dfrac{24}{18}x^{6+3}y^{-2} \\ &= -\dfrac{4}{3}x^9y^{-2} = -\dfrac{4x^9}{3y^2}\end{aligned}$

39. $\begin{aligned}\dfrac{-18x^{-2}y^3}{-12x^{-5}y^5} &= \dfrac{-18}{-12}x^{-2-(-5)}y^{3-5} \\ &= \dfrac{18}{12}x^{-2+5}y^{-2} \\ &= \dfrac{3}{2}x^3y^{-2} = \dfrac{3x^3}{2y^2}\end{aligned}$

41. $(4^3)^2 = 4^{3 \cdot 2} = 4^6$

43. $(8^4)^{-3} = 8^{4(-3)} = 8^{-12} = \dfrac{1}{8^{12}}$

45. $(6^{-4})^{-3} = 6^{-4(-3)} = 6^{12}$

47. $\begin{aligned}(5a^2b^2)^3 &= 5^3(a^2)^3(b^2)^3 \\ &= 125a^{2 \cdot 3}b^{2 \cdot 3} \\ &= 125a^6b^6\end{aligned}$

49. $\begin{aligned}(-3x^3y^{-6})^{-2} &= (-3)^{-2}(x^3)^{-2}(y^{-6})^{-2} \\ &= \dfrac{1}{(-3)^2}x^{3(-2)}y^{-6(-2)} \\ &= \dfrac{1}{9}x^{-6}y^{12} \\ &= \dfrac{y^{12}}{9x^6}\end{aligned}$

51. $\begin{aligned}(-6a^{-2}b^3c)^{-2} &= (-6)^{-2}(a^{-2})^{-2}(b^3)^{-2}c^{-2} \\ &= \dfrac{1}{(-6)^2}a^{-2(-2)}b^{3(-2)}c^{-2} \\ &= \dfrac{1}{36}a^4b^{-6}c^{-2} = \dfrac{a^4}{36b^6c^2}\end{aligned}$

53. $\left(\dfrac{4^{-3}}{3^4}\right)^3 = \dfrac{(4^{-3})^3}{(3^4)^3} = \dfrac{4^{-3 \cdot 3}}{3^{4 \cdot 3}} = \dfrac{4^{-9}}{3^{12}} = \dfrac{1}{4^9 \cdot 3^{12}}$

55. $\left(\dfrac{2x^3y^{-2}}{3y^{-3}}\right)^3 = \dfrac{(2x^3y^{-2})^3}{(3y^{-3})^3}$

$\qquad = \dfrac{2^3(x^3)^3(y^{-2})^3}{3^3(y^{-3})^3}$

$\qquad = \dfrac{8x^9y^{-6}}{27y^{-9}}$

$\qquad = \dfrac{8x^9y^{-6-(-9)}}{27}$

$\qquad = \dfrac{8x^9y^3}{27}$

57. $\left(\dfrac{125a^2b^{-3}}{5a^4b^{-2}}\right)^{-5} = \left(\dfrac{5a^4b^{-2}}{125a^2b^{-3}}\right)^5$

$\qquad = \left(\dfrac{a^{4-2}b^{-2-(-3)}}{25}\right)^5$

$\qquad = \left(\dfrac{a^2b}{25}\right)^5$

$\qquad = \dfrac{(a^2)^5(b)^5}{(25)^5} = \dfrac{a^{2\cdot5}b^{1\cdot5}}{25^{1\cdot5}}$

$\qquad = \dfrac{a^{10}b^5}{25^5}$, or $\dfrac{a^{10}b^5}{5^{10}}$

$\qquad\qquad\qquad [25^5 = (5^2)^5 = 5^{10}]$

59. $\left(\dfrac{-6^5y^4z^{-5}}{2^{-2}y^{-2}z^3}\right)^6 = (-1\cdot6^5\cdot2^2y^{4-(-2)}z^{-5-3})^6$

$\qquad = (-1\cdot6^5\cdot2^2y^6z^{-8})^6$

$\qquad = (-1)^6(6^5)^6(2^2)^6(y^6)^6(z^{-8})^6$

$\qquad = 1\cdot6^{5\cdot6}2^{2\cdot6}y^{6\cdot6}z^{-8\cdot6}$

$\qquad = 6^{30}2^{12}y^{36}z^{-48}$

$\qquad = \dfrac{6^{30}2^{12}y^{36}}{z^{48}}$

61. $[(-2x^{-4}y^{-2})^{-3}]^{-2} = [(-2)^{-3}(x^{-4})^{-3}(y^{-2})^{-3}]^{-2}$

$\qquad = [(-2)^{-3}x^{12}y^6]^{-2}$

$\qquad = [(-2)^{-3}]^{-2}(x^{12})^{-2}(y^6)^{-2}$

$\qquad = (-2)^6x^{-24}y^{-12}$

$\qquad = \dfrac{64}{x^{24}y^{12}}$

63. $\left(\dfrac{3a^{-2}b}{5a^{-7}b^5}\right)^{-7} = \left(\dfrac{5a^{-7}b^5}{3a^{-2}b}\right)^7$

$\qquad = \left(\dfrac{5a^{-7-(-2)}b^{5-1}}{3}\right)^7$

$\qquad = \left(\dfrac{5a^{-5}b^4}{3}\right)^7$

$\qquad = \dfrac{5^7(a^{-5})^7(b^4)^7}{3^7}$

$\qquad = \dfrac{5^7a^{-35}b^{28}}{3^7}$

$\qquad = \dfrac{5^7b^{28}}{3^7a^{35}}$

65. $\dfrac{10^{2a+1}}{10^{a+1}} = 10^{2a+1-(a+1)} = 10^{2a+1-a-1} = 10^a$

67. $\dfrac{9a^{x-2}}{3a^{2x+2}} = \dfrac{9}{3}\cdot\dfrac{a^{x-2}}{a^{2x+2}} = 3a^{x-2-(2x+2)} =$

$\qquad 3a^{x-2-2x-2} = 3a^{-x-4}$

69. $\dfrac{45x^{2a+4}y^{b+1}}{-9x^{a+3}y^{2+b}} = \dfrac{45}{-9}\cdot\dfrac{x^{2a+4}y^{b+1}}{x^{a+3}y^{2+b}} =$

$\qquad -5x^{2a+4-(a+3)}y^{b+1-(2+b)} = -5x^{2a+4-a-3}y^{b+1-2-b} =$

$\qquad -5x^{a+1}y^{-1}$, or $\dfrac{-5x^{a+1}}{y}$

71. $(8^x)^{4y} = 8^{x\cdot4y} = 8^{4xy}$

73. $(12^{3-a})^{2b} = 12^{(3-a)(2b)} = 12^{6b-2ab}$

75. $(5x^{a-1}y^{b+1})^{2c} = 5^{2c}x^{(a-1)(2c)}y^{(b+1)(2c)} =$

$\qquad 5^{2c}x^{2ac-2c}y^{2bc+2c}$, or $(5^2)^cx^{2ac-2c}y^{2bc+2c} =$

$\qquad 25^cx^{2ac-2c}y^{2bc+2c}$

77. $\dfrac{4x^{2a+3}y^{2b-1}}{2x^{a+1}y^{b+1}} = \dfrac{4}{2}\cdot\dfrac{x^{2a+3}y^{2b-1}}{x^{a+1}y^{b+1}} =$

$\qquad 2x^{2a+3-(a+1)}y^{2b-1-(b+1)} = 2x^{2a+3-a-1}y^{2b-1-b-1} =$

$\qquad 2x^{a+2}y^{b-2}$

79. 4.7,000,000,000.

$\qquad$ 10 places

Large number, so the exponent is positive.

$47,000,000,000 = 4.7 \times 10^{10}$

81. 0.00000001.6

$\qquad$ 8 places

Small number, so the exponent is negative.

$0.000000016 = 1.6 \times 10^{-8}$

83. 2.600,000,000.

$\qquad$ 9 places

Large number, so the exponent is positive.

$2,600,000,000 = 2.6 \times 10^9$

85. 100 millionths $= 100 \times 0.000001 = 0.0001$

$\qquad$ 0.0001.

$\qquad$ 4 places

Small number, so the exponent is negative.

100 millionths $= 0.0001 = 1 \times 10^{-4}$

87. 6.73000000.

$\qquad$ 8 places

Positive exponent so the number is large.

$6.73 \times 10^8 = 673,000,000$

89. 0.00006.6 cm

    ↑   ┘

   5 places

Negative exponent, so the number is small.

6.6×10^{-5} cm $= 0.000066$ cm

91. 1.007 billion $= 1.007 \times 10^9$

  1.007000000.

  └─────────────┘ ↑

     9 places

Positive exponent, so the number is large.

1.007 billion $= 1.007 \times 10^9 = 1,007,000,000$ users

93. $\quad (2.3 \times 10^6)(4.2 \times 10^{-11})$

$= (2.3 \times 4.2)(10^6 \times 10^{-11})$

$= 9.66 \times 10^{-5}$

95. $\quad (2.34 \times 10^{-8})(5.7 \times 10^{-4})$

$= (2.34 \times 5.7)(10^{-8} \times 10^{-4})$

$= 13.338 \times 10^{-12}$

$= (1.3338 \times 10^1) \times 10^{-12}$

$= 1.3338 \times (10^1 \times 10^{-12})$

$= 1.3338 \times 10^{-11}$

97. $\dfrac{8.5 \times 10^8}{3.4 \times 10^5} = \dfrac{8.5}{3.4} \times \dfrac{10^8}{10^5}$

$\qquad\qquad = 2.5 \times 10^3$

99. $\dfrac{4.0 \times 10^{-6}}{8.0 \times 10^{-3}} = \dfrac{4.0}{8.0} \times \dfrac{10^{-6}}{10^{-3}}$

$\qquad\qquad = 0.5 \times 10^{-3}$

$\qquad\qquad = (5 \times 10^{-1}) \times 10^{-3}$

$\qquad\qquad = 5 \times (10^{-1} \times 10^{-3})$

$\qquad\qquad = 5 \times 10^{-4}$

101. $\quad$ 2000 yr

$= 2000 \text{ yr} \times \dfrac{365 \text{ days}}{1 \text{ yr}} \times \dfrac{24 \text{ hr}}{1 \text{ day}} \times \dfrac{60 \text{ min}}{1 \text{ hr}} \times \dfrac{60 \text{ sec}}{1 \text{ min}}$

$= 63,072,000,000 \text{ sec}$

$= 6.3072 \times 10^{10} \text{ sec}$

103. $\dfrac{10,000}{300,000} = \dfrac{10^4}{3 \times 10^5}$

$\qquad\qquad \approx 0.333 \times 10^{-1}$

$\qquad\qquad = (3.33 \times 10^{-1}) \times 10^{-1}$

$\qquad\qquad = 3.33 \times 10^{-2}$

The part of the total number of words in the English language that an average person knows is about 3.33×10^{-2}.

105. 105 micrometers $= 150 \times 10^{-6}$ m

$V = lwh$

$\quad = (1.2 \text{ m})(79 \text{ m})(150 \times 10^{-6} \text{ m})$

$\quad = (1.2)(79)(150 \times 10^{-6}) \text{ m}^3$

$\quad = 14,220 \times 10^{-6} \text{ m}^3$

$\quad = 1.4220 \times 10^4 \times 10^{-6} \text{ m}^3$

$\quad = 1.422 \times 10^{-2} \text{ m}^3$

107. $20,000$ trillion $= 20,000 \times 10^{12}$

$\qquad\qquad\qquad = 2 \times 10^4 \times 10^{12}$

$\qquad\qquad\qquad = 2 \times 10^{16}$

Find the number of calculations per minute.

$2 \times 10^{16} \dfrac{\text{calculations}}{\text{second}}$

$= 2 \times 10^{16} \dfrac{\text{calculations}}{1 \text{ second}} \cdot \dfrac{60 \text{ seconds}}{1 \text{ minute}}$

$= 120 \times 10^{16} \dfrac{\text{calculations}}{\text{minute}}$

$= 1.2 \times 10^2 \times 10^{16} \dfrac{\text{calculations}}{\text{minute}}$

$= 1.2 \times 10^{18} \dfrac{\text{calculations}}{\text{minute}}$

Find the number of calculations per hour.

$1.2 \times 10^{18} \dfrac{\text{calculations}}{\text{minute}}$

$= 1.2 \times 10^{18} \dfrac{\text{calculations}}{1 \text{ minute}} \cdot \dfrac{60 \text{ minutes}}{1 \text{ hour}}$

$= 72 \times 10^{18} \dfrac{\text{calculations}}{\text{hour}}$

$= 7.2 \times 10 \times 10^{18} \dfrac{\text{calculations}}{\text{hour}}$

$= 7.2 \times 10^{19} \dfrac{\text{calculations}}{\text{hour}}$

109. First we divide to find the number of \$5 bills in \$4,540,000.

$\dfrac{\$4,540,000}{\$5} = \dfrac{4.54 \times 10^6}{5} = 0.908 \times 10^6 =$

$(9.08 \times 10^{-1}) \times 10^6 = 9.08 \times 10^5$

Then we divide 1 ton, or 2000 lb, by the number of bills to find the weight of a single bill.

$\dfrac{2000}{9.08 \times 10^5} = \dfrac{2 \times 10^3}{9.08 \times 10^5} = \dfrac{2}{9.08} \times \dfrac{10^3}{10^5} \approx$

$0.220 \times 10^{-2} = (2.2 \times 10^{-1}) \times 10^{-2} = 2.2 \times 10^{-3}$

A five-dollar bill weighs about 2.2×10^{-3} lb.

111. $\quad 9x - (-4y + 8) + (10x - 12)$

$= 9x + 4y - 8 + 10x - 12$

$= 19x + 4y - 20$

113. $4^2 + 30 \cdot 10 - 7^3 + 16 = 16 + 30 \cdot 10 - 343 + 16$

$$= 16 + 300 - 343 + 16$$

$$= 316 - 343 + 16$$

$$= -27 + 16$$

$$= -11$$

115. $20 - 5 \cdot 4 - 8 = 20 - 20 - 8$

$$= 0 - 8$$

$$= -8$$

117. $\dfrac{(2^{-2})^{-4} \times (2^3)^{-2}}{(2^{-2})^2 \cdot (2^5)^{-3}} = \dfrac{2^8 \times 2^{-6}}{2^{-4} \cdot 2^{-15}}$

$$= \dfrac{2^{8+(-6)}}{2^{-4+(-15)}}$$

$$= \dfrac{2^2}{2^{-19}}$$

$$= 2^{2-(-19)}$$

$$= 2^{21}$$

119. $\left[\left(\dfrac{a^{-2}}{b^7}\right)^{-3} \cdot \left(\dfrac{a^4}{b^{-3}}\right)^2\right]^{-1} = \left[\dfrac{(a^{-2})^{-3}}{(b^7)^{-3}} \cdot \dfrac{(a^4)^2}{(b^{-3})^2}\right]^{-1}$

$$= \left[\dfrac{a^6}{b^{-21}} \cdot \dfrac{a^8}{b^{-6}}\right]^{-1}$$

$$= \left[\dfrac{a^{6+8}}{b^{-21+(-6)}}\right]^{-1}$$

$$= \left[\dfrac{a^{14}}{b^{-27}}\right]^{-1}$$

$$= \dfrac{a^{-14}}{b^{27}}$$

$$= \dfrac{1}{a^{14}b^{27}}$$

121. $\left[\dfrac{(2x^a y^b)^3}{(-2x^a y^b)^2}\right]^2 = \left[\dfrac{(2x^a y^b)^3}{(2x^a y^b)^2}\right]^2 \quad [(-2x^a y^b)^2 = (2x^a y^b)^2]$

$$= [(2x^a y^b)^{3-2}]^2$$

$$= (2x^a y^b)^2$$

$$= 2^2 (x^a)^2 (y^b)^2$$

$$= 4x^{2a} y^{2b}$$

Chapter R Vocabulary Reinforcement

1. The sentence $x > -4$ is an example of an <u>inequality</u>.

2. In the notation 7^4, the number 7 is called the <u>base</u>.

3. A <u>variable</u> is a letter that can represent various numbers.

4. In the expression $6x$, the multipliers 6 and x are <u>factors</u>.

5. The number 5.93×10^7 is written in <u>scientific notation</u>.

6. The product of <u>reciprocals</u> is 1.

7. The sum of <u>opposites</u> is 0.

8. The <u>commutative</u> law for addition states that $a+b=b+a$.

Chapter R Concept Reinforcement

1. The statement is false. For example, let $a = 1$ and $b = 2$. Then $a - b = 1 - 2 = -1$, but $b - a = 2 - 1 = 1$.

2. The statement is true. See page 2 in the text.

3. The statement is true. We have $-(-a) = a < 0$.

4. Zero is neither positive nor negative. The given statement is false.

5. The statement is false because $|0| = 0$.

6. The statement is true. See page 16 in the text.

7. The statement is true. See page 13 in the text.

8. The statement is true. See page 57 in the text.

Chapter R Review Exercises

1. The rational numbers can be named as quotients of integers with nonzero divisors. Of the given numbers, the rational numbers are 2, $-\dfrac{2}{3}$, $0.45\overline{45}$, and -23.788.

2. We specify the conditions by which we know whether a number is in the set.

   $\{x | x \text{ is a real number less than or equal to } 46\}$

3. Since -3.9 is to the left of 2.9, we have $-3.9 < 2.9$.

4. $19 > x$

   The inequality $x < 19$ has the same meaning.

5. $-13 \geq 5$ is false since neither $-13 > 5$ nor $-13 = 5$ is true.

6. $7.01 \leq 7.01$ is true since $7.01 = 7.01$ is true.

7. $x > -4$

   We shade all numbers to the right of -4 and use a parenthesis at -4 to indicate it is not a solution.

8. $x \leq 1$

   We shade all the numbers to the left of 1 and use a bracket at 1 to indicate it is also a solution.

9. The distance of -7.23 from 0 is 7.23, so $|-7.23| = 7.23$.

10. The distance of $9 - 9$, or 0, from 0 is 0, so $|9 - 9| = 0$.

11. $6 + (-8)$

We find the difference of the absolute values, $8 - 6 = 2$. Since the negative number has the larger absolute value, the answer is negative.

$6 + (-8) = -2$

12. $-3.8 + (-4.1)$

The sum of two negative numbers is negative. We add the absolute values, $3.8 + 4.1 = 7.9$, and make the answer negative.

$-3.8 + (-4.1) = -7.9$

13. $\frac{3}{4} + \left(-\frac{13}{7}\right)$

We find the difference of the absolute values,

$\frac{13}{7} - \frac{3}{4} = \frac{52}{28} - \frac{21}{28} = \frac{31}{28}$. Since the negative number has the larger absolute value, the answer is negative.

$\frac{3}{4} + \left(-\frac{13}{7}\right) = -\frac{31}{28}$

14. $-8 - (-3) = -8 + 3 = -5$

15. $-17.3 - 9.4 = -17.3 + (-9.4) = -26.7$

16. $\frac{3}{2} - \left(-\frac{13}{4}\right) = \frac{3}{2} + \frac{13}{4} = \frac{6}{4} + \frac{13}{4} = \frac{19}{4}$

17. $(-3.8)(-2.7)$

The product of two negative numbers is positive. We multiply the absolute values, $3.8(2.7) = 10.26$, and make the answer positive.

$(-3.8)(-2.7) = 10.26$

18. $-\frac{2}{3}\left(\frac{9}{14}\right)$

The product of a negative number and a positive number is negative. We multiply the absolute values and make the answer negative.

$\frac{2}{3} \cdot \frac{9}{14} = \frac{2 \cdot 9}{3 \cdot 14} = \frac{2 \cdot 3 \cdot 3}{3 \cdot 2 \cdot 7} = \frac{2 \cdot 3}{2 \cdot 3} \cdot \frac{3}{7}$

Thus, $-\frac{2}{3}\left(\frac{9}{14}\right) = -\frac{3}{7}$

19. $-6(-7)(4) = 42(4) = 168$

20. When a negative number is divided by a positive number, the answer is negative.

$-12 \div 3 = \frac{-12}{3} = -4$

21. When a negative number is divided by a negative number, the answer is positive.

$\frac{-84}{-4} = 21$

22. When a positive number is divided by a negative number, the answer is negative.

$\frac{49}{-7} = -7$

23. $\frac{5}{6} \div \left(-\frac{10}{7}\right) = \frac{5}{6} \cdot \left(-\frac{7}{10}\right) = -\frac{5 \cdot 7}{6 \cdot 10} = -\frac{5 \cdot 7}{6 \cdot 2 \cdot 5} =$

$-\frac{7}{6 \cdot 2} \cdot \frac{5}{5} = -\frac{7}{6 \cdot 2} = -\frac{7}{12}$

24. $-\frac{5}{2} \div \left(-\frac{15}{16}\right) = -\frac{5}{2} \cdot \left(-\frac{16}{15}\right) = \frac{5 \cdot 16}{2 \cdot 15} = \frac{5 \cdot 2 \cdot 8}{2 \cdot 3 \cdot 5} =$

$\frac{2 \cdot 5}{2 \cdot 5} \cdot \frac{8}{3} = \frac{8}{3}$

25. $\frac{25}{0}$ Not defined: Division by 0.

26. $-108 \div 4.5 = \frac{-108}{4.5} = -24$

27. When $a = -7$, then $-a = -(-7) = 7$.

28. When $a = 2.3$, then $-a = -2.3$.

29. When $a = 0$, then $-a = -0 = 0$.

30. $\underbrace{a \cdot a \cdot a \cdot a \cdot a}_{5 \text{ factors}} = a^5$

31. $\underbrace{\left(-\frac{7}{8}\right)\left(-\frac{7}{8}\right)\left(-\frac{7}{8}\right)}_{3 \text{ factors}} = \left(-\frac{7}{8}\right)^3$

32. $a^{-4} = \frac{1}{a^4}$

33. $\frac{1}{x^8} = x^{-8}$

34. $2^3 - 3^4 + (13 \cdot 5 + 67) = 8 - 81 + (13 \cdot 5 + 67)$

$= 8 - 81 + (65 + 67)$

$= 8 - 81 + 132$

$= -73 + 132$

$= 59$

35. $64 \div (-4) + (-5)(20) = -16 - 100 = -116$

36. Let x represent the number. Then we have $5x$.

37. Let y represent the number. Then we have $28\%y$, or $0.28y$.

38. We have $t - 9$.

39. Let a and b represent the numbers. Then we have $\frac{a}{b} - 8$.

40. Substitute -2 for x and carry out the calculations.

$5x - 7 = 5(-2) - 7 = -10 - 7 = -17$

41. Substitute 4 for x and 20 for y and carry out the calculations.

$\frac{x - y}{2} = \frac{4 - 20}{2} = \frac{-16}{2} = -8$

42. We usually consider length to be longer than width so we substitute 12 for l and 7 for w and carry out the calculation. (The result is the same if we substitute in the opposite order.)

$A = lw = 12 \cdot 7 = 84 \text{ ft}^2$

43. Substitute to find the value of each expression.

	$x^2 - 5$	$(x+5)^2$	$(x-5)^2$	$x^2 + 5$
$x = -1$	-4	16	36	6
$x = 10$	95	225	25	105
$x = 0$	-5	25	25	5

There is no pair of expressions that is the same for the given values of x, so there are no equivalent expressions.

44.

	$2x - 14$	$2x - 7$	$2(x-7)$	$2x + 14$
$x = -1$	-16	-9	-16	12
$x = 10$	6	13	6	34
$x = 0$	-14	-7	-14	14

The values of $2x-14$ and $2(x-7)$ are the same for the given values of x and, indeed, for any allowable replacement for x. Thus, they are equivalent.

There is no other pair of expressions that is the same for the given values of x. Thus, there are no other equivalent expressions.

45. Since $9x = 3 \cdot 3x$, we multiply by 1 using $\dfrac{3x}{3x}$ as a name for 1.
$$\frac{7}{3} = \frac{7}{3} \cdot \frac{3x}{3x} = \frac{21x}{9x}$$

46. $\dfrac{-84x}{7x} = \dfrac{7x(-12)}{7x \cdot 1} = \dfrac{7x}{7x} \cdot \dfrac{-12}{1} = \dfrac{-12}{1} = -12$

47. $11 + a = a + 11$ Commutative law of addition

48. $8y = y \cdot 8$ Commutative law of multiplication

49. $(9 + a) + b = 9 + (a + b)$ Associate law of addition

50. $8(xy) = (8x)y$ Associate law of multiplication

51. $-3(2x - y) = -3 \cdot 2x - 3(-y) = -6x + 3y$

52. $4ab(2c + 1) = 4ab \cdot 2c + 4ab \cdot 1 = 8abc + 4ab$

53. $5x + 10y - 5z = 5 \cdot x + 5 \cdot 2y - 5 \cdot z = 5(x + 2y - z)$

54. $ptr + pts = pt \cdot r + pt \cdot s = pt(r + s)$

55. $2x + 6y - 5x - y = (2 - 5)x + (6 - 1)y = -3x + 5y$

56. $7c - 6 + 9c + 2 - 4c = (7 + 9 - 4)c + (-6 + 2) = 12c - 4$

57. $\quad -(-9c + 4d - 3)$
$$= 9c - 4d + 3 \qquad \begin{array}{l}\text{Changing the sign of every}\\ \text{term inside parentheses}\end{array}$$

58. $4(x - 3) - 3(x - 5) = 4x - 12 - 3x + 15 = x + 3$

59. $12x - 3(2x - 5) = 12x - 6x + 15 = 6x + 15$

60. $\quad 7x - [4 - 5(3x - 2)]$
$$= 7x - [4 - 15x + 10]$$
$$= 7x - [14 - 15x]$$
$$= 7x - 14 + 15x$$
$$= 22x - 14$$

61. $\quad 4m - 3[3(4m - 2) - (5m + 2) + 12]$
$$= 4m - 3[12m - 6 - 5m - 2 + 12]$$
$$= 4m - 3[7m + 4]$$
$$= 4m - 21m - 12$$
$$= -17m - 12$$

62. $(2x^4 y^{-3})(-5x^3 y^{-2}) = 2(-5) \cdot x^4 \cdot x^3 \cdot y^{-3} \cdot y^{-2}$
$$= -10x^7 y^{-5}$$
$$= -\frac{10x^7}{y^5}$$

63. $\dfrac{-15x^2 y^{-5}}{10x^6 y^{-8}} = \dfrac{-15}{10} x^{2-6} y^{-5-(-8)}$
$$= -\frac{3}{2} x^{-4} y^{-5+8}$$
$$= -\frac{3}{2} x^{-4} y^3$$
$$= -\frac{3y^3}{2x^4}$$

64. $(-3a^{-4}bc^3)^{-2} = (-3)^{-2}(a^{-4})^{-2}b^{-2}(c^3)^{-2}$
$$= \frac{1}{(-3)^2} a^{-4(-2)} b^{-2} c^{3(-2)}$$
$$= \frac{1}{9} a^8 b^{-2} c^{-6}$$
$$= \frac{a^8}{9b^2 c^6}$$

65. $\left[\dfrac{-2x^4 y^{-4}}{3x^{-2} y^6}\right]^{-4} = \left[\dfrac{3x^{-2} y^6}{-2x^4 y^{-4}}\right]^4$
$$= \left[\frac{3x^{-2-4} y^{6-(-4)}}{-2}\right]^4$$
$$= \left[\frac{3x^{-6} y^{10}}{-2}\right]^4$$
$$= \frac{3^4 (x^{-6})^4 (y^{10})^4}{(-2)^4}$$
$$= \frac{81x^{-6\cdot4} y^{10\cdot4}}{16}$$
$$= \frac{81x^{-24} y^{40}}{16}$$
$$= \frac{81y^{40}}{16x^{24}}$$

66. $\dfrac{2.2 \times 10^7}{3.2 \times 10^{-3}} = \dfrac{2.2}{3.2} \times \dfrac{10^7}{10^{-3}}$
$$= 0.6875 \times 10^{10}$$
$$= (6.875 \times 10^{-1}) \times 10^{10}$$
$$= 6.875 \times (10^{-1} \times 10^{10})$$
$$= 6.875 \times 10^9$$

67.
$$(3.2 \times 10^4)(4.1 \times 10^{-6}) = (3.2 \times 4.1)(10^4 \times 10^{-6})$$
$$= 13.12 \times 10^{-2}$$
$$= (1.312 \times 10) \times 10^{-2}$$
$$= 1.312 \times (10 \times 10^{-2})$$
$$= 1.312 \times 10^{-1}$$

68. We divide.
$$\frac{2.4 \times 10^{13}}{5.88 \times 10^{12}} = \frac{2.4}{5.88} \times \frac{10^{13}}{10^{12}}$$
$$\approx 0.408 \times 10$$
$$= (4.08 \times 10^{-1}) \times 10$$
$$= 4.08 \text{ light-years}$$

69. First we convert 5.0 mils to dollars.

$5.0 \times \$0.001 = \$0.005$

Next we convert 0.005 and 13.4 million to scientific notation.

$0.005 = 5 \times 10^{-3}$

$13.4 \text{ million} = 13,400,000 = 1.34 \times 10^7$

Finally we multiply to find the revenue.
$$(1.34 \times 10^7)(5 \times 10^{-3})$$
$$= (1.34 \times 5)(10^7 \times 10^{-3})$$
$$= 6.7 \times 10^4$$

The revenue will be $\$6.7 \times 10^4$.

70. $\dfrac{x-4y}{3} = \dfrac{5-4(-4)}{3} = \dfrac{5+16}{3} = \dfrac{21}{3} = 7$

Answer D is correct.

71. $2x + y = x \cdot 2 + y$ by the commutative law of multiplication, so answer B is not correct.

$2x + y = y + 2x$ by the commutative law of addition, so answer C is not correct.

$2x + y = y + 2x = y + x \cdot 2$ by the commutative laws of addition and multiplication, so answer D is not correct.

We cannot express $2x + y$ as $2y + x$ using the commutative laws of addition and multiplication, so answer A is correct.

72. $(x^y \cdot x^{3y})^3 = (x^{y+3y})^3 = (x^{4y})^3 = x^{12y}$

73. $a^{-1}b = (2^x)^{-1}(2^{x+5}) = (2^{-x})(2^{x+5}) = 2^{-x+x+5} = 2^5 = 32$

74. $3x - 3y = 3(x - y)$, so (a) and (i) are equivalent.

$(x^{-2})^5 = x^{-10}$, so (d) and (f) are equivalent.

$x(y + z) = xy + xz$, so (h) and (j) are equivalent.

There are no other equivalent expressions.

Chapter R Discussion and Writing Exercises

1. Answers may vary. Five rational numbers that are not integers are $\dfrac{1}{3}$, $-\dfrac{3}{4}$, $6\dfrac{5}{8}$, -0.001, and 1.7. They are not integers because they are not whole numbers or opposites of whole numbers.

2. The quotient 7/0 is defined to be the number that gives a result of 7 when multiplied by 0. There is no such number, so we say the quotient is not defined.

3. No; the area is quadrupled. For a triangle with base b and height h, $A = \dfrac{1}{2}bh$. For a triangle with base $2b$ and height $2h$, $A = \dfrac{1}{2} \cdot 2b \cdot 2h = 2bh = 4\left(\dfrac{1}{2}bh\right)$.

4. No; the area is quadrupled. For a parallelogram with base b and height h, $A = bh$. For a parallelogram with base $2b$ and height $2h$, $A = 2b \cdot 2h = 4(bh)$.

5. \$5 million in \$20 bills contains $\dfrac{5 \times 10^6}{20} = 0.25 \times 10^6 = 2.5 \times 10^5$ bills, and 2.5×10^5 bills would weigh $2.5 \times 10^5 \times 2.20 \times 10^{-3} = 5.5 \times 10^2$, or 550 lb. Thus, it is not possible that a criminal is carrying \$5 million in \$20 bills in a briefcase.

6. For 5^n, where n is a natural number, the one's digit will be 5. Since this is not the case with the given calculator readout, we know that the readout is an approximation.

Chapter R Test

1. The irrational numbers are the numbers that cannot be named as quotients of integers with nonzero divisors. That is, the irrational numbers are the numbers that are not rational. They are $\sqrt{7}$ and π.

2. We specify the conditions by which we know whether a number is in the set.

$\{x | x \text{ is a real number greater than } 20\}$

3. Since -4.5 is to the right of -8.7, we have $-4.5 > -8.7$.

4. $a \le 5$

The inequality $5 \ge a$ has the same meaning.

5. $-6 \ge -6$ is true since $-6 = -6$ is true.

6. $-8 \le -6$ is true since $-8 < -6$ is true.

7. $x > -2$

We shade all numbers to the right of -2 and use a parenthesis at -2 to indicate that it is not a solution.

8. The distance of 0 from 0 is 0, so $|0| = 0$.

9. The distance of $-\dfrac{7}{8}$ from 0 is $\dfrac{7}{8}$, so $\left|-\dfrac{7}{8}\right| = \dfrac{7}{8}$.

10. $7 + (-9)$

We find the difference of the absolute values, $9 - 7 = 2$. Since the negative number has the larger absolute value, the answer is negative.

$7 + (-9) = -2$

11. $-5.3 + (-7.8)$

The sum of two negative numbers is negative. We add the absolute values, $5.3 + 7.8 = 13.1$, and make the answer negative.

$-5.3 + (-7.8) = -13.1$

12. $-\dfrac{5}{2} + \left(-\dfrac{7}{2}\right)$

The sum of two negative numbers is negative. We add the absolute values, $\dfrac{5}{2} + \dfrac{7}{2} = \dfrac{12}{2} = 6$ and make the answer negative.

$-\dfrac{5}{2} + \left(-\dfrac{7}{2}\right) = -6$

13. $-6 - (-5) = -6 + 5 = -1$

14. $-18.2 + (-11.5) = -29.7$

15. $\dfrac{19}{4} - \left(-\dfrac{3}{2}\right) = \dfrac{19}{4} + \dfrac{3}{2} = \dfrac{19}{4} + \dfrac{6}{4} = \dfrac{25}{4}$

16. $(-4.1)(8.2)$

The product of a negative number and a positive number is negative. We multiply the absolute values, $(4.1)(8.2) = 33.62$, and make the answer negative.

$(-4.1)(8.2) = -33.62$

17. $-\dfrac{4}{5}\left(-\dfrac{15}{16}\right)$

The product of two negative numbers is positive. We multiply the absolute values and make the answer positive.

$\dfrac{4}{5}\left(\dfrac{15}{16}\right) = \dfrac{4 \cdot 15}{5 \cdot 16} = \dfrac{4 \cdot 3 \cdot 5}{5 \cdot 4 \cdot 4} = \dfrac{4 \cdot 5}{4 \cdot 5} \cdot \dfrac{3}{4} = \dfrac{3}{4}$

Thus, $-\dfrac{4}{5}\left(-\dfrac{15}{16}\right) = \dfrac{3}{4}$.

18. $-6(-4)(-11)2 = 24(-11)2 = (-264)2 = -528$

19. When a negative number is divided by a negative number, the answer is positive.

$-75 \div (-5) = \dfrac{-75}{-5} = 15$

20. When a negative number is divided by a positive number, the answer is negative.

$\dfrac{-10}{2} = -5$

21. $-\dfrac{5}{2} \div \left(-\dfrac{15}{16}\right) = -\dfrac{5}{2} \cdot \left(-\dfrac{16}{15}\right) = \dfrac{80}{30} = \dfrac{8}{3}$

22. $-459.2 \div 5.6 = \dfrac{-459.2}{5.6} = -82$

23. $\dfrac{-3}{0}$ Not defined: Division by 0.

24. When $a = -13$, then $-a = -(-13) = 13$.

25. When $a = 0$, then $-a = -0 = 0$.

26. $\underbrace{q \cdot q \cdot q \cdot q}_{4 \text{ factors}} = q^4$

27. $\dfrac{1}{a^9} = a^{-9}$

28.
$$1 - (2-5)^2 + 5 \div 10 \cdot 4^2$$
$$= 1 - (-3)^2 + 5 \div 10 \cdot 4^2$$
$$= 1 - 9 + 5 \div 10 \cdot 16$$
$$= 1 - 9 + 0.5 \cdot 16$$
$$= 1 - 9 + 8$$
$$= -8 + 8$$
$$= 0$$

29.
$$\dfrac{7(5 - 2 \cdot 3) - 3^2}{4^2 - 3^2} = \dfrac{7(5-6) - 3^2}{16 - 9}$$
$$= \dfrac{7(-1) - 3^2}{7}$$
$$= \dfrac{7(-1) - 9}{7}$$
$$= \dfrac{-7 - 9}{7}$$
$$= \dfrac{-16}{7}, \text{ or } -\dfrac{16}{7}$$

30. $t + 9$, or $9 + t$

31. Let x and y represent the numbers. Then we have $\dfrac{x}{y} - 12$.

32. Substitute 2 for x and -4 for y and carry out the calculations.

$3x - 3y = 3 \cdot 2 - 3(-4) = 6 + 12 = 18$

33. Substitute 3 for b and 2.5 for h and carry out the calculations.

$A = \dfrac{1}{2}bh$
$= \dfrac{1}{2}(3)(2.5)$
$= (1.5)(2.5)$
$= 3.75 \text{ cm}^2$

34.

	$x(x-3)$	$x^2 - 3x$
$x = -1$	4	4
$x = 10$	70	70
$x = 0$	0	0

The values of $x(x-3)$ and $x^2 - 3x$ are the same for the given values of x and, indeed, for any allowable replacement for x. Thus, they are equivalent.

35.

	$3x + 5x^2$	$8x^2$
$x = -1$	2	8
$x = 10$	530	800
$x = 0$	0	0

Although the expressions have the same value for $x = 0$, they are not the same for all of the given values of x, so they are not equivalent.

36. Since $36x = 4 \cdot 9x$, we multiply by 1 using $\dfrac{9x}{9x}$ as a name for 1.

$$\frac{3}{4} = \frac{3}{4} \cdot \frac{9x}{9x} = \frac{27x}{36x}$$

37. $\dfrac{-54x}{-36x} = \dfrac{-18x \cdot 3}{-18x \cdot 2} = \dfrac{-18x}{-18x} \cdot \dfrac{3}{2} = \dfrac{3}{2}$

38. $pq = qp$ Commutative law of multiplication

39. $t + 4 = 4 + t$ Commutative law of addition

40. $3 + (t + w) = (3 + t) + w$ Associative law of addition

41. $(4a)b = 4(ab)$ Associative law of multiplication

42. $-2(3a - 4b) = -2 \cdot 3a - 2 \cdot (-4b) = -6a + 8b$

43. $3\pi r(s + 1) = 3\pi r \cdot s + 3\pi r \cdot 1 = 3\pi rs + 3\pi r$

44. $ab - ac + 2ad = a \cdot b - a \cdot c + a \cdot 2d = a(b - c + 2d)$

45. $2ah + h = h \cdot 2a + h \cdot 1 = h(2a + 1)$

46. $6y - 8x + 4y + 3x = (6 + 4)y + (-8 + 3)x = 10y - 5x$

47. $4a - 7 + 17a + 21 = (4 + 17)a + (-7 + 21) = 21a + 14$

48. $-(-9x + 7y - 22) = 9x - 7y + 22$ Changing the sign of every term inside parentheses

49. $-3(x + 2) - 4(x - 5) = -3x - 6 - 4x + 20 = -7x + 14$

50.
$$4x - [6 - 3(2x - 5)]$$
$$= 4x - [6 - 6x + 15]$$
$$= 4x - [21 - 6x]$$
$$= 4x - 21 + 6x$$
$$= 10x - 21$$

51.
$$\frac{-12x^3y^{-4}}{8x^7y^{-6}} = \frac{-12}{8}x^{3-7}y^{-4-(-6)}$$
$$= -\frac{3}{2}x^{-4}y^{-4+6}$$
$$= -\frac{3}{2}x^{-4}y^2$$
$$= -\frac{3y^2}{2x^4}$$

52.
$$(3a^4b^{-2})(-2a^5b^{-3}) = 3(-2) \cdot a^4 \cdot a^5 \cdot b^{-2} \cdot b^{-3}$$
$$= -6a^9b^{-5}$$
$$= -\frac{6a^9}{b^5}$$

53.
$$(5a^{4n})(-10a^{5n}) = 5(-10) \cdot a^{4n} \cdot a^{5n}$$
$$= -50a^{4n+5n}$$
$$= -50a^{9n}$$

54. $\dfrac{-60x^{3t}}{12x^{7t}} = \dfrac{-60}{12}x^{3t-7t} = -5x^{-4t}$, or $-\dfrac{5}{x^{4t}}$

55.
$$(-3a^{-3}b^2c)^{-4} = (-3)^{-4}(a^{-3})^{-4}(b^2)^{-4}c^{-4}$$
$$= \frac{1}{(-3)^4}a^{-3(-4)}b^{2(-4)}c^{-4}$$
$$= \frac{1}{81}a^{12}b^{-8}c^{-4}$$
$$= \frac{a^{12}}{81b^8c^4}$$

56.
$$\left[\frac{-5a^{-2}b^8}{10a^{10}b^{-4}}\right]^{-4} = \left[\frac{10a^{10}b^{-4}}{-5a^{-2}b^8}\right]^4$$
$$= \left[\frac{10}{-5}a^{10-(-2)}b^{-4-8}\right]^4$$
$$= [-2a^{12}b^{-12}]^4$$
$$= (-2)^4(a^{12})^4(b^{-12})^4$$
$$= 16a^{48}b^{-48}$$
$$= \frac{16a^{48}}{b^{48}}$$

57. 0.00004.37

5 places

Small number, so the exponent is negative.

$0.0000437 = 4.37 \times 10^{-5}$

58.
$$(8.7 \times 10^{-9})(4.3 \times 10^{15}) = (8.7 \times 4.3)(10^{-9} \times 10^{15})$$
$$= 37.41 \times 10^6$$
$$= (3.741 \times 10) \times 10^6$$
$$= 3.741 \times 10 \times 10^6$$
$$= 3.741 \times 10^7$$

59.
$$\frac{1.2 \times 10^{-12}}{6.4 \times 10^{-7}} = \frac{1.2}{6.4} \times \frac{10^{-12}}{10^{-7}}$$
$$= 0.1875 \times 10^{-5}$$
$$= (1.875 \times 10^{-1}) \times 10^{-5}$$
$$= 1.875 \times (10^{-1} \times 10^{-5})$$
$$= 1.875 \times 10^{-6}$$

60. First we convert 0.002 to scientific notation.

$$0.002 = 2 \times 10^{-3}$$

Now we multiply to find the mass of Pluto.

$$\begin{aligned}
(2 \times 10^{-3})(5.98 \times 10^{24}) &= (2 \times 5.98)(10^{-3} \times 10^{24}) \\
&= 11.96 \times 10^{21} \\
&= (1.196 \times 10) \times 10^{21} \\
&= 1.196 \times (10 \times 10^{21}) \\
&= 1.196 \times 10^{22} \text{ kg}
\end{aligned}$$

Answer C is correct.

61. $(x^{-3})^{-4} = x^{12}$, so (b) and (e) are equivalent.

$5 + 5x = 5x + 5 = 5(x + 1)$, so (d), (f), and (h) are equivalent.

$5(xy) = (5x)y$, so (i) and (j) are equivalent.

There are no other equivalent expressions.

Chapter 1

Solving Linear Equations and Inequalities

Exercise Set 1.1

RC1. (d)

RC3. (b)

1. $x + 23 = 40$ Writing the equation

$\overline{17 + 23} \;?\; 40$ Substituting 17 for x

$ 40 \;\bigm|$ TRUE

Since the left-hand and the right-hand sides are the same, 17 is a solution of the equation.

3. $2x - 3 = -18$ Writing the equation

$2(-8) - 3 \;?\; -18$ Substituting -8 for x

$-16 - 3 \;\bigm|$

$ -19 \;\bigm|$ FALSE

Since the left-hand and the right-hand sides are not the same, -8 is not a solution of the equation.

5. $\dfrac{-x}{9} = -2$ Writing the equation

$\dfrac{-45}{9} \;?\; -2$ Substituting

$\phantom{\dfrac{-45}{9}} -5 \;\bigm|$ FALSE

Since the left-hand and the right-hand sides are not the same, 45 is not a solution of the equation.

7. $2 - 3x = 21$

$2 - 3 \cdot 10 \;?\; 21$ Substituting

$2 - 30 \;\bigm|$

$ -28 \;\bigm|$ FALSE

Since the left-hand and the right-hand sides are not the same, 10 is not a solution of the equation.

9. $5x + 7 = 102$

$5 \cdot 19 + 7 \;?\; 102$

$95 + 7 \;\bigm|$

$ 102 \;\bigm|$ TRUE

Since the left-hand and the right-hand sides are the same, 19 is a solution of the equation.

11. $7(y - 1) = 84$

$7(-11 - 1) \;?\; 84$

$7(-12) \;\bigm|$

$ -84 \;\bigm|$ FALSE

Since the left-hand and the right-hand sides are not the same, -11 is not a solution of the equation.

13. $y + 6 = 13$

$y + 6 - 6 = 13 - 6$ Subtracting 6 on both sides

$y + 0 = 7$ Simplifying

$y = 7$ Using the identity property of 0

Check: $y + 6 = 13$

$\overline{7 + 6} \;?\; 13$

$ 13 \;\bigm|$ TRUE

The solution is 7.

15. $-20 = x - 12$

$-20 + 12 = x - 12 + 12$ Adding 12 on both sides

$-8 = x + 0$ Simplifying

$-8 = x$ Using the identity property of 0

Check: $-20 = x - 12$

$\overline{-20 \;?\; -8 - 12}$

$ \bigm|\; -20$ TRUE

The solution is -8.

17. $-8 + x = 19$

$8 - 8 + x = 8 + 19$ Adding 8

$0 + x = 27$

$x = 27$ Using the identity property of 0

Check: $-8 + x = 19$

$\overline{-8 + 27} \;?\; 19$

$ 19 \;\bigm|$ TRUE

The solution is 27.

19. $-12 + z = -51$

$12 + (-12) + z = 12 + (-51)$ Adding 12

$0 + z = -39$

$z = -39$

The number -39 checks, so it is the solution.

21. $p - 2.96 = 83.9$

$p - 2.96 + 2.96 = 83.9 + 2.96$

$p + 0 = 86.86$

$p = 86.86$

The number 86.86 checks, so it is the solution.

23.
$$-\frac{3}{8} + x = -\frac{5}{24}$$

$$\frac{3}{8} + \left(-\frac{3}{8}\right) + x = \frac{3}{8} + \left(-\frac{5}{24}\right)$$

$$0 + x = \frac{3}{8} \cdot \frac{3}{3} + \left(-\frac{5}{24}\right)$$

$$x = \frac{9}{24} + \left(-\frac{5}{24}\right)$$

$$x = \frac{4}{24}$$

$$x = \frac{1}{6}$$

The number $\frac{1}{6}$ checks, so it is the solution.

25.
$$3x = 18$$

$$\frac{3x}{3} = \frac{18}{3} \quad \text{Dividing by 3 on both sides}$$

$$1 \cdot x = \frac{18}{3} \quad \text{Simplifying}$$

$$x = 6$$

Check: $\dfrac{3x = 18}{3 \cdot 6 \ ? \ 18}$

$\qquad\qquad 18 \ \Big| \qquad$ TRUE

The solution is 6.

27.
$$-11y = 44$$

$$\frac{-11y}{-11} = \frac{44}{-11}$$

$$1 \cdot y = \frac{44}{-11}$$

$$y = -4$$

Check: $\dfrac{-11y = 44}{-11(-4) \ ? \ 44}$

$\qquad\qquad 44 \ \Big| \qquad$ TRUE

The solution is -4.

29.
$$-\frac{x}{7} = 21$$

$$-\frac{1}{7}x = 21$$

$$-7\left(-\frac{1}{7}\right)x = -7 \cdot 21 \quad \text{Multiplying by } -7$$
$$\text{on both sides}$$

$$1 \cdot x = -147$$

$$x = -147$$

Check: $\dfrac{-\frac{x}{7} = 21}{-\frac{-147}{7} \ ? \ 21}$

$\qquad -(-21) \ \Big|$

$\qquad\qquad 21 \ \Big| \qquad$ TRUE

The solution is -147.

31.
$$-96 = -3z$$

$$\frac{-96}{-3} = \frac{-3z}{-3} \quad \text{Dividing by } -3$$

$$\frac{-96}{-3} = 1 \cdot z$$

$$32 = z$$

Check: $\dfrac{-96 = -3z}{-96 \ ? \ -3 \cdot 32}$

$\qquad\qquad -96 \qquad$ TRUE

The solution is 32.

33.
$$4.8y = -28.8$$

$$\frac{4.8y}{4.8} = \frac{-28.8}{4.8}$$

$$1 \cdot y = -\frac{28.8}{4.8}$$

$$y = -6$$

The number -6 checks, so it is the solution.

35.
$$\frac{3}{2}t = -\frac{1}{4}$$

$$\frac{2}{3} \cdot \frac{3}{2}t = \frac{2}{3} \cdot \left(-\frac{1}{4}\right)$$

$$1 \cdot t = -\frac{2}{12}$$

$$t = -\frac{1}{6}$$

The number $-\frac{1}{6}$ checks, so it is the solution.

37.
$$6x - 15 = 45$$

$$6x - 15 + 15 = 45 + 15 \quad \text{Adding 15}$$

$$6x = 60$$

$$\frac{6x}{6} = \frac{60}{6} \quad \text{Dividing by 6}$$

$$x = 10$$

Check: $\dfrac{6x - 15 = 45}{6 \cdot 10 - 15 \ ? \ 45}$

$\qquad\quad 60 - 15 \ \Big|$

$\qquad\qquad 45 \ \Big| \qquad$ TRUE

The solution is 10.

39.
$$5x - 10 = 45$$

$$5x - 10 + 10 = 45 + 10$$

$$5x = 55$$

$$\frac{5x}{5} = \frac{55}{5}$$

$$x = 11$$

Check: $\dfrac{5x - 10 = 45}{5 \cdot 11 - 10 \ ? \ 45}$

$\qquad\quad 55 - 10 \ \Big|$

$\qquad\qquad 45 \ \Big| \qquad$ TRUE

The solution is 11.

41.
$$9t + 4 = -104$$
$$9t + 4 - 4 = -104 - 4$$
$$9t = -108$$
$$\frac{9t}{9} = \frac{-108}{9}$$
$$t = -12$$

Check:
$$\begin{array}{c|c} 9t + 4 = -104 \\ \hline 9(-12) + 4 \ ? \ -104 \\ -108 + 4 \ \big| \\ -104 \ \big| & \text{TRUE} \end{array}$$

The solution is -12.

43.
$$-\frac{7}{3}x + \frac{2}{3} = -18, \text{ LCM is 3}$$
$$3\left(-\frac{7}{3}x + \frac{2}{3}\right) = 3(-18) \quad \begin{array}{l}\text{Multiplying by 3 to}\\\text{clear fractions}\end{array}$$
$$-7x + 2 = -54$$
$$-7x = -56 \qquad \text{Subtracting 2}$$
$$x = \frac{-56}{-7} \qquad \text{Dividing by } -7$$
$$x = 8$$

The number 8 checks. It is the solution.

45.
$$\frac{6}{5}x + \frac{4}{10}x = \frac{32}{10}, \text{ LCM is 10}$$
$$10\left(\frac{6}{5}x + \frac{4}{10}x\right) = 10 \cdot \frac{32}{10} \quad \begin{array}{l}\text{Multiplying by 10 to}\\\text{clear fractions}\end{array}$$
$$12x + 4x = 32$$
$$16x = 32 \qquad \text{Collecting like terms}$$
$$x = \frac{32}{16} \qquad \text{Dividing by 16}$$
$$x = 2$$

The number 2 checks. It is the solution.

47.
$$0.9y - 0.7y = 4.2$$
$$10(0.9y - 0.7y) = 10(4.2) \quad \begin{array}{l}\text{Multiplying by 10 to clear}\\\text{fractions}\end{array}$$
$$9y - 7y = 42$$
$$2y = 42 \qquad \text{Collecting like terms}$$
$$y = \frac{42}{2}$$
$$y = 21$$

The number 21 checks, so it is the solution.

49.
$$8x + 48 = 3x - 12$$
$$5x + 48 = -12 \qquad \text{Subtracting } 3x$$
$$5x = -60 \qquad \text{Subtracting 48}$$
$$x = \frac{-60}{5} \qquad \text{Dividing by 5}$$
$$x = -12$$

The number -12 checks, so it is the solution.

51.
$$7y - 1 = 27 + 7y$$
$$-1 = 27 \qquad \text{Subtracting } 7y$$

The equation $-1 = 27$ is false. No matter what number we try for x we get a false sentence. Thus, the equation has no solution.

53.
$$3x - 4 = 5 + 12x$$
$$-4 = 5 + 9x \qquad \text{Subtracting } 3x$$
$$-9 = 9x \qquad \text{Subtracting 5}$$
$$-1 = x \qquad \text{Dividing by 9}$$

The number -1 checks, so it is the solution.

55.
$$5 - 4a = a - 13$$
$$5 = 5a - 13 \qquad \text{Adding } 4a$$
$$18 = 5a \qquad \text{Adding 13}$$
$$\frac{18}{5} = a \qquad \text{Dividing by 5}$$

The number $\frac{18}{5}$ checks. It is the solution.

57.
$$3m - 7 = -7 - 4m - m$$
$$3m - 7 = -7 - 5m \qquad \text{Collecting like terms}$$
$$3m = -5m \qquad \text{Adding 7}$$
$$8m = 0 \qquad \text{Adding } 5m$$
$$m = \frac{0}{8} \qquad \text{Dividing by 8}$$
$$m = 0$$

The number 0 checks, so it is the solution.

59.
$$5x + 3 = 11 - 4x + x$$
$$5x + 3 = 11 - 3x \qquad \text{Collecting like terms}$$
$$8x + 3 = 11 \qquad \text{Adding } 3x$$
$$8x = 8 \qquad \text{Subtracting 3}$$
$$x = \frac{8}{8} \qquad \text{Dividing by 8}$$
$$x = 1$$

The number 1 checks, so it is the solution.

61.
$$-7 + 9x = 9x - 7$$
$$-7 = -7 \qquad \text{Subtracting } 9x$$

The equation $-7 = -7$ is true. Replacing x by any real number gives a true sentence. Thus, all real numbers are solutions.

63.
$$6y - 8 = 9 + 6y$$
$$-8 = 9 \qquad \text{Subtracting } 6y$$

The equation $-8 = 9$ is false. No matter what number we try for x we get a false sentence. Thus, the equation has no solution.

65.
$$2(x + 7) = 4x$$
$$2x + 14 = 4x \qquad \text{Multiplying to remove parentheses}$$
$$14 = 2x \qquad \text{Subtracting } 2x$$
$$7 = x \qquad \text{Dividing by 2}$$

Check:
$$2(x + 7) = 4x$$

$2(7 + 7)$? $4 \cdot 7$	
$2 \cdot 14$	28
28	TRUE

The solution is 7.

67. $\quad 80 = 10(3t + 2)$

$80 = 30t + 20$

$60 = 30t$

$2 = t$

Check:
$$80 = 10(3t + 2)$$

80 ? $10(3 \cdot 2 + 2)$	
$10(6 + 2)$	
$10 \cdot 8$	
80	TRUE

The solution is 2.

69. $\quad 180(n - 2) = 900$

$180n - 360 = 900$

$180n = 1260$

$n = 7$

Check:
$$180(n - 2) = 900$$

$180(7 - 2)$? 900	
$180 \cdot 5$	
900	TRUE

The solution is 7.

71. $\quad 5y - (2y - 10) = 25$

$5y - 2y + 10 = 25$

$3y + 10 = 25$

$3y = 15$

$y = 5$

Check:
$$5y - (2y - 10) = 25$$

$5 \cdot 5 - (2 \cdot 5 - 10)$? 25	
$25 - (10 - 10)$	
$25 - 0$	
25	TRUE

The solution is 5.

73. $\quad 7(3x + 6) = 11 - (x + 2)$

$21x + 42 = 11 - x - 2$

$21x + 42 = 9 - x$

$22x + 42 = 9$

$22x = -33$

$x = \dfrac{-33}{22}$

$x = -\dfrac{3}{2}$

The number $-\dfrac{3}{2}$ checks, so it is the solution.

75. $\quad 2[9 - 3(-2x - 4)] = 12x + 42$

$2[9 + 6x + 12] = 12x + 42$

$2[6x + 21] = 12x + 42$

$12x + 42 = 12x + 42$

$42 = 42$

We get a true equation. Replacing x with any real number gives a true sentence. Thus, all real numbers are solutions.

77. $\quad \dfrac{1}{8}(16y + 8) - 17 = -\dfrac{1}{4}(8y - 16)$

$2y + 1 - 17 = -2y + 4$

$2y - 16 = -2y + 4$

$4y - 16 = 4$

$4y = 20$

$y = 5$

The number 5 checks, so it is the solution.

79. $\quad 3[5 - 3(4 - t)] - 2 = 5[3(5t - 4) + 8] - 26$

$3[5 - 12 + 3t] - 2 = 5[15t - 12 + 8] - 26$

$3[-7 + 3t] - 2 = 5[15t - 4] - 26$

$-21 + 9t - 2 = 75t - 20 - 26$

$9t - 23 = 75t - 46$

$-23 = 66t - 46$

$23 = 66t$

$\dfrac{23}{66} = t$

The number $\dfrac{23}{66}$ checks, so it is the solution.

81. $\frac{2}{3}\left(\frac{7}{8}+4x\right)-\frac{5}{8}=\frac{3}{8}$

$$\frac{2}{3}\left(\frac{7}{8}+4x\right)=1$$

$$\frac{7}{12}+\frac{8}{3}x=1$$

$$12\left(\frac{7}{12}+\frac{8}{3}x\right)=12\cdot1$$

$$7+32x=12$$

$$32x=5$$

$$x=\frac{5}{32}$$

The number $\frac{5}{32}$ checks, so it is the solution.

83. $5(4x-3)-2(6-8x)+10(-2x+7)=-4(9-12x)$

$$20x-15-12+16x-20x+70=-36+48x$$

$$16x+43=-36+48x$$

$$43=-36+32x$$

$$79=32x$$

$$\frac{79}{32}=x$$

The number $\frac{79}{32}$ checks, so it is the solution.

85. $u^{-9}\cdot u^{23}=u^{-9+23}=a^{14}$

87. $(6x^5y^{-4})(-3x^{-3}y^{-7})$

$$=6\cdot(-3)\cdot x^5\cdot x^{-3}\cdot y^{-4}\cdot y^{-7}$$

$$=-18x^{5+(-3)}y^{-4+(-7)}$$

$$=-18x^2y^{-11}$$

$$=-\frac{18x^2}{y^{11}}$$

89. $2(6-10x)=2\cdot6-2\cdot10x=12-20x$

91. $-4(3x-2y+z)=-4\cdot3x-4(-2y)-4\cdot z$
$$=-12x+8y-4z$$

93. $2x-6y=2\cdot x-2\cdot3y$
$$=2(x-3y)$$

95. $4x-10y+2=2\cdot2x-2\cdot5y+2\cdot1$
$$=2(2x-5y+1)$$

97. $\{1,2,3,4,5,6,7,8,9\}$;
$\{x|x$ is a positive integer less than 10$\}$

99. $4.23x-17.898=-1.65x-42.454$

$$5.88x-17.898=-42.454$$

$$5.88x=-24.556$$

$$x\approx-4.176$$

The solution is approximately -4.176.

101. $\frac{3x}{2}+\frac{5x}{3}-\frac{13x}{6}-\frac{2}{3}=\frac{5}{6}$, LCM is 6

$$6\left(\frac{3x}{2}+\frac{5x}{3}-\frac{13x}{6}-\frac{2}{3}\right)=6\cdot\frac{5}{6}$$

Multiplying by 6 to clear fractions

$$3\cdot3x+2\cdot5x-1\cdot13x-2\cdot2=1\cdot5$$

$$9x+10x-13x-4=5$$

$$6x-4=5 \quad \text{Collecting like terms}$$

$$6x=9 \quad \text{Adding 4}$$

$$x=\frac{9}{6} \quad \text{Dividing by 6}$$

$$x=\frac{3}{2}$$

The number $\frac{3}{2}$ checks, so it is the solution.

103. $x-\{3x-[2x-(5x-(7x-1))]\}=x+7$

$$x-\{3x-[2x-(5x-7x+1)]\}=x+7$$

$$x-\{3x-[2x-(-2x+1)]\}=x+7$$

$$x-\{3x-[2x+2x-1]\}=x+7$$

$$x-\{3x-[4x-1]\}=x+7$$

$$x-\{3x-4x+1\}=x+7$$

$$x-\{-x+1\}=x+7$$

$$x+x-1=x+7$$

$$2x-1=x+7$$

$$x-1=7$$

$$x=8$$

The number 8 checks, so it is the solution.

Exercise Set 1.2

RC1. $s=t+4$

$$s-4=t+4-4$$

$$s-4=t, \text{ or } t=s-4$$

Answer (d) is correct.

RC3. $r=\frac{1}{4}q-t$

$$t+r=\frac{1}{4}q \quad \text{Adding } t$$

$$t=\frac{1}{4}q-r \quad \text{Subtracting } r$$

Answer (f) is correct.

RC5. $\frac{1}{4}s=t-q$

$$4\cdot\frac{1}{4}s=4(t-q)$$

$$s=4(t-q)$$

Answer (c) is correct.

1. $d = rt$

$\dfrac{d}{t} = r$ Dividing by t

3. $A = bh$

$\dfrac{A}{b} = h$ Dividing by b

5. $P = 2l + 2w$

$P - 2l = 2w$ Subtracting $2l$

$\dfrac{P - 2l}{2} = w$ Dividing by 2

or $\dfrac{P}{2} - l = w$

7. $A = \dfrac{1}{2}bh$

$2A = bh$ Multiplying by 2

$\dfrac{2A}{h} = b$ Dividing by h

9. $A = \dfrac{a + b}{2}$

$2A = a + b$ Multiplying by 2

$2A - b = a$ Subtracting b

11. $F = ma$

$\dfrac{F}{a} = m$ Dividing by a

13. $I = Prt$

$\dfrac{I}{Pr} = t$ Dividing by Pr

15. $E = mc^2$

$\dfrac{E}{m} = c^2$ Dividing by m

17. $Q = \dfrac{p - q}{2}$

$2Q = p - q$ Multiplying by 2

$2Q + q = p$ Adding q

19. $Ax + By = c$

$By = c - Ax$ Subtracting Ax

$y = \dfrac{c - Ax}{B}$ Dividing by B

21. $I = 1.08\dfrac{T}{N}$

$IN = 1.08T$ Multiplying by N

$N = \dfrac{1.08T}{I}$ Dividing by I

23. $C = \dfrac{3}{4}(m + 5)$

$\dfrac{4}{3} \cdot C = \dfrac{4}{3} \cdot \dfrac{3}{4}(m + 5)$ Multiplying by $\dfrac{4}{3}$

$\dfrac{4}{3}C = m + 5$

$\dfrac{4}{3}C - 5 = m$, or   Subtracting 5

$\dfrac{4C - 15}{3} = m$

25. $n = \dfrac{1}{3}(a + b - c)$

$3n = a + b - c$ Multiplying by 3

$3n - a + c = b$ Subtracting a and adding c

27. $d = R - Rst$

$d = R(1 - st)$ Factoring out R

$\dfrac{d}{1 - st} = R$ Dividing by $1 - st$

29. $T = B + Bqt$

$T = B(1 + qt)$ Factoring

$\dfrac{T}{1 + qt} = B$ Dividing by $1 + qt$

31. a) 5 ft 11 in. $= 5 \times 12$ in. $+ 11$ in. $= 60$ in. $+ 11$ in. $= 71$ in.

Substitute 185 for w, 71 for h, and 28 for a and carry out the calculation.

$R = 66 + 6.23w + 12.7h - 6.8a$

$R = 66 + 6.23(185) + 12.7(71) - 6.8(28)$

$= 66 + 1152.55 + 901.7 - 190.4$

≈ 1930 calories

b) $R = 66 + 6.23w + 12.7h - 6.8a$

$R - 66 - 12.7h + 6.8a = 6.23w$

$\dfrac{R - 66 - 12.7h + 6.8a}{6.23} = w$

33. a) 5 ft 8 in. $= 5 \times 12$ in. $+ 8$ in. $= 60$ in. $+ 8$ in. $= 68$ in.

Substitute 150 for w, 68 for h, and 25 for a and carry out the calculation.

$K = 1015.25 + 6.74w + 7.29h - 7.29a$

$K = 1015.25 + 6.74(150) + 7.29(68) - 7.29(25)$

$= 1015.25 + 1011 + 495.72 - 182.25$

≈ 2340 calories

b) $K = 1015.25 + 6.74w + 7.29h - 7.29a$

$K - 1015.25 - 6.74w - 7.29h = -7.29a$

$\dfrac{K - 1015.25 - 6.74w - 7.29h}{-7.29} = a$, or

$\dfrac{1015.25 + 6.74w + 7.29h - K}{7.29} = a$

35. a) We substitute 8.5 for d and 24.1 for a and calculate P.

$P = 9.337da - 299$

$P = 9.337(8.5)(24.1) - 299$

$P = 1912.68445 - 299$

$P = 1613.68445$

The projected birth weight is about 1614 g.

b) $P = 9.337da - 299$

$P + 299 = 9.337da$

$\dfrac{P + 299}{9.337d} = a$

37. a) We substitute 3 for a and 250 for d and calculate c.

$$c = \frac{ad}{a + 12}$$

$$c = \frac{3 \cdot 250}{3 + 12}$$

$$c = \frac{750}{15}$$

$$c = 50$$

The child's dosage is 50 mg.

b)

$$c = \frac{ad}{a + 12}$$

$$c(a + 12) = ad$$

$$\frac{c(a + 12)}{a} = d$$

This result can also be expressed as follows:

$$d = \frac{c(a + 12)}{a} = \frac{ac + 12c}{a} = c + \frac{12c}{a}$$

39. $\dfrac{80}{-16} = -\dfrac{80}{16} = -5$

41. $-\dfrac{1}{2} \div \dfrac{1}{4} = -\dfrac{1}{2} \cdot \dfrac{4}{1} = -\dfrac{4}{2} = -2$

43. $-\dfrac{2}{3} \div \left(-\dfrac{5}{6}\right) = -\dfrac{2}{3} \cdot \left(-\dfrac{6}{5}\right) = \dfrac{2 \cdot 6}{3 \cdot 5} = \dfrac{2 \cdot 2 \cdot 3}{3 \cdot 5} =$

$\dfrac{2 \cdot 2 \cdot \cancel{3}}{\cancel{3} \cdot 5} = \dfrac{4}{5}$

45. $\dfrac{-90}{15} = -\dfrac{90}{15} = -6$

47.
$$A = \pi r s + \pi r^2$$

$$A - \pi r^2 = \pi r s$$

$$\frac{A - \pi r^2}{\pi r} = s, \text{ or}$$

$$\frac{A}{\pi r} - r = s$$

49. Solve for V_1:

$$\frac{P_1 V_1}{T_1} = \frac{P_2 V_2}{T_2}$$

$$V_1 = \frac{T_1 P_2 V_2}{P_1 T_2} \quad \text{Multiplying by } \frac{T_1}{P_1}$$

Solve for P_2:

$$\frac{P_1 V_1}{T_1} = \frac{P_2 V_2}{T_2}$$

$$\frac{P_1 V_1 T_2}{T_1 V_2} = P_2 \quad \text{Multiplying by } \frac{T_2}{V_2}$$

51. We substitute \$75 for P, \$3 for I, and 5%, or 0.05, for r in the formula $t = \dfrac{I}{Pr}$.

$$t = \frac{I}{Pr}$$

$$t = \frac{\$3}{\$75(0.05)} \quad \text{Substituting}$$

$$t = 0.8$$

It will take 0.8 yr.

53. $H = W\left(\dfrac{v}{234}\right)^3$

a) $H = 2700\left(\dfrac{83}{234}\right)^3$

$H \approx 120.5$ horsepower

b) $H = 3100\left(\dfrac{73}{234}\right)^3$

$H \approx 94.1$ horsepower

Exercise Set 1.3

RC1. <u>Familiarize</u> yourself with the problem situation.

RC3. <u>Solve</u> the equation.

RC5. <u>State</u> the answer to the problem clearly.

1. *Familiarize.* Let d = Nyad's distance from the Florida coast, in miles, at 11:00 P.M. Then $3d$ = her distance from Cuba.

Translate.

Distance from Florida	plus	Distance from Cuba	is	Total distance.
d	$+$	$3d$	$=$	110

Solve. We solve the equation.

$$d + 3d = 110$$

$$4d = 110$$

$$d = 27.5$$

Check. If Nyad is 27.5 mi from Florida, then she is 3(27.5), or 82.5 mi from Cuba. Since $27.5 + 82.5 = 110$ mi, the answer checks.

State. Nyad was 27.5 mi from the Florida coast.

3. *Familiarize.* Let x = the measure of the first angle. Then $x + 7$ = the measure of the second angle, and $2x - 7$ = the measure of the third angle.

Translate. The sum of the measures of the angles of a triangle is 180°, so we have

$$x + (x + 7) + (2x - 7) = 180.$$

Solve. We solve the equation.

$$x + (x + 7) + (2x - 7) = 180$$

$$4x = 180$$

$$x = 45$$

If $x = 45$, then $x + 7 = 45 + 7 = 52$ and $2x - 7 = 2 \cdot 45 - 7 = 90 - 7 = 83$.

Check. The second angle, 52°, is 7° more than the first angle, and the third angle, 83°, is 7° less than twice the first angle. Also, $45° + 52° + 83° = 180°$. The answer checks.

State. The first angle is 45°, the second is 52°, and the third is 83°.

5. Familiarize. Let $c =$ the number of climbers who died of exposure/frostbite.

Translate.

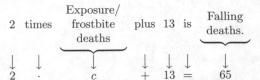

2	times	Exposure/frostbite deaths	plus	13	is	Falling deaths.
$\downarrow$	$\downarrow$	$\downarrow$	$\downarrow$	$\downarrow$	$\downarrow$	$\downarrow$
2	$\cdot$	c	$+$	13	$=$	65

Solve. We solve the equation.

$$2 \cdot c + 13 = 65$$
$$2c = 52$$
$$c = 26$$

Check. $2 \cdot 26 + 13 = 52 + 13 = 65$, the number of deaths from falling, so the answer checks.

State. 26 climbers died of exposure/frostbite.

7. Familiarize. Let $p =$ the original price. Then the sale price is $p - 30\%p$, or $p - 0.3p$, or $0.7p$.

Translate.

The sale price	is	$176.40
$\downarrow$	$\downarrow$	$\downarrow$
$0.7p$	$=$	176.40

Solve. We solve the equation.

$$0.7p = 176.40$$
$$p = 252 \quad \text{Dividing by } 0.7$$

Check. 30% of $252 is 0.3($252), or $75.60, and $252 - 75.60 = 176.40, so the answer checks.

State. The original price was $252.

9. Familiarize. Let $w =$ the width of the court. Then $w + 44 =$ the length. Recall that the perimeter of a rectangle with length l and width w is $2l + 2w$.

Translate.

Perimeter	is	288 ft.
$\downarrow$	$\downarrow$	$\downarrow$
$2(w + 44) + 2w$	$=$	288

Solve. We solve the equation.

$$2(w + 44) + 2w = 288$$
$$2w + 88 + 2w = 288$$
$$4w + 88 = 288$$
$$4w = 200$$
$$w = 50$$

If $w = 50$, then $w + 44 = 50 + 44 = 94$.

Check. The length, 94 ft, is 44 ft longer than the width. Also $2 \cdot 94 + 2 \cdot 50 = 188 + 100 = 288$. The numbers check.

State. The length of the court is 94 ft and the width is 50 ft.

11. Familiarize. Let $l =$ the length of the longest piece, in feet. Then $3l - 6 =$ the length of the second piece, and $\frac{2}{3}(3l - 6) + 2 =$ the length of the third piece.

Translate. The sum of the lengths is 168 ft, so we have

$$l + (3l - 6) + \left[\frac{2}{3}(3l - 6) + 2\right] = 168.$$

Solve. We solve the equation.

$$l + (3l - 6) + \left[\frac{2}{3}(3l - 6) + 2\right] = 168$$
$$l + 3l - 6 + 2l - 4 + 2 = 168$$
$$6l - 8 = 168$$
$$6l = 176$$
$$l = \frac{88}{3}, \text{ or } 29\frac{1}{3}$$

Then $3l - 6 = 3 \cdot \frac{88}{3} - 6 = 88 - 6 = 82$ and

$$\frac{2}{3} \cdot 82 + 2 = \frac{164}{3} + 2 = \frac{170}{3}, \text{ or } 56\frac{2}{3}.$$

Check. 82 ft is 6 less than three times $29\frac{1}{3}$ ft, and $56\frac{2}{3}$ is 2 ft more than $\frac{2}{3}$ of 82 ft. Also, $29\frac{1}{3} + 82 + 56\frac{2}{3} = 168$ ft, so the answer checks.

State. The length of the longest piece is 82 ft.

13. Familiarize. Let $p =$ the selling price of the house. Then $p - 100,000$ is the amount that exceeds $100,000. Also 7% of $100,000 is $7000.

Translate.

7% of $100,000		+	5%	of	the amount that exceeds $100,000		is	$15,250.
$\downarrow$		$\downarrow$	$\downarrow$	$\downarrow$	$\downarrow$		$\downarrow$	$\downarrow$
7000		$+$	0.05	$\cdot$	$(p - 100,000)$		$=$	15,250

Solve. We solve the equation.

$$7000 + 0.05 \cdot (p - 100,000) = 15,250$$
$$7000 + 0.05p - 5000 = 15,250$$
$$0.05p + 2000 = 15,250$$
$$0.05p = 13,250$$
$$p = 265,000$$

Check. If the selling price is $265,000, then the amount that exceeds $100,000 is $265,000 - $100,000$, or $165,000. Then the commission would be $7000 + 0.05($165,000)$, or $15,250. The answer checks.

State. The selling price of the house was $265,000.

15. Familiarize. Let $x =$ the first odd integer. Then $x + 2$ and $x + 4$ are the next two odd integers.

Translate.

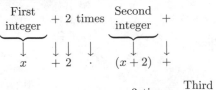

First integer	$+$	2	times	Second integer	$+$
$\downarrow$	$\downarrow$	$\downarrow$	$\downarrow$	$\downarrow$	$\downarrow$
x	$+$	2	$\cdot$	$(x + 2)$	$+$

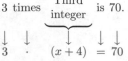

	3	times	Third integer	is	70.
	$\downarrow$	$\downarrow$	$\downarrow$	$\downarrow$	$\downarrow$
	3	$\cdot$	$(x + 4)$	$=$	70

Solve. We solve the equation.

$$x + 2 \cdot (x + 2) + 3 \cdot (x + 4) = 70$$
$$x + 2x + 4 + 3x + 12 = 70$$
$$6x + 16 = 70$$
$$6x = 54$$
$$x = 9$$

If $x = 9$, then $x + 2 = 9 + 2$, or 11, and $x + 4 = 9 + 4$, or 13.

Check. The numbers 9, 11, and 13 are consecutive odd integers. Also $9 + 2 \cdot 11 + 3 \cdot 13 = 9 + 22 + 39 = 70$. The numbers check.

State. The numbers are 9, 11, and 13.

17. *Familiarize*. Let n and $n + 1$ represent the numbers.

Translate.

$$\underbrace{\text{The sum of the numbers}}_{n + (n + 1)} \underset{=}{\text{is}} \underset{459}{459}.$$

Solve. We solve the equation.

$$n + (n + 1) = 459$$
$$2n + 1 = 459$$
$$2n = 458$$
$$n = 229$$

Then $n + 1 = 229 + 1 = 230$.

Check. 229 and 230 are consecutive numbers, and their sum is $229 + 230$, or 459. The answer checks.

State. The numbers are 229 and 230.

19. *Familiarize*. Let $w =$ the number of sheets of wallet-size photos the family bought. Each basic package includes 12 wallet-size photos, and each sheet of extra wallet-size photos contains 6 photos, so the total number of wallet-size photos is $3 \cdot 12 + 6 \cdot w$, or $36 + 6w$. The cost of 3 basic packages is $3(\$14.95)$, or $\$44.85$, and the cost of w sheets of wallet-size photos is $\$1.35w$.

Translate.

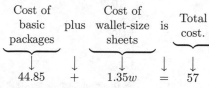

Solve. We solve the equation.

$$44.85 + 1.35w = 57$$
$$1.35w = 12.15$$
$$w = 9$$

If $w = 9$, then $36 + 6w = 36 + 6 \cdot 9 = 36 + 54 = 90$.

Check. The cost of 9 sheets of wallet-size photos is $\$1.35(9)$, or $\$12.15$, so the total cost of the basic packages and the extra photos is $\$44.85 + \$12.15$ or $\$57$. The answer checks.

State. The family bought a total of 90 wallet-size photos.

21. *Familiarize*. Let $s =$ the original salary. Then the amount of the raise is 8% of s, or $0.08s$.

Translate.

$$\underbrace{\text{Original salary}}_{s} \underset{+}{\text{plus}} \underbrace{\text{Raise}}_{0.08s} \underset{=}{\text{is}} \underset{42,066}{\$42,066}.$$

Solve. We solve the equation.

$$s + 0.08s = 42,066$$
$$1s + 0.08s = 42,066$$
$$1.08s = 42,066$$
$$s = 38,950$$

Check. 8% of $\$38,950$ is $\$3116$, and $\$38,950 + \$3116$ is $\$42,066$. The answer checks.

State. The salary before the raise was $\$38,950$.

23. *Familiarize*. Let $d =$ the number of cases of diabetes in 1973, in millions. Then an increase of 402% over this number is $d + 402\%d$, or $d + 4.02d$, or $5.02d$. This represents the number of cases of diabetes in 2010.

Translate.

$$\underbrace{\begin{array}{c}\text{Number of cases} \\ \text{of diabetes in 2010}\end{array}}_{5.02d} \underset{=}{\text{is}} \underbrace{21.1 \text{ million.}}_{21.1}$$

Solve. We solve the equation.

$$5.02d = 21.1$$
$$d \approx 4.2 \quad \text{Dividing by 5.02}$$

Check. 402% of 4.2 is $4.02(4.2)$, or 16.884, and $4.2 + 16.884 = 21.084 \approx 21.1$, so the answer checks.

State. There were approximately 4.2 million cases of diabetes in 1973.

25. a) $x = 2014 - 2012 = 2$.

Then we have

$$y = 6.5x + 41.6 = 6.5(2) + 41.6 = 13 + 41.6 = 54.6.$$

We estimate that spending on Internet search ads was $\$54.6$ billion in 2014.

b) Substitute 75 for y and solve for x.

$$y = 6.5x + 41.6$$
$$75 = 6.5x + 41.6$$
$$33.4 = 6.5x$$
$$5 \approx x$$

We estimate that spending on Internet search ads will reach $\$75$ billion about 5 years after 2012, or in 2017.

27. *Familiarize*. Let $t =$ the number of minutes it will take the plane to reach the cruising altitude. The plane needs to travel a distance of $29,000 - 8000$, or $21,000$ ft.

Translate. We use the motion formula.

$$d = rt$$
$$21,000 = 3500t$$

Solve. We solve the equation.

$$21,000 = 3500t$$

$$\frac{21,000}{3500} = t$$

$$6 = t$$

Check. The distance the plane travels at a speed of 3500 ft/min for 6 min is $3500 \cdot 6$, or 21,000 ft. The answer checks.

State. It will take 6 min for the plane to reach the cruising altitude.

29. First we will find how long it takes for Jen to travel 15 mi downstream.

Familiarize. We will use the formula $d = rt$. Let $t = $ the time, in hours, it will take Jen to travel 15 mi downstream. The speed of the boat traveling downstream is $10 + 2$, or 12 mph.

Translate.

$$d = rt$$

$$15 = 12t$$

Solve. We solve the equation.

$$15 = 12t$$

$$1.25 = t$$

Check. If the boat travels at a speed of 12 mph for 1.25 hr, it travels $12(1.25)$, or 15 mi. The answer checks.

State. It will take Jen 1.25 hr to travel 15 mi downstream.

Now we find how long it will take Jen to travel 15 mi upstream.

Familiarize. We will use the formula $d = rt$. Let $t = $ the time, in hours, it will take Jen to travel 15 mi upstream. The speed of the boat traveling upstream is $10 - 2$, or 8 mph.

Translate.

$$d = rt$$

$$15 = 8t$$

Solve. We solve the equation.

$$15 = 8t$$

$$1.875 = t$$

Check. If the boat travels at a speed of 8 mph for 1.875 hr, it travels $8(1.875)$, or 15 mi. The answer checks.

State. It will take Jen 1.875 hr to travel 15 mi upstream.

31. First we find how long it will take Fran to swim 1.8 mi upstream.

Familiarize. We will use the formula $d = rt$. Let $t = $ the time, in hours, it will take Fran to swim 1.8 mi upstream. Her speed upstream is $5 - 2.3$, or 2.7 mph.

Translate.

$$d = rt$$

$$1.8 = 2.7t$$

Solve. We solve the equation.

$$1.8 = 2.7t$$

$$\frac{2}{3} = t$$

Check. Swimming at a speed of 2.7 mph for $\frac{2}{3}$ hr, Fran swims $2.7\left(\frac{2}{3}\right)$, or 1.8 mi. The answer checks.

State. It takes Fran $\frac{2}{3}$ hr to swim 1.8 mi upstream.

Now we find how long it will take Fran to swim 1.8 mi downstream.

Familiarize. We will use the formula $d = rt$. Let $t = $ the time, in hours, it will take Fran to swim 1.8 mi downstream. Her speed downstream is $5 + 2.3$, or 7.3 mph.

Translate.

$$d = rt$$

$$1.8 = 7.3t$$

Solve. We solve the equation.

$$1.8 = 7.3t$$

$$\frac{1.8}{7.3} = t$$

$$\frac{1.8}{7.3} \cdot \frac{10}{10} = t$$

$$\frac{18}{73} = t$$

Check. Swimming at a speed of 7.3 mph for $\frac{18}{73}$ hr, Fran swims $7.3\left(\frac{18}{73}\right)$, or 1.8 mi. The answer checks.

State. It takes Fran $\frac{18}{73}$ hr to swim 1.8 mi downstream.

33.
$$5^2 - 2 \cdot 5 \cdot 12 + 12^2 = 25 - 2 \cdot 5 \cdot 12 + 144$$
$$= 25 - 10 \cdot 12 + 144$$
$$= 25 - 120 + 144$$
$$= -95 + 144$$
$$= 49$$

35.
$$\frac{12|8 - 10| + 9 \cdot 6}{5^4 + 4^5} = \frac{12|-2| + 9 \cdot 6}{625 + 1024}$$
$$= \frac{12 \cdot 2 + 9 \cdot 6}{1649}$$
$$= \frac{24 + 54}{1649}$$
$$= \frac{78}{1649}$$

37. *Familiarize*. Let $p = $ the price of the house in 2007. From 2007 to 2008, prices increased 1%, so the house was worth $p + 0.01p$, or $1.01p$. From 2008 to 2009, prices dropped 3%, so the house was then worth $1.01p - 0.03(1.01p)$, or $0.97(1.01p)$. From 2009 to 2010, prices dropped another 7%, so the value of the house became $0.97(1.01p) - 0.07(0.97)(1.01p)$, or $0.93(0.97)(1.01p)$.

Translate.

The price of the house in 2010 was $105,000.

$$0.93(0.97)(1.01p) = 105,000$$

Solve. We solve the equation.

$$0.93(0.97)(1.01p) = 105,000$$

$$p = \frac{105,000}{0.93(0.97)(1.01)}$$

$$p \approx 115,243$$

Check. If the price of the house in 2007 was \$115,243, then in 2008 it was worth 1.01(\$115,243), or about \$116,395. In 2009 it was worth 0.97(\$116,395), or about \$112,903 and in 2010 it was worth 0.93(\$112,903), or about \$105,000. The answer checks.

State. The house was worth \$115,243 in 2007.

39. *Familiarize*. Let p = the total percent change, represented in decimal notation. We represent the original population as 1 (100% of the population). After a 20% increase the population is $1 + 0.2 \cdot 1$, or 1.2. When this new population is increased by 30%, we have $1.2 + 0.3(1.2)$, or $1.3(1.2)$. When this population is decreased by 20% we have $1.3(1.2) - 0.2[1.3(1.2)]$, or $0.8(1.3)(1.2)$.

Translate.

Percent of original population after changes	minus	100% of original population	is	total percent change
↓	↓	↓	↓	↓
0.8(1.3)(1.2)	−	1	=	p

Solve.

$$0.8(1.3)(1.2) - 1 - p$$

$$1.248 - 1 = p \qquad \text{Multiplying}$$

$$0.248 = p$$

$$0.25 \approx p \qquad \text{Rounding}$$

Since p is a positive number, it represents an increase in population.

Check. We repeat the computations. The result checks.

State. The total percent change is a 25% increase.

41. *Familiarize*. Using the properties of parallel lines intersected by a transversal, we know that $m\angle 8 = m\angle 4$. Also $m\angle 4 + m\angle 2 = 180°$ and $m\angle 4 = m\angle 1$.

Translate. First we find x and use it to find $m\angle 4$.

$$\begin{array}{ccc} m\angle 8 & = & m\angle 4 \\ \downarrow & \downarrow & \downarrow \\ 5x + 25 & = & 8x + 4 \end{array}$$

Solve.

$$5x + 25 = 8x + 4$$

$$21 = 3x \qquad \text{Subtracting } 5x \text{ and } 4$$

$$7 = x$$

If $x = 7$, we have $m\angle 4 = 8 \cdot 7 + 4 = 60°$.

Then $m\angle 4 + m\angle 2 = 180°$, so

$$60° + m\angle 2 = 180°$$

$$m\angle 2 = 120°. \qquad \text{Subtracting } 60°$$

Also, $m\angle 4 = m\angle 1 = 60°$.

Check. We go over the computations. The results check.

State. $m\angle 2 = 120°$ and $m\angle 1 = 60°$.

Chapter 1 Mid-Chapter Review

1. The statement is true as shown by the following steps.

$$2x + 3 = 7$$

$$2x = 4 \qquad \text{Subtracting 3}$$

$$x = 2 \qquad \text{Dividing by 2}$$

2. The statement is true. See Example 17 on page 79 in the text.

3. The statement is false. See Example 17 on page 79 in the text.

4. When we solve an applied problem, we check the possible solution in the *original problem*. The given statement is false.

5.

$$2x - 5 = 1 - 4x$$

$$2x - 5 + 4x = 1 - 4x + 4x$$

$$6x - 5 = 1 \qquad \text{Collecting like terms}$$

$$6x - 5 + 5 = 1 + 5$$

$$6x = 6 \qquad \text{Collecting like terms}$$

$$\frac{6x}{6} = 6$$

$$x = 1 \qquad \text{Simplifying}$$

6.

$$Mx + Ny = T$$

$$Mx + Ny - Mx = T - Mx$$

$$Ny = T - Mx$$

$$y = \frac{T - Mx}{N}$$

7.

$$x + 5 = 12$$

$$\begin{array}{c|c} 7 + 5 \; ? \; 12 & \\ 12 & \text{TRUE} \end{array}$$

The number 7 is a solution of the equation.

8.

$$3x - 4 = 5$$

$$\begin{array}{c|c} 3 \cdot \dfrac{1}{3} - 4 \; ? \; 5 & \\ 1 - 4 & \\ -3 & \text{FALSE} \end{array}$$

The number $\frac{1}{3}$ is not a solution of the equation.

9.

$$\frac{-x}{8} = -3$$

$$\begin{array}{c|c} \dfrac{-(-24)}{8} \; ? \; -3 & \\ \dfrac{24}{8} & \\ 3 & \text{FALSE} \end{array}$$

The number -24 is not a solution of the equation.

10. $6(x - 3) = 36$

$$6(9 - 3) \ ? \ 36$$
$$6(6)$$
$$36 \ \Big| \ \text{TRUE}$$

The number 9 is a solution of the equation.

11. $x - 7 = -10$
$$x - 7 + 7 = -10 + 7$$
$$x = -3$$

The number -3 checks, so it is the solution.

12. $-7x = 56$
$$\frac{-7x}{-7} = \frac{56}{-7}$$
$$x = -8$$

The number -8 checks, so it is the solution.

13. $8x - 9 = 23$

$8x = 32$ Adding 9

$x = 4$ Dividing by 8

The number 4 checks, so it is the solution.

14. $1 - x = 3x - 7$

$1 = 4x - 7$ Adding x

$8 = 4x$ Adding 7

$2 = x$ Dividing by 4

The number 2 checks, so it is the solution.

15. $2 - 4y = -4y + 2$

$2 = 2$ Adding $4y$

We get an equation that is true for all real numbers, so all real numbers are solutions.

16. $\frac{3}{4}y + 2 = \frac{7}{2}$

$\frac{3}{4}y = \frac{3}{2}$ Subtracting 2

$\frac{4}{3} \cdot \frac{3}{4}y = \frac{4}{3} \cdot \frac{3}{2}$

$y = 2$ Simplifying

The number 2 checks, so it is the solution.

17. $5t - 9 = 7t - 4$

$-9 = 2t - 4$ Subtracting $5t$

$-5 = 2t$ Adding 4

$-\frac{5}{2} = t$ Dividing by 2

The number $-\frac{5}{2}$ checks, so it is the solution.

18. $4x - 11 = 11 + 4x$

$-11 = 11$ Subtracting $4x$

We get a false equation. The equation has no solution.

19. $2(y - 4) = 8y$

$2y - 8 = 8y$

$-8 = 6y$ Subtracting $2y$

$-\frac{4}{3} = y$ Dividing by 6

The number $-\frac{4}{3}$ checks, so it is the solution.

20. $4y - (y - 1) = 16$

$4y - y + 1 = 16$

$3y + 1 = 16$ Collecting like terms

$3y = 15$ Subtracting 1

$y = 5$ Dividing by 3

The number 5 checks, so it is the solution.

21. $t - 3(t - 4) = 9$

$t - 3t + 12 = 9$

$-2t + 12 = 9$ Collecting like terms

$-2t = -3$ Subtracting 12

$t = \frac{3}{2}$ Dividing by -2

The number $\frac{3}{2}$ checks, so it is the solution.

22. $6(2x + 3) = 10 - (4x - 5)$

$12x + 18 = 10 - 4x + 5$

$12x + 18 = 15 - 4x$ Collecting like terms

$16x + 18 = 15$ Adding $4x$

$16x = -3$ Subtracting 18

$x = -\frac{3}{16}$ Dividing by 16

The number $-\frac{3}{16}$ checks, so it is the solution.

23. $P = mn$

$\frac{P}{m} = n$ Dividing by m

24. $z = 3t + 3w$

$z - 3w = 3t$ Subtracting $3w$

$\frac{z - 3w}{3} = t,$ or   Dividing by 3

$\frac{z}{3} - w = t$

25. $N = \frac{r + s}{4}$

$4N = r + s$ Multiplying by 4

$4N - r = s$ Subtracting r

26. $T = 1.5\dfrac{A}{B}$

$BT = 1.5A$ Multiplying by B

$B = \dfrac{1.5A}{T},$ or $1.5\dfrac{A}{T}$

27.
$$H = \frac{2}{3}(t - 5)$$

$$\frac{3}{2}H = t - 5 \qquad \text{Multiplying by } \frac{3}{2}$$

$$\frac{3}{2}H + 5 = t, \text{ or} \qquad \text{Adding 5}$$

$$\frac{3H + 10}{2} = t$$

28.
$$f = g + ghm$$

$$f = g(1 + hm) \qquad \text{Factoring}$$

$$\frac{f}{1 + hm} = g \qquad \text{Dividing by } 1 + hm$$

29. Familiarize. Let f = number of female medical school graduates in 2002. Then an increase of 21.3% of this number is $f + 21.3\%f$, or $f + 0.213f$, or $1.213f$. This is the number of female medical school graduates in 2011.

Translate.

The number of female medical school graduates in 2011 was 8396.

$$1.213f = 8396$$

Solve. We solve the equation.
$$1.213f = 8396$$
$$f \approx 6922$$

Check. 21.3% of 6922 is $0.213(6922) \approx 1474$ and $6922 + 1474 = 8396$. The answer checks.

State. There were 6922 female medical school graduates in 2002.

30. Familiarize. Let c = the number of calories a 154-lb person would burn walking 3.5 mph for 30 min. Then 50 calories less than twice this number is $2c - 50$.

Translate. We know that $2c - 50$ represents 230 calories, so we have
$$2c - 50 = 230.$$

Solve. We solve the equation.
$$2c - 50 = 230$$
$$2c = 280$$
$$c = 140$$

Check. $2 \cdot 140 - 50 = 280 - 50 = 230$, so the answer checks.

State. A 154-lb person would burn 140 calories walking 3.5 mph for 30 min.

31. Familiarize. Let l = the length of the carpet, in feet. Then $l - 2$ = the width.

Translate. We substitute in the formula for the perimeter of a rectangle, $P = 2l + 2w$.
$$24 = 2l + 2(l - 2)$$

Solve.
$$24 = 2l + 2l - 4$$
$$24 = 4l - 4$$
$$28 = 4l$$
$$7 = l$$

If $l = 7$, then $l - 2 = 7 - 2 = 5$.

Check. The width, 5 ft, is 2 ft less than the length, 7 ft. The perimeter is $2 \cdot 7$ ft $+ 2 \cdot 5$ ft, or 14 ft $+ 10$ ft, or 24 ft. The answer checks.

State. The length of the carpet is 7 ft, and the width is 5 ft.

32. First we will find how long it will take Frederick to travel 18 mi downstream.

Familiarize. Let t = the time, in hours, it will take Frederick to travel 18 mi downstream. The speed of the boat traveling downstream is $9 + 3$, or 12 mph.

Translate. We will substitute in the formula $d = rt$.
$$18 = 12t$$

Solve.
$$18 = 12t$$
$$\frac{18}{12} = t$$
$$1.5 = t \qquad \text{Simplifying}$$

Check. At a speed of 12 mph, in 1.5 hr the boat travels $12(1.5)$, or 18 mi. The answer checks.

State. It will take Frederick 1.5 hr to travel 18 mi downstream.

Now we will find how long it will take Frederick to travel 18 mi upstream.

Familiarize. Let t = the time, in hours, it will take Frederick to travel 18 mi upstream. The speed of the boat traveling upstream is $9 - 3$, or 6 mph.

Translate. We will substitute in the formula $d = rt$.
$$18 = 6t$$

Solve.
$$18 = 6t$$
$$3 = t$$

Check. At a speed of 6 mph, in 3 hr the boat travels $6 \cdot 3$, or 18 mi. The answer checks.

State. It will take Frederick 3 hr to travel 18 mi upstream.

33. Equivalent expressions have the same value for all possible replacements. Any replacement that does not make any of the expressions undefined can be substituted for the variable. Equivalent equations have the same solution(s). True equations result only when a solution is substituted for the variable.

34. Answers may vary. A walker who knows how far and how long she walks each day wants to know her average speed each day.

35. Answers may vary. A decorator wants to have a carpet cut for a bedroom. The perimeter of the room is 54 ft and its length is 15 ft. How wide should the carpet be?

36. We can subtract by adding an opposite, so we can use the addition principle to subtract the same number on both sides of an equation. Similarly, we can divide by multiplying by a reciprocal, so we can use the multiplication principle to divide both sides of an equation by the same number.

37. The manner in which a guess or estimate is manipulated can give insight into the form of the equation to which the problem will be translated.

38. Labeling the variable clearly makes the Translate step more accurate. It also allows us to determine if the solution of the equation we translated to provides the information asked for in the original problem.

Exercise Set 1.4

RC1. (b)

RC3. (c)

RC5. (g)

1. $x - 2 \geq 6$

    -4 : We substitute and get $-4 - 2 \geq 6$, or $-6 \geq 6$, a false sentence. Therefore, -4 is not a solution.

    0 : We substitute and get $0 - 2 \geq 6$, or $-2 \geq 6$, a false sentence. Therefore, 0 is not a solution.

    4 : We substitute and get $4 - 2 \geq 6$, or $2 \geq 6$, a false sentence. Therefore, 4 is not a solution.

    8 : We substitute and get $8 - 2 \geq 6$, or $6 \geq 6$, a true sentence. Therefore, 8 is a solution.

3. $t - 8 > 2t - 3$

    0 : We substitute and get $0 - 8 > 2 \cdot 0 - 3$, or $-8 > -3$, a false sentence. Therefore, 0 is not a solution.

    -8 : We substitute and get $-8 - 8 > 2(-8) - 3$, or $-16 > -19$, a true sentence. Therefore, -8 is a solution.

    -9 : We substitute and get $-9 - 8 > 2(-9) - 3$, or $-17 > -21$, a true sentence. Therefore, -9 is a solution.

    -3 : We substitute and get $-3 - 8 > 2(-3) - 3$, or $-11 > -9$, a false sentence. Therefore, -3 is not a solution.

    $-\dfrac{7}{8}$: We substitute and get $-\dfrac{7}{8} - 8 > 2\left(-\dfrac{7}{8}\right) - 3$, or $-\dfrac{71}{8} > -\dfrac{38}{8}$, a false sentence. Therefore, $-\dfrac{7}{8}$ is not a solution.

5. Interval notation for $\{x | x < 5\}$ is $(-\infty, 5)$.

7. Interval notation for $\{x | -3 \leq x \leq 3\}$ is $[-3, 3]$.

9. $\{x | -4 > x > -8\} = \{x | -8 < x < -4\} = (-8, -4)$

11. Interval notation for the given graph is $(-2, 5)$.

13. Interval notation for the given graph is $(-\sqrt{2}, \infty)$.

15. $x + 2 > 1$

    $x + 2 - 2 > 1 - 2$ Subtracting 2

        $x > -1$

The solution set is $\{x | x > -1\}$, or $(-1, \infty)$.

17. $y + 3 < 9$

    $y + 3 - 3 < 9 - 3$ Subtracting 3

        $y < 6$

The solution set is $\{y | y < 6\}$, or $(-\infty, 6)$.

19. $a - 9 \leq -31$

    $a - 9 + 9 \leq -31 + 9$ Adding 9

        $a \leq -22$

The solution set is $\{a | a \leq -22\}$, or $(-\infty, -22]$.

21. $t + 13 \geq 9$

    $t + 13 - 13 \geq 9 - 13$ Subtracting 13

        $t \geq -4$

The solution set is $\{t | t \geq -4\}$, or $[-4, \infty)$.

23. $y - 8 > -14$

    $y - 8 + 8 > -14 + 8$ Adding 8

        $y > -6$

The solution set is $\{y | y > -6\}$, or $(-6, \infty)$.

25. $x - 11 \leq -2$

    $x - 11 + 11 \leq -2 + 11$ Adding 11

        $x \leq 9$

The solution set is $\{x | x \leq 9\}$, or $(-\infty, 9]$.

27. $8x \geq 24$

    $\dfrac{8x}{8} \geq \dfrac{24}{8}$ Dividing by 8

        $x \geq 3$

The solution set is $\{x | x \geq 3\}$, or $[3, \infty)$.

29. $0.3x < -18$

$\dfrac{0.3x}{0.3} < \dfrac{-18}{0.3}$ Dividing by 0.3

$x < -60$

The solution set is $\{x | x < -60\}$, or $(-\infty, -60)$.

31. $\dfrac{2}{3}x > 2$

$\dfrac{3}{2} \cdot \dfrac{2}{3}x > \dfrac{3}{2} \cdot 2$ Multiplying by $\dfrac{3}{2}$

$x > 3$

The solution set is $\{x | x > 3\}$, or $(3, \infty)$.

33. $-9x \geq -8.1$

$\dfrac{-9x}{-9} \leq \dfrac{-8.1}{-9}$ Dividing by -9 and reversing the inequality symbol

$x \leq 0.9$

The solution set is $\{x | x \leq 0.9\}$, or $(-\infty, 0.9]$.

35. $-\dfrac{3}{4}x \geq -\dfrac{5}{8}$

$-\dfrac{4}{3}\left(-\dfrac{3}{4}x\right) \leq -\dfrac{4}{3}\left(-\dfrac{5}{8}\right)$ Multiplying by $-\dfrac{4}{3}$ and reversing the inequality symbol

$x \leq \dfrac{20}{24}$

$x \leq \dfrac{5}{6}$

The solution set is $\left\{x \middle| x \leq \dfrac{5}{6}\right\}$, or $\left(-\infty, \dfrac{5}{6}\right]$.

37. $2x + 7 < 19$

$2x + 7 - 7 < 19 - 7$ Subtracting 7

$2x < 12$

$\dfrac{2x}{2} < \dfrac{12}{2}$ Dividing by 2

$x < 6$

The solution set is $\{x | x < 6\}$, or $(-\infty, 6)$.

39. $5y + 2y \leq -21$

$7y \leq -21$ Collecting like terms

$\dfrac{7y}{7} \leq \dfrac{-21}{7}$ Dividing by 7

$y \leq -3$

The solution set is $\{y | y \leq -3\}$, or $(-\infty, -3]$.

41. $2y - 7 < 5y - 9$

$-5y + 2y - 7 < -5y + 5y - 9$ Adding $-5y$

$-3y - 7 < -9$

$-3y - 7 + 7 < -9 + 7$ Adding 7

$-3y < -2$

$\dfrac{-3y}{-3} > \dfrac{-2}{-3}$ Dividing by -3 and reversing the inequality symbol

$y > \dfrac{2}{3}$

The solution set is $\left\{y \middle| y > \dfrac{2}{3}\right\}$, or $\left(\dfrac{2}{3}, \infty\right)$.

43. $0.4x + 5 \leq 1.2x - 4$

$-1.2x + 0.4x + 5 \leq -1.2x + 1.2x - 4$ Adding $-1.2x$

$-0.8x + 5 \leq -4$

$-0.8x + 5 - 5 \leq -4 - 5$ Subtracting 5

$-0.8x \leq -9$

$\dfrac{-0.8x}{-0.8} \geq \dfrac{-9}{-0.8}$ Dividing by -0.8 and reversing the inequality symbol

$x \geq 11.25$

The solution set is $\{x | x \geq 11.25\}$, or $[11.25, \infty)$.

45. $5x - \dfrac{1}{12} \leq \dfrac{5}{12} + 4x$

$12\left(5x - \dfrac{1}{12}\right) \leq 12\left(\dfrac{5}{12} + 4x\right)$ Clearing fractions

$60x - 1 \leq 5 + 48x$

$60x - 1 - 48x \leq 5 + 48x - 48x$ Subtracting $48x$

$12x - 1 \leq 5$

$12x - 1 + 1 \leq 5 + 1$ Adding 1

$12x \leq 6$

$\dfrac{12x}{12} \leq \dfrac{6}{12}$ Dividing by 12

$x \leq \dfrac{1}{2}$

The solution set is $\left\{x \middle| x \leq \dfrac{1}{2}\right\}$, or $\left(-\infty, \dfrac{1}{2}\right]$.

47. $4(4y - 3) \geq 9(2y + 7)$

$16y - 12 \geq 18y + 63$ Removing parentheses

$16y - 12 - 18y \geq 18y + 63 - 18y$ Subtracting $18y$

$-2y - 12 \geq 63$

$-2y - 12 + 12 \geq 63 + 12$ Adding 12

$-2y \geq 75$

$\dfrac{-2y}{-2} \leq \dfrac{75}{-2}$ Dividing by -2 and reversing the inequality symbol

$y \leq -\dfrac{75}{2}$

The solution set is $\left\{y \middle| y \leq -\dfrac{75}{2}\right\}$, or $\left(-\infty, -\dfrac{75}{2}\right]$.

49. $3(2 - 5x) + 2x < 2(4 + 2x)$

$6 - 15x + 2x < 8 + 4x$

$\quad 6 - 13x < 8 + 4x \qquad$ Collecting like terms

$\quad 6 - 17x < 8 \qquad$ Subtracting $4x$

$\quad -17x < 2 \qquad$ Subtracting 6

$\qquad x > -\dfrac{2}{17} \qquad$ Dividing by -17 and reversing the inequality symbol

The solution set is $\left\{x \middle| x > -\dfrac{2}{17}\right\}$, or $\left(-\dfrac{2}{17}, \infty\right)$.

51. $5[3m - (m + 4)] > -2(m - 4)$

$5(3m - m - 4) > -2(m - 4)$

$5(2m - 4) > -2(m - 4)$

$10m - 20 > -2m + 8$

$12m - 20 > 8 \qquad$ Adding $2m$

$12m > 28 \qquad$ Adding 20

$m > \dfrac{28}{12}$

$m > \dfrac{7}{3}$

The solution set is $\left\{m \middle| m > \dfrac{7}{3}\right\}$, or $\left(\dfrac{7}{3}, \infty\right)$.

53. $3(r - 6) + 2 > 4(r + 2) - 21$

$3r - 18 + 2 > 4r + 8 - 21$

$3r - 16 > 4r - 13 \qquad$ Collecting like terms

$-r - 16 > -13 \qquad$ Subtracting $4r$

$-r > 3 \qquad$ Adding 16

$r < -3 \qquad$ Multiplying by -1 and reversing the inequality symbol

The solution set is $\{r | r < -3\}$, or $(-\infty, -3)$.

55. $19 - (2x + 3) \le 2(x + 3) + x$

$19 - 2x - 3 \le 2x + 6 + x$

$16 - 2x \le 3x + 6 \qquad$ Collecting like terms

$16 - 5x \le 6 \qquad$ Subtracting $3x$

$-5x \le -10 \qquad$ Subtracting 16

$x \ge 2 \qquad$ Dividing by -5 and reversing the inequality symbol

The solution set is $\{x | x \ge 2\}$, or $[2, \infty)$.

57. $\dfrac{1}{4}(8y + 4) - 17 < -\dfrac{1}{2}(4y - 8)$

$2y + 1 - 17 < -2y + 4$

$2y - 16 < -2y + 4 \qquad$ Collecting like terms

$4y - 16 < 4 \qquad$ Adding $2y$

$4y < 20 \qquad$ Adding 16

$y < 5$

The solution set is $\{y | y < 5\}$, or $(-\infty, 5)$.

59. $2[4 - 2(3 - x)] - 1 \ge 4[2(4x - 3) + 7] - 25$

$2[4 - 6 + 2x] - 1 \ge 4[8x - 6 + 7] - 25$

$2[-2 + 2x] - 1 \ge 4[8x + 1] - 25$

$-4 + 4x - 1 \ge 32x + 4 - 25$

$4x - 5 \ge 32x - 21$

$-28x - 5 \ge -21$

$-28x \ge -16$

$x \le \dfrac{-16}{-28} \qquad$ Dividing by -28 and reversing the inequality symbol

$x \le \dfrac{4}{7}$

The solution set is $\left\{x \middle| x \le \dfrac{4}{7}\right\}$, or $\left(-\infty, \dfrac{4}{7}\right]$.

61. $\dfrac{4}{5}(7x - 6) < 40$

$5 \cdot \dfrac{4}{5}(7x - 6) < 5 \cdot 40 \qquad$ Clearing the fraction

$4(7x - 6) < 200$

$28x - 24 < 200$

$28x < 224$

$x < 8$

The solution set is $\{x | x < 8\}$, or $(-\infty, 8)$.

63. $\dfrac{3}{4}(3 + 2x) + 1 \ge 13$

$4\left[\dfrac{3}{4}(3 + 2x) + 1\right] \ge 4 \cdot 13 \qquad$ Clearing the fraction

$3(3 + 2x) + 4 \ge 52$

$9 + 6x + 4 \ge 52$

$6x + 13 \ge 52$

$6x \ge 39$

$x \ge \dfrac{39}{6}$, or $\dfrac{13}{2}$

The solution set is $\left\{x \middle| x \ge \dfrac{13}{2}\right\}$, or $\left[\dfrac{13}{2}, \infty\right)$.

65. $\dfrac{3}{4}\left(3x - \dfrac{1}{2}\right) - \dfrac{2}{3} < \dfrac{1}{3}$

$\dfrac{9x}{4} - \dfrac{3}{8} - \dfrac{2}{3} < \dfrac{1}{3}$

$24\left(\dfrac{9x}{4} - \dfrac{3}{8} - \dfrac{2}{3}\right) < 24 \cdot \dfrac{1}{3} \qquad$ Clearing fractions

$54x - 9 - 16 < 8$

$54x - 25 < 8$

$54x < 33$

$x < \dfrac{33}{54}$, or $\dfrac{11}{18}$

The solution set is $\left\{x \middle| x < \dfrac{11}{18}\right\}$, or $\left(-\infty, \dfrac{11}{18}\right)$.

67.
$$0.7(3x + 6) \geq 1.1 - (x + 2)$$
$$10[0.7(3x + 6)] \geq 10[1.1 - (x + 2)] \quad \text{Clearing decimals}$$
$$7(3x + 6) \geq 11 - 10(x + 2)$$
$$21x + 42 \geq 11 - 10x - 20$$
$$21x + 42 \geq -9 - 10x$$
$$31x + 42 \geq -9$$
$$31x \geq -51$$
$$x \geq -\frac{51}{31}$$

The solution set is $\left\{x \middle| x \geq -\dfrac{51}{31}\right\}$, or $\left[-\dfrac{51}{31}, \infty\right)$.

69.
$$a + (a - 3) \leq (a + 2) - (a + 1)$$
$$a + a - 3 \leq a + 2 - a - 1$$
$$2a - 3 \leq 1$$
$$2a \leq 4$$
$$a \leq 2$$

The solution set is $\{a | a \leq 2\}$, or $(-\infty, 2]$.

71. *Familiarize*. We will use the formula $I = \dfrac{703W}{H^2}$. Recall that $H = 62$ in.

***Translate*.**
$$I < 25, \text{ or } \frac{703W}{H^2} < 25$$

We replace H with 62.
$$\frac{703W}{62^2} < 25$$

***Solve*.** We solve the inequality.
$$\frac{703W}{62^2} < 25$$
$$\frac{703W}{3844} < 25$$
$$703W < 96,100$$
$$W < 136.7 \qquad \text{Rounding}$$

***Check*.** As a partial check we can substitute a value of W less than 136.7 and a value greater than 136.7 in the formula.

For $W = 136$: $I = \dfrac{703(136)}{62^2} \approx 24.9$

For $W = 137$: $I = \dfrac{703(137)}{62^2} \approx 25.1$

Since a value of W less than 136.7 gives a body mass index less than 25 and a value of W greater than 136.7 gives an index greater than 25, we have a partial check.

***State*.** Weights of approximately 136.7 lb or less will keep Alexandra's body mass index below 25. In terms of an inequality we write $\{W | W < \text{ (approximately) } 136.7 \text{ lb}\}$.

73. *Familiarize*. List the information in a table. Let $x =$ the score on the fourth test.

Test	Score
Test 1	89
Test 2	92
Test 3	95
Test 4	x
Total	360 or more

***Translate*.** We can easily get an inequality from the table.
$$89 + 92 + 95 + x \geq 360$$

***Solve*.**
$$276 + x \geq 360 \qquad \text{Collecting like terms}$$
$$x \geq 84 \qquad \text{Adding } -276$$

***Check*.** If you get 84 on the fourth test, your total score will be $89 + 92 + 95 + 84$, or 360. Any higher score will also give you an A.

***State*.** A score of 84 or better will give David an A. In terms of an inequality we write $\{x | x \geq 84\}$.

75. *Familiarize*. Let $v =$ the blue book value of the car. Since the car was not replaced, we know that \$9200 does not exceed 80% of the blue book value.

***Translate*.** We write an inequality stating that \$9200 does not exceed 80% of the blue book value.
$$9200 \leq 0.8v$$

***Solve*.**
$$9200 \leq 0.8v$$
$$11,500 < v \qquad \text{Multiplying by } \frac{1}{0.8}$$

***Check*.** We can do a partial check by substituting a value for v greater than 11,500. When $v = 11,525$, then 80% of v is $0.8(11,525)$, or \$9220. This is greater than \$9200; that is, \$9200 does not exceed this amount. We cannot check all possible values for v, so we stop here.

***State*.** The blue book value of the car is \$11,500 or more. In terms of an inequality we write $\{v | v \geq \$11,500\}$.

77. *Familiarize*. We make a table of information.

Plan A: Monthly Income	Plan B: Monthly Income
\$400 salary	\$610 salary
8% of sales	5% of sales
Total: $400 + 8\%$ of sales	Total: $610 + 5\%$ of sales

***Translate*.** We write an inequality stating that the income from Plan A is greater than the income from Plan B. We let $S =$ gross sales.
$$400 + 8\%S > 610 + 5\%S$$

***Solve*.**
$$400 + 0.08S > 610 + 0.05S$$
$$400 + 0.03S > 610$$
$$0.03S > 210$$
$$S > 7000$$

Check. We calculate for $S = \$7000$ and for some amount greater than \$7000 and some amount less than \$7000.

Plan A: Plan B:

$400 + 8\%(7000)$ $610 + 5\%(7000)$

$400 + 0.08(7000)$ $610 + 0.05(7000)$

$400 + 560$ $610 + 350$

$\$960$ $\$960$

When $S = \$7000$, the income from Plan A is equal to the income from Plan B.

Plan A: Plan B:

$400 + 8\%(8000)$ $610 + 5\%(8000)$

$400 + 0.08(8000)$ $610 + 0.05(8000)$

$400 + 640$ $610 + 400$

$\$1040$ $\$1010$

When $S = \$8000$, the income from Plan A is greater than the income from Plan B.

Plan A: Plan B:

$400 + 8\%(6000)$ $610 + 5\%(6000)$

$400 + 0.08(6000)$ $610 + 0.05(6000)$

$400 + 480$ $610 + 300$

$\$880$ $\$910$

When $S = \$6000$, the income from Plan A is less than the income from Plan B.

State. Plan A is better than Plan B when gross sales are greater than \$7000. In terms of an inequality we write $\{S | S > \$7000\}$.

79. **Familiarize**. Let $c =$ the amount of prescription costs for which plan 2 will save James money. With plan 1 he pays $\$150 + 30\%(c - \$150)$, and with plan 2 he pays $\$280 + 10\%(c - \$280)$.

Translate. We write an inequality stating that the amount paid using plan 2 is less than the amount paid using plan 1.

$$280 + 10\%(c - 280) < 150 + 30\%(c - 150)$$

Solve.
$$280 + 0.1(c - 280) < 150 + 0.3(c - 150)$$
$$280 + 0.1c - 28 < 150 + 0.3c - 45$$
$$252 + 0.1c < 105 + 0.3c$$
$$252 < 105 + 0.2c$$
$$147 < 0.2c$$
$$735 < c$$

Check. When we calculate the amount paid using each plan for $c = 735$, for some value of c less than 735, and for some value of p greater than 735, we find the following:

For $c = 735$, James pays the same amount under both plans.

For a value of c less than 735, James pays less using plan 1.

For a value of c greater than 735, James pays less using plan 2.

This gives us a partial check.

State. Plan 2 will save James money. In terms of an inequality we have $\{c | c > \$735\}$.

81. **Familiarize**. Let $p =$ the number of guests at the wedding party. Then the number of guests in excess of 25 is $p - 25$. The cost under plan A is $30p$, and the cost under plan B is $1300 + 20(p - 25)$.

Translate. We write an inequality stating that plan B costs less than plan A.

$$1300 + 20(p - 25) < 30p$$

Solve. We solve the inequality.
$$1300 + 20(p - 25) < 30p$$
$$1300 + 20p - 500 < 30p$$
$$800 + 20p < 30p$$
$$800 < 10p$$
$$80 < p$$

Check. We calculate for $p = 80$ and for some number less than 80 and some number greater than 80.

Plan A: Plan B:

$30 \cdot 80$ $1300 + 20(80 - 25)$

$\$2400$ $\$2400$

When 80 people attend, plan B costs the same as plan A.

Plan A: Plan B:

$30 \cdot 79$ $1300 + 20(79 - 25)$

$\$2370$ $\$2380$

When fewer than 80 people attend, plan B costs more than plan A.

Plan A: Plan B:

$30 \cdot 81$ $1300 + 20(81 - 25)$

$\$2430$ $\$2420$

When more than 80 people attend, plan B costs less than plan A.

State. For parties of more than 80 people, plan B will cost less. In terms of an inequality we write $\{p | p > 80\}$.

83. **Familiarize and Translate**. We use the formula $R = 2(s + 70)$ and write an inequality stating that R is less than 2100.

$$2(s + 70) < 2100$$

Solve.
$$2(s + 70) < 2100$$
$$2s + 140 < 2100$$
$$2s < 1960$$
$$s < 980$$

Check. We find R for a value of s less than 980 and for a value of s greater than 980.

When $s = 979$, $R = 2(979 + 70) = 2098 < 2100$.

When $s = 981$, $R = 2(981 + 70) = 2102 > 2100$.

We have a partial check.

State. The rent will be less than \$2100 for a square footage of less than 980 ft^2. In terms of an inequality we have $\{s | s < 980 \text{ ft}^2\}$.

85. a) In 2010, $t = 2010 - 2005 = 5$.

$$C = 82 \cdot 5 + 1923 = 410 + 1923 = 2333$$

The average cost of tuition and fees was \$2333 in 2010.

In 2014, $t = 2014 - 2005 = 9$.

$$C = 82 \cdot 9 + 1923 = 738 + 1923 = 2661$$

The average cost of tuition and fees was \$2661 in 2014.

b) *Familiarize and Translate.* We use the formula $C = 82t + 1923$ and write an inequality stating that $C > 3000$.

$$82t + 1923 > 3000$$

Solve.

$$82t + 1923 > 3000$$
$$82t > 1077$$
$$t > 13.13 \quad \text{Rounding}$$

Check. We find C for a value of t less than 13.13 and for a value of t greater than 13.13.

When $t = 13$, $C = 82 \cdot 13 + 1923 = 2989 < 3000$.

When $t = 14$, $C = 82 \cdot 14 + 1923 = 3071 > 3000$.

We have a partial check.

State. The cost of tuition and fees will be more than \$3000 for years that are more than 13.13 years since 2005. In terms of an inequality we have $\{t | t > 13.13\}$.

87. $3a - 6(2a - 5b) = 3a - 12a + 30b$
$$= -9a + 30b$$

89. $\quad 4(a - 2b) - 6(2a - 5b)$
$$= 4a - 8b - 12a + 30b$$
$$= -8a + 22b$$

91. $30x - 70y - 40 = 10 \cdot 3x - 10 \cdot 7y - 10 \cdot 4$
$$= 10(3x - 7y - 4)$$

93. $-8x + 24y - 4 = -4 \cdot 2x - 4(-6y) - 4 \cdot 1$
$$= -4(2x - 6y + 1)$$

95. $-2.3 - 8.9 = -2.3 + (-8.9) = -11.2$

97. $-2.3 + (-8.9) = -11.2$

99. a) *Familiarize.* We will use

$$S = 460 + 94p \quad \text{and} \quad D = 2000 - 60p.$$

Translate. Supply is to exceed demand, so we have

$$S > D, \text{ or}$$
$$460 + 94p > 2000 - 60p.$$

Solve. We solve the inequality.

$$460 + 94p > 2000 - 60p$$

$460 + 154p > 2000$	Adding $60p$
$154p > 1540$	Subtracting 460
$p > 10$	Dividing by 154

Check. We calculate for $p = 10$, for some value of p less than 10, and for some value of p greater than 10.

For $p = 10$: $\quad S = 460 + 94 \cdot 10 = 1400$
$$D = 2000 - 60 \cdot 10 = 1400$$

For $p = 9$: $\quad S = 460 + 94 \cdot 9 = 1306$
$$D = 2000 - 60 \cdot 9 = 1460$$

For $p = 11$: $\quad S = 460 + 94 \cdot 11 = 1494$
$$D = 2000 - 60 \cdot 11 = 1340$$

For a value of p greater than 10, supply exceeds demand. We cannot check all possible values of p, so we stop here.

State. Supply exceeds demand for values of p greater than 10. In terms of an inequality we write $\{p | p > 10\}$.

b) We have seen in part (a) that $D = S$ for $p = 10$, $S < D$ for a value of p less than 10, and $S > D$ for a value of p greater than 10. Since we cannot check all possible values of p, we stop here. Supply is less than demand for values of p less than 10. In terms of an inequality we write $\{p | p < 10\}$.

101. True

103. $\quad x + 5 \leq 5 + x$

$\quad\quad 5 \leq 5 \quad$ Subtracting x

We get a true inequality, so all real numbers are solutions.

105. $x^2 + 1 > 0$

$x^2 \geq 0$ for all real numbers, so $x^2 + 1 \geq 1 > 0$ for all real numbers.

Exercise Set 1.5

RC1. True; see page 127 in the text.

RC3. True; a number cannot be less than -4 *and* greater than 4.

1. $\{9, 10, 11\} \cap \{9, 11, 13\}$

The numbers 9 and 11 are common to the two sets, so the intersection is $\{9, 11\}$.

3. $\{a, b, c, d\} \cap \{b, f, g\}$

Only the letter b is common to the two sets. The intersection is $\{b\}$.

5. $\{9, 10, 11\} \cup \{9, 11, 13\}$

The numbers in either or both sets are 9, 10, 11, and 13, so the union is $\{9, 10, 11, 13\}$.

7. $\{a, b, c, d\} \cup \{b, f, g\}$

The letters in either or both sets are a, b, c, d, f, and g, so the union is $\{a, b, c, d, f, g\}$.

9. $\{2, 5, 7, 9\} \cap \{1, 3, 4\}$

There are no numbers common to the two sets. The intersection is the empty set, $\emptyset$.

11. $\{3, 5, 7\} \cup \emptyset$

The numbers in either or both sets are 3, 5, and 7, so the union is $\{3, 5, 7\}$.

13. $-4 < a$ *and* $a \leq 1$ can be written $-4 < a \leq 1$. In interval notation we have $(-4, 1]$.

The graph is the intersection of the graphs of $a > -4$ and $a \leq 1$.

15. We can write $1 < x < 6$ in interval notation as $(1, 6)$.

The graph is the intersection of the graphs of $x > 1$ and $x < 6$.

17. $-10 \leq 3x + 2$ *and* $3x + 2 < 17$

$\qquad -12 \leq 3x \qquad$ *and* $\qquad 3x < 15$

$\qquad -4 \leq x \qquad$ *and* $\qquad x < 5$

The solution set is the intersection of the solution sets of the individual inequalities. The numbers common to both sets are those that are greater than or equal to -4 *and* less than 5. Thus the solution set is $\{x| -4 \leq x < 5\}$, or $[-4, 5)$.

19. $3x + 7 \geq 4 \qquad$ *and* $\qquad 2x - 5 \geq -1$

$\qquad 3x \geq -3 \quad$ *and* $\qquad 2x \geq 4$

$\qquad x \geq -1 \quad$ *and* $\qquad x \geq 2$

The solution set is $\{x| x \geq -1\} \cap \{x| x \geq 2\} = \{x| x \geq 2\}$, or $[2, \infty)$.

21. $4 - 3x \geq 10 \quad$ *and* $\quad 5x - 2 > 13$

$\qquad -3x \geq 6 \quad$ *and* $\qquad 5x > 15$

$\qquad x \leq -2 \quad$ *and* $\qquad x > 3$

The solution set is $\{x| x \leq -2\} \cap \{x| x > 3\} = \emptyset$.

23. $\qquad -4 < x + 4 < 10$

$-4 - 4 < x + 4 - 4 < 10 - 4 \qquad$ Subtracting 4

$\qquad -8 < x < 6$

The solution set is $\{x| -8 < x < 6\}$, or $(-8, 6)$.

25. $\qquad 6 > -x \geq -2$

$\qquad -6 < x \leq 2 \qquad$ Multiplying by -1

The solution set is $\{x| -6 < x \leq 2\}$, or $(-6, 2]$.

27. $\qquad 2 < x + 3 \leq 9$

$2 - 3 < x + 3 - 3 \leq 9 - 3 \quad$ Subtracting 3

$\qquad -1 < x \leq 6$

The solution set is $\{x| -1 < x \leq 6\}$, or $(-1, 6]$.

29. $\qquad 1 < 3y + 4 \leq 19$

$1 - 4 < 3y + 4 - 4 \leq 19 - 4 \quad$ Subtracting 4

$\qquad -3 < 3y \leq 15$

$\qquad \dfrac{-3}{3} < \dfrac{3y}{3} \leq \dfrac{15}{3} \quad$ Dividing by 3

$\qquad -1 < y \leq 5$

The solution set is $\{y| -1 < y \leq 5\}$, or $(-1, 5]$.

31. $\qquad -10 \leq 3x - 5 \leq -1$

$-10 + 5 \leq 3x - 5 + 5 \leq -1 + 5 \quad$ Adding 5

$\qquad -5 \leq 3x \leq 4$

$\qquad \dfrac{-5}{3} \leq \dfrac{3x}{3} \leq \dfrac{4}{3} \quad$ Dividing by 3

$\qquad -\dfrac{5}{3} \leq x \leq \dfrac{4}{3}$

The solution set is $\left\{x \left| -\dfrac{5}{3} \leq x \leq \dfrac{4}{3}\right.\right\}$, or $\left[-\dfrac{5}{3}, \dfrac{4}{3}\right]$.

33. $\qquad -18 \leq -2x - 7 < 0$

$-18 + 7 \leq -2x - 7 + 7 < 0 + 7$

$\qquad -11 \leq -2x < 7$

$\qquad \dfrac{-11}{-2} \geq \dfrac{-2x}{-2} > \dfrac{7}{2}$

$\qquad \dfrac{11}{2} \geq x > -\dfrac{7}{2}$

The solution set is $\left\{x \left| -\dfrac{7}{2} < x \leq \dfrac{11}{2}\right.\right\}$, or $\left(-\dfrac{7}{2}, \dfrac{11}{2}\right]$.

35. $\qquad -\dfrac{1}{2} < \dfrac{1}{4}x - 3 \leq \dfrac{1}{2}$

$\qquad -\dfrac{1}{2} + 3 < \dfrac{1}{4}x - 3 + 3 \leq \dfrac{1}{2} + 3$

$\qquad \dfrac{5}{2} < \dfrac{1}{4}x \leq \dfrac{7}{2}$

$\qquad 4 \cdot \dfrac{5}{2} < 4 \cdot \dfrac{1}{4}x \leq 4 \cdot \dfrac{7}{2}$

$\qquad 10 < x \leq 14$

The solution set is $\{x| 10 < x \leq 14\}$, or $(10, 14]$.

37. $\qquad -4 \leq \dfrac{7 - 3x}{5} \leq 4$

$\qquad 5(-4) \leq 5 \cdot \dfrac{7 - 3x}{5} \leq 5 \cdot 4$

$\qquad -20 \leq 7 - 3x \leq 20$

$-20 - 7 \leq 7 - 3x - 7 \leq 20 - 7$

$\qquad -27 \leq -3x \leq 13$

$\qquad \dfrac{-27}{-3} \geq x \geq \dfrac{13}{-3}$

$\qquad 9 \geq x \geq -\dfrac{13}{3}$

The solution set is $\left\{x \left| -\dfrac{13}{3} \leq x \leq 9\right.\right\}$, or $\left[-\dfrac{13}{3}, 9\right]$.

39. $x < -2$ *or* $x > 1$ can be written in interval notation as $(-\infty, -2) \cup (1, \infty)$.

The graph is the union of the graphs of $x < -2$ and $x > 1$.

41. $x \le -3$ or $x > 1$ can be written in interval notation as $(-\infty, -3] \cup (1, \infty)$.

The graph is the union of the graphs of $x \le -3$ and $x > 1$.

43.
$$x + 3 < -2 \quad or \quad x + 3 > 2$$
$$x + 3 - 3 < -2 - 3 \quad or \quad x + 3 - 3 > 2 - 3$$
$$x < -5 \quad or \quad x > -1$$

The solution set is $\{x | x < -5 \text{ or } x > -1\}$, or $(-\infty, -5) \cup (-1, \infty)$.

The graph of the solution set is shown below.

45.
$$2x - 8 \le -3 \quad or \quad x - 1 \ge 3$$
$$2x - 8 + 8 \le -3 + 8 \quad or \quad x - 1 + 1 \ge 3 + 1$$
$$2x \le 5 \quad or \quad x \ge 4$$
$$\frac{2x}{2} \le \frac{5}{2} \quad or \quad x \ge 4$$
$$x \le \frac{5}{2} \quad or \quad x \ge 4$$

The solution set is $\left\{ x \middle| x \le \frac{5}{2} \text{ or } x \ge 4 \right\}$, or $\left(-\infty, \frac{5}{2} \right] \cup [4, \infty)$.

The graph of the solution set is shown below.

47. $7x + 4 \ge -17$ or $6x + 5 \ge -7$
$$7x \ge -21 \quad or \quad 6x \ge -12$$
$$x \ge -3 \quad or \quad x \ge -2$$

The solution set is $\{x | x \ge -3\}$, or $[-3, \infty)$.

The graph of the solution set is shown below.

49.
$$7 > -4x + 5 \quad or \quad 10 \le -4x + 5$$
$$7 - 5 > -4x + 5 - 5 \quad or \quad 10 - 5 \le -4x + 5 - 5$$
$$2 > -4x \quad or \quad 5 \le -4x$$
$$\frac{2}{-4} < \frac{-4x}{-4} \quad or \quad \frac{5}{-4} \ge \frac{-4x}{-4}$$
$$-\frac{1}{2} < x \quad or \quad -\frac{5}{4} \ge x$$

The solution set is $\left\{ x \middle| x \le -\frac{5}{4} \text{ or } x > -\frac{1}{2} \right\}$, or $\left(-\infty, -\frac{5}{4} \right] \cup \left(-\frac{1}{2}, \infty \right)$.

51. $3x - 7 > -10$ or $5x + 2 \le 22$
$$3x > -3 \quad or \quad 5x \le 20$$
$$x > -1 \quad or \quad x \le 4$$

All real numbers are solutions. In interval notation, the solution set is $(-\infty, \infty)$.

53.
$$-2x - 2 < -6 \quad or \quad -2x - 2 > 6$$
$$-2x - 2 + 2 < -6 + 2 \quad or \quad -2x - 2 + 2 > 6 + 2$$
$$-2x < -4 \quad or \quad -2x > 8$$
$$\frac{-2x}{-2} > \frac{-4}{-2} \quad or \quad \frac{-2x}{-2} < \frac{8}{-2}$$
$$x > 2 \quad or \quad x < -4$$

The solution set is $\{x | x < -4 \text{ or } x > 2\}$, or $(-\infty, -4) \cup (2, \infty)$.

55.
$$\frac{2}{3}x - 14 < -\frac{5}{6} \quad or \quad \frac{2}{3}x - 14 > \frac{5}{6}$$
$$6\left(\frac{2}{3}x - 14 \right) < 6\left(-\frac{5}{6} \right) \quad or \quad 6\left(\frac{2}{3}x - 14 \right) > 6 \cdot \frac{5}{6}$$
$$4x - 84 < -5 \quad or \quad 4x - 84 > 5$$
$$4x - 84 + 84 < -5 + 84 \quad or \quad 4x - 84 + 84 > 5 + 84$$
$$4x < 79 \quad or \quad 4x > 89$$
$$\frac{4x}{4} < \frac{79}{4} \quad or \quad \frac{4x}{4} > \frac{89}{4}$$
$$x < \frac{79}{4} \quad or \quad x > \frac{89}{4}$$

The solution set is $\left\{ x \middle| x < \frac{79}{4} \text{ or } x > \frac{89}{4} \right\}$, or $\left(-\infty, \frac{79}{4} \right) \cup \left(\frac{89}{4}, \infty \right)$.

57.
$$\frac{2x - 5}{6} \le -3 \quad or \quad \frac{2x - 5}{6} \ge 4$$
$$6\left(\frac{2x - 5}{6} \right) \le 6(-3) \quad or \quad 6\left(\frac{2x - 5}{6} \right) \ge 6 \cdot 4$$
$$2x - 5 \le -18 \quad or \quad 2x - 5 \ge 24$$
$$2x - 5 + 5 \le -18 + 5 \quad or \quad 2x - 5 + 5 \ge 24 + 5$$
$$2x \le -13 \quad or \quad 2x \ge 29$$
$$\frac{2x}{2} \le \frac{-13}{2} \quad or \quad \frac{2x}{2} \ge \frac{29}{2}$$
$$x \le -\frac{13}{2} \quad or \quad x \ge \frac{29}{2}$$

The solution set is $\left\{ x \middle| x \le -\frac{13}{2} \text{ or } x \ge \frac{29}{2} \right\}$, or $\left(-\infty, -\frac{13}{2} \right] \cup \left[\frac{29}{2}, \infty \right)$.

59. *Familiarize*. We will use the formula $P = 1 + \frac{d}{33}$.

Translate. We want to find those values of P for which
$$1 \le P \le 7$$
or
$$1 \le 1 + \frac{d}{33} \le 7.$$

Solve. We solve the inequality.

$$1 \le 1 + \frac{d}{33} \le 7$$

$$0 \le \frac{d}{33} \le 6$$

$$0 \le d \le 198$$

Check. We could do a partial check by substituting some values for d in the formula. The result checks.

State. The pressure is at least 1 atm and at most 7 atm for depths d in the set $\{d | 0 \text{ ft} \le d \le 198 \text{ ft}\}$.

61. *Familiarize*. Let b = the number of beats per minute. Note that $10 \text{ sec} = 10 \text{ sec} \times \frac{1 \text{ min}}{60 \text{ sec}} = \frac{10}{60} \times \frac{\text{sec}}{\text{sec}} \times 1 \text{ min} = \frac{1}{6} \text{ min}$. Then in 10 sec, or $\frac{1}{6}$ min, the woman should have between $\frac{1}{6} \cdot 138$ and $\frac{1}{6} \cdot 162$ beats.

Translate. We want to find the value of b for which

$$\frac{1}{6} \cdot 138 < b < \frac{1}{6} \cdot 162$$

Solve. We solve the inequality.

$$\frac{1}{6} \cdot 138 < b < \frac{1}{6} \cdot 162$$

$$23 < b < 27$$

Check. If the number of beats in 10 sec, or $\frac{1}{6}$ min, is between 23 and 27, then the number of beats per minute is between $6 \cdot 23$ and $6 \cdot 27$, or between 138 and 162. The answer checks.

State. The number of beats should be between 23 and 27.

63. *Familiarize*. We will use the formula $I = \frac{703W}{H^2}$, where $H = 62$. That is, $I = \frac{703W}{62^2}$.

Translate. We want to find those values of W for which

$$18.5 < I < 24.9$$

or

$$18.5 < \frac{703W}{62^2} < 24.9.$$

Solve. We solve the inequality.

$$18.5 < \frac{703W}{62^2} < 24.9$$

$$18.5 < \frac{703W}{3844} < 24.9$$

$$71,114 < 703W < 95,715.6$$

$$101.2 < W < 136.2 \qquad \text{Rounding}$$

Check. We could do a partial check by substituting some values for W in the formula. The result checks.

State. Weights between approximately 101.2 lb and 136.2 lb will allow Alexandra to keep her body mass index between 18.5 and 24.9. These weights are the values of W in the set $\{W | 101.2 \text{ lb} < W < 136.2 \text{ lb}\}$.

65. *Familiarize*. We will use the formula $c = \frac{ad}{a + 12}$, where $a = 8$. That is, $c = \frac{8d}{8 + 12}$, or $c = \frac{8d}{20} = \frac{2d}{5}$.

Translate. We want to find the values of d for which

$$100 < c < 200$$

or

$$100 < \frac{2d}{5} < 200.$$

Solve. We solve the inequality.

$$100 < \frac{2d}{5} < 200$$

$$500 < 2d < 1000$$

$$250 < d < 500$$

Check. We could do a partial check by substituting some values for d in the formula. The result checks.

State. The equivalent adult dosage is between 250 mg and 500 mg. The dosages are the values of d in the set $\{d | 250 \text{ mg} < d < 500 \text{ mg}\}$.

67.
$$8y - 3 = 3 + 8y$$
$$-3 = 3 \qquad \text{Subtracting } 8y$$

We get a false equation. Thus, the equation has no solution.

69.
$$20 = 4(3y - 7)$$
$$20 = 12y - 28$$
$$48 = 12y$$
$$4 = y$$

The solution is 4.

71.
$$-3 + 2x = 2x - 3$$
$$-3 = -3 \qquad \text{Subtracting } 2x$$

We get a true equation. Thus, all real numbers are solutions.

73.
$$x - 10 < 5x + 6 \le x + 10$$
$$-10 < 4x + 6 \le 10 \qquad \text{Subtracting } x$$
$$-16 < 4x \le 4$$
$$-4 < x \le 1$$

The solution set is $\{x | -4 < x \le 1\}$, or $(-4, 1]$.

75.
$$-\frac{2}{15} \le \frac{2}{3}x - \frac{2}{5} \le \frac{2}{15}$$

$$-\frac{2}{15} \le \frac{2}{3}x - \frac{6}{15} \le \frac{2}{15}$$

$$\frac{4}{15} \le \frac{2}{3}x \le \frac{8}{15}$$

$$\frac{3}{2} \cdot \frac{4}{15} \le \frac{3}{2} \cdot \frac{2}{3}x \le \frac{3}{2} \cdot \frac{8}{15}$$

$$\frac{2}{5} \le x \le \frac{4}{5}$$

The solution set is $\left\{x \big| \frac{2}{5} \le x \le \frac{4}{5}\right\}$, or $\left[\frac{2}{5}, \frac{4}{5}\right]$.

77. $3x < 4 - 5x < 5 + 3x$

$\qquad 0 < 4 - 8x < 5 \qquad$ Subtracting $3x$

$\qquad -4 < -8x < 1$

$\qquad \dfrac{1}{2} > x > -\dfrac{1}{8}$

The solution set is $\left\{x \middle| -\dfrac{1}{8} < x < \dfrac{1}{2}\right\}$, or $\left(-\dfrac{1}{8}, \dfrac{1}{2}\right)$.

79. $x + 4 < 2x - 6 \le x + 12$

$\qquad 4 < x - 6 \le 12 \qquad$ Subtracting x

$\qquad 10 < x \le 18$

The solution set is $\{x | 10 < x \le 18\}$, or $(10, 18]$.

81. If $-b < -a$, then $-1(-b) > -1(-a)$, or $b > a$, or $a < b$. The statement is true.

83. Let $a = 5$, $c = 12$, and $b = 2$. Then $a < c$ and $b < c$, but $a \not< b$. The given statement is false.

85. The numbers in either the set of all rational numbers or the set of all irrational numbers are all real numbers, so the union is all real numbers.

There are no numbers common to the set of all rational numbers and the set of all irrational numbers, so the intersection is $\emptyset$.

Exercise Set 1.6

RC1. $|x| > 3$

$\qquad x < -3 \quad or \quad x > 3$

The correct answer is (f).

RC3. $|x| < 3$

$\qquad -3 < x < 3$

The correct answer is (e).

RC5. $|x| \le 3$

$\qquad -3 \le x \le 3$

The correct answer is (a).

1. $|9x| = |9| \cdot |x| = 9|x|$

3. $|2x^2| = |2| \cdot |x^2|$

$\qquad = 2|x^2|$

$\qquad = 2x^2 \qquad$ Since x^2 is never negative

5. $|-2x^2| = |-2| \cdot |x^2|$

$\qquad = 2|x^2|$

$\qquad = 2x^2 \qquad$ Since x^2 is never negative

7. $|-6y| = |-6| \cdot |y| = 6|y|$

9. $\left|\dfrac{-2}{x}\right| = \dfrac{|-2|}{|x|} = \dfrac{2}{|x|}$

11. $\left|\dfrac{x^2}{-y}\right| = \dfrac{|x^2|}{|-y|}$

$\qquad = \dfrac{x^2}{|-y|}$

$\qquad = \dfrac{x^2}{|y|} \qquad$ The absolute value of the opposite of a number is the same as the absolute value of the number.

13. $\left|\dfrac{-8x^2}{2x}\right| = |-4x| = |-4| \cdot |x| = 4|x|$

15. $\left|\dfrac{4y^3}{-12y}\right| = \left|\dfrac{y^2}{-3}\right| = \dfrac{|y^2|}{|-3|} = \dfrac{y^2}{3}$

17. $|-8 - (-46)| = |38| = 38$, or

$\qquad |-46 - (-8)| = |-38| = 38$

19. $|36 - 17| = |19| = 19$, or

$\qquad |17 - 36| = |-19| = 19$

21. $|-3.9 - 2.4| = |-6.3| = 6.3$, or

$\qquad |2.4 - (-3.9)| = |6.3| = 6.3$

23. $|-5 - 0| = |-5| = 5$, or

$\qquad |0 - (-5)| = |5| = 5$

25. $|x| = 3$

$\qquad x = -3 \quad or \quad x = 3 \quad$ Absolute-value principle

The solution set is $\{-3, 3\}$.

27. $|x| = -3$

The absolute value of a number is always nonnegative. Therefore, the solution set is $\emptyset$.

29. $|q| = 0$

The only number whose absolute value is 0 is 0. The solution set is $\{0\}$.

31. $|x - 3| = 12$

$\qquad x - 3 = -12 \quad or \quad x - 3 = 12 \quad$ Absolute-value principle

$\qquad x = -9 \quad or \qquad x = 15$

The solution set is $\{-9, 15\}$.

33. $|2x - 3| = 4$

$\qquad 2x - 3 = -4 \quad or \quad 2x - 3 = 4 \quad$ Absolute-value principle

$\qquad 2x = -1 \quad or \qquad 2x = 7$

$\qquad x = -\dfrac{1}{2} \quad or \qquad x = \dfrac{7}{2}$

The solution set is $\left\{-\dfrac{1}{2}, \dfrac{7}{2}\right\}$.

35. $|4x - 9| = 14$

$\qquad 4x - 9 = -14 \quad or \quad 4x - 9 = 14$

$\qquad 4x = -5 \quad or \qquad 4x = 23$

$\qquad x = -\dfrac{5}{4} \quad or \qquad x = \dfrac{23}{4}$

The solution set is $\left\{-\dfrac{5}{4}, \dfrac{23}{4}\right\}$.

37. $|x| + 7 = 18$

$|x| + 7 - 7 = 18 - 7$ Subtracting 7

$|x| = 11$

$x = -11$ *or* $x = 11$ Absolute-value principle

The solution set is $\{-11, 11\}$.

39. $574 = 283 + |t|$

$291 = |t|$ Subtracting 283

$t = -291$ *or* $t = 291$ Absolute-value principle

The solution set is $\{-291, 291\}$.

41. $|5x| = 40$

$5x = -40$ *or* $5x = 40$

$x = -8$ *or* $x = 8$

The solution set is $\{-8, 8\}$.

43. $|3x| - 4 = 17$

$|3x| = 21$ Adding 4

$3x = -21$ *or* $3x = 21$

$x = -7$ *or* $x = 7$

The solution set is $\{-7, 7\}$.

45. $7|w| - 3 = 11$

$7|w| = 14$ Adding 3

$|w| = 2$ Dividing by 7

$w = -2$ *or* $w = 2$ Absolute-value principle

The solution set is $\{-2, 2\}$.

47. $\left|\dfrac{2x-1}{3}\right| = 5$

$\dfrac{2x-1}{3} = -5$ *or* $\dfrac{2x-1}{3} = 5$

$2x - 1 = -15$ *or* $2x - 1 = 15$

$2x = -14$ *or* $2x = 16$

$x = -7$ *or* $x = 8$

The solution set is $\{-7, 8\}$.

49. $|m + 5| + 9 = 16$

$|m + 5| = 7$ Subtracting 9

$m + 5 = -7$ *or* $m + 5 = 7$

$m = -12$ *or* $m = 2$

The solution set is $\{-12, 2\}$.

51. $10 - |2x - 1| = 4$

$-|2x - 1| = -6$ Subtracting 10

$|2x - 1| = 6$ Multiplying by -1

$2x - 1 = -6$ *or* $2x - 1 = 6$

$2x = -5$ *or* $2x = 7$

$x = -\dfrac{5}{2}$ *or* $x = \dfrac{7}{2}$

The solution set is $\left\{-\dfrac{5}{2}, \dfrac{7}{2}\right\}$.

53. $|3x - 4| = -2$

The absolute value of a number is always nonnegative. The solution set is $\emptyset$.

55. $\left|\dfrac{5}{9} + 3x\right| = \dfrac{1}{6}$

$\dfrac{5}{9} + 3x = -\dfrac{1}{6}$ *or* $\dfrac{5}{9} + 3x = \dfrac{1}{6}$

$3x = -\dfrac{13}{18}$ *or* $3x = -\dfrac{7}{18}$

$x = -\dfrac{13}{54}$ *or* $x = -\dfrac{7}{54}$

The solution set is $\left\{-\dfrac{13}{54}, -\dfrac{7}{54}\right\}$.

57. $|3x + 4| = |x - 7|$

$3x + 4 = x - 7$ *or* $3x + 4 = -(x - 7)$

$2x + 4 = -7$ *or* $3x + 4 = -x + 7$

$2x = -11$ *or* $4x + 4 = 7$

$x = -\dfrac{11}{2}$ *or* $4x = 3$

$x = -\dfrac{11}{2}$ *or* $x = \dfrac{3}{4}$

The solution set is $\left\{-\dfrac{11}{2}, \dfrac{3}{4}\right\}$.

59. $|x + 3| = |x - 6|$

$x + 3 = x - 6$ *or* $x + 3 = -(x - 6)$

$3 = -6$ *or* $x + 3 = -x + 6$

$3 = -6$ *or* $2x = 3$

$3 = -6$ *or* $x = \dfrac{3}{2}$

The first equation has no solution. The second equation has $\dfrac{3}{2}$ as a solution. There is only one solution of the original equation. The solution set is $\left\{\dfrac{3}{2}\right\}$.

61. $|2a + 4| = |3a - 1|$

$2a + 4 = 3a - 1$ *or* $2a + 4 = -(3a - 1)$

$-a + 4 = -1$ *or* $2a + 4 = -3a + 1$

$-a = -5$ *or* $5a + 4 = 1$

$a = 5$ *or* $5a = -3$

$a = 5$ *or* $a = -\dfrac{3}{5}$

The solution set is $\left\{5, -\dfrac{3}{5}\right\}$.

63. $|y - 3| = |3 - y|$

$y - 3 = 3 - y$ *or* $y - 3 = -(3 - y)$

$2y - 3 = 3$ *or* $y - 3 = -3 + y$

$2y = 6$ *or* $-3 = -3$

$y = 3$ True for all real values of y

All real numbers are solutions.

65. $|5 - p| = |p + 8|$

$\quad 5 - p = p + 8 \quad$ or $\quad 5 - p = -(p + 8)$

$\quad 5 - 2p = 8 \quad$ or $\quad 5 - p = -p - 8$

$\quad -2p = 3 \quad$ or $\quad 5 = -8$

$\quad p = -\dfrac{3}{2} \qquad$ False

The solution set is $\left\{ -\dfrac{3}{2} \right\}$.

67. $\left| \dfrac{2x - 3}{6} \right| = \left| \dfrac{4 - 5x}{8} \right|$

$\quad \dfrac{2x - 3}{6} = \dfrac{4 - 5x}{8} \quad$ or $\quad \dfrac{2x - 3}{6} = -\left(\dfrac{4 - 5x}{8}\right)$

$\quad 24\left(\dfrac{2x - 3}{6}\right) = 24\left(\dfrac{4 - 5x}{8}\right)$ or $\quad \dfrac{2x - 3}{6} = \dfrac{-4 + 5x}{8}$

$\quad 8x - 12 = 12 - 15x \quad$ or $24\left(\dfrac{2x - 3}{6}\right) = 24\left(\dfrac{-4 + 5x}{8}\right)$

$\quad 23x - 12 = 12 \qquad$ or $\quad 8x - 12 = -12 + 15x$

$\quad 23x = 24 \qquad$ or $\quad -7x - 12 = -12$

$\quad x = \dfrac{24}{23} \qquad$ or $\quad -7x = 0$

$\qquad\qquad\qquad\qquad\qquad x = 0$

The solution set is $\left\{ \dfrac{24}{23}, 0 \right\}$.

69. $\left| \dfrac{1}{2}x - 5 \right| = \left| \dfrac{1}{4}x + 3 \right|$

$\quad \dfrac{1}{2}x - 5 = \dfrac{1}{4}x + 3 \quad$ or $\quad \dfrac{1}{2}x - 5 = -\left(\dfrac{1}{4}x + 3\right)$

$\quad \dfrac{1}{4}x - 5 = 3 \qquad$ or $\quad \dfrac{1}{2}x - 5 = -\dfrac{1}{4}x - 3$

$\quad \dfrac{1}{4}x = 8 \qquad$ or $\quad \dfrac{3}{4}x - 5 = -3$

$\quad x = 32 \qquad$ or $\quad \dfrac{3}{4}x = 2$

$\quad x = 32 \qquad$ or $\quad x = \dfrac{8}{3}$

The solution set is $\left\{ 32, \dfrac{8}{3} \right\}$.

71. $|x| < 3$

$\quad -3 < x < 3$

The solution set is $\{x| -3 < x < 3\}$, or $(-3, 3)$.

73. $|x| \geq 2$

$\quad x \leq -2$ or $x \geq 2$

The solution set is $\{x|x \leq -2 \text{ or } x \geq 2\}$, or $(-\infty, -2] \cup [2, \infty)$.

75. $|x - 1| < 1$

$\quad -1 < x - 1 < 1$

$\quad 0 < x < 2$

The solution set is $\{x|0 < x < 2\}$, or $(0, 2)$.

77. $5|x + 4| \leq 10$

$\quad |x + 4| \leq 2 \qquad$ Dividing by 5

$\quad -2 \leq x + 4 \leq 2$

$\quad -6 \leq x \leq -2 \qquad$ Subtracting 4

The solution set is $\{x| -6 \leq x \leq -2\}$, or $[-6, -2]$.

79. $|2x - 3| \leq 4$

$\quad -4 \leq 2x - 3 \leq 4$

$\quad -1 \leq 2x \leq 7 \qquad$ Adding 3

$\quad -\dfrac{1}{2} \leq x \leq \dfrac{7}{2} \qquad$ Dividing by 2

The solution set is $\left\{ x| -\dfrac{1}{2} \leq x \leq \dfrac{7}{2} \right\}$, or $\left[-\dfrac{1}{2}, \dfrac{7}{2} \right]$.

81. $|2y - 7| > 10$

$\quad 2y - 7 < -10$ or $2y - 7 > 10$

$\quad 2y < -3 \quad$ or $\quad 2y > 17 \quad$ Adding 7

$\quad y < -\dfrac{3}{2} \quad$ or $\quad y > \dfrac{17}{2} \quad$ Dividing by 2

The solution set is $\left\{ y|y < -\dfrac{3}{2} \text{ or } y > \dfrac{17}{2} \right\}$, or

$\left(-\infty, -\dfrac{3}{2} \right) \cup \left(\dfrac{17}{2}, \infty \right)$.

83. $|4x - 9| \geq 14$

$\quad 4x - 9 \leq -14$ or $4x - 9 \geq 14$

$\quad 4x \leq -5 \quad$ or $\quad 4x \geq 23$

$\quad x \leq -\dfrac{5}{4} \quad$ or $\quad x \geq \dfrac{23}{4}$

The solution set is $\left\{ x|x \leq -\dfrac{5}{4} \text{ or } x \geq \dfrac{23}{4} \right\}$, or

$\left(-\infty, -\dfrac{5}{4} \right] \cup \left[\dfrac{23}{4}, \infty \right)$.

85. $|y - 3| < 12$

$\quad -12 < y - 3 < 12$

$\quad -9 < y < 15 \qquad$ Adding 3

The solution set is $\{y| -9 < y < 15\}$, or $(-9, 15)$.

87. $|2x + 3| \leq 4$

$\quad -4 \leq 2x + 3 \leq 4$

$\quad -7 \leq 2x \leq 1 \qquad$ Subtracting 3

$\quad -\dfrac{7}{2} \leq x \leq \dfrac{1}{2} \qquad$ Dividing by 2

The solution set is $\left\{ x| -\dfrac{7}{2} \leq x \leq \dfrac{1}{2} \right\}$, or $\left[-\dfrac{7}{2}, \dfrac{1}{2} \right]$.

89. $|4 - 3y| > 8$

$\quad 4 - 3y < -8 \quad$ or $\quad 4 - 3y > 8$

$\quad -3y < -12 \quad$ or $\quad -3y > 4 \quad$ Subtracting 4

$\quad y > 4 \quad$ or $\quad y < -\dfrac{4}{3} \quad$ Dividing by -3

The solution set is $\left\{ y|y < -\dfrac{4}{3} \text{ or } y > 4 \right\}$, or

$\left(-\infty, -\dfrac{4}{3} \right) \cup (4, \infty)$.

91. $|9 - 4x| \geq 14$

$\quad 9 - 4x \leq -14 \;\; or \;\; 9 - 4x \geq 14$

$\quad\quad -4x \leq -23 \;\; or \;\;\; -4x \geq 5 \quad$ Subtracting 9

$\quad\quad\quad x \geq \dfrac{23}{4} \;\; or \;\;\;\;\; x \leq -\dfrac{5}{4} \quad$ Dividing by -4

The solution set is $\left\{ x \middle| x \leq -\dfrac{5}{4} \; or \; x \geq \dfrac{23}{4} \right\}$ or

$\left(-\infty, -\dfrac{5}{4} \right] \cup \left[\dfrac{23}{4}, \infty \right)$.

93. $|3 - 4x| < 21$

$\quad -21 < 3 - 4x < 21$

$\quad -24 < -4x < 18 \quad$ Subtracting 3

$\quad\quad 6 > x > -\dfrac{9}{2} \quad$ Dividing by -4 and simplifying

The solution set is $\left\{ x \middle| 6 > x > -\dfrac{9}{2} \right\}$, or

$\left\{ x \middle| -\dfrac{9}{2} < x < 6 \right\}$, or $\left(-\dfrac{9}{2}, 6 \right)$.

95. $\left| \dfrac{1}{2} + 3x \right| \geq 12$

$\quad \dfrac{1}{2} + 3x \leq -12 \;\; or \;\; \dfrac{1}{2} + 3x \geq 12$

$\quad\quad 3x \leq -\dfrac{25}{2} \;\; or \;\;\;\; 3x \geq \dfrac{23}{2} \quad$ Subtracting $\dfrac{1}{2}$

$\quad\quad\quad x \leq -\dfrac{25}{6} \;\; or \;\;\;\;\; x \geq \dfrac{23}{6} \quad$ Dividing by 3

The solution set is $\left\{ x \middle| x \leq -\dfrac{25}{6} \; or \; x \geq \dfrac{23}{6} \right\}$, or

$\left(-\infty, -\dfrac{25}{6} \right] \cup \left[\dfrac{23}{6}, \infty \right)$.

97. $\left| \dfrac{x - 7}{3} \right| < 4$

$\quad -4 < \dfrac{x - 7}{3} < 4$

$\quad -12 < x - 7 < 12 \quad$ Multiplying by 3

$\quad\quad -5 < x < 19 \quad\quad$ Adding 7

The solution set is $\{ x | -5 < x < 19 \}$, or $(-5, 19)$.

99. $\left| \dfrac{2 - 5x}{4} \right| \geq \dfrac{2}{3}$

$\quad \dfrac{2 - 5x}{4} \leq -\dfrac{2}{3} \;\; or \;\; \dfrac{2 - 5x}{4} \geq \dfrac{2}{3}$

$\quad 2 - 5x \leq -\dfrac{8}{3} \;\; or \;\; 2 - 5x \geq \dfrac{8}{3} \quad$ Multiplying by 4

$\quad\quad -5x \leq -\dfrac{14}{3} \;\; or \;\;\; -5x \geq \dfrac{2}{3} \quad$ Subtracting 2

$\quad\quad\quad x \geq \dfrac{14}{15} \;\; or \;\;\;\; x \leq -\dfrac{2}{15} \quad$ Dividing by -5

The solution set is $\left\{ x \middle| x \leq -\dfrac{2}{15} \; or \; x \geq \dfrac{14}{15} \right\}$, or

$\left(-\infty, -\dfrac{2}{15} \right] \cup \left[\dfrac{14}{15}, \infty \right)$.

101. $|m + 5| + 9 \leq 16$

$\quad |m + 5| \leq 7 \quad\quad\quad$ Subtracting 9

$\quad -7 \leq m + 5 \leq 7$

$\quad -12 \leq m \leq 2$

The solution set is $\{ m | -12 \leq m \leq 2 \}$, or $[-12, 2]$.

103. $7 - |3 - 2x| \geq 5$

$\quad -|3 - 2x| \geq -2 \quad\quad$ Subtracting 7

$\quad\quad |3 - 2x| \leq 2 \quad\quad$ Multiplying by -1

$\quad\quad -2 \leq 3 - 2x \leq 2$

$\quad\quad -5 \leq -2x \leq -1 \quad$ Subtracting 3

$\quad\quad \dfrac{5}{2} \geq x \geq \dfrac{1}{2} \quad\quad$ Dividing by -2

The solution set is $\left\{ x \middle| \dfrac{5}{2} \geq x \geq \dfrac{1}{2} \right\}$, or $\left\{ x \middle| \dfrac{1}{2} \leq x \leq \dfrac{5}{2} \right\}$, or

$\left[\dfrac{1}{2}, \dfrac{5}{2} \right]$.

105. $\left| \dfrac{2x - 1}{3} \right| \leq 1$

$\quad -1 \leq \dfrac{2x - 1}{3} \leq 1$

$\quad -3 \leq 2x - 1 \leq 3$

$\quad -2 \leq 2x \leq 4$

$\quad -1 \leq x \leq 2$

The solution set is $\{ x | -1 \leq x \leq 2 \}$, or $[-1, 2]$.

107. $-11x + 2x \geq -36$

$\quad -9x \geq -36$

$\quad\quad x \leq 4 \quad$ Reversing the inequality symbol

The solution set is $\{ x | x \leq 4 \}$, or $(-\infty, 4]$.

109. $2(r - 1) + 4 < 3(r - 2) - 8$

$\quad 2r - 2 + 4 < 3r - 6 - 8$

$\quad\quad 2r + 2 < 3r - 14$

$\quad\quad\quad 2 < r - 14$

$\quad\quad\quad 16 < r$

The solution set is $\{ r | r > 16 \}$, or $(16, \infty)$.

111. $-3 \leq 2x + 5 \;\; or \;\; 10 > 2x - 1$

$\quad -8 \leq 2x \quad\quad or \quad 11 > 2x$

$\quad -4 \leq x \quad\quad or \quad \dfrac{11}{2} > x$

The solution set is the set of all real numbers or $(-\infty, \infty)$.

113. $|d - 6 \text{ ft}| \leq \dfrac{1}{2} \text{ ft}$

$\quad -\dfrac{1}{2} \text{ ft} \leq d - 6 \text{ ft} \leq \dfrac{1}{2} \text{ ft}$

$\quad 5\dfrac{1}{2} \text{ ft} \leq d \leq 6\dfrac{1}{2} \text{ ft}$

The solution set is $\left\{ d \middle| 5\dfrac{1}{2} \text{ ft} \leq d \leq 6\dfrac{1}{2} \text{ ft} \right\}$.

115. $|x + 5| > x$

The inequality is true for all $x < 0$ (because absolute value must be nonnegative). The solution set in this case is $\{x | x < 0\}$. If $x = 0$, we have $|0 + 5| > 0$, which is true. The solution set in this case is $\{0\}$. If $x > 0$, we have the following:

$$x + 5 < -x \quad or \quad x + 5 > x$$
$$2x < -5 \quad or \quad 5 > 0$$
$$x < -\frac{5}{2} \quad or \quad 5 > 0$$

Although $x > 0$ and $x < -\dfrac{5}{2}$ yields no solution, $x > 0$ and $5 > 0$ (true for all x) yield the solution set $\{x | x > 0\}$ in this case. The solution set for the inequality is $\{x | x < 0\} \cup \{0\} \cup \{x | x > 0\}$, or all real numbers.

117. $|7x - 2| = x + 4$

From the definition of absolute value, we know $x + 4 \geq 0$, or $x \geq -4$. So we have $x \geq -4$ and

$$7x - 2 = x + 4 \quad or \quad 7x - 2 = -(x + 4)$$
$$6x = 6 \quad or \quad 7x - 2 = -x - 4$$
$$x = 1 \quad or \quad 8x = -2$$
$$x = 1 \quad or \quad x = -\frac{1}{4}$$

The solution set is $\left\{ x \middle| x \geq -4 \text{ and } x = 1 \text{ or } x = -\dfrac{1}{4} \right\}$, or $\left\{ 1, -\dfrac{1}{4} \right\}$.

119. $|x - 6| \leq -8$

From the definition of absolute value we know that $|x - 6| \geq 0$. Thus $|x - 6| \leq -8$ is false for all x. The solution set is $\emptyset$.

121. $-3 < x < 3$ is equivalent to $|x| < 3$.

123. $x \leq -6$ or or $x \geq 6$ is equivalent to $|x| \geq 6$.

125.
$$x < -8 \quad or \quad x > 2$$
$$x + 3 < -5 \quad or \quad x + 3 > 5 \quad \text{Adding 3}$$
$$|x + 3| > 5$$

Chapter 1 Vocabulary Reinforcement

1. An <u>inequality</u> is a sentence containing $<$, $\leq$, $>$, $\geq$, or $\neq$.

2. Using <u>set-builder</u> notation, we write the solution set for $x < 7$ as $\{x | x < 7\}$.

3. Using <u>interval</u> notation, we write the solution set of $-5 \leq y < 16$ as $[-5, 16)$.

4. The <u>intersection</u> of two sets A and B is the set of all members that are common to A and B.

5. When two or more sentences are joined by the word *and* to make a compound sentence, the new sentence is called a <u>conjunction</u> of the sentences.

6. When two sets have no elements in common, the intersection of the two sets is the <u>empty set</u>.

7. Two sets with an empty intersection are said to be <u>disjoint sets</u>.

8. The <u>union</u> of two sets A and B is the collection of elements belonging to A and/or B.

9. When two or more sentences are joined by the word *or* to make a compound sentence, the new sentence is called a <u>disjunction</u> of the sentences.

10. The <u>addition principle</u> for equations states that for any real numbers a, b, and c, $a = b$ is equivalent to $a + c = b + c$.

11. The <u>multiplication principle</u> for equations states that for any real numbers a, b, and c, $a = b$ is equivalent to $a \cdot c = b \cdot c$.

12. For any real numbers a and b, the <u>distance</u> between them is $|a - b|$.

Chapter 1 Concept Reinforcement

1. True; see page 75 in the text.

2. False; the variable t appears on both sides of the formula $t = \dfrac{3B - mt}{n}$, so the original formula has not been solved for t.

3. False; see page 114 in the text.

4. False; numbers in the interval $(1, 2)$ are solutions of $x < 2$, but they are not solutions of $x \leq 1$.

5. True; see page 138 in the text.

6. False; $|0| = 0$.

7. True; we have
$$|a - b| = |-1 \cdot (-a + b)| = |-1| \cdot |-a + b| = 1 \cdot |-a + b| = |-a + b|, \text{ or } |b - a|.$$

Chapter 1 Study Guide

1.
$$\begin{array}{c|c} \multicolumn{2}{c}{28 - 7x = 7} \\ \hline 28 - 7(-3) \;?\; 7 & \\ 28 + 21 & \\ 49 & \text{FALSE} \end{array}$$

The number -3 is not a solution of the equation.

2. $2(x + 2) = 5(x - 4)$
$$2x + 4 = 5x - 20$$
$$4 = 3x - 20$$
$$24 = 3x$$
$$8 = x$$

The solution is 8.

3. $F = \dfrac{1}{4}gh$

$4F = gh$

$\dfrac{4F}{g} = h$

4. $8 - 3x \leq 3x + 6$

-2 : We substitute and get $8 - 3(-2) \leq 3(-2) + 6$, or $8 + 6 \leq -6 + 6$, or $14 \leq 0$, a false sentence. Therefore, -2 is not a solution.

5 : We substitute and get $8 - 3 \cdot 5 \leq 3 \cdot 5 + 6$, or $8 - 15 \leq 15 + 6$, or $-7 \leq 21$, a true sentence. Therefore, 5 is a solution.

5. a) Interval notation for $\{t | t < -8\}$ is $(-\infty, -8)$.

b) Interval notation for $\{x | -7 \leq x < 10\}$ is $[-7, 10)$.

c) Interval notation for $\{b | b \geq 3\}$ is $[3, \infty)$.

6. $5y + 5 < 2y - 1$

$3y + 5 < -1$

$3y < -6$

$y < -2$

The solution set is $\{y | y < -2\}$, or $(-\infty, -2)$.

The graph of the solution set is shown below.

7. $-4 \leq 5x + 6 < 11$

$-10 \leq 5x < 5$

$-2 \leq x < 1$

The solution set is $\{x | -2 \leq x < 1\}$, or $[-2, 1)$.

The graph of the solution set is shown below.

8. $z + 4 < 3 \quad or \quad 4z + 1 \geq 5$

$z < -1 \quad or \quad 4z \geq 4$

$z < -1 \quad or \quad z \geq 1$

The solution set is $\{z | z < -1 \ or \ z \geq 1\}$, or $(-\infty, -1) \cup [1, \infty)$.

The graph of the solution set is shown below.

9. $|8y^2| = |8| \cdot |y^2|$

$= 8y^2 \qquad$ Since y^2 is never negative

10. $|8 - (-20)| = |8 + 20| = |28| = 28$

11. $|5x - 1| = 9$

$5x - 1 = -9 \quad or \quad 5x - 1 = 9$

$5x = -8 \quad or \qquad 5x = 10$

$x = -\dfrac{8}{5} \quad or \qquad x = 2$

The solution set is $\left\{ -\dfrac{8}{5}, 2 \right\}$.

12. $|z + 4| = |3z - 2|$

$z + 4 = 3z - 2 \quad or \quad z + 4 = -(3z - 2)$

$-2z + 4 = -2 \quad or \quad z + 4 = -3z + 2$

$-2z = -6 \quad or \quad 4z + 4 = 2$

$z = 3 \quad or \qquad 4z = -2$

$z = 3 \quad or \qquad z = -\dfrac{1}{2}$

The solution set is $\left\{ 3, -\dfrac{1}{2} \right\}$.

13. a) $|2x + 3| < 5$

$-5 < 2x + 3 < 5$

$-8 < 2x < 2$

$-4 < x < 1$

The solution set is $\{x | -4 < x < 1\}$, or $(-4, 1)$.

b) $|3x + 2| \geq 8$

$3x + 2 \leq -8 \quad or \quad 3x + 2 \geq 8$

$3x \leq -10 \quad or \qquad 3x \geq 6$

$x \leq -\dfrac{10}{3} \quad or \qquad x \geq 2$

The solution set is $\left\{ x \middle| x \leq -\dfrac{10}{3} \ or \ x \geq 2 \right\}$, or $\left(-\infty, -\dfrac{10}{3} \right] \cup [2, \infty)$.

Chapter 1 Review Exercises

1. $-11 + y = -3$

$-11 + y + 11 = -3 + 11$

$y = 8$

The number 8 checks, so it is the solution.

2. $-7x = -3$

$\dfrac{-7x}{-7} = \dfrac{-3}{-7}$

$x = \dfrac{3}{7}$

The number $\dfrac{3}{7}$ checks, so it is the solution.

3. $-\dfrac{5}{3}x + \dfrac{7}{3} = -5$

$3\left(-\dfrac{5}{3}x + \dfrac{7}{3} \right) = 3(-5) \quad$ Clearing fractions

$-5x + 7 = -15$

$-5x = -22$

$x = \dfrac{22}{5}$

The number $\dfrac{22}{5}$ checks, so it is the solution.

4. $6(2x - 1) = 3 - (x + 10)$

$12x - 6 = 3 - x - 10$

$12x - 6 = -7 - x$

$13x - 6 = -7$

$13x = -1$

$x = -\dfrac{1}{13}$

The number $-\dfrac{1}{13}$ checks, so it is the solution.

5. $2.4x + 1.5 = 1.02$

$100(2.4x + 1.5) = 100(1.02)$ Clearing decimals

$240x + 150 = 102$

$240x = -48$

$x = -0.2$

The number -0.2 checks, so it is the solution.

6. $2(3 - x) - 4(x + 1) = 7(1 - x)$

$6 - 2x - 4x - 4 = 7 - 7x$

$2 - 6x = 7 - 7x$

$2 + x = 7$

$x = 5$

The number 5 checks, so it is the solution.

7.
$$C = \frac{4}{11}d + 3$$

$$C - 3 = \frac{4}{11}d \qquad \text{Subtracting 3}$$

$$\frac{11}{4}(C - 3) = d \qquad \text{Multiplying by } \frac{11}{4}$$

8.
$$A = 2a - 3b$$

$$A - 2a = -3b$$

$$\frac{A - 2a}{-3} = b, \text{ or}$$

$$\frac{2a - A}{3} = b$$

9. Familiarize. Let $x =$ the smaller number. Then $x + 1 =$ the larger number.

Translate.

$\underbrace{\text{Smaller number}}$ plus $\underbrace{\text{larger number}}$ is 371.

$\downarrow \qquad\quad \downarrow \qquad \downarrow \qquad\quad \downarrow \quad \downarrow$

$x \qquad\quad + \qquad (x + 1) \quad = \quad 371$

Solve. We solve the equation.

$x + (x + 1) = 371$

$2x + 1 = 371$

$2x = 370$

$x = 185$

If $x = 185$, then $x + 1 = 185 + 1 = 186$.

Check. 185 and 186 are consecutive integers and $185 + 186 = 371$. The answer checks.

State. The numbers on the markers are 185 and 186.

10. Familiarize. Let $x =$ the length of the longer piece of rope, in meters. Then $\dfrac{4}{5}x =$ the length of the shorter piece.

Translate.

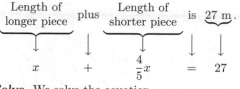

$\underbrace{\begin{array}{c}\text{Length of} \\ \text{longer piece}\end{array}}$ plus $\underbrace{\begin{array}{c}\text{Length of} \\ \text{shorter piece}\end{array}}$ is $\underbrace{27 \text{ m}}$.

$\downarrow \qquad\qquad\quad \downarrow \qquad\qquad \downarrow \qquad\quad \downarrow \quad \downarrow$

$x \qquad\qquad + \qquad \dfrac{4}{5}x \qquad = \quad 27$

Solve. We solve the equation.

$x + \dfrac{4}{5}x = 27$

$\dfrac{9}{5}x = 27$

$x = \dfrac{5}{9} \cdot 27$

$x = 15$

If $x = 15$, then $\dfrac{4}{5}x = \dfrac{4}{5} \cdot 15 = 12$.

Check. 12 m is $\dfrac{4}{5}$ of 15 m and $12 \text{ m} + 15 \text{ m} = 27$, so the answer checks.

State. The lengths of the pieces are 15 m and 12 m.

11. Familiarize. Let $p =$ the former population.

Translate.

$\underbrace{\begin{array}{c}\text{Former} \\ \text{population}\end{array}}$ plus 12% of $\underbrace{\begin{array}{c}\text{former} \\ \text{population}\end{array}}$ is 179,200

$\downarrow \qquad\quad \downarrow \quad \downarrow \;\; \downarrow \qquad \downarrow \qquad\quad \downarrow \quad \downarrow$

$p \qquad\quad + \;\; 12\% \;\cdot \qquad p \qquad = \;\; 179,200$

Solve. We solve the equation.

$p + 12\% \cdot p = 179,200$

$p + 0.12p = 179,200$

$1.12p = 179,200$

$p = 160,000$

Check. 12% of 160,000 is $0.12(160,000) = 19,200$ and $160,000 + 19,200 = 179,200$. The answer checks.

State. The former population is 160,000.

12. Familiarize. We will use the formula $d = rt$. Arnie's speed on the walkway is $3 + 6 = 9$ ft/sec.

Translate.

$d = rt$

$360 = 9t$

Solve. We solve the equation.

$360 = 9t$

$40 = t$

Check. If Arnie travels at a speed of 9 ft/sec for 40 sec, he travels $9 \cdot 40 = 360$ ft. The answer checks.

State. It will take Arnie 40 sec to walk the length of the walkway.

13. Interval is $[-8, 9)$.

14. Interval notation is $(-\infty, 40]$.

15. $x - 2 \leq -4$

$x \leq -2$

The solution set is $(-\infty, -2]$.

16. $x + 5 > 6$

$x > 1$

The solution set is $(1, \infty)$.

17. $a + 7 \leq -14$

$a \leq -21$

The solution set is $\{a | a \leq -21\}$, or $(-\infty, -21]$.

18. $y - 5 \geq -12$

$y \geq -7$

The solution set is $\{y | y \geq -7\}$, or $[-7, \infty)$.

19. $4y > -16$

$y > -4$

The solution set is $\{y | y > -4\}$, or $(-4, \infty)$.

20. $-0.3y < 9$

$y > -30$ Reversing the inequality symbol

The solution set is $\{y | y > -30\}$, or $(-30, \infty)$.

21. $-6x - 5 < 13$

$-6x < 18$

$x > -3$ Reversing the inequality symbol

The solution set is $\{x | x > -3\}$, or $(-3, \infty)$.

22. $4y + 3 \leq -6y - 9$

$10y + 3 \leq -9$

$10y \leq -12$

$y \leq -\dfrac{6}{5}$

The solution set is $\left\{y \middle| y \leq -\dfrac{6}{5}\right\}$, or $\left(-\infty, -\dfrac{6}{5}\right]$.

23. $-\dfrac{1}{2}x - \dfrac{1}{4} > \dfrac{1}{2} - \dfrac{1}{4}x$

$-\dfrac{1}{4}x - \dfrac{1}{4} > \dfrac{1}{2}$

$-\dfrac{1}{4}x > \dfrac{3}{4}$

$x < -3$ Reversing the inequality symbol

The solution set is $\{x | x < -3\}$, or $(-\infty, -3)$.

24. $0.3y - 8 < 2.6y + 15$

$-2.3y - 8 < 15$

$-2.3y < 23$

$y > -10$ Reversing the inequality symbol

The solution set is $\{y | y > -10\}$, or $(-10, \infty)$.

25. $-2(x - 5) \geq 6(x + 7) - 12$

$-2x + 10 \geq 6x + 42 - 12$

$-2x + 10 \geq 6x + 30$

$-8x + 10 \geq 30$

$-8x \geq 20$

$x \leq -\dfrac{5}{2}$ Reversing the inequality symbol

The solution set is $\left\{x \middle| x \leq -\dfrac{5}{2}\right\}$, or $\left(-\infty, -\dfrac{5}{2}\right]$.

26. *Familiarize.* Let $t =$ the length of time of the move, in hours. Then Metro Movers charges $85 + 40t$ and Champion Moving charges $60t$.

Translate.

Cost of Champion Moving	is more than	Cost of Metro Movers
$60t$	$>$	$85 + 40t$

Solve. We solve the inequality.

$60t > 85 + 40t$

$20t > 85$

$t > \dfrac{17}{4}$, or $4\dfrac{1}{4}$

Check.   When $t = \dfrac{17}{4}$ hr, Champion Moving charges $60 \cdot \dfrac{17}{4}$, or \$255, and Metro Movers charges $85 + 40 \cdot \dfrac{17}{4} = 85 + 170 = \$255$. For a value of t greater than $4\dfrac{1}{4}$, say 5, Champion Moving charges $60 \cdot 5 = \$300$, and Metro Movers charges $85 + 40 \cdot 5 = 85 + 200 = \$285$. This partial check tells us that the answer is probably correct.

State. Champion Moving is more expensive for moves taking more than $4\dfrac{1}{4}$ hr. The solution set is $\left\{t \middle| t > 4\dfrac{1}{4} \text{ hr}\right\}$.

27. *Familiarize.* Let $x =$ the amount invested at 3%. Then $30,000 - x =$ the amount invested at 4%. The interest earned on the 3% investment is $3\%x$, or $0.03x$, and the interest earned on the 4% investment is $4\%(30,000 - x)$, or $0.04(30,000 - x)$.

Translate.

Interest on 3% investment	plus	Interest on 4% investment	is at least	\$1100
$0.03x$	$+$	$0.04(30,000 - x)$	$\geq$	1100

Solve. We solve the inequality.

$0.03x + 0.04(30,000 - x) \geq 1100$

$0.03x + 1200 - 0.04x \geq 1100$

$-0.01x + 1200 \geq 1100$

$-0.01x \geq -100$

$x \leq 10,000$

Check. If \$10,000 is invested at 3%, then the amount invested at 4% is $30,000 - \$10,000$, or \$20,000. The interest earned is $0.03(\$10,000) + 0.04(\$20,000)$, or $\$300+\$800$, or $\$1100$. Then if less than $\$10,000$ is invested at 3%, the interest earned will be more than $\$1100$. This partial check shows that the answer is probably correct.

State. At most $\$10,000$ can be invested at 3% interest.

28. Interval notation for $-2 \le x < 5$ is $[-2, 5)$.

29. Interval notation for $x \le -2$ or $x > 5$ is $(-\infty, -2] \cup (5, \infty)$.

30. $\{1, 2, 5, 6, 9\} \cap \{1, 3, 5, 9\} = \{1, 5, 9\}$

31. $\{1, 2, 5, 6, 9\} \cup \{1, 3, 5, 9\} = \{1, 2, 3, 5, 6, 9\}$

32. $2x - 5 < -7$ and $3x + 8 \ge 14$

$\qquad 2x < -2$ and $\qquad 3x \ge 6$

$\qquad x < -1$ and $\qquad x \ge 2$

The intersection of $\{x | x < -1\}$ and $\{x \ge 2\}$ is $\emptyset$, so the solution set is $\emptyset$.

33. $-4 < x + 3 \le 5$

$\quad -7 < x \le 2 \qquad$ Subtracting 3

The solution set is $\{x | -7 < x \le 2\}$, or $(-7, 2]$.

34. $-15 < -4x - 5 < 0$

$\quad -10 < -4x < 5 \qquad$ Adding 5

$\quad \dfrac{5}{2} > x > -\dfrac{5}{4} \qquad$ Dividing by -4 and reversing the inequality symbol

The solution set is $\left\{ x \left| \dfrac{5}{2} > x > -\dfrac{5}{4} \right. \right\}$, or

$\left\{ x \left| -\dfrac{5}{4} < x < \dfrac{5}{2} \right. \right\}$, or $\left(-\dfrac{5}{4}, \dfrac{5}{2} \right)$.

35. $3x < -9$ or $-5x < -5$

$\quad x < -3$ or $\quad x > 1$

The solution set is $\{x | x < -3$ or $x > 1\}$, or $(-\infty, -3) \cup (1, \infty)$.

36. $2x + 5 < -17$ or $-4x + 10 \le 34$

$\quad 2x < -22$ or $\qquad -4x \le 24$

$\quad x < -11$ or $\qquad x \ge -6$

The solution set is $\{x | x < -11$ or $x \ge -6\}$, or $(-\infty, -11) \cup [-6, \infty)$.

37. $2x + 7 \le -5$ or $x + 7 \ge 15$

$\quad 2x \le -12$ or $\qquad x \ge 8$

$\quad x \le -6$ or $\qquad x \ge 8$

The solution set is $\{x | x \le -6$ or $x \ge 8\}$, or $(-\infty, -6] \cup [8, \infty)$.

38. $\left| -\dfrac{3}{x} \right| = \left| \dfrac{-3}{x} \right| = \dfrac{|-3|}{|x|} = \dfrac{3}{|x|}$

39. $\left| \dfrac{2x}{y^2} \right| = \dfrac{|2x|}{|y^2|} = \dfrac{|2| \cdot |x|}{y^2} = \dfrac{2|x|}{y^2}$

40. $\left| \dfrac{12y}{-3y^2} \right| = \left| \dfrac{-4}{y} \right| = \dfrac{|-4|}{|y|} = \dfrac{4}{|y|}$

41. $|-23 - 39| = |-62| = 62$, or

$|39 - (-23)| = |39 + 23| = |62| = 62$

42. $|x| = 6$

$\quad x = -6$ or $x = 6 \qquad$ Absolute-value principle

The solution set is $\{-6, 6\}$.

43. $|x - 2| = 7$

$\quad x - 2 = -7$ or $x - 2 = 7$

$\quad x = -5$ or $\qquad x = 9$

The solution set is $\{-5, 9\}$.

44. $|2x + 5| = |x - 9|$

$\quad 2x + 5 = x - 9$ or $2x + 5 = -(x - 9)$

$\quad x + 5 = -9$ or $2x + 5 = -x + 9$

$\quad x = -14$ or $3x + 5 = 9$

$\quad x = -14$ or $\qquad 3x = 4$

$\quad x = -14$ or $\qquad x = \dfrac{4}{3}$

The solution set is $\left\{ -14, \dfrac{4}{3} \right\}$.

45. $|5x + 6| = -8$

The absolute value of a number is always nonnegative. Thus, the solution set is $\emptyset$.

46. $|2x + 5| < 12$

$\quad -12 < 2x + 5 < 12$

$\quad -17 < 2x < 7$

$\quad -\dfrac{17}{2} < x < \dfrac{7}{2}$

The solution set is $\left\{ x \left| -\dfrac{17}{2} < x < \dfrac{7}{2} \right. \right\}$, or $\left(-\dfrac{17}{2}, \dfrac{7}{2} \right)$.

47. $|x| \ge 3.5$

$\quad x \le -3.5$ or $x \ge 3.5$

The solution set is $\{x | x \le -3.5$ or $x \ge 3.5\}$, or $(-\infty, -3.5] \cup [3.5, \infty)$.

48. $|3x - 4| \ge 15$

$\quad 3x - 4 \le -15$ or $3x - 4 \ge 15$

$\quad 3x \le -11$ or $\qquad 3x \ge 19$

$\quad x \le -\dfrac{11}{3}$ or $\qquad x \ge \dfrac{19}{3}$

The solution set is $\left\{ x \left| x \le -\dfrac{11}{3} \text{ or } x \ge \dfrac{19}{3} \right. \right\}$, or

$\left(-\infty, -\dfrac{11}{3} \right] \cup \left[\dfrac{19}{3}, \infty \right)$.

49. $|x| < 0$

The absolute value of a number is always greater than or equal to 0, so the solution set is $\emptyset$.

50. In 2010, $t = 2010 - 1980 = 30$.

$$G = 0.506t + 18.3$$
$$G = 0.506(30) + 18.3 = 15.18 + 18.3 = 33.48$$

We estimate carbon dioxide emissions to be 33.48 billion metric tons in 2010. Answer B is correct.

51. We want to find the value of t for which $35 < G < 40$. We have

$$35 < 0.506t + 18.3 < 40$$
$$16.7 < 0.506t < 21.7$$
$$33 < t < 43. \quad \text{Rounding}$$

Thus, for years between 33 yr after 1980 and 43 yr after 1980, global carbon dioxide emissions are predicted to be between 35 and 40 billion metric tons. These are the years between 2013 and 2023. Answer A is correct.

52. $|2x + 5| \le |x + 3|$

$|2x + 5| \le x + 3 \quad or \quad |2x + 5| \le -(x + 3)$

First we solve $|2x + 5| \le x + 3$.

$$-(x + 3) \le 2x + 5 \quad and \quad 2x + 5 \le x + 3$$
$$-x - 3 \le 2x + 5 \quad and \qquad x \le -2$$
$$-8 \le 3x \qquad and \qquad x \le -2$$
$$-\frac{8}{3} \le x \qquad and \qquad x \le -2$$

The solution set for this portion of the inequality is $\left\{ x \middle| -\frac{8}{3} \le x \le -2 \right\}$.

Now we solve $|2x + 5| \le -(x + 3)$.

$$-[-(x + 3)] \le 2x + 5 \quad and \quad 2x + 5 \le -(x + 3)$$
$$x + 3 \le 2x + 5 \quad and \quad 2x + 5 \le -x - 3$$
$$-2 \le x \qquad and \qquad 3x \le -8$$
$$-2 \le x \qquad and \qquad x \le -\frac{8}{3}$$

The solution set for this portion of the inequality is $\emptyset$.

Then the solution set for the original inequality is $\left\{ x \middle| -\frac{8}{3} \le x \le -2 \right\} \cup \emptyset$, or $\left\{ x \middle| -\frac{8}{3} \le x \le -2 \right\}$. This is expressed in interval notation as $\left[-\frac{8}{3}, -2 \right]$.

Chapter 1 Discussion and Writing Exercises

1. When the signs of the quantities on either side of the inequality symbol are changed, their relative positions on the number line are reversed.

2. The distance between x and -5 is $|x - (-5)|$, or $|x + 5|$. Then the solutions of the inequality $|x + 5| \le 2$ can be interpreted as "all those numbers x whose distance from -5 is at most 2 units."

3. When $b \ge c$, then $[a, b] \cup [c, d] = [a, d]$.

4. The solutions of $|x| \ge 6$ are those numbers whose distance from zero is greater than or equal to 6. In addition to the numbers in $[6, \infty)$, the distance of the numbers in $(-\infty, -6]$ from 0 is also greater than or equal to 6. Thus, $[6, \infty)$ is only part of the solution of the inequality.

5. (1) $-9(x + 2) = -9x - 18$, not $-9x + 2$. (2) This would be correct if (1) were correct except that the inequality symbol should not have been reversed. (3) If (2) were correct, the right-hand side would be -5, not 8. (4) The inequality symbol should be reversed. The correct solution is

$$7 - 9x + 6x < -9(x + 2) + 10x$$
$$7 - 9x + 6x < -9x - 18 + 10x$$
$$7 - 3x < x - 18$$
$$-4x < -25$$
$$x > \frac{25}{4}.$$

6. By definition, the notation $3 < x < 5$ indicates that $3 < x$ and $x < 5$. A solution of the disjunction $3 < x$ or $x < 5$ must be in at least one of these sets but not necessarily in both, so the disjunction cannot be written as $3 < x < 5$.

Chapter 1 Test

1.
$$x + 7 = 5$$
$$x + 7 - 7 = 5 - 7$$
$$x = -2$$

The number -2 checks, so it is the solution.

2.
$$-12x = -8$$
$$\frac{-12x}{-12} = \frac{-8}{-12}$$
$$x = \frac{2}{3}$$

The number $\frac{2}{3}$ checks, so it is the solution.

3.
$$x - \frac{3}{5} = \frac{2}{3}$$
$$x - \frac{3}{5} + \frac{3}{5} = \frac{2}{3} + \frac{3}{5}$$
$$x = \frac{10}{15} + \frac{9}{15}$$
$$x = \frac{19}{15}$$

The number $\frac{19}{15}$ checks, so it is the solution.

4.
$$3y - 4 = 8$$
$$3y = 12 \quad \text{Adding 4}$$
$$y = 4 \quad \text{Dividing by 3}$$

The number 4 checks, so it is the solution.

5. $1.7y - 0.1 = 2.1 - 0.3y$

$$2y - 0.1 = 2.1 \qquad \text{Adding } 0.3y$$
$$2y = 2.2 \qquad \text{Adding } 0.1$$
$$y = 1.1 \qquad \text{Dividing by } 2$$

The number 1.1 checks, so it is the solution.

6. $5(3x + 6) = 6 - (x + 8)$

$$15x + 30 = 6 - x - 8$$
$$15x + 30 = -2 - x$$
$$16x + 30 = -2$$
$$16x = -32$$
$$x = -2$$

The number -2 checks, so it is the solution.

7. $A = 3B - C$

$$A + C = 3B \qquad \text{Adding } C$$
$$\frac{A + C}{3} = B \qquad \text{Dividing by } 3$$

8. $m = n - nt$

$$m = n(1 - t) \qquad \text{Factoring out } n$$
$$\frac{m}{1 - t} = n \qquad \text{Dividing by } 1 - t$$

9. Familiarize. Let l = the length of the room, in feet. Then $\frac{2}{3}l$ = the width. Recall that the formula for the perimeter P of a rectangle with length l and width w is $P = 2l + 2w$.

Translate. We substitute in the formula.

$$P = 2l + 2w$$
$$48 = 2l + 2 \cdot \frac{2}{3}l$$

Solve. We solve the equation.

$$\llcorner \quad 48 = 2l + 2 \cdot \frac{2}{3}l$$
$$48 = 2l + \frac{4}{3}l$$
$$48 = \frac{10}{3}l$$
$$\frac{3}{10} \cdot 48 = l$$
$$\frac{72}{5} = l, \text{ or}$$
$$14\frac{2}{5} = l$$

If $l = \frac{72}{5}$, then $\frac{2}{3}l = \frac{2}{3} \cdot \frac{72}{5} = \frac{48}{5}$, or $9\frac{3}{5}$.

Check. $9\frac{3}{5}$ ft is two-thirds of $14\frac{2}{5}$ ft and $2 \cdot 14\frac{2}{5} + 2 \cdot 9\frac{3}{5} = 2 \cdot \frac{72}{5} + 2 \cdot \frac{48}{5} = \frac{144}{5} + \frac{96}{5} = \frac{240}{5} = 48$. The answer checks.

State. The length of the room is $14\frac{2}{5}$ ft and the width is $9\frac{3}{5}$ ft.

10. Familiarize. Let c = the number of copies the firm can make. The rental cost for 3 months is $3 \cdot \$240$, or 720, and the cost of the copies is $1.5\cent \cdot c$, or $\$0.015c$.

Translate.

Rental cost	plus	copy cost	is no more than	$1500
↓	↓	↓	↓	↓
720	+	0.015c	≤	1500

Solve. We solve the inequality.

$$720 + 0.015c \le 1500$$
$$0.015c \le 780$$
$$c \le 52{,}000$$

Check. If 52,000 copies are made, the total cost is $\$720 + \$0.015(52{,}000) = \$1500$. For more than 52,000 copies, say 52,001, the total cost is $\$720 + \$0.015(52{,}001) \approx \$1500.02$. The answer checks.

State. The law firm can make at most 52,000 copies.

11. Familiarize. Let p = the former population.

Translate.

Former population	minus	12%	of	Former population	is	158,400.
↓	↓	↓	↓	↓	↓	↓
p	−	12%	·	p	=	158,400

Solve. We solve the equation.

$$p - 12\% \cdot p = 158{,}400$$
$$p - 0.12p = 158{,}400$$
$$0.88p = 158{,}400$$
$$p = 180{,}000$$

Check. 12% of 180,000 is $0.12(180{,}000) = 21{,}600$ and $180{,}000 - 21{,}600 = 158{,}400$ so the answer checks.

State. The former population of Baytown was 180,000.

12. Familiarize. Let x = the measure of the smallest angle. Then $x + 1$ and $x + 2$ represent the measures of the other two angles. Recall that the sum of the measures of the angles in a triangle is $180°$.

Translate.

The sum of the measures	is	180°
↓	↓	↓
x + (x + 1) + (x + 2)	=	180

Solve. We solve the equation.

$$x + (x + 1) + (x + 2) = 180$$
$$3x + 3 = 180$$
$$3x = 177$$
$$x = 59$$

If $x = 59$, then $x + 1 = 59 + 1 = 60$ and $x + 2 = 59 + 2 = 61$.

Check. The numbers 59, 60, and 61 are consecutive integers and $59° + 60° + 61° = 180°$. The answer checks.

State. The measures of the angles are $59°$, $60°$, and $61°$.

13. First we will find how long it takes the boat to travel 36 mi downstream.

Familiarize. We will use the formula $d = rt$. Let $t =$ the time, in hours, it will take the boat to travel 36 mi downstream. The speed of the boat traveling downstream is $12 + 3$, or 15 mph.

Translate.
$$d = rt$$
$$36 = 15t$$

Solve. We solve the equation.
$$36 = 15t$$
$$\frac{12}{5} = t, \text{ or}$$
$$2\frac{2}{5} = t$$

Check. If the boat travels at 15 mph for $\frac{12}{5}$ hr, it travels $15 \cdot \frac{12}{5}$, or 36 mi. The answer checks.

State. It will take the boat $2\frac{2}{5}$ hr to travel 36 mi downstream.

Now we find how long it will take the boat to travel 36 mi upstream.

Familiarize. We will use the formula $d = rt$. Let $t =$ the time, in hours, it will take the boat to travel 36 mi upstream. The speed of the boat traveling upstream is $12 - 3$, or 9 mph.

Translate.
$$d = rt$$
$$36 = 9t$$

Solve. We solve the equation.
$$36 = 9t$$
$$4 = t$$

Check. If the boat travels at 9 mph for 4 hr, it travels $9 \cdot 4$, or 36 mi. The answer checks.

State. It will take the boat 4 hr to travel 36 mi upstream.

14. Interval notation for $\{x| -3 < x \le 2\}$ is $(-3, 2]$.

15. Interval notation is $(-4, \infty)$.

16. $x - 2 \le 4$
$$x \le 6 \quad \text{Adding 2}$$
The solution set is $\{x|x \le 6\}$, or $(-\infty, 6]$.

17. $-4y - 3 \ge 5$
$$-4y \ge 8$$
$$y \le -2 \quad \text{Reversing the inequality symbol}$$
The solution set is $\{y|y \le -2\}$, or $(-\infty, -2]$.

18. $x - 4 \ge 6$
$$x \ge 10 \quad \text{Adding 4}$$
The solution set is $\{x|x \ge 10\}$, or $[10, \infty)$.

19. $-0.6y < 30$
$$y > -50 \quad \text{Reversing the inequality symbol}$$
The solution set is $\{y|y > -50\}$, or $(-50, \infty)$.

20. $3a - 5 \le -2a + 6$
$$5a - 5 \le 6$$
$$5a \le 11$$
$$a \le \frac{11}{5}$$
The solution set is $\left\{a \middle| a \le \frac{11}{5}\right\}$, or $\left(-\infty, \frac{11}{5}\right]$.

21. $-5y - 1 > -9y + 3$
$$4y - 1 > 3$$
$$4y > 4$$
$$y > 1$$
The solution set is $\{y|y > 1\}$, or $(1, \infty)$.

22. $4(5 - x) < 2x + 5$
$$20 - 4x < 2x + 5$$
$$20 - 6x < 5$$
$$-6x < -15$$
$$x > \frac{5}{2}$$
The solution set is $\left\{x \middle| x > \frac{5}{2}\right\}$, or $\left(\frac{5}{2}, \infty\right)$.

23. $-8(2x + 3) + 6(4 - 5x) \ge 2(1 - 7x) - 4(4 + 6x)$
$$-16x - 24 + 24 - 30x \ge 2 - 14x - 16 - 24x$$
$$-46x \ge -14 - 38x$$
$$-8x \ge -14$$
$$x \le \frac{7}{4}$$
The solution set is $\left\{x \middle| x \le \frac{7}{4}\right\}$, or $\left(-\infty, \frac{7}{4}\right]$.

24. Familiarize. Let $t =$ the length of time of the move, in hours. Then Motivated Movers charges $105 + 30t$ and Quick-Pak Moving charges $80t$.

Translate.

Cost of Quick-Pak	is more than	Cost of Motivated Movers
$80t$	$>$	$105 + 30t$

Solve. We solve the inequality.
$$80t > 105 + 30t$$
$$50t > 105$$
$$t > \frac{21}{10}, \text{ or } 2\frac{1}{10}$$

Check. When $t = \frac{21}{10}$ hr, Motivated Movers charges $105 + 30 \cdot \frac{21}{10}$, or \$168, and Quick-Pak charges $80 \cdot \frac{21}{10}$, or \$168. For a value of t greater than $2\frac{1}{10}$, say 3, Motivated Movers charges $105 + 30 \cdot 3$, or \$195, and Quick-Pak charges $80 \cdot 3$, or \$240, so Quick-Pak is more expensive. This partial check tells us that the answer is probably correct.

State. Quick-Pak is more expensive for moves more than $2\frac{1}{10}$ hr. The solution set is $\left\{t \middle| t > 2\frac{1}{10} \text{ hr}\right\}$.

25. Familiarize. We will use the formula $P = 1 + \frac{d}{33}$.

Translate. We want to find those values of P for which

$$2 \leq P \leq 8$$

or

$$2 \leq 1 + \frac{d}{33} \leq 8.$$

Solve. We solve the inequality.

$$2 \leq 1 + \frac{d}{33} \leq 8$$

$$1 \leq \frac{d}{33} \leq 7$$

$$33 \leq d \leq 231$$

Check. We could do a partial check by substituting some values for d in the formula. The result checks.

State. The pressure is at least 2 atm and at most 8 atm for depths d in the set $\{d | 33 \text{ ft} \leq d \leq 231 \text{ ft}\}$.

26. Interval notation for $-3 \leq x \leq 4$ is $[-3, 4]$.

27. Interval notation for $x < -3 \text{ or } x > 4$ is $(-\infty, -3) \cup (4, \infty)$.

28. $5 - 2x \leq 1 \quad and \quad 3x + 2 \geq 14$

$\qquad -2x \leq -4 \quad and \quad 3x \geq 12$

$\qquad x \geq 2 \quad and \qquad x \geq 4$

The intersection of $\{x | x \geq 2\}$ and $\{x | x \geq 4\}$, is $\{x | x \geq 4\}$, or $[4, \infty)$.

29. $-3 < x - 2 < 4$

$\qquad -1 < x < 6 \qquad$ Adding 2

The solution set is $\{x | -1 < x < 6\}$, or $(-1, 6)$.

30. $-11 \leq -5x - 2 < 0$

$\qquad -9 \leq -5x < 2$

$\qquad \frac{9}{5} \geq x > -\frac{2}{5}$

The solution set is $\left\{x \middle| \frac{9}{5} \geq x > -\frac{2}{5}\right\}$, or $\left\{x \middle| -\frac{2}{5} < x \leq \frac{9}{5}\right\}$, or $\left(-\frac{2}{5}, \frac{9}{5}\right]$.

31. $-3x > 12 \quad or \quad 4x > -10$

$\qquad x < -4 \quad or \quad x > -\frac{5}{2}$

The solution set is $\left\{x \middle| x < -4 \text{ or } x > -\frac{5}{2}\right\}$, or $(-\infty, -4) \cup \left(-\frac{5}{2}, \infty\right)$.

32. $x - 7 \leq -5 \quad or \quad x - 7 \geq -10$

$\qquad x \leq 2 \quad or \qquad x \geq -3$

The union of $(-\infty, 2]$ and $[-3, \infty)$ is the set of all real numbers, or $(-\infty, \infty)$.

33. $3x - 2 < 7 \quad or \quad x - 2 > 4$

$\qquad 3x < 9 \quad or \qquad x > 6$

$\qquad x < 3 \quad or \qquad x > 6$

The solution set is $\{x | x < 3 \text{ or } x > 6\}$, or $(-\infty, 3) \cup (6, \infty)$.

34. $\left|\frac{7}{x}\right| = \frac{|7|}{|x|} = \frac{7}{|x|}$

35. $\left|\frac{-6x^2}{3x}\right| = |-2x| = |-2| \cdot |x| = 2|x|$

36. $|4.8 - (-3.6)| = |4.8 + 3.6| = |8.4| = 8.4$, or $|-3.6 - 4.8| = |-8.4| = 8.4$

37. $\{1, 3, 5, 7, 9\} \cap \{3, 5, 11, 13\} = \{3, 5\}$

38. $\{1, 3, 5, 7, 9\} \cup \{3, 5, 11, 13\} = \{1, 3, 5, 7, 9, 11, 13\}$

39. $|x| = 9$

$\qquad x = -9 \quad or \quad x = 9 \quad$ Absolute-value principle

The solution set is $\{-9, 9\}$.

40. $|x - 3| = 9$

$\qquad x - 3 = -9 \quad or \quad x - 3 = 9$

$\qquad x = -6 \quad or \qquad x = 12$

The solution set is $\{-6, 12\}$.

41. $|x + 10| = |x - 12|$

$\quad x + 10 = x - 12 \quad or \quad x + 10 = -(x - 12)$

$\qquad 10 = -12 \qquad or \quad x + 10 = -x + 12$

$\qquad 10 = -12 \qquad or \qquad 2x = 2$

$\qquad 10 = -12 \qquad or \qquad x = 1$

The first equation has no solution. The solution of the second equation is 1, so the solution set is $\{1\}$.

42. $|2 - 5x| = -10$

The absolute value of a number is always nonnegative. Thus, the solution set is $\emptyset$.

43. $|4x - 1| < 4.5$

$\qquad -4.5 < 4x - 1 < 4.5$

$\qquad -3.5 < 4x < 5.5$

$\qquad -0.875 < x < 1.375$

The solution set is $\{x | -0.875 < x < 1.375\}$, or $(-0.875, 1.375)$. This could also be expressed as $\left\{x \middle| -\frac{7}{8} < x < \frac{11}{8}\right\}$, or $\left(-\frac{7}{8}, \frac{11}{8}\right)$.

44. $|x| > 3$

$x < -3 \ \ or \ \ x > 3$

The solution set is $\{x | x < -3 \ or \ x > 3\}$, or $(-\infty, -3) \cup (3, \infty)$.

45. $\left| \dfrac{6 - x}{7} \right| \le 15$

$-15 \le \dfrac{6 - x}{7} \le 15$

$-105 \le 6 - x \le 105 \quad$ Multiplying by 7

$-111 \le -x \le 99$

$111 \ge x \ge -99$

The solution set is $\{x | 111 \ge x \ge -99\}$, or $\{x | -99 \le x \le 111\}$, or $[-99, 111]$.

46. $|-5x - 3| \ge 10$

$-5x - 3 \le -10 \ \ or \ \ -5x - 3 \ge 10$

$-5x \le -7 \quad or \qquad -5x \ge 13$

$x \ge \dfrac{7}{5} \quad or \qquad x \le -\dfrac{13}{5}$

The solution set is $\left\{ x \middle| x \le -\dfrac{13}{5} \ or \ x \ge \dfrac{7}{5} \right\}$, or

$\left(-\infty, -\dfrac{13}{5} \right] \cup \left[\dfrac{7}{5}, \infty \right)$.

47. $2(3x - 6) + 5 = 1 - (x - 6)$

$6x - 12 + 5 = 1 - x + 6$

$6x - 7 = 7 - x$

$7x - 7 = 7$

$7x = 14$

$x = 2$

The number 2 checks, so it is the solution. The solution is between 1 and 3, so answer C is correct.

48. $|3x - 4| \le -3$

The absolute value of a number is always nonnegative, so $|3x - 4|$ cannot be less than -3. Thus, the solution set is $\emptyset$.

49. $7x < 8 - 3x < 6 + 7x$

$7x < 8 - 3x \ \ and \ \ 8 - 3x < 6 + 7x$

$10x < 8 \qquad and \quad -10x < -2$

$x < \dfrac{4}{5} \qquad and \qquad x > \dfrac{1}{5}$

The intersection of $\left\{ x \middle| x < \dfrac{4}{5} \right\}$ and $\left\{ x \middle| x > \dfrac{1}{5} \right\}$ is

$\left\{ x \middle| \dfrac{1}{5} < x < \dfrac{4}{5} \right\}$, or $\left(\dfrac{1}{5}, \dfrac{4}{5} \right)$.

Chapter 2

Graphs, Functions, and Applications

RC1. False; see page 159 in the text.

RC3. True; see page 159 in the text.

RC5. True; see page 159 in the text.

RC7. $3x + 4y = 0$

$4y = -3x$

$y = -\dfrac{3}{4}x$

The answer is (d).

RC9. $4x - 3y = -4$

$-3y = -4x - 4$

$y = \dfrac{4}{3}x + \dfrac{4}{3}$

The answer is (a).

1. $A(4,1)$ is 4 units right and 1 unit up.

$B(2,5)$ is 2 units right and 5 units up.

$C(0,3)$ is 0 units left or right and 3 units up.

$D(0,-5)$ is 0 units left or right and 5 units down.

$E(6,0)$ is 6 units right and 0 units up or down.

$F(-3,0)$ is 3 units left and 0 units up or down.

$G(-2,-4)$ is 2 units left and 4 units down.

$H(-5,1)$ is 5 units left and 1 unit up.

$J(-6,6)$ is 6 units left and 6 units up.

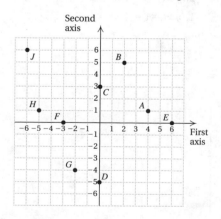

3.

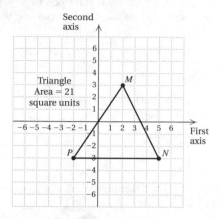

A triangle is formed. The area of a triangle is found by using the formula $A = \dfrac{1}{2}bh$. In this triangle the base and height are respectively 7 units and 6 units.

$$A = \frac{1}{2}bh = \frac{1}{2} \cdot 7 \cdot 6 = \frac{42}{2} = 21 \text{ square units}$$

5. We substitute 1 for x and -1 for y (alphabetical order of variables).

$$\begin{array}{c|c} y = 2x - 3 \\ \hline -1 \ ? \ 2 \cdot 1 - 3 \\ \ \ \ | \ -1 & \text{TRUE} \end{array}$$

Thus, $(1,-1)$ is a solution of the equation.

7. We substitute 3 for x and 5 for y (alphabetical order of variables).

$$\begin{array}{c|c} 4x - y = 7 \\ \hline 4 \cdot 3 - 5 \ ? \ 7 \\ 12 - 5 \ | \\ 7 \ | & \text{TRUE} \end{array}$$

Thus, $(3,5)$ is a solution of the equation.

9. We substitute 0 for a and $\dfrac{3}{5}$ for b (alphabetical order of variables).

$$\begin{array}{c|c} 2a + 5b = 7 \\ \hline 2 \cdot 0 + 5 \cdot \dfrac{3}{5} \ ? \ 7 \\ 0 + 3 \ | \\ 3 \ | & \text{FALSE} \end{array}$$

Thus, $\left(0, \dfrac{3}{5}\right)$ is not a solution of the equation.

11. To show that a pair is a solution, we substitute, replacing x with the first coordinate and y with the second coordinate in each pair.

$$\frac{y = 4 - x}{5\ ?\ 4 - (-1)}$$
$$\begin{array}{c|c} 4 + 1 \\ 5 & \text{TRUE} \end{array}$$

$$\frac{y = 4 - x}{1\ ?\ 4 - 3}$$
$$\begin{array}{c|c} 1 & \text{TRUE} \end{array}$$

In each case the substitution results in a true equation. Thus, $(-1, 5)$ and $(3, 1)$ are both solutions of $y = 4 - x$. We plot these points and sketch the line passing through them.

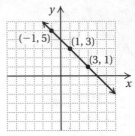

The line appears to pass through $(1, 3)$ also. We check to determine if $(1, 3)$ is a solution.

$$\frac{y = 4 - x}{3\ ?\ 4 - 1}$$
$$\begin{array}{c|c} 3 & \text{TRUE} \end{array}$$

Thus, $(1, 3)$ is another solution. There are other correct answers, including $(-2, 6)$, $(0, 4)$, $(2, 2)$, $(4, 0)$, and $(5, -1)$.

13. To show that a pair is a solution, we substitute, replacing x with the first coordinate and y with the second coordinate in each pair.

$$\frac{3x + y = 7}{3 \cdot 2 + 1\ ?\ 7}$$
$$\begin{array}{c|c} 6 + 1 \\ 7 & \text{TRUE} \end{array}$$

$$\frac{3x + y = 7}{3 \cdot 4 - 5\ ?\ 7}$$
$$\begin{array}{c|c} 12 - 5 \\ 7 & \text{TRUE} \end{array}$$

In each case the substitution results in a true equation. Thus, $(2, 1)$ and $(4, -5)$ are both solutions of $3x + y = 7$. We plot these points and sketch the line passing through them.

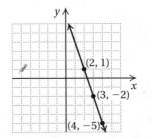

The line appears to pass through $(1, 4)$ also. We check to determine if $(1, 4)$ is a solution of $3x + y = 7$.

$$\frac{3x + y = 7}{3 \cdot 1 + 4\ ?\ 7}$$
$$\begin{array}{c|c} 3 + 4 \\ 7 & \text{TRUE} \end{array}$$

Thus, $(1, 4)$ is another solution. There are other correct answers, including $(3, -2)$.

15. To show that a pair is a solution, we substitute, replacing x with the first coordinate and y with the second coordinate in each pair.

$$\frac{6x - 3y = 3}{6 \cdot 1 - 3 \cdot 1\ ?\ 3}$$
$$\begin{array}{c|c} 6 - 3 \\ 3 & \text{TRUE} \end{array}$$

$$\frac{6x - 3y = 3}{6(-1) - 3(-3)\ ?\ 3}$$
$$\begin{array}{c|c} -6 + 9 \\ 3 & \text{TRUE} \end{array}$$

In each case the substitution results in a true equation. Thus, $(1, 1)$ and $(-1, -3)$ are both solutions of $6x - 3y = 3$. We plot these points and sketch the line passing through them.

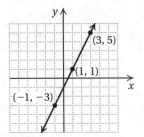

The line appears to pass through $(0, -1)$ also. We check to determine if $(0, -1)$ is a solution of $6x - 3y = 3$.

$$\frac{6x - 3y = 3}{6 \cdot 0 - 3(-1)\ ?\ 3}$$
$$\begin{array}{c|c} 0 + 3 \\ 3 & \text{TRUE} \end{array}$$

Thus, $(0, -1)$ is another solution. There are other correct answers including $(-2, -5)$, $(2, 3)$, and $(3, 5)$.

17. $y = x - 1$

We find some ordered pairs that are solutions.

When $x = -2$, $y = -2 - 1 = -3$.

When $x = -1$, $y = -1 - 1 = -2$.

When $x = 0$, $y = 0 - 1 = -1$.

When $x = 1$, $y = 1 - 1 = 0$.

When $x = 2$, $y = 2 - 1 = 1$.

When $x = 3$, $y = 3 - 1 = 2$.

x	y
-2	-3
-1	-2
0	-1
1	0
2	1
3	2

Plot these points, draw the line they determine, and label it $y = x - 1$.

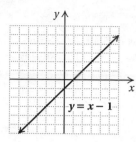

19. $y = x$

We find some ordered pairs that are solutions, plot them, and draw and label the line.

When $x = -2$, $y = -2$.

When $x = -1$, $y = -1$.

When $x = 0$, $y = 0$.

When $x = 1$, $y = 1$.

When $x = 2$, $y = 2$.

When $x = 3$, $y = 3$.

x	y
-2	-2
-1	-1
0	0
1	1
2	2
3	3

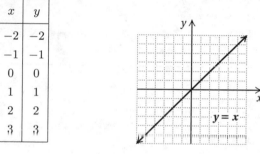

21. $y = \dfrac{1}{4}x$

We find some ordered pairs that are solutions, using multiples of 4 for x to avoid fractions. Then we plot these points and draw and label the line.

When $x = -4$, $y = \dfrac{1}{4}(-4) = -1$.

When $x = 0$, $y = \dfrac{1}{4} \cdot 0 = 0$.

When $x = 4$, $y = \dfrac{1}{4} \cdot 4 = 1$.

x	y	(x, y)
-4	-1	$(-4, -1)$
0	0	$(0, 0)$
4	1	$(4, 1)$

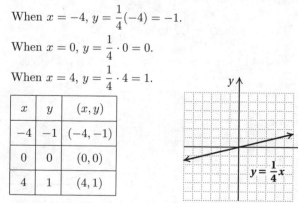

23. $y = 3 - x$

We find some ordered pairs that are solutions, plot them, and draw and label the line.

When $x = -2$, $y = 3 - (-2) = 3 + 2 = 5$.

When $x = 1$, $y = 3 - 1 = 2$.

When $x = 5$, $y = 3 - 5 = -2$.

x	y	(x, y)
-2	5	$(-2, 5)$
1	2	$(1, 2)$
5	-2	$(5, -2)$

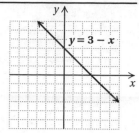

25. $y = 5x - 2$

We find some ordered pairs that are solutions, plot them, and draw and label the line.

When $x = -1$, $y = 5(-1) - 2 = -5 - 2 = -7$.

When $x = 0$, $y = 5 \cdot 0 - 2 = -2$.

When $x = 1$, $y = 5 \cdot 1 - 2 = 5 - 2 = 3$.

x	y	(x, y)
-1	-7	$(-1, -7)$
0	-2	$(0, -2)$
1	3	$(1, 3)$

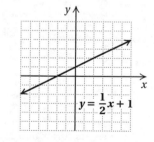

27. $y = \dfrac{1}{2}x + 1$

We find some ordered pairs that are solutions, using even numbers for x to avoid fractions. Then we plot these points and draw and label the line.

When $x = -4$, $y = \dfrac{1}{2}(-4) + 1 = -2 + 1 = -1$.

When $x = 0$, $y = \dfrac{1}{2} \cdot 0 + 1 = 1$.

When $x = 4$, $y = \dfrac{1}{2} \cdot 4 + 1 = 2 + 1 = 3$.

x	y	(x, y)
-4	-1	$(-4, -1)$
0	1	$(0, 1)$
4	3	$(4, 3)$

29. $x + y = 5$

First we solve for y.

$$x + y = 5$$
$$y = 5 - x$$

Now we find some ordered pairs that are solutions, plot them, and draw and label the line.

When $x = 0$, $y = 5 - 0 = 5$.

When $x = 2$, $y = 5 - 2 = 3$.

When $x = 5$, $y = 5 - 5 = 0$.

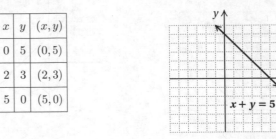

x	y	(x,y)
0	5	$(0,5)$
2	3	$(2,3)$
5	0	$(5,0)$

31. $y = -\dfrac{5}{3}x - 2$

We find some ordered pairs that are solutions, using multiples of 3 for x to avoid fractions. Then we plot these points and draw and label the line.

When $x = -3$, $y = -\dfrac{5}{3}(-3) - 2 = 5 - 2 = 3$.

When $x = 0$, $y = -\dfrac{5}{3} \cdot 0 - 2 = -2$.

When $x = 3$, $y = -\dfrac{5}{3} \cdot 3 - 2 = -5 - 2 = -7$.

x	y	(x,y)
-3	3	$(-3,3)$
0	-2	$(0,-2)$
3	-7	$(3,-7)$

33. $x + 2y = 8$

First we solve for y.

$$x + 2y = 8$$
$$2y = 8 - x$$
$$\frac{1}{2} \cdot 2y = \frac{1}{2}(8 - x)$$
$$y = \frac{1}{2} \cdot 8 - \frac{1}{2} \cdot x$$
$$y = 4 - \frac{1}{2}x$$
$$y = -\frac{1}{2}x + 4$$

Now we find some ordered pairs that are solutions, using even numbers for x to avoid fractions. Then we plot these points and draw and label the line.

When $x = -2$, $y = -\dfrac{1}{2}(-2) + 4 = 1 + 4 = 5$.

When $x = 2$, $y = -\dfrac{1}{2} \cdot 2 + 4 = -1 + 4 = 3$.

When $x = 4$, $y = -\dfrac{1}{2} \cdot 4 + 4 = -2 + 4 = 2$.

x	y	(x,y)
-2	5	$(-2,5)$
2	3	$(2,3)$
4	2	$(4,2)$

35. $y = \dfrac{3}{2}x + 1$

We find some ordered pairs that are solutions, using even numbers for x to avoid fractions. Then we plot these points and draw and label the line.

When $x = -4$, $y = \dfrac{3}{2}(-4) + 1 = -6 + 1 = -5$.

When $x = 0$, $y = \dfrac{3}{2} \cdot 0 + 1 = 1$.

When $x = 2$, $y = \dfrac{3}{2} \cdot 2 + 1 = 3 + 1 = 4$.

x	y	(x,y)
-4	-5	$(-4,-5)$
0	1	$(0,1)$
2	4	$(2,4)$

37. $8y + 2x = 4$

First we solve for y.

$$8y + 2x = 4$$
$$8y = 4 - 2x$$
$$\frac{1}{8} \cdot 8y = \frac{1}{8}(4 - 2x)$$
$$y = \frac{1}{8} \cdot 4 - \frac{1}{8} \cdot 2x$$
$$y = \frac{1}{2} - \frac{1}{4}x$$
$$y = -\frac{1}{4}x + \frac{1}{2}$$

Now we find some ordered pairs that are solutions, plot them, and draw and label the line.

When $x = -4$, $y = -\dfrac{1}{4}(-4) + \dfrac{1}{2} = 1 + \dfrac{1}{2} = \dfrac{3}{2}$.

When $x = 0$, $y = -\dfrac{1}{4} \cdot 0 + \dfrac{1}{2} = \dfrac{1}{2}$.

When $x = 4$, $y = -\dfrac{1}{4} \cdot 4 + \dfrac{1}{2} = -1 + \dfrac{1}{2} = -\dfrac{1}{2}$.

x	y	(x,y)
-4	$\dfrac{3}{2}$	$\left(-4, \dfrac{3}{2}\right)$
0	$\dfrac{1}{2}$	$\left(0, \dfrac{1}{2}\right)$
4	$-\dfrac{1}{2}$	$\left(4, -\dfrac{1}{2}\right)$

39. $8y + 2x = -4$

First we solve for y.

$$8y + 2x = -4$$
$$8y = -4 - 2x$$
$$\frac{1}{8} \cdot 8y = \frac{1}{8}(-4 - 2x)$$
$$y = \frac{1}{8}(-4) - \frac{1}{8} \cdot 2x$$
$$y = -\frac{1}{2} - \frac{1}{4}x$$
$$y = -\frac{1}{4}x - \frac{1}{2}$$

Now we find some ordered pairs that are solutions, plot them, and draw and label the line.

When $x = -4$, $y = -\frac{1}{4}(-4) - \frac{1}{2} = 1 - \frac{1}{2} = \frac{1}{2}$.

When $x = 0$, $y = -\frac{1}{4} \cdot 0 - \frac{1}{2} = -\frac{1}{2}$.

When $x = 4$, $y = -\frac{1}{4} \cdot 4 - \frac{1}{2} = -1 - \frac{1}{2} = -\frac{3}{2}$.

x	y	(x, y)
-4	$\frac{1}{2}$	$\left(-4, \frac{1}{2}\right)$
0	$-\frac{1}{2}$	$\left(0, -\frac{1}{2}\right)$
4	$-\frac{3}{2}$	$\left(4, -\frac{3}{2}\right)$

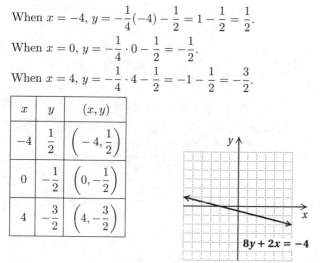

41. $y = x^2$

To find an ordered pair, we choose any number for x and then determine y. For example, if $x = 2$, then $y = 2^2 = 4$. We find several ordered pairs, plot them, and connect them with a smooth curve.

x	y
-2	4
-1	1
0	0
1	1
2	4

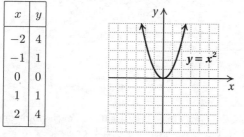

43. $y = x^2 + 2$

To find an ordered pair, we choose any number for x and then determine y. For example, if $x = 2$, then $y = 2^2 + 2 = 4 + 2 = 6$. We find several ordered pairs, plot them, and connect them with a smooth curve.

x	y
-2	6
-1	3
0	2
1	3
2	6

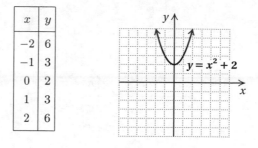

45. $y = x^2 - 3$

To find an ordered pair, we choose any number for x and then determine y. For example, if $x = 2$, then $y = 2^2 - 3 = 4 - 3 = 1$. We find several ordered pairs, plot them, and connect them with a smooth curve.

x	y
-2	1
-1	-2
0	-3
1	-2
2	1
3	6

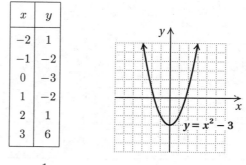

47. $y = -\frac{1}{x}$

To find an ordered pair, we choose any number for x and then determine y. For example, if $x = -4$, then $y = \frac{1}{4}$. We find several ordered pairs, plot them, and connect them with a smooth curve.

x	y
-4	$\frac{1}{4}$
-2	$\frac{1}{2}$
$-\frac{1}{2}$	2
$\frac{1}{2}$	-2
2	$-\frac{1}{2}$
4	$-\frac{1}{4}$

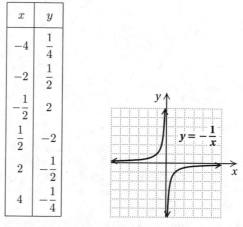

Note that we cannot use 0 as a first-coordinate, since $-1/0$ is undefined. Thus, the graph has two branches, one on each side of the y-axis.

49. $y = |x - 2|$

To find an ordered pair, we choose any number for x and then determine y. For example, if $x = 5$, then $y = |5 - 2| = |3| = 3$. We find several ordered pairs, plot them, and connect them.

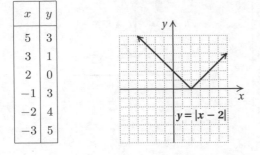

x	y
5	3
3	1
2	0
−1	3
−2	4
−3	5

$y = |x - 2|$

51. $y = x^3$

To find an ordered pair, we choose any number for x and then determine y. For example, if $x = -1$, then $y = (-1)^3 = -1$. We find several ordered pairs, plot them, and connect them with a smooth curve.

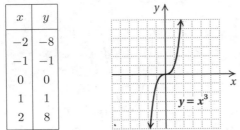

x	y
−2	−8
−1	−1
0	0
1	1
2	8

$y = x^3$

53. $-3 < 2x - 5 \le 10$

$\quad\quad 2 < 2x \le 15 \quad\quad$ Adding 5

$\quad\quad 1 < x \le \dfrac{15}{2} \quad\quad$ Dividing by 2

The solution set is $\left\{ x \middle| 1 < x \le \dfrac{15}{2} \right\}$, or $\left(1, \dfrac{15}{2} \right]$.

55. $3x - 5 \le -12 \ \ or \ \ 3x - 5 \ge 12$

$\quad\quad 3x \le -7 \quad or \quad 3x \ge 17$

$\quad\quad x \le -\dfrac{7}{3} \quad or \quad x \ge \dfrac{17}{3}$

The solution set is $\left\{ x \middle| x \le -\dfrac{7}{3} \ or \ x \ge \dfrac{17}{3} \right\}$, or

$\left(-\infty, -\dfrac{7}{3} \right] \cup \left[\dfrac{17}{3}, \infty \right)$.

57. *Familiarize*. Let $k =$ the number of people waiting for a kidney transplant. Then $k - 81,528 =$ the number waiting for a liver transplant, and the total number of people waiting for a kidney or a liver was $k + (k - 81,528)$, or $2k - 81,528$.

***Translate*.**

$$\underbrace{\text{Total number waiting for a kidney or a liver}} \ \ \underset{\downarrow}{\text{was}} \ \ \underset{\downarrow}{113,162}$$

$$\underset{\downarrow}{2k - 81,528} \quad \underset{=}{} \quad 113,162$$

***Solve*.**

$\quad 2k - 81,528 = 113,162$

$\quad\quad\quad 2k = 194,690 \quad$ Adding 81,528

$\quad\quad\quad\ k = 97,345$

If $k = 97,345$, then $k - 81,528 = 97,345 - 81,528 = 15,817$.

***Check*.** Since $97,345 + 15,817 = 113,162$ and since $15,817$ is $81,528$ fewer than $97,345$, the answer checks.

***State*.** There were $97,345$ people waiting for a kidney transplant and $15,817$ waiting for a liver transplant.

59. *Familiarize*. Let $m =$ the distance after the first $\dfrac{1}{2}$ mile, in units of $\dfrac{1}{4}$ miles. Then the taxi ride costs $\$2.00 + \$1.05m$ and the total distance is $\dfrac{1}{2} + \dfrac{m}{4}$ miles.

***Translate*.** The taxi ride costs $\$19.85$, so we have

$\quad 2.00 + 1.05m = 19.85.$

***Solve*.** We solve the equation.

$\quad\quad 2.00 + 1.05m = 19.85$

$\quad\quad 200 + 105m = 1985 \quad$ Multiplying by 100

$\quad\quad\quad\quad 105m = 1785$

$\quad\quad\quad\quad\quad\ m = 17$

When $m = 17$, the total distance is $\dfrac{1}{2} + \dfrac{17}{4} = \dfrac{19}{4} = 4\dfrac{3}{4}$ mi.

***Check*.** If the distance is $\dfrac{1}{2}$ mi plus 17 additional $\dfrac{1}{4}$ mi segments, then the fare is $\$2.00 + \$1.05(17) = \$2.00 + \$17.85 = \$19.85$. The answer checks.

***State*.** It is $4\dfrac{3}{4}$ mi from Jen's office to the South Bay Health Center.

61. $y = x^3 - 3x + 2$

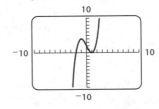

63. $y = 1/(x - 2)$

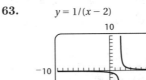

65. Note that the sum of the coordinates of each point on the graph is 4. Thus, we have $x + y = 4$, or $y = -x + 4$.

67. This is the graph of $y = |x|$ (see Example 9) with 3 subtracted from each x-coordinate. Thus, we have $y = |x| - 3$.

Exercise Set 2.2

RC1. $f(\hat{2}) = 3$

RC3. $f(-2) = -5$

1. Yes; each member of the domain is matched to only one member of the range.

3. Yes; each member of the domain is matched to only one member of the range.

5. No; a member of the domain is matched to more than one member of the range. In fact, each member of the domain is matched to 2 members of the range.

7. No; a member of the domain is matched to more than one member of the range. In fact, each member of the domain is matched to 3 members of the range.

9. The correspondence is not a function, since a number can be the area of more than one triangle.

11. Yes; each member of the domain is matched to only one member of the range.

13. $f(x) = x + 5$

   a) $f(4) = 4 + 5 = 9$

   b) $f(7) = 7 + 5 = 12$

   c) $f(-3) = -3 + 5 = 2$

   d) $f(0) = 0 + 5 = 5$

   e) $f(2.4) = 2.4 + 5 = 7.4$

   f) $f\left(\dfrac{2}{3}\right) = \dfrac{2}{3} + 5 = 5\dfrac{2}{3}$

15. $h(p) = 3p$

   a) $h(-7) = 3(-7) = -21$

   b) $h(5) = 3 \cdot 5 = 15$

   c) $h\left(\dfrac{2}{3}\right) = 3 \cdot \dfrac{2}{3} = \dfrac{6}{3} = 2$

   d) $h(0) = 3 \cdot 0 = 0$

   e) $h(6a) = 3 \cdot 6a = 18a$

   f) $h(a+1) = 3(a+1) = 3a + 3$

17. $g(s) = 3s + 4$

   a) $g(1) = 3 \cdot 1 + 4 = 3 + 4 = 7$

   b) $g(-7) = 3(-7) + 4 = -21 + 4 = -17$

   c) $g\left(\dfrac{2}{3}\right) = 3 \cdot \dfrac{2}{3} + 4 = 2 + 4 = 6$

   d) $g(0) = 3 \cdot 0 + 4 = 0 + 4 = 4$

   e) $g(a-2) = 3(a-2) + 4 = 3a - 6 + 4 = 3a - 2$

   f) $g(a+h) = 3(a+h) + 4 = 3a + 3h + 4$

19. $f(x) = 2x^2 - 3x$

   a) $f(0) = 2 \cdot 0^2 - 3 \cdot 0 = 0 - 0 = 0$

   b) $f(-1) = 2(-1)^2 - 3(-1) = 2 + 3 = 5$

   c) $f(2) = 2 \cdot 2^2 - 3 \cdot 2 = 8 - 6 = 2$

   d) $f(10) = 2 \cdot 10^2 - 3 \cdot 10 = 200 - 30 = 170$

   e) $f(-5) = 2(-5)^2 - 3(-5) = 50 + 15 = 65$

   f) $f(4a) = 2(4a)^2 - 3(4a) = 32a^2 - 12a$

21. $f(x) = |x| + 1$

   a) $f(0) = |0| + 1 = 0 + 1 = 1$

   b) $f(-2) = |-2| + 1 = 2 + 1 = 3$

   c) $f(2) = |2| + 1 = 2 + 1 = 3$

   d) $f(-10) = |-10| + 1 = 10 + 1 = 11$

   e) $f(a-1) = |a-1| + 1$

   f) $f(a+h) = |a+h| + 1$

23. $f(x) = x^3$

   a) $f(0) = 0^3 = 0$

   b) $f(-1) = (-1)^3 = -1$

   c) $f(2) = 2^3 = 8$

   d) $f(10) = 10^3 = 1000$

   e) $f(-5) = (-5)^3 = -125$

   f) $f(-3a) = (-3a)^3 = -27a^3$

25. In 1980, $s = 1980 - 1945 = 35$. We find $A(35)$.

$A(35) = 0.044(35) + 59 = 1.54 + 59 \approx 60.5$

The average age of U.S. senators was about 60.5 years in 1980.

In 2013, $s = 2013 - 1945 = 68$. We find $A(68)$.

$A(68) = 0.044(68) + 59 = 2.992 + 59 \approx 62$

The average age of U.S. senators was about 62 years in 2013.

27. $P(d) = 1 + \dfrac{d}{33}$

$P(20) = 1 + \dfrac{20}{33} = 1\dfrac{20}{33}$ atm

$P(30) = 1 + \dfrac{30}{33} = 1\dfrac{10}{11}$ atm

$P(100) = 1 + \dfrac{100}{33} = 1 + 3\dfrac{1}{33} = 4\dfrac{1}{33}$ atm

29. $W(d) = 0.112d$

$W(16) = 0.112(16) = 1.792$ cm

$W(25) = 0.112(25) = 2.8$ cm

$W(100) = 0.112(100) = 11.2$ cm

31. Graph $f(x) = -2x$.

Make a list of function values in a table.

$f(-2) = -2(-2) = 4$

$f(-1) = -2(-1) = 2$

$f(0) = -2 \cdot 0 = 0$

$f(2) = -2 \cdot 2 = -4$

x	$f(x)$
-2	4
-1	2
0	0
2	-4

Plot these points and connect them.

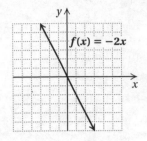

$f(x) = -2x$

33. Graph $f(x) = 3x - 1$.

Make a list of function values in a table.

$f(-1) = 3(-1) - 1 = -3 - 1 = -4$

$f(0) = 3 \cdot 0 - 1 = 0 - 1 = -1$

$f(1) = 3 \cdot 1 - 1 = 3 - 1 = 2$

$f(2) = 3 \cdot 2 - 1 = 6 - 1 = 5$

x	$f(x)$
-1	-4
0	-1
1	2
2	5

Plot these points and connect them.

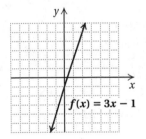

$f(x) = 3x - 1$

35. Graph $g(x) = -2x + 3$.

Make a list of function values in a table.

$g(-1) = -2(-1) + 3 = 2 + 3 = 5$

$g(0) = -2 \cdot 0 + 3 = 0 + 3 = 3$

$g(3) = -2 \cdot 3 + 3 = -6 + 3 = -3$

x	$g(x)$
-1	5
0	3
3	-3

Plot these points and connect them.

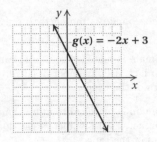

$g(x) = -2x + 3$

37. Graph $f(x) = \dfrac{1}{2}x + 1$.

Make a list of function values in a table.

$f(-2) = \dfrac{1}{2}(-2) + 1 = -1 + 1 = 0$

$f(0) = \dfrac{1}{2} \cdot 0 + 1 = 0 + 1 = 1$

$f(4) = \dfrac{1}{2} \cdot 4 + 1 = 2 + 1 = 3$

x	$f(x)$
-2	0
0	1
4	3

Plot these points and connect them.

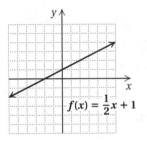

$f(x) = \dfrac{1}{2}x + 1$

39. Graph $f(x) = 2 - |x|$.

Make a list of function values in a table.

$f(-3) = 2 - |-3| = 2 - 3 = -1$

$f(-2) = 2 - |-2| = 2 - 2 = 0$

$f(-1) = 2 - |-1| = 2 - 1 = 1$

$f(0) = 2 - |0| = 2 - 0 = 2$

$f(1) = 2 - |1| = 2 - 1 = 1$

$f(2) = 2 - |2| = 2 - 2 = 0$

$f(3) = 2 - |3| = 2 - 3 = -1$

x	$f(x)$
-3	-1
-2	0
-1	1
0	2
1	1
2	0
3	-1

Plot these points and connect them.

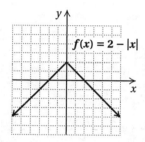

41. Graph $g(x) = |x - 1|$.

Make a list of function values in a table.

$f(-3) = |-3 - 1| = |-4| = 4$

$f(-1) = |-1 - 1| = |-2| = 2$

$f(0) = |0 - 1| = |-1| = 1$

$f(1) = |1 - 1| = |0| = 0$

$f(2) = |2 - 1| = |1| = 1$

$f(4) = |4 - 1| = |3| = 3$

x	$f(x)$
-3	4
-1	2
0	1
1	0
2	1
4	3

Plot these points and connect them.

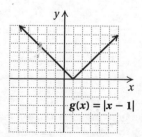

43. Graph $g(x) = x^2 + 2$.

Make a list of function values in a table.

$g(-2) = (-2)^2 + 2 = 4 + 2 = 6$

$g(-1) = (-1)^2 + 2 = 1 + 2 = 3$

$g(0) = 0^2 + 2 = 0 + 2 = 2$

$g(1) = 1^2 + 2 = 1 + 2 = 3$

$g(2) = 2^2 + 2 = 4 + 2 = 6$

x	$g(x)$
-2	6
-1	3
0	2
1	3
2	6

Plot these points and connect them.

45. Graph $f(x) = x^2 - 2x - 3$.

Make a list of function values in a table.

$f(-2) = (-2)^2 - 2(-2) - 3 = 4 + 4 - 3 = 5$

$f(-1) = (-1)^2 - 2(-1) - 3 = 1 + 2 - 3 = 0$

$f(0) = 0^2 - 2 \cdot 0 - 3 = 0 - 0 - 3 = -3$

$f(1) = 1^2 - 2 \cdot 1 - 3 = 1 - 2 - 3 = -4$

$f(2) = 2^2 - 2 \cdot 2 - 3 = 4 - 4 - 3 = -3$

$f(3) = 3^2 - 2 \cdot 3 = 3 = 9 - 6 - 3 = 0$

$f(4) = 4^2 - 2 \cdot 4 - 3 = 16 - 8 - 3 = 5$

x	$f(x)$
-2	5
-1	0
0	-3
1	-4
2	-3
3	0
4	5

Plot these points and connect them.

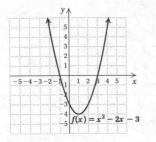

47. Graph $f(x) = -x^2 + 1$.

Make a list of function values in a table.

$f(-2) = -(-2)^2 + 1 = -4 + 1 = -3$

$f(-1) = -(-1)^2 + 1 = -1 + 1 = 0$

$f(0) = -0^2 + 1 = 0 + 1 = 1$

$f(1) = -1^2 + 1 = -1 + 1 = 0$

$f(2) = -2^2 + 1 = -4 + 1 = -3$

x	$f(x)$
-2	-3
-1	0
0	1
1	0
2	-3

Plot these points and connect them.

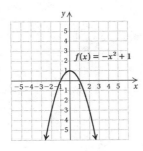

49. Graph $f(x) = x^3 + 1$.

Make a list of function values in a table.

$f(-2) = (-2)^3 + 1 = -8 + 1 = -7$

$f(-1) = (-1)^3 + 1 = -1 + 1 = 0$

$f(0) = 0^3 + 1 = 0 + 1 = 1$

$f(1) = 1^3 + 1 = 1 + 1 = 2$

$f(2) = 2^3 + 1 = 8 + 1 = 9$

x	$f(x)$
-2	-7
-1	0
0	1
1	2
2	9

Plot these points and connect them.

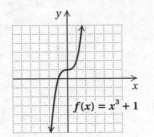

51. We can use the vertical line test:

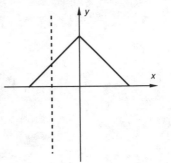

Visualize moving this vertical line across the graph. No vertical line will intersect the graph more than once. Thus, the graph is a graph of a function.

53. We can use the vertical line test.

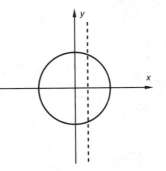

It is possible for a vertical line to intersect the graph more than once. Thus this is not a graph of a function.

55. We can use the vertical line test.

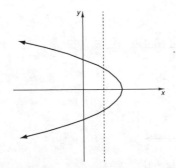

It is possible for a vertical line to intersect the graph more than once. Thus this is not the graph of a function.

57. We can use the vertical line test:

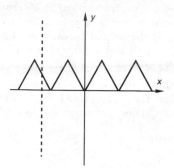

Visualize moving this vertical line across the graph. No vertical line will intersect the graph more than once. Thus, the graph is a graph of a function.

59. Locate 2009 on the horizontal axis and then move directly up to the graph. Next move across to the vertical axis. We come to about 1.8, so there were about 1.8 million children living with only grandparents in 2009.

61. Locate 2005 on the horizontal axis and then move directly up to the graph. Next move across to the vertical axis. We come to about 230,000, so there were about 230,000 pharmacists in 2005.

63.
$$-\frac{5}{3} + y = -\frac{1}{12} - \frac{5}{6}$$
$$-\frac{20}{12} + y = -\frac{1}{12} - \frac{10}{12}$$
$$-\frac{20}{12} + y = -\frac{11}{12}$$
$$y = -\frac{11}{12} + \frac{20}{12}$$
$$y = \frac{9}{12} = \frac{3}{4}$$

The solution is $\frac{3}{4}$.

65.
$$4 - 7y > 2y - 32$$
$$4 > 9y - 32$$
$$36 > 9y$$
$$4 > y$$

The solution set is $\{y|y < 4\}$, or $(-\infty, 4)$.

67.
$$7y - 2 = 3 + 7y$$
$$-2 = 3 \quad \text{Subtracting } 7y$$

We get a false equation. Thus, the equation has no solution.

69.
$$-9w \geq -99.9$$
$$w \leq 11.1 \quad \text{Reversing the inequality symbol}$$

The solution set is $\{w|w \leq 11.1\}$, or $(-\infty, 11.1]$.

71.
$$13x - 5 - x = 2(x + 5)$$
$$12x - 5 = 2x + 10$$
$$10x - 5 = 10$$
$$10x = 15$$
$$x = \frac{3}{2}, \text{ or } 1.5$$

The solution is $\frac{3}{2}$, or 1.5.

73. We first find the value of x for which $x - 6 = -2$.
$$x - 6 = -2$$
$$x = 4$$

Then we have
$$g(-2) = g(4 - 6) = 10 \cdot 4 - 1 = 40 - 1 = 39.$$

75. To find $f(g(-4))$, we first find $g(-4)$:
$$g(-4) = 2(-4) + 5 = -8 + 5 = -3.$$
Then $f(g(-4)) = f(-3) = 3(-3)^2 - 1 = 3 \cdot 9 - 1 = 27 - 1 = 26.$

To find $g(f(-4))$, we first find $f(-4)$:
$$f(-4) = 3(-4)^2 - 1 = 3 \cdot 16 - 1 = 48 - 1 = 47.$$
Then $g(f(-4)) = g(47) = 2 \cdot 47 + 5 = 94 + 5 = 99.$

77. We know that $(-1, -7)$ and $(3, 8)$ are both solutions of $g(x) = mx + b$. Substituting, we have
$$-7 = m(-1) + b, \text{ or } -7 = -m + b,$$
$$\text{and } 8 = m(3) + b, \text{ or } 8 = 3m + b.$$

Solve the first equation for b and substitute that expression into the second equation.

$-7 = -m + b$	First equation
$m - 7 = b$	Solving for b
$8 = 3m + b$	Second equation
$8 = 3m + (m - 7)$	Substituting
$8 = 3m + m - 7$	
$8 = 4m - 7$	
$15 = 4m$	
$\frac{15}{4} = m$	

We know that $m - 7 = b$, so $\frac{15}{4} - 7 = b$, or $-\frac{13}{4} = b$.
We have $m = \frac{15}{4}$ and $b = -\frac{13}{4}$, so $g(x) = \frac{15}{4}x - \frac{13}{4}$.

Exercise Set 2.3

RC1. We can calculate $5 - x$ for any value of x, so the domain of f is the set of all real numbers. The answer is (a).

RC3. We can calculate $|5 - x|$ for any value of x, so the domain of f is the set of all real numbers. The answer is (a).

RC5. We can calculate $5 - |x|$ for any value of x, so the domain of f is the set of all real numbers. The answer is (a).

1. a) Locate 1 on the horizontal axis and then find the point on the graph for which 1 is the first coordinate. From that point, look to the vertical axis to find the corresponding y-coordinate, 3. Thus, $f(1) = 3$.

b) The domain is the set of all x-values in the graph. It is $\{-4, -3, -2, -1, 0, 1, 2\}$.

c) To determine which member(s) of the domain are paired with 2, locate 2 on the vertical axis. From there look left and right to the graph to find any points for which 2 is the second coordinate. Two such points exist, $(-2, 2)$ and $(0, 2)$. Thus, the x-values for which $f(x) = 2$ are -2 and 0.

d) The range is the set of all y-values in the graph. It is $\{1, 2, 3, 4\}$.

3. a) Locate 1 on the horizontal axis and then find the point on the graph for which 1 is the first coordinate. From that point, look to the vertical axis to find the corresponding y-coordinate, about $2\frac{1}{2}$. Thus, $f(1) \approx 2\frac{1}{2}$.

b) The set of all x-values in the graph extends from -3 to 5, so the domain is $\{x| -3 \le x \le 5\}$, or $[-3, 5]$.

c) To determine which member(s) of the domain are paired with 2, locate 2 on the vertical axis. From there look left and right to the graph to find any points for which 2 is the second coordinate. One such point exists. Its first coordinate appears to be about $2\frac{1}{4}$. Thus, the x-value for which $f(x) = 2$ is about $2\frac{1}{4}$.

d) The set of all y-values in the graph extends from 1 to 4, so the range is $\{y|1 \le y \le 4\}$, or $[1, 4]$.

5. a) Locate 1 on the horizontal axis and the find the point on the graph for which 1 is the first coordinate. From that point, look to the vertical axis to find the corresponding y-coordinate. It is 1. Thus, $f(1) = 1$.

b) No endpoints are indicated and we see that the graph extends indefinitely both horizontally and vertically. Thus, the domain is the set of all real numbers.

c) To determine which member(s) of the domain are paired with 2, locate 2 on the vertical axis. From there look left and right to the graph to find any points for which 2 is the second coordinate. One such point exists. Its first coordinate is 3, so the x-value for which $f(x) = 2$ is 3.

d) The range is the set of all real numbers. (See part (b) above.)

7. a) Locate 1 on the horizontal axis and the find the point on the graph for which 1 is the first coordinate. From that point, look to the vertical axis to find the corresponding y-coordinate. It is 1. Thus, $f(1) = 1$.

b) No endpoints are indicated, so we see that the graph extends indefinitely horizontally. Thus, the domain is the set of all real numbers.

c) To determine which member(s) of the domain are paired with 2, locate 2 on the vertical axis. From there look left and right to the graph to find any points for which 2 is the second coordinate. Two such points exist, $(-2, 2)$ and $(2, 2)$. Thus, the x-values for which $f(x) = 2$ are -2 and 2.

d) The smallest y-value is 0. No endpoints are indicated, so we see that the graph extends upward indefinitely from $(0, 0)$. Thus, the range is $\{y|y \ge 0\}$, or $[0, \infty)$.

9. $f(x) = \dfrac{2}{x + 3}$

Since $\dfrac{2}{x + 3}$ cannot be calculated when the denominator is 0, we find the x-value that causes $x + 3$ to be 0:

$$x + 3 = 0$$
$$x = -3 \quad \text{Subtracting 3 on both sides}$$

Thus, -3 is not in the domain of f, while all other real numbers are. The domain of f is

$\{x|x \text{ is a real number } and \ x \ne -3\}$, or

$(-\infty, -3) \cup (-3, \infty)$.

11. $f(x) = 2x + 1$

Since we can calculate $2x + 1$ for any real number x, the domain is the set of all real numbers.

13. $f(x) = x^2 + 3$

Since we can calculate $x^2 + 3$ for any real number x, the domain is the set of all real numbers.

15. $f(x) = \dfrac{8}{5x - 14}$

Since $\dfrac{8}{5x - 14}$ cannot be calculated when the denominator is 0, we find the x-value that causes $5x - 14$ to be 0:

$$5x - 14 = 0$$
$$5x = 14$$
$$x = \frac{14}{5}$$

Thus, $\dfrac{14}{5}$ is not in the domain of f, while all other real numbers are. The domain of f is

$\left\{x \middle| x \text{ is a real number } and \ x \ne \dfrac{14}{5}\right\}$, or

$\left(-\infty, \dfrac{14}{5}\right) \cup \left(\dfrac{14}{5}, \infty\right)$.

17. $f(x) = |x| - 4$

Since we can calculate $|x| - 4$ for any real number x, the domain is the set of all real numbers.

19. $f(x) = \dfrac{x^2 - 3x}{|4x - 7|}$

Since $\dfrac{x^2 - 3x}{|4x - 7|}$ cannot be calculated when the denominator is 0, we find the x-value that causes $|4x - 7|$ to be 0:

$$|4x - 7| = 0$$
$$4x - 7 = 0$$
$$4x = 7$$
$$x = \frac{7}{4}$$

Thus, $\frac{7}{4}$ is not in the domain of f, while all other real numbers are. The domain of f is

$$\left\{ x \,\middle|\, x \text{ is a real number } and \ x \neq \frac{7}{4} \right\}, \text{ or}$$

$$\left(-\infty, \frac{7}{4} \right) \cup \left(\frac{7}{4}, \infty \right).$$

21. $g(x) = \dfrac{1}{x - 1}$

Since $\dfrac{1}{x - 1}$ cannot be calculated when the denominator is 0, we find the x-value that causes $x - 1$ to be 0:

$$x - 1 = 0$$
$$x = 1$$

Thus, 1 is not in the domain of g, while all other real numbers are. The domain of g is

$\{ x | x \text{ is a real number } and \ x \neq 1 \}$, or $(-\infty, 1) \cup (1, \infty)$.

23. $g(x) = x^2 - 2x + 1$

Since we can calculate $x^2 - 2x + 1$ for any real number x, the domain is the set of all real numbers.

25. $g(x) = x^3 - 1$

Since we can calculate $x^3 - 1$ for any real number x, the domain is the set of all real numbers.

27. $g(x) = \dfrac{7}{20 - 8x}$

Since $\dfrac{7}{20 - 8x}$ cannot be calculated when the denominator is 0, we find the x-values that cause $20 - 8x$ to be 0:

$$20 - 8x = 0$$
$$-8x = -20$$
$$x = \frac{5}{2}$$

Thus, $\frac{5}{2}$ is not in the domain of g, while all other real numbers are. The domain of g is

$$\left\{ x \,\middle|\, x \text{ is a real number } and \ x \neq \frac{5}{2} \right\}, \text{ or}$$

$$\left(-\infty, \frac{5}{2} \right) \cup \left(\frac{5}{2}, \infty \right).$$

29. $g(x) = |x + 7|$

Since we can calculate $|x + 7|$ for any real number x, the domain is the set of all real numbers.

31. $g(x) = \dfrac{-2}{|4x + 5|}$

Since $\dfrac{-2}{|4x + 5|}$ cannot be calculated when the denominator is 0, we find the x-value that causes $|4x + 5|$ to be 0:

$$|4x + 5| = 0$$
$$4x + 5 = 0$$
$$4x = -5$$
$$x = -\frac{5}{4}$$

Thus, $-\frac{5}{4}$ is not in the domain of g, while all other real numbers are. The domain of g is

$$\left\{ x \,\middle|\, x \text{ is a real number } and \ x \neq -\frac{5}{4} \right\}, \text{ or}$$

$$\left(-\infty, -\frac{5}{4} \right) \cup \left(-\frac{5}{4}, \infty \right).$$

33. The input -1 has the output -8, so $f(-1) = -8$; the input 0 has the output 0, so $f(0) = 0$; the input 1 has the output -2, so $f(1) = -2$.

35. $|x| = 8$

$$x = -8 \ or \ x = 8$$

The solution set is $\{-8, 8\}$.

37. $|x - 7| = 11$

$$x - 7 = -11 \ or \ x - 7 = 11$$
$$x = -4 \ or \ \qquad x = 18$$

The solution set is $\{-4, 18\}$.

39. $|3x - 4| = |x + 2|$

$$3x - 4 = x + 2 \ or \ 3x - 4 = -(x + 2)$$
$$2x \ \ 4 = 2 \quad or \ 3x - 4 = -x - 2$$
$$2x = 6 \quad or \ 4x - 4 = -2$$
$$x = 3 \quad or \qquad 4x = 2$$
$$x = 3 \quad or \qquad x = \frac{1}{2}$$

The solution set is $\left\{ \dfrac{1}{2}, 3 \right\}$.

41. $|3x - 8| = -11$

Since the absolute value of a number must be nonnegative, the equation has no solution. The solution is $\{\ \}$, or $\emptyset$.

43. We graph each function and determine the range.

The range of $f(x) = \dfrac{2}{x + 3}$ is $(-\infty, 0) \cup (0, \infty)$; the range of $f(x) = x^2 - 2x + 3$ is $[2, \infty)$; the range of $f(x) = |x| - 4$ is $[-4, \infty)$; the range of $f(x) = |x - 4|$ is $[0, \infty)$.

45. $f(x) = \sqrt[3]{x - 1}$

Since we can calculate the cube root of any real number, the domain is the set of all real numbers.

Chapter 2 Mid-Chapter Review

1. True; a function is a special type of relation in which each member of the domain is paired with exactly one member of the range.

2. False; see the definition of a function on page 173 of the text.

3. True; for a function $f(x) = c$, where c is a constant, all the inputs have the output c.

4. True; see the vertical-line test on page 177 of the text.

5. False; for example, see Exercise 3 above.

6. The y-value that is paired with the input 0 is 1.

The x-value that is paired with the y-value -2 is 2.

The y-value that is paired with the x-value -2 is 4.

The y-value that is paired with the x-value 4 is -5.

7. The y-value that is paired with the x-value -2 is 0.

The x-values that are paired with the y-value 0 are -2 and 3.

The y-value that is paired with the x-value 0 is -6.

The y-value that is paired with the x-value 2 is -4.

We see above that the y-value -4 is paired with the x-value 2. In addition, we see from the graph that the y-value -4 is also paired with the x-value -1.

8. We substitute -2 for x and -1 for y (alphabetical order of variables.)

$$\begin{array}{c|c} \multicolumn{2}{c}{5y + 6 = 4x} \\ \hline 5(-1) + 6 \ ? \ 4(-2) & \\ -5 + 6 & -8 \\ 1 & \text{FALSE} \end{array}$$

Thus, $(-2, -1)$ is not a solution of the equation.

9. We substitute $\frac{1}{2}$ for a and 0 for b (alphabetical order of variables.)

$$\begin{array}{c|c} \multicolumn{2}{c}{8a = 4 - b} \\ \hline 8 \cdot \frac{1}{2} \ ? \ 4 - 0 & \\ 4 & 4 \quad \text{TRUE} \end{array}$$

Thus, $\left(\frac{1}{2}, 0\right)$ is a solution of the equation.

10. Yes; each member of the domain is matched to only one member of the range.

11. No; the number 15 in the domain is matched to 2 numbers of the range, 25 and 30.

12. The set of all x-values on the graph extends from -3 through 3, so the domain is $\{x| -3 \le x \le 3\}$, or $[-3, 3]$.

The set of all y-values on the graph extends from -2 through 1, so the range is $\{y| -2 \le y \le 1\}$, or $[-2, 1]$.

13. $g(x) = 2 + x$

$g(-5) = 2 + (-5) = -3$

14. $f(x) = x - 7$

$f(0) = 0 - 7 = -7$

15. $h(x) = 8$

$h\left(\frac{1}{2}\right) = 8$

16. $f(x) = 3x^2 - x + 5$

$f(-1) = 3(-1)^2 - (-1) + 5 = 3 \cdot 1 + 1 + 5 = 3 + 1 + 5 = 9$

17. $g(p) = p^4 - p^3$

$g(10) = 10^4 - 10^3 = 10,000 - 1000 = 9000$

18. $f(t) = \frac{1}{2}t + 3$

$f(-6) = \frac{1}{2}(-6) + 3 = -3 + 3 = 0$

19. No vertical line will intersect the graph more than once. Thus, the graph is the graph of a function.

20. It is possible for a vertical line to intersect the graph more than once. Thus, the graph is not the graph of a function.

21. No vertical line will intersect the graph more than once. Thus, the graph is the graph of a function.

22. $g(x) = \dfrac{3}{12 - 3x}$

Since $\dfrac{3}{12 - 3x}$ cannot be calculated when the denominator is 0, we find the x-value that causes $12 - 3x$ to be 0:

$$12 - 3x = 0$$
$$12 = 3x$$
$$4 = x$$

Thus, the domain of g is $\{x|x$ is a real number $and\ x \neq 4\}$, or $(-\infty, 4) \cup (4, \infty)$.

23. $f(x) = x^2 - 10x + 3$

Since we can calculate $x^2 - 10x + 3$ for any real number x, the domain is the set of all real numbers.

24. $h(x) = \dfrac{x - 2}{x + 2}$

Since $\dfrac{x - 2}{x + 2}$ cannot be calculated when the denominator is 0, we find the x-value that causes $x + 2$ to be 0:

$$x + 2 = 0$$
$$x = -2$$

Thus, the domain of g is $\{x|x$ is a real number $and\ x \neq -2\}$, or $(-\infty, -2) \cup (-2, \infty)$.

25. $f(x) = |x - 4|$

Since we can calculate $|x - 4|$ for any real number x, the domain is the set of all real numbers.

26. $y = -\dfrac{2}{3}x - 2$

We find some ordered pairs that are solutions.

When $x = -3$, $y = -\dfrac{2}{3}(-3) - 2 = 2 - 2 = 0$.

When $x = 0$, $y = -\dfrac{2}{3} \cdot 0 - 2 = 0 - 2 = -2$.

When $x = 3$, $y = -\dfrac{2}{3} \cdot 3 - 2 = -2 - 2 = -4$.

x	y
-3	0
0	-2
3	-4

Plot these points, draw the line they determine, and label it $y = -\dfrac{2}{3}x - 2$.

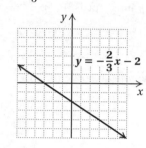

27. $f(x) = x - 1$

We find some ordered pairs that are solutions.

$f(-3) = -3 - 1 = -4.$

$f(0) = 0 - 1 = -1.$

$f(4) = 4 - 1 = 3.$

x	$f(x)$
-3	-4
0	-1
4	3

Plot these points, draw the line they determine, and label it $f(x) = x - 1$.

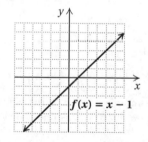

28. $h(x) = 2x + \dfrac{1}{2}$

We find some ordered pairs that are solutions.

When $x = -2$, $y = 2(-2) + \dfrac{1}{2} = -4 + \dfrac{1}{2} = -3\dfrac{1}{2}$.

When $x = 0$, $y = 2 \cdot 0 + \dfrac{1}{2} = 0 + \dfrac{1}{2} = \dfrac{1}{2}$.

When $x = 2$, $y = 2 \cdot 2 + \dfrac{1}{2} = 4 + \dfrac{1}{2} = 4\dfrac{1}{2}$.

x	$h(x)$
-2	$-3\dfrac{1}{2}$
0	$\dfrac{1}{2}$
2	$4\dfrac{1}{2}$

Plot these points, draw the line they determine, and label it $h(x) = 2x + \dfrac{1}{2}$.

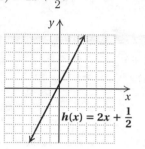

29. $g(x) = |x| - 3$

We find some ordered pairs that are solutions.

$g(-4) = |-4| - 3 = 4 - 3 = 1$

$g(-1) = |-1| - 3 = 1 - 3 = -2$

$g(0) = |0| - 3 = 0 - 3 = -3$

$g(2) = |2| - 3 = 2 - 3 = -1$

$g(3) = |3| - 3 = 3 - 3 = 0$

x	$g(x)$
-4	1
-1	-2
0	-3
2	-1
3	0

Plot these points, draw the line they determine, and label it $g(x) = |x| - 3$.

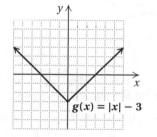

30. $y = 1 + x^2$

We find some ordered pairs that are solutions.

When $x = -2$, $y = 1 + (-2)^2 = 1 + 4 = 5$.

When $x = -1$, $y = 1 + (-1)^2 = 1 + 1 = 2$.

When $x = 0$, $y = 1 + 0^2 = 1 + 0 = 1$.

When $x = 1$, $y = 1 + 1^2 = 1 + 1 = 2$.

When $x = 2$, $y = 1 + 2^2 = 1 + 4 = 5$.

x	y
-2	5
-1	2
0	1
1	2
2	5

Plot these points, draw the line they determine, and label it $y = 1 + x^2$.

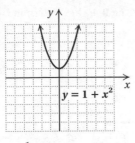

31. $f(x) = -\dfrac{1}{4}x$

We find some ordered pairs that are solutions.

$$f(-4) = -\dfrac{1}{4}(-4) = 1$$

$$f(0) = -\dfrac{1}{4} \cdot 0 = 0$$

$$f(4) = -\dfrac{1}{4} \cdot 4 = -1$$

x	$f(x)$
-4	1
0	0
4	-1

Plot these points, draw the line they determine, and label it $f(x) = -\dfrac{1}{4}x$.

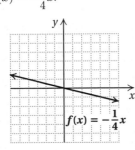

32. No; since each input has exactly one output, the number of outputs cannot exceed the number of inputs.

33. When $x < 0$, then $y < 0$ and the graph contains points in quadrant III. When $0 < x < 30$, then $y < 0$ and the graph contains points in quadrant IV. When $x > 30$, then $y > 0$ and the graph contains points in quadrant I. Thus, the graph passes through three quadrants.

34. The output -3 corresponds to the input 2. The number -3 in the range is paired with the number 2 in the domain. The point $(2, -3)$ is on the graph of the function.

35. The domain of a function is the set of all inputs, and the range is the set of all outputs.

Exercise Set 2.4

RC1. $m = \dfrac{3 - (-1)}{-1 - 2} = \dfrac{4}{-3} = -\dfrac{4}{3}$

The answer is (f).

RC3. $m = \dfrac{3 - (-1)}{-1 - 0} = \dfrac{4}{-1} = -4$

The answer is (d).

RC5. $m = \dfrac{0 - 3}{-2 - 2} = \dfrac{-3}{-4} = \dfrac{3}{4}$

The answer is (e).

1.
$$y = 4x + 5$$
$$\uparrow \qquad \uparrow$$
$$y = mx + b$$

The slope is 4, and the y-intercept is $(0, 5)$.

3.
$$f(x) = -2x - 6$$
$$\uparrow \qquad \uparrow$$
$$f(x) = mx + b$$

The slope is -2, and the y-intercept is $(0, -6)$.

5.
$$y = -\dfrac{3}{8}x - \dfrac{1}{5}$$
$$\uparrow \qquad \uparrow$$
$$y = mx + b$$

The slope is $-\dfrac{3}{8}$, and the y-intercept is $\left(0, -\dfrac{1}{5}\right)$.

7.
$$g(x) = 0.5x - 9$$
$$\uparrow \qquad \uparrow$$
$$g(x) = mx + b$$

The slope is 0.5, and the y-intercept is $(0, -9)$.

9. First we find the slope-intercept form of the equation by solving for y. This allows us to determine the slope and y-intercept easily.

$$2x - 3y = 8$$
$$-3y = -2x + 8$$
$$\dfrac{-3y}{-3} = \dfrac{-2x + 8}{-3}$$
$$y = \dfrac{2}{3}x - \dfrac{8}{3}$$

The slope is $\dfrac{2}{3}$, and the y-intercept is $\left(0, -\dfrac{8}{3}\right)$.

11. First we find the slope-intercept form of the equation by solving for y. This allows us to determine the slope and y-intercept easily.

$$9x = 3y + 6$$
$$9x - 6 = 3y$$
$$\dfrac{9x - 6}{3} = \dfrac{3y}{3}$$
$$3x - 2 = y, \text{ or}$$
$$y = 3x - 2$$

The slope is 3, and the y-intercept is $(0, -2)$.

13. First we find the slope-intercept form of the equation by solving for y. This allows us to determine the slope and y-intercept easily.

$$3 - \dfrac{1}{4}y = 2x$$
$$-\dfrac{1}{4}y = 2x - 3$$
$$-4\left(-\dfrac{1}{4}y\right) = -4(2x - 3)$$
$$y = -8x + 12$$

The slope is -8, and the y-intercept is $(0, 12)$.

15. First we find the slope-intercept form of the equation by solving for y. This allows us to determine the slope and y-intercept easily.

$$17y + 4x + 3 = 7 + 4x$$
$$17y + 3 = 7$$
$$17y = 4$$
$$y = \frac{4}{17}, \text{ or}$$
$$y = 0 \cdot x + \frac{4}{17}$$

The slope is 0, and the y-intercept is $\left(0, \frac{4}{17}\right)$.

17. We can use any two points on the line, such as $(0, 3)$ and $(4, 1)$.

$$\text{Slope} = \frac{\text{change in } y}{\text{change in } x}$$
$$= \frac{1 - 3}{4 - 0} = \frac{-2}{4} = -\frac{1}{2}$$

19. We can use any two points on the line, such as $(-3, 1)$ and $(3, 3)$.

$$\text{Slope} = \frac{\text{change in } y}{\text{change in } x}$$
$$= \frac{3 - 1}{3 - (-3)} = \frac{2}{6} = \frac{1}{3}$$

21. $\text{Slope} = \dfrac{\text{change in } y}{\text{change in } x} = \dfrac{5 - 9}{4 - 6} = \dfrac{-4}{-2} = 2$

23. $\text{Slope} = \dfrac{\text{change in } y}{\text{change in } x} = \dfrac{-8 - (-4)}{3 - 9} = \dfrac{-4}{-6} = \dfrac{2}{3}$

25. $\text{Slope} = \dfrac{\text{change in } y}{\text{change in } x} = \dfrac{8.7 - 12.4}{-5.2 - (-16.3)} = \dfrac{-3.7}{11.1} =$

$-\dfrac{37}{111} = -\dfrac{1}{3}$

27. $\text{Slope} = \dfrac{0.4}{5} = \dfrac{0.4}{5} \cdot \dfrac{10}{10} = \dfrac{4}{50} = \dfrac{2 \cdot 2}{2 \cdot 25} = \dfrac{2}{25} = 0.08 = 8\%$

The slope is $\dfrac{2}{25}$, or 8%.

29. $\text{Slope} = \dfrac{2.6}{8.2} = \dfrac{2.6}{8.2} \cdot \dfrac{10}{10} = \dfrac{26}{82} = \dfrac{2 \cdot 13}{2 \cdot 41} = \dfrac{13}{41} \approx 0.317 =$

31.7%

The slope is $\dfrac{13}{41}$, or about 31.7%.

31. We use the points $(2008, 8.2)$ and $(2015, 27.4)$.

$$\text{Slope} = \frac{\text{change in amount of purchases}}{\text{corresponding change in time}}$$
$$= \frac{27.4 - 8.2}{2015 - 2008}$$
$$= \frac{19.2}{7}$$
$$\approx 2.74$$

The rate of change is about $2.74 billion per year.

33. We can use the coordinates of any two points on the line. We'll use $(0, 30)$ and $(3, 3)$.

$$\text{Slope} = \frac{\text{change in } y}{\text{change in } x} = \frac{3 - 30}{3 - 0} = \frac{-27}{3} = -9$$

The rate of change is $-$900 per year. That is, the value is decreasing at a rate of $900 per year.

35. We can use the points $(2004, 2100)$ and $(2011, 2913)$.

$$\text{Slope} = \frac{\text{change in average amount of check}}{\text{corresponding change in time}}$$
$$= \frac{2913 - 2100}{2011 - 2004}$$
$$= \frac{813}{7}$$
$$\approx 116.14$$

The rate of change is about $116.14 per year.

37.
$$3^2 - 24 \cdot 56 + 144 \div 12$$
$$= 9 - 24 \cdot 56 + 144 \div 12$$
$$= 9 - 1344 + 144 \div 12$$
$$= 9 - 1344 + 12$$
$$= -1335 + 12$$
$$= -1323$$

39.
$$10\{2x + 3[5x - 2(-3x + y^1 - 2)]\}$$
$$= 10\{2x + 3[5x - 2(-3x + y - 2)]\}$$
$$= 10\{2x + 3[5x + 6x - 2y + 4]\}$$
$$= 10\{2x + 3[11x - 2y + 4]\}$$
$$= 10\{2x + 33x - 6y + 12\}$$
$$= 10\{35x - 6y + 12\}$$
$$= 350x - 60y + 120$$

41. *Familiarize.* Let t represent the length of a side of the triangle. Then $t - 5$ represents the length of a side of the square.

Translate.

Perimeter of the square	is the same as	perimeter of the triangle
$4(t - 5)$	$=$	$3t$

Solve.
$$4(t - 5) = 3t$$
$$4t - 20 = 3t$$
$$t - 20 = 0$$
$$t = 20$$

Check. If 20 is the length of a side of the triangle, then the length of a side of the square is $20 - 5$, or 15. The perimeter of the square is $4 \cdot 15$, or 60, and the perimeter of the triangle is $3 \cdot 20$, or 60. The numbers check.

State. The square and triangle have sides of length 15 yd and 20 yd, respectively.

43. $|5x - 8| < 32$

$$-32 < 5x - 8 < 32$$

$$-24 < 5x < 40 \qquad \text{Adding 8}$$

$$-\frac{24}{5} < x < 8 \qquad \text{Dividing by 5}$$

The solution set is $\left\{x \mid -\dfrac{24}{5} < x < 8\right\}$, or $\left(-\dfrac{24}{5}, 8\right)$.

45. $|5x - 8| = -32$

Since the absolute value of a number is nonnegative the equation has no solution. The solution is $\{\ \ \}$, or $\emptyset$.

Exercise Set 2.5

RC1. The graph of $x = -4$ is a vertical line and the graph of $y = 5$ is a horizontal line, so the graphs are perpendicular. The given statement is true. (See page 210 in the text.)

RC3. False; see page 210 in the text.

RC5. True; see page 208 in the text.

1. $x - 2 = y$

To find the x-intercept we let $y = 0$ and solve for x. We have $x - 2 = 0$, or $x = 2$. The x-intercept is $(2, 0)$.

To find the y-intercept we let $x = 0$ and solve for y.

$$x - 2 = y$$
$$0 - 2 = y$$
$$-2 = y$$

The y-intercept is $(0, -2)$. We plot these points and draw the line.

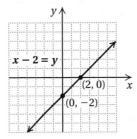

We use a third point as a check. We choose $x = 5$ and solve for y.

$$5 - 2 = y$$
$$3 = y$$

We plot $(5, 3)$ and note that it is on the line.

3. $x + 3y = 6$

To find the x-intercept we let $y = 0$ and solve for x.

$$x + 3y = 6$$
$$x + 3 \cdot 0 = 6$$
$$x = 6$$

The x-intercept is $(6, 0)$.

To find the y-intercept we let $x = 0$ and solve for y.

$$x + 3y = 6$$
$$0 + 3y = 6$$
$$3y = 6$$
$$y = 2$$

The y-intercept is $(0, 2)$.

We plot these points and draw the line.

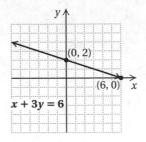

We use a third point as a check. We choose $x = 3$ and solve for y.

$$3 + 3y = 6$$
$$3y = 3$$
$$y = 1$$

We plot $(3, 1)$ and note that it is on the line.

5. $2x + 3y = 6$

To find the x-intercept we let $y = 0$ and solve for x.

$$2x + 3y = 6$$
$$2x + 3 \cdot 0 = 6$$
$$2x = 6$$
$$x = 3$$

The x-intercept is $(3, 0)$.

To find the y-intercept we let $x = 0$ and solve for y.

$$2x + 3y = 6$$
$$2 \cdot 0 + 3y = 6$$
$$3y = 6$$
$$y = 2$$

The y-intercept is $(0, 2)$.

We plot these points and draw the line.

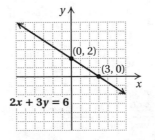

We use a third point as a check. We choose $x = -3$ and solve for y.

$$2(-3) + 3y = 6$$
$$-6 + 3y = 6$$
$$3y = 12$$
$$y = 4$$

We plot $(-3, 4)$ and note that it is on the line.

7. $f(x) = -2 - 2x$

We can think of this equation as $y = -2 - 2x$.

To find the x-intercept we let $f(x) = 0$ and solve for x. We have $0 = -2 - 2x$, or $2x = -2$, or $x = -1$. The x-intercept is $(-1, 0)$.

To find the y-intercept we let $x = 0$ and solve for $f(x)$, or y.

$$y = -2 - 2x$$
$$y = -2 - 2 \cdot 0$$
$$y = -2$$

The y-intercept is $(0, -2)$.

We plot these points and draw the line.

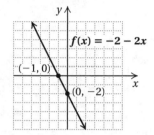

We use a third point as a check. We choose $x = -3$ and calculate y.

$$y = -2 - 2(-3) = -2 + 6 = 4$$

We plot $(-3, 4)$ and note that it is on the line.

9. $5y = -15 + 3x$

To find the x-intercept we let $y = 0$ and solve for x. We have $0 = -15 + 3x$, or $15 = 3x$, or $5 = x$. The x-intercept is $(5, 0)$.

To find the y-intercept we let $x = 0$ and solve for y.

$$5y = -15 + 3x$$
$$5y = -15 + 3 \cdot 0$$
$$5y = -15$$
$$y = -3$$

The y-intercept is $(0, -3)$.

We plot these points and draw the line.

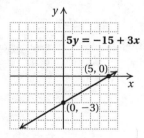

We use a third point as a check. We choose $x = -5$ and solve for y.

$$5y = -15 + 3(-5)$$
$$5y = -15 - 15$$
$$5y = -30$$
$$y = -6$$

We plot $(-5, -6)$ and note that it is on the line.

11. $2x - 3y = 6$

To find the x-intercept we let $y = 0$ and solve for x.

$$2x - 3y = 6$$
$$2x - 3 \cdot 0 = 6$$
$$2x = 6$$
$$x = 3$$

The x-intercept is $(3, 0)$.

To find the y-intercept we let $x = 0$ and solve for y.

$$2x - 3y = 6$$
$$2 \cdot 0 - 3y = 6$$
$$-3y = 6$$
$$y = -2$$

The y-intercept is $(0, -2)$.

We plot these points and draw the line.

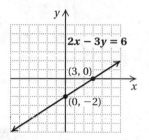

We use a third point as a check. We choose $x = -3$ and solve for y.

$$2(-3) - 3y = 6$$
$$-6 - 3y = 6$$
$$-3y = 12$$
$$y = -4$$

We plot $(-3, -4)$ and note that it is on the line.

13. $2.8y - 3.5x = -9.8$

To find the x-intercept we let $y = 0$ and solve for x.

$$2.8y - 3.5x = -9.8$$
$$2.8(0) - 3.5x = -9.8$$
$$-3.5x = -9.8$$
$$x = 2.8$$

The x-intercept is $(2.8, 0)$.

To find the y-intercept we let $x = 0$ and solve for y.

$$2.8y - 3.5x = -9.8$$
$$2.8y - 3.5(0) = -9.8$$
$$2.8y = -9.8$$
$$y = -3.5$$

The y-intercept is $(0, -3.5)$.

We plot these points and draw the line.

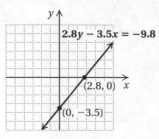

We use a third point as a check. We choose $x = 5$ and solve for y.

$$2.8y - 3.5(5) = -9.8$$
$$2.8y - 17.5 = -9.8$$
$$2.8y = 7.7$$
$$y = 2.75$$

We plot $(5, 2.75)$ and note that it is on the line.

15. $5x + 2y = 7$

To find the x-intercept we let $y = 0$ and solve for x.

$$5x + 2y = 7$$
$$5x + 2 \cdot 0 = 7$$
$$5x = 7$$
$$x = \frac{7}{5}$$

The x-intercept is $\left(\frac{7}{5}, 0\right)$.

To find the y-intercept we let $x = 0$ and solve for y.

$$5x + 2y = 7$$
$$5 \cdot 0 + 2y = 7$$
$$2y = 7$$
$$y = \frac{7}{2}$$

The y-intercept is $\left(0, \frac{7}{2}\right)$.

We plot these points and draw the line.

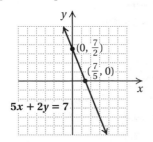

We use a third point as a check. We choose $x = 3$ and solve for y.

$$5 \cdot 3 + 2y = 7$$
$$15 + 2y = 7$$
$$2y = -8$$
$$y = -4$$

We plot $(3, -4)$ and note that it is on the line.

17. $y = \frac{5}{2}x + 1$

First we plot the y-intercept $(0, 1)$. Then we consider the slope $\frac{5}{2}$. Starting at the y-intercept and using the slope, we find another point by moving 5 units up and 2 units to the right. We get to a new point $(2, 6)$.

We can also think of the slope as $\frac{-5}{-2}$. We again start at the y-intercept $(0, 1)$. We move 5 units down and 2 units to the left. We get to another new point $(-2, -4)$. We plot the points and draw the line.

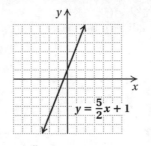

19. $f(x) = -\frac{5}{2}x - 4$

First we plot the y-intercept $(0, -4)$. We can think of the slope as $\frac{-5}{2}$. Starting at the y-intercept and using the slope, we find another point by moving 5 units down and 2 units to the right. We get to a new point $(2, -9)$.

We can also think of the slope as $\frac{5}{-2}$. We again start at the y-intercept $(0, -4)$. We move 5 units up and 2 units to the left. We get to another new point $(-2, 1)$. We plot the points and draw the line.

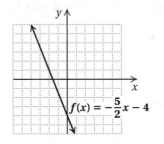

21. $x + 2y = 4$

First we write the equation in slope-intercept form by solving for y.

$$x + 2y = 4$$
$$2y = -x + 4$$
$$\frac{2y}{2} = \frac{-x + 4}{2}$$
$$y = -\frac{1}{2}x + 2$$

Now we plot the y-intercept $(0, 2)$. We can think of the slope as $\frac{-1}{2}$. Starting at the y-intercept and using the slope, we find another point by moving 1 unit down and 2 units to the right. We get to a new point $(2, 1)$.

We can also think of the slope as $\dfrac{1}{-2}$. We again start at the y-intercept $(0, 2)$. We move 1 unit up and 2 units to the left. We get to another new point $(-2, 3)$. We plot the points and draw the line.

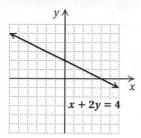

23. $4x - 3y = 12$

First we write the equation in slope-intercept form by solving for y.

$$4x - 3y = 12$$
$$-3y = -4x + 12$$
$$\frac{-3y}{-3} = \frac{-4x + 12}{-3}$$
$$y = \frac{4}{3}x - 4$$

Now we plot the y-intercept $(0, -4)$ and consider the slope $\dfrac{4}{3}$. Starting at the y-intercept and using the slope, we find another point by moving 4 units up and 3 units to the right. We get to a new point $(3, 0)$. In a similar manner we can move from the point $(3, 0)$ to find another point $(6, 4)$. We plot these points and draw the line.

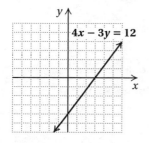

25. $f(x) = \dfrac{1}{3}x - 4$

First we plot the y-intercept $(0, -4)$. Then we consider the slope $\dfrac{1}{3}$. Starting at the y-intercept and using the slope, we find another point by moving 1 unit up and 3 units to the right. We get to a new point $(3, -3)$.

We can also think of the slope as $\dfrac{-1}{-3}$. We again start at the y-intercept $(0, -4)$. We move 1 unit down and 3 units to the left. We get to another new point $(-3, -5)$. We plot these points and draw the line.

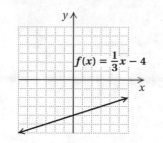

27. $5x + 4 \cdot f(x) = 4$

First we solve for $f(x)$.

$$5x + 4 \cdot f(x) = 4$$
$$4 \cdot f(x) = -5x + 4$$
$$\frac{4 \cdot f(x)}{4} = \frac{-5x + 4}{4}$$
$$f(x) = -\frac{5}{4}x + 1$$

Now we plot the y-intercept $(0, 1)$. We can think of the slope as $\dfrac{-5}{4}$. Starting at the y-intercept and using the slope, we find another point by moving 5 units down and 4 units to the right. We get to a new point $(4, -4)$.

We can also think of the slope as $\dfrac{5}{-4}$. We again start at the y-intercept $(0, 1)$. We move 5 units up and 4 units to the left. We get to another new point $(-4, 6)$. We plot these points and draw the line.

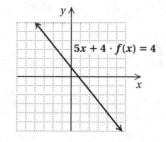

29. $x = 1$

Since y is missing, any number for y will do. Thus all ordered pairs $(1, y)$ are solutions. The graph is parallel to the y-axis.

x	y
1	-2
1	0	$\longleftarrow$ x-intercept
1	3

$\uparrow$ $\quad$ $\llcorner$ Choose any

x must $\qquad$ number for y.

be 1.

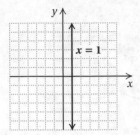

This is a vertical line, so the slope is not defined.

31. $y = -1$

Since x is missing, any number for x will do. Thus all ordered pairs $(x, -1)$ are solutions. The graph is parallel to the x-axis.

x	y	
-2	-1	
0	-1	← y-intercept
3	-1	

↑   └── y must be -1.
Choose
any number
for x.

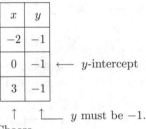

This is a horizontal line, so the slope is 0.

33. $f(x) = -6$

Since x is missing all ordered pairs $(x, 6)$ are solutions. The graph is parallel to the x-axis.

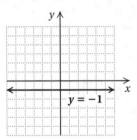

This is a horizontal line, so the slope is 0.

35. $y = 0$

Since x is missing, all ordered pairs $(x, 0)$ are solutions. The graph is the x-axis.

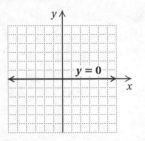

This is a horizontal line, so the slope is 0.

37. $2 \cdot f(x) + 5 = 0$

$$2 \cdot f(x) = -5$$

$$f(x) = -\frac{5}{2}$$

Since x is missing, all ordered pairs $\left(x, -\frac{5}{2}\right)$ are solutions. The graph is parallel to the x-axis.

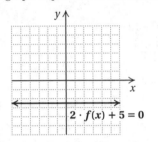

This is a horizontal line, so the slope is 0.

39. $7 - 3x = 4 + 2x$

$$7 - 5x = 4$$

$$-5x = -3$$

$$x = \frac{3}{5}$$

Since y is missing, all ordered pairs $\left(\frac{3}{5}, y\right)$ are solutions. The graph is parallel to the y-axis.

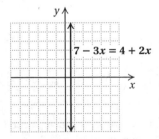

This is a vertical line, so the slope is not defined.

41. We first solve for y and determine the slope of each line.

$$x + 6 = y$$

$$y = x + 6 \quad \text{Reversing the order}$$

The slope of $y = x + 6$ is 1.

$$y - x = -2$$

$$y = x - 2$$

The slope of $y = x - 2$ is 1.

The slopes are the same, and the y-intercepts are different. The lines are parallel.

43. We first solve for y and determine the slope of each line.

$$y + 3 = 5x$$
$$y = 5x - 3$$

The slope of $y = 5x - 3$ is 5.

$$3x - y = -2$$
$$3x + 2 = y$$
$$y = 3x + 2 \quad \text{Reversing the order}$$

The slope of $y = 3x + 2$ is 3.

The slopes are not the same; the lines are not parallel.

45. We determine the slope of each line.

The slope of $y = 3x + 9$ is 3.

$$2y = 6x - 2$$
$$y = 3x - 1$$

The slope of $y = 3x - 1$ is 3.

The slopes are the same, and the y-intercepts are different. The lines are parallel.

47. We solve each equation for x.

$$12x = 3 \qquad\qquad -7x = 10$$
$$x = \frac{1}{4} \qquad\qquad x = -\frac{10}{7}$$

We have two vertical lines, so they are parallel.

49. We determine the slope of each line.

The slope of $y = 4x - 5$ is 4.

$$4y = 8 - x$$
$$4y = -x + 8$$
$$y = -\frac{1}{4}x + 2$$

The slope of $4y = 8 - x$ is $-\frac{1}{4}$.

The product of their slopes is $4\left(-\frac{1}{4}\right)$, or -1; the lines are perpendicular.

51. We determine the slope of each line.

$$x + 2y = 5$$
$$2y = -x + 5$$
$$y = -\frac{1}{2}x + \frac{5}{2}$$

The slope of $x + 2y = 5$ is $-\frac{1}{2}$.

$$2x + 4y = 8$$
$$4y = -2x + 8$$
$$y = -\frac{1}{2}x + 2$$

The slope of $2x + 4y = 8$ is $-\frac{1}{2}$.

The product of their slopes is $\left(-\frac{1}{2}\right)\left(-\frac{1}{2}\right)$, or $\frac{1}{4}$; the lines are not perpendicular. For the lines to be perpendicular, the product must be -1.

53. We determine the slope of each line.

$$2x - 3y = 7$$
$$-3y = -2x + 7$$
$$y = \frac{2}{3}x - \frac{7}{3}$$

The slope of $2x - 3y = 7$ is $\frac{2}{3}$.

$$2y - 3x = 10$$
$$2y = 3x + 10$$
$$y = \frac{3}{2}x + 5$$

The slope of $2y - 3x = 10$ is $\frac{3}{2}$.

The product of their slopes is $\frac{2}{3} \cdot \frac{3}{2} = 1$; the lines are not perpendicular. For the lines to be perpendicular, the product must be -1.

55. Solving the first equation for x and the second for y, we have $x = \frac{3}{2}$ and $y = -2$. The graph of $x = \frac{3}{2}$ is a vertical line, and the graph of $y = -2$ is a horizontal line. Since one line is vertical and the other is horizontal, the lines are perpendicular.

57. 5.3,000,000,000.

10 places

Large number, so the exponent is positive.

$53{,}000{,}000{,}000 = 5.3 \times 10^{10}$

59. 0.01. 8

2 places

Small number, so the exponent is negative.

$0.018 = 1.8 \times 10^{-2}$

61. Negative exponent, so the number is small.

0.00002. 13

5 places

$2.13 \times 10^{-5} = 0.0000213$

63. Positive exponent, so the number is large.

2.0000.

4 places

$2 \times 10^4 = 20{,}000$

65. $9x - 15y = 3 \cdot 3x - 3 \cdot 5y = 3(3x - 5y)$

67. $21p - 7pq + 14p = 7p \cdot 3 - 7p \cdot q + 7p \cdot 2$
$$= 7p(3 - q + 2)$$

69. *Familiarize*. Let $w =$ the record weight set in 2012.

Translate.

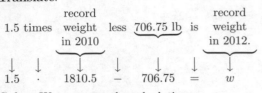

$$1.5 \cdot 1810.5 - 706.75 = w$$

Solve. We carry out the calculation.

$$1.5 \cdot 1810.5 - 706.75 = w$$
$$2715.75 - 706.75 = w$$
$$2009 = w$$

Check. We can repeat the calculation. The answer checks.

State. The record weight set in 2012 was 2009 lb.

71. Find the slope of each line.

$$5y = ax + 5$$
$$y = \frac{a}{5}x + 1$$

The slope of $5y = ax + 5$ is $\frac{a}{5}$.

$$\frac{1}{4}y = \frac{1}{10}x - 1$$
$$4 \cdot \frac{1}{4}y = 4\left(\frac{1}{10}x - 1\right)$$
$$y = \frac{2}{5}x - 4$$

The slope of $\frac{1}{4}y = \frac{1}{10}x - 1$ is $\frac{2}{5}$.

In order for the graphs to be parallel, their slopes must be the same. (Note that the y-intercepts are different.)

$$\frac{a}{5} = \frac{2}{5}$$
$$a = 2 \qquad \text{Multiplying by 5}$$

73. The y-intercept is $\left(0, \frac{2}{5}\right)$, so the equation is of the form $y = mx + \frac{2}{5}$. We substitute -3 for x and 0 for y in this equation to find m.

$$y = mx + \frac{2}{5}$$
$$0 = m(-3) + \frac{2}{5} \qquad \text{Substituting}$$
$$0 = -3m + \frac{2}{5}$$
$$3m = \frac{2}{5} \qquad \text{Adding } 3m$$
$$m = \frac{2}{15} \qquad \text{Multiplying by } \frac{1}{3}$$

The equation is $y = \frac{2}{15}x + \frac{2}{5}$.

(We could also have found the slope as follows:

$$m = \frac{\frac{2}{5} - 0}{0 - (-3)} = \frac{\frac{2}{5}}{3} = \frac{2}{15})$$

75. All points on the x-axis are pairs of the form $(x, 0)$. Thus any number for x will do and y must be 0. The equation is $y = 0$. This equation is a function because its graph passes the vertical-line test.

77. We substitute 4 for x and 0 for y.

$$y = mx + 3$$
$$0 = m(4) + 3$$
$$-3 = 4m$$
$$-\frac{3}{4} = m$$

Exercise Set 2.6

RC1. The slope of $y = \frac{4}{11}x - 2$ is $\frac{4}{11}$.

  a) A line parallel to the given line has the same slope, $\frac{4}{11}$.

  b) The slope of a line perpendicular to the given line is the opposite of the reciprocal of $\frac{4}{11}$, or $-\frac{11}{4}$.

RC3. $2x - y = -4$

$$-y = -2x - 4$$
$$y = 2x + 4$$

The slope is 2.

  a) A line parallel to the given line has the same slope, 2.

  b) The slope of a line perpendicular to the given line is the opposite of the reciprocal of 2, or $-\frac{1}{2}$.

RC5. The graph of $x = 3$ is a vertical line, so its slope is not defined.

  a) A line parallel to the given line has the same slope, so its slope is not defined.

  b) A line perpendicular to $x = 3$ is a horizontal line, so its slope is 0.

1. We use the slope-intercept equation and substitute -8 for m and 4 for b.

$$y = mx + b$$
$$y = -8x + 4$$

3. We use the slope-intercept equation and substitute 2.3 for m and -1 for b.

$$y = mx + b$$
$$y = 2.3x - 1$$

5. We use the slope-intercept equation and substitute $-\frac{7}{3}$ for m and -5 for b.

$$y = mx + b$$
$$y = -\frac{7}{3}x - 5$$

7. We use the slope-intercept equation and substitute $\frac{2}{3}$ for m and $\frac{5}{8}$ for b.

$$y = mx + b$$
$$y = \frac{2}{3}x + \frac{5}{8}$$

9. Using the point-slope equation:

Substitute 4 for x_1, 3 for y_1, and 5 for m.

$$y - y_1 = m(x - x_1)$$
$$y - 3 = 5(x - 4)$$
$$y - 3 = 5x - 20$$
$$y = 5x - 17$$

Using the slope-intercept equation:

Substitute 4 for x, 3 for y, and 5 for m in $y = mx + b$ and solve for b.

$$y = mx + b$$
$$3 = 5 \cdot 4 + b$$
$$3 = 20 + b$$
$$-17 = b$$

Then we use the equation $y = mx + b$ and substitute 5 for m and -17 for b.

$$y = 5x - 17$$

11. Using the point-slope equation:

Substitute 9 for x_1, 6 for y_1, and -3 for m.

$$y - y_1 = m(x - x_1)$$
$$y - 6 = -3(x - 9)$$
$$y - 6 = -3x + 27$$
$$y = -3x + 33$$

Using the slope-intercept equation:

Substitute 9 for x, 6 for y, and -3 for m in $y = mx + b$ and solve for b.

$$y = mx + b$$
$$6 = -3 \cdot 9 + b$$
$$6 = -27 + b$$
$$33 = b$$

Then we use the equation $y = mx + b$ and substitute -3 for m and 33 for b.

$$y = -3x + 33$$

13. Using the point-slope equation:

Substitute -1 for x_1, -7 for y_1, and 1 for m.

$$y - y_1 = m(x - x_1)$$
$$y - (-7) = 1(x - (-1))$$
$$y + 7 = 1(x + 1)$$
$$y + 7 = x + 1$$
$$y = x - 6$$

Using the slope-intercept equation:

Substitute -1 for x, -7 for y, and 1 for m in $y = mx + b$ and solve for b.

$$y = mx + b$$
$$-7 = 1(-1) + b$$
$$-7 = -1 + b$$
$$-6 = b$$

Then we use the equation $y = mx + b$ and substitute 1 for m and -6 for b.

$$y = 1x - 6, \text{ or } y = x - 6$$

15. Using the point-slope equation:

Substitute 8 for x_1, 0 for y_1, and -2 for m.

$$y - y_1 = m(x - x_1)$$
$$y - 0 = -2(x - 8)$$
$$y = -2x + 16$$

Using the slope-intercept equation:

Substitute 8 for x, 0 for y, and -2 for m in $y = mx + b$ and solve for b.

$$y = mx + b$$
$$0 = -2 \cdot 8 + b$$
$$0 = -16 + b$$
$$16 = b$$

Then we use the equation $y = mx + b$ and substitute -2 for m and 16 for b.

$$y = -2x + 16$$

17. Using the point-slope equation:

Substitute 0 for x_1, -7 for y_1, and 0 for m.

$$y - y_1 = m(x - x_1)$$
$$y - (-7) = 0(x - 0)$$
$$y + 7 = 0$$
$$y = -7$$

Using the slope-intercept equation:

Substitute 0 for x, -7 for y, and 0 for m in $y = mx + b$ and solve for b.

$$y = mx + b$$
$$-7 = 0 \cdot 0 + b$$
$$-7 = b$$

Then we use the equation $y = mx + b$ and substitute 0 for m and -7 for b.

$$y = 0x - 7, \text{ or } y = -7$$

19. Using the point-slope equation:

Substitute 1 for x_1, -2 for y_1, and $\frac{2}{3}$ for m.

$$y - y_1 = m(x - x_1)$$
$$y - (-2) = \frac{2}{3}(x - 1)$$
$$y + 2 = \frac{2}{3}x - \frac{2}{3}$$
$$y = \frac{2}{3}x - \frac{8}{3}$$

Using the slope-intercept equation:

Substitute 1 for x, -2 for y and $\frac{2}{3}$ for m in $y = mx + b$ and solve for b.

$$y = mx + b$$
$$-2 = \frac{2}{3} \cdot 1 + b$$
$$-2 = \frac{2}{3} + b$$
$$-\frac{8}{3} = b$$

Then we use the equation $y = mx + b$ and substitute $\frac{2}{3}$ for m and $-\frac{8}{3}$ for b.

$$y = \frac{2}{3}x - \frac{8}{3}$$

21. First find the slope of the line:

$$m = \frac{6-4}{5-1} = \frac{2}{4} = \frac{1}{2}$$

Using the point-slope equation:

We choose to use the point $(1, 4)$ and substitute 1 for x_1, 4 for y_1, and $\frac{1}{2}$ for m.

$$y - y_1 = m(x - x_1)$$
$$y - 4 = \frac{1}{2}(x - 1)$$
$$y - 4 = \frac{1}{2}x - \frac{1}{2}$$
$$y = \frac{1}{2}x + \frac{7}{2}$$

Using the slope-intercept equation:

We choose $(1, 4)$ and substitute 1 for x, 4 for y, and $\frac{1}{2}$ for m in $y = mx + b$. Then we solve for b.

$$y = mx + b$$
$$4 = \frac{1}{2} \cdot 1 + b$$
$$4 = \frac{1}{2} + b$$
$$\frac{7}{2} = b$$

Finally, we use the equation $y = mx + b$ and substitute $\frac{1}{2}$ for m and $\frac{7}{2}$ for b.

$$y = \frac{1}{2}x + \frac{7}{2}$$

23. First find the slope of the line:

$$m = \frac{-3-2}{-3-2} = \frac{-5}{-5} = 1$$

Using the point-slope equation:

We choose to use the point $(2, 2)$ and substitute 2 for x_1, 2 for y_1, and 1 for m.

$$y - y_1 = m(x - x_1)$$
$$y - 2 = 1(x - 2)$$
$$y - 2 = x - 2$$
$$y = x$$

Using the slope-intercept equation:

We choose $(2, 2)$ and substitute 2 for x, 2 for y, and 1 for m in $y = mx + b$. Then we solve for b.

$$y = mx + b$$
$$2 = 1 \cdot 2 + b$$
$$2 = 2 + b$$
$$0 = b$$

Finally, we use the equation $y = mx + b$ and substitute 1 for m and 0 for b.

$$y = 1x + 0, \text{ or } y = x$$

25. First find the slope of the line:

$$m = \frac{0-7}{-4-0} = \frac{-7}{-4} = \frac{7}{4}$$

Using the point-slope equation:

We choose $(0, 7)$ and substitute 0 for x_1, 7 for y_1, and $\frac{7}{4}$ for m.

$$y - y_1 = m(x - x_1)$$
$$y - 7 = \frac{7}{4}(x - 0)$$
$$y - 7 = \frac{7}{4}x$$
$$y = \frac{7}{4}x + 7$$

Using the slope-intercept equation:

We choose $(0, 7)$ and substitute 0 for x, 7 for y, and $\frac{7}{4}$ for m in $y = mx + b$. Then we solve for b.

$$y = mx + b$$
$$7 = \frac{7}{4} \cdot 0 + b$$
$$7 = b$$

Finally, we use the equation $y = mx + b$ and substitute $\frac{7}{4}$ for m and 7 for b.

$$y = \frac{7}{4}x + 7$$

27. First find the slope of the line:

$$m = \frac{-6-(-3)}{-4-(-2)} = \frac{-6+3}{-4+2} = \frac{-3}{-2} = \frac{3}{2}$$

Using the point-slope equation:

We choose $(-2, -3)$ and substitute -2 for x_1, -3 for y_1, and $\frac{3}{2}$ for m.

$$y - y_1 = m(x - x_1)$$
$$y - (-3) = \frac{3}{2}(x - (-2))$$
$$y + 3 = \frac{3}{2}(x + 2)$$
$$y + 3 = \frac{3}{2}x + 3$$
$$y = \frac{3}{2}x$$

Using the slope-intercept equation:

We choose $(-2, -3)$ and substitute -2 for x, -3 for y, and $\frac{3}{2}$ for m in $y = mx + b$. Then we solve for b.

$$y = mx + b$$
$$-3 = \frac{3}{2}(-2) + b$$
$$-3 = -3 + b$$
$$0 = b$$

Finally, we use the equation $y = mx + b$ and substitute $\frac{3}{2}$ for m and 0 for b.

$$y = \frac{3}{2}x + 0, \text{ or } y = \frac{3}{2}x$$

29. First find the slope of the line:

$$m = \frac{1 - 0}{6 - 0} = \frac{1}{6}$$

Using the point-slope equation:

We choose $(0, 0)$ and substitute 0 for x_1, 0 for y_1, and $\frac{1}{6}$ for m.

$$y - y_1 = m(x - x_1)$$
$$y - 0 = \frac{1}{6}(x - 0)$$
$$y = \frac{1}{6}x$$

Using the slope-intercept equation:

We choose $(0, 0)$ and substitute 0 for x, 0 for y, and $\frac{1}{6}$ for m in $y = mx + b$. Then we solve for b.

$$y = mx + b$$
$$0 = \frac{1}{6} \cdot 0 + b$$
$$0 = b$$

Finally, we use the equation $y = mx + b$ and substitute $\frac{1}{6}$ for m and 0 for b.

$$y = \frac{1}{6}x + 0, \text{ or } y = \frac{1}{6}x$$

31. First find the slope of the line:

$$m = \frac{-\frac{1}{2} - 6}{\frac{1}{4} - \frac{3}{4}} = \frac{-\frac{13}{2}}{-\frac{1}{2}} = 13$$

Using the point-slope equation:

We choose $\left(\frac{3}{4}, 6\right)$ and substitute $\frac{3}{4}$ for x_1, 6 for y_1, and 13 for m.

$$y - y_1 = m(x - x_1)$$
$$y - 6 = 13\left(x - \frac{3}{4}\right)$$
$$y - 6 = 13x - \frac{39}{4}$$
$$y = 13x - \frac{15}{4}$$

Using the slope-intercept equation:

We choose $\left(\frac{3}{4}, 6\right)$ and substitute $\frac{3}{4}$ for x, 6 for y, and 13 for m in $y = mx + b$. Then we solve for b.

$$y = mx + b$$
$$6 = 13 \cdot \frac{3}{4} + b$$
$$6 = \frac{39}{4} + b$$
$$-\frac{15}{4} = b$$

Finally, we use the equation $y = mx + b$ and substitute 13 for m and $-\frac{15}{4}$ for b.

$$y = 13x - \frac{15}{4}$$

33. First solve the equation for y and determine the slope of the given line.

$$x + 2y = 6 \qquad \text{Given line}$$
$$2y = -x + 6$$
$$y = -\frac{1}{2}x + 3$$

The slope of the given line is $-\frac{1}{2}$. The line through $(3, 7)$ must have slope $-\frac{1}{2}$.

Using the point-slope equation:

Substitute 3 for x_1, 7 for y_1, and $-\frac{1}{2}$ for m.

$$y - y_1 - m(x - x_1)$$
$$y - 7 = -\frac{1}{2}(x - 3)$$
$$y - 7 = -\frac{1}{2}x + \frac{3}{2}$$
$$y = -\frac{1}{2}x + \frac{17}{2}$$

Using the slope-intercept equation:

Substitute 3 for x, 7 for y, and $-\frac{1}{2}$ for m and solve for b.

$$y = mx + b$$
$$7 = -\frac{1}{2} \cdot 3 + b$$
$$7 = -\frac{3}{2} + b$$
$$\frac{17}{2} = b$$

Then we use the equation $y = mx + b$ and substitute $-\frac{1}{2}$ for m and $\frac{17}{2}$ for b.

$$y = -\frac{1}{2}x + \frac{17}{2}$$

35. First solve the equation for y and determine the slope of the given line.

$$5x - 7y = 8 \qquad \text{Given line}$$

$$5x - 8 = 7y$$

$$\frac{5}{7}x - \frac{8}{7} = y$$

$$y = \frac{5}{7}x - \frac{8}{7}$$

The slope of the given line is $\frac{5}{7}$. The line through $(2, -1)$ must have slope $\frac{5}{7}$.

Using the point-slope equation:

Substitute 2 for x_1, -1 for y_1, and $\frac{5}{7}$ for m.

$$y - y_1 = m(x - x_1)$$

$$y - (-1) = \frac{5}{7}(x - 2)$$

$$y + 1 = \frac{5}{7}x - \frac{10}{7}$$

$$y = \frac{5}{7}x - \frac{17}{7}$$

Using the slope-intercept equation:

Substitute 2 for x, -1 for y, and $\frac{5}{7}$ for m and solve for b.

$$y = mx + b$$

$$-1 = \frac{5}{7} \cdot 2 + b$$

$$-1 = \frac{10}{7} + b$$

$$-\frac{17}{7} = b$$

Then we use the equation $y = mx + b$ and substitute $\frac{5}{7}$ for m and $-\frac{17}{7}$ for b.

$$y = \frac{5}{7}x - \frac{17}{7}$$

37. First solve the equation for y and determine the slope of the given line.

$$3x - 9y = 2 \quad \text{Given line}$$

$$3x - 2 = 9y$$

$$\frac{1}{3}x - \frac{2}{9} = y$$

The slope of the given line is $\frac{1}{3}$. The line through $(-6, 2)$ must have slope $\frac{1}{3}$.

Using the point-slope equation:

Substitute -6 for x_1, 2 for y_1, and $\frac{1}{3}$ for m.

$$y - y_1 = m(x - x_1)$$

$$y - 2 = \frac{1}{3}(x - (-6))$$

$$y - 2 = \frac{1}{3}(x + 6)$$

$$y - 2 = \frac{1}{3}x + 2$$

$$y = \frac{1}{3}x + 4$$

Using the slope-intercept equation:

Substitute -6 for x, 2 for y, and $\frac{1}{3}$ for m and solve for b.

$$y = mx + b$$

$$2 = \frac{1}{3}(-6) + b$$

$$2 = -2 + b$$

$$4 = b$$

Then we use the equation $y = mx + b$ and substitute $\frac{1}{3}$ for m and 4 for b.

$$y = \frac{1}{3}x + 4$$

39. First solve the equation for y and determine the slope of the given line.

$$2x + y = -3 \qquad \text{Given line}$$

$$y = -2x - 3$$

The slope of the given line is -2. The slope of the perpendicular line is the opposite of the reciprocal of -2. Thus, the line through $(2, 5)$ must have slope $\frac{1}{2}$.

Using the point-slope equation:

Substitute 2 for x_1, 5 for y_1, and $\frac{1}{2}$ for m.

$$y - y_1 = m(x - x_1)$$

$$y - 5 = \frac{1}{2}(x - 2)$$

$$y - 5 = \frac{1}{2}x - 1$$

$$y = \frac{1}{2}x + 4$$

Using the slope-intercept equation:

Substitute 2 for x, 5 for y, and $\frac{1}{2}$ for m and solve for b.

$$y = mx + b$$

$$5 = \frac{1}{2} \cdot 2 + b$$

$$5 = 1 + b$$

$$4 = b$$

Then we use the equation $y = mx + b$ and substitute $\frac{1}{2}$ for m and 4 for b.

$$y = \frac{1}{2}x + 4$$

41. First solve the equation for y and determine the slope of the given line.

$$3x + 4y = 5 \qquad \text{Given line}$$
$$4y = -3x + 5$$
$$y = -\frac{3}{4}x + \frac{5}{4}$$

The slope of the given line is $-\frac{3}{4}$. The slope of the perpendicular line is the opposite of the reciprocal of $-\frac{3}{4}$. Thus, the line through $(3, -2)$ must have slope $\frac{4}{3}$.

Using the point-slope equation:

Substitute 3 for x_1, -2 for y_1, and $\frac{4}{3}$ for m.

$$y - y_1 = m(x - x_1)$$
$$y - (-2) = \frac{4}{3}(x - 3)$$
$$y + 2 = \frac{4}{3}x - 4$$
$$y = \frac{4}{3}x - 6$$

Using the slope-intercept equation:

Substitute 3 for x, -2 for y, and $\frac{4}{3}$ for m.

$$y = mx + b$$
$$-2 = \frac{4}{3} \cdot 3 + b$$
$$-2 = 4 + b$$
$$-6 = b$$

Then we use the equation $y = mx + b$ and substitute $\frac{4}{3}$ for m and -6 for b.

$$y = \frac{4}{3}x - 6$$

43. First solve the equation for y and determine the slope of the given line.

$$2x + 5y = 7 \qquad \text{Given line}$$
$$5y = -2x + 7$$
$$y = -\frac{2}{5}x + \frac{7}{5}$$

The slope of the given line is $-\frac{2}{5}$. The slope of the perpendicular line is the opposite of the reciprocal of $-\frac{2}{5}$. Thus, the line through $(0, 9)$ must have slope $\frac{5}{2}$.

Using the point-slope equation:

Substitute 0 for x_1, 9 for y_1, and $\frac{5}{2}$ for m.

$$y - y_1 = m(x - x_1)$$
$$y - 9 = \frac{5}{2}(x - 0)$$
$$y - 9 = \frac{5}{2}x$$
$$y = \frac{5}{2}x + 9$$

Using the slope-intercept equation:

Substitute 0 for x, 9 for y, and $\frac{5}{2}$ for m.

$$y = mx + b$$
$$9 = \frac{5}{2} \cdot 0 + b$$
$$9 = b$$

Then we use the equation $y = mx + b$ and substitute $\frac{5}{2}$ for m and 9 for b.

$$y = \frac{5}{2}x + 9$$

45. a) The problem describes a situation in which a fee per bag is charged after an initial charge for the first three bags. For 4 bags, the total cost is $\$10 + \$5 \cdot 1$. For 5 bags, the total cost is $\$10 + \$5 \cdot 2$. Then for x bags beyond the first 3 bags, the total cost is $C(x) - 10 + 5x$, or $C(x) = 5x + 10$. (Keep in mind that x is the number of additional bags after the first three bags.)

b) The y-intercept is $(0, 10)$ and the slope, or rate of change, is 5, or $\frac{5}{1}$. We plot $(0, 10)$ and from there we move up 5 units and right 1 unit to $(1, 15)$. Then we draw a line through these points. We can also find a third point as a check.

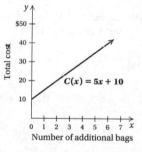

c) Since 7 bags is 4 bags in addition to the initial 3 bags, we find $C(4)$.

$$C(4) = 5 \cdot 4 + 10 = 20 + 10 = 30$$

Thus, it would cost $30 to shred 7 bags of documents.

47. a) The problem describes a situation in which the value of the lawn mower decreases at a rate of $85 per month from an initial value of $9400. After 1 month, the value is $\$9400 - \$85 \cdot 1$. After 2 months, the value is $\$9400 - \$85 \cdot 2$. Then after t months, the value is $V(t) = 9400 - 85t$.

b) For $V(t) = 9400 - 85t$, or $V(t) = -85t + 9400$, the y-intercept is $(0, 9400)$ and the slope is -85, or $-\frac{85}{1}$. Plot $(0, 9400)$ and from there move down 85 units and right 1 unit to $(1, 9315)$. Then we draw a line through these points. We can find a third point as a check.

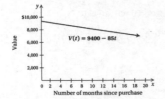

c) $C(18) = 9400 - 85 \cdot 18 = 9400 - 1530 = \$7870$

49. a) The data points are $(0, 11)$ and $(8, 27)$. First we find the slope.

$$m = \frac{27 - 11}{8 - 0} = \frac{16}{8} = 2$$

Using the slope and the y-intercept, $(0, 11)$, we write the function:

$S(x) = 2x + 11$, where x is the number of years since 2004 and $S(x)$ is in billions of dollars.

b) In 2008, $x = 2008 - 2004 = 4$.

$$S(4) = 2 \cdot 4 + 11 = 8 + 11 = 19$$

We estimate that sales of organic food in 2008 were \$19 billion.

In 2017, $x = 2017 - 2004 = 13$.

$$S(13) = 2 \cdot 13 + 11 = 26 + 11 = 37$$

We predict that sales of organic food will be \$37 billion in 2017.

51. a) One data point is $(0, 22,800)$. In 2012, $x = 2012 - 1995 = 17$, so another data point is $(17, 17,540)$.

We find the slope.

$$m = \frac{17,540 - 22,800}{17 - 0} = \frac{-5260}{17} \approx -309.41.$$

Using the slope and the y-intercept, $(0, 22,800)$, we write the function.

$D(x) = -309.41x + 22,800$, where x is the number of years since 1995.

b) In 2000, $x = 2000 - 1995 = 5$

$$D(5) = -309.41(5) + 22,800 = 21,252.95 \approx 21,253$$

We estimate that there were about 21,253 new-auto dealerships in 2000.

c) Substitute 15,500 for $D(x)$ and solve for x.

$$15,500 = -309.41x + 22,800$$
$$-7300 = -309.41x$$
$$24 \approx x$$

We estimate that there will be 15,500 new-auto dealerships about 24 years after 1995, or in 2019.

53. a) One data point is $(0, 46.56)$. In 2011, $t = 2011 - 2003 = 8$, so another data point is $(8, 49.33)$.

We find the slope.

$$m = \frac{49.33 - 46.56}{8 - 0} = \frac{2.77}{8} \approx 0.346$$

Using the slope and the y-intercept, $(0, 46.56)$, we write the function:

$$E(t) = 0.346t + 46.56.$$

b) In 2016, $t = 2016 - 2003 = 13$.

$$E(13) = 0.346(13) + 46.56 \approx 51.06$$

We estimate that the life expectancy is about 51.06 years in 2016.

55. $2x + 3 > 51$

$\quad\quad 2x > 48$ \quad\quad Subtracting 3

$\quad\quad\quad x > 24$ \quad\quad Dividing by 2

The solution set is $\{x | x > 24\}$, or $(24, \infty)$.

57. $2x + 3 \le 51$

$\quad\quad 2x \le 48$ \quad\quad Subtracting 3

$\quad\quad\quad x \le 24$ \quad\quad Dividing by 2

The solution set is $\{x | x \le 24\}$, or $(-\infty, 24]$.

59. $|2x + 3| \le 13$

$\quad -13 \le 2x + 3 \le 13$

$\quad -16 \le 2x \le 10$ \quad\quad Subtracting 3

$\quad\quad -8 \le x \le 5$ \quad\quad Dividing by 2

The solution set is $\{x | -8 \le x \le 5\}$, or $[-8, 5]$.

61. $|5x - 4| = -8$

Since the absolute value of a number must be nonnegative, the equation has no solution. The solution set is $\{\ \}$, or $\emptyset$.

63. First find the slopes.

$$m_1 = \frac{k - 8}{-3 - 4} = \frac{k - 8}{-7}$$

$$m_2 = \frac{3 - (-6)}{5 - 1} = \frac{9}{4}$$

If the lines are parallel, the slopes must be equal.

$$\frac{k - 8}{-7} = \frac{9}{4}$$
$$4(k - 8) = -63$$
$$4k - 32 = -63$$
$$4k = -31$$
$$k = -\frac{31}{4}, \text{ or } -7.75$$

Chapter 2 Vocabulary Reinforcement

1. The graph of $x = a$ is a <u>vertical</u> line with x-intercept $(a, 0)$.

2. The <u>point-slope</u> equation of a line with slope m and passing through (x_1, y_1) is $y - y_1 = m(x - x_1)$.

3. A <u>function</u> is a correspondence between a first set, called the <u>domain</u>, and a second set, called the <u>range</u>, such that each member of the <u>domain</u> corresponds to <u>exactly one</u> member of the <u>range</u>.

4. The <u>slope</u> of a line containing points (x_1, y_1) and (x_2, y_2) is given by $m = $ the change in y/the change in x, also described as rise/run.

5. Two lines are <u>perpendicular</u> if the product of their slopes is -1.

6. The equation $y = mx + b$ is called the <u>slope-intercept</u> equation of a line with slope m and y-intercept $(0, b)$.

7. Lines are <u>parallel</u> if they have the same slope and different y-intercepts.

Chapter 2 Concept Reinforcement

1. False; the slope of a vertical line is not defined. See page 208 in the text.

2. True; see page 196 in the text.

3. False; parallel lines have the same slope and *different y*-intercepts. See page 209 in the text.

Chapter 2 Study Guide

1. A member of the domain is matched to more than one member of the range, so the correspondence is not a function.

2. $g(x) = \dfrac{1}{2}x - 2$

$g(0) = \dfrac{1}{2} \cdot 0 - 2 = 0 - 2 = -2$

$g(-2) = \dfrac{1}{2}(-2) - 2 = -1 - 2 = -3$

$g(6) = \dfrac{1}{2} \cdot 6 - 2 = 3 - 2 = 1$

3. $y = \dfrac{2}{5}x - 3$

We find some ordered pairs that are solutions, plot them, and draw and label the line.

When $x = -5$, $y = \dfrac{2}{5}(-5) - 3 = -2 - 3 = -5$.

When $x = 0$, $y = \dfrac{2}{5} \cdot 0 - 3 = 0 - 3 = -3$.

When $x = 5$, $y = \dfrac{2}{5} \cdot 5 - 3 = 2 - 3 = -1$.

x	y
-5	-5
0	-3
5	-1

4. No vertical line can cross the graph at more than one point, so the graph is that of a function.

5. The set of all x-values on the graph extends from -4 through 5, so the domain is $\{x \mid -4 \le x \le 5\}$, or $[-4, 5]$.

The set of all y-values on the graph extends from -2 through 4, so the range is $\{y \mid -2 \le y \le 4\}$, or $[-2, 4]$.

6. Since $\dfrac{x - 3}{3x + 9}$ cannot be calculated when $3x + 9$ is 0, we solve $3x + 9 = 0$.

$3x + 9 = 0$

$3x = -9$

$x = -3$

Thus, the domain of g is $\{x \mid x \text{ is a real number } and$ $x \ne -3\}$, or $(-\infty, -3) \cup (-3, \infty)$.

7. $m = \dfrac{-8 - 2}{2 - (-3)} = \dfrac{-10}{5} = -2$

8. $3x = -6y + 12$

$3x - 12 = -6y$

$-\dfrac{1}{2}x + 2 = y$ Dividing by -6

The slope is $-\dfrac{1}{2}$, and the y-intercept is $(0, 2)$.

9. $3y - 3 = x$

To find the y-intercept, let $x = 0$ and solve for y.

$3y - 3 = 0$

$3y = 3$

$y = 1$

The y-intercept is $(0, 1)$.

To find the x-intercept, let $y = 0$ and solve for x.

$3 \cdot 0 - 3 = x$

$0 - 3 = x$

$-3 = x$

The x-intercept is $(-3, 0)$.

We plot these points and draw the line.

We find a third point as a check. Let $x = 3$.

$3y - 3 = 3$

$3y = 6$

$y = 2$

We see that the point $(3, 2)$ is on the line.

10. $y = \dfrac{1}{4}x - 3$

First we plot the y-intercept $(0, -3)$. Then we consider the slope $\dfrac{1}{4}$. Starting at the y-intercept, we find another point by moving 1 unit up and 4 units to the right. We get to the point $(4, -2)$. We can also think of the slope as $\dfrac{-1}{-4}$. We again start at the y-intercept and move down 1 unit and 4 units to the left. We get to a third point $(-4, -4)$. We plot the points and draw the graph.

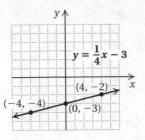

11. $y = 3$

All ordered pairs $(x, 3)$ are solutions. The graph is a horizontal line that intersects the y-axis at $(0, 3)$.

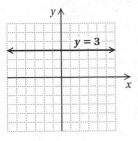

12. $x = -\dfrac{5}{2}$

All points $\left(-\dfrac{5}{2}, y\right)$ are solutions. The graph is a vertical line that intersects the x-axis at $\left(-\dfrac{5}{2}, 0\right)$.

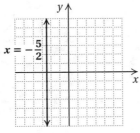

13. We first solve for y and determine the slope of each line.

$$-3x + 8y = -8$$
$$8y = 3x - 8$$
$$y = \frac{3}{8}x - 1$$

The slope of $-3x + 8y = -8$ is $\dfrac{3}{8}$.

$$8y = 3x + 40$$
$$y = \frac{3}{8}x + 5$$

The slope of $8y = 3x + 40$ is $\dfrac{3}{8}$.

The slopes are the same and the y-intercepts, $(0, -1)$ and $(0, 5)$ are different, so the lines are parallel.

14. We first solve for y and determine the slope of each line.

$$5x - 2y = -8$$
$$-2y = -5x - 8$$
$$y = \frac{5}{2}x + 4$$

The slope of $5x - 2y = -8$ is $\dfrac{5}{2}$.

$$2x + 5y = 15$$
$$5y = -2x + 15$$
$$y = -\frac{2}{5}x + 3$$

The slope of $2x + 5y = 15$ is $-\dfrac{2}{5}$.

The slopes are different, so the lines are not parallel. The product of the slopes is $\dfrac{5}{2}\left(-\dfrac{2}{5}\right) = -1$, so the lines are perpendicular.

15. $y = mx + b$ Slope-intercept equation

$y = -8x + 0.3$

16. Using the point-slope equation:

$$y - (-3) = -4\left(x - \frac{1}{2}\right)$$
$$y + 3 = -4x + 2$$
$$y = -4x - 1$$

Using the slope intercept equation:

$$-3 = -4\left(\frac{1}{2}\right) + b$$
$$-3 = -2 + b$$
$$-1 = b$$

Then, substituting in $y = mx + b$, we have $y = -4x - 1$.

17. We first find the slope:

$$m = \frac{-3 - 7}{4 - (-2)} = \frac{-10}{6} = -\frac{5}{3}$$

We use the point-slope equation.

$$y - 7 = -\frac{5}{3}[x - (-2)]$$
$$y - 7 = -\frac{5}{3}(x + 2)$$
$$y - 7 = -\frac{5}{3}x - \frac{10}{3}$$
$$y = -\frac{5}{3}x + \frac{11}{3}$$

18. First we find the slope of the given line:

$$4x - 3y = 6$$
$$-3y = -4x + 6$$
$$y = \frac{4}{3}x - 2$$

A line parallel to this line has slope $\dfrac{4}{3}$.

We use the slope-intercept equation.

$$-5 = \frac{4}{3}(2) + b$$
$$-5 = \frac{8}{3} + b$$
$$-\frac{23}{3} = b$$

Then we have $y = \dfrac{4}{3}x - \dfrac{23}{3}$.

19. From Exercise 18 above we know that the slope of the given line is $\frac{4}{3}$. The slope of a line perpendicular to this line is $-\frac{3}{4}$.

We use the point-slope equation.

$$y - (-5) = -\frac{3}{4}(x - 2)$$

$$y + 5 = -\frac{3}{4}x + \frac{3}{2}$$

$$y = -\frac{3}{4}x - \frac{7}{2}$$

Chapter 2 Review Exercises

1. No; a member of the domain, 3, is matched to more than one member of the range.

2. Yes; each member of the domain is matched to only one member of the range.

3. $g(x) = -2x + 5$

$g(0) = -2 \cdot 0 + 5 = 0 + 5 = 5$

$g(-1) = -2(-1) + 5 = 2 + 5 = 7$

4. $f(x) = 3x^2 - 2x + 7$

$f(0) = 3 \cdot 0^2 - 2 \cdot 0 + 7 = 0 - 0 + 7 = 7$

$f(-1) = 3(-1)^2 - 2(-1) + 7 = 3 \cdot 1 - 2(-1) + 7 = 3 + 2 + 7 = 12$

5. $C(t) = 309.2t + 3717.7$

$C(10) = 309.2(10) + 3717.7 = 3092 + 3717.7 = 6809.7 \approx 6810$

We estimate that the average cost of tuition and fees will be about $6810 in 2010.

6. $y = -3x + 2$

We find some ordered pairs that are solutions, plot them, and draw and label the line.

When $x = -1$, $y = -3(-1) + 2 = 3 + 2 = 5$.

When $x = 1$, $y = -3 \cdot 1 + 2 = -3 + 2 = -1$

When $x = 2$, $y = -3 \cdot 2 + 2 = -6 + 2 = -4$.

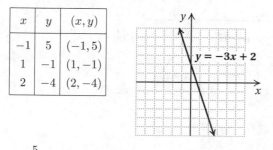

x	y	(x, y)
-1	5	$(-1, 5)$
1	-1	$(1, -1)$
2	-4	$(2, -4)$

7. $y = \frac{5}{2}x - 3$

We find some ordered pairs that are solutions, using multiples of 2 to avoid fractions. Then we plot these points and draw and label the line.

When $x = 0$, $y = \frac{5}{2} \cdot 0 - 3 = 0 - 3 = -3$.

When $x = 2$, $y = \frac{5}{2} \cdot 2 - 3 = 5 - 3 = 2$.

When $x = 4$, $y = \frac{5}{2} \cdot 4 - 3 = 10 - 3 = 7$.

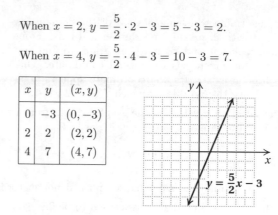

x	y	(x, y)
0	-3	$(0, -3)$
2	2	$(2, 2)$
4	7	$(4, 7)$

8. $y = |x - 3|$

To find an ordered pair, we choose any number for x and then determine y. For example, if $x = 5$, then $y = |5 - 3| = |2| = 2$. We find several ordered pairs, plot them, and connect them.

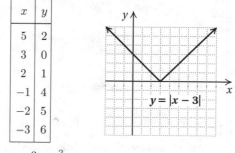

x	y
5	2
3	0
2	1
-1	4
-2	5
-3	6

9. $y = 3 - x^2$

To find an ordered pair, we choose any number for x and then determine y. For example, if $x = 2$, then $3 - 2^2 = 3 - 4 = -1$. We find several ordered pairs, plot them, and connect them with a smooth curve.

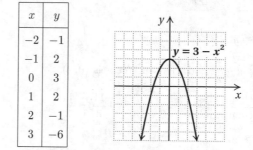

x	y
-2	-1
-1	2
0	3
1	2
2	-1
3	-6

10. No vertical line will intersect the graph more than once. Thus, the graph is the graph of a function.

11. It is possible for a vertical line to intersect the graph more than once. Thus, this is not the graph of a function.

12. a) Locate 2 on the horizontal axis and then find the point on the graph for which 2 is the first coordinate. From that point, look to the vertical axis to find the corresponding y-coordinate, 3. Thus, $f(2) = 3$.

b) The set of all x-values in the graph extends from -2 to 4, so the domain is $\{x | -2 \le x \le 4\}$, or $[-2, 4]$.

c) To determine which member(s) of the domain are paired with 2, locate 2 on the vertical axis. From there look left and right to the graph to find any points for which 2 is the second coordinate. One such point exists. Its first coordinate appears to be -1. Thus, the x-value for which $f(x) = 2$ is -1.

d) The set of all y-values in the graph extends from 1 to 5, so the range is $\{y|1 \le y \le 5\}$, or $[1, 5]$.

13. $f(x) = \dfrac{5}{x-4}$

Since $\dfrac{5}{x-4}$ cannot be calculated when the denominator is 0, we find the x-value that causes $x - 4$ to be 0:

$$x - 4 = 0$$
$$x = 4 \quad \text{Adding 4 on both sides}$$

Thus, 4 is not in the domain of f, while all other real numbers are. The domain of f is

$\{x|x \text{ is a real number } and \ x \ne 4\}$, or $(-\infty, 4) \cup (4, \infty)$.

14. $g(x) = x - x^2$

Since we can calculate $x - x^2$ for any real number x, the domain is the set of all real numbers.

15. $f(x) = -3x + 2$

$$f(x) = mx + b$$

The slope is -3, and the y-intercept is $(0, 2)$.

16. First we find the slope-intercept form of the equation by solving for y. This allows us to determine the slope and y-intercept easily.

$$4y + 2x = 8$$
$$4y = -2x + 8$$
$$\frac{4y}{4} = \frac{-2x + 8}{4}$$
$$y = -\frac{1}{2}x + 2$$

The slope is $-\dfrac{1}{2}$, and the y-intercept is $(0, 2)$.

17. Slope $= \dfrac{\text{change in } y}{\text{change in } x} = \dfrac{-4 - 7}{10 - 13} = \dfrac{-11}{-3} = \dfrac{11}{3}$

18. $2y + x = 4$

To find the x-intercept we let $y = 0$ and solve for x.

$$2y + x = 4$$
$$2 \cdot 0 + x = 4$$
$$x = 4$$

The x-intercept is $(4, 0)$.

To find the y-intercept we let $x = 0$ and solve for y.

$$2y + x = 4$$
$$2y + 0 = 4$$
$$2y = 4$$
$$y = 2$$

The y-intercept is $(0, 2)$.

We plot these points and draw the line.

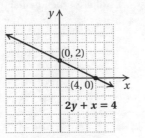

We use a third point as a check. We choose $x = -2$ and solve for y.

$$2y + (-2) = 4$$
$$2y = 6$$
$$y = 3$$

We plot $(-2, 3)$ and note that it is on the line.

19. $2y = 6 - 3x$

To find the x-intercept we let $y = 0$ and solve for x.

$$2y = 6 - 3x$$
$$2 \cdot 0 = 6 - 3x$$
$$0 = 6 - 3x$$
$$3x = 6$$
$$x = 2$$

The x-intercept is $(2, 0)$.

To find the y-intercept we let $x = 0$ and solve for y.

$$2y = 6 - 3x$$
$$2y = 6 - 3 \cdot 0$$
$$2y = 6$$
$$y = 3$$

The y-intercept is $(0, 3)$.

We plot these points and draw the line.

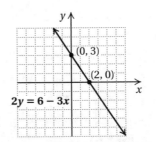

We use a third point as a check. We choose $x = 4$ and solve for y.

$$2y = 6 - 3 \cdot 4$$
$$2y = 6 - 12$$
$$2y = -6$$
$$y = -3$$

We plot $(4, -3)$ and note that it is on the line.

20. $g(x) = -\dfrac{2}{3}x - 4$

First we plot the y-intercept $(0, -4)$. We can think of the slope as $\dfrac{-2}{3}$. Starting at the y-intercept and using the slope, we find another point by moving 2 units down and 3 units to the right. We get to a new point $(3, -6)$.

We can also think of the slope as $\dfrac{2}{-3}$. We again start at the y-intercept $(0, -4)$. We move 2 units up and 3 units to the left. We get to another new point $(-3, -2)$. We plot the points and draw the line.

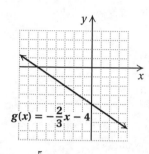

21. $f(x) = \dfrac{5}{2}x + 3$

First we plot the y-intercept $(0, 3)$. Then we consider the slope $\dfrac{5}{2}$. Starting at the y-intercept and using the slope, we find another point by moving 5 units up and 2 units to the right. We get to a new point $(2, 8)$.

We can also think of the slope as $\dfrac{-5}{-2}$. We again start at the y-intercept $(0, 3)$. We move 5 units down and 2 units to the left. We get to another new point $(-2, -2)$. We plot the points and draw the line.

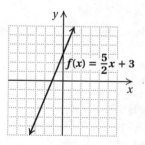

22. $x = -3$

Since y is missing, all ordered pairs $(-3, y)$ are solutions. The graph is parallel to the y-axis.

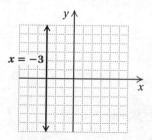

23. $f(x) = 4$

Since x is missing, all ordered pairs $(x, 4)$ are solutions. The graph is parallel to the x-axis.

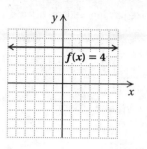

24. We first solve each equation for y and determine the slope of each line.
$$y + 5 = -x$$
$$y = -x - 5$$
The slope of $y + 5 = -x$ is -1.
$$x - y - 2$$
$$x = y + 2$$
$$x - 2 = y$$
The slope of $x - y = 2$ is 1.

The slopes are not the same, so the lines are not parallel. The product of the slopes is $-1 \cdot 1$, or -1, so the lines are perpendicular.

25. We first solve each equation for y and determine the slope of each line.
$$3x - 5 = 7y$$
$$\frac{3}{7}x - \frac{5}{7} = y$$
The slope of $3x - 5 = 7y$ is $\dfrac{3}{7}$.
$$7y - 3x = 7$$
$$7y = 3x + 7$$
$$y = \frac{3}{7}x + 1$$
The slope of $7y - 3x = 7$ is $\dfrac{3}{7}$.

The slopes are the same and the y-intercepts are different, so the lines are parallel.

26. We first solve each equation for y and determine the slope of each line.
$$4y + x = 3$$
$$4y = -x + 3$$
$$y = -\frac{1}{4}x + \frac{3}{4}$$
The slope of $4y + x = 3$ is $-\dfrac{1}{4}$.
$$2x + 8y = 5$$
$$8y = -2x + 5$$
$$y = -\frac{1}{4}x + \frac{5}{8}$$
The slope of $2x + 8y = 5$ is $-\dfrac{1}{4}$.

The slopes are the same and the y-intercepts are different, so the lines are parallel.

27. $x = 4$ is a vertical line and $y = -3$ is a horizontal line, so the lines are perpendicular.

28. We use the slope-intercept equation and substitute 4.7 for m and -23 for b..
$$y = mx + b$$
$$y = 4.7x - 23$$

29. Using the point-slope equation:

Substitute 3 for x_1, -5 for y_1, and -3 for m.
$$y - y_1 = m(x - x_1)$$
$$y - (-5) = -3(x - 3)$$
$$y + 5 = -3x + 9$$
$$y = -3x + 4$$

Using the slope-intercept equation:

Substitute 3 for x, -5 for y, and -3 for m in $y = mx + b$ and solve for b.
$$y = mx + b$$
$$-5 = -3 \cdot 3 + b$$
$$-5 = -9 + b$$
$$4 = b$$

Then we use the equation $y = mx + b$ and substitute -3 for m and 4 for b.
$$y = -3x + 4$$

30. First find the slope of the line:
$$m = \frac{6 - 3}{-4 - (-2)} = \frac{3}{-2} = -\frac{3}{2}$$

Using the point-slope equation:

We choose to use the point $(-2, 3)$ and substitute -2 for x_1, 3 for y_1, and $-\frac{3}{2}$ for m.
$$y - y_1 = m(x - x_1)$$
$$y - 3 = -\frac{3}{2}(x - (-2))$$
$$y - 3 = -\frac{3}{2}(x + 2)$$
$$y - 3 = -\frac{3}{2}x - 3$$
$$y = -\frac{3}{2}x$$

Using the slope-intercept equation:

We choose $(-2, 3)$ and substitute -2 for x, 3 for y, and $-\frac{3}{2}$ for m in $y = mx + b$. Then we solve for b.
$$3 = -\frac{3}{2}(-2) + b$$
$$3 = 3 + b$$
$$0 = b$$

Finally, we use the equation $y = mx + b$ and substitute $-\frac{3}{2}$ for m and 0 for b.
$$y = -\frac{3}{2}x + 0, \text{ or } y = -\frac{3}{2}x$$

31. First solve the equation for y and determine the slope of the given line.
$$5x + 7y = 8 \qquad \text{Given line}$$
$$7y = -5x + 8$$
$$y = -\frac{5}{7}x + \frac{8}{7}$$

The slope of the given line is $-\frac{5}{7}$. The line through $(14, -1)$ must have slope $-\frac{5}{7}$.

Using the point-slope equation:

Substitute 14 for x_1, -1 for y_1, and $-\frac{5}{7}$ for m.
$$y - y_1 = m(x - x_1)$$
$$y - (-1) = -\frac{5}{7}(x - 14)$$
$$y + 1 = -\frac{5}{7}x + 10$$
$$y = -\frac{5}{7}x + 9$$

Using the slope-intercept equation:

Substitute 14 for x, -1 for y, and $-\frac{5}{7}$ for m and solve for b.
$$y = mx + b$$
$$-1 = -\frac{5}{7} \cdot 14 + b$$
$$-1 = -10 + b$$
$$9 = b$$

Then we use the equation $y = mx + b$ and substitute $-\frac{5}{7}$ for m and 9 for b.
$$y = -\frac{5}{7}x + 9$$

32. First solve the equation for y and determine the slope of the given line.
$$3x + y = 5 \qquad \text{Given line}$$
$$y = -3x + 5$$

The slope of the given line is -3. The slope of the perpendicular line is the opposite of the reciprocal of -3. Thus, the line through $(5, 2)$ must have slope $\frac{1}{3}$.

Using the point-slope equation:

Substitute 5 for x_1, 2 for y_1, and $\frac{1}{3}$ for m.
$$y - y_1 = m(x - x_1)$$
$$y - 2 = \frac{1}{3}(x - 5)$$
$$y - 2 = \frac{1}{3}x - \frac{5}{3}$$
$$y = \frac{1}{3}x + \frac{1}{3}$$

Using the slope-intercept equation:

Substitute 5 for x, 2 for y, and $\frac{1}{3}$ for m and solve for b.

$$y = mx + b$$
$$2 = \frac{1}{3} \cdot 5 + b$$
$$2 = \frac{5}{3} + b$$
$$\frac{1}{3} = b$$

Then we use the equation $y = mx + b$ and substitute $\frac{1}{3}$ for m and $\frac{1}{3}$ for b.

$$y = \frac{1}{3}x + \frac{1}{3}$$

33. a) We form pairs of the type (x, R) where x is the number of years since 1972 and R is the record. We have two pairs, $(0, 44.66)$ and $(40, 43.94)$. These are two points on the graph of the linear function we are seeking.

First we find the slope:
$$m = \frac{43.94 - 44.66}{40 - 0} = \frac{-0.72}{40} = -0.018.$$

Using the slope and the y-intercept, $(0, 44.66)$ we write the function: $R(x) = -0.018x + 44.66$, where x is the number of years after 1972.

b) 2000 is 28 years after 1972, so to estimate the record in 2000, we find $R(28)$:
$$R(28) = -0.018(28) + 44.66$$
$$\approx 44.16$$

The estimated record was about 44.16 seconds in 2000.

2010 is 38 years after 1972, so to estimate the record in 2010, we find $R(38)$:
$$R(38) = -0.018(38) + 44.66$$
$$\approx 43.98$$

The estimated record was about 43.98 seconds in 2010.

34. $f(x) = \dfrac{x+3}{x-2}$

We cannot calculate $\dfrac{x+3}{x-2}$ when the denominator is 0, so we solve $x - 2 = 0$.
$$x - 2 = 0$$
$$x = 2$$

Thus, the domain of f is $(-\infty, 2) \cup (2, \infty)$. Answer C is correct.

35. First we find the slope of the given line.
$$3y - \frac{1}{2}x = 0$$
$$3y = \frac{1}{2}x$$
$$y = \frac{1}{6}x$$

The slope is $\frac{1}{6}$. The slope of a line perpendicular to the given line is -6. We use the point-slope equation.
$$y - 1 = -6[x - (-2)]$$
$$y - 1 = -6(x + 2)$$
$$y - 1 = -6x - 12$$
$$y = -6x - 11, \text{ or}$$
$$6x + y = -11$$

Answer A is correct.

36. The cost of x jars of preserves is $\$2.49x$, and the shipping charges are $\$3.75 + \$0.60x$. Then the total cost is $\$2.49x + \$3.75 + \$0.60x$, or $\$3.09x + \$3.75$. Thus, a linear function that can be used to determine the cost of buying and shipping x jars of preserves is $f(x) = 3.09x + 3.75$.

Chapter 2 Discussion and Writing Exercises

1. A line's x- and y-intercepts are the same only when the line passes through the origin. The equation for such a line is of the form $y = mx$.

2. The concept of slope is useful in describing how a line slants. A line with positive slope slants up from left to right. A line with negative slope slants down from left to right. The larger the absolute value of the slope, the steeper the slant.

3. Find the slope-intercept form of the equation.
$$4x + 5y = 12$$
$$5y = -4x + 12$$
$$y = -\frac{4}{5}x + \frac{12}{5}$$

This form of the equation indicates that the line has a negative slope and thus should slant down from left to right. The student apparently graphed $y = \frac{4}{5}x + \frac{12}{5}$.

4. For $R(t) = 50t + 35$, $m = 50$ and $b = 35$; 50 signifies that the cost per hour of a repair is $\$50$; 35 signifies that the minimum cost of a repair job is $\$35$.

5. $m = \dfrac{\text{change in } y}{\text{change in } x}$

As we move from one point to another on a vertical line, the y-coordinate changes but the x-coordinate does not. Thus, the change in y is a non-zero number while the change in x is 0. Since division by 0 is undefined, the slope of a vertical line is undefined.

As we move from one point to another on a horizontal line, the y-coordinate does not change but the x-coordinate does. Thus, the change in y is 0 while the change in x is a non-zero number, so the slope is 0.

6. Using algebra, we find that the slope-intercept form of the equation is $y = \frac{5}{2}x - \frac{3}{2}$. This indicates that the y-intercept is $\left(0, -\frac{3}{2}\right)$, so a mistake has been made. It appears that the student graphed $y = \frac{5}{2}x + \frac{3}{2}$.

Chapter 2 Test

1. Yes; each member of the domain is matched to only one member of the range.

2. No; a member of the domain, Lake Placid, is matched to more than one member of the range.

3. $f(x) = -3x - 4$

   $f(0) = -3 \cdot 0 - 4 = 0 - 4 = -4$

   $f(-2) = -3(-2) - 4 = 6 - 4 = 2$

4. $g(x) = x^2 + 7$

   $g(0) = 0^2 + 7 = 0 + 7 = 7$

   $g(-1) = (-1)^2 + 7 = 1 + 7 = 8$

5. $h(x) = -6$

   $h(-4) = -6$

   $h(-6) = -6$

6. $f(x) = |x + 7|$

   $f(-10) = |-10 + 7| = |-3| = 3$

   $f(-7) = |-7 + 7| = |0| = 0$

7. $y = -2x - 5$

   We find some ordered pairs that are solutions, plot them, and draw and label the line.

   When $x = 0$, $y = -2 \cdot 0 - 5 = 0 - 5 = -5$.

   When $x = -2$, $y = -2(-2) - 5 = 4 - 5 = -1$.

   When $x = -4$, $y = -2(-4) - 5 = 8 - 5 = 3$.

x	y
0	-5
-2	-1
-4	3

   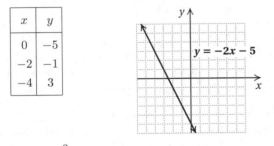

8. $f(x) = -\dfrac{3}{5}x$

   We find some function values, plot the corresponding points, and draw the graph.

   $f(-5) = -\dfrac{3}{5}(-5) = 3$

   $f(0) = -\dfrac{3}{5} \cdot 0 = 0$

   $f(5) = -\dfrac{3}{5} \cdot 5 = -3$

x	$f(x)$
-5	3
0	0
5	-3

   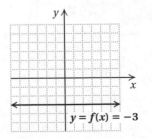

9. $g(x) = 2 - |x|$

   We find some function values, plot the corresponding points, and draw the graph.

   $g(-4) = 2 - |-4| = 2 - 4 = -2$

   $g(-2) = 2 - |-2| = 2 - 2 = 0$

   $g(0) = 2 - |0| = 2 - 0 = 2$

   $g(3) = 2 - |3| = 2 - 3 = -1$

   $g(5) = 2 - |5| = 2 - 5 = -3$

x	$g(x)$
-4	-2
-2	0
0	2
3	-1
5	-3

   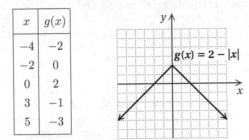

10. $f(x) = x^2 + 2x - 3$

    We find some function values, plot the corresponding points, and draw the graph.

    $f(-4) = (-4)^2 + 2(-4) - 3 = 16 - 8 - 3 = 5$

    $f(-3) = (-3)^2 + 2(-3) - 3 = 9 - 6 - 3 = 0$

    $f(-1) = (-1)^2 + 2(-1) - 3 = 1 - 2 - 3 = -4$

    $f(0) = 0^2 + 2 \cdot 0 - 3 = -3$

    $f(1) = 1^2 + 2 \cdot 1 - 3 = 1 + 2 - 3 = 0$

    $f(2) = 2^2 + 2 \cdot 2 - 3 = 4 + 4 - 3 = 5$

x	$f(x)$
-4	5
-3	0
-1	-4
0	-3
1	0
2	5

    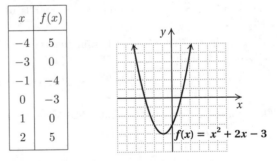

11. $y = f(x) = -3$

    Since x is missing, all ordered pairs $(x, -3)$ are solutions. The graph is parallel to the x-axis.

12. $2x = -4$

$x = -2$

Since y is missing, all ordered pairs $(-2, y)$ are solutions. The graph is parallel to the y-axis.

13. a) In 2005, $x = 2005 - 1990 = 15$. We find $A(15)$.

$A(15) = 0.233(15) + 5.87 = 9.365 \approx 9.4$

The median age of cars in 2005 was about 9.4 yr.

b) Substitute 7.734 for $A(t)$ and solve for t.

$7.734 = 0.233t + 5.87$

$1.864 = 0.233t$

$8 = t$

The median age of cars was 7.734 yr 8 years after 1990, or in 1998.

14. No vertical line will intersect the graph more than once. Thus, the graph is the graph of a function.

15. It is possible for a vertical line to intersect the graph more than once. Thus, this is not the graph of a function.

16. $f(x) = \dfrac{8}{2x + 3}$

Since $\dfrac{8}{2x + 3}$ cannot be calculated when the denominator is 0, we find the x-value that causes $2x + 3$ to be 0:

$2x + 3 = 0$

$2x = -3$

$x = -\dfrac{3}{2}$

Thus, $-\dfrac{3}{2}$ is not in the domain of f, while all other real numbers are. The domain of f is

$\left\{ x \middle| x \text{ is a real number } and \ x \neq -\dfrac{3}{2} \right\}$, or

$\left(-\infty, -\dfrac{3}{2} \right) \cup \left(-\dfrac{3}{2}, \infty \right).$

17. $g(x) = 5 - x^2$

Since we can calculate $5 - x^2$ for any real number x, the domain is the set of all real numbers.

18. a) Locate 1 on the horizontal axis and then find the point on the graph for which 1 is the first coordinate. From that point, look to the vertical axis to find the corresponding y-coordinate, 1. Thus, $f(1) = 1$.

b) The set of all x-values in the graph extends from -3 to 4, so the domain is $\{x| -3 \leq x \leq 4\}$, or $[-3, 4]$.

c) To determine which member(s) of the domain are paired with 2, locate 2 on the vertical axis. From there look left and right to the graph to find any points for which 2 is the second coordinate. One such point exists. Its first coordinate is -3, so the x-value for which $f(x) = 2$ is -3.

d) The set of all y-values in the graph extends from -1 to 2, so the range is $\{y| -1 \leq y \leq 2\}$, or $[-1, 2]$.

19. $f(x) = -\dfrac{3}{5}x + 12$

$f(x) = mx + b$

The slope is $-\dfrac{3}{5}$, and the y-intercept is $(0, 12)$.

20. First we find the slope-intercept form of the equation by solving for y. This allows us to determine the slope and y-intercept easily.

$-5y - 2x = 7$

$-5y = 2x + 7$

$\dfrac{-5y}{-5} = \dfrac{2x + 7}{-5}$

$y = -\dfrac{2}{5}x - \dfrac{7}{5}$

The slope is $-\dfrac{2}{5}$, and the y-intercept is $\left(0, -\dfrac{7}{5} \right)$.

21. Slope $= \dfrac{\text{change in } y}{\text{change in } x} = \dfrac{-2 - 3}{-2 - 6} = \dfrac{-5}{-8} = \dfrac{5}{8}$

22. Slope $= \dfrac{\text{change in } y}{\text{change in } x} = \dfrac{5.2 - 5.2}{-4.4 - (-3.1)} = \dfrac{0}{-1.3} = 0$

23. We can use the coordinates of any two points on the graph. We'll use $(10, 0)$ and $(25, 12)$.

Slope $= \dfrac{\text{change in } y}{\text{change in } x} = \dfrac{12 - 0}{25 - 10} = \dfrac{12}{15} = \dfrac{4}{5}$

The slope, or rate of change is $\dfrac{4}{5}$ km/min.

24. $2x + 3y = 6$

To find the x-intercept we let $y = 0$ and solve for x.

$2x + 3y = 6$

$2x + 3 \cdot 0 = 6$

$2x = 6$

$x = 3$

The x-intercept is $(3, 0)$.

To find the y-intercept we let $x = 0$ and solve for y.

$2x + 3y = 6$

$2 \cdot 0 + 3y = 6$

$3y = 6$

$y = 2$

The y-intercept is $(0, 2)$.

We plot these points and draw the line.

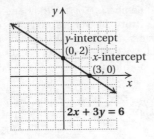

We use a third point as a check. We choose $x = -3$ and solve for y.

$$2(-3) + 3y = 6$$
$$-6 + 3y = 6$$
$$3y = 12$$
$$y = 4$$

We plot $(-3, 4)$ and note that it is on the line.

25. $f(x) = -\dfrac{2}{3}x - 1$

First we plot the y-intercept $(0, -1)$. We can think of the slope as $\dfrac{-2}{3}$. Starting at the y-intercept and using the slope, we find another point by moving 2 units down and 3 units to the right. We get to a new point $(3, -3)$.

We can also think of the slope as $\dfrac{2}{-3}$. We again start at the y-intercept $(0, -1)$. We move 2 units up and 3 units to the left. We get to another new point $(-3, 1)$. We plot the points and draw the line.

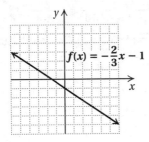

26. We first solve each equation for y and determine the slope of each line.

$$4y + 2 = 3x$$
$$4y = 3x - 2$$
$$y = \dfrac{3}{4}x - \dfrac{1}{2}$$

The slope of $4y + 2 = 3x$ is $\dfrac{3}{4}$.

$$-3x + 4y = -12$$
$$4y = 3x - 12$$
$$y = \dfrac{3}{4}x - 3$$

The slope of $-3x + 4y = -12$ is $\dfrac{3}{4}$.

The slopes are the same and the y-intercepts are different, so the lines are parallel.

27. The slope of $y = -2x + 5$ is -2.

We solve the second equation for y and determine the slope.

$$2y - x = 6$$
$$2y = x + 6$$
$$y = \dfrac{1}{2}x + 3$$

The slopes are not the same, so the lines are not parallel. The product of the slopes is $-2 \cdot \dfrac{1}{2}$, or -1, so the lines are perpendicular.

28. We use the slope-intercept equation and substitute -3 for m and 4.8 for b.

$$y = mx + b$$
$$y = -3x + 4.8$$

29. $y = f(x) = mx + b$
$$f(x) = 5.2x - \dfrac{5}{8}$$

30. Using the point-slope equation:

Substitute 1 for x_1, -2 for y_1, and -4 for m.

$$y - y_1 = m(x - x_1)$$
$$y - (-2) = -4(x - 1)$$
$$y + 2 = -4x + 4$$
$$y = -4x + 2$$

Using the slope-intercept equation:

Substitute 1 for x, -2 for y, and -4 for m in $y = mx + b$ and solve for b.

$$y = mx + b$$
$$-2 = -4 \cdot 1 + b$$
$$-2 = -4 + b$$
$$2 = b$$

Then we use the equation $y = mx + b$ and substitute -4 for m and 2 for b.

$$y = -4x + 2$$

31. First find the slope of the line:
$$m = \dfrac{-6 - 15}{4 - (-10)} = \dfrac{-21}{14} = -\dfrac{3}{2}$$

Using the point-slope equation:

We choose to use the point $(4, -6)$ and substitute 4 for x_1, 6 for y_1, and $-\dfrac{3}{2}$ for m.

$$y - y_1 = m(x - x_1)$$
$$y - (-6) = -\dfrac{3}{2}(x - 4)$$
$$y + 6 = -\dfrac{3}{2}x + 6$$
$$y = -\dfrac{3}{2}x$$

Using the slope-intercept equation:

We choose $(4, -6)$ and substitute 4 for x, -6 for y, and $-\dfrac{3}{2}$ for m in $y = mx + b$. Then we solve for b.

$$y = mx + b$$
$$-6 = -\frac{3}{2} \cdot 4 + b$$
$$-6 = -6 + b$$
$$0 = b$$

Finally, we use the equation $y = mx + b$ and substitute $-\frac{3}{2}$ for m and 0 for b.

$$y = -\frac{3}{2}x + 0, \text{ or } y = -\frac{3}{2}x$$

32. First solve the equation for y and determine the slope of the given line.

$$x - 2y = 5 \qquad \text{Given line}$$
$$-2y = -x + 5$$
$$y = \frac{1}{2}x - \frac{5}{2}$$

The slope of the given line is $\frac{1}{2}$. The line through $(4, -1)$ must have slope $\frac{1}{2}$.

Using the point-slope equation:

Substitute 4 for x_1, -1 for y_1, and $\frac{1}{2}$ for m.

$$y - y_1 = m(x - x_1)$$
$$y - (-1) = \frac{1}{2}(x - 4)$$
$$y + 1 = \frac{1}{2}x - 2$$
$$y = \frac{1}{2}x - 3$$

Using the slope-intercept equation:

Substitute 4 for x, -1 for y, and $\frac{1}{2}$ for m and solve for b.

$$y = mx + b$$
$$-1 = \frac{1}{2}(4) + b$$
$$-1 = 2 + b$$
$$-3 = b$$

Then we use the equation $y = mx + b$ and substitute $\frac{1}{2}$ for m and -3 for b.

$$y = \frac{1}{2}x - 3$$

33. First solve the equation for y and determine the slope of the given line.

$$x + 3y = 2 \qquad \text{Given line}$$
$$3y = -x + 2$$
$$y = -\frac{1}{3}x + \frac{2}{3}$$

The slope of the given line is $-\frac{1}{3}$. The slope of the perpendicular line is the opposite of the reciprocal of $-\frac{1}{3}$. Thus, the line through $(2, 5)$ must have slope 3.

Using the point-slope equation:

Substitute 2 for x_1, 5 for y_1, and 3 for m.

$$y - y_1 = m(x - x_1)$$
$$y - 5 = 3(x - 2)$$
$$y - 5 = 3x - 6$$
$$y = 3x - 1$$

Using the slope-intercept equation:

Substitute 2 for x, 5 for y, and 3 for m and solve for b.

$$y = mx + b$$
$$5 = 3 \cdot 2 + b$$
$$5 = 6 + b$$
$$-1 = b$$

Then we use the equation $y = mx + b$ and substitute 3 for m and -1 for b.

$$y = 3x - 1$$

34. a) Note that $2010 - 1970 = 40$. Thus, the data points are $(0, 23.2)$ and $(40, 28.2)$. We find the slope.

$$m = \frac{28.2 - 23.2}{40 - 0} = \frac{5}{40} = 0.125$$

Using the slope and the y-intercept, $(0, 23.2)$, we write the function: $A(x) = 0.125x + 23.2$

b) In 2008, $x = 2008 - 1970 = 38$.

$A(38) = 0.125(38) + 23.2 = 27.95$ years

In 2015, $x = 2015 - 1970 = 45$.

$A(45) = 0.125(45) + 23.2 = 28.825$ years

35. Using the point-slope equation, $y - y_1 = m(x - x_1)$, with $x_1 = 3$, $y_1 = 1$, and $m = -2$ we have $y - 1 = -2(x - 3)$. Thus, answer B is correct.

36. First solve each equation for y and determine the slopes.

$$3x + ky = 17$$
$$ky = -3x + 17$$
$$y = -\frac{3}{k}x + \frac{17}{k}$$

The slope of $3x + ky = 17$ is $-\frac{3}{k}$.

$$8x - 5y = 26$$
$$-5y = -8x + 26$$
$$y = \frac{8}{5}x - \frac{26}{5}$$

The slope of $8x - 5y = 26$ is $\frac{8}{5}$.

If the lines are perpendicular, the product of their slopes is -1.

$$-\frac{3}{k} \cdot \frac{8}{5} = -1$$
$$-\frac{24}{5k} = -1$$
$$24 = 5k \qquad \text{Multiplying by } -5k$$
$$\frac{24}{5} = k$$

37. Answers may vary. One such function is $f(x) = 3$.

Cumulative Review Chapters 1 - 2

1. a) Note that $2004 - 1950 = 54$, so 2004 is 54 yr after 1950. Then the data points are $(0, 3.85)$ and $(54, 3.50)$.

First we find the slope.

$$m = \frac{3.50 - 3.85}{54 - 0} = \frac{-0.35}{54} \approx -0.006$$

Using the slope and the y-intercept, $(0, 3.85)$, we write the function: $R(x) = -0.006x + 3.85$.

b) In 2008, $x = 2008 - 1950 = 58$.

$R(58) = -0.006(58) + 3.85 \approx 3.50$ min

In 2010, $x = 2010 - 1950 = 60$.

$R(60) = -0.006(60) + 3.85 \approx 3.49$ min

2. a) Locate 15 on the x-axis and the find the point on the graph for which 15 is the first coordinate. From that point, look to the vertical axis to find the corresponding y-coordinate, 6. Thus, $f(15) = 6$.

b) The set of all x-values in the graph extends from 0 to 30, so the domain is $\{x | 0 \le x \le 30\}$, or $[0, 30]$.

c) To determine which member(s) of the domain are paired with 14, locate 14 on the vertical axis. From there look left and right to the graph to find any points for which 14 is the second coordinate. One such point exists. Its first coordinate is 25. Thus, the x-value for which $f(x) = 14$ is 25.

d) The set of all y-values in the graph extends from 0 to 15, so the range is $\{y | 0 \le y \le 15\}$, or $[0, 15]$.

3.
$$x + 9.4 = -12.6$$
$$x + 9.4 - 9.4 = -12.6 - 9.4$$
$$x = -22$$

The solution is -22.

4.
$$\frac{2}{3}x - \frac{1}{4} = -\frac{4}{5}x$$
$$60\left(\frac{2}{3}x - \frac{1}{4}\right) = 60\left(-\frac{4}{5}x\right) \quad \text{Clearing fractions}$$
$$60 \cdot \frac{2}{3}x - 60 \cdot \frac{1}{4} = -48x$$
$$40x - 15 = -48x$$
$$40x - 15 - 40x = -48x - 40x$$
$$-15 = -88x$$
$$\frac{-15}{-88} = \frac{-88x}{-88}$$
$$\frac{15}{88} = x$$

The solution is $\frac{15}{88}$.

5.
$$-2.4t = -48$$
$$\frac{-2.4t}{-2.4} = \frac{-48}{-2.4}$$
$$t = 20$$

The solution is 20.

6.
$$4x + 7 = -14$$
$$4x = -21 \quad \text{Subtracting 7}$$
$$x = -\frac{21}{4} \quad \text{Dividing by 4}$$

The solution is $-\frac{21}{4}$.

7.
$$3n - (4n - 2) = 7$$
$$3n - 4n + 2 = 7$$
$$-n + 2 = 7$$
$$-n = 5$$
$$n = -5 \quad \text{Multiplying by } -1$$

The solution is -5.

8.
$$5y - 10 = 10 + 5y$$
$$-10 = 10 \quad \text{Subtracting } 5y$$

We get a false equation, so the original equation has no solution.

9.
$$W = Ax + By$$
$$W - By = Ax \quad \text{Subtracting } By$$
$$\frac{W - By}{A} = x \quad \text{Dividing by } A$$

10.
$$M = A + 4AB$$
$$M = A(1 + 4B) \quad \text{Factoring out } A$$
$$\frac{M}{1 + 4B} = A \quad \text{Dividing by } 1 + 4B$$

11.
$$y - 12 \le -5$$
$$y \le 7 \quad \text{Adding 12}$$

The solution set is $\{y | y \le 7\}$, or $(-\infty, 7]$.

12.
$$6x - 7 < 2x - 13$$
$$4x - 7 < -13$$
$$4x < -6$$
$$x < -\frac{3}{2}$$

The solution set is $\left\{x \middle| x < -\frac{3}{2}\right\}$, or $\left(-\infty, -\frac{3}{2}\right)$.

13.
$$5(1 - 2x) + x < 2(3 + x)$$
$$5 - 10x + x < 6 + 2x$$
$$5 - 9x < 6 + 2x$$
$$5 - 11x < 6$$
$$-11x < 1$$
$$x > -\frac{1}{11} \quad \text{Reversing the inequality symbol}$$

The solution set is $\left\{x \middle| x > -\frac{1}{11}\right\}$, or $\left(-\frac{1}{11}, \infty\right)$.

14. $x + 3 < -1$ *or* $x + 9 \ge 1$
$$x < -4 \quad or \quad x \ge -8$$

The intersection of $\{x | x < -4\}$ and $\{x | x \ge -8\}$ is the set of all real numbers. This is the solution set.

15. $-3 < x + 4 \le 8$

$-7 < x \le 4$

The solution set is $\{x | -7 < x \le 4\}$, or $(-7, 4]$.

16. $-8 \le 2x - 4 \le -1$

$-4 \le 2x \le 3$

$-2 \le x \le \dfrac{3}{2}$

The solution set is $\left\{x \middle| -2 \le x \le \dfrac{3}{2}\right\}$, or $\left[-2, \dfrac{3}{2}\right]$.

17. $|x| = 8$

$x = -8 \ \ or \ \ x = 8$

The solution set is $\{-8, 8\}$.

18. $|y| > 4$

$y < -4 \ \ or \ \ y > 4$

The solution set is $\{y | y < -4 \ or \ y > 4\}$, or $(-\infty, -4) \cup (4, \infty)$.

19. $|4x - 1| \le 7$

$-7 \le 4x - 1 \le 7$

$-6 \le 4x \le 8$

$-\dfrac{3}{2} \le x \le 2$

The solution set is $\left\{x \middle| -\dfrac{3}{2} \le x \le 2\right\}$, or $\left[-\dfrac{3}{2}, 2\right]$.

20. First solve the equation for y and determine the slope of the given line.

$4y - x = 3$ Given line

$4y = x + 3$

$y = \dfrac{1}{4}x + \dfrac{3}{4}$

The slope of the given line is $\dfrac{1}{4}$. The slope of the perpendicular line is the opposite of the reciprocal of $\dfrac{1}{4}$. Thus, the line through $(-4, -6)$ must have slope -4.

Using the point-slope equation:

Substitute -4 for x_1, -6 for y_1, and -4 for m.

$y - y_1 = m(x - x_1)$

$y - (-6) = -4(x - (-4))$

$y + 6 = -4(x + 4)$

$y + 6 = -4x - 16$

$y = -4x - 22$

Using the slope-intercept equation:

Substitute -4 for x, -6 for y, and -4 for m.

$y = mx + b$

$-6 = -4(-4) + b$

$-6 = 16 + b$

$-22 = b$

Then we use the equation $y = mx + b$ and substitute -4 for m and -22 for b.

$y = -4x - 22$

21. First solve the equation for y and determine the slope of the given line.

$4y - x = 3$ Given line

$4y = x + 3$

$y = \dfrac{1}{4}x + \dfrac{3}{4}$

The slope of the given line is $\dfrac{1}{4}$. The line through $(-4, -6)$ must have slope $\dfrac{1}{4}$.

Using the point-slope equation:

Substitute -4 for x_1, -6 for y_1, and $\dfrac{1}{4}$ for m.

$y - y_1 = m(x - x_1)$

$y - (-6) = \dfrac{1}{4}(x - (-4))$

$y + 6 = \dfrac{1}{4}(x + 4)$

$y + 6 = \dfrac{1}{4}x + 1$

$y = \dfrac{1}{4}x - 5$

Using the slope-intercept equation:

Substitute -4 for x, -6 for y, and $\dfrac{1}{4}$ for m and solve for b.

$y = mx + b$

$-6 = \dfrac{1}{4}(-4) + b$

$-6 = -1 + b$

$-5 - b$

Then we use the equation $y = mx + b$ and substitute $\dfrac{1}{4}$ for m and -5 for b.

$y = \dfrac{1}{4}x - 5$

22. $y = -2x + 3$

We find some ordered pairs that are solutions, plot them, and draw and label the graph.

When $x = -1$, $y = -2(-1) + 3 = 2 + 3 = 5$.

When $x = 1$, $y = -2 \cdot 1 + 3 = -2 + 3 = 1$.

When $x = 3$, $y = -2 \cdot 3 + 3 = -6 + 3 = -3$.

x	y
-1	5
1	1
3	-3

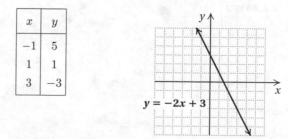

$y = -2x + 3$

23. $3x = 2y + 6$

To find the x-intercept we let $y = 0$ and solve for x.

$$3x = 2y + 6$$
$$3x = 2 \cdot 0 + 6$$
$$3x = 6$$
$$x = 2$$

The x-intercept is $(2, 0)$.

To find the y-intercept we let $x = 0$ and solve for y.

$$3x = 2y + 6$$
$$3 \cdot 0 = 2y + 6$$
$$0 = 2y + 6$$
$$-2y = 6$$
$$y = -3$$

The y-intercept is $(0, -3)$.

We plot these points and draw the line.

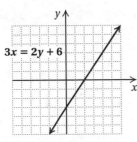

We use a third point as a check. We choose $x = 4$ and solve for y.

$$3 \cdot 4 = 2y + 6$$
$$12 = 2y + 6$$
$$6 = 2y$$
$$3 = y$$

We plot $(4, 3)$ and note that it is on the line.

24. $4x + 16 = 0$
$$4x = -16$$
$$x = -4$$

Since y is missing, all ordered pairs $(-4, y)$ are solutions. The graph is parallel to the y-axis.

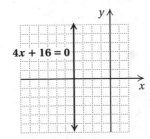

25. $-2y = -6$
$$y = 3$$

Since x is missing, all ordered pairs $(x, 3)$ are solutions. The graph is parallel to the x-axis.

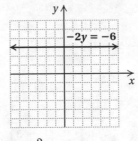

26. $f(x) = \dfrac{2}{3}x + 1$

We calculate some function values, plot the corresponding points, and connect them.

$$f(-3) = \frac{2}{3}(-3) + 1 = -2 + 1 = -1$$
$$f(0) = \frac{2}{3} \cdot 0 + 1 = 0 + 1 = 1$$
$$f(3) = \frac{2}{3} \cdot 3 + 1 = 2 + 1 = 3$$

x	$f(x)$
-3	-1
0	1
3	3

27. $g(x) = 5 - |x|$

We calculate some function values, plot the corresponding points, and connect them.

$$g(-2) = 5 - |-2| = 5 - 2 = 3$$
$$g(-1) = 5 - |-1| = 5 - 1 = 4$$
$$g(0) = 5 - |0| = 5 - 0 = 5$$
$$g(1) = 5 - |1| = 5 - 1 = 4$$
$$g(2) = 5 - |2| = 5 - 2 = 3$$
$$g(3) = 5 - |3| = 5 - 3 = 2$$

x	$g(x)$
-2	3
-1	4
0	5
1	4
2	3
3	2

28. First we find the slope-intercept form of the equation by solving for y. This allows us to determine the slope and y-intercept easily.

$$-4y + 9x = 12$$
$$-4y = -9x + 12$$
$$\frac{-4y}{-4} = \frac{-9x + 12}{-4}$$
$$y = \frac{9}{4}x - 3$$

The slope is $\dfrac{9}{4}$, and the y-intercept is $(0, -3)$.

29. Slope $= \dfrac{\text{change in } y}{\text{change in } x} = \dfrac{3 - 7}{-1 - 2} = \dfrac{-4}{-3} = \dfrac{4}{3}$

30. Using the point-slope equation:

Substitute 2 for x_1, -11 for y_1, and -3 for m.

$$y - y_1 = m(x - x_1)$$
$$y - (-11) = -3(x - 2)$$
$$y + 11 = -3x + 6$$
$$y = -3x - 5$$

Using the slope-intercept equation:

Substitute 2 for x, -11 for y, and -3 for m in $y = mx + b$ and solve for b.

$$y = mx + b$$
$$-11 = -3 \cdot 2 + b$$
$$-11 = -6 + b$$
$$-5 = b$$

Then use the equation $y = mx + b$ and substitute -3 for m and -5 for b.

$$y = -3x - 5$$

31. First find the slope of the line:

$$m = \dfrac{3 - 2}{-6 - 4} = \dfrac{1}{-10} = -\dfrac{1}{10}$$

Using the point-slope equation:

We choose to use the point $(4, 2)$ and substitute 4 for x_1, 2 for y_1, and $-\dfrac{1}{10}$ for m.

$$y - y_1 = m(x - x_1)$$
$$y - 2 = -\dfrac{1}{10}(x - 4)$$
$$y - 2 = -\dfrac{1}{10}x + \dfrac{2}{5}$$
$$y = -\dfrac{1}{10}x + \dfrac{12}{5}$$

Using the slope-intercept equation:

We choose $(4, 2)$ and substitute 4 for x, 2 for y, and $-\dfrac{1}{10}$ for m in $y = mx + b$. Then we solve for b.

$$y = mx + b$$
$$2 = -\dfrac{1}{10} \cdot 4 + b$$
$$2 = -\dfrac{2}{5} + b$$
$$\dfrac{12}{5} = b$$

Finally, we use the equation $y = mx + b$ and substitute $-\dfrac{1}{10}$ for m and $\dfrac{12}{5}$ for b.

$$y = -\dfrac{1}{10}x + \dfrac{12}{5}$$

32. **Familiarize.** Let $w =$ the width, in meters. Then $w + 6 =$ the length. Recall that the formula for the perimeter of a rectangle is $P = 2l + 2w$.

Translate. We use the formula for perimeter.

$$80 = 2(w + 6) + 2w$$

Solve. We solve the equation.

$$80 = 2(w + 6) + 2w$$
$$80 = 2w + 12 + 2w$$
$$80 = 4w + 12$$
$$68 = 4w$$
$$17 = w$$

If $w = 17$, then $w + 6 = 17 + 6 = 23$.

Check. 23 m is 6 m more than 17 m, and $2 \cdot 23 + 2 \cdot 17 = 46 + 34 = 80$ m. The answer checks.

State. The length is 23 m and the width is 17 m.

33. **Familiarize.** Let $s =$ David's old salary. Then his new salary is $s + 20\%s$, or $s + 0.2s$, or $1.2s$.

Translate.

$$\underbrace{\text{New salary}} \text{ is } \$27{,}000$$
$$\downarrow \qquad \downarrow \quad \downarrow$$
$$1.2s \quad = \quad 27{,}000$$

Solve. We solve the equation.

$$1.2s = 27{,}000$$
$$s = 22{,}500$$

Check. 20% of $\$22{,}500$ is $0.2(\$22{,}500)$, or $\$4500$, and $\$22{,}500 + \$4500 = \$27{,}000$. The answer checks.

State. David's old salary was $\$22{,}500$.

34. First we solve each equation for y and determine the slopes.

a) $7y - 3x = 21$
$$7y = 3x + 21$$
$$y = \dfrac{3}{7}x + 3$$

The slope is $\dfrac{3}{7}$.

b) $-3x - 7y = 12$
$$-7y = 3x + 12$$
$$y = -\dfrac{3}{7}x - \dfrac{12}{7}$$

The slope is $-\dfrac{3}{7}$.

c) $7y + 3x = 21$
$$7y = -3x + 21$$
$$y = -\dfrac{3}{7}x + 3$$

The slope is $-\dfrac{3}{7}$.

d) $3y + 7x = 12$
$$3y = -7x + 12$$
$$y = -\dfrac{7}{3}x + 4$$

The slope is $-\dfrac{7}{3}$.

The only pair of slopes whose product is -1 is $\dfrac{3}{7}$ and $-\dfrac{7}{3}$. Thus, equations (1) and (4) represent perpendicular lines.

35. We have two data points, $(1000, 101{,}000)$ and $(1250, 126{,}000)$. We find the slope of the line containing these points.

$$m = \frac{126{,}000 - 101{,}000}{1250 - 1000} = \frac{25{,}000}{250} = 100$$

We will use the point-slope equation with $x_1 = 1000$, $y_1 = 101{,}000$ and $m = 100$.

$$y - y_1 = m(x - x_1)$$
$$y - 101{,}000 = 100(x - 1000)$$
$$y - 101{,}000 = 100x - 100{,}000$$
$$y = 100x + 1000$$

Now we find the value of y when $x = 1500$.

$$y = 100 \cdot 1500 + 1000 = 150{,}000 + 1000 = 151{,}000$$

Thus, when \$1500 is spent on advertising, weekly sales increase by \$151,000.

36. $x + 5 < 3x - 7 \leq x + 13$

$$x + 5 < 3x - 7 \quad and \quad 3x - 7 \leq x + 13$$
$$5 < 2x - 7 \quad and \quad 2x - 7 \leq 13$$
$$12 < 2x \quad\quad and \quad\quad 2x \leq 20$$
$$6 < x \quad\quad and \quad\quad x \leq 10$$

The solution set is $\{x | 6 < x \ and \ x \leq 10\}$, or $\{x | 6 < x \leq 10\}$, or $(6, 10]$.

Chapter 3

Systems of Equations

RC1. False; see page 245 in the text.

RC3. True; see page 245 in the text.

1. Graph both lines on the same set of axes.

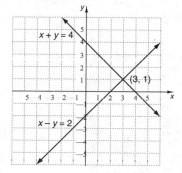

The solution (point of intersection) seems to be the point $(3, 1)$.

Check:

$x + y = 4$		$x - y = 2$	
$3 + 1$? 4		$3 - 1$? 2	
4	TRUE	2	TRUE

The solution is $(3, 1)$.

Since the system of equations has a solution it is consistent. Since there is exactly one solution, the equations are independent.

3. Graph both lines on the same set of axes.

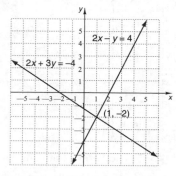

The solution (point of intersection) seems to be the point $(1, -2)$.

Check:

$2x - y = 4$		$2x + 3y = -4$	
$2 \cdot 1 - (-2)$? 4		$2 \cdot 1 + 3(-2)$? -4	
$2 + 2$		$2 - 6$	
4	TRUE	-4	TRUE

The solution is $(1, -2)$.

Since the system of equations has a solution, it is consistent. Since there is exactly one solution, the equations are independent.

5. Graph both lines on the same set of axes.

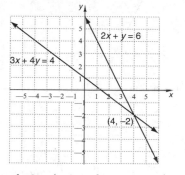

The solution (point of intersection) seems to be the point $(4, -2)$.

Check:

$2x + y = 6$		$3x + 4y = 4$	
$2 \cdot 4 + (-2)$? 6		$3 \cdot 4 + 4(-2)$? 4	
$8 - 2$		$12 - 8$	
6	TRUE	4	TRUE

The solution is $(4, -2)$.

Since the system of equations has a solution, it is consistent. Since there is exactly one solution, the equations are independent.

7. Graph both lines on the same set of axes.

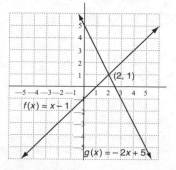

The solution seems to be the point $(2, 1)$.

Check:

$$f(x) = x - 1$$

1 ? 2 − 1

    1   TRUE

$$g(x) = -2x + 5$$

1 ? −2 · 2 + 5

    −4 + 5

    1   TRUE

The solution is $(2, 1)$.

Since the system of equations has a solution, it is consistent. Since there is exactly one solution, the equations are independent.

9. Graph both lines on the same set of axes.

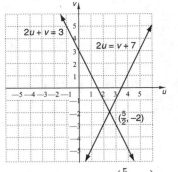

The solution seems to be $\left(\dfrac{5}{2}, -2\right)$.

Check:

$$2u + v = 3$$

$2 \cdot \dfrac{5}{2} + (-2)$? 3

    5 − 2

    3   TRUE

$$2u = v + 7$$

$2 \cdot \dfrac{5}{2}$? −2 + 7

    5   5   TRUE

The solution is $\left(\dfrac{5}{2}, -2\right)$.

Since the system of equations has a solution, it is consistent. Since there is exactly one solution, the equations are independent.

11. Graph both lines on the same set of axes.

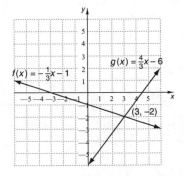

The ordered pair $(3, -2)$ checks in both equations. It is the solution.

Since the system of equations has a solution, it is consistent. Since there is exactly one solution, the equations are independent.

13. Graph both lines on the same set of axes.

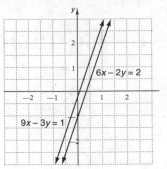

The lines are parallel. There is no solution.

Since the system of equations has no solution, it is inconsistent. Since there is no solution, the equations are independent.

15. Graph both lines on the same set of axes.

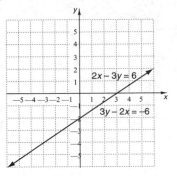

The graphs are the same. Any solution of one of the equations is also a solution of the other. Each equation has an infinite number of solutions. Thus the system of equations has an infinite number of solutions. Since the system of equations has a solution, it is consistent. Since there are infinitely many solutions, the equations are dependent.

17. Graph both lines on the same set of axes.

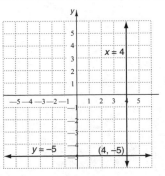

The ordered pair $(4, -5)$ checks in both equations. It is the solution.

Since the system of equations has a solution, it is consistent. Since there is exactly one solution, the equations are independent.

19. Graph both lines on the same set of axes.

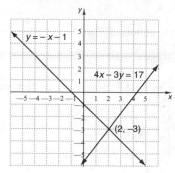

The ordered pair $(2, -3)$ checks in both equations. It is the solution.

Since the system of equations has a solution, it is consistent. Since there is exactly one solution, the equations are independent.

21. Since the system of equations has a solution, it is consistent. Since there is exactly one solution, the equations are independent. The graph of the system consists of a vertical line and a horizontal line, each passing through $(3, 3)$. Thus, system **F** corresponds to this graph.

23. Since the system of equations has a solution, it is consistent. Since there are infinitely many solutions, the equations are dependent. The equations in system **B** are equivalent, so their graphs are the same. In addition the graph corresponds to the one shown, so system **B** corresponds to this graph.

25. Since the system of equations has no solution, it is inconsistent. Since there is no solution, the equations are independent. The equations in system **D** have the same slope and different y-intercepts and have the graphs shown, so this system corresponds to the given graph.

27. $3x + 4 = x - 2$

$2x + 4 = -2$ Adding $-x$ on both sides

$2x = -6$ Adding -4 on both sides

$x = -3$ Multiplying by $\dfrac{1}{2}$ on both sides

The solution is -3.

29. $4x - 5x = 8x - 9 + 11x$

$-x = 19x - 9$ Collecting like terms

$-20x = -9$ Adding $-19x$ on both sides

$x = \dfrac{9}{20}$ Multiplying by $-\dfrac{1}{20}$ on both sides

The solution is $\dfrac{9}{20}$.

31. Graph these equations, solving each equation for y first, if necessary. We get $y = \dfrac{13.78 - 2.18x}{7.81}$ and $y = \dfrac{5.79x - 8.94}{3.45}$. Using the INTERSECT feature, we find that the point of intersection is $(2.23, 1.14)$.

33. Graph both lines on the same set of axes.

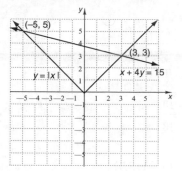

The solutions appear to be $(-5, 5)$ and $(3, 3)$.

Check:

For $(-5, 5)$:

$$\frac{y = |x|}{5 \;\; ? \;\; |-5|}$$
$$\begin{array}{c|c} & 5 \quad \text{TRUE} \end{array}$$

$$\frac{x + 4y = 15}{-5 + 4 \cdot 5 \;\; ? \;\; 15}$$
$$\begin{array}{c|c} -5 + 20 & \\ 15 & \text{TRUE} \end{array}$$

For $(3, 3)$:

$$\frac{y = |x|}{3 \;\; ? \;\; |3|}$$
$$\begin{array}{c|c} & 3 \quad \text{TRUE} \end{array}$$

$$\frac{x + 4y = 15}{3 + 4 \cdot 3 \;\; ? \;\; 15}$$
$$\begin{array}{c|c} 3 + 12 & \\ 15 & \text{TRUE} \end{array}$$

Both pairs check. The solutions are $(-5, 5)$ and $(3, 3)$.

Exercise Set 3.2

RC1. True; see page 251 in the text.

RC3. False; see Example 1 on page 251 in the text.

1. $y = 5 - 4x$,   (1)

$2x - 3y = 13$ (2)

We substitute $5 - 4x$ for y in the second equation and solve for x.

$$2x - 3y = 13 \quad (2)$$
$$2x - 3(5 - 4x) = 13 \quad \text{Substituting}$$
$$2x - 15 + 12x = 13$$
$$14x - 15 = 13$$
$$14x = 28$$
$$x = 2$$

Next we substitute 2 for x in either equation of the original system and solve for y.

$$y = 5 - 4x \quad (1)$$
$$y = 5 - 4 \cdot 2 \quad \text{Substituting}$$
$$y = 5 - 8$$
$$y = -3$$

We check the ordered pair $(2, -3)$.

$$\frac{y = 5 - 4x}{}$$

$$\begin{array}{c|l}
-3 \ ? \ 5 - 4 \cdot 2 & \\
\quad\ \ \ 5 - 8 & \\
\hline
-3 \ \Big| \ -3 & \text{TRUE}
\end{array}$$

$$\frac{2x - 3y = 13}{}$$

$$\begin{array}{c|l}
2 \cdot 2 - 3(-3) \ ? \ 13 & \\
\quad\ \ \ 4 + 9 & \\
\hline
\qquad\quad 13 \ \Big| \ 13 & \text{TRUE}
\end{array}$$

Since $(2, -3)$ checks, it is the solution.

3. $2y + x = 9$,   (1)

$\quad\ x = 3y - 3$ (2)

We substitute $3y - 3$ for x in the first equation and solve for y.

$$2y + x = 9 \qquad (1)$$
$$2y + (3y - 3) = 9 \qquad \text{Substituting}$$
$$5y - 3 = 9$$
$$5y = 12$$
$$y = \frac{12}{5}$$

Next we substitute $\dfrac{12}{5}$ for y in either equation of the original system and solve for x.

$$x = 3y - 3 \qquad\qquad\qquad (2)$$
$$x = 3 \cdot \frac{12}{5} - 3 = \frac{36}{5} - \frac{15}{5} = \frac{21}{5}$$

We check the ordered pair $\left(\dfrac{21}{5}, \dfrac{12}{5}\right)$.

$$\frac{2y + x = 9}{}$$

$$\begin{array}{c|l}
2 \cdot \dfrac{12}{5} + \dfrac{21}{15} \ ? \ 9 & \\[2mm]
\quad \dfrac{24}{5} + \dfrac{21}{5} & \\[2mm]
\qquad\ \ \dfrac{45}{5} & \\[2mm]
\hline
\qquad\quad 9 \ \Big| \ 9 & \text{TRUE}
\end{array}$$

$$\frac{x = 3y - 3}{}$$

$$\begin{array}{c|l}
\dfrac{21}{5} \ ? \ 3 \cdot \dfrac{12}{5} - 3 & \\[2mm]
\qquad \dfrac{36}{5} - \dfrac{15}{5} & \\[2mm]
\hline
\dfrac{21}{5} \ \Big| \ \dfrac{21}{5} & \text{TRUE}
\end{array}$$

Since $\left(\dfrac{21}{5}, \dfrac{12}{5}\right)$ checks, it is the solution.

5. $3s - 4t = 14$,   (1)

$\quad\ 5s + t = 8$ (2)

We solve the second equation for t.

$$5s + t = 8 \qquad\qquad (2)$$
$$t = 8 - 5s \qquad (3)$$

We substitute $8 - 5s$ for t in the first equation and solve for s.

$$3s - 4t = 14 \qquad (1)$$
$$3s - 4(8 - 5s) = 14 \qquad \text{Substituting}$$
$$3s - 32 + 20s = 14$$
$$23s - 32 = 14$$
$$23s = 46$$
$$s = 2$$

Next we substitute 2 for s in Equation (1), (2), or (3). It is easiest to use Equation (3) since it is already solved for t.

$$t = 8 - 5 \cdot 2 = 8 - 10 = -2$$

We check the ordered pair $(2, -2)$.

$$\frac{3s - 4t = 14}{}$$

$$\begin{array}{c|l}
3 \cdot 2 - 4(-2) \ ? \ 14 & \\
\quad\ \ \ 6 + 8 & \\
\hline
\qquad\quad 14 \ \Big| \ 14 & \text{TRUE}
\end{array}$$

$$\frac{5s + t = 8}{}$$

$$\begin{array}{c|l}
5 \cdot 2 + (-2) \ ? \ 8 & \\
\quad\ \ \ 10 - 2 & \\
\hline
\qquad\quad 8 \ \Big| \ 8 & \text{TRUE}
\end{array}$$

Since $(2, -2)$ checks, it is the solution.

7. $9x - 2y = -6$,   (1)

$\quad\ 7x + 8 = y$ (2)

We substitute $7x + 8$ for y in the first equation and solve for x.

$$9x - 2y = -6 \qquad (1)$$
$$9x - 2(7x + 8) = -6 \qquad \text{Substituting}$$
$$9x - 14x - 16 = -6$$
$$-5x - 16 = -6$$
$$-5x = 10$$
$$x = -2$$

Next we substitute -2 for x in either equation of the original system and solve for y.

$$7x + 8 = y \qquad (2)$$
$$7(-2) + 8 = y$$
$$-14 + 8 = y$$
$$-6 = y$$

We check the ordered pair $(-2, -6)$.

$$\frac{9x - 2y = -6}{}$$

$9(-2) - 2(-6)$? -6

$-18 + 12$

-6 | TRUE

$$\frac{7x + 8 = y}{}$$

$7(-2) + 8$? -6

$-14 + 8$

-6 | TRUE

Since $(-2, -6)$ checks, it is the solution.

9. $-5s + t = 11,$ (1)

$4s + 12t = 4$ (2)

We solve the first equation for t.

$-5s + t = 11$ (1)

$t = 5s + 11$ (3)

We substitute $5s + 11$ for t in the second equation and solve for s.

$4s + 12t = 4$ (2)

$4s + 12(5s + 11) = 4$

$4s + 60s + 132 = 4$

$64s + 132 = 4$

$64s = -128$

$s = -2$

Next we substitute -2 for s in Equation (3).

$t = 5s + 11 = 5(-2) + 11 = -10 + 11 = 1$

We check the ordered pair $(-2, 1)$.

$$\frac{-5s + t = 11}{}$$

$-5(-2) + 1$? 11

$10 + 1$

11 | 11 TRUE

$$\frac{4s + 12t = 4}{}$$

$4(-2) + 12 \cdot 1$? 4

$-8 + 12$

4 | 4 TRUE

Since $(-2, 1)$ checks, it is the solution.

11. $2x - 3 = y$ (1)

$y - 2x = 1,$ (2)

We substitute $2x - 3$ for y in the second equation and solve for x.

$y - 2x = 1$ (2)

$2x - 3 - 2x = 1$ Substituting

$-3 = 1$ Collecting like terms

We have a false equation. Therefore, there is no solution.

13. $3a - b = 7,$ (1)

$2a + 2b = 5$ (2)

We solve the first equation for b.

$3a - b = 7$ (1)

$-b = -3a + 7$

$b = 3a - 7$ (3)

We substitute $3a - 7$ for b in the second equation and solve for a.

$2a + 2b = 5$ (2)

$2a + 2(3a - 7) = 5$

$2a + 6a - 14 = 5$

$8a - 14 = 5$

$8a = 19$

$a = \dfrac{19}{8}$

We substitute $\dfrac{19}{8}$ for a in Equation (3).

$b = 3a - 7 = 3 \cdot \dfrac{19}{8} - 7 = \dfrac{57}{8} - \dfrac{56}{8} = \dfrac{1}{8}$

The ordered pair $\left(\dfrac{19}{8}, \dfrac{1}{8} \right)$ checks in both equations. It is the solution.

15. $2x - 6y = 4,$ (1)

$3y + 2 = x$ (2)

We substitute $3y + 2$ for x in the first equation and solve for y.

$2x - 6y = 4$

$2(3y + 2) - 6y = 4$

$6y + 4 - 6y = 4$

$4 = 4$

We have a true equation. Any value of y will make this equation true. Thus, the system of equations has infinitely many solutions.

17. $2x + 2y = 2,$ (1)

$3x - y = 1$ (2)

We solve the second equation for y.

$3x - y = 1$ (2)

$-y = -3x + 1$

$y = 3x - 1$ (3)

We substitute $3x - 1$ for y in the first equation and solve for x.

$2x + 2y = 2$ (1)

$2x + 2(3x - 1) = 2$

$2x + 6x - 2 = 2$

$8x - 2 = 2$

$8x = 4$

$x = \dfrac{1}{2}$

Next we substitute $\dfrac{1}{2}$ for x in Equation (3).

$$y = 3x - 1 = 3 \cdot \frac{1}{2} - 1 = \frac{3}{2} - 1 = \frac{1}{2}$$

The ordered pair $\left(\frac{1}{2}, \frac{1}{2}\right)$ checks in both equations. It is the solution.

19. **Familiarize**. Let l = the length of the court and w = the width. Recall that the perimeter P of a rectangle with length l and width w is given by $P = 2l + 2w$.

Translate.

$$\underbrace{\text{The perimeter}}_{\downarrow \atop 2l + 2w} \;\; \underset{\downarrow \atop =}{\text{is}} \;\; \underbrace{60 \text{ m.}}_{\downarrow \atop 60}$$

$$\underbrace{\text{The length}}_{\downarrow \atop l} \;\; \underset{\downarrow \atop =}{\text{is}} \;\; \underset{\downarrow \atop 5}{5} \;\; \underset{\downarrow \atop \cdot}{\text{times}} \;\; \underbrace{\text{the width.}}_{\downarrow \atop w}$$

We have system of equations.

$$2l + 2w = 60, \quad (1)$$
$$l = 5w \qquad\quad (2)$$

Solve. We substitute $5w$ for l in Equation (1) and solve for w.

$$2l + 2w = 60$$
$$2 \cdot 5w + 2w = 60$$
$$10w + 2w = 60$$
$$12w = 60$$
$$w = 5$$

Now substitute 5 for w in Equation (2) and find l.

$$l = 5w = 5 \cdot 5 = 25$$

Check. If the length is 25 m and the width is 5 m, then the perimeter is $2 \cdot 25 + 2 \cdot 5$, or 60 m. Also, the length is five times the width. The answer checks.

State. The length of the court is 25 m, and the width is 5 m.

21. **Familiarize**. Using the drawing in the text, we let x and y represent the measures of the angles.

Translate.

$$\underbrace{\text{The sum of the measures}}_{\downarrow \atop x + y} \;\; \underset{\downarrow \atop =}{\text{is}} \;\; \underbrace{180°.}_{\downarrow \atop 180}$$

$$\underbrace{\text{One angle}}_{\downarrow \atop x} \;\; \underset{\downarrow \atop = 3}{\text{is}} \;\; \underset{\downarrow}{3} \;\; \underset{\downarrow \atop \cdot}{\text{times}} \;\; \underbrace{\text{the other}}_{\downarrow \atop y} \;\; \underset{\downarrow \atop -}{\text{less}} \;\; \underset{\downarrow \atop 12}{12°.}$$

We have a system of equations.

$$x + y = 180, \quad (1)$$
$$x = 3y - 12 \quad (2)$$

Solve. Substitute $3y - 12$ for x in Equation (1) and solve for y.

$$x + y = 180$$
$$(3y - 12) + y = 180$$
$$4y - 12 = 180$$
$$4y = 192$$
$$y = 48$$

Now substitute 48 for y in Equation (2) and find x.

$$x = 3y - 12 = 3 \cdot 48 - 12 = 132$$

Check. The sum of the measures is $48° + 132°$, or $180°$. Also, $132°$ is $12°$ less than three times $48°$. The answer checks.

State. The measures of the angles are $48°$ and $132°$.

23. **Familiarize**. Let x = number of games won and y = number of games tied. The total points earned in x wins is $2x$; the total points earned in y ties is $1 \cdot y$, or y.

Translate.

$$\underbrace{\text{Points from wins}}_{\downarrow \atop 2x} \;\; \underset{\downarrow \atop +}{\text{plus}} \;\; \underbrace{\text{points from ties}}_{\downarrow \atop y} \;\; \underset{\downarrow \downarrow \atop = \;\; 60}{\text{is } 60.}$$

$$\underbrace{\text{Number of wins}}_{\downarrow \atop x} \;\; \underset{\downarrow \atop =}{\text{is}} \;\; \underbrace{9 \text{ more than the number of ties.}}_{\downarrow \atop 9 + y}$$

We have a system of equations:

$$2x + y = 60,$$
$$x = 9 + y$$

Solve. We solve the system of equations. We use substitution.

$$2(9 + y) + y = 60 \quad \text{Substituting } 9 + y \text{ for } x \text{ in (1)}$$
$$18 + 2y + y = 60$$
$$18 + 3y = 60$$
$$3y = 42$$
$$y = 14$$

$$x = 9 + 14 \quad \text{Substituting 14 for } y \text{ in (2)}$$
$$x = 23$$

Check. The number of wins, 23, is 9 more than the number of ties, 14.

Points from wins: $\qquad 23 \times 2 = 46$

Points from ties: $\qquad\;\; 14 \times 1 = \underline{14}$

$$\text{Total} \quad 60$$

The numbers check.

State. The team had 23 wins and 14 ties.

25. $y = 1.3x - 7$

The equation is in slope-intercept form, $y = mx + b$. The slope is 1.3.

27. $A = \dfrac{pq}{7}$

$$7A = pq \quad \text{Multiplying by 7}$$
$$\frac{7A}{q} = p \quad \text{Dividing by } q$$

29. $-4x + 5(x - 7) = 8x - 6(x + 2)$

$$-4x + 5x - 35 = 8x - 6x - 12 \quad \text{Removing parentheses}$$
$$x - 35 = 2x - 12 \qquad\quad \text{Collecting like terms}$$
$$-35 = x - 12 \qquad\qquad \text{Subtracting } x$$
$$-23 = x \qquad\qquad\quad\;\; \text{Adding 12}$$

The solution is -23.

31. $2 = m + b,$ (1)   Substituting $(1, 2)$

     $4 = -3m + b$ (2)   Substituting $(-3, 4)$

$$2 = m + b \quad (1)$$
$$\underline{-4 = 3m - b} \quad \text{Multiplying (2) by } -1$$
$$-2 = 4m$$
$$-\frac{1}{2} = m$$

Substitute $-\dfrac{1}{2}$ for m in (1).

$$2 = -\frac{1}{2} + b$$
$$\frac{5}{2} = b$$

Thus, $m = -\dfrac{1}{2}$ and $b = \dfrac{5}{2}$.

33. ***Familiarize***. Let $l =$ the original length, in inches, and $w =$ the original width, in inches. Then $w - 6 =$ the width after 6 in. is cut off.

Translate.

$\underbrace{\text{The original perimeter}}$ is $\underbrace{156 \text{ in}}$.
$$\downarrow \qquad\qquad\qquad\quad \downarrow \quad \downarrow$$
$$2l + 2w \qquad\qquad = \quad 156$$

$\underbrace{\text{The length}}$ becomes 4 times $\underbrace{\text{the new width}}$.
$$\downarrow \qquad\quad \downarrow \quad \downarrow \quad \downarrow \qquad \downarrow$$
$$l \qquad\quad = \quad 4 \quad \cdot \qquad (w-6)$$

We have a system of equations:

$$2l + 2w = 156, \quad (1)$$
$$l - 4 \cdot (w - 6) \quad (2)$$

Solve. Substitute $4(w-6)$ for l in Equation (1) and solve for w.

$$2l + 2w = 156$$
$$2 \cdot 4(w - 6) + 2w = 156$$
$$8w - 48 + 2w = 156$$
$$10w - 48 = 156$$
$$10w = 204$$
$$w = 20.4$$

Now substitute 20.4 for w in Equation (2) and find l.

$$l = 4 \cdot (w - 6) = 4(20.4 - 6) = 4(14.4) = 57.6$$

Check. The original perimeter is $2(57.6) + 2(20.4)$, or $115.2 + 40.8$, or 156 in. If 6 in. is cut off the width, then the width becomes $20.4 - 6$, or 14.4 in., and the length is 4 times the width, or 57.6 in. The answer checks.

State. The length is 57.6 in., and the width is 20.4 in.

Exercise Set 3.3

RC1. If a system of equations has a solution, then it is <u>consistent</u>.

RC3. If a system of equations has infinitely many solutions, then it is <u>consistent</u>.

RC5. If the graphs of the equations in a system of two equations in two variables are parallel, then the system is <u>inconsistent</u>.

1. $x + 3y = 7$ (1)

$$\underline{-x + 4y = 7} \quad (2)$$
$$0 + 7y = 14 \quad \text{Adding}$$
$$7y = 14$$
$$y = 2$$

Substitute 2 for y in one of the original equations and solve for x.

$$x + 3y = 7 \quad \text{Equation (1)}$$
$$x + 3 \cdot 2 = 7 \quad \text{Substituting}$$
$$x + 6 = 7$$
$$x = 1$$

Check:

$x + 3y = 7$		$-x + 4y = 7$	
$1 + 3 \cdot 2 \; ? \; 7$		$-1 + 4 \cdot 2 \; ? \; 7$	
$1 + 6$		$-1 + 8$	
7	TRUE	7	TRUE

Since $(1, 2)$ checks, it is the solution.

3. $9x + 5y = 6$ (1)

$$\underline{2x - 5y = -17} \quad (2)$$
$$11x + 0 = -11 \quad \text{Adding}$$
$$11x = -11$$
$$x = -1$$

Substitute -1 for x in one of the original equations and solve for y.

$$9x + 5y = 6 \quad \text{Equation (1)}$$
$$9(-1) + 5y = 6 \quad \text{Substituting}$$
$$-9 + 5y = 6$$
$$5y = 15$$
$$y = 3$$

We obtain $(-1, 3)$. This checks, so it is the solution.

5. $5x + 3y = -11,$ (1)

     $3x - y = -1$ (2)

We multiply by 3 on both sides of the second equation and then add.

$$5x + 3y = -11$$
$$\underline{9x - 3y = -3}$$
$$14x + 0 = -14$$
$$14x = -14$$
$$x = -1$$

Substitute -1 for x in one of the original equations and solve for y.

$$3x - y = -1 \quad \text{Equation (2)}$$
$$3(-1) - y = -1$$
$$-3 - y = -1$$
$$-y = 2$$
$$y = -2$$

We obtain $(-1, -2)$. This checks, so it is the solution.

7. $5r - 3s = 19, \quad (1)$
$\quad 2r - 6s = -2 \quad (2)$

We multiply by -2 on both sides of the first equation and then add.

$$\begin{array}{r} -10r + 6s = -38 \\ \underline{2r - 6s = -2} \\ -8r + 0 = -40 \end{array}$$
$$-8r = -40$$
$$r = 5$$

Substitute 5 for r in one of the original equations and solve for s.

$$5r - 3s = 19 \quad \text{Equation (1)}$$
$$5 \cdot 5 - 3s = 19$$
$$25 - 3s = 19$$
$$-3s = -6$$
$$s = 2$$

We obtain $(5, 2)$. This checks, so it is the solution.

9. $2x + 3y = 1,$
$\quad 4x + 6y = 2$

Multiply the first equation by -2 and then add.

$$\begin{array}{r} -4x - 6y = -2 \\ \underline{4x + 6y = 2} \\ 0 = 0 \quad \text{Adding} \end{array}$$

We have an equation that is true for all numbers x and y. The system is dependent and has an infinite number of solutions.

11. $5x - 9y = 7,$
$\quad 7y - 3x = -5$

We first write the second equation in the form $Ax + By = C$.

$$5x - 9y = 7 \quad (1)$$
$$-3x + 7y = -5 \quad (2)$$

We use the multiplication principle with both equations and then add.

$$\begin{array}{rl} 15x - 27y = 21 & \text{Multiplying by 3} \\ \underline{-15x + 35y = -25} & \text{Multiplying by 5} \\ 0 + 8y = -4 & \text{Adding} \end{array}$$
$$8y = -4$$
$$y = -\frac{1}{2}$$

Substitute $-\dfrac{1}{2}$ for y in one of the original equations and solve for x.

$$5x - 9y = 7 \quad \text{Equation (1)}$$
$$5x - 9\left(-\frac{1}{2}\right) = 7 \quad \text{Substituting}$$
$$5x + \frac{9}{2} = \frac{14}{2}$$
$$5x = \frac{5}{2}$$
$$x = \frac{1}{2}$$

We obtain $\left(\dfrac{1}{2}, -\dfrac{1}{2}\right)$. This checks, so it is the solution.

13. $3x + 2y = 24, \quad (1)$
$\quad 2x + 3y = 26 \quad (2)$

We use the multiplication principle with both equations and then add.

$$\begin{array}{rl} 9x + 6y = 72 & \text{Multiplying by 3} \\ \underline{-4x - 6y = -52} & \text{Multiplying by } -2 \\ 5x + 0 = 20 \end{array}$$
$$5x = 20$$
$$x = 4$$

Substitute 4 for x in one of the original equations and solve for y.

$$3x + 2y = 24 \quad \text{Equation (1)}$$
$$3 \cdot 4 + 2y = 24$$
$$12 + 2y = 24$$
$$2y = 12$$
$$y = 6$$

We obtain $(4, 6)$. This checks, so it is the solution.

15. $2x - 4y = 5,$
$\quad 2x - 4y = 6$

Multiply the first equation by -1 and then add.

$$\begin{array}{r} -2x + 4y = -5 \\ \underline{2x - 4y = 6} \\ 0 = 1 \end{array}$$

We have a false equation. The system has no solution.

17. $2a + b = 12, \quad (1)$
$\quad a + 2b = -6 \quad (2)$

Multiply both sides of the first equation by -2 and then add.

$$\begin{array}{r} -4a - 2b = -24 \\ \underline{a + 2b = -6} \\ -3a + 0 = -30 \end{array}$$
$$-3a = -30$$
$$a = 10$$

Substitute 10 for a in one of the original equations and then solve for b.

$$2a + b = 12 \quad \text{Equation (1)}$$
$$2 \cdot 10 + b = 12$$
$$20 + b = 12$$
$$b = -8$$

We obtain $(10, -8)$. This checks, so it is the solution.

19. $\dfrac{1}{3}x + \dfrac{1}{5}y = 7, \quad (1)$

$\dfrac{1}{6}x - \dfrac{2}{5}y = -4 \quad (2)$

We use the multiplication principle with both equations to clear the fractions.

$5x + 3y = 105, \quad (3) \quad \text{Multiplying (1) by 15}$

$5x - 12y = -120 \quad (4) \quad \text{Multiplying (2) by 30}$

Now we multiply both sides of Equation (4) by -1 and then add.

$$\begin{array}{r} 5x + 3y = 105 \quad (3) \\ -5x + 12y = 120 \\ \hline 0 + 15y = 225 \\ 15y = 225 \\ y = 15 \end{array}$$

Substitute 15 for y in one of the equations in which the fractions were cleared and then solve for x.

$$5x + 3y = 105 \quad \text{Equation (3)}$$
$$5x + 3 \cdot 15 = 105$$
$$5x + 45 = 105$$
$$5x = 60$$
$$x = 12$$

We obtain $(12, 15)$. This checks, so it is the solution.

21. $\dfrac{1}{5}x + \dfrac{1}{2}y = 6, \quad (1)$

$\dfrac{2}{5}x - \dfrac{3}{2}y = -8 \quad (2)$

We could clear fractions first, but instead we will multiply both sides of the first equation by 3 and then add.

$$\begin{array}{r} \dfrac{3}{5}x + \dfrac{3}{2}y = 18 \\ \dfrac{2}{5}x - \dfrac{3}{2}y = -8 \quad \text{Equation (2)} \\ \hline x + 0 = 10 \\ x = 10 \end{array}$$

Substitute 10 for x in one of the original equations and then solve for y.

$\dfrac{1}{5}x + \dfrac{1}{2}y = 6 \quad \text{Equation (1)}$

$\dfrac{1}{5} \cdot 10 + \dfrac{1}{2}y = 6$

$2 + \dfrac{1}{2}y = 6$

$\dfrac{1}{2}y = 4$

$y = 8$

We obtain $(10, 8)$. This checks, so it is the solution.

23. $\dfrac{1}{2}x - \dfrac{1}{3}y = -4, \quad (1)$

$\dfrac{1}{4}x + \dfrac{5}{6}y = 4 \quad (2)$

First we use the multiplication principle with both equations to clear the fractions.

$$3x - 2y = -24 \quad (3) \quad \text{Multiplying (1) by 6}$$
$$3x + 10y = 48 \quad (4) \quad \text{Multiplying (2) by 12}$$

Now we multiply both sides of the first equation by -1 and then add.

$$\begin{array}{r} -3x + 2y = 24 \\ 3x + 10y = 48 \\ \hline 0 + 12y = 72 \\ 12y = 72 \\ y = 6 \end{array}$$

Substitute 6 for y in one of the equations in which fractions were cleared and then solve for x.

$$3x - 2y = -24 \quad \text{Equation (3)}$$
$$3x - 2 \cdot 6 = -24$$
$$3x - 12 = -24$$
$$3x = -12$$
$$x = -4$$

We obtain $(-4, 6)$. This checks, so it is the solution.

25. $0.3x - 0.2y = 4,$

$0.2x + 0.3y = 0.5$

We first multiply each equation by 10 to clear decimals.

$3x - 2y = 40 \quad (1)$

$2x + 3y = 5 \quad (2)$

We use the multiplication principle with both equations of the resulting system.

$$\begin{array}{ll} \text{From (1):} & 9x - 6y = 120 \quad \text{Multiplying by 3} \\ \text{From (2):} & \underline{4x + 6y = 10} \quad \text{Multiplying by 2} \\ & 13x + 0 = 130 \quad \text{Adding} \\ & 13x = 130 \\ & x = 10 \end{array}$$

Substitute 10 for x in one of the equations in which the decimals were cleared and solve for y.

$$2x + 3y = 5 \quad \text{Equation (2)}$$
$$2 \cdot 10 + 3y = 10 \quad \text{Substituting}$$
$$20 + 3y = 5$$
$$3y = -15$$
$$y = -5$$

We obtain $(10, -5)$. This checks, so it is the solution.

27. $0.05x + 0.25y = 22,$
$0.15x + 0.05y = 24$

We first multiply each equation by 100 to clear decimals.

$$5x + 25y = 2200 \quad (1)$$
$$15x + 5y = 2400 \quad (2)$$

We multiply by -5 on both sides of the second equation and add.

$$
\begin{array}{r}
5x + 25y = 2200 \\
-75x - 25y = -12{,}000 \quad \text{Multiplying by } -5 \\
\hline
-70x + 0 = -9800 \quad \text{Adding}
\end{array}
$$

$$-70x = -9800$$
$$x = \frac{-9800}{-70}$$
$$x = 140$$

Substitute 140 for x in one of the equations in which the decimals were cleared and solve for y.

$$5x + 25y = 2200 \quad \text{Equation (1)}$$
$$5 \cdot 140 + 25y = 2200 \quad \text{Substituting}$$
$$700 + 25y = 2200$$
$$25y = 1500$$
$$y = 60$$

We obtain $(140, 60)$. This checks, so it is the solution.

29. **Familiarize.** Let x = the larger number and y = the smaller number.

Translate.

The sum of the numbers is 63.

$$x + y \qquad\qquad = 63$$

The larger number minus the smaller number is 9.

$$x \qquad\quad - \qquad\quad y \qquad\quad = 9$$

We have a system of equations.

$$x + y = 63, \quad (1)$$
$$x - y = 9 \quad (2)$$

Solve. We add the equations.

$$
\begin{array}{r}
x + y = 63 \\
x - y = 9 \\
\hline
2x = 72
\end{array}
$$

$$x = 36$$

Substitute 36 for x in Equation (1) and solve for y.

$$36 + y = 63$$
$$y = 27$$

Check. $36 + 27 = 63$ and $36 - 27 = 9$, so the answer checks.

State. The numbers are 36 and 27.

31. **Familiarize.** Let x = the larger number and y = the smaller number.

Translate.

The sum of the numbers is 3.

$$x + y \qquad\qquad = 3$$

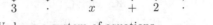

Three times the larger number plus two times the smaller number is 24.

$$3 \cdot x + 2 \cdot y = 24$$

We have a system of equations.

$$x + y = 3, \quad (1)$$
$$3x + 2y = 24 \quad (2)$$

Solve. First we multiply by -2 on both sides of Equation (1) and then add.

$$
\begin{array}{r}
-2x - 2y = -6 \\
3x + 2y = 24 \\
\hline
x = 18
\end{array}
$$

Substitute 18 for x in Equation (1) and solve for y.

$$18 + y = 3$$
$$y = -15$$

Check. $18 + (-15) = 3$ and $3 \cdot 18 + 2(-15) = 54 - 30 = 24$, so the answer checks.

State. The numbers are 18 and -15.

33. **Familiarize.** Let x = the measure of the larger angle and y = the measure of the smaller angle.

Translate.

The sum of the measures is $90°$.

$$x + y \qquad\qquad = 90$$

The difference of the measures is $6°$.

$$x - y \qquad\qquad = 6$$

We have a system of equations.

$$x + y = 90, \quad (1)$$
$$x - y = 6 \quad (2)$$

Solve. We add the equations.

$$
\begin{array}{r}
x + y = 90 \\
x - y = 6 \\
\hline
2x = 96
\end{array}
$$

$$x = 48$$

Substitute 48 for x in Equation (1) and solve for y.

$$48 + y = 90$$
$$y = 42$$

Check. $48° + 42° = 90°$ and $48° - 42° = 6°$, so the answer checks.

State. The measures of the angles are 48° and 42°.

35. Familiarize. Let $x =$ the number of two-point shots made and $y =$ the number of three-point shots made. Then $2x$ represents the number of points scored on two-point shots and $3y$ represents the number of points scored on three-point shots.

Translate.

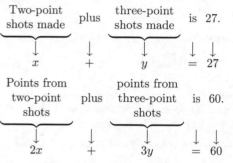

We have a system of equations.

$$x + y = 27, \quad (1)$$
$$2x + 3y = 60 \quad (2)$$

Solve. First we multiply Equation (1) by -2 and then add.

$$-2x - 2y = -54$$
$$\underline{2x + 3y = 60}$$
$$y = 6$$

Substitute 6 for y in Equation (1) and solve for x.

$$x + 6 = 27$$
$$x = 21$$

Check. $21 + 6 = 27$ and $2 \cdot 21 + 3 \cdot 6 = 42 + 18 = 60$, so the answer checks.

State. 21 two-point shots and 6 three-point shots were made.

37. Familiarize. Let x and y represent the number of 3-credit and 4-credit courses, respectively. Then $3x$ and $4y$ represent the number of credits from the courses.

Translate.

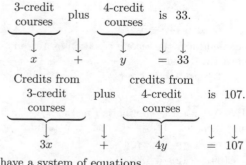

We have a system of equations.

$$x + y = 33, \quad (1)$$
$$3x + 4y = 107 \quad (2)$$

Solve. First we multiply Equation (1) by -3 and then add.

$$-3x - 3y = -99$$
$$\underline{3x + 4y = 107}$$
$$y = 8$$

Now substitute 8 for y in Equation (2) and solve for x.

$$x + 8 = 33$$
$$x = 25$$

Check. $25 + 8 = 33$ and $3 \cdot 25 + 4 \cdot 8 = 75 + 32 = 107$, so the answer checks.

State. The students took 25 3-credit courses and 8 4-credit courses.

39. $f(x) = 3x^2 - x + 1$

$f(0) = 3 \cdot 0^2 - 0 + 1 = 0 - 0 + 1 = 1$

41. $f(x) = 3x^2 - x + 1$

$f(-2) = 3(-2)^2 - (-2) + 1 = 12 + 2 + 1 = 15$

43. We cannot calculate $f(x)$ when the denominator is 0. We set the denominator equal to 0 and solve for x.

$$x + 7 = 0$$
$$x = -7$$

The domain is $\{x | x$ is a real number *and* $x \neq -7\}$, or $(-\infty, -7) \cup (-7, \infty)$.

45. Substitute $-\dfrac{3}{5}$ for m and -7 for b in the slope-intercept equation.

$$y = mx + b$$
$$y = -\frac{3}{5}x - 7$$

47. Graph these equations, solving each equation for y first, if necessary. We get $y = \dfrac{3.5x - 106.2}{2.1}$ and $y = \dfrac{-4.1x - 106.28}{16.7}$. Using the INTERSECT feature, we find that the point of intersection is $(23.12, -12.04)$.

49. Substitute -5 for x and -1 for y in the first equation.

$$A(-5) - 7(-1) = -3$$
$$-5A + 7 = -3$$
$$-5A = -10$$
$$A = 2$$

Then substitute -5 for x and -1 for y in the second equation.

$$-5 - B(-1) = -1$$
$$-5 + B = -1$$
$$B = 4$$

We have $A = 2$, $B = 4$.

51. $(0, -3)$ and $\left(-\dfrac{3}{2}, 6\right)$ are two solutions of $px - qy = -1$.

Substitute 0 for x and -3 for y.

$$p \cdot 0 - q \cdot (-3) = -1$$
$$3q = -1$$
$$q = -\dfrac{1}{3}$$

Substitute $-\dfrac{3}{2}$ for x and 6 for y.

$$p \cdot \left(-\dfrac{3}{2}\right) - q \cdot 6 = -1$$
$$-\dfrac{3}{2}p - 6q = -1$$

Substitute $-\dfrac{1}{3}$ for q and solve for p.

$$-\dfrac{3}{2}p - 6 \cdot \left(-\dfrac{1}{3}\right) = -1$$
$$-\dfrac{3}{2}p + 2 = -1$$
$$-\dfrac{3}{2}p = -3$$
$$-\dfrac{2}{3} \cdot \left(-\dfrac{3}{2}p\right) = -\dfrac{2}{3} \cdot (-3)$$
$$p = 2$$

Thus, $p = 2$ and $q = -\dfrac{1}{3}$.

Exercise Set 3.4

RC1. The mixture has a total of <u>10</u> gal.

RC3. Berry Choice is 15% fruit juice, so the number of gallons of fruit juice in y gal of Berry Choice is <u>0.15y</u>.

1. Familiarize. Let $x = $ the number of e-books purchased and $y = $ the number of games purchased.

Translate. We organize the information in a table.

Type of purchase	e-book	Game	Total
Number purchased	x	y	68
Price	\$3.99	\$1.99	
Amount spent	$3.99x$	$1.99y$	225.32

The "Number purchased" row of the table gives us one equation:

$$x + y = 68.$$

The "Amount spent" row of the table gives us a second equation:

$$3.99x + 1.99y = 225.32.$$

We have a system of equations.

$$x + y = 68,$$
$$3.99x + 1.99y = 225.32$$

We can multiply the second equation by 100 on both sides to clear the decimals.

$$x + y = 68, \qquad (1)$$
$$399x + 199y = 22{,}532 \quad (2)$$

Solve. We use the elimination method. Begin by multiplying Equation (1) by -199 and then adding.

$$-199x - 199y = -13{,}532$$
$$\underline{399x + 199y = 22{,}532}$$
$$200x \qquad\qquad = 9000$$
$$x = 45$$

Substitute 45 for x in Equation (1) and solve for y.

$$45 + y = 68$$
$$y = 23$$

Check. The number of purchases is $45 + 23$, or 68. The amount spent was $\$3.99(45) + \$1.99(23) = \$179.55 + \$45.77 = \$225.32$. The answer checks.

State. Laura bought 45 e-books and 23 games.

3. Familiarize. If the 18 bouquets were identical and a total of \$86.76 was spent for the bouquets, then the cost of each bouquet was \$86.76/18, or \$4.82. Let $x = $ the number of foil balloons and $y = $ the number of latex balloons in each bouquet.

Translate. We organize the information in a table.

Type of balloon	Foil	Latex	Total
Number in bouquet	x	y	9
Price	\$1.99	\$0.12	
Amount spent	$1.99x$	$0.12y$	4.82

The "Number in bouquet" row of the table gives us one equation:

$$x + y = 9.$$

The "Amount spent" row of the table gives us a second equation:

$$1.99x + 0.12y = 4.82.$$

We have a system of equations.

$$x + y = 9,$$
$$1.99x + 0.12y = 4.82$$

We can multiply the second equation by 100 on both sides to clear the decimals.

$$x + y = 9, \qquad (1)$$
$$199x + 12y = 482 \quad (2)$$

Solve. We use the elimination method. Begin by multiplying Equation (1) by -12 and then adding.

$$-12x - 12y = -108$$
$$\underline{199x + 12y = 482}$$
$$187x \qquad\qquad = 374$$
$$x = 2$$

Substitute 2 for x in Equation (1) and solve for y.

$$2 + y = 9$$
$$y = 7$$

Check. The number of balloons in each bouquet was $2+7$, or 9. The cost of each bouquet was $\$1.99(2) + \$0.12(7) = \$3.98 + \$0.84 = \$4.82$. The answer checks.

State. Each bouquet contained 2 foil balloons and 7 latex balloons.

5. Familiarize. Let x and y represent the number of ounces of vinegar and olive oil, respectively, used in the mixture.

Translate.

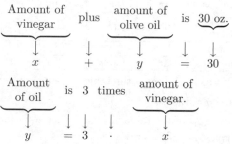

We have a system of equations.

$$x + y = 30, \quad (1)$$
$$y = 3x \quad\quad (2)$$

Solve. We use the substitution method. First we substitute $3x$ for y in Equation (1) and solve for x.

$$x + y = 30$$
$$x + 3x = 30$$
$$4x = 30$$
$$x = \frac{15}{2}, \text{ or } 7\frac{1}{2}$$

Now substitute $\frac{15}{2}$ for x in Equation (2) and find y.

$$y = 3x$$
$$y = 3 \cdot \frac{15}{2} = \frac{45}{2}, \text{ or } 22\frac{1}{2}$$

Check. $22\frac{1}{2}$ oz $+ 7\frac{1}{2}$ oz $= 30$ oz, and $22\frac{1}{2}$ oz $= 3 \cdot 7\frac{1}{2}$ oz, so the answer checks.

State. The mixture should contain $7\frac{1}{2}$ oz of vinegar and $22\frac{1}{2}$ oz of olive oil.

7. Familiarize. Let x and y represent the number of pounds of the 40% and the 10% mixture to be used. The final mixture contains 25% (10 lb), or 0.25(10 lb), or 2.5 lb of peanuts.

Translate. We organize the information in a table.

	40% mixture	10% mixture	Wedding mixture
Number of pounds	x	y	10
Percent of peanuts	40%	10%	25%
Pounds of peanuts	$0.4x$	$0.1y$	2.5

The first row of the table gives us one equation:

$$x + y = 10$$

The last row gives us a second equation:

$$0.4x + 0.1y = 2.5$$

After clearing decimals, we have the problem translated to a system of equations:

$$x + y = 10, \quad (1)$$
$$4x + y = 25 \quad (2)$$

Solve. We solve the system of equations using the elimination method.

$$-x - y = -10 \quad \text{Multiplying (1) by } -1$$
$$\underline{4x + y = 25}$$
$$3x = 15$$
$$x = 5$$

Now substitute 5 for x in Equation (1) and solve for y.

$$5 + y = 10$$
$$y = 5$$

Check. If 5 lb of each mixture is used, the total wedding mixture is $5 + 5$, or 10 lb. The amount of peanuts in the wedding mixture is $0.4(5) + 0.1(5)$, or $2 + 0.5$, or 2.5 lb. The answer checks.

State. 5 lb of each type of mixture should be used.

9. Familiarize. Let $x =$ the number of liters of 25% solution and $y =$ the number of liters of 50% solution to be used. The mixture contains 40%(10 L), or 0.4(10 L) $= 4$ L of acid.

Translate. We organize the information in a table.

	25% solution	50% solution	Mixture
Number of liters	x	y	10
Percent of acid	25%	50%	40%
Amount of acid	$0.25x$	$0.5y$	4 L

We get one equation from the "Number of liters" row of the table.

$$x + y = 10$$

The last row of the table yields a second equation.

$$0.25x + 0.5y = 4$$

After clearing decimals, we have the problem translated to a system of equations:

$$x + y = 10, \quad (1)$$
$$25x + 50y = 400 \quad (2)$$

Solve. We use the elimination method to solve the system of equations.

$$-25x - 25y = -250 \quad \text{Multiplying (1) by } -25$$
$$\underline{25x + 50y = 400}$$
$$25y = 150$$
$$y = 6$$

Substitute 6 for y in (1) and solve for x.

$$x + 6 = 10$$
$$x = 4$$

Check. The total amount of the mixture is 4 Lb + 6 L, or 10 L. The amount of acid in the mixture is 0.25(4 L) + 0.5(6 L) = 1 L + 3 L = 4 L. The answer checks.

State. 4 L of the 25% solution and 6 L of the 50% solution should be mixed.

11. Familiarize. Let x and y represent the number of packets of sweet pepper seeds and hot-pepper seeds, respectively, in the assortment.

Translate. We organize the information in a table.

Type	Sweet pepper	Hot pepper	Assortment
Number	x	y	16
Price	$2.85	$4.29	$3.30
Value	2.85x	4.29y	3.30(16), or 52.80

The "Number" row of the table gives us one equation:

$$x + y = 16.$$

The "Value" row of the table gives us a second equation:

$$2.85x + 4.29y = 52.80.$$

We have a system of equations.

$$x + y = 16,$$
$$2.85x + 4.29y = 52.80$$

We can multiply the second equation by 100 on both sides to clear the decimals.

$$x + y = 16, \quad (1)$$
$$285x + 429y = 5280 \quad (2)$$

Solve. We use the elimination method. First we multiply Equation (1) by -285 and then add.

$$-285x - 285y = -4560$$
$$\underline{285x + 429y = 5280}$$
$$144y = 720$$
$$y = 5$$

Now substitute 5 for y in Equation (1) and solve for x.

$$x + 5 = 16$$
$$x = 11$$

Check. 11 + 5 = 16 and $2.85(11) + $4.29(5) = $31.35 + $21.45 = $52.80, so the answer checks.

State. The assortment contains 11 packets of sweet-pepper seeds and 5 packets of hot-pepper seeds.

13. Familiarize. Let x = the amount of the 6% loan and y = the amount of the 9% loan. Recall that the formula for simple interest is

$$\text{Interest} = \text{Principal} \cdot \text{Rate} \cdot \text{Time}.$$

Translate. We organize the information in a table.

	6% loan	9% loan	Total
Principal	x	y	$12,000
Interest Rate	6%	9%	
Time	1 yr	1 yr	
Interest	0.06x	0.09y	$855

The "Principal" row of the table gives us one equation:

$$x + y = 12,000$$

The last row of the table yields another equation:

$$0.06x + 0.09y = 855$$

After clearing decimals, we have the problem translated to a system of equations:

$$x + y = 12,000 \quad (1)$$
$$6x + 9y = 85,500 \quad (2)$$

Solve. We use the elimination method to solve the system of equations.

$$-6x - 6y = -72,000 \quad \text{Multiplying (1) by } -6$$
$$\underline{6x + 9y = 85,500}$$
$$3y = 13,500$$
$$y = 4500$$

Substitute 4500 for y in (1) and solve for x.

$$x + 4500 = 12,000$$
$$x = 7500$$

Check. The loans total $7500 + $4500, or $12,000. The total interest is 0.06($7500) + 0.09($4500) = $450 + $405 = $855. The answer checks.

State. The 6% loan was for $7500, and the 9% loan was for $4500.

15. Familiarize. From the bar graph we see that whole milk is 4% milk fat, milk for cream cheese is 8% milk fat, and cream is 30% milk fat. Let x = the number of pounds of whole milk and y = the number of pounds of cream to be used. The mixture contains 8%(200 lb), or 0.08(200 lb) = 16 lb of milk fat.

Translate. We organize the information in a table.

	Whole milk	Cream	Mixture
Number of pounds	x	y	200
Percent of milk fat	4%	30%	8%
Amount of milk fat	0.04x	0.3y	16 lb

We get one equation from the " Number of pounds" row of the table:

$$x + y = 200$$

The last row of the table yields a second equation:

$$0.04x + 0.3y = 16$$

After clearing decimals, we have the problem translated to a system of equations:

$$x + y = 200, \quad (1)$$
$$4x + 30y = 1600 \quad (2)$$

Solve. We use the elimination method to solve the system of equations.

$$-4x - 4y = -800 \qquad \text{Multiplying (1) by } -4$$
$$\underline{4x + 30y = 1600}$$
$$26y = 800$$
$$y = \frac{400}{13}, \text{ or } 30\frac{10}{13}$$

Substitute $\frac{400}{13}$ for y in (1) and solve for x.

$$x + \frac{400}{13} = 200$$
$$x = \frac{2200}{13}, \text{ or } 169\frac{3}{13}$$

Check. The total amount of the mixture is

$$\frac{2200}{13} \text{ lb} + \frac{400}{13} \text{ lb} = \frac{2600}{13} \text{ lb} = 200 \text{ lb}. \text{ The amount}$$

of milk fat in the mixture is $0.04\left(\frac{2200}{13} \text{ lb}\right) +$

$$0.3\left(\frac{400}{13} \text{ lb}\right) = \frac{88}{13} \text{ lb} + \frac{120}{13} \text{ lb} = \frac{208}{13} \text{ lb} = 16 \text{ lb}.$$

The answer checks.

State. $169\frac{3}{13}$ lb of whole milk and $30\frac{10}{13}$ lb of cream should be mixed.

17. Familiarize. Let $x =$ the amount of the 5.5% investment and $y =$ the amount of the 4% investment. Recall that the formula for simple interest is

$$\text{Interest} = \text{Principal} \cdot \text{Rate} \cdot \text{Time}.$$

Translate. We organize the information in a table.

	5.5% investment	4% investment	Total
Amount	x	y	$3200
Interest Rate	5.5%	4%	
Time	1 yr	1 yr	
Interest	$0.055x$	$0.04y$	$155

The "Amount" row of the table gives us one equation:

$$x + y = 3200$$

The last row of the table yields another equation:

$$0.055x + 0.04y = 155$$

After clearing decimals, we have the problem translated to a system of equations:

$$x + y = 3200 \qquad (1)$$
$$55x + 40y = 155,000 \quad (2)$$

Solve. We use the elimination method to solve the system of equations.

$$-40x - 40y = -128,000 \quad \text{Multiplying (1) by } -40$$
$$\underline{55x + 40y = 155,000}$$
$$15x \quad\quad = 27,000$$
$$x = 1800$$

Substitute 1800 for x in (1) and solve for y.

$$1800 + y = 3200$$
$$y = 1400$$

Check. The total amount invested was $1800 + $1400, or $3200. The interest earned was $0.055($1800) + 0.04($1400) = $99 + $56 = 155. The answer checks.

State. $1800 was invested at 5.5% and $1400 was invested at 4%.

19. Familiarize. Let $x =$ the number of $5 bills and $y =$ the number of $1 bills. The total value of the $5 bills is $5x$, and the total value of the $1 bills is $1 \cdot y$, or y.

Translate.

The total number of bills is 22.

$$x + y \qquad\qquad = 22$$

The total value of the bills is $50.

$$5x + y \qquad\qquad = 50$$

We have a system of equations:

$$x + y = 22, \quad (1)$$
$$5x + y = 50 \quad (2)$$

Solve. We use the elimination method.

$$-x - y = -22 \quad \text{Multiplying (1) by } -1$$
$$\underline{5x + y = 50}$$
$$4x \quad\quad = 28$$
$$x = 7$$

$$7 + y = 22 \quad \text{Substituting 7 for } x \text{ in (1)}$$
$$y = 15$$

Check. Total number of bills: $7 + 15 = 22$

Total value of bills: $\$5 \cdot 7 + \$1 \cdot 15 = \$35 + \$15 = \$50$.

The numbers check.

State. There are 7 $5 bills and 15 $1 bills.

21. Familiarize. We first make a drawing.

Slow train d miles	75 mph	$(t + 2)$ hr
Fast train d miles	125 mph	t hr

From the drawing we see that the distances are the same. Now complete the chart.

$$d = r \cdot t$$

	Distance	Rate	Time	
Slow train	d	75	$t+2$	$\rightarrow d = 75(t+2)$
Fast train	d	125	t	$\rightarrow d = 125t$

Translate. Using $d = rt$ in each row of the table, we get a system of equations:

$$d = 75(t+2),$$
$$d = 125t$$

Solve. We solve the system of equations.

$$125t = 75(t+2) \quad \text{Using substitution}$$
$$125t = 75t + 150$$
$$50t = 150$$
$$t = 3$$

Then $d = 125t = 125 \cdot 3 = 375$

Check. At 125 mph, in 3 hr the fast train will travel $125 \cdot 3 = 375$ mi. At 75 mph, in $3 + 2$, or 5 hr the slow train will travel $75 \cdot 5 = 375$ mi. The numbers check.

State. The trains will meet 375 mi from the station.

23. Familiarize. We first make a drawing. Let $d = $ the distance and $r = $ the speed of the canoe in still water. Then when the canoe travels downstream its speed is $r + 6$, and its speed upstream is $r - 6$. From the drawing we see that the distances are the same.

Downstream, 6 km/h current

d km, $r + 6$, 4 hr

Upstream, 6 km/h current

d km, $r - 6$, 10 hr

Organize the information in a table.

	Distance	Rate	Time
With current	d	$r+6$	4
Against current	d	$r-6$	10

Translate. Using $d = rt$ in each row of the table, we get a system of equations:

$$d = 4(r+6), \qquad d = 4r + 24,$$
$$\text{or}$$
$$d = 10(r-6) \qquad d = 10r - 60$$

Solve. Solve the system of equations.

$$4r + 24 = 10r - 60 \quad \text{Using substitution}$$
$$24 = 6r - 60$$
$$84 = 6r$$
$$14 = r$$

Check. When $r = 14$, then $r + 6 = 14 + 6 = 20$, and the distance traveled in 4 hr is $4 \cdot 20 = 80$ km. Also, $r - 6 = 14 - 6 = 8$, and the distance traveled in 10 hr is $8 \cdot 10 = 80$ km. The answer checks.

State. The speed of the canoe in still water is 14 km/h.

25. Familiarize. Let $s = $ the speed of the plane in still air and $w = $ the wind speed. Then the speed of the plane against the headwind is $s - w$, and the speed with the tailwind is $s + w$. Note that 48 min $= 48$ min $\times \dfrac{1 \text{ hr}}{60 \text{ min}} = 0.8$ hr.

Translate. We use the formula $d = rt$ to write two equations.

Against headwind: $270 = (s - w)3$, or $270 = 3s - 3w$

With tailwind: $270 = (s + w)1.8$, or $270 = 1.8s + 1.8w$

After clearing decimals, we have the following system of equations.

$$3s - 3w = 270, \quad (1)$$
$$18s + 18w = 2700 \quad (2)$$

Solve. Multiply Equation (1) by 6 and then add.

$$18s - 18w = 1620$$
$$\underline{18x + 18w = 2700}$$
$$36s \qquad = 4320$$
$$s = \quad 120$$

Now substitute 120 for s in Equation (1) and solve for w.

$$3(120) - 3w = 270$$
$$360 - 3w = 270$$
$$-3w = -90$$
$$w = 30$$

Check. The plane's speed against the headwind is $120 - 30$, or 90, mph. Flying at 90 mph for 3 hr, the plane would travel $90 \cdot 3$, or 270 mi. The plane's speed with the tailwind is $120 + 30$, or 150, mph. Flying at 150 mph for 1.8 hr, the plane would travel $150 \cdot 1.8$, or 270 mi. The answer checks.

State. The wind speed is 30 mph, and the speed of the plane in still air is 120 mph.

27. Familiarize. We first make a drawing. Let $d = $ the distance traveled at 420 km/h and $t = $ the time traveled. Then $1000 - d = $ the distance traveled at 330 km/h.

d km, 420 km/h, t hr $1000 - d$ km, 330 km/h, t hr

$\longmapsto$ 1000 km $\longmapsto$

We list the information in a table.

$$d = r \cdot t$$

	Distance	Rate	Time	
Faster airplane	d	420	t	$\rightarrow d = 420t$
Slower airplane	$1000 - d$	330	t	$\rightarrow 1000 - d = 330t$

Translate. Using $d = rt$ in each row of the table, we get a system of equations:

$$d = 420t, \quad (1)$$
$$1000 - d = 330t \quad (2)$$

Solve. We use substitution.

$$1000 - 420t = 330t \quad \text{Substituting } 420t \text{ for } d \text{ in (2)}$$
$$1000 = 750t$$
$$\frac{4}{3} = t$$

Check. If $t = \frac{4}{3}$, then $420 \cdot \frac{4}{3} = 560$, the distance traveled by the faster airplane. Also, $330 \cdot \frac{4}{3} = 440$, the distance traveled by the slower plane. The sum of the distances is $560 + 440$, or 1000 km. The values check.

State. The airplanes will meet after $\frac{4}{3}$ hr, or $1\frac{1}{3}$ hr.

29. Familiarize. We make a drawing. Note that the plane's speed traveling toward London is $360 + 50$, or 410 mph, and the speed traveling toward New York City is $360 - 50$, or 310 mph. Also, when the plane is d mi from New York City, it is $3458 - d$ mi from London.

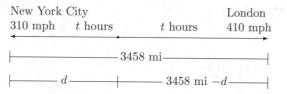

Organize the information in a table.

	Distance	Rate	Time
Toward NYC	d	310	t
Toward London	$3458 - d$	410	t

Translate. Using $d = rt$ in each row of the table, we get a system of equations:

$$d = 310t, \quad (1)$$
$$3458 - d = 410t \quad (2)$$

Solve. We solve the system of equations.

$$3458 - 310t = 410t \quad \text{Using substitution}$$
$$3458 = 720t$$
$$4.8028 \approx t$$

Substitute 4.8028 for t in (1).

$$d \approx 310(4.8028) \approx 1489$$

Check. If the plane is 1489 mi from New York City, it can return to New York City, flying at 310 mph, in $1489/310 \approx 4.8$ hr. If the plane is $3458 - 1489$, or 1969 mi from London, it can fly to London, traveling at 410 mph, in $1969/410 \approx 4.8$ hr. Since the times are the same, the answer checks.

State. The point of no return is about 1489 mi from New York City.

31. The numbers that are in both sets are 6, 8, and 10. We have

$$\{2, 4, 6, 8, 10\} \cap \{6, 7, 8, 9, 10\} = \{6, 8, 10\}.$$

33. $|3a| = |3| \cdot |a| = 3|a|$

35. $\left| \dfrac{-3}{y} \right| = \dfrac{|-3|}{|y|} = \dfrac{3}{|y|}$

37. Familiarize. Let $x =$ the amount of the original solution that remains after some of the original solution is drained and replaced with pure antifreeze. Let $y =$ the amount of the original solution that is drained and replaced with pure antifreeze.

Translate. We organize the information in a table. Keep in mind that the table contains information regarding the solution *after* some of the original solution is drained and replaced with pure antifreeze.

	Original Solution	Pure Anti-freeze	New Mixture
Amount of solution	x	y	16 L
Percent of antifreeze	30%	100%	50%
Amount of antifreeze in solution	$0.3x$	$1 \cdot y$, or y	$0.5(16)$, or 8

The "Amount of solution" row gives us one equation:
$$x + y = 16$$

The last row gives us a second equation:
$$0.3x + y = 8$$

After clearing the decimal we have the following system of equations:

$$x + y = 16, \quad (1)$$
$$3x + 10y = 80 \quad (2)$$

Solve. We use the elimination method.

$$\begin{array}{rl} -3x - 3y = -48 & \text{Multiplying (1) by } -3 \\ \underline{3x + 10y = 80} & \\ 7y = 32 & \\ y = \dfrac{32}{7}, \text{ or } 4\dfrac{4}{7} & \end{array}$$

Although the problem only asks for the amount of pure antifreeze added, we will also find x in order to check.

$$x + 4\frac{4}{7} = 16 \quad \text{Substituting } 4\frac{4}{7} \text{ for } y \text{ in (1)}$$
$$x = 11\frac{3}{7}$$

Check. Total amount of new mixture: $11\frac{3}{7} + 4\frac{4}{7} = 16$ L

Amount of antifreeze in new mixture:
$$0.3\left(11\frac{3}{7}\right) + 4\frac{4}{7} = \frac{3}{10} \cdot \frac{80}{7} + \frac{32}{7} = \frac{56}{7} = 8 \text{ L}$$
The numbers check.

State. Michelle should drain $4\frac{4}{7}$ L of the original solution and replace it with pure antifreeze.

39. *Familiarize*. Let x and y represent the number of city miles and highway miles that were driven, respectively. Then in city driving, $\dfrac{x}{18}$ gallons of gasoline are used; in highway driving, $\dfrac{y}{24}$ gallons are used.

Translate. We organize the information in a table.

Type of driving	City	Highway	Total
Number of miles	x	y	465
Gallons of gasoline used	$\dfrac{x}{18}$	$\dfrac{y}{24}$	23

The first row of the table gives us one equation:

$$x + y = 465$$

The second row gives us another equation:

$$\frac{x}{18} + \frac{y}{24} = 23$$

After clearing fractions, we have the following system of equations:

$$x + y = 465, \qquad (1)$$
$$24x + 18y = 9936 \quad (2)$$

Solve. We solve the system of equations using the elimination method.

$$-18x - 18y = -8370 \quad \text{Multiplying (1) by } -18$$
$$\underline{24x + 18y = \quad 9936}$$
$$6x \qquad\quad = \quad 1566$$
$$x = \quad 261$$

Now substitute 261 for x in Equation (1) and solve for y.

$$261 + y = 465$$
$$y = 204$$

Check. The total mileage is $261 + 204$, or 465. In 216 city miles, $261/18$, or 14.5 gal of gasoline are used; in 204 highway miles, $204/24$, or 8.5 gal are used. Then a total of $14.5 + 8.5$ or 23 gal of gasoline are used. The answer checks.

State. 261 miles were driven in the city, and 204 miles were driven on the highway.

Chapter 3 Mid-Chapter Review

1. False; see pages 253 and 260 in the text.

2. False; see page 245 in the text.

3. True; see page 245 in the text.

4. True; a vertical line $x = a$ and a horizontal line $y = b$ intersect at exactly one point, (a, b).

5. $x + 2y = 3,$

$\quad y = x - 6$

$\quad x + 2(x - 6) = 3$

$\quad x + 2x - 12 = 3$

$\quad 3x - 12 = 3$

$\quad\quad 3x = 15$

$\quad\quad\quad x = 5$

$y = 5 - 6$

$y = -1$

The solution is $(5, -1)$.

6. $3x - 2y = 5,$

$\quad 2x + 4y = 14$

$\quad 6x - 4y = 10$

$\quad \underline{2x + 4y = 14}$

$\quad 8x \qquad = 24$

$\quad\quad x = 3$

$\quad 2 \cdot 3 + 4y = 14$

$\quad\quad 6 + 4y = 14$

$\quad\quad\quad 4y = 8$

$\quad\quad\quad y = 2$

The solution is $(3, 2)$.

7. Graph the lines on the same set of axes.

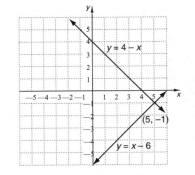

The solution appears to be $(5, -1)$.

Check:

$y = x - 6$		$y = 4 - x$	
$-1 \;?\; 5 - 6$		$-1 \;?\; 4 - 5$	
$\quad\; -1$	TRUE	$\quad\; -1$	TRUE

The solution is $(5, -1)$.

Since the system of equations has a solution, it is consistent. Since there is exactly one solution, the equations are independent.

8. Graph the lines on the same set of axes.

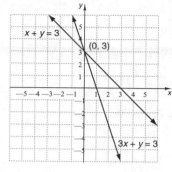

The solution appears to be $(0, 3)$.

Check:

$$\begin{array}{c|c} x + y = 3 \\ \hline 0 + 3 \ ? \ 3 \\ 3 \ \Big| \ \text{TRUE} \end{array} \qquad \begin{array}{c|c} 3x + y = 3 \\ \hline 3 \cdot 0 + 3 \ ? \ 3 \\ 0 + 3 \ \Big| \\ 3 \ \Big| \ \text{TRUE} \end{array}$$

The solution is $(0, 3)$.

Since the system of equations has a solution, it is consistent. Since there is exactly one solution, the equations are independent.

9. Graph the lines on the same set of axes.

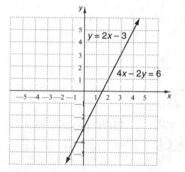

The graphs are the same. There is an infinite number of solutions. Since the system of equations has a solution, it is consistent. Since there are infinitely many solutions, the equations are dependent.

10. Graph the lines on the same set of axes.

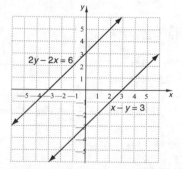

The lines are parallel. There is no solution. Since the system of equations has no solution, it is inconsistent. Since there is no solution, the equations are independent.

11. $x = y + 2,$ (1)

    $2x - 3y = -2$ (2)

Substitute $y + 2$ for x in the Equation (2) and solve for y.

$$2(y + 2) - 3y = -2$$
$$2y + 4 - 3y = -2$$
$$-y + 4 = -2$$
$$-y = -6$$
$$y = 6$$

Substitute 6 for y in Equation (1) and find x.

$$x = 6 + 2 = 8$$

The solution is $(8, 6)$.

12. $y = x - 5,$ (1)

    $x - 2y = 8$ (2)

Substitute $x - 5$ for y in the Equation (2) and solve for x.

$$x - 2(x - 5) = 8$$
$$x - 2x + 10 = 8$$
$$-x + 10 = 8$$
$$-x = -2$$
$$x = 2$$

Substitute 2 for x in Equation (1) and find y.

$$y = 2 - 5 = -3$$

The solution is $(2, -3)$.

13. $4x + 3y = 3,$ (1)

    $y = x + 8$ (2)

Substitute $x + 8$ for y in the Equation (1) and solve for x.

$$4x + 3(x + 8) = 3$$
$$4x + 3x + 24 = 3$$
$$7x + 24 = 3$$
$$7x = -21$$
$$x = -3$$

Substitute -3 for x in Equation (2) and find y.

$$y = -3 + 8 = 5$$

The solution is $(-3, 5)$.

14. $3x - 2y = 1,$ (1)

    $x = y + 1$ (2)

Substitute $y + 1$ for x in the Equation (1) and solve for y.

$$3(y + 1) - 2y = 1$$
$$3y + 3 - 2y = 1$$
$$y + 3 = 1$$
$$y = -2$$

Substitute -2 for y in Equation (2) and find x.

$$x = -2 + 1 = -1$$

The solution is $(-1, -2)$.

15. $2x + y = 2,$ (1)

$\quad\dfrac{x - y = 4}{3x \quad\ = 6}$ (2)

$3x \quad\ = 6$ Adding

$\qquad\ x = 2$

Substitute 2 for x in Equation (1) and solve for y.

$2 \cdot 2 + y = 2$

$4 + y = 2$

$y = -2$

The solution is $(2, -2)$.

16. $x - 2y = 13,$ (1)

$\quad\dfrac{x + 2y = -3}{2x \qquad = 10}$ (2)

$2x \qquad = 10$ Adding

$\qquad x = 5$

Substitute 5 for x in Equation (2) and solve for y.

$5 + 2y = -3$

$2y = -8$

$y = -4$

The solution is $(5, -4)$.

17. $3x - 4y = 5,$ (1)

$5x - 2y = -1$ (2)

First multiply by -2 on both sides of Equation (2) and then add.

$3x - 4y = 5$

$\dfrac{-10x + 4y = 2}{-7x \qquad = 7}$

$-7x \qquad = 7$

$x = -1$

Substitute -1 for x in Equation (1) and solve for y.

$3(-1) - 4y = 5$

$-3 - 4y = 5$

$-4y = 8$

$y = -2$

The solution is $(-1, -2)$.

18. $3x + 2y = 11,$ (1)

$2x + 3y = 9$ (2)

We use the multiplication principle on both equations and then add.

$9x + 6y = 33$ Multiplying (1) by 3

$\dfrac{-4x - 6y = -18}{5x \qquad = 15}$ Multiplying (2) by -2

$5x \qquad = 15$

$x = 3$

Substitute 3 for x in Equation (1) and solve for y.

$3 \cdot 3 + 2y = 11$

$9 + 2y = 11$

$2y = 2$

$y = 1$

The solution is $(3, 1)$.

19. $x - 2y = 5,$ (1)

$3x - 6y = 10$ (2)

First we multiply by -3 on both sides of Equation (1) and then add.

$-3x + 6y = -15$

$\dfrac{3x - 6y = 10}{0 = -5}$

$0 = -5$

We get a false equation. There is no solution.

20. $4x - 6y = 2,$ (1)

$-2x + 3y = -1$ (2)

First we multiply by 2 on both sides of Equation (2) and then add.

$4x - 6y = 2$

$\dfrac{-4x + 6y = -2}{0 = 0}$

$0 = 0$

We obtain an equation that is true for all values of x and y. Thus, if an ordered pair is a solution of one of the original equations, it is a solution of the other equation also. The system of equations has an infinite number of solutions.

21. $\dfrac{1}{2}x + \dfrac{1}{3}y = 1,$ (1)

$\dfrac{1}{5}x - \dfrac{3}{4}y = 11$ (2)

We multiply both sides of Equation 1 by 6 and both sides of Equation (2) by 20 to clear fractions.

$3x + 2y = 6$ (3)

$4x - 15y = 220$ (4)

We use the multiplication principle on both equations again and then add.

$45x + 30y = 90$ Multiplying (3) by 15

$\dfrac{8x - 30y = 440}{53x \qquad = 530}$ Multiplying (4) by 2

$53x \qquad = 530$

$x = 10$

Substitute 10 for x in Equation (3) and solve for y.

$3 \cdot 10 + 2y = 6$

$30 + 2y = 6$

$2y = -24$

$y = -12$

The solution is $(10, -12)$.

22. $0.2x + 0.3y = 0.6,$ (1)

$0.1x - 0.2y = -2.5$ (2)

We could begin by clearing the decimals. Instead, we will multiply by -2 on both sides of Equation (2) and then add.

$0.2x + 0.3y = 0.6$

$\dfrac{-0.2x + 0.4y = 5}{0.7y = 5.6}$

$0.7y = 5.6$

$y = 8$

Substitute 8 for y in Equation (2) and solve for x.

$$0.1x - 0.2(8) = -2.5$$
$$0.1x - 1.6 = -2.5$$
$$0.1x = -0.9$$
$$x = -9$$

The solution is $(-9, 8)$.

23. Familiarize. Let $l =$ the length of the garden, in feet, and let $w =$ the width.

Translate.

Perimeter is 44 ft.

$$2l + 2w = 44$$

Width is length less 2 ft.

$$w = l - 2$$

We have a system of equations.

$$2l + 2w = 44, \quad (1)$$
$$w = l - 2 \quad (2)$$

Solve. We substitute $l - 2$ for w in Equation (1) and solve for l.

$$2l + 2(l - 2) = 44$$
$$2l + 2l - 4 = 44$$
$$4l - 4 = 44$$
$$4l = 48$$
$$l = 12$$

Now substitute 12 for l in Equation (2) and find w.

$$w = 12 - 2 = 10$$

Check. The perimeter is $2 \cdot 12$ ft $+ 2 \cdot 10$ ft $= 24$ ft $+ 20$ ft $= 44$ ft. Also, the width, 10 ft, is 2 ft less than the length, 12 ft. The answer checks.

State. The length of the garden is 12 ft, and the width is 10 ft.

24. Familiarize. Let x and y represent the amounts invested at 2% and at 3%, respectively. Then the 2% investment earns $0.02x$, and the 3% investment earns $0.03y$.

Translate.

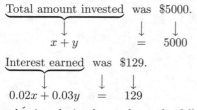

Total amount invested was $5000.

$$x + y = 5000$$

Interest earned was $129.

$$0.02x + 0.03y = 129$$

After clearing decimals, we have the following system of equations.

$$x + y = 5000, \quad (1)$$
$$2x + 3y = 12{,}900 \quad (2)$$

Solve. Multiply both sides of Equation (1) by -2 and then add.

$$-2x - 2y = -10{,}000$$
$$\underline{2x + 3y = 12{,}900}$$
$$y = 2900$$

Substitute 2900 for y in Equation (1) and solve for x.

$$x + 2900 = 5000$$
$$x = 2100$$

Check.

Total investment: $\$2100 + \$2900 = \$5000$

Amount earned at 2%: $0.02(\$2100) = \$42$

Amount earned at 3%: $0.03(\$2900) = \$87$

Total earnings: $\$42 + \$87 = \$129$

The answer checks.

State. \$2100 was invested at 2%, and \$2900 was invested at 3%.

25. Familiarize. Let $x =$ the number of liters of 20% solution and $y =$ the number of liters of 50% solution to be used. The mixture contains 30% (84 L), or 0.3(84 L), or 25.2 L of acid.

Translate. We organize the information in a table.

	20% solution	50% solution	Mixture
Number of liters	x	y	84
Percent of acid	20%	50%	30%
Amount of acid	$0.2x$	$0.5y$	25.2 L

We get one equation from the "Number of liters" row of the table.

$$x + y = 84$$

The last row of the table yields a second equation.

$$0.2x + 0.5y = 25.2$$

After clearing decimals, we have the following system of equations.

$$x + y = 84, \quad (1)$$
$$2x + 5y = 252 \quad (2)$$

Solve. Multiply both sides of Equation (1) by -2 and then add.

$$-2x - 2y = -168$$
$$\underline{2x + 5y = 252}$$
$$3y = 84$$
$$y = 28$$

Substitute 28 for y in Equation (1) and solve for x.

$$x + 28 = 84$$
$$x = 56$$

Check. The total amount of the mixture is 56 L $+$ 28 L, or 84 L. The amount of acid in the mixture is $0.2(56$ L$) + 0.5(28$ L$) = 11.2$ L $+ 14$ L $= 25.2$ L. The answer checks.

State. 56 L of the 20% solution and 28 L of the 50% solution should be used.

26. Familiarize. Let $d =$ the distance traveled and $r =$ the speed of the boat in still water, in mph. Then when the boat travels downstream its speed is $r + 6$, and its speed upstream is $r - 6$. The distances are the same.

We organize the information in a table.

	Distance	Rate	Time
Downstream	d	$r+6$	5
Upstream	d	$r-6$	8

Translate. Using $d = rt$ in each row of the table, we get a system of equations.

$$d = (r+6)5, \qquad d = 5r + 30, \quad (1)$$
$$\text{or}$$
$$d = (r-6)8 \qquad d = 8r - 48 \quad (2)$$

Solve. Substitute $8r - 48$ for d in Equation (1) and solve for r.

$$8r - 48 = 5r + 30$$
$$3r - 48 = 30$$
$$3r = 78$$
$$r = 26$$

Check. When $r = 26$, then $r + 6 = 26 + 6 = 32$, and the distance traveled in 5 hr is $32 \cdot 5$, or 160 mi. Also, $r - 6 = 26 - 6 = 20$, and the distance traveled in 8 hr is $20 \cdot 8$, or 160 mi. Since the distances are the same, the answer checks.

State. The speed of the boat in still water is 26 mph.

27. Graphically: 1.   Graph $y = \dfrac{3}{4}x + 2$ and $y = \dfrac{2}{5}x - 5$ and find the point of intersection. The first coordinate of this point is the solution of the original equation.

2. Rewrite the equation as $\dfrac{7}{20}x + 7 = 0$. Then graph $y = \dfrac{7}{20}x + 7$ and find the x-intercept. The first coordinate of this point is the solution of the original equation.

Algebraically: 1. Use the addition and multiplication principles for equations.

2. Multiply by 20 to clear the fractions and then use the addition and multiplication principles for equations.

28. a) Answers may vary.
$$x + y = 1,$$
$$x - y = 7$$

b) Answers may vary
$$x + 2y = 5,$$
$$3x + 6y = 10$$

c) Answers may vary.
$$x - 2y = 3,$$
$$3x - 6y = 9$$

29. Answers may vary. Form a linear expression in two variables and set it equal to two different constants. See Exercises 10 and 19 in this review for examples.

30. Answers may vary. Let any linear equation be one equation in the system. Multiply by a constant on both sides of that equation to get the second equation in the system. See Exercises 9 and 20 in this review for examples.

Exercise Set 3.5

RC1. (b)

RC3. (a)

1.
$$x + y + z = 2, \quad (1)$$
$$2x - y + 5z = -5, \quad (2)$$
$$-x + 2y + 2z = 1 \quad (3)$$

Add Equations (1) and (2) to eliminate y:

$$x + y + z = 2 \quad (1)$$
$$\underline{2x - y + 5z = -5} \quad (2)$$
$$3x \qquad + 6z = -3 \quad (4) \quad \text{Adding}$$

Use a different pair of equations and eliminate y:

$$4x - 2y + 10z = -10 \quad \text{Multiplying (2) by 2}$$
$$\underline{-x + 2y + 2z = 1} \quad (3)$$
$$3x \qquad + 12z = -9 \quad (5) \quad \text{Adding}$$

Now solve the system of Equations (4) and (5).

$$3x + 6z = -3 \quad (4)$$
$$3x + 12z = -9 \quad (5)$$

$$-3x - 6z = 3 \quad \text{Multiplying (4) by } -1$$
$$\underline{3x + 12z = -9} \quad (5)$$
$$6z = -6 \quad \text{Adding}$$
$$z = -1$$

$$3x + 6(-1) = -3 \quad \text{Substituting } -1 \text{ for } z \text{ in (4)}$$
$$3x - 6 = -3$$
$$3x = 3$$
$$x = 1$$

$$1 + y + (-1) = 2 \quad \text{Substituting 1 for } x \text{ and } -1$$
$$\text{for } z \text{ in (1)}$$
$$y = 2 \quad \text{Simplifying}$$

We obtain $(1, 2, -1)$. This checks, so it is the solution.

3.
$$2x - y + z = 5, \quad (1)$$
$$6x + 3y - 2z = 10, \quad (2)$$
$$x - 2y + 3z = 5 \quad (3)$$

We start by eliminating z from two different pairs of equations.

$$4x - 2y + 2z = 10 \quad \text{Multiplying (1) by 2}$$
$$\underline{6x + 3y - 2z = 10} \quad (2)$$
$$10x + y = 20 \quad (4) \quad \text{Adding}$$

$$-6x + 3y - 3z = -15 \quad \text{Multiplying (1) by } -3$$
$$\underline{x - 2y + 3z = 5} \quad (3)$$
$$-5x + y = -10 \quad (5) \quad \text{Adding}$$

Now solve the system of Equations (4) and (5).

$$10x + y = 20 \quad (4)$$
$$\underline{5x - y = 10} \quad \text{Multiplying (5) by } -1$$
$$15x = 30 \quad \text{Adding}$$
$$x = 2$$

$$10 \cdot 2 + y = 20 \quad \text{Substituting 2 for } x \text{ in (4)}$$
$$20 + y = 20$$
$$y = 0$$

$$2 \cdot 2 - 0 + z = 5 \quad \text{Substituting 2 for } x \text{ and 0}$$
$$ \quad \text{for } y \text{ in (1)}$$
$$4 + z = 5$$
$$z = 1$$

We obtain $(2, 0, 1)$. This checks, so it is the solution.

5. $2x - 3y + z = 5, \quad (1)$
$x + 3y + 8z = 22, \quad (2)$
$3x - y + 2z = 12 \quad (3)$

We start by eliminating y from two different pairs of equations.

$$2x - 3y + z = 5 \quad (1)$$
$$\underline{x + 3y + 8z = 22} \quad (2)$$
$$3x + 9z = 27 \quad (4) \quad \text{Adding}$$

$$x + 3y + 8z = 22 \quad (2)$$
$$\underline{9x - 3y + 6z = 36} \quad \text{Multiplying (3) by 3}$$
$$10x + 14z = 58 \quad (5) \quad \text{Adding}$$

Solve the system of Equations (4) and (5).

$$3x + 9z = 27 \quad (4)$$
$$10x + 14z = 58 \quad (5)$$

$$30x + 90z = 270 \quad \text{Multiplying (4) by 10}$$
$$\underline{-30x - 42z = -174} \quad \text{Multiplying (5) by } -3$$
$$48z = 96 \quad \text{Adding}$$
$$z = 2$$

$$3x + 9 \cdot 2 = 27 \quad \text{Substituting 2 for } z \text{ in (4)}$$
$$3x + 18 = 27$$
$$3x = 9$$
$$x = 3$$

$$2 \cdot 3 - 3y + 2 = 5 \quad \text{Substituting 3 for } x \text{ and 2}$$
$$ \quad \text{for } z \text{ in (1)}$$
$$-3y + 8 = 5$$
$$-3y = -3$$
$$y = 1$$

We obtain $(3, 1, 2)$. This checks, so it is the solution.

7. $3a - 2b + 7c = 13, \quad (1)$
$a + 8b - 6c = -47, \quad (2)$
$7a - 9b - 9c = -3 \quad (3)$

We start by eliminating a from two different pairs of equations.

$$3a - 2b + 7c = 13 \quad (1)$$
$$\underline{-3a - 24b + 18c = 141} \quad \text{Multiplying (2) by } -3$$
$$-26b + 25c = 154 \quad (4) \quad \text{Adding}$$

$$-7a - 56b + 42c = 329 \quad \text{Multiplying (2) by } -7$$
$$\underline{7a - 9b - 9c = -3} \quad (3)$$
$$-65b + 33c = 326 \quad (5) \quad \text{Adding}$$

Now solve the system of Equations (4) and (5).

$$-26b + 25c = 154 \quad (4)$$
$$-65b + 33c = 326 \quad (5)$$

$$-130b + 125c = 770 \quad \text{Multiplying (4) by 5}$$
$$\underline{130b - 66c = -652} \quad \text{Multiplying (5) by } -2$$
$$59c = 118$$
$$c = 2$$

$$-26b + 25 \cdot 2 = 154 \quad \text{Substituting 2 for } c \text{ in (4)}$$
$$-26b + 50 = 154$$
$$-26b = 104$$
$$b - -4$$

$$a + 8(-4) - 6(2) = -47 \quad \text{Substituting } -4 \text{ for } b \text{ and}$$
$$ \quad 2 \text{ for } c \text{ in (2)}$$
$$a - 32 - 12 = -47$$
$$a - 44 = -47$$
$$a = -3$$

We obtain $(-3, -4, 2)$. This checks, so it is the solution.

9. $2x + 3y + z = 17, \quad (1)$
$x - 3y + 2z = -8, \quad (2)$
$5x - 2y + 3z = 5 \quad (3)$

We start by eliminating y from two different pairs of equations.

$$2x + 3y + z = 17 \quad (1)$$
$$\underline{x - 3y + 2z = -8} \quad (2)$$
$$3x + 3z = 9 \quad (4) \quad \text{Adding}$$

$$4x + 6y + 2z = 34 \quad \text{Multiplying (1) by 2}$$
$$\underline{15x - 6y + 9z = 15} \quad \text{Multiplying (3) by 3}$$
$$19x + 11z = 49 \quad (5) \quad \text{Adding}$$

Now solve the system of Equations (4) and (5).

$$3x + 3z = 9 \quad (4)$$
$$19x + 11z = 49 \quad (5)$$

$$33x + 33z = 99 \quad \text{Multiplying (4) by 11}$$
$$\underline{-57x - 33z = -147} \quad \text{Multiplying (5) by } -3$$
$$-24x = -48$$
$$x = 2$$

$3 \cdot 2 + 3z = 9$ Substituting 2 for x in (4)

$6 + 3z = 9$

$3z = 3$

$z = 1$

$2 \cdot 2 + 3y + 1 = 17$ Substituting 2 for x and 1 for z in (1)

$3y + 5 = 17$

$3y = 12$

$y = 4$

We obtain $(2, 4, 1)$. This checks, so it is the solution.

11. $\quad 2x + y + z = -2, \quad (1)$

$\quad 2x - y + 3z = 6, \quad (2)$

$\quad 3x - 5y + 4z = 7 \quad (3)$

We start by eliminating y from two different pairs of equations.

$$2x + y + z = -2 \quad (1)$$
$$\underline{2x - y + 3z = 6} \quad (2)$$
$$4x + 4z = 4 \quad (4) \quad \text{Adding}$$

$$10x + 5y + 5z = -10 \quad \text{Multiplying (1) by 5}$$
$$\underline{3x - 5y + 4z = 7} \quad (3)$$
$$13x + 9z = -3 \quad (5) \quad \text{Adding}$$

Now solve the system of Equations (4) and (5).

$\quad 4x + 4z = 4 \quad (4)$

$\quad 13x + 9z = -3 \quad (5)$

$$36x + 36z = 36 \quad \text{Multiplying (4) by 9}$$
$$\underline{-52x - 36z = 12} \quad \text{Multiplying (5) by } -4$$
$$-16x = 48 \quad \text{Adding}$$
$$x = -3$$

$4(-3) + 4z = 4$ Substituting -3 for x in (4)

$-12 + 4z = 4$

$4z = 16$

$z = 4$

$2(-3) + y + 4 = -2$ Substituting -3 for x and 4 for z in (1)

$y - 2 = -2$

$y = 0$

We obtain $(-3, 0, 4)$. This checks, so it is the solution.

13. $\quad x - y + z = 4, \quad (1)$

$\quad 5x + 2y - 3z = 2, \quad (2)$

$\quad 3x - 7y + 4z = 8 \quad (3)$

We start by eliminating z from two different pairs of equations.

$$3x - 3y + 3z = 12 \quad \text{Multiplying (1) by 3}$$
$$\underline{5x + 2y - 3z = 2} \quad (2)$$
$$8x - y = 14 \quad (4) \quad \text{Adding}$$

$-4x + 4y - 4z = -16$ Multiplying (1) by -4

$\underline{3x - 7y + 4z = 8} \quad (3)$

$-x - 3y = -8 \quad (5) \quad$ Adding

Now solve the system of Equations (4) and (5).

$\quad 8x - y = 14 \quad (4)$

$\quad -x - 3y = -8 \quad (5)$

$$8x - y = 14 \quad (4)$$
$$\underline{-8x - 24y = -64} \quad \text{Multiplying (5) by 8}$$
$$-25y = -50$$
$$y = 2$$

$8x - 2 = 14$ Substituting 2 for y in (4)

$8x = 16$

$x = 2$

$2 - 2 + z = 4$ Substituting 2 for x and 2 for y in (1)

$z = 4$

We obtain $(2, 2, 4)$. This checks, so it is the solution.

15. $\quad 4x - y - z = 4, \quad (1)$

$\quad 2x + y + z = -1, \quad (2)$

$\quad 6x - 3y - 2z = 3 \quad (3)$

We start by eliminating y from two different pairs of equations.

$$4x - y - z = 4 \quad (1)$$
$$\underline{2x + y + z = -1} \quad (2)$$
$$6x = 3 \quad (4) \quad \text{Adding}$$

At this point we can either continue by eliminating y from a second pair of equations or we can solve (4) for x and substitute that value in a different pair of the original equations to obtain a system of two equations in two variables. We take the second option.

$6x = 3 \quad (4)$

$x = \dfrac{1}{2}$

Substitute $\dfrac{1}{2}$ for x in (1):

$4\left(\dfrac{1}{2}\right) - y - z = 4$

$2 - y - z = 4$

$-y - z = 2 \quad (5)$

Substitute $\dfrac{1}{2}$ for x in (3):

$6\left(\dfrac{1}{2}\right) - 3y - 2z = 3$

$3 - 3y - 2z = 3$

$-3y - 2z = 0 \quad (6)$

Solve the system of Equations (5) and (6).

$$2y + 2z = -4 \quad \text{Multiplying (5) by } -2$$
$$\underline{-3y - 2z = 0} \quad (6)$$
$$-y = -4$$
$$y = 4$$

$$-4 - z = 2 \qquad \text{Substituting 4 for } y \text{ in (5)}$$
$$-z = 6$$
$$z = -6$$

We obtain $\left(\dfrac{1}{2}, 4, -6\right)$. This checks, so it is the solution.

17. $\quad a - 2b - 5c = -3, \quad (1)$
$\qquad 3a + b - 2c = -1, \quad (2)$
$\qquad 2a + 3b + c = 4 \qquad (3)$

We start by eliminating b from two different pairs of equations.

$$a - 2b - 5c = -3 \quad (1)$$
$$\underline{6a + 2b - 4c = -2} \quad \text{Multiplying (2) by 2}$$
$$7a \qquad - 9c = -5 \quad (4) \text{ Adding}$$

$$-9a - 3b + 6c = 3 \quad \text{Multiplying (2) by } -3$$
$$\underline{2a + 3b + c = 4}$$
$$-7a \qquad + 7c = 7 \quad (5) \text{ Adding}$$

Now solve the system of Equations (4) and (5).

$$7a - 9c = -5$$
$$\underline{-7a + 7c = 7}$$
$$-2c = 2$$
$$c = -1$$

$$7a - 9(-1) = -5 \qquad \text{Substituting } -1 \text{ for } c \text{ in (4)}$$
$$7a + 9 = -5$$
$$7a = -14$$
$$a = -2$$

$$3(-2) + b - 2(-1) = -1 \qquad \text{Substituting } -2 \text{ for } a \text{ and}$$
$$\qquad\qquad\qquad\qquad -1 \text{ for } c \text{ in (2)}$$
$$-6 + b + 2 = -1$$
$$b - 4 = -1$$
$$b = 3$$

We obtain $(-2, 3, -1)$. This checks, so it is the solution.

19. $\quad 2r + 3s + 12t = 4, \quad (1)$
$\qquad 4r - 6s + 6t = 1, \quad (2)$
$\qquad r + s + t = 1 \quad (3)$

We start by eliminating s from two different pairs of equations.

$$4r + 6s + 24t = 8 \quad \text{Multiplying (1) by 2}$$
$$\underline{4r - 6s + 6t = 1} \quad (2)$$
$$8r \qquad + 30t = 9 \quad (4) \quad \text{Adding}$$

$$4r - 6s + 6t = 1 \quad (2)$$
$$\underline{6r + 6s + 6t = 6} \quad \text{Multiplying (3) by 6}$$
$$10r \qquad + 12t = 7 \quad (5) \quad \text{Adding}$$

Solve the system of Equations (4) and (5).

$$40r + 150t = 45 \quad \text{Multiplying (4) by 5}$$
$$\underline{-40r - 48t = -28} \quad \text{Multiplying (5) by } -4$$
$$102t = 17$$
$$t = \frac{17}{102}$$
$$t = \frac{1}{6}$$

$$8r + 30\left(\frac{1}{6}\right) = 9 \quad \text{Substituting } \frac{1}{6} \text{ for } t \text{ in (4)}$$
$$8r + 5 = 9$$
$$8r = 4$$
$$r = \frac{1}{2}$$

$$\frac{1}{2} + s + \frac{1}{6} = 1 \quad \text{Substituting } \frac{1}{2} \text{ for } r \text{ and}$$
$$\qquad\qquad\qquad \frac{1}{6} \text{ for } t \text{ in (3)}$$
$$s + \frac{2}{3} = 1$$
$$s = \frac{1}{3}$$

We obtain $\left(\dfrac{1}{2}, \dfrac{1}{3}, \dfrac{1}{6}\right)$. This checks, so it is the solution.

21. $\quad a + 2b + c = 1, \quad (1)$
$\qquad 7a + 3b - c = -2, \quad (2)$
$\qquad a + 5b + 3c = 2 \qquad (3)$

We start by eliminating c from two different pairs of equations.

$$a + 2b + c = 1 \quad (1)$$
$$\underline{7a + 3b - c = -2} \quad (2)$$
$$8a + 5b = -1 \quad (4)$$

$$21a + 9b - 3c = -6 \quad \text{Multiplying (2) by 3}$$
$$\underline{a + 5b + 3c = 2}$$
$$22a + 14b = -4 \quad (5)$$

Now solve the system of Equations (4) and (5).

$$112a + 70b = -14 \quad \text{Multiplying (4) by 14}$$
$$\underline{-110a - 70b = 20} \quad \text{Multiplying (5) by } -5$$
$$2a = 6$$
$$a = 3$$

$$8 \cdot 3 + 5b = -1 \quad \text{Substituting 3 for } a \text{ in (4)}$$
$$24 + 5b = -1$$
$$5b = -25$$
$$b = -5$$

$$3 + 2(-5) + c = 1 \quad \text{Substituting 3 for } a \text{ and}$$
$$\qquad\qquad\qquad -5 \text{ for } b \text{ in (1)}$$
$$3 - 10 + c = 1$$
$$-7 + c = 1$$
$$c = 8$$

We obtain $(3, -5, 8)$. This checks, so it is the solution.

23. $x + y + z = 57,$ (1)
             $-2x + y \quad\; = 3,$ (2)
             $x \quad\quad - z = 6$ (3)

We will use the substitution method. Solve Equations (2) and (3) for y and z, respectively. Then substitute in Equation (1) to solve for x.

$$-2x + y = 3 \qquad \text{Solving (2) for } y$$
$$y = 2x + 3$$
$$x - z = 6 \qquad \text{Solving (3) for } z$$
$$-z = -x + 6$$
$$z = x - 6$$
$$x + (2x + 3) + (x - 6) = 57 \quad \text{Substituting in (1)}$$
$$4x - 3 = 57$$
$$4x = 60$$
$$x = 15$$

To find y, substitute 15 for x in $y = 2x + 3$:

$$y = 2 \cdot 15 + 3 = 33$$

To find z, substitute 15 for x in $z = x - 6$:

$$z = 15 - 6 = 9$$

We obtain $(15, 33, 9)$. This checks, so it is the solution.

25. $r + s \quad\quad = 5,$ (1)
             $3s + 2t = -1,$ (2)
             $4r \quad\; + t = 14$ (3)

We will use the elimination method. Note that there is no t in Equation (1). We will use Equations (2) and (3) to obtain another equation with no t-term.

$$3s + 2t = -1 \quad (2)$$
$$\underline{-8r \quad\quad - 2t = -28} \quad \text{Multiplying (3) by } -2$$
$$-8r + 3s \quad\quad = -29 \quad (4) \quad \text{Adding}$$

Now solve the system of Equations (1) and (4).

$$r + s = 5 \quad (1)$$
$$-8r + 3s = -29 \quad (4)$$

$$8r + 8s = 40 \quad \text{Multiplying (1) by 8}$$
$$\underline{-8r + 3s = -29} \quad (4)$$
$$11s = 11 \quad \text{Adding}$$
$$s = 1$$

$$r + 1 = 5 \quad \text{Substituting 1 for } s \text{ in (1)}$$
$$r = 4$$

$$4 \cdot 4 + t = 14 \quad \text{Substituting 4 for } r \text{ in (3)}$$
$$16 + t = 14$$
$$t = -2$$

We obtain $(4, 1, -2)$. This checks, so it is the solution.

27. $x + y + z = 105,$ (1)
             $10y - z = 11,$ (2)
             $2x - 3y \quad\; = 7$ (3)

We will use the elimination method. Note that there is no z in Equation (3). We will use Equations (1) and (2) to obtain another equation with no z-term.

$$x + y + z = 105 \quad (1)$$
$$\underline{10y - z = 11} \quad (2)$$
$$x + 11y \quad\quad = 116 \quad (4)$$

Now solve the system of Equations (4) and (3).

$$-2x - 22y = -232 \quad \text{Multiplying (4) by } -2$$
$$\underline{2x - 3y = 7} \quad (3)$$
$$-25y = -225$$
$$y = 9$$

$$x + 11 \cdot 9 = 116 \quad \text{Substituting 9 for } y \text{ in (4)}$$
$$x + 99 = 116$$
$$x = 17$$

$$17 + 9 + z = 105 \quad \text{Substituting 14 for } x \text{ and}$$
$$\qquad\qquad\qquad\qquad 9 \text{ for } y \text{ in (1)}$$
$$26 + z = 105$$
$$z = 79$$

We obtain $(17, 9, 79)$. This checks, so it is the solution.

29. $Q = 4(a + b)$

$$\frac{Q}{4} = a + b \qquad \text{Dividing by 4}$$
$$\frac{Q}{4} - b = a, \text{ or} \qquad \text{Subtracting } b$$
$$\frac{Q - 4b}{4} = a$$

31. $F = \dfrac{1}{2}t(c - d)$

$$2F = t(c - d) \quad \text{Multiplying by 2}$$
$$2F = tc - td \quad \text{Removing parentheses}$$
$$2F + td = tc \qquad \text{Adding } td$$
$$\frac{2F + td}{t} = c, \text{ or} \qquad \text{Dividing by } t$$
$$\frac{2F}{t} + d = c$$

33. $Ax - By = c$

$$Ax = By + c \qquad \text{Adding } By$$
$$Ax - c = By \qquad\quad \text{Subtracting } c$$
$$\frac{Ax - c}{B} = y \qquad\quad \text{Dividing by } B$$

35. $y = 5 - 4x$, or $y = -4x + 5$

The equation is in slope-intercept form, $y = mx + b$. The slope is -4, and the y-intercept is $(0, 5)$.

37. $7x - 6.4y = 20$

$$-6.4y = -7x + 20$$
$$y = 1.09375x - 3.125 \quad \text{Dividing by } -6.4$$

The equation is now in slope-intercept form, $y = mx + b$. The slope is 1.09375, and the y-intercept is $(0, -3.125)$.

39.
$$w + x + y + z = 2, \quad (1)$$
$$w + 2x + 2y + 4z = 1, \quad (3)$$
$$w - x + y + z = 6, \quad (3)$$
$$w - 3x - y + z = 2 \quad (4)$$

Start by eliminating w from three different pairs of equations.

$$w + x + y + z = 2 \quad (1)$$
$$\underline{-w - 2x - 2y - 4z = -1} \quad \text{Multiplying (2) by } -1$$
$$-x - y - 3z = 1 \quad (5) \quad \text{Adding}$$

$$w + x + y + z = 2 \quad (1)$$
$$\underline{-w + x - y - z = -6} \quad \text{Multiplying (3) by } -1$$
$$2x = -4 \quad (6) \quad \text{Adding}$$

$$w + x + y + z = 2 \quad (1)$$
$$\underline{-w + 3x + y - z = -2} \quad \text{Multiplying (4) by } -1$$
$$4x + 2y = 0 \quad (7) \quad \text{Adding}$$

We can solve (6) for x:
$$2x = -4$$
$$x = -2$$

Substitute -2 for x in (7):
$$4(-2) + 2y = 0$$
$$-8 + 2y = 0$$
$$2y = 8$$
$$y = 4$$

Substitute -2 for x and 4 for y in (5):
$$-(-2) - 4 - 3z = 1$$
$$-2 - 3z = 1$$
$$-3z = 3$$
$$z = -1$$

Substitute -2 for x, 4 for y, and -1 for z in (1):
$$w - 2 + 4 - 1 = 2$$
$$w + 1 = 2$$
$$w = 1$$

We obtain $(1, -2, 4, -1)$. This checks, so it is the solution.

Exercise Set 3.6

RC1. (c)

RC3. (a)

1. Familiarize. Let x, y, and z represent the average reading, math, and writing scores, respectively.

Translate.

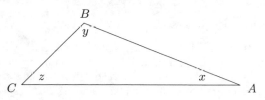

Sum of the scores was 1498.
$$x + y + z = 1498$$

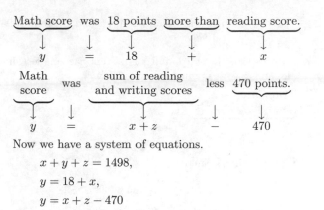

Math score was 18 points more than reading score.
$$y = 18 + x$$

Math score was sum of reading and writing scores less 470 points.
$$y = x + z - 470$$

Now we have a system of equations.
$$x + y + z = 1498,$$
$$y = 18 + x,$$
$$y = x + z - 470$$

Solve. Solving the system, we get $(496, 514, 488)$.

Check. The total score is $496 + 514 + 488$, or 1498. The math score, 514, is 18 more than the reading score, 496. The sum of the reading and writing scores is $496 + 488$, or 984, and 470 points less than this is $984 - 470$, or 514, the math score. The answer checks.

State. The average reading, math, and writing scores were 496, 514, and 488, respectively.

3. Familiarize. We first make a drawing.

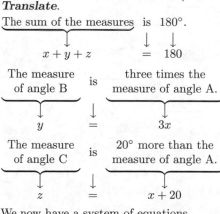

We let x, y, and z represent the measures of angles A, B, and C, respectively. The measures of the angles of a triangle add up to $180°$.

Translate.

The sum of the measures is $180°$.
$$x + y + z = 180$$

The measure of angle B is three times the measure of angle A.
$$y = 3x$$

The measure of angle C is $20°$ more than the measure of angle A.
$$z = x + 20$$

We now have a system of equations.
$$x + y + z = 180,$$
$$y = 3x,$$
$$z = x + 20$$

Solve. Solving the system we get $(32, 96, 52)$.

Check. The sum of the measures is $32° + 96° + 52°$, or $180°$. Three times the measure of angle A is $3 \cdot 32°$, or $96°$, the measure of angle B. $20°$ more than the measure of angle A is $32° + 20°$, or $52°$, the measure of angle C. The numbers check.

State. The measures of angles A, B, and C are $32°$, $96°$, and $52°$, respectively.

5. *Familiarize*. Let x, y, and z represent the smallest, middle, and largest numbers, respectively.

Translate.

$\underbrace{\text{Sum of the numbers}}$ $\underset{\downarrow}{\text{is}}$ $\underset{\downarrow}{55}$.

$\underset{\downarrow}{x+y+z}\underset{\downarrow}{=}\underset{\downarrow}{55}$

$\underset{\downarrow}{\text{Largest}}\underset{\downarrow}{\text{minus}}\underset{\downarrow}{\text{smallest}}\text{is}49$.

$\underset{\downarrow}{z}\underset{\downarrow}{-}\underset{\downarrow}{x}=49$

$\underset{\downarrow}{\text{Smallest}}\underset{\downarrow}{\text{plus}}\underset{\downarrow}{\text{middle}}\text{is}13$.

$\underset{\downarrow}{x}\underset{\downarrow}{+}\underset{\downarrow}{y}=13$

We now have a system of equations.

$$x + y + z = 55,$$
$$z - x = 49,$$
$$x + y = 13$$

Solve. Solving the system we get $(-7, 20, 42)$.

Check. $-7 + 20 + 42 = 55$; $42 - (-7) = 49$; and $-7 + 20 = 13$. The answer checks.

State. The numbers are -7, 20, and 42.

7. *Familiarize*. Let $x =$ the number of Sixteens, $y =$ the number of Originals, and $z =$ the number of Powers sold. Then the amounts collected from selling each size of smoothie are $\$3.90x$, $\$4.90y$, and $\$5.70z$, respectively. The number of ounces of each type of drink sold are $16x$, $22y$ and $30z$, respectively.

Translate.

$\underbrace{\text{Number of drinks sold}}$ $\underset{\downarrow}{\text{was}}$ $\underset{\downarrow}{34}$.

$\underset{\downarrow}{x+y+z}\underset{\downarrow}{=}\underset{\downarrow}{34}$

$\underbrace{\text{Amount collected}}$ $\underset{\downarrow}{\text{was}}$ $\underset{\downarrow}{\$163}$.

$\underset{\downarrow}{3.90x + 4.90y + 5.70z}\underset{\downarrow}{=}\underset{\downarrow}{163}$

$\underbrace{\text{Number of ounces sold}}$ $\underset{\downarrow}{\text{was}}$ $\underset{\downarrow}{752}$.

$\underset{\downarrow}{16x + 22y + 30z}\underset{\downarrow}{=}\underset{\downarrow}{752}$

We have a system of equations.

$$x + y + z = 34,$$
$$3.90x + 4.90y + 5.70z = 163,$$
$$16x + 22y + 30z = 752.$$

Solve. Solving the system, we get $(10, 16, 8)$.

Check. The number of drinks served was $10 + 16 + 8$, or 34. The amount collected was $\$3.90(10) + \$4.90(16) + \$5.70(8) = \$39 + \$78.40 + \$45.60 = \$163$. The number of ounces sold was $16(10) + 22(16) + 30(8) = 160 + 352 + 240 = 752$. The answer checks.

State. Elliot sold 10 Sixteens, 16 Originals and 8 Powers.

9. *Familiarize*. Let x, y, and z represent the number of milligrams of cholesterol in 1 egg, 1 cupcake, and 1 slice of pizza, respectively.

Translate.

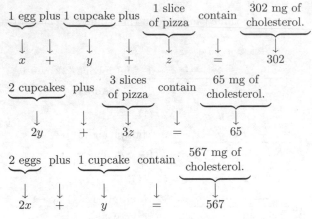

We have a system of equations.

$$x + y + z = 302,$$
$$2y + 3z = 65,$$
$$2x + y = 567$$

Solve. Solving the system, we get $(274, 19, 9)$.

Check. $274 \text{ mg} + 19 \text{ mg} + 9 \text{ mg} = 302 \text{ mg}$; $2 \cdot 19 \text{ mg} + 3 \cdot 9 \text{ mg} = 65 \text{ mg}$; $2 \cdot 274 \text{ mg} + 19 \text{ mg} = 567 \text{ mg}$. The answer checks

State. An egg contains 274 mg of cholesterol, a cupcake contains 19 mg, and a slice of pizza contains 9 mg.

11. *Familiarize*. Let $x =$ the cost of automatic transmission, $y =$ the cost of power door locks, and $z =$ the cost of air conditioning. The prices of the options are added to the basic price of $\$14,685$.

Translate.

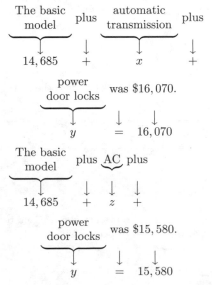

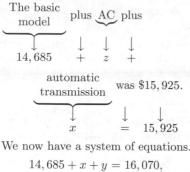

$$14,685 + x + y = 16,070,$$
$$14,685 + z + y = 15,580,$$
$$14,685 + z + x = 15,925$$

Solve. Solving the system we get $(865, 520, 375)$.

Check. The basic model with automatic transmission and power door locks costs $14,685 + $865 + 520, or $16,070$. The basic model with AC and power door locks costs $14,685 + $375 + 520, or $15,580$. The basic model with AC and automatic transmission costs $14,685 + $375 + 865, or $15,925$. The numbers check.

State. Automatic transmission costs $865, power door locks cost $520, and AC costs $375.

13. **Familiarize.** Let x, y, and z represent the average amounts spent for veterinary expenses per dog, cat, and bird, respectively.

Translate.

Sum of expenses was $290.

$$x + y + z = 290$$

Amount spent per dog was $110 more than amounts spent per cat and bird.

$$x = 110 + y + z$$

Amount spent per cat was 9 times amount spent per bird.

$$y = 9 \cdot z$$

Now we have a system of equations.

$$x + y + z = 290,$$
$$x = 110 + y + z,$$
$$y = 9z$$

Solve. Solving the system, we get $(200, 81, 9)$.

Check. The sum of the expenses is $200 + $81 + 9, or 290. The average amount spent per dog, $200, is $110 more than $81 + 9, or 90, the sum of the amounts spent per cat and bird. The average amount spent per cat, $81, is 9 times the average amount spent per bird, $9. The answer checks.

State. The average amounts spent for veterinary expenses per dog, cat, and bird were $200, $81, and $9, respectively.

15. **Familiarize.** Let $r =$ the number of servings of roast beef, $p =$ the number of baked potatoes, and $b =$ the number of servings of broccoli. Then r servings of roast beef contain $300r$ Calories, $20r$ g of protein, and no vitamin C. In p baked potatoes there are $100p$ Calories, $5p$ g of protein, and $20p$ mg of vitamin C. And b servings of broccoli contain $50b$ Calories, $5b$ g of protein, and $100b$ mg of vitamin C. The patient requires 800 Calories, 55 g of protein, and 220 mg of vitamin C.

Translate. Write equations for the total number of calories, the total amount of protein, and the total amount of vitamin C.

$$300r + 100p + 50b = 800 \quad \text{(Calories)}$$
$$20r + 5p + 5b = 55 \quad \text{(protein)}$$
$$20p + 100b = 220 \quad \text{(vitamin C)}$$

We now have a system of equations.

Solve. Solving the system we get $(2, 1, 2)$.

Check. Two servings of roast beef provide 600 Calories, 40 g of protein, and no vitamin C. One baked potato provides 100 Calories, 5 g of protein, and 20 mg of vitamin C. And 2 servings of broccoli provide 100 Calories, 10 g of protein, and 200 mg of vitamin C. Together, then, they provide 800 Calories, 55 g of protein, and 220 mg of vitamin C. The values check.

State. The dietician should prepare 2 servings of roast beef, 1 baked potato, and 2 servings of broccoli.

17. **Familiarize.** Let $x =$ the amount invested in the first fund, $y =$ the amount invested in the second fund, and $z =$ the amount invested in the third fund. Then the earnings from the investments were $0.02x$, $0.06y$, and $0.03z$.

Translate.

The total amount invested was $80,000.

$$x + y + z = 80,000$$

The total earnings were $2250.

$$0.02x + 0.06y + 0.03z = 2250$$

The earnings from the first fund were $150 more than the earnings from the third fund.

$$0.02x = 150 + 0.03z$$

Now we have a system of equations.

$$x + y + z = 80,000,$$
$$0.02x + 0.06y + 0.03z = 2250,$$
$$0.02x = 150 + 0.03z$$

Solve. Solving the system we get $(45,000, 10,000, 25,000)$.

Check. The total investment was $45,000 + $10,000 + $25,000$, or $80,000$. The total earnings were $0.02(\$45,000) + 0.06(10,000) + 0.03(25,000) = \$900 + \$600 + \$750 = \$2250$. The earnings from the first fund,

$900, were $150 more than the earnings from the third fund, $750.

State. $45,000 was invested in the first fund, $10,000 in the second fund, and $25,000 in the third fund.

19. **Familiarize**. Let x, y, and z represent the number of par-3, par-4, and par-5 holes, respectively. Then a par golfer shoots $3x$ on the par-3 holes, $4x$ on the par-4 holes, and $5x$ on the par-5 holes.

Translate.

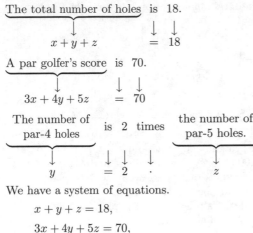

We have a system of equations.

$$x + y + z = 18,$$
$$3x + 4y + 5z = 70,$$
$$y = 2z$$

Solve. Solving the system we get $(6, 8, 4)$.

Check. The numbers add up to 18. A par golfer would shoot $3 \cdot 6 + 4 \cdot 8 + 5 \cdot 4$, or 70. The number of par-4 holes, 8, is twice the number of par-5 holes, 4. The numbers check.

State. There are 6 par-3 holes, 8 par-4 holes, and 4 par-5 holes.

21. **Familiarize**. It helps to organize the information in a table. We let x, y, and z represent the weekly productions of the individual machines.

Machines Working	A	B	C
Weekly Production	x	y	z

Machines Working	A & B	B & C	A, B, & C
Weekly Production	3400	4200	5700

Translate. From the table, we obtain three equations.

$$x + y + z = 5700 \quad \text{(All three machines working)}$$
$$x + y \quad\quad = 3400 \quad \text{(A and B working)}$$
$$\quad y + z = 4200 \quad \text{(B and C working)}$$

Solve. Solving the system we get $(1500, 1900, 2300)$.

Check. The sum of the weekly productions of machines A, B & C is $1500 + 1900 + 2300$, or 5700. The sum of the weekly productions of machines A and B is $1500 + 1900$, or 3400. The sum of the weekly productions of machines B and C is $1900 + 2300$, or 4200. The numbers check.

State. In a week Machine A can polish 1500 lenses, Machine B can polish 1900 lenses, and Machine C can polish 2300 lenses.

23. Graph $f(x) = 2x - 3$.

We find some points on the graph, plot them, and draw the line containing them.

For $x = -1$, $f(-1) = 2(-1) - 3 = -2 - 3 = -5$.

For $x = 1$, $f(1) = 2 \cdot 1 - 3 = 2 - 3 = -1$.

For $x = 3$, $f(3) = 2 \cdot 3 - 3 = 6 - 3 = 3$.

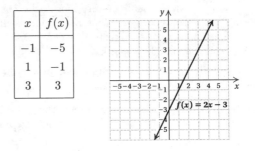

x	$f(x)$
-1	-5
1	-1
3	3

25. Graph $h(x) = x^2 - 2$.

We find some points on the graph, plot them, and draw the curve through them.

For $x = -2$, $h(-2) = (-2)^2 - 2 = 4 - 2 = 2$.

For $x = -1$, $h(-1) = (-1)^2 - 2 = 1 - 2 = -1$.

For $x = 0$, $h(0) = 0^2 - 2 = 0 - 2 = -2$.

For $x = 1$, $h(1) = 1^2 - 2 = 1 - 2 = -1$.

For $x = 2$, $h(2) = 2^2 - 2 = 4 - 2 = 2$.

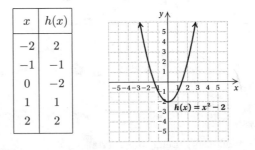

x	$h(x)$
-2	2
-1	-1
0	-2
1	1
2	2

27. There is no vertical line that can intersect the graph more than one time, so this is the graph of a function.

29. **Familiarize**. We first make a drawing with additional labels.

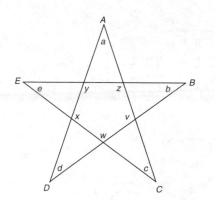

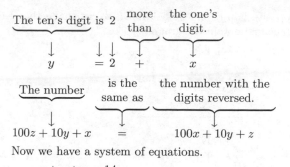

We let a, b, c, d, and e represent the angle measures at the tips of the star. We also label the interior angles of the pentagon v, w, x, y, and z. We recall the following geometric fact:

The sum of the measures of the interior angles of a polygon of n sides is given by $(n-2)180°$.

Using this fact we know:

1. The sum of the angle measures of a triangle is $(3-2)180°$, or $180°$.

2. The sum of the angle measures of a pentagon is $(5-2)180°$, or $3(180°)$.

Translate. Using fact (1) listed above we obtain a system of 5 equations.

$$a + v + d = 180$$
$$b + w + e = 180$$
$$c + x + a = 180$$
$$d + y + b = 180$$
$$e + z + c = 180$$

Solve. Adding we obtain

$$2a + 2b + 2c + 2d + 2e + v + w + x + y + z = 5(180)$$

$$2(a+b+c+d+e) + (v+w+x+y+z) = 5(180)$$

Using fact (2) listed above we substitute $3(180)$ for $(v+w+x+y+z)$ and solve for $(a+b+c+d+e)$.

$$2(a+b+c+d+e) + 3(180) = 5(180)$$
$$2(a+b+c+d+e) = 2(180)$$
$$a+b+c+d+e = 180$$

Check. We should repeat the above calculations.

State. The sum of the angle measures at the tips of the star is $180°$.

31. Familiarize. Let $x =$ the one's digit, $y =$ the ten's digit, and $z =$ the hundred's digit. Then the number is represented by $100z + 10y + x$. When the digits are reversed, the resulting number is represented by $100x + 10y + z$.

Translate.

The sum of the digits is 14.

$$x + y + z = 14$$

The ten's digit is 2 more than the one's digit.

$$y = 2 + x$$

The number is the same as the number with the digits reversed.

$$100z + 10y + x = 100x + 10y + z$$

Now we have a system of equations.

$$x + y + z = 14,$$
$$y = 2 + x,$$
$$100z + 10y + x = 100x + 10y + z$$

Solve. Solving the system we get $(4, 6, 4)$.

Check. If the number is 464, then the sum of the digits is $4+6+4$, or 14. The ten's digit, 6, is 2 more than the one's digit, 4. If the digits are reversed the number is unchanged. The result checks.

State. The number is 464.

Exercise Set 3.7

RC1. A graph of an inequality is a drawing that represents its solutions.

RC3. The graph of $4x - y < 3$ is a half-plane.

RC5. For $4x - y < 3$, the related equation is $4x - y = 3$.

1. We use alphabetical order to replace x by -3 and y by 3.

$$\frac{3x + y < -5}{}$$
$$3(-3) + 3 \ ? \ -5$$
$$-9 + 3 \ \big|$$
$$-6 \ \big| \quad \text{TRUE}$$

Since $-6 < -5$ is true, $(-3, 3)$ is a solution.

3. We use alphabetical order to replace x by 5 and y by 9.

$$\frac{2x - y > -1}{}$$
$$2 \cdot 5 - 9 \ ? \ -1$$
$$10 - 9 \ \big|$$
$$1 \ \big| \quad \text{TRUE}$$

Since $1 > -1$ is true, $(5, 9)$ is a solution.

5. Graph: $y > 2x$

We first graph the line $y = 2x$. We draw the line dashed since the inequality symbol is $>$. To determine which half-plane to shade, test a point not on the line. We try $(1, 1)$ and substitute:

$$\frac{y > 2x}{}$$
$$1 \ ? \ 2 \cdot 1$$
$$\big| \ 2 \quad \text{FALSE}$$

Since $1 > 2$ is false, $(1, 1)$ is not a solution, nor are any points in the half-plane containing $(1, 1)$. The points in the opposite half-plane are solutions, so we shade that half-plane and obtain the graph.

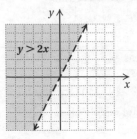

7. Graph: $y < x + 1$

First graph the line $y = x + 1$. Draw it dashed since the inequality symbol is $<$. Test the point $(0,0)$ to determine if it is a solution.

$$\frac{y < x + 1}{0 \;?\; 0 + 1}$$
$$\qquad \Big|\; 1 \qquad \text{TRUE}$$

Since $0 < 1$ is true, we shade the half-plane containing $(0,0)$ and obtain the graph.

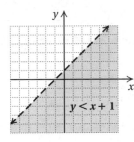

9. Graph: $y > x - 2$

First graph the line $y = x - 2$. Draw a dashed line since the inequality symbol is $>$. Test the point $(0,0)$ to determine if it is a solution.

$$\frac{y > x - 2}{0 \;?\; 0 - 2}$$
$$\qquad \Big|\; -2 \qquad \text{TRUE}$$

Since $0 > -2$ is true, we shade the half-plane containing $(0,0)$ and obtain the graph.

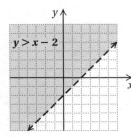

11. Graph: $x + y < 4$

First graph $x + y = 4$. Draw the line dashed since the inequality symbol is $<$. Test the point $(0,0)$ to determine if it is a solution.

$$\frac{x + y < 4}{0 + 0 \;?\; 4}$$
$$\qquad 0 \quad\Big|\quad \text{TRUE}$$

Since $0 < 4$ is true, we shade the half-plane containing $(0,0)$ and obtain the graph.

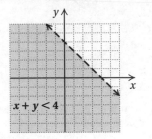

13. Graph: $3x + 4y \le 12$

We first graph $3x + 4y = 12$. Draw the line solid since the inequality symbol is $\le$. Test the point $(0,0)$ to determine if it is a solution.

$$\frac{3x + 4y \le 12}{3 \cdot 0 + 4 \cdot 0 \;?\; 12}$$
$$\qquad\qquad 0 \quad\Big|\quad \text{TRUE}$$

Since $0 \le 12$ is true, we shade the half-plane containing $(0,0)$ and obtain the graph.

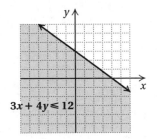

15. Graph: $2y - 3x > 6$

We first graph $2y - 3x = 6$. Draw the line dashed since the inequality symbol is $>$. Test the point $(0,0)$ to determine if it is a solution.

$$\frac{2y - 3x > 6}{2 \cdot 0 - 3 \cdot 0 \;?\; 6}$$
$$\qquad\qquad 0 \quad\Big|\quad \text{FALSE}$$

Since $0 > 6$ is false, we shade the half-plane that does not contain $(0,0)$ and obtain the graph.

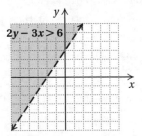

17. Graph: $3x - 2 \le 5x + y$

$$-2 \le 2x + y$$

We first graph $-2 = 2x + y$. Draw the line solid since the inequality symbol is $\le$. Test the point $(0,0)$ to determine if it is a solution.

$$\frac{-2 \le 2x + y}{-2 \ ? \ 2 \cdot 0 + 0}$$
$$\begin{array}{c|c} & 0 \quad\quad \text{TRUE} \end{array}$$

Since $-2 \le 0$ is true, we shade the half-plane containing $(0,0)$ and obtain the graph.

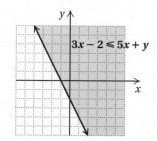

19. Graph: $x < 5$

We first graph $x = 5$. Draw the line dashed since the inequality symbol is $<$. Test the point $(0,0)$ to determine if it is a solution.

$$\frac{x < 5}{0 \ ? \ 5} \quad \text{TRUE}$$

Since $0 < 5$ is true, we shade the half-plane containing $(0,0)$ and obtain the graph.

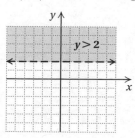

21. Graph: $y > 2$

We first graph $y = 2$. We draw the line dashed since the inequality symbol is $>$. Test the point $(0,0)$ to determine if it is a solution.

$$\frac{y > 2}{0 \ ? \ 2} \quad \text{FALSE}$$

Since $0 > 2$ is false, we shade the half-plane that does not contain $(0,0)$ and obtain the graph.

23. Graph: $2x + 3y \le 6$

We first graph $2x + 3y = 6$. We draw the line solid since the inequality symbol is $\le$. Test the point $(0,0)$ to determine if it is a solution.

$$\frac{2x + 3y \le 6}{2 \cdot 0 + 3 \cdot 0 \ ? \ 6}$$
$$\begin{array}{c|c} 0 & \text{TRUE} \end{array}$$

Since $0 \le 6$ is true, we shade the half-plane containing $(0,0)$ and obtain the graph.

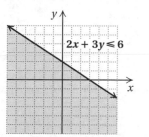

25. The intercepts of the graph of the related equation are $(0,-2)$ and $(3,0)$, so inequality **F** could be the correct one. Since $(0,0)$ is in the solution set of this inequality and the half-plane containing $(0,0)$ is shaded, we know that inequality **F** corresponds to this graph.

27. The intercepts of the graph of the related equation are $(-5,0)$ and $(0,3)$, so inequality **B** could be the correct one. Since $(0,0)$ is in the solution set of this inequality and the half-plane containing $(0,0)$ is shaded, we know that inequality **B** corresponds to this graph.

29. The intercepts of the graph of the related equation are $(-3,0)$ and $(0,-3)$, so inequality **C** could be the correct one. Since $(0,0)$ is not in the solution set of the inequality and the half-plane that does not contain $(0,0)$ is shaded, we know that inequality **C** corresponds to this graph.

31. Graph: $y \ge x$,

$$y \le -x + 2$$

We graph the lines $y = x$ and $y = -x + 2$, using solid lines. We indicate the region for each inequality by arrows at the ends of the lines. Note where the regions overlap, and shade the region of solutions.

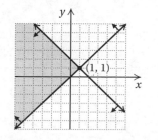

To find the vertex we solve the system of related equations:

$$y = x,$$
$$y = -x + 2$$

Solving, we obtain the vertex $(1,1)$.

33. Graph: $y > x$,
$$y < -x + 1$$

We graph the lines $y = x$ and $y = -x + 1$, using dashed lines. We indicate the region for each inequality by arrows at the ends of the lines. Note where the regions overlap, and shade the region of solutions.

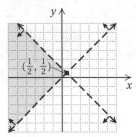

To find the vertex we solve the system of related equations:
$$y = x,$$
$$y = -x + 1$$

Solving, we obtain the vertex $\left(\dfrac{1}{2}, \dfrac{1}{2}\right)$.

35. Graph: $x \leq 3$,
$$y \geq -3x + 2$$

Graph the lines $x = 3$ and $y = -3x + 2$, using solid lines. Indicate the region for each inequality by arrows, and shade the region where they overlap.

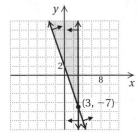

To find the vertex we solve the system of related equations:
$$x = 3,$$
$$y = -3x + 2$$

Solving, we obtain the vertex $(3, -7)$.

37. Graph: $x + y \leq 1$,
$$x - y \leq 2$$

Graph the lines $x + y = 1$ and $x - y = 2$, using solid lines. Indicate the region for each inequality by arrows, and shade the region where they overlap.

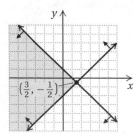

To find the vertex we solve the system of related equations:
$$x + y = 1,$$
$$x - y = 2$$

The vertex is $\left(\dfrac{3}{2}, -\dfrac{1}{2}\right)$.

39. Graph: $y \leq 2x + 1$,   (1)
$$y \geq -2x + 1, \quad (2)$$
$$x \leq 2 \qquad\quad (3)$$

Shade the intersection of the graphs of $y \leq 2x + 1$, $y \geq -2x + 1$, and $x \leq 2$.

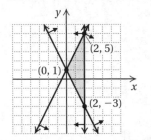

To find the vertices we solve three different systems of equations. From (1) and (2) we obtain the vertex $(0, 1)$. From (1) and (3) we obtain the vertex $(2, 5)$. From (2) and (3) we obtain the vertex $(2, -3)$.

41. Graph: $x + 2y \leq 12$,   (1)
$$2x + y \leq 12, \quad (2)$$
$$x \geq 0, \qquad\ (3)$$
$$y \geq 0 \qquad\ (4)$$

Shade the intersection of the graphs of the four inequalities above.

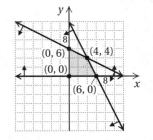

To find the vertices we solve four different systems of equations, as follows:

System of equations	Vertex
From (1) and (2)	$(4, 4)$
From (1) and (3)	$(0, 6)$
From (2) and (4)	$(6, 0)$
From (3) and (4)	$(0, 0)$

43.
$$5(3x - 4) = -2(x + 5)$$
$$15x - 20 = -2x - 10$$
$$17x - 20 = -10$$
$$17x = 10$$
$$x = \dfrac{10}{17}$$

The solution is $\dfrac{10}{17}$.

45. $2(x-1) + 3(x-2) - 4(x-5) = 10$

$2x - 2 + 3x - 6 - 4x + 20 = 10$

$x + 12 = 10$

$x = -2$

The solution is -2.

47. $5x + 7x = -144$

$12x = -144$

$x = -12$

The solution is -12.

49. $f(x) = |2 - x|$

$f(0) = |2 - 0| = |2| = 2$

51. $f(x) = |2 - x|$

$f(10) = |2 - 10| = |-8| = 8$

53. The height must be less than the crest width, so we have

$h < 2w$.

The crest width cannot exceed one-and-a-half times the height, so we have

$w \leq 1.5h$.

The height cannot exceed 3200 ft, so we have

$h \leq 3200$.

The height and width must be nonnegative, so we have

$h > 0$,

$w > 0$.

We have a system of inequalities.

$h < 2w$,

$w \leq 1.5h$,

$h \leq 3200$,

$h > 0$,

$w > 0$

We could also express this as

$h < 2w$,

$0 < w \leq 1.5h$,

$0 < h \leq 3200$.

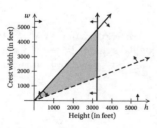

Chapter 3 Vocabulary Reinforcement

1. A solution of a system of two equations in two variables is an ordered <u>pair</u> that makes both equations true.

2. A <u>consistent</u> system of equations has at least one solution.

3. A solution of a system of three equations in three variables is an ordered <u>triple</u> that makes all three equations true.

4. If, for a system of two equations in two variables, the graphs of the equations are different lines, then the equations are <u>independent</u>.

5. The graph of an inequality like $x > 2y$ is a <u>half-plane</u>.

Chapter 3 Concept Reinforcement

1. False; see page 245 in the text.

2. True; see page 245 in the text.

3. The point $(0, b)$ is the y-intercept of a linear equation. If both equations contain the point $(0, b)$, then the equations have the same y-intercept.

4. The graph of $x - 4$ is a vertical line and the graph of $y = -4$ is a horizontal line, so the lines have a point of intersection. Thus the system of equations $x = 4$ and $y = -4$ is consistent. The given statement is false.

Chapter 3 Study Guide

1. Graph both lines on the same set of axes.

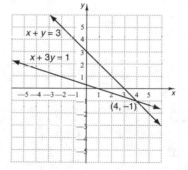

The solution appears to be $(4, -1)$.

Check:

$x + 3y = 1$		$x + y = 3$	
$4 + 3(-1)$? 1		$4 + (-1)$? 3	
$4 - 3$		3	TRUE
1	TRUE		

The solution is $(4, -1)$.

Since the system of equations has a solution, it is consistent. Since there is exactly one solution, the equations are independent.

2. $2x + y = 2$, (1)

$3x + 2y = 5$ (2)

We will use the substitution method. First solve Equation (1) for y.

$2x + y = 2$

$y = -2x + 2$ (3)

Now substitute $-2x+2$ for y in Equation (2) and solve for x.

$$3x + 2(-2x + 2) = 5$$
$$3x - 4x + 4 = 5$$
$$-x + 4 = 5$$
$$-x = 1$$
$$x = -1$$

Substitute -1 for x in Equation (3) and find y.

$$y = -2(-1) + 2 = 2 + 2 = 4$$

The solution is $(-1, 4)$.

3. $2x + 3y = 5,$ (1)

    $3x + 4y = 6$ (2)

We will use the elimination method. Use the multiplication principle on both equations and then add.

$$
\begin{array}{ll}
8x + 12y = 20 & \text{Multiplying (1) by 4} \\
\underline{-9x - 12y = -18} & \text{Multiplying (2) by } -3 \\
-x \quad\quad = 2 & \\
x = -2 &
\end{array}
$$

Substitute -2 for x in one of the original equations and solve for y.

$$2(-2) + 3y = 5 \quad \text{Using Equation (1)}$$
$$-4 + 3y = 5$$
$$3y = 9$$
$$y = 3$$

The solution is $(-2, 3)$.

4. *Familiarize*. We let x and y represent the amounts of the investments. We will use the simple interest formula, $I = Prt$.

Translate. We organize the information in a table.

	6% loan	5% loan	Total
Principal	x	y	$23,000
Interest rate	6%	5%	
Time	1 yr	1 yr	
Interest	$0.06x$	$0.05y$	$1237

The "Principal" row of the table gives us one equation:

$$x + y = 23,000.$$

The "Interest" row of the table gives us a second equation:

$$0.06x + 0.05y = 1237.$$

After clearing decinmals, we have the following system of equations.

$$x + y = 23,000 \quad (1)$$
$$6x + 5y = 123,700 \quad (2)$$

Solve. We multiply Equation (1) by -5 and then add.

$$
\begin{array}{l}
-5x - 5y = -115,000 \\
\underline{6x + 5y = 123,700} \\
x = 8700
\end{array}
$$

Substitute 8700 for x in Equation (1) and solve for y.

$$8700 + y = 23,000$$
$$y = 14,300$$

Check. $\$8700 + \$14,300 = \$23,000$ and $0.06(\$8700) + 0.05(\$14,300) = \$522 + \$715 = \$1237$, so the answer checks.

State. $8700 was invested at 6%, and $14,300 was invested at 5%.

5. $x - y + z = 9,$ (1)

    $2x + y + 2z = 3,$ (2)

    $4x + 2y - 3z = -1$ (3)

We will start by eliminating y from two different pairs of equations.

$$
\begin{array}{ll}
x - y + z = 9 & (1) \\
\underline{2x + y + 2z = 3} & (2) \\
3x \quad\quad + 3z = 12 & (4)
\end{array}
$$

$$
\begin{array}{ll}
2x - 2y + 2z = 18 & \text{Multiplying (1) by 2} \\
\underline{4x + 2y - 3z = -1} & (3) \\
6x \quad\quad - z = 17 & (5)
\end{array}
$$

Now solve the system of Equations (4) and (5).

$$
\begin{array}{ll}
3x + 3z = 12 & (4) \\
\underline{18x - 3z = 51} & \text{Multiplying (5) by 3} \\
21x \quad\quad = 63 & \\
x = 3 &
\end{array}
$$

$$3 \cdot 3 + 3z = 12 \quad \text{Substituting 3 for } x \text{ in (4)}$$
$$9 + 3z = 12$$
$$3z = 3$$
$$z = 1$$

$$
\begin{array}{ll}
2 \cdot 3 + y + 2 \cdot 1 = 3 & \text{Substituting 3 for } x \\
& \text{and 1 for } z \text{ in (2)} \\
6 + y + 2 = 3 & \\
8 + y = 3 & \\
y = -5 &
\end{array}
$$

The solution is $(3, -5, 1)$.

6. Graph: $3x - 2y > 6$

We first graph the line $3x - 2y = 6$. We draw the line dashed since the inequality symbol is $>$. To determine which half-plane to shade, test a point not on the line. We try $(0, 0)$.

$$
\begin{array}{c|c}
\multicolumn{2}{c}{3x - 2y > 6} \\
\hline
3 \cdot 0 - 2 \cdot 0 \;?\; 6 & \\
0 - 0 & \\
0 & \text{FALSE}
\end{array}
$$

Since $0 > 6$ is false, we shade the half-plane that does not contain $(0, 0)$.

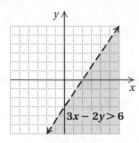

7. Graph: $x - 2y \leq 4$, (1)

$\qquad x + y \leq 4$, (2)

$\qquad x - 1 \geq 0$ (3)

Graph the lines $x - 2y = 4$, $x + y = 4$, and $x - 1 = 0$ (or $x = 1$) using solid lines. Indicate the region for each inequality by arrows and shade the region where they overlap.

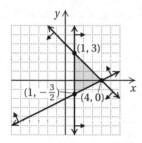

To find the vertices we solve three different systems of equations. From (1) and (2) we obtain the vertex $(4, 0)$. From (1) and (3) we obtain the vertex $\left(1, -\dfrac{3}{2}\right)$. From (2) and (3) we obtain the vertex $(1, 3)$.

Chapter 3 Review Exercises

1. Graph both lines on the same set of axes.

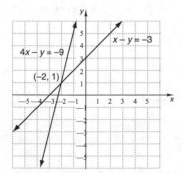

The solution (point of intersection) seems to be the point $(-2, 1)$.

Check:

$4x - y = -9$		$x - y = -3$	
$4(-2) - 1 \ ? \ -9$		$-2 - 1 \ ? \ -3$	
$-8 - 1$		-3	TRUE
-9	TRUE		

The solution is $(-2, 1)$.

Since the system of equations has a solution it is consistent. Since there is exactly one solution, the equations are independent.

2. Graph both lines on the same set of axes.

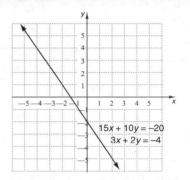

The graphs are the same. Any solution of one of the equations is also a solution of the other. Each equation has an infinite number of solutions. Thus the system of equations has an infinite number of solutions.

Since the system of equations has a solution, it is consistent. Since there are infinitely many solutions, the equations are dependent.

3. Graph both lines on the same set of axes.

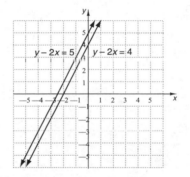

The lines are parallel. There is no solution.

Since the system of equations has no solution, it is inconsistent. Since there is no solution, the equations are independent.

4. $2x - 3y = 5$, (1)

$\qquad x = 4y + 5$ (2)

Substitute $4y + 5$ for x in Equation (1) and solve for y.

$$2(4y + 5) - 3y = 5$$
$$8y + 10 - 3y = 5$$
$$5y + 10 = 5$$
$$5y = -5$$
$$y = -1$$

Substitute -1 for y in Equation (2) and find x.

$$x = 4(-1) + 5 = -4 + 5 = 1$$

The solution is $(1, -1)$.

5. $y = x + 2,$ (1)

  $y - x = 8$ (2)

Substitute $x + 2$ for y in the second equation and solve for x.

$$y - x = 8 \quad (2)$$
$$x + 2 - x = 8$$
$$2 = 8$$

We get a false equation. There is no solution.

6. $7x - 4y = 6,$ (1)

  $y - 3x = -2$ (2)

First solve the second equation for y.

$$y - 3x = -2 \quad (2)$$
$$y = 3x - 2 \quad (3)$$

Now substitute $3x - 2$ for y in the first equation and solve for x.

$$7x - 4y = 6 \quad (1)$$
$$7x - 4(3x - 2) = 6$$
$$7x - 12x + 8 = 6$$
$$-5x + 8 = 6$$
$$-5x = -2$$
$$x = \frac{2}{5}$$

Next substitute $\frac{2}{5}$ for x in Equation (3) and find y.

$$y = 3 \cdot \frac{2}{5} - 2 = \frac{6}{5} - \frac{10}{5} = -\frac{4}{5}$$

Since $\left(\frac{2}{5}, -\frac{4}{5}\right)$ checks, it is the solution.

7. $x + 3y = -3,$ (1)

  $\dfrac{2x - 3y = 21}{3x \qquad = 18}$ (2)

  $3x \qquad = 18$ Adding

  $x = 6$

Substitute 6 for x in Equation (1) and solve for y.

$$6 + 3y = -3$$
$$3y = -9$$
$$y = -3$$

The solution is $(6, -3)$.

8. $3x - 5y = -4,$ (1)

  $5x - 3y = \quad 4$ (2)

We multiply Equation (1) by 3 and Equation (2) by -5 and then add.

$$9x - 15y = -12$$
$$\underline{-25x + 15y = -20}$$
$$-16x \qquad = -32$$
$$x = 2$$

Now substitute 2 for x in one of the equations and solve for y.

$$3x - 5y = -4 \quad (1)$$
$$3 \cdot 2 - 5y = -4$$
$$6 - 5y = -4$$
$$-5y = -10$$
$$y = 2$$

Since $(2, 2)$ checks, it is the solution.

9. $\dfrac{1}{3}x + \dfrac{2}{9}y = 1,$

  $\dfrac{3}{2}x + \dfrac{1}{2}y = 6$

We multiply by 9 on both sides of the first equation and by 2 on both sides of the second equation to clear the fractions.

$$3x + 2y = 9, \quad (1)$$
$$3x + \ y = 12 \quad (2)$$

Now we multiply both sides of Equation (2) by -1 and then add.

$$3x + 2y = 9$$
$$\underline{-3x - \ y = -12}$$
$$y = -3$$

Substitute -3 for y in Equation (2) and solve for x.

$$3x + (-3) = 12$$
$$3x = 15$$
$$x = 5$$

The solution is $(5, -3)$.

10. $1.5x - 3 = -2y,$

  $3x + 4y = 6$

First we rewrite the first equation in the form $Ax + By = C$.

$$1.5x + 2y = 3, \ (1)$$
$$3x + 4y = \ 6 \ (2)$$

Now multiply Equation (1) by -2 and then add.

$$-3x - 4y = -6$$
$$\underline{3x + 4y = 6}$$
$$0 = 0$$

We get an equation that is true for all numbers x and y. The system has an infinite number of solutions.

11. _Familiarize._ Let $x = $ the number of less expensive brushes sold and $y = $ the number of more expensive brushes sold.

Translate. We organize the information in a table.

Kind of brush	Less expensive	More expensive	Total
Number sold	x	y	45
Price	\$8.50	\$9.75	
Amount taken in	$8.50x$	$9.75y$	398.75

The "Number sold" row of the table gives us one equation:

$$x + y = 45$$

The "Amount taken in" row gives us a second equation:

$$8.50x + 9.75y = 398.75$$

We have a system of equations:

$$x + y = 45,$$
$$8.50x + 9.75y = 398.75$$

We can multiply the second equation on both sides by 100 to clear the decimals:

$$x + y = 45, \quad (1)$$
$$850x + 975y = 39,875 \quad (2)$$

Solve. We solve the system of equations using the elimination method. Begin by multiplying Equation (1) by -850.

$$-850x - 850y = -38,250 \quad \text{Multiplying (1)}$$
$$\underline{850x + 975y = 39,875}$$
$$125y = 1625$$
$$y = 13$$

Substitute 13 for y in (1) and solve for x.

$$x + 13 = 45$$
$$x = 32$$

Check. The number of brushes sold is $32 + 13$, or 45. The amount taken in was $\$8.50(32) + \$9.75(13) = \$272 + \$126.75 = \$398.75$. The answer checks.

State. 32 of the less expensive brushes were sold, and 13 of the more expensive brushes were sold.

12. **Familiarize.** Let x and y represent the number of liters of Orange Thirst and Quencho that should be used, respectively. The amount of orange juice in the mixture is $10\%(10 \text{ L})$, or $0.1(10 \text{ L})$, or 1 L.

Translate. We organize the information in a table.

	Orange Thirst	Quencho	Mixture
Number of liters	x	y	10
Percent of juice	15%	5%	10%
Liters of juice	$0.15x$	$0.05y$	1

The first row of the table gives us one equation.

$$x + y = 10$$

The last row yields a second equation.

$$0.15x + 0.05y = 1$$

After clearing decimals we have the following system of equations.

$$x + y = 10, \quad (1)$$
$$15x + 5y = 100 \quad (2)$$

Solve. We use the elimination method. First we multiply Equation (1) by -5 and then add.

$$-5x - 5y = -50$$
$$\underline{15x + 5y = 100}$$
$$10x = 50$$
$$x = 5$$

Now substitute 5 for x in one of the equations and solve for y.

$$x + y = 10 \quad (1)$$
$$5 + y = 10$$
$$y = 5$$

Check. If 5 L of each type of juice are used, then the mixture contains $5 + 5$, or 10 L. The amount of orange juice in the mixture is $0.15(5) + 0.05(5)$, or $0.75 + 0.25$, or 1 L. The answer checks.

State. 5 L of Orange Thirst and 5 L of Quencho should be used.

13. **Familiarize.** We first make a drawing.

Slow train			
d miles	44 mph	$(t + 1)$ hr	
Fast train			
d miles	52 mph	t hr	

From the drawing we see that the distances are the same. We organize the information in a table.

d	$=$	r	$\cdot$	t

	Distance	Rate	Time	
Slow train	d	44	$t + 1$	$\rightarrow d = 44(t+1)$
Fast train	d	52	t	$\rightarrow d = 52t$

Translate. Using $d = rt$ in each row of the table, we get a system of equations:

$$d = 44(t + 1),$$
$$d = 52t$$

Solve. We solve the system of equations.

$$52t = 44(t + 1) \quad \text{Using substitution}$$
$$52t = 44t + 44$$
$$8t = 44$$
$$t = \frac{11}{2}, \text{ or } 5\frac{1}{2}$$

Check. At 52 mph, in $5\frac{1}{2}$ hr the fast train will travel $52 \cdot \frac{11}{2}$, or 286 mi. At 44 mph, in $5\frac{1}{2} + 1$, or $6\frac{1}{2}$ hr, the slow train travels $44 \cdot \frac{13}{2}$, or 286 mi. Since the distances are the same, the answer checks.

State. The second train will travel $5\frac{1}{2}$ hr before it overtakes the first train.

14. $x + 2y + \ z = 10,$ (1)

$2x - \ y + \ z = \ 8,$ (2)

$3x + \ y + 4z = \ \ 2$ (3)

We start by eliminating y from two different pairs of equations.

$x + 2y + \ z = 10$ (1)

$\underline{4x - 2y + 2z = 16}$ Multiplying (2) by 2

$5x \qquad + 3z = 26$ (4)

$2x - y + \ z = 8$ (2)

$\underline{3x + y + 4z = 2}$ (3)

$5x \qquad + 5z = 10$ (5)

Now solve the system of Equations (4) and (5). We multiply Equation (5) by -1 and then add.

$5x + 3z = 26 \qquad$ (4)

$\underline{-5x - 5z = -10}$

$\qquad - 2z = 16$

$\qquad z = -8$

$5x + 3(-8) = 26$ Substituting -8 for z in (4)

$5x - 24 = 26$

$5x = 50$

$x = 10$

$3 \cdot 10 + y + 4(-8) = 2$ Substituting 10 for x and -8 for z in (3)

$30 + y - 32 = 2$

$y - 2 = 2$

$y = 4$

We obtain $(10, 4, -8)$. This checks, so it is the solution.

15. $3x + 2y + \ z = 1,$ (1)

$2x - \ y - 3z = 1,$ (2)

$-x + 3y + 2z = 6$ (3)

We start by eliminating x from two different pairs of equations.

$3x + \ 2y + \ z = 1 \qquad$ (1)

$\underline{-3x + \ 9y + 6z = 18}$ Multiplying (3) by 3

$\qquad 11y + 7z = 19 \qquad$ (4)

$2x - \ y - 3z = 1 \qquad$ (2)

$\underline{-2x + 6y + 4z = 12}$ Multiplying (3) by 2

$\qquad 5y + \ z = 13 \qquad$ (5)

Now we solve the system of Equations (4) and (5).

$11y + 7z = 19 \qquad$ (4)

$\underline{-35y - 7z = -91}$ Multiplying (5) by -7

$-24y \qquad = -72$

$y = 3$

$5 \cdot 3 + z = 13$ Substituting 3 for y in (5)

$15 + z = 13$

$z = -2$

$-x + 3 \cdot 3 + 2(-2) = 6$ Substituting 3 for y and -2 for z in (3)

$-x + 9 - 4 = 6$

$-x + 5 = 6$

$-x = 1$

$x = -1$

We obtain $(-1, 3, -2)$. This checks, so it is the solution.

16. $2x - 5y - 2z = -4,$ (1)

$7x + 2y - 5z = -6,$ (2)

$-2x + 3y + 2z = \ \ 4$ (3)

We start by eliminating x from two different pairs of equations.

$14x - 35y - 14z = -28$ Multiplying (1) by 7

$\underline{-14x - \ 4y + 10z = 12}$ Multiplying (2) by -2

$- 39y - \ 4z = -16$ (4)

$2x - 5y - 2z = -4$ (1)

$\underline{-2x + 3y + 2z = 4}$ (3)

$- 2y \qquad = 0$

$y = 0$

Substitute 0 for y in Equation (4) and solve for z.

$-39 \cdot 0 - 4z = -16$

$-4z = -16$

$z = 4$

Now substitute 0 for y and 4 for z in one of the original equations and solve for x. We use Equation (1).

$2x - 5 \cdot 0 - 2 \cdot 4 = -4$

$2x - 8 = -4$

$2x = 4$

$x = 2$

We obtain $(2, 0, 4)$. This checks, so it is the solution.

17. $x + \ y + 2z = 1,$ (1)

$x - \ y + \ z = 1,$ (2)

$x + 2y + \ z = 2$ (3)

We start by eliminating y from two different pairs of equations.

$x + y + 2z = 1$ (1)

$\underline{x - y + \ z = 1}$ (2)

$2x \qquad + 3z = 2$ (4)

$2x - 2y + 2z = 2$ Multiplying (2) by 2

$\underline{x + 2y + \ z = 2}$

$3x \qquad + 3z = 4$ (5)

Now solve the system of Equations (4) and (5).

$2x + 3z = 2 \qquad$ (4)

$\underline{-3x - 3z = -4}$ Multiplying (5) by -1

$-x \qquad = -2$

$x = 2$

$$2 \cdot 2 + 3z = 2 \qquad \text{Substituting 2 for } x \text{ in (4)}$$
$$4 + 3z = 2$$
$$3z = -2$$
$$z = -\frac{2}{3}$$
$$2 + y + 2\left(-\frac{2}{3}\right) = 1 \qquad \text{Substituting 2 for } x \text{ and}$$
$$-\frac{2}{3} \text{ for } z \text{ in (1)}$$
$$2 + y - \frac{4}{3} = 1$$
$$y + \frac{2}{3} = 1$$
$$y = \frac{1}{3}$$

We obtain $\left(2, \frac{1}{3}, -\frac{2}{3}\right)$. This checks, so it is the solution.

18. Familiarize. Let a, b, and c represent the measures of angles A, B, and C, respectively. Recall that the sum of the measures of the angles of a triangle is 180°.

Translate.

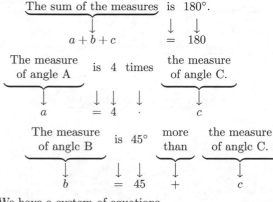

We have a system of equations.

$$a + b + c = 180,$$
$$a = 4c,$$
$$b = 45 + c$$

Solve. Solving the system we get $\left(90, 67\frac{1}{2}, 22\frac{1}{2}\right)$.

Check. The sum of the measures is $90° + 67\frac{1}{2}° + 22\frac{1}{2}°$, or 180°. Four times the measure of angle C is $4\left(22\frac{1}{2}°\right)$, or 90°, the measure of angle A; 45° more than the measure of angle C is $22\frac{1}{2}° + 45°$, or $67\frac{1}{2}°$, the measure of angle B. The answer checks.

State. The measures of angles A, B, and C are 90°, $67\frac{1}{2}°$, and $22\frac{1}{2}°$, respectively.

19. Familiarize. Let x, y, and z represent the prices of 1 bag of caramel nut crunch popcorn, 1 bag of plain popcorn, and 1 bag of mocha choco latte popcorn, respectively.

Translate. The total cost of 1 bag of each type of popcorn was $49, so we have

$$x + y + z = 49.$$

The price of the caramel nut crunch popcorn was six times the price of the plain popcorn and $16 more than the price of the mocha choco latte popcorn, so we have

$$x = 6y \text{ and } x = z + 16.$$

We have a system of equations.

$$x + y + z = 49,$$
$$x = 6y,$$
$$x = z + 16$$

Solve. Solving the equation we have $(30, 5, 14)$.

Check. The total cost was $30 + $5 + $14, or $90. $30 is six times $5 and is $16 more than $14. The answer checks.

State. One bag of caramel nut crunch popcorn costs $30, one bag of plain popcorn costs $5, and one bag of mocha choco latte popcorn costs $14.

20. Graph: $2x + 3y < 12$

First graph the line $2x + 3y = 12$. Draw it dashed since the inequality symbol is $<$. Test the point $(0, 0)$ to determine if it is a solution.

$$\frac{2x + 3y < 12}{2 \cdot 0 + 3 \cdot 0 \ ? \ 12}$$
$$0 \ \bigm| \ \text{TRUE}$$

Since $0 < 12$ is true, we shade the half-plane containing $(0, 0)$ and draw the graph.

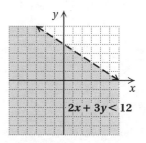

21. Graph: $y \leq 0$

First graph the line $y = 0$ (the x-axis). Draw it solid since the inequality symbol is $\leq$. Test the point $(1, 2)$ to determine if it is a solution.

$$\frac{y \leq 0}{2 \ ? \ 0 \ \text{FALSE}}$$

Since $2 \leq 0$ is false, we shade the half-plane that does not contain $(1, 2)$ and obtain the graph.

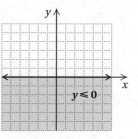

22. Graph: $x + y \geq 1$

First graph the line $x + y = 1$. Draw it solid since the inequality symbol is $\geq$. Test the point $(0, 0)$ to determine if it is a solution.

$$\frac{x + y \geq 1}{0 + 0 ~?~ 1}$$
$$\begin{array}{c|c} 0 & \text{FALSE} \end{array}$$

Since $0 \geq 1$ is false, we shade the half-plane that does not contain $(0, 0)$ and obtain the graph.

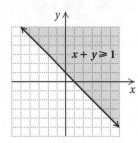

23. Graph: $y \geq -3$,

$x \geq 2$

We graph the lines $y = -3$ and $x = 2$ using solid lines. We indicate the region for each inequality by arrows at the ends of the lines. Note where the regions overlap, and shade the region of solutions.

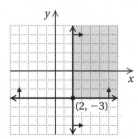

To find the vertex we solve the system of related equations:

$$y = -3,$$
$$x = 2$$

We see that the vertex is $(2, -3)$.

24. Graph: $x + 3y \geq -1$,

$x + 3y \leq 4$

We graph the lines $x + 3y = -1$ and $x + 3y = 4$ using solid lines. We indicate the region for each inequality by arrows at the ends of the lines. Note where the regions overlap, and shade the region of solutions.

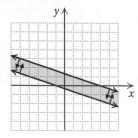

25. Graph: $x - y \leq 3$,   (1)

$x + y \geq -1$,   (2)

$y \leq 2$ (3)

We graph the lines $x - y = 3$, $x + y = -1$, and $y = 2$ using solid lines. We indicate the region for each inequality by arrows at the ends of the lines. Note where the regions overlap, and shade the region of solutions.

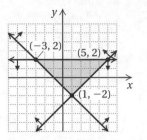

To find the vertices we solve three different systems of equations. From (1) and (2) we obtain the vertex $(1, -2)$. From (1) and (3) we obtain the vertex $(5, 2)$. From (2) and (3) we obtain the vertex $(-3, 2)$.

26. ***Familiarize***. Let x and y represent the numbers.

Translate.

$$\underbrace{\text{The sum of the numbers}}_{\downarrow} \underbrace{\text{is}}_{\downarrow} \underbrace{-2.}_{\downarrow}$$
$$\qquad\qquad x + y \qquad\qquad = \quad -2$$

$$\text{Twice } \underbrace{\text{one number}}_{\downarrow} \underbrace{\text{plus}}_{\downarrow} \underbrace{\text{the other}}_{\downarrow} \underbrace{\text{is}}_{\downarrow} \underbrace{4.}_{\downarrow}$$
$$2\cdot \qquad x \qquad + \qquad y \qquad = \quad 4$$

We have a system of equations.

$$x + y = -2, \quad (1)$$
$$2x + y = \quad 4 \quad (2)$$

Solve. We multiply both sides of Equation (1) by -1 and then add.

$$-x - y = 2$$
$$\underline{2x + y = 4}$$
$$x \qquad = 6$$

Substitute 6 for x in Equation (1) and solve for y.

$$6 + y = -2$$
$$y = -8$$

Check. $6 + (-8) = -2$, and $2 \cdot 6 + (-8) = 12 - 8 = 4$. The answer checks.

State. The numbers are 6 and -8.

Since one number is 6, C is the correct answer.

27. ***Familiarize***. Let $t =$ the time the cars travel, and let $d =$ the distance the first car travels. Then the distance the second car travels is $275 - d$. We organize the information in a table.

	Distance	Rate	Time
First car	d	50	t
Second car	$275 - d$	60	t

Translate. Using $d = rt$ in each row of the table, we get a system of equations.

$$d = 50t,$$
$$275 - d = 60t$$

Solve.

$$275 - 50t = 60t \quad \text{Using substitution}$$
$$275 = 110t$$
$$2.5 = t$$

Check. In 2.5 hr, the first car travels $50(2.5)$, or 125 mi, and the second car travels $60(2.5)$, or 150 mi. Then the cars are 125 mi + 150 mi, or 275 mi, apart after 2.5 hr. The answer checks.

State. The cars will be 275 mi apart after 2.5 hr.

A is the correct answer.

28. We graph the equations and find the points of intersection.

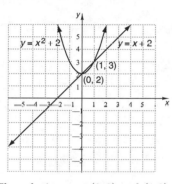

The solutions are $(0, 2)$ and $(1, 3)$.

Chapter 3 Discussion and Writing Exercises

1. Answers may vary. One day a florist sold a total of 23 hanging baskets and flats of petunias. Hanging baskets cost $10.95 each and flats of petunias cost $12.95 each. The sales totaled $269.85. How many of each were sold?

2. We know that machines A and B can polish 3400 lenses, so machine C can polish $5700 - 3400$, or 2300 lenses. Machines B and C can polish 4200 lenses, so machine B can polish $4200 - 2300$, or 1900 lenses. Then machine A can polish $5700 - 1900 - 2300$, or 1500 lenses.

3. Let $x =$ the number of adults in the audience, $y =$ the number of senior citizens, and $z =$ the number of children. The total attendance is 100, so we have Equation (1), $x + y + z = 100$. The amount taken in was $100, so Equation (2) is $10x + 3y + 0.5z = 100$. There is no other information that can be translated to an equation. Clearing decimals in Equation (2) and then eliminating z gives us Equation (3), $95 + 25y = 500$. Dividing by 5 on both sides, we have Equation (4), $19x + 5y = 100$.

Since we have only two equations, it is not possible to eliminate z from another pair of equations. However, in Equation (4), note that 5 is a factor of both $5y$ and 100. Therefore, 5 must also be a factor of $19x$, and hence of x, since 5 is not a factor of 19. Then for some positive integer n, $x = 5n$. (We require n to be positive, since the number of adults clearly cannot be negative and must also be nonzero since the exercise states that the audience consists of *adults*, senior citizens, and children.) We have:

$$19 \cdot 5n + 5y = 100$$
$$19n + y = 20 \quad \text{Dividing by 5}$$

Since n and y must both be positive, $n = 1$. (If $n > 1$, then $19n + y > 20$.) Then $x = 5 \cdot 1$, or 5.

$$19 \cdot 5 + 5y = 100 \quad \text{Substituting in (4)}$$
$$y = 1$$

$$5 + 1 + z = 100 \quad \text{Substituting in (1)}$$
$$z = 94$$

There were 5 adults, 1 senior citizen, and 94 children in the audience.

4. No; the symbol $\geq$ does not always yield a graph in which the half-plane above the line is shaded. For the inequality $-y \geq 3$, for example, the half-plane below the line $y = -3$ is shaded.

Chapter 3 Test

1. Graph both lines on the same set of axes.

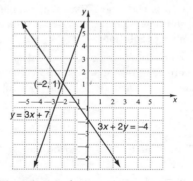

The solution (point of intersection) seems to be the point $(-2, 1)$.

Check:

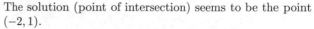

$y = 3x + 7$		$3x + 2y = -4$	
$1 \; ? \; 3(-2) + 7$		$3(-2) + 2 \cdot 1 \; ? \; -4$	
$-6 + 7$		$-6 + 2$	
1	TRUE	-4	TRUE

The solution is $(-2, 1)$.

Since the system of equations has a solution, it is consistent. Since there is exactly one solution, the equations are independent.

2. Graph both lines on the same set of axes.

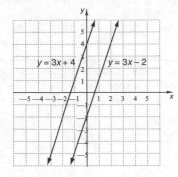

The lines are parallel. There is no solution.

Since the system of equations has no solution, it is inconsistent. Since there is no solution, the equations are independent.

3. Graph both lines on the same set of axes.

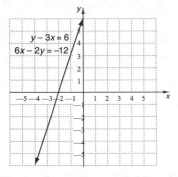

The graphs are the same. Any solution of one of the equations is also a solution of the other. Each equation has an infinite number of solutions. Thus the system of equations has an infinite number of solutions.

Since the system of equations has a solution, it is consistent. Since there are infinitely many solutions, the equations are dependent.

4. $4x + 3y = -1$,   (1)

   $y = 2x - 7$ (2)

Substitute $2x - 7$ for y in the first equation and solve for x.

$$4x + 3(2x - 7) = -1$$
$$4x + 6x - 21 = -1$$
$$10x - 21 = -1$$
$$10x = 20$$
$$x = 2$$

Next we substitute 2 for x in Equation (2) to find y.

$$y = 2x - 7 = 2 \cdot 2 - 7 = 4 - 7 = -3$$

The ordered pair $(2, -3)$ checks in both equations. It is the solution.

5. $x = 3y + 2$,   (1)

   $2x - 6y = 4$ (2)

Substitute $3y + 2$ for x in the second equation and solve for y.

$$2(3y + 2) - 6y = 4$$
$$6y + 4 - 6y = 4$$
$$4 = 4$$

We have an equation that is true for all values of x and y. The system of equations has infinitely many solutions.

6. $x + 2y = 6$,   (1)

   $2x + 3y = 7$ (2)

Solve the first equation for x.

$$x + 2y = 6$$
$$x = -2y + 6   (3)$$

Now substitute $-2y + 6$ for x in the second equation and solve for y.

$$2(-2y + 6) + 3y = 7$$
$$-4y + 12 + 3y = 7$$
$$-y + 12 = 7$$
$$-y = -5$$
$$y = 5$$

Next we substitute 5 for y in Equation 3 and find x.

$$x = -2 \cdot 5 + 6 = -10 + 6 = -4$$

The ordered pair $(-4, 5)$ checks in both equations. It is the solution.

7. $2x + 5y = 3$,   (1)

   $\underline{-2x + 3y = 5}$ (2)

        $8y = 8$ Adding

        $y = 1$

Now substitute 1 for y in one of the equations and solve for x.

$$2x + 5y = 3   \text{Equation (1)}$$
$$2x + 5 \cdot 1 = 3$$
$$2x + 5 = 3$$
$$2x = -2$$
$$x = -1$$

The ordered pair $(-1, 1)$ checks in both equations. It is the solution.

8. $x + y = -2$,   (1)

   $4x - 6y = -3$ (2)

Multiply Equation (1) by 6 and then add.

$$6x + 6y = -12$$
$$\underline{4x - 6y = -3}$$
$$10x \qquad = -15$$
$$x = -\frac{3}{2}$$

Now substitute $-\dfrac{3}{2}$ for x in one of the original equations and solve for y.

$$x + y = -2 \quad \text{Equation (1)}$$
$$-\frac{3}{2} + y = -2$$
$$y = -\frac{1}{2}$$

The ordered pair $\left(-\dfrac{3}{2}, -\dfrac{1}{2}\right)$ checks in both equations. It is the solution.

9.
$$\frac{2}{3}x - \frac{4}{5}y = 1, \quad (1)$$
$$\frac{1}{3} - \frac{2}{5}y = 2 \quad (2)$$

We first multiply each equation by 15 to clear the fractions.

$$15\left(\frac{2}{3}x - \frac{4}{5}y\right) = 15 \cdot 1 \rightarrow 10x - 12y = 15 \quad (3)$$
$$15\left(\frac{1}{3}x - \frac{2}{5}y\right) = 15 \cdot 2 \rightarrow 5x - 6y = 30 \quad (4)$$

Now we multiply Equation (4) by -2 and then add.

$$\begin{aligned} 10x - 12y &= 15 \\ -10x + 12y &= -60 \\ \hline 0 &= -45 \end{aligned}$$

We get a false equation. The system of equations has no solution.

10. Familiarize. Let $l = $ the length and $w = $ the width, in feet.

Translate. The width is 42 ft less than the length, so we have one equation:

$$w = l - 42.$$

We use the formula $P = 2l + 2w$ to get a second equation:

$$288 = 2l + 2w.$$

We have a system of equations.

$$w = l - 42, \quad (1)$$
$$288 = 2l + 2w \quad (2)$$

Solve. We use the substitution method.

$$288 = 2l + 2(l - 42)$$
$$288 = 2l + 2l - 84$$
$$288 = 4l - 84$$
$$372 = 4l$$
$$93 = l$$

Substitute 93 for l in Equation (1) and find w.

$$w = 93 - 42 = 51$$

Check. The length, 93 ft, is 42 ft more than the width, 51 ft. Also, $2 \cdot 93 + 2 \cdot 51 = 288$ ft. The answer checks.

State. The length is 93 ft, and the width is 51 ft.

11. Familiarize. Let $d = $ the distance traveled and $r = $ the speed of the plane in still air, in km/h. Then with a 20-km/h tailwind the plane's speed is $r + 20$. The speed of

the plane traveling against the wind is $r - 20$. We organize the information in a table.

	Distance	Rate	Time
With wind	d	$r + 20$	5
Against wind	d	$r - 20$	7

Translate. Using $d = rt$ in each row of the table, we get a system of equations.

$$d = (r + 20)5,$$
$$d = (r - 20)7, \quad \text{or}$$

$$d = 5r + 100,$$
$$d = 7r - 140$$

Solve. We use the substitution method.

$$5r + 100 = 7r - 140$$
$$100 = 2r - 140$$
$$240 = 2r$$
$$120 = r$$

Check. If $r = 120$, then $r + 20 = 120 + 20 = 140$, and the distance traveled in 5 hr is $140 \cdot 5$, or 700 km. Also, $r - 20 = 120 - 20 = 100$, and the distance traveled in 7 hr is $100 \cdot 7$, or 700 km. The distances are the same, so the answer checks.

State. The speed of the plane in still air is 120 km/h.

12. Familiarize. Let $b = $ the number of buckets of wings and $d = $ the number of chicken dinners sold. We organize the information in a table.

	Buckets	Dinners	Total
Number sold	b	d	28
Price	\$12	\$7	
Amount collected	$12b$	$7d$	281

Translate. The first and last rows of the table give us two equations.

$$b + d = 28, \quad (1)$$
$$12b + 7d = 281 \quad (2)$$

Solve. We use the elimination method.

$$\begin{aligned} -7b - 7d &= -196 \quad \text{Multiplying (1) by } -7 \\ 12b + 7d &= 281 \\ \hline 5b &= 85 \\ b &= 17 \end{aligned}$$

Substitute 17 for b in Equation (1) and solve for d.

$$17 + d = 28$$
$$d = 11$$

Check. The number of orders filled was $17 + 11$, or 28. The amount taken in was $\$12 \cdot 17 + \$7 \cdot 11$, or \$281. The answer checks.

State. 17 buckets of wings and 11 chicken dinners were sold.

13. *Familiarize*. Let $x =$ the number of liters of 20% solution and $y =$ the number of liters of 45% solution to be used. We organize the information in a table.

	20% solution	45% solution	Mixture
Number of liters	x	y	20
Percent of salt	20%	45%	30%
Amount of salt	$0.2x$	$0.45y$	$0.3(20)$, or 6

***Translate*.** We get one equation from the first row of the table.

$$x + y = 20$$

The last row of the table yields a second equation.

$$0.2x + 0.45y = 6$$

After clearing decimals, we have the following system of equations.

$$x + y = 20, \quad (1)$$
$$20x + 45y = 600 \quad (2)$$

***Solve*.** We use the elimination method.

$$
\begin{array}{ll}
-20x - 20y = -400 & \text{Multiplying (1) by } -20 \\
\underline{20x + 45y = 600} & \\
25y = 200 & \\
y = 8 &
\end{array}
$$

Substitute 8 for y in Equation (1) and solve for x.

$$x + 8 = 20$$
$$x = 12$$

***Check*.** 12 L + 8 L = 20 L. The amount of salt in the mixture is $0.2(12) + 0.45(8) = 2.4 + 3.6$, or 6 L. The answer checks.

***State*.** 12 L of 20% solution and 8 L of 45% solution should be used.

14.
$$6x + 2y - 4z = 15, \quad (1)$$
$$-3x - 4y + 2z = -6, \quad (2)$$
$$4x - 6y + 3z = 8 \quad (3)$$

We start by eliminating y from two different pairs of equations.

$$
\begin{array}{ll}
12x + 4y - 8z = 30 & \text{Multiplying (1) by 2} \\
\underline{-3x - 4y + 2z = -6} & (2) \\
9x \qquad - 6z = 24 & (4)
\end{array}
$$

$$
\begin{array}{ll}
18x + 6y - 12z = 45 & \text{Multiplying (1) by 3} \\
\underline{4x - 6y + 3z = 8} & (3) \\
22x \qquad - 9z = 53 & (5)
\end{array}
$$

Now solve the system of Equations (4) and (5).

$$9x - 6z = 24 \quad (4)$$
$$22x - 9z = 53 \quad (5)$$

$$
\begin{array}{ll}
27x - 18z = 72 & \text{Multiplying (4) by 3} \\
\underline{-44x + 18z = -106} & \text{Multiplying (5) by } -2 \\
-17x \qquad = -34 & \\
x = 2 &
\end{array}
$$

$$
\begin{array}{ll}
9 \cdot 2 - 6z = 24 & \text{Substituting 2 for } x \text{ in (4)} \\
18 - 6z = 24 & \\
-6z = 6 & \\
z = -1 &
\end{array}
$$

$$
\begin{array}{ll}
6 \cdot 2 + 2y - 4(-1) = 15 & \text{Substituting 2 for } x \\
& \text{and } -1 \text{ for } z \text{ in (1)} \\
12 + 2y + 4 = 15 & \\
16 + 2y = 15 & \\
2y = -1 & \\
y = -\dfrac{1}{2} &
\end{array}
$$

We obtain $\left(2, -\dfrac{1}{2}, -1\right)$. This checks, so it is the solution.

15. *Familiarize*. Let x, y, and z represent the number of hours worked by the electrician, the carpenter, and the plumber, respectively. In x hours the carpenter earns $21x$; in y hours the carpenter earns $19.50y$; and in z hours the plumber earns $24z$.

***Translate*.**

$$
\begin{array}{ccc}
\underbrace{\text{Total time worked}} & \text{is} & \underbrace{\text{21.5 hr.}} \\
\downarrow & \downarrow & \downarrow \\
x + y + z & = & 21.5
\end{array}
$$

$$
\begin{array}{ccc}
\underbrace{\text{Total amount earned}} & \text{is} & \text{\$469.50.} \\
\downarrow & \downarrow & \downarrow \\
21x + 19.50y + 24z & = & 469.50
\end{array}
$$

$$
\begin{array}{cccccc}
\underbrace{\text{Plumber's hours}} & \text{are} & 2 & \underbrace{\text{more than}} & \underbrace{\text{carpenter's hours.}} \\
\downarrow & \downarrow & \downarrow & \downarrow & \downarrow \\
z & = & 2 & + & y
\end{array}
$$

We have a system of equations.

$$x + y + z = 21.5,$$
$$21x + 19.50y + 24z = 469.50,$$
$$z = 2 + y$$

***Solve*.** Solving the system, we get $(3.5, 8, 10)$.

***Check*.** The total time worked was $3.5 + 8 + 10$, or 21.5 hr. The amount earned was $\$21(3.5) + \$19.50(8) + \$24(10)$, or $\$469.50$. The time worked by the plumber, 10 hr, is 2 more than the time worked by the carpenter, 8 hr. The answer checks.

***State*.** The electrician worked 3.5 hr.

16. Graph: $y \geq x - 2$

We first graph the line $y = x - 2$. We draw a solid line because the inequality symbol is $\geq$. Test the point $(0, 0)$ to determine if it is a solution.

$$
\begin{array}{c}
y \geq x - 2 \\
\hline
0 \ ? \ 0 - 2 \\
\bigg| \ -2 \qquad \text{TRUE}
\end{array}
$$

Since $0 \geq -2$ is true, we shade the half-plane that contains $(0, 0)$.

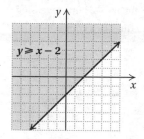

17. Graph: $x - 6y < -6$

We first graph the line $x - 6y = 6$. We draw the line dashed since the inequality symbol is $<$. We test the point $(0, 0)$ to determine if it is a solution.

$$\frac{x - 6y < -6}{0 - 6 \cdot 0 \ ? \ -6}$$
$$0 \ \bigg| \ \text{FALSE}$$

Since $0 < -6$ is false, we shade the half-plane that does not contain $(0, 0)$.

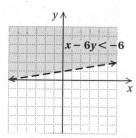

18. Graph: $x + y \geq 3$,
$\qquad\qquad x - y \geq 5$

Graph the lines $x + y = 3$ and $x - y = 5$ using solid lines. Indicate the region for each inequality by arrows and shade the region where they overlap.

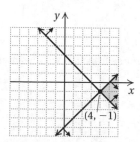

To find the vertex we solve the system of related equations:

$$x + y = 3,$$
$$x - y = 5$$

Solving, we obtain the vertex $(4, -1)$.

19. Graph: $2y - x \geq -4$, \quad (1)
$\qquad\qquad 2y + 3x \leq -6$, \quad (2)
$\qquad\qquad\qquad\ \ y \leq 0$, \quad (3)
$\qquad\qquad\qquad\ \ x \leq 0$ \quad (4)

Shade the intersection of the graphs of the four inequalities above.

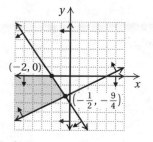

To find the two vertices we solve two systems of equations, as follows:

System of equations	Vertex
From (1) and (2)	$\left(-\dfrac{1}{2}, -\dfrac{9}{4}\right)$
From (2) and (3)	$(-2, 0)$

20. *Familiarize.* Let x, y, and z represent the amounts invested at 2%, 3%, and 5%, respectively. The amounts earned by the funds are $0.02x$, $0.03y$, and $0.05z$.

Translate.

$$\underbrace{\text{Amount invested}} \text{ was } \$30,000.$$
$$x + y + z = 30,000$$

$$\underbrace{\text{Total earnings}} \text{ were } \$990.$$
$$0.02x + 0.03y + 0.05z = 990$$

$$\underbrace{\text{Earnings at 5\%}} \text{ were } \$280 \ \underbrace{\text{more than}} \ \underbrace{\text{earnings at 2\%.}}$$
$$0.05z = 280 + 0.02x$$

We have a system of equations.

$$x + y + z = 30,000,$$
$$0.02x + 0.03y + 0.05z = 990,$$
$$0.05z = 280 + 0.02x$$

Solve. Solving the system, we get $(11,000, 9000, 10,000)$.

Check. The total amount invested was $\$11,000 + \$9000 + \$10,000$, or $\$30,000$. The amounts earned at 2%, 3%, and 5% are $0.02(\$11,000)$, $0.03(\$9000)$, and $0.05(\$10,000)$, or $\$220$, $\$270$, and $\$500$, respectively. Thus the total earnings were $\$220 + \$270 + \$500$, or $\$990$. The amount earned at 5%, $\$500$, is $\$280$ more than $\$220$, the amount earned at 2%. The answer checks.

State. $\$10,000$ was invested at 5%.

Answer B is correct.

21. Substituting -1 for x and 3 for $f(x)$, we have

$$3 = m(-1) + b, \text{ or } 3 = -m + b.$$

Substituting -2 for x and -4 for $f(x)$, we have

$$-4 = m(-2) + b, \text{ or } -4 = -2m + b.$$

Then we have a system of equations:

$$3 = -m + b, \quad (1)$$
$$-4 = -2m + b \quad (2)$$

We solve the system of equations.

$$\begin{aligned} -3 = m - b & \quad \text{Multiplying (1) by } -1 \\ \underline{-4 = -2m + b} & \quad (2) \\ -7 = -m & \\ 7 = m & \end{aligned}$$

Substitute 7 for m in (1) and solve for b.

$$3 = -7 + b$$
$$10 = b$$

Thus, we have $m = 7$ and $b = 10$.

Cumulative Review Chapters 1 - 3

1. $6y - 5(3y - 4) = 10$
    $6y - 15y + 20 = 10$
    $-9y + 20 = 10$
    $-9y = -10$
    $y = \dfrac{10}{9}$

    The solution is $\dfrac{10}{9}$.

2. $-3 + 5x = 2x + 15$
    $-3 + 3x = 15$
    $3x = 18$
    $x = 6$

    The solution is 6.

3. $A = \pi r^2 h$
    $\dfrac{A}{\pi r^2} = h$

4. $L = \dfrac{1}{3}m(k + p)$
    $3L = m(k + p)$
    $\dfrac{3L}{m} = k + p$
    $\dfrac{3L}{m} - k = p, \text{ or}$
    $\dfrac{3L - km}{m} = p$

5. $5x + 8 > 2x + 5$
    $3x + 8 > 5$
    $3x > -3$
    $x > -1$

    The solution set is $\{x | x > -1\}$, or $(-1, \infty)$.

6. $-12 \le -3x + 1 < 0$
    $-13 \le -3x < -1$
    $\dfrac{13}{3} \ge x > \dfrac{1}{3} \qquad$ Dividing by -3 and reversing the inequality symbol

    The solution set is $\left\{ x \left| \dfrac{1}{3} < x \le \dfrac{13}{3} \right. \right\}$, or $\left(\dfrac{1}{3}, \dfrac{13}{3} \right]$.

7. $2x - 10 \le -4 \quad or \quad x - 4 \ge 3$
    $2x \le 6 \qquad or \qquad x \ge 7$
    $x \le 3 \qquad or \qquad x \ge 7$

    The solution set is $\{x | x \le 3 \ or \ x \ge 7\}$, or $(-\infty, 3] \cup [7, \infty)$.

8. $|x + 1| = 4$
    $x + 1 = -4 \quad or \quad x + 1 = 4$
    $x = -5 \quad or \qquad x = 3$

    The solution set is $\{-5, 3\}$.

9. $|8y - 3| \ge 15$
    $8y - 3 \le -15 \quad or \quad 8y - 3 \ge 15$
    $8y \le -12 \quad or \qquad 8y \ge 18$
    $y \le -\dfrac{3}{2} \quad or \qquad y \ge \dfrac{9}{4}$

    The solution set is $\left\{ y \left| y \le -\dfrac{3}{2} \ or \ y \ge \dfrac{9}{4} \right. \right\}$, or
    $\left(-\infty, -\dfrac{3}{2} \right] \cup \left[\dfrac{9}{4}, \infty \right)$.

10. $|2x + 1| = |x - 4|$
    $2x + 1 = -(x - 4) \quad or \quad 2x + 1 = x - 4$
    $2x + 1 = -x + 4 \quad or \quad x + 1 = -4$
    $3x + 1 = 4 \qquad or \qquad x = -5$
    $3x = 3 \qquad or \qquad x = -5$
    $x = 1 \qquad or \qquad x = -5$

    The solutions set is $\{1, -5\}$.

11. $|-18 - (-7)| = |-18 + 7| = |-11| = 11$

12. $3y = 9$
    $y = 3$

    This is the equation of a horizontal line with y-intercept $(0, 3)$.

13. $f(x) = -\dfrac{1}{2}x - 3$

Find some function values. Then plot points and connect them.

$f(-4) = -\dfrac{1}{2}(-4) - 3 = 2 - 3 = -1$

$f(0) = -\dfrac{1}{2} \cdot 0 - 3 = 0 - 3 = -3$

$f(2) = -\dfrac{1}{2} \cdot 2 - 3 = -1 - 3 = -4$

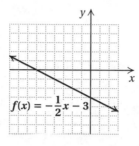

14. $3x - 1 = y$

Find some ordered pairs that are solutions, plot these points, and draw the graph.

When $x = -1$, $y = 3(-1) - 1 = -3 - 1 = -4$.

When $x = 0$, $y = 3 \cdot 0 - 1 = 0 - 1 = -1$.

When $x = 2$, $y = 3 \cdot 2 - 1 = 6 - 1 = 5$.

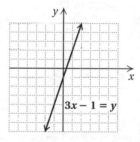

15. $3x + 5y = 15$

Find and plot the intercepts and then draw the graph.

y-intercept: $3 \cdot 0 + 5y = 15$

$5y = 15$

$y = 3$

The y-intercept is $(0, 3)$.

x-intercept: $3x + 5 \cdot 0 = 15$

$3x = 15$

$x = 5$

The x-intercept is $(5, 0)$.

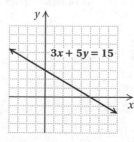

16. $y > 3x - 4$

First graph the line $y = 3x - 4$. We draw the line dashed because the inequality symbol is $>$. To determine which half-plane to shade, test a point not on the line. We try $(0, 0)$.

$$\begin{array}{c|l} y > 3x - 4 \\ \hline 0 \ ? \ 3 \cdot 0 - 4 \\ \quad\ \ 0 - 4 \\ \quad\ \ -4 & \text{TRUE} \end{array}$$

Since $0 > -4$ is true, we shade the half-plane that contains $(0, 0)$.

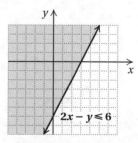

17. $2x - y \le 6$

First graph the line $2x - y = 6$. We draw the line solid because the inequality symbol is $\le$. To determine which half-plane to shade, test a point not on the line. We try $(0, 0)$.

$$\begin{array}{c|l} 2x - y \le 6 \\ \hline 2 \cdot 0 - 0 \ ? \ 6 \\ \quad 0 - 0 \\ \quad\ \ 0 & \text{TRUE} \end{array}$$

Since $0 \le 6$ is true, we shade the half-plane that contains $(0, 0)$.

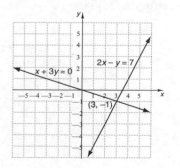

18. Graph both equations on the same set of axes.

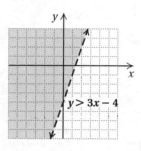

The solution appears to be $(3, -1)$.

Check:

$$\begin{array}{c|c}
2x - y = 7 & x + 3y = 0 \\
\hline
2 \cdot 3 - (-1) \ ? \ 7 & 3 + 3(-1) \ ? \ 0 \\
6 + 1 \ \Big| & 3 - 3 \ \Big| \\
7 \ \Big| \ \text{TRUE} & 0 \ \Big| \ \text{TRUE}
\end{array}$$

The solution is $(3, -1)$.

Since the system of equations has a solution, it is consistent. Since there is exactly one solution, the equations are independent.

19. $3x + 4y = 4$, (1)

$x = 2y + 2$ (2)

Substitute $2y + 2$ for x in Equation (1) and solve for y.

$$3(2y + 2) + 4y = 4$$
$$6y + 6 + 4y = 4$$
$$10y + 6 = 4$$
$$10y = -2$$
$$y = -\frac{1}{5}$$

Substitute $-\dfrac{1}{5}$ for y in Equation (2) and find x.

$$x = 2\left(-\frac{1}{5}\right) + 2 = -\frac{2}{5} + \frac{10}{5} = \frac{8}{5}$$

The solution is $\left(\dfrac{8}{5}, -\dfrac{1}{5}\right)$.

20. $3x + y = 2$, (1)

$\underline{6x - y = 7}$ (2)

$9x \quad\ = 9$ Adding

$x = 1$

Substitute 1 for x in Equation (1) and solve for y.

$$3 \cdot 1 + y = 2$$
$$3 + y = 2$$
$$y = -1$$

The solution is $(1, -1)$.

21. $4x + 3y = 5$, (1)

$3x + 2y = 3$ (2)

We use the multiplication principle with both equations and then add.

$8x + 6y = 10$ Multiplying (1) by 2

$\underline{-9x - 6y = -9}$ Multiplying (2) by -3

$-x \quad\ = 1$

$x = -1$

Substitute -1 for x in Equation (2) and solve for y.

$$3(-1) + 2y = 3$$
$$-3 + 2y = 3$$
$$2y = 6$$
$$y = 3$$

The solution is $(-1, 3)$.

22. $x - y + z = 1$, (1)

$2x + y + z = 3$, (2)

$x + y - 2z = 4$ (3)

We start by eliminating y from two different pairs of equations.

$$\begin{array}{ll}
x - y + z = 1 & (1) \\
\underline{2x + y + z = 3} & (2) \\
3x \quad\ + 2z = 4 & (4)
\end{array}$$

$$\begin{array}{ll}
x - y + z = 1 & (1) \\
\underline{x + y - 2z = 4} & (3) \\
2x \quad\ - z = 5 & (5)
\end{array}$$

Now solve the system of equations (4) and (5).

$$\begin{array}{ll}
3x + 2z = 4 & (4) \\
\underline{4x - 2z = 10} & \text{Multiplying (5) by 2} \\
7x \quad\ = 14 & \\
x = 2 &
\end{array}$$

$2 \cdot 2 - z = 5$ Substituting 2 for x in (5)

$4 - z = 5$

$-z = 1$

$z = -1$

$2 \cdot 2 + y + (-1) = 3$ Substituting 2 for x and -1 for z in (2)

$4 + y - 1 = 3$

$y + 3 = 3$

$y = 0$

The solution is $(2, 0, -1)$.

23. Graph: $x + y \le -3$,

$x - y \le 1$

Graph the lines $x + y = -3$ and $x - y = 1$ using solid lines. We indicate the region for each inequality by arrows and shade the region where they overlap.

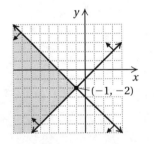

To find the vertex we solve the system of related equations:

$x + y = -3$,

$x - y = 1$

Solving, we obtain the vertex $(-1, -2)$.

24. Graph: $4y - 3x \geq -12$, (1)

$4y + 3x \geq -36$, (2)

$y \leq 0$, (3)

$x \leq 0$ (4)

Shade the intersection of the graphs of the four inequalities.

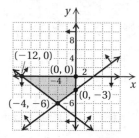

To find the vertices we solve four systems of inequalities. From (1) and (2) we obtain the vertex $(-4, -6)$. From (1) and (4) we obtain the vertex $(0, -3)$. From (2) and (3) we obtain the vertex $(-12, 0)$. From (3) and (4) we obtain the vertex $(0, 0)$.

25. a) The domain is the set of all x-values in the graph, $\{-5, -3, -1, 1, 3\}$.

b) The range is the set of all y-values in the graph, $\{-3, -2, 1, 4, 5\}$.

c) The y-value that corresponds to $x = -3$ is -2, so $f(-3) = -2$.

d) The function value -5 corresponds to the input 3.

26. $f(x) = \dfrac{7}{2x - 1}$

We cannot calculate $\dfrac{7}{2x - 1}$ when the denominator is 0.

We find the value of x for which $2x - 1 = 0$.

$2x - 1 = 0$

$2x = 1$

$x = \dfrac{1}{2}$

The domain is $\left\{ x \middle| x \text{ is a real number } and \ x \neq \dfrac{1}{2} \right\}$ or $\left(-\infty, \dfrac{1}{2} \right) \cup \left(\dfrac{1}{2}, \infty \right)$.

27. $g(x) = 1 - 2x^2$

$g(-1) = 1 - 2(-1)^2 = 1 - 2 \cdot 1 = 1 - 2 = -1$

$g(0) = 1 - 2 \cdot 0^2 = 1 - 2 \cdot 0 = 1 - 0 = 1$

$g(3) = 1 - 2 \cdot 3^2 = 1 - 2 \cdot 9 = 1 - 18 = -17$

28. $5y - 4x = 20$

$5y = 4x + 20$

$y = \dfrac{4}{5}x + 4$

The equation is now in slope-intercept form, $y = mx + b$. The slope is $\dfrac{4}{5}$, and the y-intercept is $(0, 4)$.

29. $y = mx + b$

$2 = -3 \cdot 5 + b$ Substituting

$2 = -15 + b$

$17 = b$

We have $y = -3x + 17$.

30. First we find the slope of the line

$$m = \dfrac{5 - (-3)}{-3 - (-1)} = \dfrac{5 + 3}{-3 + 1} = \dfrac{8}{-2} = -4$$

Now we use the point-slope equation.

$y - (-3) = -4[x - (-1)]$

$y + 3 = -4(x + 1)$

$y + 3 = -4x - 4$

$y = -4x - 7$

31. First we write each equation in slope-intercept form.

$x - 2y - 4 \qquad\qquad 4x + 2y = 1$

$-2y = -x + 4 \qquad\qquad 2y = -4x + 1$

$y = \dfrac{1}{2}x - 2 \qquad\qquad y = -2x + \dfrac{1}{2}$

The slopes, $\dfrac{1}{2}$ and -2, are different, so the lines are not parallel. The product of the slopes is $\dfrac{1}{2}(-2)$, or -1, so the lines are perpendicular.

32. First we find the slope of the given line.

$3x - 9y = 2$

$-9y = -3x + 2$

$y = \dfrac{1}{3}x - \dfrac{2}{9}$

The slope of the given line is $\dfrac{1}{3}$. A line parallel to this line will also have slope $\dfrac{1}{3}$. We use the point-slope equation.

$y - 2 = \dfrac{1}{3}[x - (-6)]$

$y - 2 = \dfrac{1}{3}(x + 6)$

$y - 2 = \dfrac{1}{3}x + 2$

$y = \dfrac{1}{3}x + 4$

33. *Familiarize.* Let l = the length of one piece of wire, in meters. Then $\dfrac{2}{3}l$ = the length of the other piece.

Translate. The total length of the wire is 10 m, so we have

$$l + \dfrac{2}{3}l = 10.$$

Solve. We solve the equation.

$l + \dfrac{2}{3}l = 10$

$\dfrac{5}{3}l = 10$

$l = \dfrac{3}{5} \cdot 10 = 6$

If $l = 6$, then $\frac{2}{3}l = \frac{2}{3} \cdot 6 = 4$.

Check. 6 m $+ 4$ m $= 10$ m, and 4 m is $\frac{2}{3}$ of 6 m. The answer checks.

State. The wire should be cut so that one piece is 6 m and the other is 4 m.

34. Familiarize. Let $s =$ the score on the last test.

Translate. The sum of the four scores must be at least 360 so we have

$$87 + 94 + 91 + s \geq 360.$$

Solve. We solve the equation

$$87 + 94 + 91 + s \geq 360$$
$$272 + s \geq 360$$
$$s \geq 88$$

Check. If the last score is 88, then the total of the four scores is $97 + 94 + 91 + 88$, or 360. A score of 88 or higher will yield a total score of 360 or more. The answer checks.

State. A score of 88 or higher will give Adam an A. In terms of an inequality, we have $\{s | s \geq 88\}$.

35. Familiarize. Let $x =$ the number of scientific calculators ordered and $y =$ the number of graphing calculators. Then the scientific calculators cost $9x$, and the graphing calculators cost $78y$.

$$\underbrace{\text{Total number of calculators}}_{x + y} \underset{=}{\text{ is }} \underset{45}{\text{ 45.}}$$

$$\underbrace{\text{Total cost of order}}_{9x + 78y} \underset{=}{\text{ is }} \underset{2268}{\$2268.}$$

We have a system of equations.

$$x + y = 45, \quad (1)$$
$$9x + 78y = 2268 \quad (2)$$

Solve. We begin by multiplying both sides of Equation (1) by -9 and then adding.

$$-9x - 9y = -405$$
$$\underline{9x + 78y = 2268}$$
$$69y = 1863$$
$$y = 27$$

Substitute 27 for y in Equation (1) and solve for x.

$$x + 27 = 45$$
$$x = 18$$

Check. The total number of calculators is $18 + 27$, or 45. The total cost of the order is $\$9 \cdot 18 + \$78 \cdot 27$, or $\$162 + \$2106$, or $\$2268$. The answer checks.

State. 18 scientific calculators and 27 graphing calculators were ordered.

36. Familiarize. Let $x =$ the number of liters of 15% solution and $y =$ the number of liters of 25% solution to be used. The mixture contains 18%(30 L), or 0.18(30 L), or 5.4 L of alcohol.

Translate. We organize the information in a table.

	15% solution	25% solution	Mixture
Number of liters	x	y	30
Percent of alcohol	15%	25%	18%
Amount of alcohol	$0.15x$	$0.25y$	5.4 L

We get one equation from the "Number of liters" row of the table.

$$x + y = 30$$

The last row of the table yields a second equation.

$$0.15x + 0.25y = 5.4$$

After clearing decimals, we have the following system of equations.

$$x + y = 30, \quad (1)$$
$$15x + 25y = 540 \quad (2)$$

Solve. We begin by multiplying both sides of Equation (1) by -15 and then adding.

$$-15x - 15y = -450$$
$$\underline{15x + 25y = 540}$$
$$10y = 90$$
$$y = 9$$

Substitute 9 for y in Equation (1) and solve for x.

$$x + 9 = 30$$
$$x = 21$$

Check. The mixture contains 21 L $+ 9$ L, or 30 L. The amount of alcohol in the mixture is $0.15(21) + 0.25(9)$, or 3.15 L $+ 2.25$, or 5.4 L. The answer checks.

State. 21 L of 15% solution and 9 L of 25% solution should be used.

37. Familiarize. Let $t =$ the time, in hours, the first train travels. Then $t - 3 =$ the time the second train travels. The trains will travel the same distance, d km. We organize the information in a table.

	Distance	Rate	Time
First train	d	80	t
Second train	d	120	$t - 3$

Translate. Using $d = rt$ in each row of the table, we have a system of equations.

$$d = 80t, \quad (1)$$
$$d = 120(t - 3) \quad (2)$$

Solve. Substitute $80t$ for d in Equation (2) and solve for t.

$$80t = 120(t - 3)$$
$$80t = 120t - 360$$
$$-40t = -360$$
$$t = 9$$

In 9 hr, the first train travels $80 \cdot 9$, or 720 km.

Check. We have seen that the first train will travel 720 km. In $9 - 3$ or 6 hr, the second train will travel $120 \cdot 6$, or 720 km. The distances are the same, so the answer checks.

State. The second train will overtake the first train 720 km from the station.

38. Familiarize. Let x, y, and z represent the amounts spent for electricity, rent, and telephone, respectively.

Translate.

$$\underbrace{\text{Total amount spent}}_{x+y+z} \; \underbrace{\text{was}}_{=} \; \underbrace{\$680.}_{680}$$

$$\underbrace{\text{Electric bill}}_{x} \; \underbrace{\text{was}}_{=} \; \underbrace{\text{one-fourth}}_{\frac{1}{4}} \; \underbrace{\text{of}}_{\cdot} \; \underbrace{\text{the rent.}}_{y}$$

$$\underbrace{\text{Rent}}_{y} \; \underbrace{\text{was}}_{=} \; \underbrace{\$400}_{400} \; \underbrace{\text{more than}}_{+} \; \underbrace{\text{the phone bill.}}_{z}$$

We have a system of equations.

$$x + y + z = 680,$$
$$x = \frac{1}{4}y,$$
$$y = 400 + z$$

Solve. Solving the system of equations, we get $(120, 480, 80)$.

Check. The total amount spent was $\$120 + \$480 + \$80$, or $\$680$. The electric bill, $\$120$, is one-fourth of the rent, $\$480$. The rent was $\$400$ more than the phone bill, $\$80$. The answer checks.

State. The electric bill was $\$120$.

39. We find a function that relates the amount spent on radio advertising to weekly sales increase. We have the data points $(1000, \, 101{,}000)$ and $(1250, \, 126{,}000)$.

$$m = \frac{126{,}000 - 101{,}000}{1250 - 1000} = \frac{25{,}000}{250} = 100$$

Now we can use the slope-intercept equation.

$$101{,}000 = 100(1000) + b$$
$$101{,}000 = 100{,}000 + b$$
$$1000 = b$$

Then we have the function

$$S(a) = 100a + 1000$$

Now we find $S(1500)$.

$$S(1500) = 100 \cdot 1500 + 1000 = 150{,}000 + 1000 = 151{,}000$$

Sales would increase about $\$151{,}000$ when $\$1500$ is spent on radio advertising.

40. We have the points $(5 - 3)$ and $(-4, 2)$.

$$m = \frac{-3 - 2}{5 - (-4)} = \frac{-5}{9} = -\frac{5}{9}$$

We use the slope-intercept equation to find b.

$$-3 = -\frac{5}{9} \cdot 5 + b$$
$$-3 = -\frac{25}{9} + b$$
$$-\frac{27}{9} + \frac{25}{9} = b$$
$$-\frac{2}{9} = b$$

Chapter 4

Polynomials and Polynomial Functions

RC1. (c)

RC3. (a)

RC5. (h)

RC7. (g)

1. $-9x^4 - x^3 + 7x^2 + 6x - 8$

Term	$-9x^4$	$-x^3$	$7x^2$	$6x$	-8
Degree	4	3	2	1	0
Degree of polynomial	4				
Leading term	$-9x^4$				
Leading coefficient	-9				
Constant term	-8				

3. $t^3 + 4t^7 + s^2t^4 - 2$

Term	t^3	$4t^7$	s^2t^4	-2
Degree	3	7	6	0
Degree of polynomial	7			
Leading term	$4t^7$			
Leading coefficient	4			
Constant term	-2			

5. $u^7 + 8u^2v^6 + 3uv + 4u - 1$

Term	u^7	$8u^2v^6$	$3uv$	$4u$	-1
Degree	7	8	2	1	0
Degree of polynomial	8				
Leading term	$8u^2v^6$				
Leading coefficient	8				
Constant term	-1				

7. $-4y^3 - 6y^2 + 7y + 23$

9. $-xy^3 + x^2y^2 + x^3y + 1$

11. $-9b^5y^5 - 8b^2y^3 + 2by$

13. $5 + 12x - 4x^3 + 8x^5$

15. $3xy^3 + x^2y^2 - 9x^3y + 2x^4$

17. $-7ab + 4ax - 7ax^2 + 4x^6$

19. $P(x) = 3x^2 - 2x + 5$

$P(4) = 3 \cdot 4^2 - 2 \cdot 4 + 5$

$= 48 - 8 + 5$

$= 45$

$P(-2) = 3(-2)^2 - 2(-2) + 5$

$= 12 + 4 + 5$

$= 21$

$P(0) = 3 \cdot 0^2 - 2 \cdot 0 + 5$

$= 0 - 0 + 5$

$- 5$

21. $p(x) = 9x^3 + 8x^2 - 4x - 9$

$p(-3) = 9(-3)^3 + 8(-3)^2 - 4(-3) - 9$

$= -243 + 72 + 12 - 9$

$= -168$

$p(0) = 9 \cdot 0^3 + 8 \cdot 0^2 - 4 \cdot 0 - 9$

$= 0 + 0 - 0 - 9$

$= -9$

$p(1) = 9 \cdot 1^3 + 8 \cdot 1^2 - 4 \cdot 1 - 9$

$= 9 + 8 - 4 - 9$

$= 4$

$p\left(\frac{1}{2}\right) = 9\left(\frac{1}{2}\right)^3 + 8\left(\frac{1}{2}\right)^2 - 4 \cdot \frac{1}{2} - 9$

$= \frac{9}{8} + 2 - 2 - 9$

$= -\frac{63}{8},$ or $-7\frac{7}{8}$

23. $P(x) = 0.0157x^3 + 0.1163x^2 - 1.3396x + 3.7063$

$P(25) = 0.0157 \cdot 25^3 + 0.1163 \cdot 25^2 - 1.3396 \cdot 25 + 3.7063$

≈ 288 watts

25. a) Locate 2 on the horizontal axis. From there move vertically to the graph and then horizontally to the $M(t)$-axis. This locates a value of about 340. Thus, about 340 mg of ibuprofen is in the the bloodstream 2 hr after 400 mg have been swallowed.

b) Locate 4 on the horizontal axis. From there move vertically to the graph and then horizontally to the $M(t)$-axis. This locates a value of about 190. Thus, about 190 mg of ibuprofen is in the the bloodstream 4 hr after 400 mg have been swallowed.

c) Locate 5 on the horizontal axis. From there move vertically to the graph and then horizontally to the $M(t)$-axis. This locates a value of about 65. Thus, $M(5) \approx 65$.

d) Locate 3 on the horizontal axis. From there move vertically to the graph and then horizontally to the $M(t)$-axis. This locates a value of about 300. Thus, $M(3) \approx 300$.

27. $R(x) = 240x - 0.5x^2$

a) $\begin{aligned} R(50) &= 240(50) - 0.5(50)^2 \\ &= 240(50) - 0.5(2500) \\ &= 12,000 - 1250 \\ &= \$10,750 \end{aligned}$

b) $\begin{aligned} R(95) &= 240(95) - 0.5(95)^2 \\ &= 240(95) - 0.5(9025) \\ &= 22,800 - 4512.5 \\ &= \$18,287.50 \end{aligned}$

29. We subtract:
$$\begin{aligned} P(x) &= R(x) - C(x) \\ &= 280x - 0.4x^2 - (7000 + 0.6x^2) \\ &= 280x - 0.4x^2 + (-7000 - 0.6x^2) \quad \text{Adding the} \\ &\qquad\qquad\qquad\qquad\qquad\qquad\qquad\quad \text{opposite} \\ &= -x^2 + 280x - 7000 \end{aligned}$$
The total profit is given by $P(x) = -x^2 + 280x - 7000$.

31. Substitute 162 for G, 78 for W_1, and 68 for L_2.
$$\begin{aligned} M &= G - W_1 - L_2 + 1 \\ &= 162 - 78 - 68 + 1 \\ &= 17 \end{aligned}$$
The magic number is 17.

33. Substitute 162 for G, 86 for W_1, and 69 for L_2.
$$\begin{aligned} M &= G - W_1 - L_2 + 1 \\ &= 162 - 86 - 69 + 1 \\ &= 8 \end{aligned}$$
The magic number is 8.

35. $6x^2 - 7x^2 + 3x^2 = (6 - 7 + 3)x^2 = 2x^2$

37. $\begin{aligned} & 7x - 2y - 4x + 6y \\ &= (7 - 4)x + (-2 + 6)y \\ &= 3x + 4y \end{aligned}$

39. $\begin{aligned} & 3a + 9 - 2 + 8a - 4a + 7 \\ &= (3 + 8 - 4)a + (9 - 2 + 7) \\ &= 7a + 14 \end{aligned}$

41. $\begin{aligned} & 3a^2b + 4b^2 - 9a^2b - 6b^2 \\ &= (3 - 9)a^2b + (4 - 6)b^2 \\ &= -6a^2b - 2b^2 \end{aligned}$

43. $\begin{aligned} & 8x^2 - 3xy + 12y^2 + x^2 - y^2 + 5xy + 4y^2 \\ &= (8 + 1)x^2 + (-3 + 5)xy + (12 - 1 + 4)y^2 \\ &= 9x^2 + 2xy + 15y^2 \end{aligned}$

45. $\begin{aligned} & 4x^2y - 3y + 2xy^2 - 5x^2y + 7y + 7xy^2 \\ &= (4 - 5)x^2y + (-3 + 7)y + (2 + 7)xy^2 \\ &= -x^2y + 4y + 9xy^2 \end{aligned}$

47. $\begin{aligned} & (3x^2 + 5y^2 + 6) + (2x^2 - 3y^2 - 1) \\ &= (3 + 2)x^2 + (5 - 3)y^2 + (6 - 1) \\ &= 5x^2 + 2y^2 + 5 \end{aligned}$

49. $\begin{aligned} & (2a + 3b - c) + (4a - 2b + 2c) \\ &= (2 + 4)a + (3 - 2)b + (-1 + 2)c \\ &= 6a + b + c \end{aligned}$

51. $\begin{aligned} & (a^2 - 3b^2 + 4c^2) + (-5a^2 + 2b^2 - c^2) \\ &= (1 - 5)a^2 + (-3 + 2)b^2 + (4 - 1)c^2 \\ &= -4a^2 - b^2 + 3c^2 \end{aligned}$

53. $\begin{aligned} & (x^2 + 3x - 2xy - 3) + (-4x^2 - x + 3xy + 2) \\ &= (1 - 4)x^2 + (3 - 1)x + (-2 + 3)xy + (-3 + 2) \\ &= -3x^2 + 2x + xy - 1 \end{aligned}$

55. $\begin{aligned} & (7x^2y - 3xy^2 + 4xy) + (-2x^2y - xy^2 + xy) \\ &= (7 - 2)x^2y + (-3 - 1)xy^2 + (4 + 1)xy \\ &= 5x^2y - 4xy^2 + 5xy \end{aligned}$

57. $\begin{aligned} & (2r^2 + 12r - 11) + (6r^2 - 2r + 4) + (r^2 - r - 2) \\ &= (2 + 6 + 1)r^2 + (12 - 2 - 1)r + (-11 + 4 - 2) \\ &= 9r^2 + 9r - 9 \end{aligned}$

59. $\begin{aligned} & \left(\frac{2}{3}xy + \frac{5}{6}xy^2 + 5.1x^2y\right) + \left(-\frac{4}{5}xy + \frac{3}{4}xy^2 - 3.4x^2y\right) \\ &= \left(\frac{2}{3} - \frac{4}{5}\right)xy + \left(\frac{5}{6} + \frac{3}{4}\right)xy^2 + (5.1 - 3.4)x^2y \\ &= \left(\frac{10}{15} - \frac{12}{15}\right)xy + \left(\frac{10}{12} + \frac{9}{12}\right)xy^2 + 1.7x^2y \\ &= -\frac{2}{15}xy + \frac{19}{12}xy^2 + 1.7x^2y \end{aligned}$

61. $5x^3 - 7x^2 + 3x - 6$

a) $-(5x^3 - 7x^2 + 3x - 6)$ Writing an inverse sign in front

b) $-5x^3 + 7x^2 - 3x + 6$ Writing the opposite of each term

63. $-13y^2 + 6ay^4 - 5by^2$

a) $-(-13y^2 + 6ay^4 - 5by^2)$

b) $13y^2 - 6ay^4 + 5by^2$

65. $\begin{aligned} & (7x - 2) - (-4x + 5) \\ &= (7x - 2) + (4x - 5) \quad \text{Adding the opposite} \\ &= 11x - 7 \end{aligned}$

67. $\quad (-3x^2 + 2x + 9) - (x^2 + 5x - 4)$

$= (-3x^2 + 2x + 9) + (-x^2 - 5x + 4)$ Adding the opposite

$= -4x^2 - 3x + 13$

69. $\quad (5a + c - 2b) - (3a + 2b - 2c)$

$= (5a + c - 2b) + (-3a - 2b + 2c)$

$= 2a + 3c - 4b$

71. $\quad (3x^2 - 2x - x^3) - (5x^2 - 8x - x^3)$

$= (3x^2 - 2x - x^3) + (-5x^2 + 8x + x^3)$

$= -2x^2 + 6x$

73. $\quad (5a^2 + 4ab - 3b^2) - (9a^2 - 4ab + 2b^2)$

$= (5a^2 + 4ab - 3b^2) + (-9a^2 + 4ab - 2b^2)$

$= -4a^2 + 8ab - 5b^2$

75. $\quad (6ab - 4a^2b + 6ab^2) - (3ab^2 - 10ab - 12a^2b)$

$= (6ab - 4a^2b + 6ab^2) + (-3ab^2 + 10ab + 12a^2b)$

$= 16ab + 8a^2b + 3ab^2$

77. $\quad (0.09y^4 - 0.052y^3 + 0.93) -$
$\qquad (0.03y^4 - 0.084y^3 + 0.94y^2)$

$= (0.09y^4 - 0.052y^3 + 0.93) +$
$\qquad (-0.03y^4 + 0.084y^3 - 0.94y^2)$

$= 0.06y^4 + 0.032y^3 - 0.94y^2 + 0.93$

79. $\quad \left(\dfrac{5}{8}x^4 - \dfrac{1}{4}x^2 - \dfrac{1}{2}\right) - \left(-\dfrac{3}{8}x^4 + \dfrac{3}{4}x^2 + \dfrac{1}{2}\right)$

$= \left(\dfrac{5}{8}x^4 - \dfrac{1}{4}x^2 - \dfrac{1}{2}\right) + \left(\dfrac{3}{8}x^4 - \dfrac{3}{4}x^2 - \dfrac{1}{2}\right)$

$= x^4 - x^2 - 1$

81. Graph: $f(x) = \dfrac{2}{3}x - 1$.

We find some ordered pairs that are solutions. By choosing multiples of 3 for x we can avoid fractional values when calculating $f(x)$.

For $x = -3$, $f(-3) = \dfrac{2}{3}(-3) - 1 = -2 - 1 = -3$.

For $x = 0$, $f(0) = \dfrac{2}{3} \cdot 0 - 1 = 0 - 1 = -1$.

For $x = 3$, $f(3) = \dfrac{2}{3} \cdot 3 - 1 = 2 - 1 = 1$.

x	$f(x)$	$(x, f(x))$
-3	-3	$(-3, -3)$
0	-1	$(0, -1)$
3	1	$(3, 1)$

We plot these points and draw the graph.

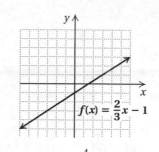

$f(x) = \dfrac{2}{3}x - 1$

83. Graph: $g(x) = \dfrac{4}{x - 3}$.

We choose x-values and find the corresponding values of $g(x)$. We list the results in a table.

x	$f(x)$	$(x, f(x))$
-5	$-\dfrac{1}{2}$	$\left(-5, -\dfrac{1}{2}\right)$
-3	$-\dfrac{2}{3}$	$\left(-3, -\dfrac{2}{3}\right)$
-1	-1	$(-1, -1)$
0	$-\dfrac{4}{3}$	$\left(0, -\dfrac{4}{3}\right)$
1	-2	$(1, -2)$
2	-4	$(2, -4)$
4	4	$(4, 4)$
5	2	$(5, 2)$
6	$\dfrac{4}{3}$	$\left(6, \dfrac{4}{3}\right)$
7	1	$(7, 1)$
9	$\dfrac{2}{3}$	$\left(9, \dfrac{2}{3}\right)$

Since $4/0$ is undefined, we cannot use 3 as a first coordinate. The graph has two branches, one on each side of the line $x = 3$.

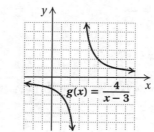

$g(x) = \dfrac{4}{x - 3}$

85. $\quad -3x - 7 = x - 5$

$\qquad -7 = 4x - 5$ Adding $3x$

$\qquad -2 = 4x$ Adding 5

$\qquad -\dfrac{1}{2} = x$ Dividing by 4

The solution is $-\dfrac{1}{2}$.

87. $x - (7 - x) = 2(x + 3)$

$x - 7 + x = 2x + 6$

$2x - 7 = 2x + 6$

$-7 = 6$ Subtracting $2x$

We get a false equation, so there is no solution.

89. Graph: $y = \frac{4}{3}x + 2$.

First we plot the y-intercept $(0, 2)$. Starting at $(0, 2)$ and using the slope, $\frac{4}{3}$, we find another point by moving 4 units up and 3 units to the right. We get to the point $(3, 6)$. We can also think of the slope as $\frac{-4}{-3}$. Starting again at $(0, 2)$, move 4 units down and 3 units to the left to the point $(-3, -2)$. Plot the points and draw the line.

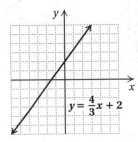

91. Graph: $y = 0.4x - 3$

We write the equation as $y = \frac{4}{10}x - 3$, or $y = \frac{2}{5}x - 3$. First we plot the y-intercept $(0, -3)$. Starting at $(0, -3)$ and using the slope, $\frac{2}{5}$, we find another point by moving 2 units up and 5 units to the right. We get to the point $(5, -1)$. We can also think of the slope as $\frac{-2}{-5}$. Starting again at $(0, -3)$, move 2 units down and 5 units to the left to the point $(-5, -5)$. Plot the points and draw the line.

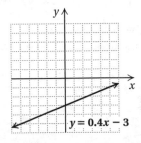

93. First we find the number of truffles in the display.

$$N(x) = \frac{1}{6}x^3 + \frac{1}{2}x^2 + \frac{1}{3}x$$

$$N(5) = \frac{1}{6} \cdot 5^3 + \frac{1}{2} \cdot 5^2 + \frac{1}{3} \cdot 5$$

$$= \frac{1}{6} \cdot 125 + \frac{1}{2} \cdot 25 + \frac{5}{3}$$

$$= \frac{125}{6} + \frac{25}{2} + \frac{5}{3}$$

$$= \frac{125}{6} + \frac{75}{6} + \frac{10}{6}$$

$$= \frac{210}{6} = 35$$

There are 35 truffles in the display. Now find the volume of one truffle. Each truffle's diameter is 3 cm, so the radius is $\frac{3}{2}$, or 1.5 cm.

$$V(r) = \frac{4}{3}\pi r^3$$

$$V(1.5) \approx \frac{4}{3}(3.14)(1.5)^3 \approx 14.13 \text{ cm}^3$$

Finally, multiply the number of truffles and the volume of a truffle to find the total volume of chocolate.

$$35(14.13 \text{ cm}^3) = 494.55 \text{ cm}^3$$

The display contains about 494.55 cm^3 of chocolate.

95. $f(0) = 0.05 \cdot 0^4 - 0^2 + 5 = 5$, but the graph shows that $f(0) = -5$.

97. $(47x^{4a} + 3x^{3a} + 22x^{2a} + x^a + 1) +$

$(37x^{3a} + 8x^{2a} + 3)$

$= 47x^{4a} + (3 + 37)x^{3a} + (22 + 8)x^{2a} + x^a + (1 + 3)$

$= 47x^{4a} + 40x^{3a} + 30x^{2a} + x^a + 4$

Exercise Set 4.2

RC1. True; see page 331 in the text.

RC3. True; see page 333 in the text.

1. $8y^2 \cdot 3y = (8 \cdot 3)(y^2 \cdot y) = 24y^3$

3. $2x(-10x^2y) = [2(-10)](x \cdot x^2)(y) = -20x^3y$

5. $(5x^5y^4)(-2xy^3) = [5(-2)](x^5 \cdot x)(y^4 \cdot y^3) = -10x^6y^7$

7. $2z(7 - x)$

$= 2z \cdot 7 - 2z \cdot x$ Using a distributive law

$= 14z - 2zx$ Multiplying monomials

9. $6ab(a + b)$

$= 6ab \cdot a + 6ab \cdot b$ Using a distributive law

$= 6a^2b + 6ab^2$ Multiplying monomials

11. $5cd(3c^2d - 5cd^2)$

$= 5cd \cdot 3c^2d - 5cd \cdot 5cd^2$

$= 15c^3d^2 - 25c^2d^3$

13. $(5x + 2)(3x - 1)$

$= 15x^2 - 5x + 6x - 2$ FOIL

$= 15x^2 + x - 2$

15. $(s + 3t)(s - 3t)$

$= s^2 - (3t)^2$ $(A + B)(A - B) = A^2 - B^2$

$= s^2 - 9t^2$

17. $(x - y)(x - y)$

$= x^2 - 2xy + y^2$ $(A - B)^2 = A^2 - 2AB + B^2$

19. $(x^3 + 8)(x^3 - 5)$

$= x^6 - 5x^3 + 8x^3 - 40$ FOIL

$= x^6 + 3x^3 - 40$

21. $\quad (a^2 - 2b^2)(a^2 - 3b^2)$

$= a^4 - 3a^2b^2 - 2a^2b^2 + 6b^4 \qquad \text{FOIL}$

$= a^4 - 5a^2b^2 + 6b^4$

23. $\quad (x - 4)(x^2 + 4x + 16)$

$= (x - 4)(x^2) + (x - 4)(4x) + (x - 4)(16)$

$\qquad\qquad\qquad\qquad \text{Using a distributive law}$

$= x(x^2) - 4(x^2) + x(4x) - 4(4x) + x(16) - 4(16)$

$\qquad\qquad\qquad\qquad \text{Using a distributive law}$

$= x^3 - 4x^2 + 4x^2 - 16x + 16x - 64$

$\qquad\qquad\qquad \text{Multiplying monomials}$

$= x^3 - 64 \qquad \text{Collecting like terms}$

25. $\quad (x + y)(x^2 - xy + y^2)$

$= (x + y)x^2 + (x + y)(-xy) + (x + y)(y^2)$

$= x(x^2) + y(x^2) + x(-xy) + y(-xy) + x(y^2) + y(y^2)$

$= x^3 + x^2y - x^2y - xy^2 + xy^2 + y^3$

$= x^3 + y^3$

27.
$$
\begin{array}{r}
a^2 + a - 1 \\
a^2 + 4a - 5 \\
\hline
-5a^2 - 5a + 5 \quad \text{Multiplying by } -5 \\
4a^3 + 4a^2 - 4a \qquad \text{Multiplying by } 4a \\
a^4 + a^3 - a^2 \qquad\quad \text{Multiplying by } a^2 \\
\hline
a^4 + 5a^3 - 2a^2 - 9a + 5 \quad \text{Adding}
\end{array}
$$

29.
$$
\begin{array}{r}
4a^2b - 2ab + 3b^2 \\
ab - 2b + a \\
\hline
4a^3b - 2a^2b + 3ab^2 \qquad (1) \\
-6b^3 \qquad\quad +4ab^2 - 8a^2b^2 \qquad (2) \\
3ab^3 \qquad\qquad\qquad - 2a^2b^2 + 4a^3b^2 \quad (3) \\
\hline
3ab^3 - 6b^3 + 4a^3b - 2a^2b + 7ab^2 - 10a^2b^2 + 4a^3b^2 \quad (4)
\end{array}
$$

(1) Multiplying by a

(2) Multiplying by $-2b$

(3) Multiplying by ab

(4) Adding

31. $\quad \left(x + \dfrac{1}{4}\right)\left(x + \dfrac{1}{4}\right)$

$= x^2 + \dfrac{1}{4}x + \dfrac{1}{4}x + \dfrac{1}{16} \qquad \text{FOIL}$

$= x^2 + \dfrac{1}{2}x + \dfrac{1}{16}$

33. $\quad \left(\dfrac{1}{2}x - \dfrac{2}{3}\right)\left(\dfrac{1}{4}x + \dfrac{1}{3}\right)$

$= \dfrac{1}{8}x^2 + \dfrac{1}{6}x - \dfrac{1}{6}x - \dfrac{2}{9} \qquad \text{FOIL}$

$= \dfrac{1}{8}x^2 - \dfrac{2}{9}$

35. $\quad (1.3x - 4y)(2.5x + 7y)$

$= 3.25x^2 + 9.1xy - 10xy - 28y^2 \qquad \text{FOIL}$

$= 3.25x^2 - 0.9xy - 28y^2$

37. $\quad (a + 8)(a + 5)$

$= a^2 + 5a + 8a + 40 \qquad \text{FOIL}$

$= a^2 + 13a + 40$

39. $\quad (y + 7)(y - 4)$

$= y^2 - 4y + 7y - 28 \qquad \text{FOIL}$

$= y^2 + 3y - 28$

41. $\quad \left(3a + \dfrac{1}{2}\right)^2$

$= (3a)^2 + 2(3a)\left(\dfrac{1}{2}\right) + \left(\dfrac{1}{2}\right)^2$

$\qquad\qquad (A + B)^2 = A^2 + 2AB + B^2$

$= 9a^2 + 3a + \dfrac{1}{4}$

43. $\quad (x - 2y)^2$

$= x^2 - 2(x)(2y) + (2y)^2$

$\qquad\qquad (A - B)^2 = A^2 - 2AB + B^2$

$= x^2 - 4xy + 4y^2$

45. $\quad \left(b - \dfrac{1}{3}\right)\left(b - \dfrac{1}{2}\right)$

$= b^2 - \dfrac{1}{2}b - \dfrac{1}{3}b + \dfrac{1}{6} \qquad \text{FOIL}$

$= b^2 - \dfrac{3}{6}b - \dfrac{2}{6}b + \dfrac{1}{6}$

$= b^2 - \dfrac{5}{6}b + \dfrac{1}{6}$

47. $\quad (2x + 9)(x + 2)$

$= 2x^2 + 4x + 9x + 18 \qquad \text{FOIL}$

$= 2x^2 + 13x + 18$

49. $\quad (20a - 0.16b)^2$

$= (20a)^2 - 2(20a)(0.16b) + (0.16b)^2$

$\qquad\qquad (A - B)^2 = A^2 - 2AB + B^2$

$= 400a^2 - 6.4ab + 0.0256b^2$

51. $\quad (2x - 3y)(2x + y)$

$= 4x^2 + 2xy - 6xy - 3y^2 \qquad \text{FOIL}$

$= 4x^2 - 4xy - 3y^2$

53. $\quad (x^3 + 2)^2$

$= (x^3)^2 + 2 \cdot x^3 \cdot 2 + 2^2 \quad (A + B)^2 = A^2 + 2AB + B^2$

$= x^6 + 4x^3 + 4$

55. $\quad (2x^2 - 3y^2)^2$

$= (2x^2)^2 - 2(2x^2)(3y^2) + (3y^2)^2$

$\qquad\qquad (A - B)^2 = A^2 - 2AB + B^2$

$= 4x^4 - 12x^2y^2 + 9y^4$

57. $\quad (a^3b^2 + 1)^2$

$= (a^3b^2)^2 + 2 \cdot a^3b^2 \cdot 1 + 1^2$

$\qquad\qquad (A + B)^2 = A^2 + 2AB + B^2$

$= a^6b^4 + 2a^3b^2 + 1$

59. $\quad (0.1a^2 - 5b)^2$

$\quad = (0.1a^2)^2 - 2(0.1a^2)(5b) + (5b)^2$

$\qquad\qquad\qquad (A - B)^2 = A^2 - 2AB + B^2$

$\quad = 0.01a^4 - a^2 b + 25b^2$

61. $A = P(1 + i)^2$

$\quad A = P(1 + 2i + i^2) \qquad$ FOIL

$\quad A = P + 2Pi + Pi^2 \qquad$ Multiplying by P

63. $\quad (d + 8)(d - 8)$

$\quad = d^2 - 8^2 \qquad (A + B)(A - B) = A^2 - B^2$

$\quad = d^2 - 64$

65. $\quad (2c + 3)(2c - 3)$

$\quad = (2c)^2 - 3^2 \qquad (A + B)(A - B) = A^2 - B^2$

$\quad = 4c^2 - 9$

67. $\quad (6m - 5n)(6m + 5n)$

$\quad = (6m)^2 - (5n)^2 \qquad (A + B)(A - B) = A^2 - B^2$

$\quad = 36m^2 - 25n^2$

69. $\quad (x^2 + yz)(x^2 - yz)$

$\quad = (x^2)^2 - (yz)^2 \qquad (A + B)(A - B) = A^2 - B^2$

$\quad = x^4 - y^2 z^2$

71. $\quad (-mn + m^2)(mn + m^2)$

$\quad = (m^2 - mn)(m^2 + mn)$

$\quad = (m^2)^2 - (mn)^2 \qquad (A + B)(A - B) = A^2 - B^2$

$\quad = m^4 - m^2 n^2$

73. $\quad (-3pt + 4p^2)(4p^2 + 3pt)$

$\quad = (4p^2 - 3pt)(4p^2 + 3pt)$

$\quad = (4p^2)^2 - (3pt)^2$

$\quad = 16p^4 - 9p^2 t^2$

75. $\quad \left(\dfrac{1}{2}p - \dfrac{2}{3}q\right)\left(\dfrac{1}{2}p + \dfrac{2}{3}q\right)$

$\quad = \left(\dfrac{1}{2}p\right)^2 - \left(\dfrac{2}{3}q\right)^2$

$\quad = \dfrac{1}{4}p^2 - \dfrac{4}{9}q^2$

77. $\quad (x + 1)(x - 1)(x^2 + 1)$

$\quad = (x^2 - 1^2)(x^2 + 1)$

$\quad = (x^2 - 1)(x^2 + 1)$

$\quad = (x^2)^2 - 1^2$

$\quad = x^4 - 1$

79. $\quad (a - b)(a + b)(a^2 - b^2)$

$\quad = (a^2 - b^2)(a^2 - b^2)$

$\quad = (a^2 - b^2)^2$

$\quad = (a^2)^2 - 2(a^2)(b^2) + (b^2)^2$

$\quad = a^4 - 2a^2 b^2 + b^4$

81. $\quad (a + b + 1)(a + b - 1)$

$\quad = [(a + b) + 1][(a + b) - 1]$

$\quad = (a + b)^2 - 1^2$

$\quad = a^2 + 2ab + b^2 - 1$

83. $\quad (2x + 3y + 4)(2x + 3y - 4)$

$\quad = [(2x + 3y) + 4][(2x + 3y) - 4]$

$\quad = (2x + 3y)^2 - 4^2$

$\quad = 4x^2 + 12xy + 9y^2 - 16$

85. $f(x) = 5x + x^2$

$\quad f(t - 1) = 5(t - 1) + (t - 1)^2$

$\qquad\quad = 5t - 5 + t^2 - 2t + 1$

$\qquad\quad = t^2 + 3t - 4$

$\quad f(p + 1) = 5(p + 1) + (p + 1)^2$

$\qquad\quad = 5p + 5 + p^2 + 2p + 1$

$\qquad\quad = p^2 + 7p + 6$

$\quad\; f(a + h) - f(a)$

$\quad = [5(a + h) + (a + h)^2] - [5a + a^2]$

$\quad = 5a + 5h + a^2 + 2ah + h^2 - 5a - a^2$

$\quad = h^2 + 5h + 2ah$

$\quad f(t - 2) + c = 5(t - 2) + (t - 2)^2 + c$

$\qquad\quad = 5t - 10 + t^2 - 4t + 4 + c$

$\qquad\quad = t^2 + t - 6 + c$

$\quad f(a) + 5 = 5a + a^2 + 5$

87. $f(x) = 3x^2 - 7x + 8$

$\quad\; f(t - 1)$

$\quad = 3(t - 1)^2 - 7(t - 1) + 8$

$\quad = 3(t^2 - 2t + 1) - 7t + 7 + 8$

$\quad = 3t^2 - 6t + 3 - 7t + 7 + 8$

$\quad = 3t^2 - 13t + 18$

$\quad\; f(p + 1)$

$\quad = 3(p + 1)^2 - 7(p + 1) + 8$

$\quad = 3(p^2 + 2p + 1) - 7p - 7 + 8$

$\quad = 3p^2 + 6p + 3 - 7p - 7 + 8$

$\quad = 3p^2 - p + 4$

$\quad\; f(a + h) - f(a)$

$\quad = [3(a + h)^2 - 7(a + h) + 8] - [3a^2 - 7a + 8]$

$\quad = 3(a^2 + 2ah + h^2) - 7a - 7h + 8 - 3a^2 + 7a - 8$

$\quad = 3a^2 + 6ah + 3h^2 - 7a - 7h + 8 - 3a^2 + 7a - 8$

$\quad = 3h^2 + 6ah - 7h$

$\quad\; f(t - 2) + c$

$\quad = 3(t - 2)^2 - 7(t - 2) + 8 + c$

$\quad = 3(t^2 - 4t + 4) - 7t + 14 + 8 + c$

$\quad = 3t^2 - 12t + 12 - 7t + 14 + 8 + c$

$\quad = 3t^2 - 19t + 34 + c$

$\quad f(a) + 5 = 3a^2 - 7a + 8 + 5$

$\qquad\quad = 3a^2 - 7a + 13$

89. $f(x) = 5x - x^2$

$f(t-1) = 5(t-1) - (t-1)^2$

$\quad = 5t - 5 - (t^2 - 2t + 1)$

$\quad = 5t - 5 - t^2 + 2t - 1$

$\quad = -t^2 + 7t - 6$

$f(p+1) = 5(p+1) - (p+1)^2$

$\quad = 5p + 5 - (p^2 + 2p + 1)$

$\quad = 5p + 5 - p^2 - 2p - 1$

$\quad = -p^2 + 3p + 4$

$f(a+h) - f(a)$

$= [5(a+h) - (a+h)^2] - [5a - a^2]$

$= 5a + 5h - (a^2 + 2ah + h^2) - 5a + a^2$

$= 5a + 5h - a^2 - 2ah - h^2 - 5a + a^2$

$= -h^2 + 5h - 2ah$

$f(t-2) + c = 5(t-2) - (t-2)^2 + c$

$\quad = 5t - 10 - (t^2 - 4t + 4) + c$

$\quad = 5t - 10 - t^2 + 4t - 4 + c$

$\quad = -t^2 + 9t - 14 + c$

$f(a) + 5 = 5a - a^2 + 5$

91. $f(x) = 4 + 3x - x^2$

$f(t-1) = 4 + 3(t-1) - (t-1)^2$

$\quad = 4 + 3t - 3 - (t^2 - 2t + 1)$

$\quad = 4 + 3t - 3 - t^2 + 2t - 1$

$\quad = -t^2 + 5t$

$f(p+1) = 4 + 3(p+1) - (p+1)^2$

$\quad = 4 + 3p + 3 - (p^2 + 2p + 1)$

$\quad = 4 + 3p + 3 - p^2 - 2p - 1$

$\quad = -p^2 + p + 6$

$f(a+h) - f(a)$

$= [4 + 3(a+h) - (a+h)^2] - [4 + 3a - a^2]$

$= 4 + 3a + 3h - (a^2 + 2ah + h^2) - 4 - 3a + a^2$

$= 4 + 3a + 3h - a^2 - 2ah - h^2 - 4 - 3a + a^2$

$= -h^2 + 3h - 2ah$

$f(t-2) + c = 4 + 3(t-2) - (t-2)^2 + c$

$\quad = 4 + 3t - 6 - (t^2 - 4t + 4) + c$

$\quad = 4 + 3t - 6 - t^2 + 4t - 4 + c$

$\quad = -t^2 + 7t - 6 + c$

$f(a) + 5 = 4 + 3a - a^2 + 5$

$\quad = -a^2 + 3a + 9$

93. Familiarize. When Rachel's sister catches up with her, their distances traveled are the same. Let $d =$ this distance. Also let $t =$ the time for Rachel's sister to catch up with her. Then $t + 2 =$ the time Rachel travels before her sister catches up with her. Organize the information in a table.

	Distance	Rate	Time
	d	$=$	$r \cdot t$
Rachel	d	55	$t + 2$
Sister	d	75	t

Translate. Using $d = r \cdot t$ in each row of the table, we get a system of equations.

$$d = 55(t+2), \quad (1)$$
$$d = 75t \qquad (2)$$

Solve. We use the substitution method. First, substitute $75t$ for d in Equation (1) and solve for t.

$$75t = 55(t+2)$$
$$75t = 55t + 110$$
$$20t = 110$$
$$t = 5.5$$

If the sister's time is 5.5 hr, then Rachel's time is $5.5 + 2$, or 7.5 hr.

Check. At 55 mph, in 7.5 hr Rachel travels $55(7.5)$, or 412.5 mi. At 75 mph, in 5.5 hr her sister travels $75(5.5)$, or 412.5 mi. Since the distances are the same, the answer checks.

State. Rachel's sister will catch up with her 5.5 hr after the sister leaves (or 7.5 hr after Rachel leaves).

95.
$$5x + 9y = 2 \quad (1)$$
$$\underline{4x - 9y = 10 \quad (2)}$$
$$9x \quad\quad = 12 \quad \text{Adding}$$
$$x = \frac{4}{3}$$

Substitute $\frac{4}{3}$ for x in one of the equations and solve for y.

$$5x + 9y = 2 \qquad (1)$$
$$5 \cdot \frac{4}{3} + 9y = 2$$
$$\frac{20}{3} + 9y = 2$$
$$9y = -\frac{14}{3}$$
$$y = -\frac{14}{27}$$

The ordered pair $\left(\frac{4}{3}, -\frac{14}{27}\right)$ checks. It is the solution.

97.
$$2x - 3y = 1, \quad (1)$$
$$4x - 6y = 2 \quad (2)$$

We multiply Equation (1) by -2 and add.

$$-4x + 6y = -2$$
$$\underline{4x - 6y = 2}$$
$$0 = 0$$

We get an equation that is true for all values of x and y. There are infinitely many solutions.

99. Left to the student

101. $(z^{n^2})^{n^3}(z^{4n^3})^{n^2} = z^{n^5} \cdot z^{4n^5} = z^{n^5 + 4n^5} = z^{5n^5}$

103. $(r^2 + s^2)^2(r^2 + 2rs + s^2)(r^2 - 2rs + s^2)$

$= (r^2 + s^2)^2(r + s)^2(r - s)^2$

$= (r^2 + s^2)^2[(r + s)(r - s)]^2$

$= (r^2 + s^2)^2(r^2 - s^2)^2$

$= [(r^2 + s^2)(r^2 - s^2)]^2$

$= (r^4 - s^4)^2$

$= r^8 - 2r^4 s^4 + s^8$

105. $\left(3x^5 - \dfrac{5}{11}\right)^2$

$= (3x^5)^2 - 2(3x^5)\left(\dfrac{5}{11}\right) + \left(\dfrac{5}{11}\right)^2$

$= 9x^{10} - \dfrac{30}{11}x^5 + \dfrac{25}{121}$

107. $(x^a + y^b)(x^a - y^b)(x^{2a} + y^{2b})$

$= (x^{2a} - y^{2b})(x^{2a} + y^{2b})$

$= x^{4a} - y^{4b}$

109. $(x - 1)(x^2 + x + 1)(x^3 + 1)$

$= (x^3 + x^2 + x - x^2 - x - 1)(x^3 + 1)$

$= (x^3 - 1)(x^3 + 1)$

$= x^6 - 1$

Exercise Set 4.3

RC1. To factor a polynomial is to express it as a <u>product</u>.

RC3. The expression $x(x - 2)$ is a <u>factorization</u> of $x^2 - 2x$.

RC5. The expression $4x$ is a <u>common</u> factor of the terms of the polynomial $8x + 12x^3$.

1. $6a^2 + 3a$

$= 3a \cdot 2a + 3a \cdot 1$

$= 3a(2a + 1)$

3. $x^3 + 9x^2$

$= x^2 \cdot x + x^2 \cdot 9$

$= x^2(x + 9)$

5. $8x^2 - 4x^4$

$= 4x^2 \cdot 2 - 4x^2 \cdot x^2$

$= 4x^2(2 - x^2)$

7. $4x^2 y - 12xy^2$

$= 4xy \cdot x - 4xy \cdot 3y$

$= 4xy(x - 3y)$

9. $3y^2 - 3y - 9$

$= 3 \cdot y^2 - 3 \cdot y - 3 \cdot 3$

$= 3(y^2 - y - 3)$

11. $4ab - 6ac + 12ad$

$= 2a \cdot 2b - 2a \cdot 3c + 2a \cdot 6d$

$= 2a(2b - 3c + 6d)$

13. $10a^4 + 15a^2 - 25a - 30$

$= 5 \cdot 2a^4 + 5 \cdot 3a^2 - 5 \cdot 5a - 5 \cdot 6$

$= 5(2a^4 + 3a^2 - 5a - 6)$

15. $15x^2 y^5 z^3 - 12x^4 y^4 z^7$

$= 3x^2 y^4 z^3 \cdot 5y - 3x^2 y^4 z^3 \cdot 4x^2 z^4$

$= 3x^2 y^4 z^3(5y - 4x^2 z^4)$

17. $14a^4 b^3 c^5 + 21a^3 b^5 c^4 - 35a^4 b^4 c^3$

$= 7a^3 b^3 c^3 \cdot 2ac^2 + 7a^3 b^3 c^3 \cdot 3b^2 c - 7a^3 b^3 c^3 \cdot 5ab$

$= 7a^3 b^3 c^3(2ac^2 + 3b^2 c - 5ab)$

19. $-5x - 45 = -5(x + 9)$

21. $-6a - 84 = -6(a + 14)$

23. $-2x^2 + 2x - 24 = -2(x^2 - x + 12)$

25. $-3y^2 + 24y = -3y(y - 8)$

27. $-a^4 + 2a^3 - 13a^2 - 1$

$= -1(a^4 - 2a^3 + 13a^2 + 1)$, or

$= -(a^4 - 2a^3 + 13a^2 + 1)$

29. $-3y^3 + 12y^2 - 15y + 24 = -3(y^3 - 4y^2 + 5y - 8)$

31. $\pi r^2 h + \dfrac{4}{3}\pi r^3 = \pi r^2 \cdot h + \pi r^2 \cdot \dfrac{4}{3}r = \pi r^2\left(h + \dfrac{4}{3}r\right)$

We could also factor as follows:

$\pi r^2 h + \dfrac{4}{3}\pi r^3 = 1 \cdot \pi r^2 h + \dfrac{4}{3}\pi r^3$

$= \dfrac{3}{3}\pi r^2 h + \dfrac{4}{3}\pi r^3$

$= \dfrac{1}{3}\pi r^2 \cdot 3h + \dfrac{1}{3}\pi r^2 \cdot 4r$

$= \dfrac{1}{3}\pi r^2(3h + 4r)$

33. a) $h(t) = -16t^2 + 72t$

$h(t) = -8t(2t - 9)$

b) $h(2) = -16 \cdot 2^2 + 72 \cdot 2 = 80$

$h(2) = -8(2)(2 \cdot 2 - 9) = -8(2)(-5) = 80$

The expressions have the same value for $t = 2$, so the factorization is probably correct.

35. $R(x) = 280x - 0.4x^2$

$R(x) = 0.4x(700 - x)$

37. $a(b - 2) + c(b - 2) = (b - 2)(a + c)$

39. $(x - 2)(x + 5) + (x - 2)(x + 8)$

$= (x - 2)[(x + 5) + (x + 8)]$

$= (x - 2)(2x + 13)$

41. $y^8 - 7y^7 + y - 7$

$= y^7(y - 7) + 1 \cdot (y - 7)$

$= (y - 7)(y^7 + 1)$

43. $\quad ac + ad + bc + bd$

$= a(c + d) + b(c + d)$

$= (c + d)(a + b)$

45. $\quad b^3 - b^2 + 2b - 2$

$= b^2(b - 1) + 2(b - 1)$

$= (b - 1)(b^2 + 2)$

47. $\quad y^3 + 8y^2 - 5y - 40$

$= y^2(y + 8) - 5(y + 8)$

$= (y + 8)(y^2 - 5)$

49. $\quad 24x^3 + 72x - 36x^2 - 108$

$= 12(2x^3 + 6x - 3x^2 - 9)$

$= 12[2x(x^2 + 3) - 3(x^2 + 3)]$

$= 12(x^2 + 3)(2x - 3)$

51. $a^4 - a^3 + a^2 + a = a(a^3 - a^2 + a + 1)$

53. $\quad 2y^4 + 6y^2 - 5y^2 - 15$

$= 2y^2(y^2 + 3) - 5(y^2 + 3)$

$= (y^2 + 3)(2y^2 - 5)$

55. $|x - 3| = 10$

$x - 3 = -10 \ \ or \ \ x - 3 = 10$

$x = -7 \ \ or \ \ \ \ \ \ \ x = 13$

The solution set is $\{-7, 13\}$.

57. $|2 - x| \le 12$

$-12 \le 2 - x \le 12$

$-14 \le -x \le 10$

$14 \ge x \ge -10 \quad$ Reversing the inequality symbol

The solution set is $\{x| -10 \le x \le 14\}$, or $[-10, 14]$.

59. $8 \le x - 7 \le 10$

$15 \le x \le 17$

The solution set is $\{x|15 \le x \le 17\}$, or $[15, 17]$.

61. $2x - 7 > 6 \quad or \quad 3x + 1 < 2$

$2x > 13 \quad or \quad \quad 3x < 1$

$x > \dfrac{13}{2} \quad or \quad \quad x < \dfrac{1}{3}$

The solution set is $\left\{ x \Big| x < \dfrac{1}{3} \ or \ x > \dfrac{13}{2} \right\}$, or

$\left(-\infty, \dfrac{1}{3} \right) \cup \left(\dfrac{13}{2}, \infty \right)$.

63. $x^5 y^4 + \underline{\quad} = x^3 y(\underline{\quad} + xy^5)$

The term that goes in the first blank is the product of $x^3 y$ and xy^5, or $x^4 y^6$.

The term that goes in the second blank is the expression that is multiplied with $x^3 y$ to obtain $x^5 y^4$, or $x^2 y^3$. Thus, we have

$x^5 y^4 + x^4 y^6 = x^3 y(x^2 y^3 + xy^5)$.

65. $\quad rx^2 - rx + 5r + sx^2 - sx + 5s$

$= r(x^2 - x + 5) + s(x^2 - x + 5)$

$= (x^2 - x + 5)(r + s)$

67. $\quad a^4 x^4 + a^4 x^2 + 5a^4 + a^2 x^4 + a^2 x^2 + 5a^2 +$

$\quad\quad\quad\quad\quad\quad\quad\quad\quad 5x^4 + 5x^2 + 25$

$= a^4(x^4 + x^2 + 5) + a^2(x^4 + x^2 + 5) + 5(x^4 + x^2 + 5)$

$= (x^4 + x^2 + 5)(a^4 + a^2 + 5)$

69. $x^{1/3} - 7x^{4/3} = x^{1/3} \cdot 1 - 7x \cdot x^{1/3} = x^{1/3}(1 - 7x)$

71. $\quad x^{1/3} - 5x^{1/2} + 3x^{3/4}$

$= x^{4/12} - 5x^{6/12} + 3x^{9/12}$

$= x^{4/12}(1 - 5x^{2/12} + 3x^{5/12})$

$= x^{1/3}(1 - 5x^{1/6} + 3x^{5/12})$

73. $\quad 3a^{n+1} + 6a^n - 15a^{n+2}$

$= 3a^n \cdot a + 3a^n \cdot 2 - 3a^n(5a^2)$

$= 3a^n(a + 2 - 5a^2)$

75. $\quad 7y^{2a+b} - 5y^{a+b} + 3y^{a+2b}$

$= y^{a+b} \cdot 7y^a - y^{a+b}(5) + y^{a+b} \cdot 3y^b$

$= y^{a+b}(7y^a - 5 + 3y^b)$

Exercise Set 4.4

RC1. To factor $x^2 + 19x - 20$, we look for a factorization of -20 in which the positive factor has the greater absolute value. This is due to the fact that the coefficient of the middle term is positive. The answer is (c).

RC3. To factor $x^2 + 7x + 12$, we look for a factorization of 12 in which both factors are positive. This is due to the fact that the coefficient of the middle term is positive. The answer is (a).

1. $x^2 + 13x + 36$

We look for two numbers whose product is 36 and whose sum is 13. Since both 36 and 13 are positive, we need consider only positive factors.

Pairs of Factors	Sums of Factors
1, 36	37
2, 18	20
3, 12	15
4, 9	13
6, 6	12

The numbers we need are 4 and 9. The factorization is $(x + 4)(x + 9)$.

3. $t^2 - 8t + 15$

Since the constant term, 15, is positive and the coefficient of the middle term, -8, is negative, we look for a factorization of 15 in which both factors are negative. Their sum must be -8.

Pairs of Factors	Sums of Factors
$-1, -15$	-16
$-3, -5$	-8

The numbers we need are -3 and -5. The factorization is $(t-3)(t-5)$.

5. $x^2 - 8x - 33$

Since the constant term, -33, is negative, we look for a factorization of -33 in which one factor is positive and one factor is negative. The sum of the factors must be -8, so the negative factor must have the larger absolute value. Thus we consider only pairs of factors in which the negative factor has the larger absolute value.

Pairs of Factors	Sums of Factors
$1, -33$	-32
$3, -11$	-8

The numbers we want are 3 and -11. The factorization is $(x+3)(x-11)$.

7. $2y^2 - 16y + 32$

   $= 2(y^2 - 8y + 16)$ Removing the common factor

We now factor $y^2 - 8y + 16$. We look for two numbers whose product is 16 and whose sum is -8. Since the constant term is positive and the coefficient of the middle term is negative, we look for a factorization of 16 in which both factors are negative.

Pairs of Factors	Sums of Factors
$-1, -16$	-17
$-2, -8$	-10
$-4, -4$	-8

The numbers we need are -4 and -4.

$$y^2 - 8y + 16 = (y-4)(y-4)$$

We must not forget to include the common factor 2.

$$2y^2 - 16y + 32 = 2(y-4)(y-4).$$

9. $p^2 + 3p - 54$

Since the constant term is negative, we look for a factorization of -54 in which one factor is positive and one factor is negative. We consider only pairs of factors in which the positive factor has the larger absolute value, since the sum of the factors, 3, is positive.

Pairs of Factors	Sums of Factors
$54, -1$	53
$27, -2$	25
$18, -3$	15
$9, -6$	3

The numbers we need are 9 and -6. The factorization is $(p+9)(p-6)$.

11. $12x + x^2 + 27 = x^2 + 12x + 27$

We look for two numbers whose product is 27 and whose sum is 12. Since both 27 and 12 are positive, we need consider only positive factors.

Pairs of Factors	Sums of Factors
$1, 27$	28
$3, 9$	12

The numbers we want are 3 and 9. The factorization is $(x+3)(x+9)$.

13. $y^2 - \dfrac{2}{3}y + \dfrac{1}{9}$

Since the constant term, $\dfrac{1}{9}$, is positive and the coefficient of the middle term, $-\dfrac{2}{3}$, is negative, we look for a factorization of $\dfrac{1}{9}$ in which both factors are negative. Their sum must be $-\dfrac{2}{3}$.

Pairs of Factors	Sums of Factors
$-1, -\dfrac{1}{9}$	$-\dfrac{10}{9}$
$-\dfrac{1}{3}, -\dfrac{1}{3}$	$-\dfrac{2}{3}$

The numbers we need are $-\dfrac{1}{3}$ and $-\dfrac{1}{3}$. The factorization is $\left(y - \dfrac{1}{3}\right)\left(y - \dfrac{1}{3}\right)$.

15. $t^2 - 4t + 3$

Since the constant term, 3, is positive and the coefficient of the middle term, -4, is negative, we look for a factorization of 3 in which both factors are negative. Their sum must be -4. The only possibility is $-1, -3$. These are the numbers we need. The factorization is $(t-1)(t-3)$.

17. $5x + x^2 - 14 = x^2 + 5x - 14$

Since the constant term, -14, is negative, we look for a factorization of -14 in which one factor is positive and one factor is negative. Their sum must be 5, so the positive factor must have the larger absolute value. We consider only pairs of factors in which the positive factor has the larger absolute value.

Pairs of Factors	Sums of Factors
$-1, 14$	13
$-2, 7$	5

The numbers we need are -2 and 7. The factorization is $(x-2)(x+7)$.

19. $x^2 + 5x + 6$

We look for two numbers whose product is 6 and whose sum is 5. Since 6 and 5 are both positive, we need consider only positive factors.

Pairs of Factors	Sums of Factors
$1, 6$	7
$2, 3$	5

The numbers we need are 2 and 3. The factorization is $(x+2)(x+3)$.

21. $56 + x - x^2 = -x^2 + x + 56 = -1(x^2 - x - 56)$

We now factor $x^2 - x - 56$. Since the constant term, -56, is negative, we look for a factorization of -56 in which one factor is positive and one factor is negative. We consider only pairs of factors in which the negative factor has the larger absolute value, since the sum of the factors, -1, is negative.

Pairs of Factors	Sums of Factors
$-56,\ 1$	-55
$-28,\ 2$	-26
$-14,\ 4$	-10
$-8,\ 7$	-1

The numbers we need are -8 and 7. Thus, $x^2 - x - 56 = (x - 8)(x + 7)$. We must not forget to include the factor that was factored out earlier:

$$-x^2 + x + 56$$
$$= -1(x^2 - x - 56)$$
$$= -1(x - 8)(x + 7)$$
$$= (-x + 8)(x + 7) \qquad \text{Multiplying } x - 8 \text{ by } -1$$
$$= (x - 8)(-x - 7) \qquad \text{Multiplying } x + 7 \text{ by } -1$$

23. $32y + 4y^2 - y^3$

There is a common factor, y. We also factor out -1 in order to make the leading coefficient positive.

$$32y + 4y^2 - y^3 = -y(-32 - 4y + y^2)$$
$$= -y(y^2 - 4y - 32)$$

Now we factor $y^2 - 4y - 32$. Since the constant term, -32, is negative, we look for a factorization of -32 in which one factor is positive and one factor is negative. We consider only pairs of factors in which the negative factor has the larger absolute value, since the sum of the factors, -4, is negative.

Pairs of Factors	Sums of Factors
$-32,\ 1$	-31
$-16,\ 2$	-14
$-8,\ 4$	-4

The numbers we need are -8 and 4. Thus, $y^2 - 4y - 32 = (y - 8)(y + 4)$. We must not forget to include the common factor:

$$32y + 4y^2 - y^3$$
$$= -y(y^2 - 4y - 32)$$
$$= -y(y - 8)(y + 4)$$
$$= -1 \cdot y(y - 8)(y + 4)$$
$$= y(-y + 8)(y + 4) \qquad \text{Multiplying } y - 8 \text{ by } -1$$
$$= y(y - 8)(-y - 4) \qquad \text{Multiplying } y + 4 \text{ by } -1$$

25. $x^4 + 11x^2 - 80$

First make a substitution. We let $u = x^2$, so $u^2 = x^4$. Then we consider $u^2 + 11u - 80$. We look for pairs of factors of -80, one positive and one negative, such that the positive factor has the larger absolute value and the sum of the factors is 11.

Pairs of Factors	Sums of Factors
$80,\ -1$	79
$40,\ -2$	38
$20,\ -4$	16
$16,\ -5$	11
$10,\ -8$	2

The numbers we need are 16 and -5. Then $u^2 + 11u - 80 = (u + 16)(u - 5)$. Replacing u by x^2 we obtain the factorization of the original trinomial: $(x^2 + 16)(x^2 - 5)$.

27. $x^2 - 3x + 7$

There are no factors of 7 whose sum is -3. This trinomial is not factorable into binomials with integer coefficients.

29. $x^2 + 12xy + 27y^2$

We look for numbers p and q such that $x^2 + 12xy + 27y^2 = (x + py)(x + qy)$. Our thinking is much the same as if we were factoring $x^2 + 12x + 27$. We look for factors of 27 whose sum is 12. Those factors are 9 and 3. Then

$$x^2 + 12xy + 27y^2 = (x + 9y)(x + 3y).$$

31. $2x^2 - 8x - 90 = 2(x^2 - 4x - 45)$

Now we factor $x^2 - 4x - 45$. We look for two numbers whose product is -45 and whose sum is -4. The numbers we need are -9 and 5. We have: $2x^2 - 8x - 90 = 2(x^2 - 4x - 45) = 2(x - 9)(x + 5)$

33. $-z^2 + 36 - 9z = -z^2 - 9z + 36 = -1(z^2 + 9z - 36)$

Now we factor $z^2 + 9z - 36$. We look for two numbers whose product is -36 and whose sum is 9. The numbers we need are 12 and -3. We have:

$$-z^2 - 9z + 36$$
$$= -1(z^2 + 9z - 36)$$
$$= -1(z + 12)(z - 3)$$
$$= (-z - 12)(z - 3) \qquad \text{Multiplying } z + 12 \text{ by } -1$$
$$= (z + 12)(-z + 3) \qquad \text{Multiplying } z - 3 \text{ by } -1$$

35. $x^4 + 50x^2 + 49$

Substitute u for x^2 (and hence u^2 for x^4). Consider $u^2 + 50u + 49$. We look for a pair of positive factors of 49 whose sum is 50.

Pairs of Factors	Sums of Factors
$7,\ 7$	14
$1,\ 49$	50

The numbers we need are 1 and 49. Then $u^2 + 50u + 49 = (u + 1)(u + 49)$. Replacing u by x^2 we have

$$x^4 + 50x^2 + 49 = (x^2 + 1)(x^2 + 49).$$

37. $x^6 + 11x^3 + 18$

Substitute u for x^3 (and hence u^2 for x^6). Consider $u^2 + 11u + 18$. We look for two numbers whose product is 18 and whose sum is 11. Since both 18 and 11 are positive, we need consider only positive factors.

Pairs of Factors	Sums of Factors
1, 18	19
2, 9	11
3, 6	9

The numbers we need are 2 and 9. Then $u^2 + 11u + 18 = (u+2)(u+9)$. Replacing u by x^3 we obtain the factorization of the original trinomial: $(x^3 + 2)(x^3 + 9)$.

39. $x^8 - 11x^4 + 24$

Substitute u for x^4 (and hence u^2 for x^8). Consider $u^2 - 11u + 24$. Since the constant term, 24, is positive and the coefficient of the middle term, -11, is negative, we look for a factorization of 24 in which both factors are negative. Their sum must be -11.

Pairs of Factors	Sums of Factors
$-1, -24$	-25
$-2, -12$	-14
$-3, -8$	-11
$-4, -6$	-10

The numbers we need are -3 and -8. Then $u^2 - 11u + 24 = (u-3)(u-8)$. Replacing u by x^4 we obtain the factorization of the original trinomial: $(x^4 - 3)(x^4 - 8)$.

41. $y^2 - 0.8y + 0.16$

We look for two numbers whose product is 0.16 and whose sum is -0.8. The numbers we need are -0.4 and -0.4. Then
$$y^2 - 0.8y + 0.16 = (y - 0.4)(y - 0.4)$$

43. $12 - b^{10} - b^{20} = -1(-12 + b^{10} + b^{20}) = -1(b^{20} + b^{10} - 12)$

We factor $b^{20} + b^{10} - 12$. We look for two numbers whose product is -12 and whose sum is 1. The numbers we need are 4 and -3. We have:
$$12 - b^{10} - b^{20}$$
$$= -1(b^{20} + b^{10} - 12)$$
$$= -1(b^{10} + 4)(b^{10} - 3)$$
$$= (-b^{10} - 4)(b^{10} - 3) \quad \text{Multiplying } b^{10} + 4 \text{ by } -1$$
$$= (b^{10} + 4)(-b^{10} + 3) \quad \text{Multiplying } b^{10} - 3 \text{ by } -1$$

45. *Familiarize.* Let $x =$ the number of pounds of Countryside rice and $y =$ the number of pounds of Mystic rice to be used in the mixture.

Translate. We organize the information in a table.

	Countryside	Mystic	Mixture
Number of pounds	x	y	25
Percent of wild rice	10%	50%	35%
Pounds of wild rice	$0.1x$	$0.5x$	$0.35(25)$

From the "Number of pounds" row of the table we get one equation.
$$x + y = 25$$
We get a second equation from the last row of the table.

$0.1x + 0.5x = 0.35(25)$, or
$0.1x + 0.5x = 8.75$

Clearing decimals, we have the following system of equations:
$$x + y = 25, \quad (1)$$
$$10x + 50y = 875 \quad (2)$$

Solve. We use the elimination method to solve the system of equations. We multiply Equation (1) by -10 and add.
$$-10x - 10y = -250$$
$$\underline{10x + 50y = \quad 875}$$
$$40y = \quad 625$$
$$y = 15.625$$

Now substitute 15.625 for y in Equation (1) and solve for x.
$$x + 15.625 = 25$$
$$x = 9.375$$

Check. The total weight of the mixture is $9.375 + 15.625$, or 25 lb. The amount of wild rice in the mixture is $0.1(9.375) + 0.5(15.625) = 0.9375 + 7.8125 = 8.75$ lb. This is 35% of 25 lb. The numbers check.

State. 9.375 lb, or $9\frac{3}{8}$ lb, of Countryside Rice and 15.625 lb, or $15\frac{5}{8}$ lb of Mystic Rice should be used.

47. The graph is that of a function, because no vertical line can cross the graph at more than one point.

49. The graph is not that of a function, because a vertical line can cross the graph at more than one point.

51. $f(x) = x^2 - 2$

$f(x)$ can be calculated for any number x, so the domain is the set of all real numbers.

53. $f(x) = \dfrac{3}{4x - 7}$

We cannot calculate $\dfrac{3}{4x - 7}$ when the denominator is 0.
$$4x - 7 = 0$$
$$4x = 7$$
$$x = \frac{7}{4}$$

The domain is $\left\{ x \middle| x \text{ is a real number and } x \neq \dfrac{7}{4} \right\}$, or $\left(-\infty, \dfrac{7}{4} \right) \cup \left(\dfrac{7}{4}, \infty \right)$.

55. All such m are the sums of the factors of 75.

Pair of Factors	Sum of Factors
75, 1	76
$-75, -1$	-76
25, 3	28
$-25, -3$	-28
15, 5	20
$-15, -5$	-20

m can be 76, -76, 28, -28, 20, or -20.

57. $20(-365) = -7300$ and $20 + (-365) = -345$ so the other factor is $(x - 365)$.

Chapter 4 Mid-Chapter Review

1. True; $5x + 2x^2 - 4x^3 = x(5 + 2x - 4x^2)$.

2. The statement is false because one of the variables has a negative exponent. See page 320 in the text.

3. True; see page 321 in the text.

4. False; $-(-x^2 + x) = x^2 - x \neq x - x^2$.

5. True; $144 - x^2 = 12^2 - x^2$.

6. $(8w - 3)(w - 5) = (8w)(w) + (8w)(-5) + (-3)(w) + (-3)(-5)$
$$= 8w^2 - 40w - 3w + 15$$
$$= 8w^2 - 43w + 15$$

7. $c^3 - 8c^2 - 48c = c \cdot c^2 - c \cdot 8c - c \cdot 48$
$$= c(c^2 - 8c - 48)$$
$$= c(c + 4)(c - 12)$$

8. $x^{20} + 8x^{10} - 9 = (x^{10})^2 + 8(x^{10}) - 9$
$$= (x^{10} + 9)(x^{10} - 1)$$

9. $5y^3 + 20y^2 - y - 4 = 5y^2(y + 4) + (-1)(y + 4)$
$$= (y + 4)(5y^2 - 1)$$

10. $-a^7 + a^4 - a + 8$

Terms: $-a^7,\ a^4,\ -a,\ 8$

Degree of each term: 7, 4, 1$(-a = -a^1)$, 0$(8 = 8a^0)$

Degree of the polynomial: 7

Leading term: $-a^7$

Leading coefficient: -1 $(-a^7 = -1 \cdot a^7)$

Constant term: 8

11. $3x^4 + 2x^3w^5 - 12x^2w + 4x^2 - 1$

Terms: $3x^4,\ 2x^3w^5,\ -12x^2w,\ 4x^2,\ -1$

Degree of each term: 4, 8$(3 + 5 = 8)$, 3$(2 + 1 = 3)$, 2, 0

Degree of the polynomial: 8

Leading term: $2x^3w^5$

Leading coefficient: 2

Constant term: -1

12. $-2y + 5 - y^3 + y^9 - 2y^4 = 5 - 2y - y^3 - 2y^4 + y^9$

13. $2qx - 9qr + 2x^5 - 4qx^2 = 2x^5 - 4qx^2 + 2qx - 9qr$

14. $h(x) = -x^3 - 4x + 5$

$h(0) = -0^3 - 4 \cdot 0 + 5 = 5$

$h(-2) = -(-2)^3 - 4(-2) + 5 = -(-8) - 4(-2) + 5 = 8 + 8 + 5 = 21$

$h\left(\dfrac{1}{2}\right) = -\left(\dfrac{1}{2}\right)^3 - 4 \cdot \dfrac{1}{2} + 5 = -\dfrac{1}{8} - 2 + 5 = 2\dfrac{7}{8}$, or $\dfrac{23}{8}$

15. $f(x) = \dfrac{1}{2}x^4 - x^3$

$f(-1) = \dfrac{1}{2}(-1)^4 - (-1)^3 = \dfrac{1}{2} \cdot 1 - (-1) = \dfrac{1}{2} + 1 = 1\dfrac{1}{2}$, or $\dfrac{3}{2}$

$f(1) = \dfrac{1}{2}(1)^4 - 1^3 = \dfrac{1}{2} \cdot 1 - 1 = \dfrac{1}{2} - 1 = -\dfrac{1}{2}$

$f(0) = \dfrac{1}{2} \cdot 0^4 - 0^3 = 0 - 0 = 0$

16. $f(x) = x^2 + 2x - 9$

$f(a - 2) = (a - 2)^2 + 2(a - 2) - 9$
$$= a^2 - 4a + 4 + 2a - 4 - 9$$
$$= a^2 - 2a - 9$$

$f(a + h) - f(a) = (a + h)^2 + 2(a + h) - 9 - (a^2 + 2a - 9)$
$$= a^2 + 2ah + h^2 + 2a + 2h - 9 - a^2 - 2a + 9$$
$$= 2ah + h^2 + 2h$$

17. $(3a^2 - 7b + ab + 2) + (-5a^2 + 4b - 5ab - 3)$
$$= (3a^2 - 5a^2) + (-7b + 4b) + (ab - 5ab) + (2 - 3)$$
$$= -2a^2 - 3b - 4ab - 1$$

18. $(x^2 + 10x - 4) + (9x^2 - 2x + 1) + (x^2 - x - 5)$
$$= (x^2 + 9x^2 + x^2) + (10x - 2x - x) + (-4 + 1 - 5)$$
$$= 11x^2 + 7x - 8$$

19. $(b - 12)(b + 1) = b^2 + b - 12b - 12 = b^2 - 11b - 12$

20. $c^2(3c^2 - c^3) = c^2 \cdot 3c^2 - c^2 \cdot c^3 = 3c^4 - c^5$

21. $(y^4 - 6)(y^4 + 3) = y^8 + 3y^4 - 6y^4 - 18 = y^8 - 3y^4 - 18$

22. $(7y^2 - 2y^3 - 5y) - (y^2 - 3y - 6y^3)$
$$= 7y^2 - 2y^3 - 5y - y^2 + 3y + 6y^3$$
$$= 4y^3 + 6y^2 - 2y$$

23. $(8x - 11) - (-x + 1) = 8x - 11 + x - 1 = 9x - 12$

24. $(4x - 5)^2 = (4x)^2 - 2 \cdot 4x \cdot 5 + 5^2 = 16x^2 - 40x + 25$

25. $(2x + 5)^2 = (2x)^2 + 2 \cdot 2x \cdot 5 + 5^2 = 4x^2 + 20x + 25$

26. $(0.01x - 0.5y) - (2.5y - 0.1x) = 0.01x - 0.5y - 2.5y + 0.1x = 0.11x - 3y$

27. $-13x^2 \cdot 10xy = -130x^3y$

28.
$$
\begin{array}{r}
x^2\ -\ 2xy\ +\ 3y^2 \\
x\ +\ y \\
\hline
x^2y\ -\ 2xy^2\ +\ 3y^3 \\
x^3\ -\ 2x^2y\ +\ 3xy^2 \\
\hline
x^3\ -\ x^2y\ +\ xy^2\ +\ 3y^3
\end{array}
$$

29. $(5x - 7)(2x + 9) = 10x^2 + 45x - 14x - 63 = 10x^2 + 31x - 63$

30. $(9x - 4)(9x + 4) = (9x)^2 - 4^2 = 81x^2 - 16$

31. $5h^2 + 7h = h \cdot 5h + h \cdot 7 = h(5h + 7)$

32. $x^2 + 8x - 20$

We look for two numbers whose product is -20 and whose sum is 8. The numbers we need are 10 and -2.

$$x^2 + 8x - 20 = (x + 10)(x - 2)$$

33. $21 - 4b - b^2 = -b^2 - 4b + 21 = -1(b^2 + 4b - 21)$

Now we factor $b^2 + 4b - 21$. We look for two numbers whose product is -21 and whose sum is 4. The numbers we need are 7 and -3.

$$-b^2 - 4b + 21$$
$$= -1(b^2 + 4b - 21)$$
$$= -1(b + 7)(b - 3)$$
$$= (-b - 7)(b - 3) \qquad \text{Multiplying } b - 7 \text{ by } -1$$
$$= (b + 7)(-b + 3) \qquad \text{Multiplying } b + 3 \text{ by } -1$$

We could also express the last form of the answer as $(7 + b)(3 - b)$.

34. $m^2 + \dfrac{2}{7}m + \dfrac{1}{49} = m^2 + 2 \cdot m \cdot \dfrac{1}{7} + \left(\dfrac{1}{7}\right)^2 = \left(m + \dfrac{1}{7}\right)^2$

35. $2xy - x^2y - 5x + 10 = xy(2 - x) + 5(-x + 2)$
$$= xy(2 - x) + 5(2 - x)$$
$$= (2 - x)(xy + 5)$$

36. $3w^2 - 6w + 3 = 3(w^2 - 2w + 1) = 3(w - 1)^2$

37. $t^3 + 3t^2 + t + 3 = t^2(t + 3) + (t + 3)$
$$= (t + 3)(t^2 + 1)$$

38. $24xy^6z^4 - 16x^4y^3z = 8xy^3z \cdot 3y^3z^3 - 8xy^3z \cdot 2x^3$
$$= 8xy^3z(3y^3z^3 - 2x^3)$$

39. $x^2 + 8x + 6$

There is no pair of factors of 6 whose sum is 8. This trinomial is not factorable into binomials.

40. One explanation is as follows. The expression $-(a - b)$ is the opposite of $a - b$. Since $(a - b) + (b - a) = 0$, then $-(a - b) = b - a$.

41. No; if the coefficients of at least one pair of like terms are opposites, then the sum is a monomial. For example, $(2x + 3) + (-2x + 1) = 4$, a monomial.

42. No; consider the polynomial $3x^{11} + 5x^7$. All of the coefficients and exponents are prime numbers, yet the polynomial can be factored so it is not prime.

43. When coefficients and/or exponents are large, a polynomial is more easily evaluated after it has been factored.

44. a) The middle term, $2 \cdot a \cdot 3$, is missing from the right-hand side.

$$(a + 3)^2 = a^2 + 6a + 9$$

b) The middle term, $-2ab$, is missing from the right-hand side and the sign preceding b^2 is incorrect.

$$(a - b)(a - b) = a^2 - 2ab + b^2$$

c) The product of the outside terms and the product of the inside terms are missing from the right-hand side.

$$(x + 3)(x - 4) = x^2 - x - 12$$

d) There should be a minus sign between the terms of the product.

$$(p + 7)(p - 7) = p^2 - 49$$

e) The middle term, $-2 \cdot t \cdot 3$, is missing from the right-hand side and the sign preceding 9 is incorrect.

$$(t - 3)^2 = t^2 - 6t + 9$$

45. Answer may vary. For the polynomial $4a^3 - 12a$, an incorrect factorization is $4a(a - 3)$. Evaluating both the polynomial and the factorization for $a = 0$, we get 0 in each case. Thus, the evaluation does not catch the mistake.

Exercise Set 4.5

RC1. The product of the First terms must be $\underline{2x^2}$.

RC3. The sum of the Outside and the Inside products must be $\underline{-3x}$ (the middle term).

RC5. Multiply the <u>leading</u> coefficient 10 and the <u>constant</u> 2.

RC7. Split the middle term, $\underline{21x}$, writing it as the sum of $20x$ and x.

1. $3x^2 - 14x - 5$

We will use the FOIL method.

1) There is no common factor (other than 1 or -1).

2) We factor the first term, $3x^2$. The factors are $3x$ and x. We have this possibility:

$$(3x + \quad)(x + \quad)$$

3) Next we factor the last term, -5. The possibilities are $-1 \cdot 5$ and $1(-5)$.

4) We look for combinations of factors from steps (2) and (3) such that the sum of their products is the middle term, $-14x$. We try the possibilities:

$$(3x - 1)(x + 5) = 3x^2 + 14x - 5$$
$$(3x + 1)(x - 5) = 3x^2 - 14x - 5$$

The factorization is $(3x + 1)(x - 5)$.

3. $10y^3 + y^2 - 21y$

We will use the ac-method.

1) Look for a common factor. We factor out y:

$$y(10y^2 + y - 21)$$

2) Factor the trinomial $10y^2 + y - 21$. Multiply the leading coefficient, 10, and the constant, -21.

$$10(-21) = -210$$

3) Look for a factorization of -210 in which the sum of the factors is the coefficient of the middle term, 1.

Pairs of Factors	Sums of Factors
$-1,\ \ \ 210$	209
$1,\ -210$	-209
$-2,\ \ \ 105$	103
$2,\ -105$	-103
$-3,\ \ \ 70$	67
$3,\ -70$	-67
$-5,\ \ \ 42$	37
$5,\ -42$	-37
$-6,\ \ \ 35$	29
$6,\ -35$	-29
$-7,\ \ \ 30$	23
$7,\ -30$	-23
$-10,\ \ \ 21$	11
$10,\ -21$	-11
$-14,\ \ \ 15$	1 $\leftarrow -14+15=1$
$14,\ -15$	-1

4) Next, split the middle term, y, as follows:

$$y = -14y + 15y$$

5) Factor by grouping:

$$10y^2 + y - 21 = 10y^2 - 14y + 15y - 21$$
$$= 2y(5y - 7) + 3(5y - 7)$$
$$= (5y - 7)(2y + 3)$$

We must include the common factor to get a factorization of the original trinomial:

$$10y^3 + y^2 - 21y = y(5y - 7)(2y + 3)$$

5. $3c^2 - 20c + 32$

We will use the FOIL method.

1) There is no common factor(other than 1 or -1).

2) Factor the first term, $3c^2$. The factors are $3c$ and c. We have this possibility:

$$(3c+\ \ )(c+\ \ )$$

3) Next we factor the last term, 32. The possibilities are $1 \cdot 32$, $-1(-32)$, $2 \cdot 16$, $(-2)(-16)$, $4 \cdot 8$, and $-4(-8)$.

4) We look for a combination of factors from steps (2) and (3) such that the sum of their products is the middle term, $-20c$. Trial and error leads us to the correct factorization, $(3c - 8)(c - 4)$.

7. $35y^2 + 34y + 8$

We will use the *ac*-method.

1) There is no common factor (other than 1 or -1).

2) Multiply the leading coefficient, 35, and the constant, 8: $35(8) = 280$

3) Try to factor 280 so the sum of the factors is 34. We need only consider pairs of positive factors since 280 and 34 are both positive.

Pairs of Factors	Sums of Factors
$280,\ \ 1$	281
$140,\ \ 2$	142
$70,\ \ 4$	74
$56,\ \ 5$	61
$40,\ \ 7$	47
$28,\ 10$	38
$20,\ 14$	34

4) Split $34y$ as follows:

$$34y = 20y + 14y$$

5) Factor by grouping:

$$35y^2 + 34y + 8 = 35y^2 + 20y + 14y + 8$$
$$= 5y(7y + 4) + 2(7y + 4)$$
$$= (7y + 4)(5y + 2)$$

9. $4t + 10t^2 - 6 = 10t^2 + 4t - 6$

We will use the FOIL method.

1) Factor out the common factor, 2:

$$2(5t^2 + 2t - 3)$$

2) Now we factor out the trinomial $5t^2 + 2t - 3$.

Factor the first term, $5t^2$. The factors are $5t$ and t. We have this possibility:

$$(5t+\ \ )(t+\ \ )$$

3) Factor the last term, -3. The possibilities are $1(-3)$ and $-1 \cdot 3$.

4) Look for factors in steps (2) and (3) such that the sum of the products is the middle term, $2t$. Trial and error leads us to the correct factorization:
$5t^2 + 2t - 3 = (5t - 3)(t + 1)$

We must include the common factor to get a factorization of the original trinomial:

$$4t + 10t^2 - 6 = 2(5t - 3)(t + 1)$$

11. $8x^2 - 16 - 28x = 8x^2 - 28x - 16$

We will use the *ac*-method.

1) Factor out the common factor, 4:

$$4(2x^2 - 7x - 4)$$

2) Now we factor the trinomial $2x^2 - 7x - 4$. Multiply the leading coefficient, 2, and the constant, -4: $2(-4) = -8$

3) Factor -8 so the sum of the factors is -7. We need only consider pairs of factors in which the negative factor has the larger absolute value, since their sum is negative.

Pairs of Factors	Sums of Factors
$-4,\ \ \ 2$	-2
$-8,\ \ \ 1$	-7

4) Split $-7x$ as follows:

$$-7x = -8x + x$$

5) Factor by grouping:
$$2x^2 - 7x - 4 = 2x^2 - 8x + x - 4$$
$$= 2x(x-4) + (x-4)$$
$$= (x-4)(2x+1)$$

We must include the common factor to get a factorization of the original trinomial:
$$8x^2 - 16 - 28x = 4(x-4)(2x+1)$$

13. $18a^2 - 51a + 15$

We will use the FOIL method.

1) Factor out the common factor, 3:
$$3(6a^2 - 17a + 5)$$

2) We now factor the trinomial $6a^2 - 17a + 5$.

Factor the first term, $6a^2$. The factors are $6a$, a and $3a$, $2a$. We have these possibilities:
$$(6a+\ \ )(a+\ \ ) \text{ and } (3a+\ \ )(2a+\ \ )$$

3) Factor the last term, 5. The possibilities are $5 \cdot 1$ and $-5(-1)$.

4) Look for factors in steps (2) and (3) such that the sum of the products is the middle term, $-17a$. Trial and error leads us to the correct factorization:
$6a^2 - 17a + 5 = (3a-1)(2a-5)$

We must include the common factor to get a factorization of the original trinomial:
$$18a^2 - 51a + 15 = 3(3a-1)(2a-5)$$

15. $30t^2 + 85t + 25$

We will use the ac-method.

1) Factor out the common factor, 5:
$$5(6t^2 + 17t + 5)$$

2) Now we factor the trinomial $6t^2 + 17t + 5$. Multiply the leading coefficient, 6, and the constant, 5:
$6 \cdot 5 = 30$

3) Factor 30 so the sum of the factors is 17. We need to consider only positive pairs of factors since the middle term and the constant are both positive.

Pairs of Factors	Sums of Factors
30, 1	31
15, 2	17
10, 3	13
6, 5	11

4) Split $17t$ as follows:
$$17t = 15t + 2t$$

5) Factor by grouping:
$$6t^2 + 17t + 5 = 6t^2 + 15t + 2t + 5$$
$$= 3t(2t+5) + (2t+5)$$
$$= (2t+5)(3t+1)$$

We must include the common factor to get a factorization of the original trinomial:
$$30t^2 + 85t + 25 = 5(2t+5)(3t+1)$$

17. $12x^3 - 31x^2 + 20x$

We will use the FOIL method.

1) Factor out the common factor, x:
$$x(12x^2 - 31x + 20)$$

2) We now factor the trinomial $12x^2 - 31x + 20$. Factor the first term, $12x^2$. The factors are $12x$, x and $6x$, $2x$ and $4x$, $3x$. We have these possibilities:
$(12x+\ )(x+\ ), (6x+\ )(2x+\ ), (4x+\ )(3x+\ )$

3) Factor the last term, 20. The possibilities are $20 \cdot 1$, $-20(-1)$, $10 \cdot 2$, $-10(-2)$, $5 \cdot 4$, and $-5(-4)$.

4) Look for factors in steps (2) and (3) such that the sum of the products is the middle term, $-31x$. Trial and error leads us to the correct factorization:
$12x^2 - 31x + 20 = (4x-5)(3x-4)$

We must include the common factor to get a factorization of the original trinomial:
$$12x^3 - 31x^2 + 20x = x(4x-5)(3x-4)$$

19. $14x^4 - 19x^3 - 3x^2$

We will use the ac-method.

1) Factor out the common factor, x^2:
$$x^2(14x^2 - 19x - 3)$$

2) Now we factor the trinomial $14x^2 - 19x - 3$. Multiply the leading coefficient, 14, and the constant, -3:
$14(-3) = -42$

3) Factor -42 so the sum of the factors is -19. We need only consider pairs of factors in which the negative factor has the larger absolute value, since the sum is negative.

Pairs of Factors	Sums of Factors
-42, 1	-41
-21, 2	-19
-14, 3	-11
-7, 6	-1

4) Split $-19x$ as follows:
$$-19x = -21x + 2x$$

5) Factor by grouping:
$$14x^2 - 19x - 3 = 14x^2 - 21x + 2x - 3$$
$$= 7x(2x-3) + 2x - 3$$
$$= (2x-3)(7x+1)$$

We must include the common factor to get a factorization of the original trinomial:
$$14x^4 - 19x^3 - 3x^2 = x^2(2x-3)(7x+1)$$

21. $3a^2 - a - 4$

We will use the FOIL method.

1) There is no common factor (other than 1 or -1).

2) Factor the first term, $3a^2$. The factors are $3a$ and a. We have this possibility: $(3a+\ )(a+\ )$

3) Factor the last term, -4. The possibilities are $4(-1)$, $-4 \cdot 1$, and $2(-2)$.

4) Look for factors in steps (2) and (3) such that the sum of the products is the middle term, $-a$. Trial and error leads us to the correct factorization: $(3a - 4)(a + 1)$

23. $9x^2 + 15x + 4$

We will use the ac-method.

1) There is no common factor (other than 1 or -1).

2) Multiply the leading coefficient and the constant: $9(4) = 36$

3) Factor 36 so the sum of the factors is 15. We need only consider pairs of positive factors since 36 and 15 are both positive.

Pairs of Factors	Sums of Factors
36, 1	37
18, 2	20
12, 3	15
9, 4	13
6, 6	12

4) Split $15x$ as follows: $15x = 12x + 3x$

5) Factor by grouping:
$$9x^2 + 15x + 4 = 9x^2 + 12x + 3x + 4$$
$$= 3x(3x + 4) + 3x + 4$$
$$= (3x + 4)(3x + 1)$$

25. $3 + 35z - 12z^2 = -12z^2 + 35z + 3$

We will use the FOIL method.

1) Factor out -1 so the leading coefficient is positive: $-1(12z^2 - 35z - 3)$

2) Now we factor the trinomial $12z^2 - 35z - 3$. Factor the first term, $12z^2$. The factors are $12z$, z and $6z$, $2z$ and $4z$, $3z$. We have these possibilities: $(12z+\ )(z+\ )$, $(6z+\ )(2z+\ )$, $(4z+\ )(3z+\ )$

3) Factor the last term, -3. The possibilities are $3(-1)$ and $-3 \cdot 1$.

4) Look for factors in steps (2) and (3) such that the sum of the products is the middle term, $-35z$. Trial and error leads us to the correct factorization: $(12z + 1)(z - 3)$

We must include the common factor to get a factorization of the original trinomial:
$$3 + 35z - 12z^2 = -1(12z + 1)(z - 3), \text{ or}$$
$$(-12z - 1)(z - 3), \text{ or } (12z + 1)(-z + 3)$$

27. $-4t^2 - 4t + 15$

We will use the ac-method.

1) Factor out -1 so the leading coefficient is positive: $-1(4t^2 + 4t - 15)$

2) Now we factor the trinomial $4t^2 + 4t - 15$. Multiply the leading coefficient and the constant: $4(-15) = -60$

3) Factor -60 so the sum of the factors is 4. The desired factorization is $10(-6)$.

4) Split $4t$ as follows:
$$4t = 10t - 6t$$

5) Factor by grouping:
$$4t^2 + 4t - 15 = 4t^2 + 10t - 6t - 15$$
$$= 2t(2t + 5) - 3(2t + 5)$$
$$= (2t + 5)(2t - 3)$$

We must include the common factor to get a factorization of the original trinomial:
$$-4t^2 - 4t + 15 = -1(2t + 5)(2t - 3), \text{ or}$$
$$(-2t - 5)(2t - 3), \text{ or } (2t + 5)(-2t + 3)$$

29. $3x^3 - 5x^2 - 2x$

We will use the FOIL method.

1) Factor out the common factor, x: $x(3x^2 - 5x - 2)$

2) Now we factor the trinomial $3x^2 - 5x - 2$. Factor the first term, $3x^2$. The factors are $3x$ and x. We have this possibility: $(3x+\ )(x+\ )$

3) Factor the last term, -2. The possibilities are $2(-1)$ and $-2 \cdot 1$.

4) Look for factors in steps (2) and (3) such that the sum of the products is the middle term, $-5x$. Trial and error leads us to the correct factorization: $(3x + 1)(x - 2)$

We must include the common factor to get a factorization of the original trinomial:
$$3x^3 - 5x^2 - 2x = x(3x + 1)(x - 2)$$

31. $24x^2 - 2 - 47x = 24x^2 - 47x - 2$

We will use the ac-method.

1) There is no common factor (other than 1 or -1).

2) Multiply the leading coefficient and the constant: $24(-2) = -48$

3) Factor -48 so the sum of the factors is -47. The desired factorization is $-48 \cdot 1$.

4) Split $-47x$ as follows:
$$-47x = -48x + x$$

5) Factor by grouping:
$$24x^2 - 47x - 2 = 24x^2 - 48x + x - 2$$
$$= 24x(x - 2) + (x - 2)$$
$$= (x - 2)(24x + 1)$$

33. $-8t^3 - 8t^2 + 30t$

We will use the FOIL method.

1) Factor out the common factor, $-2t$: $-2t(4t^2 + 4t - 15)$

2) Now factor the trinomial $4t^2 + 4t - 15$. Factor the first term, $4t^2$. The possibilities are $4t \cdot t$ and $2t \cdot 2t$. The possible factorizations are of the form:

$(4t+\ )(t+\ )$ and $(2t+\ )(2t+\ )$

3) Factor the last term, -15. The possibilities are $-15 \cdot 1$, $15(-1)$, $-5 \cdot 3$, and $5(-3)$.

4) Look for factors in steps (2) and (3) such that the sum of the products is the middle term, $4t$. Trial and error leads us to the correct factorization: $(2t + 5)(2t - 3)$

We must include the common factor to get a factorization of the original trinomial:
$$-8t^3 - 8t^2 + 30t = -2t(2t + 5)(2t - 3)$$

35. $-24x^3 + 2x + 47x^2$, or $-24x^3 + 47x^2 + 2x$

We will use the ac-method.

1) Factor out the common factor, $-x$:
$$-x(24x^2 - 47x - 2)$$

Now factor the trinomial $24x^2 - 47x - 2$. From Exercise 31 we know that the factorization is $(x - 2)(24x + 1)$. We must include the common factor to get a factorization of the original trinomial:
$$-24x^3 + 2x + 47x^2 = -x(x - 2)(24x + 1)$$

37. $21x^2 + 37x + 12$

We will use the FOIL method.

1) There is no common factor (other than 1 or -1).

2) Factor the first term $21x^2$. The factors are $21x$, x and $7x$, $3x$. We have these possibilities: $(21x+ \quad)(x+ \quad)$ and $(7x+ \quad)(3x+ \quad)$.

3) Factor the last term, 12. The possibilities are $12 \cdot 1$, $-12 \cdot (-1)$, $6 \cdot 2$, $-6 \cdot (-2)$, $4 \cdot 3$, and $-4 \cdot (-3)$.

4) Look for factors in steps (2) and (3) such that the sum of the products is the middle term, $37x$. Trial and error leads us to the correct factorization: $(7x + 3)(3x + 4)$

39. $40x^4 + 16x^2 - 12$

We will use the ac-method.

1) Factor out the common factor, 4.
$$4(10x^4 + 4x^2 - 3)$$

Now we will factor the trinomial $10x^4 + 4x^2 - 3$. Substitute u for x^2 (and u^2 for x^4), and factor $10u^2 + 4u - 3$.

2) Multiply the leading coefficient and the constant: $10(-3) = -30$

3) Factor -30 so the sum of the factors is 4. This cannot be done. The trinomial $10u^2 + 4u - 3$ cannot be factored into binomials with integer coefficients. We have
$$40x^4 + 16x^2 - 12 = 4(10x^4 + 4x^2 - 3)$$

41. $12a^2 - 17ab + 6b^2$

We will use the FOIL method. (Our thinking is much the same as if we were factoring $12a^2 - 17a + 6$.)

1) There is no common factor (other than 1 or -1).

2) Factor the first term, $12a^2$. The factors are $12a$, a and $6a$, $2a$ and $4a$, $3a$. We have these possibilities: $(12a+ \quad)(a+ \quad)$ and $(6a+ \quad)(2a+ \quad)$ and $(4a+ \quad)(3a+ \quad)$.

3) Factor the last term, $6b^2$. The possibilities are $6b \cdot b$, $-6b \cdot (-b)$, $3b \cdot 2b$, and $-3b \cdot (-2b)$.

4) Look for factors in steps (2) and (3) such that the sum of the products is the middle term, $-17ab$. Trial and error leads us to the correct factorization: $(4a - 3b)(3a - 2b)$

43. $2x^2 + xy - 6y^2$

We will use the ac-method.

1) There is no common factor (other than 1 or -1).

2) Multiply the coefficients of the first and last terms: $2(-6) = -12$

3) Factor -12 so the sum of the factors is 1. The desired factorization is $4(-3)$.

4) Split xy as follows:
$$xy = 4xy - 3xy$$

5) Factor by grouping:
$$\begin{aligned} 2x^2 + xy - 6y^2 &= 2x^2 + 4xy - 3xy - 6y^2 \\ &= 2x(x + 2y) - 3y(x + 2y) \\ &= (x + 2y)(2x - 3y) \end{aligned}$$

45. $12x^2 - 58xy + 56y^2$

We will use the FOIL method.

1) Factor out the common factor, 2:
$$2(6x^2 - 29xy + 28y^2)$$

2) Now we factor the trinomial $6x^2 - 29xy + 28y^2$. Factor the first term, $6x^2$. The factors are $6x$, x and $3x$, $2x$. We have these possibilities: $(6x+ \quad)(x+ \quad)$ and $(3x+ \quad)(2x+ \quad)$.

3) Factor the last term, $28y^2$. The possibilities are $28y \cdot y$, $-28y \cdot (-y)$, $14y \cdot 2y$, $-14 \cdot (-2y)$, $7y \cdot 4y$, and $-7y \cdot (-4y)$.

4) Look for factors in steps (2) and (3) such that the sum of the products is the middle term, $-29xy$. Trial and error leads us to the correct factorization: $(3x - 4y)(2x - 7y)$

We must include the common factor to get a factorization of the original trinomial:
$$12x^2 - 58xy + 56y^2 = 2(3x - 4y)(2x - 7y)$$

47. $9x^2 - 30xy + 25y^2$

We will use the ac-method.

1) There is no common factor (other than 1 or -1).

2) Multiply the coefficients of the first and last terms: $9(25) = 225$

3) Factor 225 so the sum of the factors is -30. The desired factorization is $-15(-15)$.

4) Split $-30xy$ as follows:

$$-30xy = -15xy - 15xy$$

5) Factor by grouping:

$$9x^2 - 30xy + 25y^2 = 9x^2 - 15xy - 15xy + 25y^2$$
$$= 3x(3x - 5y) - 5y(3x - 5y)$$
$$= (3x - 5y)(3x - 5y)$$

49. $3x^6 + 4x^3 - 4$

We will use the FOIL method.

1) There is no common factor (other than 1 or -1). Substitute u for x^3 (and hence u^2 for x^6). We factor $3u^2 + 4u - 4$.

2) Factor the first term, $3u^2$. The factors are $3u$ and u. We have this possibility:
$(3u+ \ )(u+ \ )$

3) Factor the last term, -4. The possibilities are $-1 \cdot 4$, $1 \cdot (-4)$, and $-2 \cdot 2$.

4) Look for factors in steps (2) and (3) such that the sum of the products is the middle term, $4u$. Trial and error leads us to the correct factorization of $3u^2 + 4u - 4$: $(3u - 2)(u + 2)$. Replacing u with x^3 we have the factorization of the original trinomial: $(3x^3 - 2)(x^3 + 2)$.

51. a) $h(10) = -16(0)^2 + 80(0) + 224 = 224$ ft

$h(1) = -16(1)^2 + 80(1) + 224 = 288$ ft

$h(3) = -16(3)^2 + 80(3) + 224 = 320$ ft

$h(4) = -16(4)^2 + 80(4) + 224 = 288$ ft

$h(6) = -16(6)^2 + 80(6) + 224 = 128$ ft

b) $h(t) = -16t^2 + 80t + 224$

We will use the grouping method.

1) Factor out -16 so the leading coefficient is positive: $-16(t^2 - 5t - 14)$

2) Factor the trinomial $t^2 - 5t - 14$. Multiply the leading coefficient and the constant: $1(-14) = -14$

3) Factor -14 so the sum of the factors is -5. The desired factorization is $-7 \cdot 2$.

4) Split $-5t$ as follows:

$$-5t = -7t + 2t$$

5) Factor by grouping:

$$t^2 - 5t - 14 = t^2 - 7t + 2t - 14$$
$$= t(t - 7) + 2(t - 7)$$
$$= (t - 7)(t + 2)$$

We must include the common factor to get a factorization of the original trinomial.

$$h(t) = -16(t - 7)(t + 2)$$

53. $\quad x + 2y - \ z = 0, \quad (1)$
$\quad 4x + 2y + 5z = 6, \quad (2)$
$\quad 2x - \ y + \ z = 5 \quad (3)$

First we will eliminate z from Equations (1) and (2).

$\quad 5x + 10y - 5z = 0 \quad$ Multiplying (1) by 5
$\quad \underline{4x + \ 2y + 5z = 6} \quad (2)$
$\quad 9x + 12y \quad\quad = 6 \quad (4)$

Now add Equations (1) and (3) to eliminate z.

$\quad x + 2y - z = 0$
$\quad \underline{2x - \ y + z = 5}$
$\quad 3x + \ y \quad\quad = 5 \quad\quad (5)$

Now solve the system composed of Equations (4) and (5). We multiply Equation (5) by -3 and add.

$\quad\quad 9x + 12y = \quad 6$
$\quad \underline{-9x - \ 3y = -15}$
$\quad\quad\quad\quad 9y = \ -9$
$\quad\quad\quad\quad\ y = \ -1$

Substitute -1 for y in Equation (4) or (5) and solve for x. We use Equation (5) here.

$\quad\quad 3x - 1 = 5$
$\quad\quad\quad 3x = 6$
$\quad\quad\quad\ x = 2$

Now substitute 2 for x and -1 for y in one of the original equations and solve for z. We use Equation (3).

$\quad\quad 2 \cdot 2 - (-1) + z = 5$
$\quad\quad\quad 4 + 1 + z = 5$
$\quad\quad\quad\quad 5 + z = 5$
$\quad\quad\quad\quad\quad\ z = 0$

The triple $(2, -1, 0)$ checks, so it is the solution.

55. $\quad 2x + 9y + 6z = 5, \quad (1)$
$\quad\ x - \ y + \ z = 4, \quad (2)$
$\quad 3x + 2y + 3z = 7 \quad (3)$

First we will eliminate x from Equations (1) and (2).

$\quad 2x + \ 9y + 6z = \quad 5 \quad (1)$
$\quad \underline{-2x + \ 2y - 2z = -8} \quad$ Multiplying (2) by -2
$\quad\quad\quad 11y + 4z = -3 \quad (4)$

Now eliminate x from Equations (2) and (3).

$\quad -3x + 3y - 3z = -12 \quad$ Multiplying (2) by -3
$\quad \underline{\ 3x + 2y + 3z = \quad\ 7} \quad (3)$
$\quad\quad\quad 5y \quad\quad\quad = -5$
$\quad\quad\quad\quad\ y = \ -1$

Note that the last step eliminated z as well as x, allowing us to solve for y. Now substitute -1 for y in Equation (4) and solve for z.

$\quad\quad 11(-1) + 4z = -3$
$\quad\quad\quad -11 + 4z = -3$
$\quad\quad\quad\quad\quad 4z = 8$
$\quad\quad\quad\quad\quad\ z = 2$

Substitute -1 for y and 2 for z in one of the original equations and solve for x. We use Equation (2).

$$x - (-1) + 2 = 4$$
$$x + 1 + 2 = 4$$
$$x + 3 = 4$$
$$x = 1$$

The triple $(1, -1, 2)$ checks, so it is the solution.

57. Write the first equation in slope-intercept form.

$$y - 2x = 18$$
$$y = 2x + 18$$

The second equation, $2x - 7 = y$, or $y = 2x - 7$, is in slope-intercept form.

Since the equations have the same slope, 2, and different y-intercepts, $(0, 8)$ and $(0, -7)$, the graphs of the equations are parallel lines.

59. Write each equation in slope-intercept form.

$$2x + 5y = 4 \qquad\qquad 2x - 5y = -3$$
$$5y = -2x + 4 \qquad\qquad -5y = -2x - 3$$
$$y = -\frac{2}{5}x + \frac{4}{5} \qquad\qquad y = \frac{2}{5}x + \frac{3}{5}$$

The slope of the first line is $-\frac{2}{5}$, and the slope of the second is $\frac{2}{5}$. The slopes are not the same nor is their product -1, so the graphs of the equations are neither parallel nor perpendicular.

61. First we find the slope.

$$m = \frac{-4 - (-3)}{5 - (-2)} = \frac{-1}{7} = -\frac{1}{7}$$

Now substitute $-\frac{1}{7}$ for m and the coordinates of either point in the slope-intercept equation $y = mx + b$ and solve for b. We will use the point $(5, -4)$.

$$y = mx + b$$
$$-4 = -\frac{1}{7} \cdot 5 + b$$
$$-4 = -\frac{5}{7} + b$$
$$-\frac{23}{7} = b$$

Now use the equation $y = mx + b$ again, substituting $-\frac{1}{7}$ for m and $-\frac{23}{7}$ for b.

$$y = -\frac{1}{7}x - \frac{23}{7}$$

63. First we find the slope.

$$m = \frac{-4 - 3}{7 - (-10)} = \frac{-7}{17} = -\frac{7}{17}$$

Now substitute $-\frac{7}{17}$ for m and the coordinates of either point in the slope-intercept equation $y = mx + b$ and solve for b. We will use the point $(-10, 3)$.

$$y = mx + b$$
$$3 = -\frac{7}{17}(-10) + b$$
$$3 = \frac{70}{17} + b$$
$$-\frac{19}{17} = b$$

Now use the equation $y = mx + b$ again, substituting $-\frac{7}{17}$ for m and $-\frac{19}{17}$ for b.

$$y = -\frac{7}{17}x - \frac{19}{17}$$

65. Left to the student

67. $7a^2b^2 + 6 + 13ab = 7a^2b^2 + 13ab + 6$

We will use the ac-method. There is no common factor (other than 1 or -1). Multiply the leading coefficient and the constant: $7(6) = 42$. Factor 42 so the sum of the factors is 13. The desired factorization is $6 \cdot 7$. Split the middle term and factor by grouping.

$$7a^2b^2 + 13ab + 6 = 7a^2b^2 + 6ab + 7ab + 6$$
$$= ab(7ab + 6) + 7ab + 6$$
$$= (7ab + 6)(ab + 1)$$

69. $9x^2y^2 - 4 + 5xy = 9x^2y^2 + 5xy - 4$

We will use the ac-method. There is no common factor (other than 1 or -1). Multiply the leading coefficient and the constant: $9(-4) = -36$. Factor -36 so the sum of the factors is 5. The desired factorization is $9(-4)$. Split the middle term and factor by grouping.

$$9x^2y^2 + 5xy - 4 = 9x^2y^2 + 9xy - 4xy - 4$$
$$= 9xy(xy + 1) - 4(xy + 1)$$
$$= (xy + 1)(9xy - 4)$$

71. $x^{2a} + 5x^a - 24$

$x^{2a} = (x^a)^2$, so the factorization is of the form $(x^a + \quad)(x^a + \quad)$.

Look for factors of -24 whose sum is 5. The factors are 8 and -3. Then the factorization is $(x^a + 8)(x^a - 3)$.

Exercise Set 4.6

RC1. $x^2 - 100 = x^2 - 10^2$; this is a difference of squares.

RC3. $x^3 - 1000 = x^3 - 10^3$; this is a difference of cubes.

RC5. $16x^2 + 40x + 25 = (4x)^2 + 2 \cdot 4x \cdot 5 + 5^2$; this is a trinomial square.

RC7. $27x^3 + 1 = (3x)^3 + 1^3$; this is a sum of cubes.

1. $x^2 - 4x + 4 = (x - 2)^2$ \qquad Find the square terms and write their square roots with a minus sign between them.

3. $y^2 + 18y + 81 = (y + 9)^2$ Find the square terms and write their square roots with a minus sign between them.

5. $x^2 + 1 + 2x = x^2 + 2x + 1$ Writing in descending order
$= (x + 1)^2$ Factoring the trinomial square

7. $9y^2 + 12y + 4 = (3y + 2)^2$ Find the square terms and write their square roots with a minus sign between them.

9. $-18y^2 + y^3 + 81y = y^3 - 18y^2 + 81y$ Writing in descending order
$= y(y^2 - 18y + 81)$ Removing the common factor
$= y(y - 9)^2$ Factoring the trinomial square

11. $12a^2 + 36a + 27 = 3(4a^2 + 12a + 9)$ Removing the common factor
$= 3(2a + 3)^2$ Factoring the trinomial square

13. $2x^2 - 40x + 200 = 2(x^2 - 20x + 100)$
$= 2(x - 10)^2$

15. $1 - 8d + 16d^2 = (1 - 4d)^2$,   Find the square terms
or $(4d - 1)^2$ and write their square roots with a minus sign between them.

17. $3a^3 - 6a^2 + 3a = 3a(a^2 - 2a + 1)$ Removing the common factor
$= 3a(a - 1)^2$ Factoring the trinomial square

19. $0.25x^2 + 0.30x + 0.09 = (0.5x + 0.3)^2$ Find the square terms and write their square roots with a plus sign between them.

21. $p^2 - 2pq + q^2 = (p - q)^2$

23. $a^2 + 4ab + 4b^2 = (a + 2b)^2$

25. $25a^2 - 30ab + 9b^2 = (5a - 3b)^2$

27. $y^6 + 26y^3 + 169 = (y^3 + 13)^2$ Find the square terms and write their square roots with a plus sign between them

29. $16x^{10} - 8x^5 + 1 = (4x^5 - 1)^2$ $[16x^{10} = (4x^5)^2]$

31. $x^4 + 2x^2y^2 + y^4 = (x^2 + y^2)^2$

33. $p^2 - 49 = p^2 - 7^2 = (p + 7)(p - 7)$

35. $y^4 - 8y^2 + 16 = (y^2 - 4)^2$ Find the square terms and write their square roots with a minus sign between them.
$= [(y + 2)(y - 2)]^2$ Factoring the difference of squares
$= (y + 2)^2(y - 2)^2$

37. $p^2q^2 - 25 = (pq)^2 - 5^2 = (pq + 5)(pq - 5)$

39. $6x^2 - 6y^2$
$= 6(x^2 - y^2)$ Removing the common factor
$= 6(x + y)(x - y)$ Factoring the difference of squares

41. $4xy^4 - 4xz^4$
$= 4x(y^4 - z^4)$ Removing the common factor
$= 4x[(y^2)^2 - (z^2)^2]$
$= 4x(y^2 + z^2)(y^2 - z^2)$ Factoring the difference of squares
$= 4x(y^2 + z^2)(y + z)(y - z)$ Factoring $y^2 - z^2$

43. $4a^3 - 49a = a(4a^2 - 49)$
$= a[(2a)^2 - 7^2]$
$= a(2a + 7)(2a - 7)$

45. $3x^8 - 3y^8 = 3(x^8 - y^8)$
$= 3[(x^4)^2 - (y^4)^2)]$
$= 3(x^4 + y^4)(x^4 - y^4)$
$= 3(x^4 + y^4)[(x^2)^2 - (y^2)^2]$
$= 3(x^4 + y^4)(x^2 + y^2)(x^2 - y^2)$
$= 3(x^4 + y^4)(x^2 + y^2)(x + y)(x - y)$

47. $9a^4 - 25a^2b^4 = a^2(9a^2 - 25b^4)$
$= a^2[(3a)^2 - (5b^2)^2]$
$= a^2(3a + 5b^2)(3a - 5b^2)$

49. $\dfrac{1}{36} - z^2 = \left(\dfrac{1}{6}\right)^2 - z^2 = \left(\dfrac{1}{6} + z\right)\left(\dfrac{1}{6} - z\right)$

51. $0.04x^2 - 0.09y^2 = (0.2x)^2 - (0.3y)^2$
$= (0.2x + 0.3y)(0.2x - 0.3y)$

53. $m^3 - 7m^2 - 4m + 28$
$= m^2(m - 7) - 4(m - 7)$ Factoring by grouping
$= (m - 7)(m^2 - 4)$
$= (m - 7)(m + 2)(m - 2)$ Factoring the difference of squares

55. $a^3 - ab^2 - 2a^2 + 2b^2$
$= a(a^2 - b^2) - 2(a^2 - b^2)$ Factoring by grouping
$= (a^2 - b^2)(a - 2)$
$= (a + b)(a - b)(a - 2)$ Factoring the difference of squares

57. $(a + b)^2 - 100 = (a + b)^2 - 10^2$
$= (a + b + 10)(a + b - 10)$

59. $144 - (p - 8)^2$ Difference of squares

$\quad = [12 + (p - 8)][12 - (p - 8)]$

$\quad = (12 + p - 8)(12 - p + 8)$

$\quad = (4 + p)(20 - p)$

61. $a^2 + 2ab + b^2 - 9$

$\quad = (a^2 + 2ab + b^2) - 9$ Grouping as a difference of squares

$\quad = (a + b)^2 - 3^2$

$\quad = (a + b + 3)(a + b - 3)$

63. $r^2 - 2r + 1 - 4s^2$

$\quad = (r^2 - 2r + 1) - 4s^2$ Grouping as a difference of squares

$\quad = (r - 1)^2 - (2s)^2$

$\quad = (r - 1 + 2s)(r - 1 - 2s)$

65. $2m^2 + 4mn + 2n^2 - 50b^2$

$\quad = 2(m^2 + 2mn + n^2 - 25b^2)$ Removing the common factor

$\quad = 2[(m^2 + 2mn + n^2) - 25b^2]$ Grouping as a difference of squares

$\quad = 2[(m + n)^2 - (5b)^2]$

$\quad = 2(m + n + 5b)(m + n - 5b)$

67. $9 - (a^2 + 2ab + b^2) = 9 - (a + b)^2$

$\qquad\qquad\qquad\qquad = [3 + (a + b)][3 - (a + b)]$, or

$\qquad\qquad\qquad\qquad (3 + a + b)(3 - a - b)$

69. $z^3 + 27 = z^3 + 3^3$

$\qquad\qquad = (z + 3)(z^2 - 3z + 9)$

$\qquad\qquad A^3 + B^3 = (A + B)(A^2 - AB + B^2)$

71. $x^3 - 1 = x^3 - 1^3$

$\qquad\qquad = (x - 1)(x^2 + x + 1)$

$\qquad\qquad A^3 - B^3 = (A - B)(A^2 + AB + B^2)$

73. $8 - 27b^3 = 2^3 - (3b)^3$

$\qquad\qquad = (2 - 3b)(4 + 6b + 9b^2)$

75. $8a^3 + 1 = (2a)^3 + 1^3$

$\qquad\qquad = (2a + 1)(4a^2 - 2a + 1)$

$\qquad\qquad A^3 + B^3 = (A + B)(A^2 - AB + B^2)$

77. $8x^3 + 27 = (2x)^3 + 3^3$

$\qquad\qquad = (2x + 3)(4x^2 - 6x + 9)$

79. $a^3 - b^3 = (a - b)(a^2 + ab + b^2)$

81. $a^3 + \dfrac{1}{8} = a^3 + \left(\dfrac{1}{2}\right)^3$

$\qquad\qquad = \left(a + \dfrac{1}{2}\right)\left(a^2 - \dfrac{1}{2}a + \dfrac{1}{4}\right)$

83. $x^3 + 0.001 = x^3 + (0.1)^3$

$\qquad\qquad = (x + 0.1)(x^2 - 0.1x + 0.01)$

85. $2y^3 - 128 = 2(y^3 - 64)$

$\qquad\qquad = 2(y^3 - 4^3)$

$\qquad\qquad = 2(y - 4)(y^2 + 4y + 16)$

87. $24a^3 + 3 = 3(8a^3 + 1)$

$\qquad\qquad = 3[(2a)^3 + 1^3]$

$\qquad\qquad = 3(2a + 1)(4a^2 - 2a + 1)$

89. $rs^3 + 64r = r(s^3 + 64)$

$\qquad\qquad = r(s^3 + 4^3)$

$\qquad\qquad = r(s + 4)(s^2 - 4s + 16)$

91. $5x^5 - 40x^2z^3 = 5x^2(x^3 - 8z^3)$

$\qquad\qquad = 5x^2[x^3 - (2z)^3]$

$\qquad\qquad = 5x^2(x - 2z)(x^2 + 2xz + 4z^2)$

93. $64x^6 - 8t^6 = 8(8x^6 - t^6)$

$\qquad\qquad = 8[(2x^2)^3 - (t^2)^3]$

$\qquad\qquad = 8(2x^2 - t^2)(4x^4 + 2x^2t^2 + t^4)$

95. $z^6 - 1$

$\quad = (z^3)^2 - 1^2$ Writing as a difference of squares

$\quad = (z^3 + 1)(z^3 - 1)$ Factoring a difference of squares

$\quad = (z + 1)(z^2 - z + 1)(z - 1)(z^2 + z + 1)$

$\qquad\qquad$ Factoring a sum and a difference of cubes

97. $t^6 + 64y^6 = (t^2)^3 + (4y^2)^3$

$\qquad\qquad = (t^2 + 4y^2)(t^4 - 4t^2y^2 + 16y^4)$

99. $8w^9 - z^9 = (2w^3)^3 - (z^3)^3$

$\qquad\qquad = (2w^3 - z^3)(4w^6 + 2w^3z^3 + z^6)$

101. $\dfrac{1}{8}c^3 + d^3 = \left(\dfrac{1}{2}c\right)^3 + d^3$

$\qquad\qquad = \left(\dfrac{1}{2}c + d\right)\left(\dfrac{1}{4}c^2 - \dfrac{1}{2}cd + d^2\right)$

103. $0.001x^3 - 0.008y^3 = (0.1x)^3 - (0.2y)^3$

$\qquad\qquad = (0.1x - 0.2y)(0.01x^2 + 0.02xy + 0.04y^2)$

105. $7x - 2y = -11$,   (1)

$\qquad 2x + 7y = 18$ (2)

To eliminate y, multiply Equation (1) by 7 and Equation (2) by 2 and add.

$49x - 14y = -77$

$\underline{4x + 14y = 36}$

$53x = -41$

$\qquad\qquad x = -\dfrac{41}{53}$

Now substitute $-\dfrac{41}{53}$ for x in one of the original equations and solve for y.

$$2x + 7y = 18 \quad (2)$$

$$2\left(-\frac{41}{53}\right) + 7y = 18$$

$$-\frac{82}{53} + 7y = 18$$

$$7y = \frac{1036}{53}$$

$$y = \frac{148}{53}$$

The pair $\left(-\frac{41}{53}, \frac{148}{53}\right)$ checks, so it is the solution.

107. $x - y = -12, \quad (1)$

$$\underline{x + y = 14 \quad (2)}$$

$$2x = 2 \quad \text{Adding}$$

$$x = 1$$

Now substitute 1 for x in one of the original equations and solve for y.

$$x + y = 14 \quad (2)$$

$$1 + y = 14$$

$$y = 13$$

The pair $(1, 13)$ checks, so it is the solution.

109. $x - y \le 5,$

$x + y \ge 3$

Graph the lines $x - y = 5$ and $x + y = 3$ using solid lines and then shade the region where the solution sets overlap.

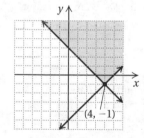

To find the vertex, solve the system of related equations:

$$x - y = 5$$

$$x + y = 3$$

Solving, we obtain the vertex $(4, -1)$.

111. $x - y \ge 5, \quad (1)$

$x + y \le 3, \quad (2)$

$x \ge 1 \quad (3)$

Shade the intersection of the graphs of $x - y \ge 5$, $x + y \le 3$, and $x \ge 1$.

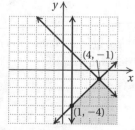

To find the vertices we solve two different systems of equations. From (1) and (2) we obtain the vertex $(4, -1)$. From (1) and (3) we obtain the vertex $(1, -4)$.

113. To find an equation of the line through $(-2, -4)$ parallel to $x - y = 5$, we first write $x - y = 5$ in slope-intercept form.

$$x - y = 5$$

$$-y = -x + 5$$

$$y = x - 5$$

The slope of $y = x - 5$ is 1, so we want to find an equation with slope 1 containing the point $(-2, -4)$. First substitute 1 for m, -2 for x, and -4 for y in $y = mx + b$.

$$y = mx + b$$

$$-4 = 1(-2) + b$$

$$-4 = -2 + b$$

$$-2 = b$$

Now use $y = mx + b$ again, substituting 1 for m and -2 for b.

$$y = 1 \cdot x - 2, \text{ or } y = x - 2$$

We have seen above that the slope of $x - y = 5$ is 1. Then the slope of a line perpendicular to this line is the opposite of the reciprocal of 1, or -1. Thus, we want to find an equation of the line with slope -1 containing $(-2, -4)$. First substitute -1 for m, -2 for x, and -4 for y in $y = mx + b$.

$$y = mx + b$$

$$-4 = -1(-2) + b$$

$$-4 = 2 + b$$

$$-6 = b$$

Now use $y = mx + b$ again, substituting -1 for m and -6 for b.

$$y = -1 \cdot x - 6, \text{ or } y = -x - 6$$

115. To find an equation of the line through $(4, 5)$ parallel to $y = -\frac{1}{2}x + 3$, we first observe that the slope of $y = -\frac{1}{2}x + 3$ is $-\frac{1}{2}$. Thus, we want to find an equation of the line with slope $-\frac{1}{2}$ containing $(4, 5)$. Begin by substituting $-\frac{1}{2}$ for m, 4 for x, and 5 for y in $y = mx + b$.

$$y = mx + b$$

$$5 = -\frac{1}{2} \cdot 4 + b$$

$$5 = -2 + b$$

$$7 = b$$

Now use $y = mx + b$ again, substituting $-\frac{1}{2}$ for m and 7 for b.

$$y = -\frac{1}{2}x + 7$$

The slope of a line perpendicular to $y = -\dfrac{1}{2}x + 3$ is the opposite of the reciprocal of $-\dfrac{1}{2}$, or 2. Thus, we want to find an equation of the line with slope 2 containing $(4, 5)$. First substitute 2 for m, 4 for x, and 5 for y in $y = mx + b$.

$$y = mx + b$$
$$5 = 2 \cdot 4 + b$$
$$5 = 8 + b$$
$$-3 = b$$

Now use $y = mx + b$ again, substituting 2 for m and -3 for b.

$$y = 2x - 3$$

117. If $P(x) = x^3$, then

$$P(a + h) - P(a)$$
$$= (a + h)^3 - a^3$$
$$= (a + h - a)[(a + h)^2 + (a + h)(a) + a^2]$$
$$= h(a^2 + 2ah + h^2 + a^2 + ah + a^2)$$
$$= h(3a^2 + 3ah + h^2)$$

119. a) $\pi R^2 h - \pi r^2 h = \pi h(R^2 - r^2)$
$$= \pi h(R + r)(R - r)$$

b) Note that 4 m = 400 cm.

$$\pi R^2 h - \pi r^2 h$$
$$= \pi (50)^2 (400) - \pi (10)^2 (400)$$
$$= 1,000,000\pi - 40,000\pi$$
$$= 960,000\pi \text{ cm}^3 \quad \text{(or } 0.96\pi \text{ m}^3)$$
$$\approx 3,014,400 \text{ cm}^3 \quad \text{Using 3.14 for } \pi$$

$$\pi h(R + r)(R - r)$$
$$= \pi (400)(50 + 10)(50 - 10)$$
$$= \pi (400)(60)(40)$$
$$= 960,000\pi \text{ cm}^3 \quad \text{(or } 0.96\pi \text{ m}^3)$$
$$\approx 3,014,400 \text{ cm}^3 \quad \text{Using 3.14 for } \pi$$

If we use the π key on a calculator, the result is approximately $3,015,929$ cm^3.

121. $5c^{100} - 80d^{100}$
$$= 5(c^{100} - 16d^{100})$$
$$= 5(c^{50} + 4d^{50})(c^{50} - 4d^{50})$$
$$= 5(c^{50} + 4d^{50})(c^{25} + 2d^{25})(c^{25} - 2d^{25})$$

123. $x^{6a} + y^{3b} = (x^{2a})^3 + (y^b)^3$
$$= (x^{2a} + y^b)(x^{4a} - x^{2a}y^b + y^{2b})$$

125. $3x^{3a} + 24y^{3b} = 3(x^{3a} + 8y^{3b})$
$$= 3[(x^a)^3 + (2y^b)^3]$$
$$= 3(x^a + 2y^b)(x^{2a} - 2x^a y^b + 4y^{2b})$$

127. $\dfrac{1}{24}x^3 y^3 + \dfrac{1}{3}z^3 = \dfrac{1}{3}\left(\dfrac{1}{8}x^3 y^3 + z^3\right)$
$$= \dfrac{1}{3}\left[\left(\dfrac{1}{2}xy\right)^3 + z^3\right]$$
$$= \dfrac{1}{3}\left(\dfrac{1}{2}xy + z\right)\left(\dfrac{1}{4}x^2 y^2 - \dfrac{1}{2}xyz + z^2\right)$$

129. $(x + y)^3 - x^3$
$$= [(x + y) - x][(x + y)^2 + x(x + y) + x^2]$$
$$= (x + y - x)(x^2 + 2xy + y^2 + x^2 + xy + x^2)$$
$$= y(3x^2 + 3xy + y^2)$$

131. $(a + 2)^3 - (a - 2)^3$
$$= [(a + 2) - (a - 2)][(a + 2)^2 + (a + 2)(a - 2) + (a - 2)^2]$$
$$= (a + 2 - a + 2)(a^2 + 4a + 4 + a^2 - 4 + a^2 - 4a + 4)$$
$$= 4(3a^2 + 4)$$

Exercise Set 4.7

RC1. Always look first for a <u>common</u> factor.

RC3. If there are three terms, determine whether the trinomial is a <u>square</u>.

RC5. Always factor <u>completely</u>.

1. $y^2 - 225$
$$= y^2 - 15^2 \quad \text{Difference of squares}$$
$$= (y + 15)(y - 15)$$

3. $2x^2 + 11x + 12$
$$= (2x + 3)(x + 4) \qquad \text{FOIL or } ac\text{-method}$$

5. $5x^4 - 20$
$$= 5(x^4 - 4) \quad \text{Removing the common factor}$$
$$= 5[(x^2)^2 - 2^2] \quad \text{Difference of squares}$$
$$= 5(x^2 + 2)(x^2 - 2)$$

7. $p^2 + 36 + 12p$
$$= p^2 + 12p + 36 \quad \text{Trinomial square}$$
$$= (p + 6)^2$$

9. $2x^2 - 10x - 132$
$$= 2(x^2 - 5x - 66)$$
$$= 2(x - 11)(x + 6) \quad \text{Trial and error}$$

11. $9x^2 - 25y^2$
$$= (3x)^2 - (5y)^2 \quad \text{Difference of squares}$$
$$= (3x + 5y)(3x - 5y)$$

13. $4m^4 - 100$
$$= 4(m^4 - 25) \quad \text{Removing the common factor}$$
$$= 4[(m^2)^2 - 5^2] \quad \text{Difference of squares}$$
$$= 4(m^2 + 5)(m^2 - 5)$$

15. $6w^2 + 12w - 18$
$$= 6(w^2 + 2w - 3) \quad \text{Removing the common factor}$$
$$= 6(w + 3)(w - 1) \quad \text{Trial and error}$$

17. $2xy^2 - 50x$
$$= 2x(y^2 - 25)$$
$$= 2x(y + 5)(y - 5)$$

19. $225 - (a - 3)^2$ Difference of squares

$= [15 + (a - 3)][15 - (a - 3)]$

$= [15 + a - 3][15 - a + 3]$

$= (12 + a)(18 - a)$

21. $m^6 - 1$

$= (m^3)^2 - 1^2$ Difference of squares

$= (m^3 + 1)(m^3 - 1)$ Sum and difference of cubes

$= (m + 1)(m^2 - m + 1)(m - 1)(m^2 + m + 1)$

23. $x^2 + 6x - y^2 + 9$

$= x^2 + 6x + 9 - y^2$

$= (x + 3)^2 - y^2$ Difference of squares

$= [(x + 3) + y][(x + 3) - y]$

$= (x + 3 + y)(x + 3 - y)$

25. $250x^3 - 128y^3$

$= 2(125x^3 - 64y^3)$

$= 2[(5x)^3 - (4y)^3]$ Difference of cubes

$= 2(5x - 4y)(25x^2 + 20xy + 16y^2)$

27. $8m^3 + m^6 - 20$

$= m^6 + 8m^3 - 20$

$= (m^3)^2 + 8m^3 - 20$

$= (m^3 - 2)(m^3 + 10)$ Trial and error

29. $ac + cd - ab - bd$

$= c(a + d) - b(a + d)$ Factoring by grouping

$= (a + d)(c - b)$

31. $50b^2 - 5ab - a^2$

$= (5b - a)(10b + a)$ FOIL or ac-method

33. $-7x^2 + 2x^3 + 4x - 14$

$= 2x^3 - 7x^2 + 4x - 14$

$= x^2(2x - 7) + 2(2x - 7)$ Factoring by grouping

$= (2x - 7)(x^2 + 2)$

35. $2x^3 + 6x^2 - 8x - 24$

$= 2(x^3 + 3x^2 - 4x - 12)$

$= 2[x^2(x + 3) - 4(x + 3)]$ Factoring by grouping

$= 2(x + 3)(x^2 - 4)$ Difference of squares

$= 2(x + 3)(x + 2)(x - 2)$

37. $16x^3 + 54y^3$

$= 2(8x^3 + 27y^3)$

$= 2[(2x)^3 + (3y)^3]$ Sum of cubes

$= 2(2x + 3y)(4x^2 - 6xy + 9y^2)$

39. $6y - 60x^2y - 9xy$

$= 3y(2 - 20x^2 - 3x)$

$= 3y(-20x^2 - 3x + 2)$ Rearranging

$= 3y(-1)(20x^2 + 3x - 2)$ Factoring out -1

$= -3y(5x + 2)(4x - 1)$ FOIL or ac-method

$= 3y(-5x - 2)(4x - 1)$ Multiplying $5x + 2$ by -1

$= 3y(5x + 2)(-4x + 1)$ Multiplying $4x - 1$ by -1

41. $a^8 - b^8$ Difference of squares

$= (a^4 + b^4)(a^4 - b^4)$ Difference of squares

$= (a^4 + b^4)(a^2 + b^2)(a^2 - b^2)$ Difference of squares

$= (a^4 + b^4)(a^2 + b^2)(a + b)(a - b)$

43. $a^3b - 16ab^3$

$= ab(a^2 - 16b^2)$ Difference of squares

$= ab(a + 4b)(a - 4b)$

45. $\dfrac{1}{16}x^2 - \dfrac{1}{6}xy^2 + \dfrac{1}{9}y^4$

$= \dfrac{1}{16}x^2 - \dfrac{1}{6}xy^2 + \dfrac{1}{9}(y^2)^2$ Trinomial square

$= \left(\dfrac{1}{4}x - \dfrac{1}{3}y^2\right)^2$

47. $5x^3 - 5x^2y - 5xy^2 + 5y^3$

$= 5(x^3 - x^2y - xy^2 + y^3)$

$= 5[x^2(x - y) - y^2(x - y)]$ Factoring by grouping

$= 5(x - y)(x^2 - y^2)$

$= 5(x - y)(x + y)(x - y)$ Factoring the difference of squares

$= 5(x - y)^2(x + y)$

49. $42ab + 27a^2b^2 + 8$

$= 27a^2b^2 + 42ab + 8$

$= (9ab + 2)(3ab + 4)$ FOIL or ac-method

51. $8y^4 - 125y$

$= y(8y^3 - 125)$

$= y[(2y)^3 - 5^3]$ Difference of cubes

$= y(2y - 5)(4y^2 + 10y + 25)$

53. $a^2 - b^2 - 6b - 9$

$= a^2 - (b^2 + 6b + 9)$ Factoring out -1

$= a^2 - (b + 3)^2$ Difference of squares

$= [a + (b + 3)][a - (b + 3)]$

$= (a + b + 3)(a - b - 3)$

55. $q^2 - 10q + 25 - r^2$

$= (q - 5)^2 - r^2$ Difference of squares

$= (q - 5 + r)(q - 5 - r)$

57. *Familiarize*. Let $x =$ the number of correct answers and $y =$ the number of incorrect answers. Then the total number of points awarded for the correct answers is $2x$, the total number of points deducted for the incorrect answers is $\frac{1}{2}y$, and the total score is $2x - \frac{1}{2}y$. Assume that all the questions were answered.

Translate.

$\underbrace{\text{The number of questions}}_{x + y}$ is 75.

$$x + y = 75$$

$\underbrace{\text{The total score}}_{2x - \frac{1}{2}y}$ is 100.

$$2x - \frac{1}{2}y = 100$$

After clearing the fraction in the second equation, we have a system of equations.

$$x + y = 75,$$
$$4x - y = 200$$

Solve. We use the elimination method. Begin by adding the equations.

$$
\begin{aligned}
x + y &= 75 \\
4x - y &= 200 \\
\hline
5x &= 275 \\
x &= 55
\end{aligned}
$$

Now substitute 55 for x in the first equation and solve for y.

$$55 + y = 75$$
$$y = 20$$

Check. If there are 55 correct answers and 20 incorrect answers, the total number of questions is $55+20$, or 75. For the 55 correct answers, $2 \cdot 55$ or 110 points are awarded; for the 20 incorrect answers, $\frac{1}{2} \cdot 20$ or 10, points are deducted. Then the total score is $110-10$ or 100. The answer checks.

State. There were 55 correct answers and 20 incorrect answers.

59. $30y^4 - 97xy^2 + 60x^2$

$= (5y^2 - 12x)(6y^2 - 5x)$ FOIL or ac-method

61. $5x^3 - \dfrac{5}{27}$

$= 5\left(x^3 - \dfrac{1}{27}\right)$

$= 5\left[x^3 - \left(\dfrac{1}{3}\right)^3\right]$ Difference of cubes

$= 5\left(x - \dfrac{1}{3}\right)\left(x^2 + \dfrac{1}{3}x + \dfrac{1}{9}\right)$

63. $(x - p)^2 - p^2$ Difference of squares

$= (x - p + p)(x - p - p)$

$= x(x - 2p)$

65. $(y - 1)^4 - (y - 1)^2$

$= (y - 1)^2[(y - 1)^2 - 1]$
            Removing the common factor

$= (y - 1)^2[(y - 1) + 1][(y - 1) - 1]$
            Factoring the difference of squares

$= (y - 1)^2(y)(y - 2)$, or $y(y - 1)^2(y - 2)$

67. $4x^2 + 4xy + y^2 - r^2 + 6rs - 9s^2$

$= (4x^2 + 4xy + y^2) - 1(r^2 - 6rs + 9s^2)$ Grouping

$= (2x + y)^2 - (r - 3s)^2$ Difference of squares

$= [(2x + y) + (r - 3s)][(2x + y) - (r - 3s)]$

$= (2x + y + r - 3s)(2x + y - r + 3s)$

69. $c^{2w+1} + 2c^{w+1} + c$

$= c^{2w} \cdot c + 2c^w \cdot c + c$

$= c(c^{2w} + 2c^w + 1)$

$= c[(c^w)^2 + 2(c^w) + 1]$ Trinomial square

$= c(c^w + 1)^2$

71. $3(x + 1)^2 + 9(x + 1) - 12$

$= 3[(x + 1)^2 + 3(x + 1) - 4]$

$= 3[(x + 1) + 4][(x + 1) - 1]$ Factor u^2+3u-4
                                where $u=x+1$

$= 3(x + 5)(x)$, or

$3x(x + 5)$

73. $x^6 - 2x^5 + x^4 - x^2 + 2x - 1$

$= x^4(x^2 - 2x + 1) - (x^2 - 2x + 1)$

$= (x^2 - 2x + 1)(x^4 - 1)$

$= (x - 1)^2(x^2 + 1)(x + 1)(x - 1)$

$= (x - 1)^3(x^2 + 1)(x + 1)$

75. $y^9 - y$

$= y(y^8 - 1)$

$= y(y^4 + 1)(y^4 - 1)$

$= y(y^4 + 1)(y^2 + 1)(y^2 - 1)$

$= y(y^4 + 1)(y^2 + 1)(y + 1)(y - 1)$

Exercise Set 4.8

RC1. False; see page 381 in the text.

RC3. False; see Example 3 on page 382 in the text.

1. $x^2 + 3x = 28$

$x^2 + 3x - 28 = 0$ Getting 0 on one side

$(x + 7)(x - 4) = 0$ Factoring

$x + 7 = 0$ *or* $x - 4 = 0$ Principle of zero
                                products

$x = -7$ *or* $x = 4$

The solutions are -7 and 4.

3. $y^2 + 9 = 6y$

$y^2 - 6y + 9 = 0$ Getting 0 on one side

$(y - 3)(y - 3) = 0$ Factoring

$y - 3 = 0$ *or* $y - 3 = 0$ Principle of zero
                                products

$y = 3$ *or* $y = 3$

There is only one solution, 3.

5. $x^2 + 20x + 100 = 0$

$(x + 10)(x + 10) = 0$ Factoring

$x + 10 = 0 \quad or \quad x + 10 = 0$ Principle of zero products

$\qquad x = -10 \quad or \qquad x = -10$

There is only one solution, -10.

7. $9x + x^2 + 20 = 0$

$x^2 + 9x + 20 = 0$ Changing order

$(x + 5)(x + 4) = 0$ Factoring

$x + 5 = 0 \quad or \quad x + 4 = 0$ Principle of zero products

$\quad x = -5 \quad or \qquad x = -4$

The solutions are -5 and -4.

9. $x^2 + 8x = 0$

$x(x + 8) = 0$ Factoring

$x = 0 \quad or \quad x + 8 = 0$ Principle of zero products

$x = 0 \quad or \qquad x = 8$

The solutions are 0 and -8.

11. $x^2 - 25 = 0$

$(x + 5)(x - 5) = 0$ Factoring

$x + 5 = 0 \quad or \quad x - 5 = 0$ Principle of zero products

$\quad x = -5 \quad or \qquad x = 5$

The solutions are -5 and 5.

13. $z^2 = 144$

$z^2 - 144 = 0$ Getting 0 on one side

$(z + 12)(z - 12) = 0$ Factoring

$z + 12 = 0 \quad or \quad z - 12 = 0$ Principle of zero products

$\qquad z = -12 \quad or \qquad z = 12$

The solutions are -12 and 12.

15. $y^2 + 2y = 63$

$y^2 + 2y - 63 = 0$

$(y + 9)(y - 7) = 0$

$y + 9 = 0 \quad or \quad y - 7 = 0$

$\quad y = -9 \quad or \qquad y = 7$

The solutions are -9 and 7.

17. $32 + 4x - x^2 = 0$

$0 = x^2 - 4x - 32$

$0 = (x - 8)(x + 4)$

$x - 8 = 0 \quad or \quad x + 4 = 0$

$\quad x = 8 \quad or \qquad x = -4$

The solutions are 8 and -4.

19. $3b^2 + 8b + 4 = 0$

$(3b + 2)(b + 2) = 0$

$3b + 2 = 0 \quad or \quad b + 2 = 0$

$\quad 3b = -2 \quad or \qquad b = -2$

$\quad b = -\dfrac{2}{3} \quad or \qquad b = -2$

The solutions are $-\dfrac{2}{3}$ and -2.

21. $8y^2 - 10y + 3 = 0$

$(4y - 3)(2y - 1) = 0$

$4y - 3 = 0 \quad or \quad 2y - 1 = 0$

$\quad 4y = 3 \quad or \qquad 2y = 1$

$\quad y = \dfrac{3}{4} \quad or \qquad y = \dfrac{1}{2}$

The solutions are $\dfrac{3}{4}$ and $\dfrac{1}{2}$.

23. $6z - z^2 = 0$

$0 = z^2 - 6z$

$0 = z(z - 6)$

$z = 0 \quad or \quad z - 6 = 0$

$z = 0 \quad or \qquad z = 6$

The solutions are 0 and 6.

25. $12z^2 + z = 6$

$12z^2 + z - 6 = 0$

$(4z + 3)(3z - 2) = 0$

$4z + 3 = 0 \quad or \quad 3z - 2 = 0$

$\quad 4z = -3 \quad or \qquad 3z = 2$

$\quad z = -\dfrac{3}{4} \quad or \qquad z = \dfrac{2}{3}$

The solutions are $-\dfrac{3}{4}$ and $\dfrac{2}{3}$.

27. $7x^2 - 7 = 0$

$7(x^2 - 1) = 0$

$7(x + 1)(x - 1) = 0$

$x + 1 = 0 \quad or \quad x - 1 = 0$

$\quad x = -1 \quad or \qquad x = 1$

The solutions are -1 and 1.

29. $10 - r - 21r^2 = 0$

$21r^2 + r - 10 = 0$ Multiplying by -1 and rearranging

$(3r - 2)(7r + 5) = 0$

$3r - 2 = 0 \quad or \quad 7r + 5 = 0$

$\quad 3r = 2 \quad or \qquad 7r = -5$

$\quad r = \dfrac{2}{3} \quad or \qquad r = -\dfrac{5}{7}$

The solutions are $\dfrac{2}{3}$ and $-\dfrac{5}{7}$.

31. $15y^2 = 3y$

$15y^2 - 3y = 0$

$3y(5y - 1) = 0$

$3y = 0 \quad or \quad 5y - 1 = 0$
$y = 0 \quad or \qquad\quad 5y = 1$

$y = 0 \quad or \qquad\quad y = \dfrac{1}{5}$

The solutions are 0 and $\dfrac{1}{5}$.

33. $14 = x(x - 5)$

$14 = x^2 - 5x$

$0 = x^2 - 5x - 14 \quad$ Getting 0 on one side

$0 = (x - 7)(x + 2)$

$x - 7 = 0 \quad or \quad x + 2 = 0$
$x = 7 \quad or \qquad x = -2$

The solutions are 7 and -2.

35. $2x^3 - 2x^2 = 12x$

$2x^3 - 2x^2 - 12x = 0$

$2x(x^2 - x - 6) = 0$

$2x(x - 3)(x + 2) = 0$

$2x = 0 \quad or \quad x - 3 = 0 \quad or \quad x + 2 = 0$
$x = 0 \quad or \qquad x = 3 \quad or \qquad x = -2$

The solutions are 0, 3, and -2.

37. $2x^3 = 128x$

$2x^3 - 128x = 0$

$2x(x^2 - 64) = 0$

$2x(x + 8)(x - 8) = 0$

$2x = 0 \quad or \quad x + 8 = 0 \quad or \quad x - 8 = 0$
$x = 0 \quad or \qquad x = -8 \quad or \qquad x = 8$

The solutions are 0, -8, and 8.

39. $t^4 - 26t^2 + 25 = 0$

$(t^2 - 1)(t^2 - 25) = 0$

$(t + 1)(t - 1)(t + 5)(t - 5) = 0$

$t + 1 = 0 \quad or \quad t - 1 = 0 \quad or \quad t + 5 = 0 \quad or \quad t - 5 = 0$
$t = -1 \quad or \qquad t = 1 \quad or \qquad t = -5 \quad or \qquad t = 5$

The solutions are -1, 1, -5, and 5.

41. $(a - 4)(a + 4) = 20$

$a^2 - 16 = 20$

$a^2 - 36 = 0$

$(a + 6)(a - 6) = 0$

$a + 6 = 0 \quad or \quad a - 6 = 0$
$a = -6 \quad or \qquad a = 6$

The solutions are -6 and 6.

43. $x(5 + 12x) = 28$

$5x + 12x^2 = 28$

$12x^2 + 5x - 28 = 0$

$(4x + 7)(3x - 4) = 0$

$4x + 7 = 0 \quad or \quad 3x - 4 = 0$
$4x = -7 \quad or \qquad 3x = 4$

$x = -\dfrac{7}{4} \quad or \qquad x = \dfrac{4}{3}$

The solutions are $-\dfrac{7}{4}$ and $\dfrac{4}{3}$.

45. We set $f(x)$ equal to 8.

$x^2 + 12x + 40 = 8$

$x^2 + 12x + 32 = 0$

$(x + 8)(x + 4) = 0$

$x + 8 = 0 \quad or \quad x + 4 = 0$
$x = -8 \quad or \qquad x = -4$

The values of x for which $f(x) = 8$ are -8 and -4.

47. We set $g(x)$ equal to 12.

$2x^2 + 5x = 12$

$2x^2 + 5x - 12 = 0$

$(2x - 3)(x + 4) = 0$

$2x - 3 = 0 \quad or \quad x + 4 = 0$
$2x = 3 \quad or \qquad x = -4$

$x = \dfrac{3}{2} \quad or \qquad x = -4$

The values of x for which $g(x) = 12$ are $\dfrac{3}{2}$ and -4.

49. We set $h(x)$ equal to -27.

$12x + x^2 = -27$

$12x + x^2 + 27 = 0$

$x^2 + 12x + 27 = 0 \quad$ Rearranging

$(x + 3)(x + 9) = 0$

$x + 3 = 0 \quad or \quad x + 9 = 0$
$x = -3 \quad or \qquad x = -9$

The values of x for which $h(x) = -27$ are -3 and -9.

51. $f(x) = \dfrac{3}{x^2 - 4x - 5}$

$f(x)$ cannot be calculated for any x-value for which the denominator, $x^2 - 4x - 5$, is 0. To find the excluded values, we solve:

$x^2 - 4x - 5 = 0$

$(x - 5)(x + 1) = 0$

$x - 5 = 0 \quad or \quad x + 1 = 0$
$x = 5 \quad or \qquad x = -1$

The domain of f is $\{x | x$ is a real number *and* $x \neq 5$ *and* $x \neq -1\}$.

53. $f(x) = \dfrac{x}{6x^2 - 54}$

$f(x)$ cannot be calculated for any x-value for which the denominator, $6x^2 - 54$, is 0. To find the excluded values, we solve:

$6x^2 - 54 = 0$

$6(x^2 - 9) = 0$

$6(x + 3)(x - 3) = 0$

$x + 3 = 0 \quad or \quad x - 3 = 0$

$x = -3 \quad or \qquad x = 3$

The domain of f is $\{x | x$ is a real number *and*

$x \neq -3$ *and* $x \neq 3\}$.

55. $f(x) = \dfrac{x - 5}{25x^2 - 10x + 1}$

$f(x)$ cannot be calculated for any x-value for which the denominator, $25x^2 - 10x + 1$, is 0. To find the excluded values, we solve:

$25x^2 - 10x + 1 = 0$

$(5x - 1)(5x - 1) = 0$

$5x - 1 = 0 \quad or \quad 5x - 1 = 0$

$5x = 1 \quad or \qquad 5x = 1$

$x = \dfrac{1}{5} \quad or \qquad x = \dfrac{1}{5}$

The domain of f is $\left\{ x | x \text{ is a real number } and \ x \neq \dfrac{1}{5} \right\}$.

57. $f(x) = \dfrac{7}{5x^3 - 35x^2 + 50x}$

$f(x)$ cannot be calculated for any x-value for which the denominator, $5x^3 - 35x^2 + 50x$, is 0. To find the excluded values, we solve:

$5x^3 - 35x^2 + 50x = 0$

$5x(x^2 - 7x + 10) = 0$

$5x(x - 2)(x - 5) = 0$

$5x = 0 \quad or \quad x - 2 = 0 \quad or \quad x - 5 = 0$

$x = 0 \quad or \qquad x = 2 \quad or \qquad x = 5$

The domain of f is $\{x | x$ is a real number *and*

$x \neq 0$ *and* $x \neq 2$ *and* $x \neq 5\}$.

59. From the graph we see that the x-intercepts are $(-5, 0)$ and $(9, 0)$. The solutions of the equation are the first coordinates of the x-intercepts, -5 and 9.

61. From the graph we see that the x-intercepts are $(-4, 0)$ and $(8, 0)$. The solutions of the equation are the first coordinates of the x-intercepts, -4 and 8.

63. *Familiarize*. Let $w =$ the width of the book, in cm. Then $w + 5 =$ the length. Recall that the area of a rectangle is given by length $\times$ width.

***Translate*.**

$\underbrace{\text{The area}}_{} \quad \text{is} \quad \underbrace{84 \text{ cm}^2}_{}.$

$\qquad \downarrow \qquad \quad \downarrow \qquad \downarrow$

$w(w + 5) \quad = \qquad 84$

***Solve*.** We solve the equation.

$w(w + 5) = 84$

$w^2 + 5w = 84$

$w^2 + 5w - 84 = 0$

$(w + 12)(w - 7) = 0$

$w + 12 = 0 \qquad or \quad w - 7 = 0$

$w = -12 \quad or \qquad w = 7$

***Check*.** The width cannot be negative so we check only 7. If the width is 7 cm, then the length is $7 + 5$, or 12 cm, and the area is $7 \cdot 12$, or 84 cm^2. The answer checks.

***State*.** The length is 12 cm, and the width is 7 cm.

65. *Familiarize*. Using the labels on the drawing in the text, we let x represent the base of the triangle and $x + 2$ represent the height, in feet. Recall that the formula for the area of the triangle with base b and height h is $\dfrac{1}{2}bh$.

***Translate*.**

$\underbrace{\text{The area}}_{} \quad \text{is} \quad \underbrace{12 \text{ ft}^2}_{}.$

$\quad \downarrow \qquad \qquad \downarrow \qquad \downarrow$

$\dfrac{1}{2}x(x + 2) \quad = \qquad 12$

***Solve*.** We solve the equation:

$\dfrac{1}{2}x(x + 2) = 12$

$x(x + 2) = 24 \quad \text{Multiplying by 2}$

$x^2 + 2x = 24$

$x^2 + 2x - 24 = 0$

$(x + 6)(x - 4) = 0$

$x + 6 = 0 \quad or \quad x - 4 = 0$

$x = -6 \quad or \qquad x = 4$

***Check*.** We check only 4 since the length of the base cannot be negative. If the base is 4 ft, then the height is $4 + 2$, or 6 ft, and the area is $\dfrac{1}{2} \cdot 4 \cdot 6$, or 12 ft^2. The answer checks.

***State*.** The height is 6 ft, and the base is 4 ft.

67. *Familiarize*. We make a drawing and label it. We let x represent the length of a side of the original square, in cm.

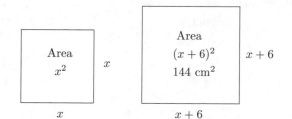

***Translate*.**

$\underbrace{\text{Area of new square}}_{} \quad \text{is} \quad \underbrace{144 \text{ cm}^2}_{}.$

$\qquad \downarrow \qquad \qquad \quad \downarrow \qquad \downarrow$

$\quad (x + 6)^2 \qquad \quad = \qquad 144$

***Solve*.** We solve the equation:

$(x + 6)^2 = 144$

$x^2 + 12x + 36 = 144$

$x^2 + 12x - 108 = 0$

$(x - 6)(x + 18) = 0$

$x - 6 = 0 \quad or \quad x + 18 = 0$

$x = 6 \quad or \qquad x = -18$

***Check*.** We only check 6 since the length of a side cannot be negative. If we increase the length by 6, the new length is $6+6$, or 12 cm. Then the new area is $12 \cdot 12$, or 144 cm^2. We have a solution.

State. The length of a side of the original square is 6 cm.

69. **Familiarize.** Let x represent the first integer, $x+2$ the second, and $x+4$ the third.

Translate.

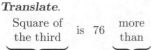

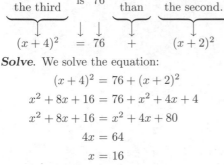

Solve. We solve the equation:
$$(x+4)^2 = 76 + (x+2)^2$$
$$x^2 + 8x + 16 = 76 + x^2 + 4x + 4$$
$$x^2 + 8x + 16 = x^2 + 4x + 80$$
$$4x = 64$$
$$x = 16$$

Check. We check the integers 16, 18, and 20. The square of 20, or 400, is 76 more than 324, the square of 18. The answer checks.

State. The integers are 16, 18, and 20.

71. **Familiarize.** Using the labels in the text, we let $x =$ the width of the frame, in centimeters. Then we see that $20 - 2x =$ the length of the picture and $12 - 2x =$ the width of the picture. Recall that the area A of a rectangle with length l and width w is given by $A = lw$.

Translate.

$$\underbrace{\text{The area of the picture}}_{(20-2x)(12-2x)} \text{ is } \underbrace{84 \text{ cm}^2.}_{84}$$

Solve. We solve the equation.
$$(20-2x)(12-2x) = 84$$
$$240 - 64x + 4x^2 = 84$$
$$4x^2 - 64x + 156 = 0$$
$$4(x^2 - 16x + 39) = 0$$
$$x^2 - 16x + 39 = 0$$
$$(x-13)(x-3) = 0$$
$$x - 13 = 0 \quad or \quad x - 3 = 0$$
$$x = 13 \quad or \qquad x = 3$$

Check. If the width of the frame were 13 cm, then the width of the picture would be $12 - 2 \cdot 13$, or -14 cm. This is not possible, so 13 cannot be a solution. If the width of the frame is 3 cm, then the length and width of the picture are $20 - 2 \cdot 3$, or 14 cm, and $12 - 2 \cdot 3$, or 6 cm, respectively. The area of the picture is 14 cm $\cdot$ 6 cm, or 84 cm^2. The number 3 checks.

State. The frame is 3 cm wide.

73. **Familiarize.** Using the labels in the drawing in the text, we let $w =$ the width, in feet, and $w + 4 =$ the length. Recall that the area of a rectangle is given by length $\times$ width.

Translate.

$$\underbrace{\text{The area}}_{} \text{ is } \underbrace{96 \text{ ft}^2.}_{}$$
$$w(w+4) = 96$$

Solve. We solve the equation.
$$w(w+4) = 96$$
$$w^2 + 4w = 96$$
$$w^2 + 4w - 96 = 0$$
$$(w+12)(w-8) = 0$$
$$w + 12 = 0 \quad or \quad w - 8 = 0$$
$$w = -12 \quad or \qquad w = 8$$

Check. The width cannot be negative so we check only 8. If the width is 8 ft, then the length is $8 + 4$, or 12 ft, and the area is $8 \cdot 12$, or 96 ft^2. The answer checks.

State. The length is 12 ft, and the width is 8 ft.

75. **Familiarize.** If $d =$ the distance from the base of the tower to the end of the wire, then $d + 4 =$ the height of the tower.

Translate. We use the Pythagorean theorem.
$$d^2 + (d+4)^2 = 20^2$$

Solve. We solve the equation.
$$d^2 + (d^2 + 8d + 16) = 400$$
$$2d^2 + 8d + 16 = 400$$
$$2d^2 + 8d - 384 = 0$$
$$2(d+16)(d-12) = 0$$
$$d + 16 = 0 \quad or \quad d - 12 = 0$$
$$d = -16 \quad or \qquad d = 12$$

Check. The distance cannot be negative, so we check only 12. If $d = 12$, then $d + 4 = 16$ and $12^2 + 16^2 = 400 = 20^2$. The answer checks.

State. The distance d is 12 ft, and the height of the tower is 16 ft.

77. **Familiarize.** Let $x =$ the height that the ladder reaches on the wall. Then $x + 1 =$ the length of the ladder.

Translate. We use the Pythagorean theorem.
$$9^2 + x^2 = (x+1)^2$$

Solve. We solve the equation.
$$9^2 + x^2 = (x+1)^2$$
$$81 + x^2 = x^2 + 2x + 1$$
$$81 = 2x + 1$$
$$80 = 2x$$
$$40 = x$$

If $x = 40$, then $x + 1 = 40 + 1 = 41$.

Check. 40 and 41 are consecutive integers, and $9^2 + 40^2 = 81 + 1600 = 1681 = (41)^2$, so the answer checks.

State. The ladder is 41 ft long.

79. **Familiarize.** Let x and $x + 2$ represent the lengths of the legs and $x + 4$ represent the length of the hypotenuse.

Translate. We use the Pythagorean theorem.

$$x^2 + (x+2)^2 = (x+4)^2$$

Solve. We solve the equation.

$$x^2 + (x+2)^2 = (x+4)^2$$
$$x^2 + x^2 + 4x + 4 = x^2 + 8x + 16$$
$$2x^2 + 4x + 4 = x^2 + 8x + 16$$
$$x^2 - 4x - 12 = 0$$
$$(x-6)(x+2) = 0$$
$$x - 6 = 0 \ \ or \ \ x + 2 = 0$$
$$x = 6 \ \ or \ \ \ \ \ \ x = -2$$

Check. Since the length of a side cannot be negative, -2 cannot be a solution. We check 6. When $x = 6$, then $x + 2 = 6 + 2 = 8$ and $x + 4 = 6 + 4 = 10$. The numbers 6, 8, and 10 are consecutive even integers and $6^2 + 8^2 = 36 + 64 = 100 = 10^2$. The answer checks.

State. The lengths of the sides are 6, 8, and 10.

81. *Familiarize*. We will use the equation
$h(t) = -16t^2 + 96t + 880$.

Translate.

$$\underbrace{\text{Height}} \quad \text{is} \quad \underbrace{\text{0 ft.}}$$
$$\downarrow \quad\quad\quad \downarrow \quad\quad \downarrow$$
$$-16t^2 + 96t + 880 \ = \ \ 0$$

Solve. We solve the equation:

$$-16(t^2 - 6t - 55) = 0$$
$$-16(t-11)(t+5) = 0$$
$$t - 11 = 0 \ \ or \ \ t + 5 = 0$$
$$t = 11 \ \ or \ \ \ \ \ \ t = -5$$

Check. The number -5 is not a solution, since time cannot be negative in this application. When $t = 11$, $h(t) = -16 \cdot 11^2 + 96 \cdot 11 + 880 = 0$. We have a solution.

State. The object reaches the ground after 11 sec.

83. $|-3 - (-4)| = |-3 + 4| = |1| = 1$

85. $\left|-\dfrac{3}{5} - \dfrac{2}{3}\right| = \left|-\dfrac{9}{15} - \dfrac{10}{15}\right| = \left|-\dfrac{19}{15}\right| = \dfrac{19}{15}$

87. First find the slope.

$$m = \frac{-4 - 7}{-8 - (-2)} = \frac{-11}{-6} = \frac{11}{6}$$

Now substitute $\dfrac{11}{6}$ for m and the coordinates of one of the given points for x and y in the equation $y = mx + b$ and solve for b. We will use the point $(-2, 7)$.

$$y = mx + b$$
$$7 = \frac{11}{6}(-2) + b$$
$$7 = -\frac{11}{3} + b$$
$$\frac{32}{3} = b$$

Now use $y = mx + b$ again, substituting $\dfrac{11}{6}$ for m and $\dfrac{32}{3}$ for b.

$$y = \frac{11}{6}x + \frac{32}{3}$$

89. First find the slope.

$$m = \frac{4 - 7}{8 - (-2)} = \frac{-3}{10} = -\frac{3}{10}$$

Now substitute $-\dfrac{3}{10}$ for m and the coordinates of one of the given points for x and y in the equation $y = mx + b$ and solve for b. We will use the point $(-2, 7)$.

$$y = mx + b$$
$$7 = -\frac{3}{10}(-2) + b$$
$$7 = \frac{3}{5} + b$$
$$\frac{32}{5} = b$$

Now we use $y = mx + b$ again, substituting $-\dfrac{3}{10}$ for m and $\dfrac{32}{5}$ for b.

$$y = -\frac{3}{10}x + \frac{32}{5}$$

91. The solutions of $-x^2 - 2x + 3 = 0$ are the first coordinates of the x-intercepts. From the graph we see that these are -3 and 1. The solution set is $\{-3, 1\}$.

To solve $-x^2 - 2x + 3 \geq -5$ we find the x-values for which $f(x) \geq -5$. From the graph we see that these are the values in the interval $[-4, 2]$. The solution set can also be expressed as $\{x| -4 \leq x \leq 2\}$.

93. a) One method is to graph $y = x^4 - 3x^3 - x^2 + 5$ and then use the ZERO feature. The solutions are 1.2522305 and 3.1578935.

b) One method is to graph $y_1 = x^4 - 3x^3 - x^2 + 5$ and $y_2 = 5$. Then use the INTERSECT feature to find the first coordinates of the points of intersection of the graphs. The solutions are -0.3027756, 0, and 3.3027756.

c) One method is to graph $y_1 = x^4 - 3x^3 - x^2 + 5$ and $y_2 = -8$. Then use the INTERSECT feature to find the first coordinates of the points of intersection of the graphs. The solutions are 2.1387475 and 2.7238657.

d) One method is to graph $y_1 = x^4$ and $y_2 = 1 + 3x^3 + x^2$. Then use the INTERSECT feature to find the first coordinates of the points of intersection of the graphs. The solutions are -0.7462555 and 3.3276509.

Chapter 4 Vocabulary Reinforcement

1. When the terms of a polynomial are written such that the exponents increase from left to right, we say that the polynomial is written in ascending order.

2. To factor a polynomial is to express it as a product.

3. A _factor_ of a polynomial P is a polynomial that can be used to express P as a product.

4. A _factorization_ of a polynomial is an expression that names that polynomial as a product.

5. When factoring a polynomial with four terms, try factoring by _grouping_.

6. A trinomial square is the square of a _binomial_.

7. The principle of _zero_ products states that if $ab = 0$, then $a = 0$ or $b = 0$.

8. The factorization of a _difference_ of squares is the product of the sum and difference of two terms.

Chapter 4 Concept Reinforcement

1. False; if $ab = 0$ then $a = 0$ _or_ $b = 0$.

2. True; $27 - t^3 = 3^3 - t^3$.

3. False; the expression is not a binomial because y has a negative exponent.

Chapter 4 Study Guide

1. $-6x^4 + 5x^3 - x^2 + 10x - 1$

Terms: $-6x^4$, $5x^3$, $-x^2$, $10x$, -1

Degree of each term: 4, 3, 2, 1, 0

Degree of the polynomial: 4

Leading term: $-6x^4$

Leading coefficient: -6

Constant term: -1

2. $(3y^2 - 6y^3 + 7y) - (y^2 - 10y - 8y^3 + 8)$
$$= 3y^2 - 6y^3 + 7y - y^2 + 10y + 8y^3 - 8$$
$$= 2y^3 + 2y^2 + 17y - 8$$

3. $(3x - 5y)(x + 2y) = 3x^2 + 6xy - 5xy - 10y^2$
$$= 3x^2 + xy - 10y^2$$

4. $(2y + 7)^2 = (2y)^2 + 2 \cdot 2y \cdot 7 + 7^2 = 4y^2 + 28y + 49$

5. $(5d + 10)(5d - 10) = (5d)^2 - 10^2 = 25d^2 - 100$

6. $f(x) = 3x^2 - x + 2$
$$f(x + 1) = 3(x + 1)^2 - (x + 1) + 2$$
$$= 3(x^2 + 2x + 1) - (x + 1) + 2$$
$$= 3x^2 + 6x + 3 - x - 1 + 2$$
$$= 3x^2 + 5x + 4$$
$$f(a+h) - f(a) = 3(a + h)^2 - (a + h) + 2 - (3a^2 - a + 2)$$
$$= 3(a^2 + 2ah + h^2) - (a+h) + 2 - (3a^2 - a + 2)$$
$$= 3a^2 + 6ah + 3h^2 - a - h + 2 - 3a^2 + a - 2$$
$$= 6ah + 3h^2 - h$$

7. $y^3 + 3y^2 - 8y - 24 = y^2(y + 3) - 8(y + 3)$
$$= (y + 3)(y^2 - 8)$$

8. $3x^2 + 19x - 72$

There is no common factor (other than 1 or -1). The factorization will be of the form
$$(3x + \quad)(x + \quad)$$
Look for a pair of factors of -72 that yield the desired middle term. We have
$$3x^2 + 19x - 72 = (3x - 8)(x + 9)$$

9. $10x^2 - 33x - 7$

There is no common factor (other than 1 or -1). Multiply the leading coefficient and the constant: $10(-7) = -70$. Look for a pair of factors of -70 whose sum is -33. The numbers we need are -35 and 2. Split the middle term and factor by grouping.
$$10x^2 - 33x - 7 = 10x^2 - 35x + 2x - 7$$
$$= 5x(2x - 7) + (2x - 7)$$
$$= (2x - 7)(5x + 1)$$

10. $81x^2 - 72x + 16 = (9x)^2 - 2 \cdot 9x \cdot 4 + 4^2 = (9x - 4)^2$

11. $100t^2 - 1 = (10t)^2 - 1 = (10t + 1)(10t - 1)$

12. $216x^3 + 1 = (6x)^3 + 1^3 = (6x + 1)(36x^2 - 6x + 1)$

13. $1000y^3 - 27 = (10y)^3 - 3^3 = (10y - 3)(100y^2 + 30y + 9)$

14. $3x^2 - x = 14$
$$3x^2 - x - 14 = 0$$
$$(3x - 7)(x + 2) = 0$$
$$3x - 7 = 0 \quad or \quad x + 2 = 0$$
$$3x = 7 \quad or \qquad x = -2$$
$$x = \frac{7}{3} \quad or \qquad x = -2$$
The solutions are $\frac{7}{3}$ and -2.

Chapter 4 Review Exercises

1. $3x^6y - 7x^8y^3 + 2x^3 - 3x^2$

a) The degrees of the terms are $6 + 1$, or 7; $8 + 3$, or 11; 3; and 2. The degree of the polynomial is 11.

b) The leading term is the term of highest degree, $-7x^8y^3$. The leading coefficient is -7.

c) $-3x^2 + 2x^3 + 3x^6y - 7x^8y^3$

d) $-7x^8y^3 + 3x^6y + 2x^3 - 3x^2$, or
   $-7x^8y^3 + 3x^6y - 3x^2 + 2x^3$

2. $P(x) = x^3 - x^2 + 4x$
$$P(0) = 0^3 - 0^2 + 4 \cdot 0 = 0$$
$$P(-1) = (-1)^3 - (-1)^2 + 4(-1)$$
$$= -1 - 1 - 4$$
$$= -6$$

3. $P(x) = 4 - 2x - x^2$

$P(-2) = 4 - 2(-2) - (-2)^2$

$\qquad = 4 + 4 - 4$

$\qquad = 4$

$P(5) = 4 - 2 \cdot 5 - 5^2$

$\qquad = 4 - 10 - 25$

$\qquad = -31$

4. $8x + 13y - 15x + 10y = 8x - 15x + 13y + 10y$

$\qquad\qquad\qquad\qquad\quad = -7x^2 + 23y$

5. $\quad 3ab - 10 + 5ab^2 - 2ab + 7ab^2 + 14$

$\quad = (3 - 2)ab + (5 + 7)ab^2 + (-10 + 14)$

$\quad = ab + 12ab^2 + 4$

6. In 2010, $t = 2010 - 2008 = 2$. Locate 2 on the t-axis, and then move up to the graph. Now move horizontally to the $f(t)$-axis. This locates a value of about 4.9, so we estimate that the number of children participating in football in 2010 was about 4.9 million.

7. $\quad (-6x^3 - 4x^2 + 3x + 1) + (5x^3 + 2x + 6x^2 + 1)$

$\quad = (-6 + 5)x^3 + (-4 + 6)x^2 + (3 + 2)x + (1 + 1)$

$\quad = -x^3 + 2x^2 + 5x + 2$

8. $\quad (4x^3 - 2x^2 - 7x + 5) + (8x^2 - 3x^3 - 9 + 6x)$

$\quad = (4 - 3)x^3 + (-2 + 8)x^2 + (-7 + 6)x + (5 - 9)$

$\quad = x^3 + 6x^2 - x - 4$

9.
$$-9xy^2 - xy + 6x^2y$$
$$4xy^2 - xy - 5x^2y$$
$$\underline{-3xy^2 + 6xy + 12x^2y}$$
$$-8xy^2 + 4xy + 13x^2y$$

10. $\quad (3x - 5) - (-6x + 2)$

$\quad = (3x - 5) + (6x - 2)$

$\quad = 9x - 7$

11. $\quad (4a - b + 3c) - (6a - 7b - 4c)$

$\quad = (4a - b + 3c) + (-6a + 7b + 4c)$

$\quad = -2a + 6b + 7c$

12. $\quad (9p^2 - 4p + 4) - (-7p^2 + 4p + 4)$

$\quad = (9p^2 - 4p + 4) + (7p^2 - 4p - 4)$

$\quad = 16p^2 - 8p$

13. $\quad (6x^2 - 4xy + y^2) - (2x^2 + 3xy - 2y^2)$

$\quad = (6x^2 - 4xy + y^2) + (-2x^2 - 3xy + 2y^2)$

$\quad = 4x^2 - 7xy + 3y^2$

14. $(3x^2y)(-6xy^3) = [3(-6)](x^2 \cdot x)(y \cdot y^3) = -18x^3y^4$

15.
$$x^4 - 2x^2 + 3$$
$$x^4 + x^2 - 1$$
$$-x^4 + 2x^2 - 3$$
$$x^6 - 2x^4 + 3x^2$$
$$\underline{x^8 - 2x^6 + 3x^4}$$
$$x^8 - x^6 \qquad + 5x^2 - 3$$

16. $\quad (4ab + 3c)(2ab - c)$

$\quad = 8a^2b^2 - 4abc + 6abc - 3c^2 \quad$ FOIL

$\quad = 8a^2b^2 + 2abc - 3c^2$

17. $(2x + 5y)(2x - 5y) = (2x)^2 - (5y)^2 = 4x^2 - 25y^2$

18. $(2x - 5y)^2 = (2x)^2 - 2 \cdot 2x \cdot 5y + (5y)^2 = 4x^2 - 20xy + 25y^2$

19.
$$5x^2 - 7x + 3$$
$$\underline{4x^2 + 2x - 9}$$
$$-45x^2 + 63x - 27$$
$$10x^3 - 14x^2 + 6x$$
$$\underline{20x^4 - 28x^3 + 12x^2}$$
$$20x^4 - 18x^3 - 47x^2 + 69x - 27$$

20. $(x^2 + 4y^3)^2 = (x^2)^2 + 2 \cdot x^2 \cdot 4y^3 + (4y^3)^2 = x^4 + 8x^2y^3 + 16y^6$

21. $\quad (x - 5)(x^2 + 5x + 25)$

$\quad = (x - 5)(x^2) + (x - 5)(5x) + (x - 5)(25)$

$\quad = x^3 - 5x^2 + 5x^2 - 25x + 25x - 125$

$\quad = x^3 - 125$

22. $\quad \left(x - \dfrac{1}{3}\right)\left(x - \dfrac{1}{6}\right)$

$\quad = x^2 - \dfrac{1}{6}x - \dfrac{1}{3}x + \dfrac{1}{18} \quad$ FOIL

$\quad = x^2 - \dfrac{1}{2}x + \dfrac{1}{18}$

23. $f(x) = x^2 - 2x - 7$

$f(a - 1) = (a - 1)^2 - 2(a - 1) - 7$

$\qquad\quad = a^2 - 2a + 1 - 2a + 2 - 7$

$\qquad\quad = a^2 - 4a - 4$

$f(a + h) - f(a)$

$= (a + h)^2 - 2(a + h) - 7 - (a^2 - 2a - 7)$

$= a^2 + 2ah + h^2 - 2a - 2h - 7 - a^2 + 2a + 7$

$= 2ah + h^2 - 2h$

24. $9y^4 - 3y^2 = 3y^2 \cdot 3y^2 - 3y^2 \cdot 1$

$\qquad\qquad = 3y^2(3y^2 - 1)$

25. $15x^4 - 18x^3 + 21x^2 - 9x = 3x(5x^3 - 6x^2 + 7x - 3)$

26. $\quad a^2 - 12a + 27$

$\quad = (a - 3)(a - 9) \quad$ Trial and error

27. $\quad 3m^2 + 14m + 8$

$\quad = (3m + 2)(m + 4) \quad$ FOIL or ac-method

28. $\quad 25x^2 + 20x + 4$

$\quad = (5x)^2 + 2 \cdot 5x \cdot 2 + 2^2 \quad$ Trinomial square

$\quad = (5x + 2)^2$

29. $\quad 4y^2 - 16$

$\quad = 4(y^2 - 4) \qquad$ Difference of squares

$\quad = 4(y + 2)(y - 2)$

30. $ax + 2bx - ay - 2by$

$= x(a + 2b) - y(a + 2b)$ Factoring by grouping

$= (a + 2b)(x - y)$

31. $4x^4 + 4x^2 + 20 = 4(x^4 + x^2 + 5)$

32. $27x^3 - 8$

$= (3x)^3 - 2^3$ Difference of cubes

$= (3x - 2)(9x^2 + 6x + 4)$

33. $0.064b^3 - 0.125c^3$

$= (0.4b)^3 - (0.5c)^3$ Difference of cubes

$= (0.4b - 0.5c)(0.16b^2 + 0.2bc + 0.25c^2)$

34. $y^5 - y$

$= y(y^4 - 1)$

$= y(y^2 + 1)(y^2 - 1)$

$= y(y^2 + 1)(y + 1)(y - 1)$

35. $2z^8 - 16z^6$

$= 2z^6(z^2 - 8)$

36. $54x^6y - 2y$

$= 2y(27z^6 - 1)$ Difference of cubes

$= 2y(3x^2 - 1)(9x^4 + 3x^2 + 1)$

37. $1 + a^3$ Sum of cubes

$= (1 + a)(1 - a + a^2)$

38. $36x^2 - 120x + 100$

$= 4(9x^2 - 30x + 25)$ Trinomial square

$= 4(3x - 5)^2$

39. $6t^2 + 17pt + 5p^2$

$= (3t + p)(2t + 5p)$ FOIL or ac-method

40. $x^3 + 2x^2 - 9x - 18$

$= x^2(x + 2) - 9(x + 2)$

$= (x + 2)(x^2 - 9)$

$= (x + 2)(x + 3)(x - 3)$

41. $a^2 - 2ab + b^2 - 4t^2$

$= (a - b)^2 - 4t^2$

$= (a - b + 2t)(a - b - 2t)$

42. $x^2 - 20x = -100$

$x^2 - 20x + 100 = 0$

$(x - 10)(x - 10) = 0$

$x - 10 = 0$ or $x - 10 = 0$

$x = 10$ or $x = 10$

The solution is 10.

43. $6b^2 - 13b + 6 = 0$

$(2b - 3)(3b - 2) = 0$

$2b - 3 = 0$ or $3b - 2 = 0$

$2b = 3$ or $3b = 2$

$b = \dfrac{3}{2}$ or $b = \dfrac{2}{3}$

The solutions are $\dfrac{3}{2}$ and $\dfrac{2}{3}$.

44. $8y^2 = 14y$

$8y^2 - 14y = 0$

$2y(4y - 7) = 0$

$2y = 0$ or $4y - 7 = 0$

$y = 0$ or $4y = 7$

$y = 0$ or $y = \dfrac{7}{4}$

The solutions are 0 and $\dfrac{7}{4}$.

45. $r^2 = 16$

$r^2 - 16 = 0$

$(r + 4)(r - 4) = 0$

$r + 4 = 0$ or $r - 4 = 0$

$r = -4$ or $r = 4$

The solutions are -4 and 4.

46. We set $f(x)$ equal to 4.

$x^2 - 7x - 40 = 4$

$x^2 - 7x - 44 = 0$

$(x + 4)(x - 11) = 0$

$x + 4 = 0$ or $x - 11 = 0$

$x = -4$ or $x = 11$

The values of x for which $f(x) = 4$ are -4 and 11.

47. $f(x) = \dfrac{x - 3}{3x^2 + 19x - 14}$

$f(x)$ cannot be calculated for any x-value for which the denominator, $3x^2 + 19x - 14$, is 0. To find the excluded values, we solve:

$3x^2 + 19x - 14 = 0$

$(3x - 2)(x + 7) = 0$

$3x - 2 = 0$ or $x + 7 = 0$

$3x = 2$ or $x = -7$

$x = \dfrac{2}{3}$ or $x = -7$

The domain of f is $\left\{ x \middle| x \text{ is a real number } and\ x \neq \dfrac{2}{3} \ and \right.$

$\left. x \neq -7 \right\}$.

48. *Familiarize.* Using the labels on the drawing in the text, we let w = the width of the photograph and $w + 3$ = the length, in inches. Then the dimensions with the border added are $w + 2 + 2$ and $w + 3 + 2 + 2$, or $w + 4$ and $w + 7$.

Translate. We use the formula for the area of a rectangle, $A = lw$.

$$(w + 7)(w + 4) = 108$$

Solve. We solve the equation.

$$(w + 7)(w + 4) = 108$$
$$w^2 + 11w + 28 = 108$$
$$w^2 + 11w - 80 = 0$$
$$(w + 16)(w - 5) = 0$$
$$w + 16 = 0 \quad or \quad w - 5 = 0$$
$$w = -16 \quad or \quad w = 5$$

Check. The width cannot be negative, so we check only 5. If the width of the photograph is 5 in., then the length is $5 + 3$, or 8 in. With the border added, the dimensions are $5 + 2 + 2$ and $8 + 2 + 2$, or 9 in. and 12 in. The area is $9 \cdot 12$, or 108 in^2. The answer checks.

State. The length of the photograph is 8 in., and the width is 5 in.

49. Familiarize. Let x, $x+2$, and $x+4$ represent the integers.

Translate. The sum of the squares of the integers is 83, so we have

$$x^2 + (x + 2)^2 + (x + 4)^2 = 83.$$

Solve. We solve the equation.

$$x^2 + (x + 2)^2 + (x + 4)^2 = 83$$
$$x^2 + x^2 + 4x + 4 + x^2 + 8x + 16 = 83$$
$$3x^2 + 12x + 20 = 83$$
$$3x^2 + 12x - 63 = 0$$
$$3(x^2 + 4x - 21) = 0$$
$$3(x + 7)(x - 3) = 0$$
$$x + 7 = 0 \quad or \quad x - 3 = 0$$
$$x = -7 \quad or \quad x = 3$$

If $x = -7$, then $x+2 = -7+2$, or -5, and $x+4 = -7+4 = -3$. If $x = 3$, then $x + 2 = 3 + 2$, or 5, and $x + 4 = 3 + 4$, or 7.

Check. -7, -5, and -3 are consecutive odd integers and $(-7)^2 + (-5)^2 + (-3)^2 = 49 + 25 + 9 = 83$. Also, 3, 5, and 7 are consecutive odd integers and $3^2 + 5^2 + 7^2 = 9 + 25 + 49 = 83$. Both answers check.

State. The integers are -7, -5, and -3 or 3, 5, and 7.

50. Familiarize. Let s = the length of a side of the square. Then the area is $s \cdot s$, or s^2.

Translate.

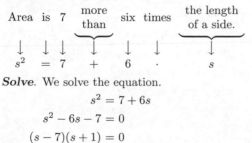

Solve. We solve the equation.

$$s^2 = 7 + 6s$$
$$s^2 - 6s - 7 = 0$$
$$(s - 7)(s + 1) = 0$$

$$s - 7 = 0 \quad or \quad s + 1 = 0$$
$$s = 7 \quad or \quad s = -1$$

Check. The length of a side cannot be negative, so we check only 7. The area is 7^2, or 49, and 7 more than six times 7 is $7 + 6 \cdot 7$, or $7 + 42$, or 49. The answer checks.

State. The length of a side of the square is 7.

51. $t^3 - 64 = t^3 - 4^3 = (t - 4)(t^2 + 4t + 16)$

The correct choice is A.

52. $hm + 5hn - gm - 5gn = h(m + 5n) - g(m + 5n)$
$$= (m + 5n)(h - g)$$

The correct choice is C.

53. $128x^6 - 2y^6$
$$= 2(64x^6 - y^6) \qquad \text{Difference of squares}$$
$$= 2(8x^3 + y^3)(8x^3 - y^3) \quad \text{Sum of cubes and}$$
$$\qquad\qquad\qquad\qquad\qquad \text{difference of cubes}$$
$$= 2(2x + y)(4x^2 - 2xy + y^2)(2x - y)(4x^2 + 2xy + y^2)$$

54. $(x + 1)^3 - (x - 1)^3 \qquad \text{Difference of cubes}$
$$= [(x + 1) - (x - 1)][(x + 1)^2 + (x + 1)(x - 1) + (x - 1)^2]$$
$$= (x + 1 - x + 1)(x^2 + 2x + 1 + x^2 - 1 + x^2 - 2x + 1)$$
$$= 2(3x^2 + 1)$$

55. $[a - (b - 1)][(b - 1)^2 + a(b - 1) + a^2] =$
$$[a - (b - 1)][a^2 + a(b - 1) + (b - 1)^2]$$

This product is of the form $(A - B)(A^2 + AB + B^2)$ where $A = a$ and $B = b - 1$. We know that this product is the factorization of $A^3 - B^3$, so we have $a^3 - (b - 1)^3$.

56. $64x^3 = x$
$$64x^3 - x = 0$$
$$x(64x^2 - 1) = 0$$
$$x(8x + 1)(8x - 1) = 0$$
$$x = 0 \quad or \quad 8x + 1 = 0 \quad or \quad 8x - 1 = 0$$
$$x = 0 \quad or \quad 8x = -1 \quad or \quad 8x = 1$$
$$x = 0 \quad or \quad x = -\frac{1}{8} \quad or \quad x = \frac{1}{8}$$

The solutions are 0, $-\dfrac{1}{8}$, and $\dfrac{1}{8}$.

Chapter 4 Discussion and Writing Exercises

1. A sum of two squares can be factored when there is a common factor that is a perfect square. For example, consider $4 + 4x^2$:

$$4 + 4x^2 = 2^2 + (2x)^2, \text{ and}$$
$$4 + 4x^2 = 4(1 + x^2)$$

2. See the procedure on page 359 of the text.

3. Add the opposite of the polynomial being subtracted.

4. To solve $P(x) = 0$, find the first coordinate(s) of the x-intercept(s) of $y = P(x)$.

To solve $P(x) = 4$, find the first coordinate(s) of the points of intersection of the graphs of $y_1 = P(x)$ and $y_2 = 4$.

5. To use factoring, write $x^3 - 8 = (x - 2)(x^2 + 2x + 4)$ and $(x-2)^3 = (x-2)(x-2)(x-2)$. Since $(x-2)(x^2+2x+4) \neq (x-2)(x-2)(x-2)$, then $x^3-8 \neq (x-2)^3$. To use graphing, enter $y_1 = x^3 - 8$ and $y_2 = (x - 2)^3$, and show that the graphs are different.

6. Both are correct. The factorizations are equivalent:

$$(a - b)(x - y)$$
$$= -1(b - a)(-1)(y - x)$$
$$= (-1)(-1)(b - a)(y - x)$$
$$= (b - a)(y - x)$$

7. $x = 5 \ or \ \ \ \ \ x = -3$

$x - 5 = 0 \ or \ x + 3 = 0$

$(x - 5)(x + 3) = 0$

$x^2 - 2x - 15 = 0$

There cannot be more than two solutions of a quadratic equation. This is because a quadratic equation is factorable into at most two different linear factors. Each of these has one solution when set equal to zero as required by the principle of zero products.

8. The discussion could include the following points:

a) We can now solve certain polynomial equations.

b) Whereas most linear equations have exactly one solution, non-linear polynomial equations can have more than one solution.

c) We used factoring and the principle of zero products to solve polynomial equations.

Chapter 4 Test

1. $3xy^3 - 4x^2y + 5x^5y^4 - 2x^4y$

a), b)

Term	$3xy^3$	$-4x^2y$	$5x^5y^4$	$-2x^4y$
Degree	4	3	9	5
Degree of polynomial	9			
Leading term	$5x^5y^4$			
Leading coefficient	5			

c) $3xy^3 - 4x^2y - 2x^4y + 5x^5y^4$

d) $5x^5y^4 + 3xy^3 - 4x^2y - 2x^4y$ or

$5x^5y^4 + 3xy^3 - 2x^4y - 4x^2y$

2. $P(x) = 2x^3 + 3x^2 - x + 4$

$P(0) = 2 \cdot 0^3 + 3 \cdot 0^2 - 0 + 4$

$\ \ \ \ \ = 0 + 0 - 0 + 4$

$\ \ \ \ \ = 4$

$P(-2) = 2(-2)^3 + 3(-2)^2 - (-2) + 4$

$\ \ \ \ \ \ \ = 2(-8) + 3(4) - (-2) + 4$

$\ \ \ \ \ \ \ = -16 + 12 + 2 + 4$

$\ \ \ \ \ \ \ = 2$

3. In 2010, $t = 2010 - 1980 = 30$. Locate 30 on the t-axis and then move up to the graph. Now move horizontally to the $m(t)$-axis. This locates a value of about 250, so we estimate that about 250 million tons of municipal solid waste was generated in 2010.

4. $5xy - 2xy^2 - 2xy + 5xy^2$

$= (5 - 2)xy + (-2 + 5)xy^2$

$= 3xy + 3xy^2$

5. $(-6x^3 + 3x^2 - 4y) + (3x^3 - 2y - 7y^2)$

$= (-6 + 3)x^3 + 3x^2 + (-4 - 2)y - 7y^2$

$= -3x^3 + 3x^2 - 6y - 7y^2$

6. $(4a^3 - 2a^2 + 6a - 5) + (3a^3 - 3a + 2 - 4a^2)$

$= (4 + 3)a^3 + (-2 - 4)a^2 + (6 - 3)a + (-5 + 2)$

$= 7a^3 - 6a^2 + 3a - 3$

7. $(5m^3 - 4m^2n - 6mn^2 - 3n^3) +$

$\ \ \ \ \ \ \ \ \ (9mn^2 - 4n^3 + 2m^3 + 6m^2n)$

$= (5 + 2)m^3 + (-4 + 6)m^2n + (-6 + 9)mn^2 + (-3 - 4)n^3$

$= 7m^3 + 2m^2n + 3mn^2 - 7n^3$

8. $(9a - 4b) - (3a + 4b) = (9a - 4b) + (-3a - 4b)$

$\ \ \ \ \ \ \ \ \ \ \ \ \ \ \ \ = 6a - 8b$

9. $(4x^2 - 3x + 7) - (-3x^2 + 4x - 6)$

$= (4x^2 - 3x + 7) + (3x^2 - 4x + 6)$

$= 7x^2 - 7x + 13$

10. $(6y^2 - 2y - 5y^3) - (4y^2 - 7y - 6y^3)$

$= (6y^2 - 2y - 5y^3) + (-4y^2 + 7y + 6y^3)$

$= 2y^2 + 5y + y^3$

11. $(-4x^2y)(-16xy^2) = [-4(-16)](x^2 \cdot x)(y \cdot y^2) = 64x^3y^3$

12. $(6a - 5b)(2a + b)$

$= 12a^2 + 6ab - 10ab - 5b^2 \ \ \ \ \ \ $ FOIL

$= 12a^2 - 4ab - 5b^2$

13. $(x - y)(x^2 - xy - y^2)$

$= (x - y)(x^2) - (x - y)(xy) - (x - y)(y^2)$

$= x \cdot x^2 - y \cdot x^2 - x \cdot xy - (-y)(xy) - x \cdot y^2 - (-y)(y^2)$

$= x^3 - x^2y - x^2y + xy^2 - xy^2 + y^3$

$= x^3 - 2x^2y + y^3$

14.

$$3m^2 \ + \ 4m \ - \ 2$$
$$\underline{- \ \ m^2 \ - \ 3m \ + \ 5}$$
$$15m^2 + 20m - 10$$
$$- \ 9m^3 \ - 12m^2 \ + \ 6m$$
$$\underline{-3m^4 \ - \ 4m^3 \ + \ 2m^2 }$$
$$-3m^4 - 13m^3 \ + \ 5m^2 \ + 26m - 10$$

15. $(4y - 9)^2$
$$= (4y)^2 - 2 \cdot 4y \cdot 9 + 9^2$$
$$ \qquad (A - B)^2 = A^2 - 2AB + B^2$$
$$= 16y^2 - 72y + 81$$

16. $(x - 2y)(x + 2y)$
$$= x^2 - (2y)^2 \quad (A + B)(A - B) = A^2 - B^2$$
$$= x^2 - 4y^2$$

17. $f(x) = x^2 - 5x$
$$f(a + 10) = (a + 10)^2 - 5(a + 10)$$
$$= a^2 + 20a + 100 - 5a - 50$$
$$= a^2 + 15a + 50$$
$$f(a + h) - f(a) = (a + h)^2 - 5(a + h) - (a^2 - 5a)$$
$$= a^2 + 2ah + h^2 - 5a - 5h - a^2 + 5a$$
$$= 2ah + h^2 - 5h$$

18. $9x^2 + 7x = x \cdot 9x + x \cdot 7 = x(9x + 7)$

19. $24y^3 + 16y^2 = 8y^2 \cdot 3y + 8y^2 \cdot 2 = 8y^2(3y + 2)$

20. $y^3 + 5y^2 - 4y - 20 = y^2(y + 5) - 4(y + 5)$
$$= (y + 5)(y^2 - 4)$$
$$= (y + 5)(y + 2)(y - 2)$$

21. $p^2 - 12p - 28$

We look for a pair of factors of -28 whose sum is -12. The numbers we need are -14 and 2.
$$p^2 - 12p - 28 = (p - 14)(p + 2)$$

22. $12m^2 + 20m + 3$

We will use the FOIL method.

1) There are no common factors (other than 1 or -1).

2) Factor the first term, $12m^2$. The possibilities are $(12m+ \quad)(m+ \quad)$ and $(6m+ \quad)(2m+ \quad)$ and $(4m+ \quad)(3m+ \quad)$.

3) Factor the last term, 3. We need to consider only positive factors because both the middle term and the last term are positive. The factors are 3 and 1.

4) Look for factors in steps (2) and (3) such that the sum of the products is the middle term, $20m$. Trial and error leads us to the correct factorization: $(6m + 1)(2m + 3)$.

23. $9y^2 - 25 = (3y)^2 - 5^2 = (3y + 5)(3y - 5)$

24. $3r^3 - 3 = 3(r^3 - 1)$
$$= 3(r - 1)(r^2 + r + 1)$$
$$A^3 - B^3 = (A - B)(A^2 + AB + B^2)$$

25. $9x^2 + 25 - 30x = 9x^2 - 30x + 25$
$$= (3x)^2 - 2 \cdot 3x \cdot 5 + 5^2$$
$$= (3x - 5)^2$$

26. $(z + 1)^2 - b^2 = (z + 1 + b)(z + 1 - b)$

27. $x^8 - y^8 = (x^4)^2 - (y^4)^2$
$$= (x^4 + y^4)(x^4 - y^4)$$
$$= (x^4 + y^4)[(x^2)^2 - (y^2)^2]$$
$$= (x^4 + y^4)(x^2 + y^2)(x^2 - y^2)$$
$$= (x^4 + y^4)(x^2 + y^2)(x + y)(x - y)$$

28. $y^2 + 8y + 16 - 100t^2$
$$= (y + 4)^2 - (10t)^2$$
$$= (y + 4 + 10t)(y + 4 - 10t)$$

29. $20a^2 - 5b^2 = 5(4a^2 - b^2) = 5(2a + b)(2a - b)$

30. $24x^2 - 46x + 10$

We will use the ac-method.

1) We factor out the common factor, 2.
$$2(12x^2 - 23x + 5)$$

2) Now we factor the trinomial $12x^2 - 23x + 5$. Multiply the leading coefficient, 12, and the constant, 5.
$$12 \cdot 5 = 60$$

3) Look for a factorization of 60 in which the sum of the factors is the coefficient of the middle term, -23. The factors we need are -20 and -3.

4) Split the middle term as follows:
$$-23x = \quad 20x - 3x$$

5) Factor by grouping.
$$12x^2 - 23x + 5 = 12x^2 - 20x - 3x + 5$$
$$= 4x(3x - 5) - (3x - 5)$$
$$= (3x - 5)(4x - 1)$$

We must include the common factor to get a factorization of the original trinomial.
$$24x^2 - 46x + 10 = 2(3x - 5)(4x - 1)$$

31. $16a^7b + 54ab^7$
$$= 2ab(8a^6 + 27b^6)$$
$$= 2ab[(2a^2)^3 + (3b^2)^3]$$
$$= 2ab(2a^2 + 3b^2)[(2a^2)^2 - 2a^2 \cdot 3b^2 + (3b^2)^2]$$
$$ \qquad A^3 + B^3 = (A + B)(A^2 - AB + B^2)$$
$$= 2ab(2a^2 + 3b^2)(4a^4 - 6a^2b^2 + 9b^4)$$

32.
$$x^2 - 18 = 3x$$
$$x^2 - 3x - 18 = 0$$
$$(x - 6)(x + 3) = 0$$
$$x - 6 = 0 \ \ or \ \ x + 3 = 0$$
$$x = 6 \ \ or \qquad x = -3$$

The solutions are 6 and -3.

33.
$$5y^2 - 125 = 0$$
$$5(y^2 - 25) = 0$$
$$5(y + 5)(y - 5) = 0$$
$$y + 5 = 0 \quad or \quad y - 5 = 0$$
$$y = -5 \quad or \qquad y = 5$$
The solutions are -5 and 5.

34.
$$2x^2 + 21 = -17x$$
$$2x^2 + 17x + 21 = 0$$
$$(2x + 3)(x + 7) = 0$$
$$2x + 3 = 0 \quad or \quad x + 7 = 0$$
$$2x = -3 \quad or \qquad x = -7$$
$$x = -\frac{3}{2} \quad or \qquad x = -7$$
The solutions are $-\dfrac{3}{2}$ and -7.

35. We set $f(x)$ equal to 11.
$$3x^2 - 15x + 11 = 11$$
$$3x^2 - 15x = 0$$
$$3x(x - 5) = 0$$
$$3x = 0 \quad or \quad x - 5 = 0$$
$$x = 0 \quad or \qquad x = 5$$
The values of x for which $f(x) = 11$ are 0 and 5.

36. $f(x) = \dfrac{3 - x}{x^2 + 2x + 1}$

$f(x)$ cannot be calculated for any x-value for which the denominator is 0. To find the excluded values, we solve:
$$x^2 + 2x + 1 = 0$$
$$(x + 1)(x + 1) = 0$$
$$x + 1 = 0 \quad or \quad x + 1 = 0$$
$$x = -1 \quad or \qquad x = -1$$
The domain of f is $\{x | x$ is a real number $and \; x \neq -1\}$, or $(-\infty, -1) \cup (-1, \infty)$.

37. *Familiarize*. Let $w =$ the width, in cm. Then $w + 3 =$ the length.

***Translate*.** We use the formula for the area of a rectangle, $A = l \cdot w$.
$$40 = (w + 3)w$$

***Solve*.** We solve the equation.
$$40 = (w + 3)w$$
$$40 = w^2 + 3w$$
$$0 = w^2 + 3w - 40$$
$$0 = (w + 8)(w - 5)$$
$$w + 8 = 0 \quad or \quad w - 5 = 0$$
$$w = -8 \quad or \qquad w = 5$$

***Check*.** The width cannot be negative, so we check only 5. When $w = 5$, then $w + 3 = 5 + 3 = 8$. If the length is 8 cm and the width is 5 cm, then the length is 3 cm more than the width and the area is $8 \cdot 5$, or 40 cm². The answer checks.

***State*.** The length is 8 cm, and the width is 5 cm.

38. *Familiarize*. Let $d =$ the distance the ladder reaches up the wall, in feet. Then $d + 2 =$ the length of the ladder. We make a drawing.

***Translate*.** We use the Pythagorean theorem.
$$a^2 + b^2 = c^2$$
$$10^2 + d^2 = (d + 2)^2$$

***Solve*.** We solve the equation.
$$10^2 + d^2 = (d + 2)^2$$
$$100 + d^2 = d^2 + 4d + 4$$
$$100 = 4d + 4$$
$$96 = 4d$$
$$24 = d$$
If $d = 24$, then $d + 2 = 24 + 2 = 26$.

***Check*.** 26 ft is 2 ft more than 24 ft. Also, $10^2 + 24^2 = 100 + 576 = 676 = 26^2$, so the answer checks.

***State*.** The ladder reaches 24 ft up the wall.

39. $f(n) = \dfrac{1}{2}n^2 - \dfrac{1}{2}n$
$$f(n) = \frac{1}{2}n \cdot n - \frac{1}{2}n \cdot 1$$
$$f(n) = \frac{1}{2}n(n - 1)$$

40. $8x^3 - 1 = (2x - 1)(4x^2 + 2x + 1)$

Answer C is correct.

41. $6x^{2n} - 7x^n - 20 = 6(x^n)^2 - 7x^n - 20 = (3x^n + 4)(2x^n - 5)$

42.
$$(p + q)^2 = p^2 + 2pq + q^2$$
$$29 = p^2 + 2 \cdot 5 + q^2 \quad \text{Substituting 29 for } (p+q)^2$$
$$\text{and 5 for } pq$$
$$29 = p^2 + 10 + q^2$$
$$19 = p^2 + q^2$$

Cumulative Review Chapters 1 - 4

1.
$$(x^2 + 4x - xy - 9) + (-3x^2 - 3x + 8)$$
$$= (1 - 3)x^2 + (4 - 3)x - xy + (-9 + 8)$$
$$= -2x^2 + x - xy - 1$$

2.
$$(6x^2 - 3x + 2x^3) - (8x^2 - 9x + 2x^3)$$
$$= (6x^2 - 3x + 2x^3) + (-8x^2 + 9x - 2x^3)$$
$$= (6 - 8)x^2 + (-3 + 9)x + (2 - 2)x^3$$
$$= -2x^2 + 6x$$

3.

$$
\begin{array}{r}
a^2 \ - \ a \ - \ 3 \\
a^2 \ + 2a \ - \ 3 \\
\hline
- \ 3a^2 \ + 3a \ + 9 \\
2a^3 \ - 2a^2 \ - 6a \\
a^4 \ - \ a^3 \ - 3a^2 \\
\hline
a^4 \ + \ a^3 \ - 8a^2 \ - 3a \ + 9
\end{array}
$$

4 $(x+4)(x+9)$

$= x^2 + 9x + 4x + 36$ FOIL

$= x^2 + 13x + 36$

5. $8 - 3x = 6x - 10$

$8 - 9x = -10$ Subtracting $6x$

$-9x = -18$ Subtracting 8

$x = 2$ Dividing by -9

The solution is 2.

6. $\dfrac{1}{2}x - 3 = \dfrac{7}{2}$

$2\left(\dfrac{1}{2}x - 3\right) = 2 \cdot \dfrac{7}{2}$ Clearing fractions

$x - 6 = 7$

$x = 13$

The solution is 13.

7. $A = \dfrac{1}{2}h(a+b)$

$\dfrac{2A}{h} = a + b$ Multiplying by $\dfrac{2}{h}$

$\dfrac{2A}{h} - a = b$, or

$\dfrac{2A - ah}{h} = b$

8. $6x - 1 \le 3(5x + 2)$

$6x - 1 \le 15x + 6$

$-9x - 1 \le 6$

$-9x \le 7$

$x \ge -\dfrac{7}{9}$ Dividing by -9 and reversing the inequality symbol

The solution set is $\left\{x \middle| x \ge -\dfrac{7}{9}\right\}$, or $\left[-\dfrac{7}{9}, \infty\right)$.

9. $4x - 3 < 2$ or $x - 3 > 1$

$4x < 5$ or $x > 4$

$x < \dfrac{5}{4}$ or $x > 4$

The solution set is $\left\{x \middle| x < \dfrac{5}{4} \text{ or } x > 4\right\}$, or

$\left(-\infty, \dfrac{5}{4}\right) \cup (4, \infty).$

10. $|2x - 3| < 7$

$-7 < 2x - 3 < 7$

$-4 < 2x < 10$

$-2 < x < 5$

The solution set is $\{x | -2 < x < 5\}$, or $(-2, 5)$.

11. $x + y + z = -5,$ (1)

$\quad x - z = 10,$ (2)

$\quad y - z = 12$ (3)

Since Equation (2) does not have a y-term, we eliminate y from a pair of equations.

$$
\begin{array}{r}
x + y + \ z = -5 \quad (1) \\
- y + \ z = -12 \quad \text{Multiplying (3) by } -1 \\
\hline
x \qquad + 2z = -17 \quad (5)
\end{array}
$$

Now we solve the system of Equations (2) and (5).

$x - z = 10$ (2)

$x + 2z = -17$ (5)

We multiply Equation (2) by 2 and then add.

$$
\begin{array}{r}
2x - 2z = 20 \\
x + 2z = -17 \\
\hline
3x \qquad = 3 \\
x = 1
\end{array}
$$

$1 - z = 10$ Substituting in (2)

$-z = 9$

$z = -9$

$1 + y - 9 = -5$ Substituting in (1)

$y - 8 = -5$

$y = 3$

The solution is $(1, 3, -9)$.

12. $2x + 5y = -2,$ (1)

$\quad 5x + 3y = 14$ (2)

We multiply Equation (1) by 3 and Equation (2) by -5 and then add.

$$
\begin{array}{r}
6x + 15y = -6 \\
-25x - 15y = -70 \\
\hline
-19x \qquad = -76 \\
x = 4
\end{array}
$$

$2 \cdot 4 + 5y = -2$ Substituting in (1)

$8 + 5y = -2$

$5y = -10$

$y = -2$

The solution is $(4, -2)$.

13. $3x - y = 7$, (1)

$2x + 2y = 5$ (2)

We multiply Equation (1) by 2 and then add.

$$6x - 2y = 14$$
$$\underline{2x + 2y = 5}$$
$$8x \qquad = 19$$
$$x = \frac{19}{8}$$

$2 \cdot \dfrac{19}{8} + 2y = 5$ Substituting in (2)

$$\frac{19}{4} + 2y = 5$$
$$2y = \frac{1}{4}$$
$$y = \frac{1}{8}$$

The solution is $\left(\dfrac{19}{8}, \dfrac{1}{8}\right)$.

14. $x + 2y - z = 0$, (1)

$3x + y - 2z = -1$, (2)

$x - 4y + z = -2$ (3)

First we eliminate z from two different pairs of equations.

$$-2x - 4y + 2z = 0 \qquad \text{Multiplying (1) by } -2$$
$$\underline{3x + y - 2z = -1 \quad (2)}$$
$$x - 3y \qquad = -1 \quad (4)$$

$$x + 2y - z = 0 \quad (1)$$
$$\underline{x - 4y + z = -2 \quad (3)}$$
$$2x - 2y \qquad = -2 \quad (5)$$

Now we solve the system of equations (4) and (5). We multiply Equation (4) by -2 and then add.

$$-2x + 6y = 2$$
$$\underline{2x - 2y = -2}$$
$$4y = 0$$
$$y = 0$$

$x - 3 \cdot 0 = -1$ Substituting in (4)

$$x = -1$$

$-1 - 4 \cdot 0 + z = -2$ Substituting in (3)

$$-1 + z = -2$$
$$z = -1$$

The solution is $(-1, 0, -1)$.

15. $11x + x^2 + 24 = 0$

$x^2 + 11x + 24 = 0$ Rearranging

$(x + 3)(x + 8) = 0$

$x + 3 = 0 \quad or \quad x + 8 = 0$

$x = -3 \quad or \qquad x = -8$

The solutions are -3 and -8.

16. $2x^2 - 15x = -7$

$2x^2 - 15x + 7 = 0$

$(2x - 1)(x - 7) = 0$

$2x - 1 = 0 \quad or \quad x - 7 = 0$

$2x = 1 \quad or \qquad x = 7$

$x = \dfrac{1}{2} \quad or \qquad x = 7$

The solutions are $\dfrac{1}{2}$ and 7.

17. We set $f(x)$ equal to 4.

$$3x^2 + 4x = 4$$
$$3x^2 + 4x - 4 = 0$$
$$(3x - 2)(x + 2) = 0$$

$3x - 2 = 0 \quad or \quad x + 2 = 0$

$3x = 2 \quad or \qquad x = -2$

$x = \dfrac{2}{3} \quad or \qquad x = -2$

The values of x for which $f(x) = 4$ are $\dfrac{2}{3}$ and -2.

18. $F(x) = \dfrac{x + 7}{x^2 - 2x - 15}$

The values of x excluded from the domain are those for which the denominator is 0. We find those values.

$$x^2 - 2x - 15 = 0$$
$$(x - 5)(x + 3) = 0$$

$x - 5 = 0 \quad or \quad x + 3 = 0$

$x = 5 \quad or \qquad x = -3$

The domain of F is $\{x | x$ is a real number $and \ x \neq 5$ $and \ x \neq -3\}$.

19. $3x^3 - 12x^2 = 3x^2 \cdot x - 3x^2 \cdot 4 = 3x^2(x - 4)$

20. $2x^4 + x^3 + 2x + 1$

$= x^3(2x + 1) + (2x + 1) \qquad \text{Factoring by grouping}$

$= (2x + 1)(x^3 + 1) \qquad\qquad \text{Sum of cubes}$

$= (2x + 1)(x + 1)(x^2 - x + 1)$

21. $x^2 + 5x - 14$

$= (x + 7)(x - 2) \quad \text{Trial and error}$

22. $20a^2 - 23a + 6$

$= (4a - 3)(5a - 2) \quad \text{FOIL or } ac\text{-method}$

23. $4x^2 - 25 \quad \text{Difference of squares}$

$= (2x + 5)(2x - 5)$

24. $2x^2 - 28x + 98$

$= 2(x^2 - 14x + 49) \quad \text{Trinomial square}$

$= 2(x - 7)^2$

25. $a^3 + 1000 \qquad \text{Sum of cubes}$

$= (a + 10)(a^2 - 10a + 100)$

26. $64x^3 - 1 \qquad \text{Difference of cubes}$

$= (4x - 1)(16x^2 + 4x + 1)$

27. $4a^3 + a^6 - 12$

   $= a^6 + 4a^3 - 12$ Rearranging

   $= (a^3 + 6)(a^3 - 2)$ Trial and error

28. $4x^4y^2 - x^2y^4$

   $= x^2y^2(4x^2 - y^2)$

   $= x^2y^2(2x + y)(2x - y)$

29. ***Familiarize***. Let x, y, and z represent the cost of an 8×10 print, an 11×14 print, and a 24×36 print, respectively.

   Translate. We have three equations.

   Greg's total cost: $2x + y = 41.40$

   Sara's total cost: $x + y + z = 101.20$

   Austin's total cost: $2y + 2z = 184$

   Solve. Solving the system of equations, we get $(9.20, 23, 69)$.

   Check. Greg's cost: $2(\$9.20) + \$23 = \$41.40$

   Sara's cost: $\$9.20 + \$23 + \$69 = \$101.20$

   Austin's cost: $2 \cdot \$23 + 2 \cdot \$69 = \$184$

   The answer checks.

   State. An 8×10 print costs $\$9.20$, and 11×14 print costs $\$23$, and a 24×36 print costs $\$69$.

30. $x < 1$ *or* $x \geq 2$

   We shade all points to the left of 1 and use a parentheses at 1 to show that it is not a solution. We also shade all points to the right of 2 and use a bracket at 2 to show that 2 is a solution.

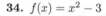

31. $y = -2x$

   We find some ordered pairs that are solutions and draw and label the line.

   When $x = -2$, $y = -2(-2) = 4$.

   When $x = 0$, $y = -2 \cdot 0 = 0$.

   When $x = 2$, $y = -2 \cdot 2 = -4$.

x	y
-2	4
0	0
2	-4

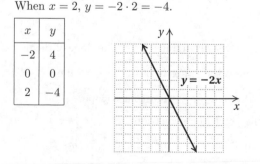

32. $6y + 24 = 0$

   $6y = -24$

   $y = -4$

   Since x is missing, all ordered pairs $(x, -4)$ are solutions. The graph is parallel to the x-axis.

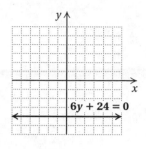

33. $y > x + 6$

   First graph $y = x + 6$. Draw the line dashed since the inequality symbol is $>$. Test the point $(0, 0)$ to determine if it is a solution.

$$\frac{y > x + 6}{0 \ ? \ 0 + 6}$$
$$\Big| \quad 6 \qquad \text{FALSE}$$

   Since $0 > 6$ is false, we shade the half-plane that does not contain $(0, 0)$.

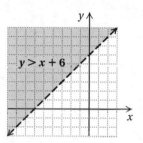

34. $f(x) = x^2 - 3$

   Make a list of function values in a table.

   $f(-2) = (-2)^2 - 3 = 4 - 3 = 1$

   $f(-1) = (-1)^2 - 3 = 1 - 3 = -2$

   $f(0) = 0^2 - 3 = -3$

   $f(1) = 1^2 - 3 = 1 - 3 = -2$

   $f(2) = 2^2 - 3 = 4 - 3 = 1$

x	$f(x)$
-2	1
-1	-2
0	-3
1	-2
2	1

   Plot these points and connect them.

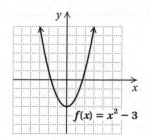

35. $g(x) = 4 - |x|$

Make a list of function values in a table.

$g(-5) = 4 - |-5| = 4 - 5 = -1$

$g(-3) = 4 - |-3| = 4 - 3 = 1$

$g(-1) = 4 - |-1| = 4 - 1 = 3$

$g(0) = 4 - |0| = 4 - 0 = 4$

$g(2) = 4 - |2| = 4 - 2 = 2$

$g(4) = 4 - |4| = 4 - 4 = 0$

x	$g(x)$
-5	-1
-3	1
-1	3
0	4
2	2
4	0

Plot these points and connect them.

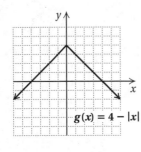

36. $2x + 3y \le 6,$ (1)

$5x - 5y \le 15,$ (2)

$x \ge 0$ (3)

Shade the intersection of the graphs of the three inequalities above.

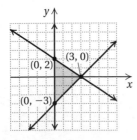

To find the vertices we solve three systems of equations, as follows:

System of equations	Vertex
From (1) and (2)	$(3, 0)$
From (1) and (3)	$(0, 2)$
From (2) and (3)	$(0, -3)$

37. First solve the equation for y and determine the slope of the given line.

$x + 2y = 6$ Given line

$2y = -x + 6$

$y = -\dfrac{1}{2}x + 3$

The slope of the given line is $-\dfrac{1}{2}$. The line through $(3, 7)$ must have slope $-\dfrac{1}{2}$.

Using the point-slope equation:

Substitute 3 for x_1, 7 for y_1, and $-\dfrac{1}{2}$ for m.

$y - y_1 = m(x - x_1)$

$y - 7 = -\dfrac{1}{2}(x - 3)$

$y - 7 = -\dfrac{1}{2}x + \dfrac{3}{2}$

$y = -\dfrac{1}{2}x + \dfrac{17}{2}$

Using the slope-intercept equation:

Substitute 3 for x, 7 for y, and $-\dfrac{1}{2}$ for m and solve for b.

$y = mx + b$

$7 = -\dfrac{1}{2} \cdot 3 + b$

$7 = -\dfrac{3}{2} + b$

$\dfrac{17}{2} = b$

Then we use the equation $y = mx + b$ and substitute $-\dfrac{1}{2}$ for m and $\dfrac{17}{2}$ for b.

$y = -\dfrac{1}{2}x + \dfrac{17}{2}$

38. First solve the equation for y and determine the slope of the given line.

$3x + 4y = 5$ Given line

$4y = -3x + 5$

$y = -\dfrac{3}{4}x + \dfrac{5}{4}$

The slope of the given line is $-\dfrac{3}{4}$. The slope of the perpendicular line is the opposite of the reciprocal of $-\dfrac{3}{4}$. Thus, the line through $(3, -2)$ must have slope $\dfrac{4}{3}$.

Using the point-slope equation:

Substitute 3 for x_1, -2 for y_1, and $\dfrac{4}{3}$ for m.

$y - y_1 = m(x - x_1)$

$y - (-2) = \dfrac{4}{3}(x - 3)$

$y + 2 = \dfrac{4}{3}x - 4$

$y = \dfrac{4}{3}x - 6$

Using the slope-intercept equation:

Substitute 3 for x, -2 for y, and $\frac{4}{3}$ for m.

$$y = mx + b$$
$$-2 = \frac{4}{3} \cdot 3 + b$$
$$-2 = 4 + b$$
$$-6 = b$$

Then we use the equation $y = mx + b$ and substitute $\frac{4}{3}$ for m and -6 for b.

$$y = \frac{4}{3}x - 6$$

39. First find the slope of the line.

$$m = \frac{0 - 4}{-2 - (-1)} = \frac{-4}{-1} = 4$$

Using the point-slope equation:

We choose $(-2, 0)$ and substitute -2 for x_1, 0 for y_1, and 4 for m.

$$y - y_1 = m(x - x_1)$$
$$y - 0 = 4(x - (-2))$$
$$y = 4(x + 2)$$
$$y = 4x + 8$$

Using the slope-intercept equation:

We choose $(-2, 0)$ and substitute -2 for x, 0 for y, and 4 for m. Then solve for b.

$$y = mx + b$$
$$0 = 4(-2) + b$$
$$0 = -8 + b$$
$$8 = b$$

Finally, we use the equation $y = mx + b$ and substitute 4 for m and 8 for b.

$$y = 4x + 8$$

40. Using the point-slope equation:

Substitute 2 for x_1, 1 for y_1, and -3 for m.

$$y - y_1 = m(x - x_1)$$
$$y - 1 = -3(x - 2)$$
$$y - 1 = -3x + 6$$
$$y = -3x + 7$$

Using the slope-intercept equation:

Substitute 2 for x, 1 for y, and -3 for m. Then solve for b.

$$y = mx + b$$
$$1 = -3 \cdot 2 + b$$
$$1 = -6 + b$$
$$7 = b$$

Finally, we use the equation $y = mx + b$ and substitute -3 for m and 7 for b.

$$y = -3x + 7$$

41. We have the data points $(2009, 123.89)$ and $(2012, 152.52)$.

$$\text{Rate of change} = \frac{152.52 - 123.89}{2012 - 2009} = \frac{28.63}{3} \approx 9.54$$

The rate of change in spending was about \$9.54 per year.

42. a) $N(6) = 6^2 - 6 = 36 - 6 = 30$ games

b) $72 = n^2 - n$
$$0 = n^2 - n - 72$$
$$0 = (n - 9)(n + 8)$$
$$n - 9 = 0 \quad or \quad n + 8 = 0$$
$$n = 9 \quad or \qquad n = -8$$

The number of teams cannot be negative, so there are 9 teams in the league.

43. *Familiarize.* Referring to the drawing in the text, we see that the dimensions of the Lucite are $5 + 2x$ cm by $4 + 2x$ cm.

Translate.

Area of 1 piece of Lucite	is	$5\frac{1}{2}$	times	area of card.
↓	↓	↓	↓	↓
$(5 + 2x)(4 + 2x)$	$=$	$5\frac{1}{2}$	$\cdot$	$5 \cdot 4$

Solve. We solve the equation.

$$(5 + 2x)(4 + 2x) = 5\frac{1}{2} \cdot 5 \cdot 4$$
$$20 + 18x + 4x^2 = 110$$
$$4x^2 + 18x - 90 = 0$$
$$2(2x^2 + 9x - 45) = 0$$
$$2(2x + 15)(x - 3) = 0$$
$$2x + 15 = 0 \qquad or \quad x - 3 = 0$$
$$2x = -15 \quad or \qquad x = 3$$
$$x = -\frac{15}{2} \quad or \qquad x = 3$$

Check. Since the width of the border cannot be negative, we check only 3. If $x = 3$, then $5 + 2x = 5 + 2 \cdot 3 = 5 + 6 = 11$ and $4 + 2x = 4 + 2 \cdot 3 = 4 + 6 = 10$.

Then the area of a piece of the Lucite is $11 \cdot 10$, or 110 cm, and the area of the card is $5 \cdot 4$, or 20 cm. Since $5\frac{1}{2} \cdot 20 = 110$, the answer checks.

State. The dimensions of the Lucite are 11 cm by 10 cm.

44. $|x + 1| \le |x - 3|$

$|x + 1| \le x - 3 \quad or \quad |x + 1| \le -(x - 3)$

First we solve $|x + 1| \le x - 3$.

$$-(x - 3) \le x + 1 \quad and \quad x + 1 \le x - 3$$
$$-x + 3 \le x + 1 \quad and \qquad 1 \le -3$$
$$2 \le 2x \qquad and \qquad 1 \le -3$$
$$1 \le x \qquad and \qquad 1 \le -3$$

Since $1 \le -3$ is false for all values of x, the solution set for this portion of the inequality is $\emptyset$.

Now we solve $|x + 1| \le -(x - 3)$.

$$-[-(x - 3)] \le x + 1 \quad and \quad x + 1 \le -(x - 3)$$
$$x - 3 \le x + 1 \quad and \quad x + 1 \le -x + 3$$
$$-3 \le 1 \qquad and \qquad 2x \le 2$$
$$-3 \le 1 \qquad and \qquad x \le 1$$

Since $-3 \leq 1$ is true for all values of x, the solution set for this portion of the inequality is $\{x | x \leq 1\}$, or $(-\infty, 1]$.

Then the solution set for the original inequality is
$\emptyset \cup \{x | x \leq 1\}$, or $\{x \leq 1\}$, or $(-\infty, 1]$.

Chapter 5

Rational Expressions, Equations, and Functions

Exercise Set 5.1

RC1. $\dfrac{1}{x} \div \dfrac{1}{8} = \dfrac{1}{x} \cdot \dfrac{8}{1} = \dfrac{8}{x}$; the answer is (c).

RC3. (e)

RC5. $\dfrac{1}{x} \div 8 = \dfrac{1}{x} \cdot \dfrac{1}{8} = \dfrac{1}{8x}$; the answer is (b).

RC7. $\dfrac{x}{8} \cdot \dfrac{8}{x} = \dfrac{8x}{8x} = 1$; the answer is (f).

1. $\dfrac{5t^2 - 64}{3t + 17}$

We set the denominator equal to 0 and solve.

$$3t + 17 = 0$$
$$3t = -17$$
$$t = -\dfrac{17}{3}$$

The expression is not defined for the number $-\dfrac{17}{3}$.

3. $\dfrac{x^3 - x^2 + x + 2}{x^2 + 12x + 35}$

We set the denominator equal to 0 and solve.

$$x^2 + 12x + 35 = 0$$
$$(x + 5)(x + 7) = 0$$
$$x + 5 = 0 \quad or \quad x + 7 = 0$$
$$x = -5 \quad or \qquad x = -7$$

The expression is not defined for the numbers -5 and -7.

5. $f(x) = \dfrac{4x - 5}{x + 7}$

We set the denominator equal to 0 and solve.

$$x + 7 = 0$$
$$x = -7$$

The domain of $f(x) = \dfrac{4x - 5}{x + 7}$ is

$\{x | x$ is a real number $and\ x \neq -7\}$, or
$(-\infty, -7) \cup (-7, \infty)$.

7. $g(x) = \dfrac{7}{3x - x^2}$

We set the denominator equal to 0 and solve.

$$3x - x^2 = 0$$
$$x(3 - x) = 0$$
$$x = 0 \quad or \quad 3 - x = 0$$
$$x = 0 \quad or \qquad 3 = x$$

The domain of $g(x) = \dfrac{7}{3x - x^2}$ is

$\{x | x$ is a real number $and\ x \neq 0\ and\ x \neq 3\}$, or
$(-\infty, 0) \cup (0, 3) \cup (3, \infty)$.

9. In Exercise 1 we found that the only replacement for which the rational expression $\dfrac{5t^2 - 64}{3t + 17}$ is not defined is

$-\dfrac{17}{3}$. Then the domain of $f(t) = \dfrac{5t^2 - 64}{3t + 17}$ is

$\left\{ x \middle| x \text{ is a real number } and\ x \neq -\dfrac{17}{3} \right\}$, or

$\left(-\infty, -\dfrac{17}{3} \right) \cup \left(-\dfrac{17}{3}, \infty \right)$.

11. In Exercise 3 we found that the replacements for which the rational expression $\dfrac{x^3 - x^2 + x + 2}{x^2 + 12x + 35}$ is not defined are -7 and -5. Then the domain of $f(x) = \dfrac{x^3 - x^2 + x + 2}{x^2 + 12x + 35}$ is

$\{x | x$ is a real number $and\ x \neq -7\ and\ x \neq -5\}$, or
$(-\infty, -7) \cup (-7, -5) \cup (-5, \infty)$.

13. $\dfrac{7x}{7x} \cdot \dfrac{x + 2}{x + 8} = \dfrac{7x(x + 2)}{7x(x + 8)}$ Multiplying numerators and multiplying denominators

15. $\dfrac{q - 5}{q + 3} \cdot \dfrac{q + 5}{q + 5} = \dfrac{(q - 5)(q + 5)}{(q + 3)(q + 5)}$ Multiplying numerators and multiplying denominators

17. $\dfrac{15y^5}{5y^4} = \dfrac{3 \cdot 5 \cdot y^4 \cdot y}{5 \cdot y^4 \cdot 1}$ Factoring the numerator and the denominator

$= \dfrac{5y^4}{5y^4} \cdot \dfrac{3y}{1}$ Factoring the rational expression

$= 1 \cdot 3y$ $\dfrac{5y^4}{5y^4} = 1$

$= 3y$ Removing a factor of 1

19. $\dfrac{16p^3}{24p^7} = \dfrac{8p^3 \cdot 2}{8p^3 \cdot 3p^4}$ Factoring the numerator and the denominator

$= \dfrac{8p^3}{8p^3} \cdot \dfrac{2}{3p^4}$ Factoring the rational expression

$= 1 \cdot \dfrac{2}{3p^4}$ $\dfrac{8p^3}{8p^3} = 1$

$= \dfrac{2}{3p^4}$ Removing a factor of 1

21. $\dfrac{9a - 27}{9} = \dfrac{9(a - 3)}{9 \cdot 1}$ Factoring the numerator and the denominator

$\qquad = \dfrac{9}{9} \cdot \dfrac{a - 3}{1}$

$\qquad = \dfrac{a - 3}{1}$ Removing a factor of 1

$\qquad = a - 3$

23. $\dfrac{12x - 15}{21} = \dfrac{3(4x - 5)}{3 \cdot 7}$ Factoring the numerator and the denominator

$\qquad = \dfrac{3}{3} \cdot \dfrac{4x - 5}{7}$

$\qquad = \dfrac{4x - 5}{7}$ Removing a factor of 1

25. $\dfrac{4y - 12}{4y + 12} = \dfrac{4(y - 3)}{4(y + 3)} = \dfrac{4}{4} \cdot \dfrac{y - 3}{y + 3} = \dfrac{y - 3}{y + 3}$

27. $\dfrac{t^2 - 16}{t^2 - 8t + 16} = \dfrac{(t + 4)(t - 4)}{(t - 4)(t - 4)} = \dfrac{t + 4}{t - 4} \cdot \dfrac{t - 4}{t - 4} = \dfrac{t + 4}{t - 4}$

29. $\dfrac{x^2 - 9x + 8}{x^2 + 3x - 4} = \dfrac{(x - 8)(x - 1)}{(x + 4)(x - 1)} = \dfrac{x - 8}{x + 4} \cdot \dfrac{x - 1}{x - 1} = \dfrac{x - 8}{x + 4}$

31. $\dfrac{w^3 - z^3}{w^2 - z^2} = \dfrac{(w - z)(w^2 + wz + z^2)}{(w + z)(w - z)} =$

$\dfrac{w - z}{w - z} \cdot \dfrac{w^2 + wz + z^2}{w + z} = \dfrac{w^2 + wz + z^2}{w + z}$

33. $\dfrac{x^4}{3x + 6} \cdot \dfrac{5x + 10}{5x^7}$

$= \dfrac{x^4(5x + 10)}{(3x + 6)(5x^7)}$ Multiplying the numerators and the denominators

$= \dfrac{x^4(5)(x + 2)}{3(x + 2)(5)(x^4)(x^3)}$ Factoring the numerator and the denominator

$= \dfrac{x^4(5)(x + 2)(1)}{3(x + 2)(5)(x^4)(x^3)}$ Removing a factor of 1: $\dfrac{(x^4)(5)(x + 2)}{(x + 2)(5)(x^4)} = 1$

$= \dfrac{1}{3x^3}$ Simplifying

35. $\dfrac{x^2 - 16}{x^2} \cdot \dfrac{x^2 - 4x}{x^2 - x - 12}$

$= \dfrac{(x^2 - 16)(x^2 - 4x)}{x^2(x^2 - x - 12)}$ Multiplying the numerators and the denominators

$= \dfrac{(x + 4)(x - 4)(x)(x - 4)}{x \cdot x(x - 4)(x + 3)}$ Factoring the numerator and the denominator

$= \dfrac{(x + 4)(x - 4)(x)(x - 4)}{x \cdot x(x - 4)(x + 3)}$ Removing a factor of 1

$= \dfrac{(x + 4)(x - 4)}{x(x + 3)}$

37. $\dfrac{y^2 - 16}{2y + 6} \cdot \dfrac{y + 3}{y - 4} = \dfrac{(y^2 - 16)(y + 3)}{(2y + 6)(y - 4)}$

$= \dfrac{(y + 4)(y - 4)(y + 3)}{2(y + 3)(y - 4)}$

$= \dfrac{(y + 4)(y - 4)(y + 3)}{2(y + 3)(y - 4)}$

$= \dfrac{y + 4}{2}$

39. $\dfrac{x^2 - 2x - 35}{2x^3 - 3x^2} \cdot \dfrac{4x^3 - 9x}{7x - 49}$

$= \dfrac{(x^2 - 2x - 35)(4x^3 - 9x)}{(2x^3 - 3x^2)(7x - 49)}$

$= \dfrac{(x - 7)(x + 5)(x)(2x + 3)(2x - 3)}{x \cdot x(2x - 3)(7)(x - 7)}$

$= \dfrac{(x - 7)(x + 5)(x)(2x + 3)(2x - 3)}{x \cdot x(2x - 3)(7)(x - 7)}$

$= \dfrac{(x + 5)(2x + 3)}{7x}$

41. $\dfrac{c^3 + 8}{c^2 - 4} \cdot \dfrac{c^2 - 4c + 4}{c^2 - 2c + 4}$

$= \dfrac{(c^3 + 8)(c^2 - 4c + 4)}{(c^2 - 4)(c^2 - 2c + 4)}$

$= \dfrac{(c + 2)(c^2 - 2c + 4)(c - 2)(c - 2)}{(c + 2)(c - 2)(c^2 - 2c + 4) \cdot 1}$

$= \dfrac{(c + 2)(c^2 - 2c + 4)(c - 2)}{(c + 2)(c^2 - 2c + 4)(c - 2)} \cdot \dfrac{c - 2}{1}$

$= \dfrac{c - 2}{1}$

$= c - 2$

43. $\dfrac{x^2 - y^2}{x^3 - y^3} \cdot \dfrac{x^2 + xy + y^2}{x^2 + 2xy + y^2}$

$= \dfrac{(x^2 - y^2)(x^2 + xy + y^2)}{(x^3 - y^3)(x^2 + 2xy + y^2)}$

$= \dfrac{(x + y)(x - y)(x^2 + xy + y^2) \cdot 1}{(x - y)(x^2 + xy + y^2)(x + y)(x + y)}$

$= \dfrac{(x + y)(x - y)(x^2 + xy + y^2)}{(x + y)(x - y)(x^2 + xy + y^2)} \cdot \dfrac{1}{x + y}$

$= \dfrac{1}{x + y}$

45. $\dfrac{12x^8}{3y^4} \div \dfrac{16x^3}{6y}$

$= \dfrac{12x^8}{3y^4} \cdot \dfrac{6y}{16x^3}$ Multiplying by the reciprocal of the divisor

$= \dfrac{12x^8(6y)}{3y^4(16x^3)}$ Multiplying the numerators and the denominators

$= \dfrac{3 \cdot 4 \cdot x^3 \cdot x^5 \cdot 2 \cdot 3 \cdot y}{3 \cdot y \cdot y^3 \cdot 4 \cdot 2 \cdot 2 \cdot x^3}$ Factoring the numerator and the denominator

$= \dfrac{\cancel{3} \cdot \cancel{4} \cdot \cancel{x^3} \cdot x^5 \cdot \cancel{2} \cdot 3 \cdot \cancel{y}}{\cancel{3} \cdot \cancel{y} \cdot y^3 \cdot \cancel{4} \cdot \cancel{2} \cdot 2 \cdot \cancel{x^3}}$ Removing a factor of 1

$= \dfrac{3x^5}{2y^3}$

47. $\dfrac{3y+15}{y} \div \dfrac{y+5}{y} = \dfrac{3y+15}{y} \cdot \dfrac{y}{y+5}$

$= \dfrac{(3y+15)(y)}{y(y+5)}$

$= \dfrac{3(y+5)(y)}{y(y+5) \cdot 1}$

$= \dfrac{3(\cancel{y+5})(\cancel{y})}{\cancel{y}(\cancel{y+5}) \cdot 1}$

$= \dfrac{3}{1}$

$= 3$

49. $\dfrac{y^2-9}{y} \div \dfrac{y+3}{y+2} = \dfrac{y^2-9}{y} \cdot \dfrac{y+2}{y+3}$

$= \dfrac{(y^2-9)(y+2)}{y(y+3)}$

$= \dfrac{(y+3)(y-3)(y+2)}{y(y+3)}$

$= \dfrac{(\cancel{y+3})(y-3)(y+2)}{y(\cancel{y+3})}$

$= \dfrac{(y-3)(y+2)}{y}$

51. $\dfrac{4a^2-1}{a^2-4} \div \dfrac{2a-1}{a-2} = \dfrac{4a^2-1}{a^2-4} \cdot \dfrac{a-2}{2a-1}$

$= \dfrac{(4a^2-1)(a-2)}{(a^2-4)(2a-1)}$

$= \dfrac{(2a+1)(2a-1)(a-2)}{(a+2)(a-2)(2a-1)}$

$= \dfrac{(2a+1)(\cancel{2a-1})(\cancel{a-2})}{(a+2)(\cancel{a-2})(\cancel{2a-1})}$

$= \dfrac{2a+1}{a+2}$

53. $\dfrac{x^2-16}{x^2-10x+25} \div \dfrac{3x-12}{x^2-3x-10}$

$= \dfrac{x^2-16}{x^2-10x+25} \cdot \dfrac{x^2-3x-10}{3x-12}$

$= \dfrac{(x^2-16)(x^2-3x-10)}{(x^2-10x+25)(3x-12)}$

$= \dfrac{(x+4)(x-4)(x-5)(x+2)}{(x-5)(x-5)(3)(x-4)}$

$= \dfrac{(x+4)(\cancel{x-4})(\cancel{x-5})(x+2)}{(\cancel{x-5})(x-5)(3)(\cancel{x-4})}$

$= \dfrac{(x+4)(x+2)}{3(x-5)}$

55. $\dfrac{y^3+3y}{y^2-9} \div \dfrac{y^2+5y-14}{y^2+4y-21}$

$= \dfrac{y^3+3y}{y^2-9} \cdot \dfrac{y^2+4y-21}{y^2+5y-14}$

$= \dfrac{(y^3+3y)(y^2+4y-21)}{(y^2-9)(y^2+5y-14)}$

$= \dfrac{y(y^2+3)(y+7)(y-3)}{(y+3)(y-3)(y+7)(y-2)}$

$= \dfrac{y(y^2+3)(\cancel{y+7})(\cancel{y-3})}{(y+3)(\cancel{y-3})(\cancel{y+7})(y-2)}$

$= \dfrac{y(y^2+3)}{(y+3)(y-2)}$

57. $\dfrac{x^3-64}{x^3+64} \div \dfrac{x^2-16}{x^2-4x+16}$

$= \dfrac{x^3-64}{x^3+64} \cdot \dfrac{x^2-4x+16}{x^2-16}$

$= \dfrac{(x^3-64)(x^2-4x+16)}{(x^3+64)(x^2-16)}$

$= \dfrac{(x-4)(x^2+4x+16)(x^2-4x+16)}{(x+4)(x^2-4x+16)(x+4)(x-4)}$

$= \dfrac{(x-4)(x^2-4x+16)}{(x-4)(x^2-4x+16)} \cdot \dfrac{x^2+4x+16}{(x+4)(x+4)}$

$= \dfrac{x^2+4x+16}{(x+4)(x+4)}, \text{ or } \dfrac{x^2+4x+16}{(x+4)^2}$

59. $\dfrac{8x^3y^3+27x^3}{64x^3y^3-x^3} \div \dfrac{4x^2y^2-9x^2}{16x^2y^2+4x^2y+x^2}$

$= \dfrac{8x^3y^3+27x^3}{64x^3y^3-x^3} \cdot \dfrac{16x^2y^2+4x^2y+x^2}{4x^2y^2-9x^2}$

$= \dfrac{(8x^3y^3+27x^3)(16x^2y^2+4x^2y+x^2)}{(64x^3y^3-x^3)(4x^2y^2-9x^2)}$

$= \dfrac{x^3(8y^3+27)(x^2)(16y^2+4y+1)}{x^3(64y^3-1)(x^2)(4y^2-9)}$

$= \dfrac{x^3(2y+3)(4y^2-6y+9)(x^2)(16y^2+4y+1)}{x^3(4y-1)(16y^2+4y+1)(x^2)(2y+3)(2y-3)}$

$= \dfrac{x^3(2y+3)(x^2)(16y^2+4y+1)}{x^3(16y^2+4y+1)(x^2)(2y+3)} \cdot \dfrac{4y^2-6y+9}{(4y-1)(2y-3)}$

$= \dfrac{4y^2-6y+9}{(4y-1)(2y-3)}$

61. $\left[\dfrac{r^2 - 4s^2}{r + 2s} \div (r + 2s)\right] \cdot \dfrac{2s}{r - 2s}$

$= \left[\dfrac{r^2 - 4s^2}{r + 2s} \cdot \dfrac{1}{r + 2s}\right] \cdot \dfrac{2s}{r - 2s}$

$= \dfrac{(r^2 - 4s^2)(1)(2s)}{(r + 2s)(r + 2s)(r - 2s)}$

$= \dfrac{(r + 2s)(r - 2s)(2s)}{(r + 2s)(r + 2s)(r - 2s)}$

$= \dfrac{(\cancel{r + 2s})(\cancel{r - 2s})(2s)}{(\cancel{r + 2s})(r + 2s)(\cancel{r - 2s})}$

$= \dfrac{2s}{r + 2s}$

63. $\dfrac{y^2 - 2y}{y^2 + y - 2} \cdot \dfrac{y - 1}{y^2 + 4y + 4} \div \dfrac{y^2 + 2y - 8}{y^4}$

$= \dfrac{y^2 - 2y}{y^2 + y - 2} \cdot \dfrac{y - 1}{y^2 + 4y + 4} \cdot \dfrac{y^4}{y^2 + 2y - 8}$

$= \dfrac{(y^2 - 2y)(y - 1)(y^4)}{(y^2 + y - 2)(y^2 + 4y + 4)(y^2 + 2y - 8)}$

$= \dfrac{y(y - 2)(y - 1)(y^4)}{(y + 2)(y - 1)(y + 2)(y + 2)(y + 4)(y - 2)}$

$= \dfrac{y(\cancel{y - 2})(\cancel{y - 1})(y^4)}{(y + 2)(\cancel{y - 1})(y + 2)(y + 2)(y + 4)(\cancel{y - 2})}$

$= \dfrac{y^5}{(y + 2)^3(y + 4)}$

65. The function can be written as a set of six ordered pairs, $\{(-4, 3), (-2, 1), (0, -3), (2, -2), (4, 0), (6, 4)\}$.

The domain is the set of all first coordinates, $\{-4, -2, 0, 2, 4, 6\}$.

The range is the set of all second coordinates, $\{-3, -2, 0, 1, 3, 4\}$.

67. The domain is the set of all x-values on the graph, $[-5, 5]$.

The range is the set of all y-values on the graph, $[-4, 4]$.

69. $6a^2 + 5ab - 25b^2$

We can factor this using the FOIL method or the grouping method. The factorization is $(3a - 5b)(2a + 5b)$.

71. $10x^2 - 80x + 70 = 10(x^2 - 8x + 7)$

To factor $x^2 - 8x + 7$ we find factors of 7 whose sum is -8. The numbers we need are -1 and -7, so $x^2 - 8x + 7 = (x - 1)(x - 7)$. Then $10x^2 - 80x + 70 = 10(x - 1)(x - 7)$.

73. $21p^2 + p - 10$

We can factor this using the FOIL method or the grouping method. The factorization is $(7p + 5)(3p - 2)$.

75. $2x^3 - 16x^2 - 66x = 2x(x^2 - 8x - 33)$

To factor $x^2 - 8x - 33$ we find factors of -33 whose sum is -8. The numbers we need are -11 and 3, so $x^2 - 8x - 33 = (x - 11)(x + 3)$. Then $2x^3 - 16x^2 - 66x = 2x(x - 11)(x + 3)$.

77. Substitute $-\dfrac{2}{3}$ for m and -5 for b in the slope-intercept equation, $y = mx + b$. The equation is $y = -\dfrac{2}{3}x - 5$.

79. $\dfrac{x(x + 1) - 2(x + 3)}{(x + 1)(x + 2)(x + 3)} = \dfrac{x^2 + x - 2x - 6}{(x + 1)(x + 2)(x + 3)}$

$= \dfrac{x^2 - x - 6}{(x + 1)(x + 2)(x + 3)}$

$= \dfrac{(x - 3)(x + 2)}{(x + 1)(x + 2)(x + 3)}$

$= \dfrac{(x - 3)(\cancel{x + 2})}{(x + 1)(\cancel{x + 2})(x + 3)}$

$= \dfrac{x - 3}{(x + 1)(x + 3)}$

81. $\dfrac{m^2 - t^2}{m^2 + t^2 + m + t + 2mt} = \dfrac{m^2 - t^2}{(m^2 + 2mt + t^2) + (m + t)}$

$= \dfrac{(m + t)(m - t)}{(m + t)^2 + (m + t)}$

$= \dfrac{(m + t)(m - t)}{(m + t)[(m + t) + 1]}$

$= \dfrac{(\cancel{m + t})(m - t)}{(\cancel{m + t})(m + t + 1)}$

$= \dfrac{m - t}{m + t + 1}$

83. $g(x) = \dfrac{2x + 3}{4x - 1}$

$g(5) = \dfrac{2 \cdot 5 + 3}{4 \cdot 5 - 1} = \dfrac{10 + 3}{20 - 1} = \dfrac{13}{19}$

$g(0) = \dfrac{2 \cdot 0 + 3}{4 \cdot 0 - 1} = \dfrac{3}{-1} = -3$

$g\left(\dfrac{1}{4}\right) = \dfrac{2 \cdot \dfrac{1}{4} + 3}{4 \cdot \dfrac{1}{4} - 1} = \dfrac{\dfrac{1}{2} + 3}{1 - 1} = \dfrac{\dfrac{7}{2}}{0}$; since division by 0 is not

defined, $g(0)$ is not defined.

$g(a + h) = \dfrac{2(a + h) + 3}{4(a + h) - 1} = \dfrac{2a + 2h + 3}{4a + 4h - 1}$

Exercise Set 5.2

RC1. $\dfrac{10x}{x - 7} - \dfrac{3x + 5}{x - 7} = \dfrac{10x - (3x + 5)}{x - 7} = \dfrac{10x - 3x - 5}{x - 7} =$

$\dfrac{7x - 5}{x - 7}$

RC3. $\dfrac{9y - 2}{y^2 - 10} - \dfrac{y + 1}{y^2 - 10} = \dfrac{9y - 2 - (y + 1)}{y^2 - 10} =$

$\dfrac{9y - 2 - y - 1}{y^2 - 10} = \dfrac{8y - 3}{y^2 - 10}$

1. $15 = 3 \cdot 5$

$40 = 2 \cdot 2 \cdot 2 \cdot 5$

LCM $= 2 \cdot 2 \cdot 2 \cdot 3 \cdot 5$, or 120

(We used each factor the greatest number of times that it occurs in any one prime factorization.)

3. $18 = 2 \cdot 3 \cdot 3$

$48 = 2 \cdot 2 \cdot 2 \cdot 2 \cdot 3$

LCM $= 2 \cdot 2 \cdot 2 \cdot 2 \cdot 3 \cdot 3$, or 144

5. $30 = 2 \cdot 3 \cdot 5$

$105 = 3 \cdot 5 \cdot 7$

LCM $= 2 \cdot 3 \cdot 5 \cdot 7$, or 210

7. $9 = 3 \cdot 3$

$15 = 3 \cdot 5$

$5 = 5$

LCM $= 3 \cdot 3 \cdot 5$, or 45

9. $\dfrac{5}{6} + \dfrac{4}{15} = \dfrac{5}{2 \cdot 3} + \dfrac{4}{3 \cdot 5}$, LCD $= 2 \cdot 3 \cdot 5$, or 30

$$= \dfrac{5}{2 \cdot 3} \cdot \dfrac{5}{5} + \dfrac{4}{3 \cdot 5} \cdot \dfrac{2}{2}$$

$$= \dfrac{25}{2 \cdot 3 \cdot 5} + \dfrac{8}{2 \cdot 3 \cdot 5}$$

$$= \dfrac{33}{2 \cdot 3 \cdot 5} = \dfrac{\cancel{3} \cdot 11}{2 \cdot \cancel{3} \cdot 5}$$

$$= \dfrac{11}{10}$$

11. $\dfrac{7}{36} + \dfrac{1}{24}$

$$= \dfrac{7}{2 \cdot 2 \cdot 3 \cdot 3} + \dfrac{1}{2 \cdot 2 \cdot 2 \cdot 3}, \quad \text{LCD} = 2 \cdot 2 \cdot 2 \cdot 3 \cdot 3, \text{ or } 72$$

$$= \dfrac{7}{2 \cdot 2 \cdot 3 \cdot 3} \cdot \dfrac{2}{2} + \dfrac{1}{2 \cdot 2 \cdot 2 \cdot 3} \cdot \dfrac{3}{3}$$

$$= \dfrac{14}{2 \cdot 2 \cdot 2 \cdot 3 \cdot 3} + \dfrac{3}{2 \cdot 2 \cdot 2 \cdot 3 \cdot 3}$$

$$= \dfrac{17}{2 \cdot 2 \cdot 2 \cdot 3 \cdot 3}$$

$$= \dfrac{17}{72}$$

13. $\dfrac{3}{4} + \dfrac{7}{30} + \dfrac{1}{16}$

$$= \dfrac{3}{2 \cdot 2} + \dfrac{7}{2 \cdot 3 \cdot 5} + \dfrac{1}{2 \cdot 2 \cdot 2 \cdot 2}, \quad \text{LCD} = 2 \cdot 2 \cdot 2 \cdot 2 \cdot 3 \cdot 5$$

$$= \dfrac{3}{2 \cdot 2} \cdot \dfrac{2 \cdot 2 \cdot 3 \cdot 5}{2 \cdot 2 \cdot 3 \cdot 5} + \dfrac{7}{2 \cdot 3 \cdot 5} \cdot \dfrac{2 \cdot 2 \cdot 2}{2 \cdot 2 \cdot 2} +$$

$$\dfrac{1}{2 \cdot 2 \cdot 2 \cdot 2} \cdot \dfrac{3 \cdot 5}{3 \cdot 5}$$

$$= \dfrac{180}{2 \cdot 2 \cdot 2 \cdot 2 \cdot 3 \cdot 5} + \dfrac{56}{2 \cdot 2 \cdot 2 \cdot 2 \cdot 3 \cdot 5} + \dfrac{15}{2 \cdot 2 \cdot 2 \cdot 2 \cdot 3 \cdot 5}$$

$$= \dfrac{251}{2 \cdot 2 \cdot 2 \cdot 2 \cdot 3 \cdot 5}$$

$$= \dfrac{251}{240}$$

15. $21x^2 y = 3 \cdot 7 \cdot x \cdot x \cdot y$

$7xy = 7 \cdot x \cdot y$

LCM $= 3 \cdot 7 \cdot x \cdot x \cdot y$, or $21x^2 y$

17. $y^2 - 100 = (y + 10)(y - 10)$

$10y + 100 = 10(y + 10)$

LCM $= 10(y + 10)(y - 10)$

19. $15ab^2 = 3 \cdot 5 \cdot a \cdot b \cdot b$

$3ab = 3 \cdot a \cdot b$

$10a^3 b = 2 \cdot 5 \cdot a \cdot a \cdot a \cdot b$

LCM $= 2 \cdot 3 \cdot 5 \cdot a \cdot a \cdot a \cdot b \cdot b$, or $30a^3 b^2$

21. $5y - 15 = 5(y - 3)$

$y^2 - 6y + 9 = (y - 3)(y - 3)$

LCM $= 5(y - 3)(y - 3)$, or $5(y - 3)^2$

23. $y^2 - 25 = (y + 5)(y - 5)$

$5 - y$

We can use $y - 5$ from the prime factorization of $y^2 - 25$ or $5 - y$ from the second expression, but not both.

LCM $= (y + 5)(y - 5)$, or $(y + 5)(5 - y)$

25. $2r^2 - 5r - 12 = (2r + 3)(r - 4)$

$3r^2 - 13r + 4 = (3r - 1)(r - 4)$

$r^2 - 16 = (r + 4)(r - 4)$

LCM $= (2r + 3)(r - 4)(3r - 1)(r + 4)$

27. $x^5 + 4x^3 = x^3(x^2 + 4) = x \cdot x \cdot x(x^2 + 4)$

$x^3 - 4x^2 + 4x = x(x^2 - 4x + 4) = x(x - 2)(x - 2)$

LCM $= x \cdot x \cdot x(x - 2)(x - 2)(x^2 + 4)$, or $x^3(x - 2)^2(x^2 + 4)$

29. $x^5 - 2x^4 + x^3 = x^3(x^2 - 2x + 1) = x \cdot x \cdot x(x - 1)(x - 1)$

$2x^3 + 2x = 2x(x^2 + 1)$

$5x + 5 = 5(x + 1)$

LCM $= 2 \cdot 5 \cdot x \cdot x \cdot x(x - 1)(x - 1)(x + 1)(x^2 + 1)$, or $10x^3(x - 1)^2(x + 1)(x^2 + 1)$

31. $\dfrac{x - 2y}{x + y} + \dfrac{x + 9y}{x + y}$

$$= \dfrac{x - 2y + x + 9y}{x + y} \qquad \text{Adding the numerators}$$

$$= \dfrac{2x + 7y}{x + y}$$

33. $\dfrac{4y + 2}{y - 2} - \dfrac{y - 3}{y - 2}$

$$= \dfrac{4y + 2 - (y - 3)}{y - 2} \qquad \text{Subtracting numerators}$$

$$= \dfrac{4y + 2 - y + 3}{y - 2}$$

$$= \dfrac{3y + 5}{y - 2}$$

35. $\dfrac{a^2}{a-b} + \dfrac{b^2}{b-a}$

$= \dfrac{a^2}{a-b} + \dfrac{b^2}{b-a} \cdot \dfrac{-1}{-1}$ Multiplying by 1, using $\dfrac{-1}{-1}$

$= \dfrac{a^2}{a-b} + \dfrac{-b^2}{a-b}$

$= \dfrac{a^2 - b^2}{a-b}$ Adding numerators

$= \dfrac{(a+b)(a-b)}{a-b}$ Factoring the numerator

$= \dfrac{(a+b)(a\!\!\!\!-b)}{1(a\!\!\!\!-b)}$ Removing a factor of 1

$= \dfrac{a+b}{1}$

$= a+b$

37. $\dfrac{6}{y} - \dfrac{7}{-y}$

$= \dfrac{6}{y} - \dfrac{7}{-y} \cdot \dfrac{-1}{-1}$ Multiplying by 1, using $\dfrac{-1}{-1}$

$= \dfrac{6}{y} - \dfrac{-7}{y}$

$= \dfrac{6-(-7)}{y}$ Subtracting numerators

$= \dfrac{13}{y}$

39. $\dfrac{4a-2}{a^2-49} + \dfrac{5+3a}{49-a^2}$

$= \dfrac{4a-2}{a^2-49} + \dfrac{5+3a}{49-a^2} \cdot \dfrac{-1}{-1}$ Multiplying by 1, using $\dfrac{-1}{-1}$

$= \dfrac{4a-2}{a^2-49} + \dfrac{-5-3a}{a^2-49}$

$= \dfrac{4a-2-5-3a}{a^2-49}$ Adding numerators

$= \dfrac{a-7}{a^2-49}$

$= \dfrac{a-7}{(a+7)(a-7)}$ Factoring

$= \dfrac{(a\!\!\!\!-7) \cdot 1}{(a+7)(a\!\!\!\!-7)}$ Removing a factor of 1

$= \dfrac{1}{a+7}$

41. $\dfrac{a^3}{a-b} + \dfrac{b^3}{b-a}$

$= \dfrac{a^3}{a-b} + \dfrac{b^3}{b-a} \cdot \dfrac{-1}{-1}$

$= \dfrac{a^3}{a-b} + \dfrac{-b^3}{a-b}$

$= \dfrac{a^3 - b^3}{a-b}$

$= \dfrac{(a-b)(a^2+ab+b^2)}{a-b}$

$= \dfrac{(a\!\!\!\!-b)(a^2+ab+b^2)}{(a\!\!\!\!-b) \cdot 1}$

$= a^2 + ab + b^2$

43. $\dfrac{y-2}{y+4} + \dfrac{y+3}{y-5}$ LCD $= (y+4)(y-5)$

$= \dfrac{y-2}{y+4} \cdot \dfrac{y-5}{y-5} + \dfrac{y+3}{y-5} \cdot \dfrac{y+4}{y+4}$

$= \dfrac{(y^2 - 7y + 10) + (y^2 + 7y + 12)}{(y+4)(y-5)}$

$= \dfrac{2y^2 + 22}{(y+4)(y-5)} = \dfrac{2(y^2+11)}{(y+4)(y-5)}$

45. $\dfrac{4xy}{x^2-y^2} + \dfrac{x-y}{x+y}$

$= \dfrac{4xy}{(x+y)(x-y)} + \dfrac{x-y}{x+y}$ LCD $= (x+y)(x-y)$

$= \dfrac{4xy}{(x+y)(x-y)} + \dfrac{x-y}{x+y} \cdot \dfrac{x-y}{x-y}$

$= \dfrac{4xy + x^2 - 2xy + y^2}{(x+y)(x-y)}$

$= \dfrac{x^2 + 2xy + y^2}{(x+y)(x-y)} = \dfrac{(x+y)(x+y)}{(x+y)(x-y)}$

$= \dfrac{(x\!\!\!\!+y)(x+y)}{(x\!\!\!\!+y)(x-y)} = \dfrac{x+y}{x-y}$

47. $\dfrac{9x+2}{3x^2-2x-8} + \dfrac{7}{3x^2+x-4}$

$= \dfrac{9x+2}{(3x+4)(x-2)} + \dfrac{7}{(3x+4)(x-1)}$

    LCD $= (3x+4)(x-2)(x-1)$

$= \dfrac{9x+2}{(3x+4)(x-2)} \cdot \dfrac{x-1}{x-1} + \dfrac{7}{(3x+4)(x-1)} \cdot \dfrac{x-2}{x-2}$

$= \dfrac{9x^2 - 7x - 2 + 7x - 14}{(3x+4)(x-2)(x-1)}$

$= \dfrac{9x^2 - 16}{(3x+4)(x-2)(x-1)} = \dfrac{(3x+4)(3x-4)}{(3x+4)(x-2)(x-1)}$

$= \dfrac{(3x\!\!\!\!+4)(3x-4)}{(3x\!\!\!\!+4)(x-2)(x-1)}$

$= \dfrac{3x-4}{(x-2)(x-1)}$

49. $\dfrac{4}{x+1} + \dfrac{x+2}{x^2-1} + \dfrac{3}{x-1}$

$= \dfrac{4}{x+1} + \dfrac{x+2}{(x+1)(x-1)} + \dfrac{3}{x-1}$

$\qquad\qquad\qquad \text{LCD} = (x+1)(x-1)$

$= \dfrac{4}{x+1} \cdot \dfrac{x-1}{x-1} + \dfrac{x+2}{(x+1)(x-1)} + \dfrac{3}{x-1} \cdot \dfrac{x+1}{x+1}$

$= \dfrac{4x-4+x+2+3x+3}{(x+1)(x-1)}$

$= \dfrac{8x+1}{(x+1)(x-1)}$

51. $\dfrac{x-1}{3x+15} - \dfrac{x+3}{5x+25}$

$= \dfrac{x-1}{3(x+5)} - \dfrac{x+3}{5(x+5)}$

$\qquad\qquad \text{LCD} = 3 \cdot 5(x+5), \text{ or } 15(x+5)$

$= \dfrac{x-1}{3(x+5)} \cdot \dfrac{5}{5} - \dfrac{x+3}{5(x+5)} \cdot \dfrac{3}{3}$

$= \dfrac{5x-5-(3x+9)}{15(x+5)}$

$= \dfrac{5x-5-3x-9}{15(x+5)}$

$= \dfrac{2x-14}{15(x+5)} = \dfrac{2(x-7)}{15(x+5)}$

53. $\dfrac{5ab}{a^2-b^2} - \dfrac{a-b}{a+b}$

$= \dfrac{5ab}{(a+b)(a-b)} - \dfrac{a-b}{a+b} \qquad \text{LCD} = (a+b)(a-b)$

$= \dfrac{5ab}{(a+b)(a-b)} - \dfrac{a-b}{a+b} \cdot \dfrac{a-b}{a-b}$

$= \dfrac{5ab-(a^2-2ab+b^2)}{(a+b)(a-b)}$

$= \dfrac{5ab-a^2+2ab-b^2}{(a+b)(a-b)}$

$= \dfrac{-a^2+7ab-b^2}{(a+b)(a-b)}$

55. $\dfrac{3y}{y^2-7y+10} - \dfrac{2y}{y^2-8y+15}$

$= \dfrac{3y}{(y-5)(y-2)} - \dfrac{2y}{(y-5)(y-3)}$

$\qquad\qquad \text{LCD} = (y-5)(y-2)(y-3)$

$= \dfrac{3y}{(y-5)(y-2)} \cdot \dfrac{y-3}{y-3} - \dfrac{2y}{(y-5)(y-3)} \cdot \dfrac{y-2}{y-2}$

$= \dfrac{3y^2-9y-(2y^2-4y)}{(y-5)(y-2)(y-3)}$

$= \dfrac{3y^2-9y-2y^2+4y}{(y-5)(y-2)(y-3)}$

$= \dfrac{y^2-5y}{(y-5)(y-2)(y-3)} = \dfrac{y(y-5)}{(y-5)(y-2)(y-3)}$

$= \dfrac{y(y-5)}{(y-5)(y-2)(y-3)}$

$= \dfrac{y}{(y-2)(y-3)}$

57. $\dfrac{y}{y^2-y-20} + \dfrac{2}{y+4}$

$= \dfrac{y}{(y-5)(y+4)} + \dfrac{2}{y+4} \qquad \text{LCD} = (y-5)(y+4)$

$= \dfrac{y}{(y-5)(y+4)} + \dfrac{2}{y+4} \cdot \dfrac{y-5}{y-5}$

$= \dfrac{y+2y-10}{(y-5)(y+4)}$

$= \dfrac{3y-10}{(y-5)(y+4)}$

59. $\dfrac{3y+2}{y^2+5y-24} + \dfrac{7}{y^2+4y-32}$

$= \dfrac{3y+2}{(y+8)(y-3)} + \dfrac{7}{(y+8)(y-4)}$

$\qquad\qquad \text{LCD} = (y+8)(y-3)(y-4)$

$= \dfrac{3y+2}{(y+8)(y-3)} \cdot \dfrac{y-4}{y-4} + \dfrac{7}{(y+8)(y-4)} \cdot \dfrac{y-3}{y-3}$

$= \dfrac{3y^2-10y-8+7y-21}{(y+8)(y-3)(y-4)}$

$= \dfrac{3y^2-3y-29}{(y+8)(y-3)(y-4)}$

61. $\dfrac{3x-1}{x^2+2x-3} - \dfrac{x+4}{x^2-9}$

$= \dfrac{3x-1}{(x+3)(x-1)} - \dfrac{x+4}{(x+3)(x-3)}$

$\qquad\qquad \text{LCD} = (x+3)(x-1)(x-3)$

$= \dfrac{3x-1}{(x+3)(x-1)} \cdot \dfrac{x-3}{x-3} - \dfrac{x+4}{(x+3)(x-3)} \cdot \dfrac{x-1}{x-1}$

$= \dfrac{3x^2-10x+3-(x^2+3x-4)}{(x+3)(x-1)(x-3)}$

$= \dfrac{3x^2-10x+3-x^2-3x+4}{(x+3)(x-1)(x-3)}$

$= \dfrac{2x^2-13x+7}{(x+3)(x-1)(x-3)}$

63. $\dfrac{1}{x+1} - \dfrac{x}{x-2} + \dfrac{x^2+2}{x^2-x-2}$

$= \dfrac{1}{x+1} - \dfrac{x}{x-2} + \dfrac{x^2+2}{(x-2)(x+1)}$

$\qquad\qquad \text{LCD} = (x+1)(x-2)$

$= \dfrac{1}{x+1} \cdot \dfrac{x-2}{x-2} - \dfrac{x}{x-2} \cdot \dfrac{x+1}{x+1} + \dfrac{x^2+2}{(x-2)(x+1)}$

$= \dfrac{x-2-(x^2+x)+x^2+2}{(x+1)(x-2)}$

$= \dfrac{x-2-x^2-x+x^2+2}{(x+1)(x-2)}$

$= \dfrac{0}{(x+1)(x-2)}$

$= 0$

65. $\dfrac{y-3}{y-4} - \dfrac{y+2}{y+4} + \dfrac{y-7}{y^2-16}$

$= \dfrac{y-3}{y-4} - \dfrac{y+2}{y+4} + \dfrac{y-7}{(y+4)(y-4)}$

$\qquad\qquad\qquad\text{LCM} = (y+4)(y-4)$

$= \dfrac{y-3}{y-4} \cdot \dfrac{y+4}{y+4} - \dfrac{y+2}{y+4} \cdot \dfrac{y+4}{y+4} + \dfrac{y-7}{(y+4)(y-4)}$

$= \dfrac{(y-3)(y+4) - (y+2)(y-4) + (y-7)}{(y+4)(y-4)}$

$= \dfrac{y^2 + y - 12 - (y^2 - 2y - 8) + y - 7}{(y+4)(y-4)}$

$= \dfrac{y^2 + y - 12 - y^2 + 2y + 8 + y - 7}{(y+4)(y-4)}$

$= \dfrac{4y - 11}{(y+4)(y-4)}$

67. $\dfrac{y+2}{y+4} + \dfrac{y-7}{y^2-16} - \dfrac{y-3}{y-4}$

$= \dfrac{y+2}{y+4} + \dfrac{y-7}{(y+4)(y-4)} - \dfrac{y-3}{y-4}$

$\qquad\qquad\qquad\text{LCM} = (y+4)(y-4)$

$= \dfrac{y+2}{y+4} \cdot \dfrac{y-4}{y-4} + \dfrac{y-7}{(y+4)(y-4)} - \dfrac{y-3}{y-4} \cdot \dfrac{y+4}{y+4}$

$= \dfrac{(y^2 - 2y - 8) + (y - 7) - (y^2 + y - 12)}{(y+4)(y-4)}$

$= \dfrac{y^2 - 2y - 8 + y - 7 - y^2 - y + 12}{(y+4)(y-4)}$

$= \dfrac{-2y - 3}{(y+4)(y-4)}$

69. $\dfrac{4x}{x^2-1} + \dfrac{3x}{1-x} - \dfrac{4}{x-1}$

$= \dfrac{4x}{x^2-1} + \dfrac{3x}{1-x} \cdot \dfrac{-1}{-1} - \dfrac{4}{x-1}$

$= \dfrac{4x}{(x+1)(x-1)} + \dfrac{-3x}{x-1} - \dfrac{4}{x-1}$

$\qquad\qquad\qquad\text{LCD} = (x+1)(x-1)$

$= \dfrac{4x}{(x+1)(x-1)} + \dfrac{-3x}{x-1} \cdot \dfrac{x+1}{x+1} - \dfrac{4}{x-1} \cdot \dfrac{x+1}{x+1}$

$= \dfrac{4x - 3x^2 - 3x - 4x - 4}{(x+1)(x-1)}$

$= \dfrac{-3x^2 - 3x - 4}{(x+1)(x-1)}$

71. $\dfrac{1}{x+y} + \dfrac{1}{y-x} - \dfrac{2x}{x^2-y^2}$

$= \dfrac{1}{x+y} + \dfrac{1}{y-x} \cdot \dfrac{-1}{-1} - \dfrac{2x}{x^2-y^2}$

$= \dfrac{1}{x+y} + \dfrac{-1}{x-y} - \dfrac{2x}{(x+y)(x-y)}$

$\qquad\qquad\qquad\text{LCD} = (x+y)(x-y)$

$= \dfrac{1}{x+y} \cdot \dfrac{x-y}{x-y} + \dfrac{-1}{x-y} \cdot \dfrac{x+y}{x+y} - \dfrac{2x}{(x+y)(x-y)}$

$= \dfrac{x - y - x - y - 2x}{(x+y)(x-y)}$

$= \dfrac{-2x - 2y}{(x+y)(x-y)} = \dfrac{-2(x+y)}{(x+y)(x-y)}$

$= \dfrac{-2\cancel{(x+y)}}{\cancel{(x+y)}(x-y)}$

$= \dfrac{-2}{x-y}, \text{ or } \dfrac{2}{y-x}$

73. $\dfrac{x+5}{x-3} - \dfrac{x+2}{x+1} - \dfrac{6x+10}{x^2-2x-3}$

$= \dfrac{x+5}{x-3} - \dfrac{x+2}{x+1} - \dfrac{6x+10}{(x-3)(x+1)}$

$\qquad\qquad\qquad\text{LCD} = (x-3)(x+1)$

$= \dfrac{x+5}{x-3} \cdot \dfrac{x+1}{x+1} - \dfrac{x+2}{x+1} \cdot \dfrac{x-3}{x-3} - \dfrac{6x+10}{(x-3)(x+1)}$

$= \dfrac{(x+5)(x+1) - (x+2)(x-3) - (6x+10)}{(x-3)(x+1)}$

$= \dfrac{x^2 + 6x + 5 - (x^2 - x - 6) - 6x - 10}{(x-3)(x+1)}$

$= \dfrac{x^2 + 6x + 5 - x^2 + x + 6 - 6x - 10}{(x-3)(x+1)}$

$= \dfrac{x+1}{(x-3)(x+1)}$

$= \dfrac{1 \cdot \cancel{(x+1)}}{(x-3)\cancel{(x+1)}}$

$= \dfrac{1}{x-3}$

75. Graph: $2x - 3y > 6$

We first graph the line $2x - 3y = 6$. The intercepts are $(0, -2)$ and $(3, 0)$. We draw the line dashed since the inequality symbol is $>$. To determine which half-plane to shade, we consider a test point not on the line. We try $(0, 0)$:

$$\begin{array}{c|c} 2x - 3y > 6 \\ \hline 2 \cdot 0 - 3 \cdot 0 \; ? \; 6 \\ 0 \; \big| \qquad \text{FALSE} \end{array}$$

Since $0 > 6$ is false, we shade the half-plane that does not contain $(0, 0)$.

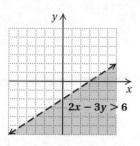

77. Graph: $5x + 3y \leq 15$

We first graph the line $5x + 3y = 15$. The intercepts are $(3,0)$ and $(0,5)$. We draw the line solid since the inequality symbol is $\leq$. To determine which half-plane to shade, we consider a test point not on the line. We try $(0,0)$:

$$\frac{5x + 3y \leq 15}{5 \cdot 0 + 3 \cdot 0 \ ? \ 15}$$
$$0 \ | \qquad \text{TRUE}$$

Since $0 \leq 15$ is true, we shade the half-plane that contains $(0,0)$.

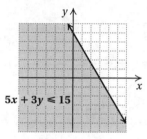

79. $t^3 - 8 = t^3 - 2^3$
$$= (t - 2)(t^2 + 2t + 4)$$

81. $23x^4 + 23x$
$$= 23x(x^3 + 1)$$
$$= 23x(x + 1)(x^2 - x + 1)$$

83. First we find the slope of the given line.
$$3y + 8x = 10$$
$$3y = -8x + 10$$
$$\frac{1}{3} \cdot 3y = \frac{1}{3}(-8x + 10)$$
$$y = -\frac{8}{3}x + \frac{10}{3}$$

The slope is $-\frac{8}{3}$. A line parallel to this line will have the same slope. We use the point-slope equation.
$$y - y_1 = m(x - x_1)$$
$$y - (-6) = -\frac{8}{3}(x - 4)$$
$$y + 6 = -\frac{8}{3}x + \frac{32}{3}$$
$$y = -\frac{8}{3}x + \frac{14}{3}$$

85. From the graph or from the equation of the function we see that the domain is $(-\infty, 2) \cup (2, \infty)$. From the graph we see that the range is $(-\infty, 0) \cup (0, \infty)$.

87. $x^8 - x^4 = x^4(x^2 + 1)(x + 1)(x - 1)$
$x^5 - x^2 = x^2(x - 1)(x^2 + x + 1)$
$x^5 - x^3 = x^3(x + 1)(x - 1)$
$x^5 + x^2 = x^2(x + 1)(x^2 - x + 1)$
LCM =
$x^4(x^2 + 1)(x + 1)(x - 1)(x^2 + x + 1)(x^2 - x + 1)$

89.
$$\frac{x + y + 1}{y - (x + 1)} + \frac{x + y - 1}{x - (y - 1)} - \frac{x - y - 1}{1 - (y - x)}$$
$$= \frac{x + y + 1}{-x + y - 1} + \frac{x + y - 1}{x - y + 1} - \frac{x - y - 1}{x - y + 1}$$
$$= \frac{x + y + 1}{-x + y - 1} \cdot \frac{-1}{-1} + \frac{x + y - 1}{x - y + 1} - \frac{x - y - 1}{x - y + 1}$$
$$= \frac{-x - y - 1}{x - y + 1} + \frac{x + y - 1}{x - y + 1} - \frac{x - y - 1}{x - y + 1}$$
$$= \frac{-x - y - 1 + x + y - 1 - (x - y - 1)}{x - y + 1}$$
$$= \frac{-x - y - 1 + x + y - 1 - x + y + 1}{x - y + 1}$$
$$= \frac{-x + y - 1}{x - y + 1}$$
$$= \frac{-(x - y + 1)}{x - y + 1}$$
$$= \frac{-1 \cdot (x - y + 1)}{1 \cdot (x - y + 1)}$$
$$= \frac{-1}{1} \cdot \frac{x - y + 1}{x - y + 1}$$
$$= -1$$

91.
$$\frac{x}{x^4 - y^4} - \frac{1}{x^2 + 2xy + y^2}$$
$$= \frac{x}{(x^2 + y^2)(x + y)(x - y)} - \frac{1}{(x + y)(x + y)}$$
$$= \frac{x}{(x^2 + y^2)(x + y)(x - y)} \cdot \frac{x + y}{x + y} -$$
$$\frac{1}{(x + y)(x + y)} \cdot \frac{(x^2 + y^2)(x - y)}{(x^2 + y^2)(x - y)}$$
$$= \frac{x^2 + xy}{(x^2 + y^2)(x + y)(x - y)(x + y)} -$$
$$\frac{x^3 - x^2y + xy^2 - y^3}{(x^2 + y^2)(x + y)(x - y)(x + y)}$$
$$= \frac{x^2 + xy - (x^3 - x^2y + xy^2 - y^3)}{(x^2 + y^2)(x + y)(x - y)(x + y)}$$
$$= \frac{x^2 + xy - x^3 + x^2y - xy^2 + y^3}{(x^2 + y^2)(x + y)(x - y)(x + y)}$$

Exercise Set 5.3

RC1. 4

RC3. 7

RC5. $x + 4 = x - (-4); \ -4$

1.
$$\frac{24x^6 + 18x^5 - 36x^2}{6x^2}$$
$$= \frac{24x^6}{6x^2} + \frac{18x^5}{6x^2} - \frac{36x^2}{6x^2}$$
$$= 4x^4 + 3x^3 - 6$$

3.
$$\frac{45y^7 - 20y^4 + 15y^2}{5y^2}$$
$$= \frac{45y^7}{5y^2} - \frac{20y^4}{5y^2} + \frac{15y^2}{5y^2}$$
$$= 9y^5 - 4y^2 + 3$$

5.
$$(32a^4b^3 + 14a^3b^2 - 22a^2b) \div 2a^2b$$
$$= \frac{32a^4b^3 + 14a^3b^2 - 22a^2b}{2a^2b}$$
$$= \frac{32a^4b^3}{2a^2b} + \frac{14a^3b^2}{2a^2b} - \frac{22a^2b}{2a^2b}$$
$$= 16a^2b^2 + 7ab - 11$$

7.
$$\begin{array}{r} x + 7 \\ x+3\overline{)\,x^2 + 10x + 21} \\ \underline{x^2 + 3x} \\ 7x + 21 \\ \underline{7x + 21} \\ 0 \end{array}$$
$(x^2 + 10x) - (x^2 + 3x) = 7x$

The answer is $x + 7$.

9.
$$\begin{array}{r} a - 12 \\ a+4\overline{)\,a^2 - 8a - 16} \\ \underline{a^2 + 4a} \\ -12a - 16 \\ \underline{-12a - 48} \\ 32 \end{array}$$
$(a^2 - 8a) - (a^2 + 4a) = -12a$
$(-12a - 16) - (-12a - 48) = 32$

The answer is $a - 12$, R 32, or $a - 12 + \dfrac{32}{a+4}$.

11.
$$\begin{array}{r} x + 2 \\ x+5\overline{)\,x^2 + 7x + 14} \\ \underline{x^2 + 5x} \\ 2x + 14 \\ \underline{2x + 10} \\ 4 \end{array}$$
$(x^2 + 7x) - (x^2 + 5x) = 2x$
$(2x + 14) - (2x + 10) = 4$

The answer is $x + 2$, R 4, or $x + 2 + \dfrac{4}{x+5}$.

13.
$$\begin{array}{r} 2y^2 - y + 2 \\ 2y+4\overline{)\,4y^3 + 6y^2 + 0y + 14} \\ \underline{4y^3 + 8y^2} \\ -2y^2 + 0y \\ \underline{-2y^2 - 4y} \\ 4y + 14 \\ \underline{4y + 8} \\ 6 \end{array}$$

The answer is $2y^2 - y + 2$, R 6, or $2y^2 - y + 2 + \dfrac{6}{2y+4}$.

15.
$$\begin{array}{r} 2y^2 + 2y - 1 \\ 5y-2\overline{)\,10y^3 + 6y^2 - 9y + 10} \\ \underline{10y^3 - 4y^2} \\ 10y^2 - 9y \\ \underline{10y^2 - 4y} \\ -5y + 10 \\ \underline{-5y + 2} \\ 8 \end{array}$$

The answer is $2y^2 + 2y - 1$, R 8, or $2y^2 + 2y - 1 + \dfrac{8}{5y-2}$.

17.
$$\begin{array}{r} 2x^2 - x - 9 \\ x^2+2\overline{)\,2x^4 - x^3 - 5x^2 + x - 6} \\ \underline{2x^4 + 4x^2} \\ -x^3 - 9x^2 + x \\ \underline{-x^3 - 2x} \\ -9x^2 + 3x - 6 \\ \underline{-9x^2 - 18} \\ 3x + 12 \end{array}$$

The answer is $2x^2 - x - 9$, R $(3x + 12)$, or
$$2x^2 - x - 9 + \frac{3x+12}{x^2+2}.$$

19.
$$\begin{array}{r} 2x^3 + 5x^2 + 17x + 51 \\ x^2-3x\overline{)\,2x^5 - x^4 + 2x^3 + 0x^2 - x} \\ \underline{2x^5 - 6x^4} \\ 5x^4 + 2x^3 \\ \underline{5x^4 - 15x^3} \\ 17x^3 + 0x^2 \\ \underline{17x^3 - 51x^2} \\ 51x^2 - x \\ \underline{51x^2 - 153x} \\ 152x \end{array}$$

The answer is $2x^3 + 5x^2 + 17x + 51$, R $152x$, or
$$2x^3 + 5x^2 + 17x + 51 + \frac{152x}{x^2-3x}.$$

21. $(x^3 - 2x^2 + 2x - 5) \div (x - 1)$
$$\begin{array}{r|rrrr} 1 & 1 & -2 & 2 & -5 \\ & & 1 & -1 & 1 \\ \hline & 1 & -1 & 1 & -4 \end{array}$$
The answer is $x^2 - x + 1$, R -4, or $x^2 - x + 1 + \dfrac{-4}{x-1}$.

23. $(a^2 + 11a - 19) \div (a + 4) =$
$(a^2 + 11a - 19) \div [a - (-4)]$
$$\begin{array}{r|rrr} -4 & 1 & 11 & -19 \\ & & -4 & -28 \\ \hline & 1 & 7 & -47 \end{array}$$
The answer is $a + 7$, R -47, or $a + 7 + \dfrac{-47}{a+4}$.

25. $(x^3 - 7x^2 - 13x + 3) \div (x - 2)$
$$\begin{array}{r|rrrr} 2 & 1 & -7 & -13 & 3 \\ & & 2 & -10 & -46 \\ \hline & 1 & -5 & -23 & -43 \end{array}$$
The answer is $x^2 - 5x - 23$, R -43, or $x^2 - 5x - 23 + \dfrac{-43}{x-2}$.

27. $(3x^3 + 7x^2 - 4x + 3) \div (x + 3) =$

$(3x^3 + 7x^2 - 4x + 3) \div [x - (-3)]$

$$\begin{array}{r|rrrr} -3 & 3 & 7 & -4 & 3 \\ & & -9 & 6 & -6 \\ \hline & 3 & -2 & 2 & \!\mid\! -3 \end{array}$$

The answer is $3x^2 - 2x + 2$, R -3, or $3x^2 - 2x + 2 + \dfrac{-3}{x+3}$.

29. $(y^3 - 3y + 10) \div (y - 2) =$

$(y^3 + 0y^2 - 3y + 10) \div (y - 2)$

$$\begin{array}{r|rrrr} 2 & 1 & 0 & -3 & 10 \\ & & 2 & 4 & 2 \\ \hline & 1 & 2 & 1 & \!\mid\! 12 \end{array}$$

The answer is $y^2 + 2y + 1$, R 12, or $y^2 + 2y + 1 + \dfrac{12}{y-2}$.

31. $(3x^4 - 25x^2 - 18) \div (x - 3) =$

$(3x^4 + 0x^3 - 25x^2 + 0x - 18) \div (x - 3)$

$$\begin{array}{r|rrrrr} 3 & 3 & 0 & -25 & 0 & -18 \\ & & 9 & 27 & 6 & 18 \\ \hline & 3 & 9 & 2 & 6 & \!\mid\! 0 \end{array}$$

The answer is $3x^3 + 9x^2 + 2x + 6$.

33. $(x^3 - 8) \div (x - 2) = (x^3 + 0x^2 + 0x - 8) \div (x - 2)$

$$\begin{array}{r|rrrr} 2 & 1 & 0 & 0 & -8 \\ & & 2 & 4 & 8 \\ \hline & 1 & 2 & 4 & \!\mid\! 0 \end{array}$$

The answer is $x^2 + 2x + 4$.

35. $(y^4 - 16) \div (y - 2) =$

$(y^4 + 0y^3 + 0y^2 + 0y - 16) \div (y - 2)$

$$\begin{array}{r|rrrrr} 2 & 1 & 0 & 0 & 0 & -16 \\ & & 2 & 4 & 8 & 16 \\ \hline & 1 & 2 & 4 & 8 & \!\mid\! 0 \end{array}$$

The answer is $y^3 + 2y^2 + 4y + 8$.

37. $(y^8 - 1) \div (y + 1) = (y^8 - 1) \div [y - (-1)]$

$$\begin{array}{r|rrrrrrrrr} -1 & 1 & 0 & 0 & 0 & 0 & 0 & 0 & 0 & -1 \\ & & -1 & 1 & -1 & 1 & -1 & 1 & -1 & 1 \\ \hline & 1 & -1 & 1 & -1 & 1 & -1 & 1 & -1 & \!\mid\! 0 \end{array}$$

The answer is $y^7 - y^6 + y^5 - y^4 + y^3 - y^2 + y - 1$.

39. Graph: $2x - 3y < 6$

We first graph the line $2x - 3y = 6$. The intercepts are $(0, -2)$ and $(3, 0)$. We draw the line dashed since the inequality symbol is $<$. To determine which half-plane to shade, we consider a test point not on the line. We try $(0, 0)$:

$$\begin{array}{c} 2x - 3y < 6 \\ \hline 2 \cdot 0 - 3 \cdot 0 \; ? \; 6 \\ 0 \; \mid \qquad \text{TRUE} \end{array}$$

Since $0 < 6$ is true, we shade the half-plane that contains $(0, 0)$.

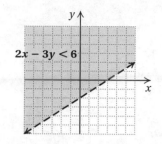

41. Graph $y > 4$.

We first graph the line $y = 4$. This is a horizontal line 4 units above the x-axis. We draw the line dashed since the inequality symbol is $>$. To determine which half-plane to shade, we write $y > 4$ as $0x + y > 4$ and consider a test point not on the line. We try $(0, 0)$:

$$\begin{array}{c} 0x + y > 4 \\ \hline 0 \cdot 0 + 0 \; ? \; 4 \\ 0 \; \mid \qquad \text{FALSE} \end{array}$$

Since $0 > 4$ is false, we shade the half-plane that does not contain $(0, 0)$.

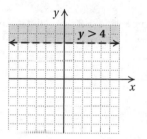

43. Graph: $f(x) = x^2$.

We select x-values and find the corresponding y-values. We plot these ordered pairs and draw the graph.

x	y
-2	4
-1	1
0	0
1	1
2	4

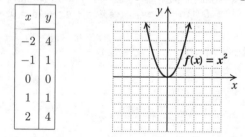

45. Graph: $f(x) = 3 - x^2$.

We select x-values and find the corresponding y-values. We plot these ordered pairs and draw the graph.

x	y
-3	-6
-2	-1
-1	2
0	3
1	2
2	-1
3	-6

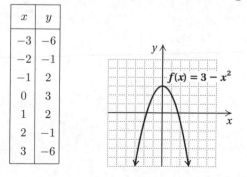

47. $x^2 - 5x = 0$

$x(x - 5) = 0$

$x = 0 \;\; or \;\; x - 5 = 0$

$x = 0 \;\; or \;\;\;\;\;\; x = 5$

The solutions are 0 and 5.

49. $12x^2 = 17x + 5$

$12x^2 - 17x - 5 = 0$

$(4x + 1)(3x - 5) = 0$

$4x + 1 = 0 \;\;\; or \;\; 3x - 5 = 0$

$4x = -1 \;\; or \;\;\;\;\;\; 3x = 5$

$x = -\dfrac{1}{4} \;\; or \;\;\;\;\;\; x = \dfrac{5}{3}$

The solutions are $-\dfrac{1}{4}$ and $\dfrac{5}{3}$.

51. $f(x) = 4x^3 + 16x^2 - 3x - 45$

$f(-3) = 4(-3)^3 + 16(-3)^2 - 3(-3) - 45$

$= -108 + 144 + 9 - 45$

$= 0$

Since $f(-3) = 0$, we know that -3 is a solution of $f(x) = 0$. That is, $f(x) = 0$ when $x = -3$, or when $x + 3 = 0$. We find $f(x) \div (x + 3)$.

$$
\begin{array}{r}
4x^2+4x-15 \\
x+3\,\overline{\smash{\big)}\,4x^3+16x^2-3x-45} \\
\underline{4x^3+12x^2} \\
4x^2-3x \\
\underline{4x^2+12x} \\
-15x-45 \\
\underline{-15x-45} \\
0
\end{array}
$$

Now we know that $(x + 3)(4x^2 + 4x - 15) = 0$. To find the other solutions of $f(x) = 0$ we solve $4x^2 + 4x - 15 = 0$.

$4x^2 + 4x - 15 = 0$

$(2x - 3)(2x + 5) = 0$

$2x - 3 = 0 \;\; or \;\; 2x + 5 = 0$

$2x = 3 \;\; or \;\;\;\;\;\; 2x = -5$

$x = \dfrac{3}{2} \;\; or \;\;\;\;\;\; x = -\dfrac{5}{2}$

Then the solutions of $f(x) = 0$ are -3, $\dfrac{3}{2}$, and $-\dfrac{5}{2}$.

53.
$$
\begin{array}{r}
x-5 \\
x+2\,\overline{\smash{\big)}\,x^2-3x+2k} \\
\underline{x^2+2x} \\
-5x+2k \\
\underline{-5x-10} \\
2k+10
\end{array}
$$

The remainder is 7. Thus, we solve the following equation for k.

$2k + 10 = 7$

$2k = -3$

$k = -\dfrac{3}{2}$

55.
$$
\begin{array}{r}
a^2+ab \\
a^2 + 3ab + 2b^2\,\overline{\smash{\big)}\,a^4+4a^3b + 5a^2b^2 + 2ab^3} \\
\underline{a^4+3a^3b + 2a^2b^2} \\
a^3b + 3a^2b^2 + 2ab^3 \\
\underline{a^3b + 3a^2b^2 + 2ab^3} \\
0
\end{array}
$$

The answer is $a^2 + ab$.

Exercise Set 5.4

RC1. The expression is a <u>complex</u> rational expression.

RC3. The <u>least common denominator</u> of the rational expressions $\dfrac{7}{x}$, $\dfrac{2}{3}$, and $\dfrac{4}{x}$ is $3x$.

1. $\dfrac{2+\dfrac{3}{5}}{4-\dfrac{1}{2}}$

The LCM of the denominators is $5 \cdot 2$, or 10. We multiply by 1 using $10/10$.

$$
\dfrac{2+\dfrac{3}{5}}{4-\dfrac{1}{2}} = \dfrac{2+\dfrac{3}{5}}{4-\dfrac{1}{2}} \cdot \dfrac{10}{10}
$$

$$
= \dfrac{\left(2+\dfrac{3}{5}\right) \cdot 10}{\left(4-\dfrac{1}{2}\right) \cdot 10}
$$

$$
= \dfrac{2\cdot 10+\dfrac{3}{5} \cdot 10}{4\cdot 10-\dfrac{1}{2} \cdot 10}
$$

$$
= \dfrac{20+6}{40-5}
$$

$$
= \dfrac{26}{35}
$$

3. $\dfrac{\dfrac{2}{3}+\dfrac{4}{5}}{\dfrac{3}{4}-\dfrac{1}{2}}$

The LCM of the denominators is $3 \cdot 5 \cdot 2 \cdot 2$, or 60. We multiply by 1 using $60/60$.

$$\frac{\frac{2}{3}+\frac{4}{5}}{\frac{3}{4}-\frac{1}{2}} = \frac{\frac{2}{3}+\frac{4}{5}}{\frac{3}{4}-\frac{1}{2}} \cdot \frac{60}{60}$$

$$= \frac{\left(\frac{2}{3}+\frac{4}{5}\right)\cdot 60}{\left(\frac{3}{4}-\frac{1}{2}\right)\cdot 60}$$

$$= \frac{\frac{2}{3}\cdot 60 + \frac{4}{5}\cdot 60}{\frac{3}{4}\cdot 60 - \frac{1}{2}\cdot 60}$$

$$= \frac{40+48}{45-30}$$

$$= \frac{88}{15}$$

5. $\dfrac{\frac{x}{y^2}}{\frac{y^3}{x^2}} = \dfrac{x}{y^2}\cdot\dfrac{x^2}{y^3}$ Multiplying by the reciprocal
of the divisor

$$= \frac{x^3}{y^5}$$

7. $\dfrac{\frac{9x^2-y^2}{xy}}{\frac{3x-y}{y}}$

$$= \frac{9x^2-y^2}{xy}\cdot\frac{y}{3x-y} \qquad \begin{array}{l}\text{Multiplying by the recip-}\\\text{rocal of the divisor}\end{array}$$

$$= \frac{(9x^2-y^2)(y)}{xy(3x-y)}$$

$$= \frac{(3x+y)(3x-y)(y)}{xy\,(3x-y)}$$

$$= \frac{3x+y}{x}$$

9. $\dfrac{\frac{1}{a}+2}{\frac{1}{a}-1}$

The LCM of all the denominators is a. We multiply by 1
using a/a.

$$\frac{\frac{1}{a}+2}{\frac{1}{a}-1} = \frac{\frac{1}{a}+2}{\frac{1}{a}-1}\cdot\frac{a}{a}$$

$$= \frac{\left(\frac{1}{a}+2\right)\cdot a}{\left(\frac{1}{a}-1\right)\cdot a}$$

$$= \frac{\frac{1}{a}\cdot a + 2a}{\frac{1}{a}\cdot a - a}$$

$$= \frac{1+2a}{1-a}$$

11. $\dfrac{x-\frac{1}{x}}{x+\frac{1}{x}}$

The LCM of all the denominators is x. We multiply by 1
using x/x.

$$\frac{x-\frac{1}{x}}{x+\frac{1}{x}} = \frac{x-\frac{1}{x}}{x+\frac{1}{x}}\cdot\frac{x}{x}$$

$$= \frac{\left(x-\frac{1}{x}\right)\cdot x}{\left(x+\frac{1}{x}\right)\cdot x}$$

$$= \frac{x\cdot x - \frac{1}{x}\cdot x}{x\cdot x + \frac{1}{x}\cdot x}$$

$$= \frac{x^2-1}{x^2+1} = \frac{(x+1)(x-1)}{x^2+1}$$

13. $\dfrac{\frac{3}{x}+\frac{4}{y}}{\frac{4}{x}-\frac{3}{y}}$

The LCM of all the denominators is xy. We multiply by 1
using $\dfrac{xy}{xy}$.

$$\frac{\frac{3}{x}+\frac{4}{y}}{\frac{4}{x}-\frac{3}{y}} = \frac{\frac{3}{x}+\frac{4}{y}}{\frac{4}{x}-\frac{3}{y}}\cdot\frac{xy}{xy}$$

$$= \frac{\left(\frac{3}{x}+\frac{4}{y}\right)\cdot xy}{\left(\frac{4}{x}-\frac{3}{y}\right)\cdot xy}$$

$$= \frac{\frac{3}{x}\cdot xy + \frac{4}{y}\cdot xy}{\frac{4}{x}\cdot xy - \frac{3}{y}\cdot xy}$$

$$= \frac{3y+4x}{4y-3x}$$

15. $\dfrac{a - \dfrac{3a}{b}}{b - \dfrac{b}{a}}$

$= \dfrac{a - \dfrac{3a}{b}}{b - \dfrac{b}{a}} \cdot \dfrac{ab}{ab}$ Using the LCM of the denominators

$= \dfrac{\left(a - \dfrac{3a}{b}\right) \cdot ab}{\left(b - \dfrac{b}{a}\right) \cdot ab}$

$= \dfrac{a(ab) - \dfrac{3a}{b} \cdot ab}{b(ab) - \dfrac{b}{a} \cdot ab}$

$= \dfrac{a^2 b - 3a^2}{ab^2 - b^2}$

$= \dfrac{a^2(b - 3)}{b^2(a - 1)}$

17. $\dfrac{\dfrac{1}{a} + \dfrac{1}{b}}{\dfrac{a^2 - b^2}{ab}}$

$= \dfrac{\dfrac{1}{a} + \dfrac{1}{b}}{\dfrac{a^2 - b^2}{ab}} \cdot \dfrac{ab}{ab}$ Using the LCM of the denominators

$= \dfrac{\left(\dfrac{1}{a} + \dfrac{1}{b}\right) \cdot ab}{\left(\dfrac{a^2 - b^2}{ab}\right) \cdot ab}$

$= \dfrac{\dfrac{1}{a} \cdot ab + \dfrac{1}{b} \cdot ab}{\dfrac{a^2 - b^2}{ab} \cdot ab}$

$= \dfrac{b + a}{a^2 - b^2} = \dfrac{b + a}{(a + b)(a - b)}$

$= \dfrac{a + b}{a + b} \cdot \dfrac{1}{a - b}$ $(b + a = a + b)$

$= \dfrac{1}{a - b}$

19. $\dfrac{\dfrac{1}{x + h} - \dfrac{1}{x}}{h}$

$= \dfrac{\dfrac{1}{x + h} \cdot \dfrac{x}{x} - \dfrac{1}{x} \cdot \dfrac{x + h}{x + h}}{h}$ Adding in the numerator

$= \dfrac{\dfrac{x - x - h}{x(x + h)}}{h} = \dfrac{\dfrac{-h}{x(x + h)}}{h}$

$= \dfrac{-h}{x(x + h)} \cdot \dfrac{1}{h}$ Multiplying by the reciprocal of the divisor

$= \dfrac{-h}{hx(x + h)}$

$= \dfrac{h}{h} \cdot \dfrac{-1}{x(x + h)}$ Removing a factor of 1

$= \dfrac{-1}{x(x + h)}, \text{ or } -\dfrac{1}{x(x + h)}$

21. $\dfrac{\dfrac{x^2 - x - 12}{x^2 - 2x - 15}}{\dfrac{x^2 + 8x + 12}{x^2 - 5x - 14}}$

$= \dfrac{x^2 - x - 12}{x^2 - 2x - 15} \cdot \dfrac{x^2 - 5x - 14}{x^2 + 8x + 12}$ Multiplying by the reciprocal of the divisor

$= \dfrac{(x^2 - x - 12)(x^2 - 5x - 14)}{(x^2 - 2x - 15)(x^2 + 8x + 12)}$

$= \dfrac{(\cancel{x + 3})(x - 4)(x - 7)(\cancel{x + 2})}{(x - 5)(\cancel{x + 3})(x + 6)(\cancel{x + 2})}$ Removing a factor of 1

$= \dfrac{(x - 4)(x - 7)}{(x - 5)(x + 6)}$

23. $\dfrac{\dfrac{1}{x + 2} + \dfrac{4}{x - 3}}{\dfrac{2}{x - 3} - \dfrac{7}{x + 2}}$

$= \dfrac{\dfrac{1}{x + 2} + \dfrac{4}{x - 3}}{\dfrac{2}{x - 3} - \dfrac{7}{x + 2}} \cdot \dfrac{(x + 2)(x - 3)}{(x + 2)(x - 3)}$ Using the LCM of the denominators

$= \dfrac{\dfrac{1}{x + 2} \cdot (x + 2)(x - 3) + \dfrac{4}{x - 3} \cdot (x + 2)(x - 3)}{\dfrac{2}{x - 3} \cdot (x + 2)(x - 3) - \dfrac{7}{x + 2} \cdot (x + 2)(x - 3)}$

$= \dfrac{x - 3 + 4(x + 2)}{2(x + 2) - 7(x - 3)}$

$= \dfrac{x - 3 + 4x + 8}{2x + 4 - 7x + 21}$

$= \dfrac{5x + 5}{-5x + 25}$

$= \dfrac{\cancel{5}(x + 1)}{\cancel{5}(-x + 5)}$ Removing a factor of 1

$= \dfrac{x + 1}{-x + 5}, \text{ or } \dfrac{x + 1}{5 - x}$

25. $\dfrac{\dfrac{6}{x^2-4}-\dfrac{5}{x+2}}{\dfrac{7}{x^2-4}-\dfrac{4}{x-2}}$

$=\dfrac{\dfrac{6}{(x+2)(x-2)}-\dfrac{5}{x+2}\cdot\dfrac{x-2}{x-2}}{\dfrac{7}{(x+2)(x-2)}-\dfrac{4}{x-2}\cdot\dfrac{(x+2)}{(x+2)}}$

         Finding the LCM of the denomina-
         tors and multiplying by 1

$=\dfrac{\dfrac{6}{(x+2)(x-2)}-\dfrac{5x-10}{(x+2)(x-2)}}{\dfrac{7}{(x+2)(x-2)}-\dfrac{4x+8}{(x+2)(x-2)}}$

$=\dfrac{\dfrac{6-5x+10}{(x+2)(x-2)}}{\dfrac{7-4x-8}{(x+2)(x-2)}}$ Subtracting in the numerator
                      and in the denominator

$=\dfrac{\dfrac{16-5x}{(x+2)(x-2)}}{\dfrac{-1-4x}{(x+2)(x-2)}}$

$=\dfrac{16-5x}{(x+2)(x-2)}\cdot\dfrac{(x+2)(x-2)}{-1-4x}$

         Multiplying by the reciprocal of the denominator

$=\dfrac{(16-5x)(x+2)(x-2)}{(x+2)(x-2)(-1-4x)}$ Removing a factor of 1

$=\dfrac{16-5x}{-1-4x}$

$=\dfrac{-1\cdot(5x-16)}{-1(4x+1)}=\dfrac{5x-16}{4x+1}$

27. $\dfrac{\dfrac{1}{z^2}-\dfrac{1}{w^2}}{\dfrac{1}{z^3}+\dfrac{1}{w^3}}$

$=\dfrac{\dfrac{1}{z^2}-\dfrac{1}{w^2}}{\dfrac{1}{z^3}+\dfrac{1}{w^3}}\cdot\dfrac{z^3w^3}{z^3w^3}$ Multiplying by the LCM of
                     the denominators

$=\dfrac{\left(\dfrac{1}{z^2}-\dfrac{1}{w^2}\right)\cdot z^3w^3}{\left(\dfrac{1}{z^3}+\dfrac{1}{w^3}\right)\cdot z^3w^3}$

$=\dfrac{zw^3-z^3w}{w^3+z^3}$

$=\dfrac{zw(w^2-z^2)}{(w+z)(w^2-wz+z^2)}$

$=\dfrac{zw(w+z)(w-z)}{(w+z)(w^2-wz+z^2)}$

$=\dfrac{zw(w-z)}{w^2-wz+z^2}$

29. $\dfrac{\dfrac{3}{x^2+2x-3}-\dfrac{1}{x^2-3x-10}}{\dfrac{3}{x^2-6x+5}-\dfrac{1}{x^2+5x+6}}$

$=\dfrac{\dfrac{3}{(x+3)(x-1)}-\dfrac{1}{(x-5)(x+2)}}{\dfrac{3}{(x-5)(x-1)}-\dfrac{1}{(x+3)(x+2)}}$

$=\dfrac{\dfrac{3}{(x+3)(x-1)}-\dfrac{1}{(x-5)(x+2)}}{\dfrac{3}{(x-5)(x-1)}-\dfrac{1}{(x+3)(x+2)}}\cdot$

         $\dfrac{(x+3)(x-1)(x-5)(x+2)}{(x+3)(x-1)(x-5)(x+2)}$

         Multiplying by 1, using the LCD

$=\dfrac{3(x-5)(x+2)-(x+3)(x-1)}{3(x+3)(x+2)-(x-1)(x-5)}$

$=\dfrac{3(x^2-3x-10)-(x^2+2x-3)}{3(x^2+5x+6)-(x^2-6x+5)}$

$=\dfrac{3x^2-9x-30-x^2-2x+3}{3x^2+15x+18-x^2+6x-5}$

$=\dfrac{2x^2-11x-27}{2x^2+21x+13}$

31. *Familiarize.* Let $p=$ the number of pages in the 2012 tax code.

     Translate.

		pages for 1984	less	5292 pages	is	pages for 2012.
3	times					
↓	↓	↓	↓	↓	↓	↓
3	·	26,300	−	5292	=	p

     Solve. We carry out the calculation.

         $3\cdot 26,300-5292=p$

           $78,900-5292=p$

               $73,608=p$

     Check. We can repeat the calculation. The answer checks.

     State. The tax code for 2012 contained 73,608 pages.

33. $4x^3+20x^2+6x$

     $=2x(2x^2+10x+3)$ Factoring out the largest
                               common factor

     The trinomial $2x^2+10x+3$ cannot be factored as a product of binomials so we have the complete factorization.

35. $y^3-8=y^3-2^3=(y-2)(y^2+2y+4)$

37. $1000x^3+1=(10x)^3+1^3=(10x+1)(100x^2-10x+1)$

39. $y^3-64x^3=y^3-(4x)^3=(y-4x)(y^2+4xy+16x^2)$

41. $T=\dfrac{r+s}{3}$

      $3T=r+s$ Multiplying by 3

   $3T-r=s$ Subtracting r

43. $f(x) = x^2 - 3$

$f(-5) = (-5)^2 - 3 = 25 - 3 = 22$

45. $f(a) = \dfrac{3}{a^2}$, $f(a+h) = \dfrac{3}{(a+h)^2}$

$\dfrac{f(a+h) - f(a)}{h} = \dfrac{\dfrac{3}{(a+h)^2} - \dfrac{3}{a^2}}{h}$

$= \dfrac{3a^2 - 3(a+h)^2}{a^2(a+h)^2} \cdot \dfrac{1}{h}$

$= \dfrac{3a^2 - 3a^2 - 6ah - 3h^2}{a^2(a+h)^2} \cdot \dfrac{1}{h}$

$= \dfrac{-6ah - 3h^2}{a^2(a+h)^2 h}$

$= \dfrac{-3h(2a + h)}{a^2(a+h)^2 h}$

$= \dfrac{-3\cancel{h}(2a + h)}{a^2(a+h)^2 \cancel{h}}$

$= \dfrac{-3(2a + h)}{a^2(a+h)^2}$

47. $f(a) = \dfrac{1}{1-a}$, $f(a+h) = \dfrac{1}{1-(a+h)} = \dfrac{1}{1-a-h}$

$\dfrac{f(a+h) - f(a)}{h}$

$= \dfrac{\dfrac{1}{1-a-h} - \dfrac{1}{1-a}}{h}$

$= \dfrac{1 - a - (1 - a - h)}{(1-a-h)(1-a)} \cdot \dfrac{1}{h}$

$= \dfrac{1 - a - 1 + a + h}{(1-a-h)(1-a)} \cdot \dfrac{1}{h}$

$= \dfrac{h}{(1-a-h)(1-a)h}$

$= \dfrac{\cancel{h} \cdot 1}{(1-a-h)(1-a)\cancel{h}}$

$= \dfrac{1}{(1-a-h)(1-a)}$

49. $\dfrac{5x^{-1} - 5y^{-1} + 10x^{-1}y^{-1}}{6x^{-1} - 6y^{-1} + 12x^{-1}y^{-1}}$

$= \dfrac{\dfrac{5}{x} - \dfrac{5}{y} + \dfrac{10}{xy}}{\dfrac{6}{x} - \dfrac{6}{y} + \dfrac{12}{xy}}$

$= \dfrac{\dfrac{5}{x} - \dfrac{5}{y} + \dfrac{10}{xy}}{\dfrac{6}{x} - \dfrac{6}{y} + \dfrac{12}{xy}} \cdot \dfrac{xy}{xy}$

$= \dfrac{5y - 5x + 10}{6y - 6x + 12}$

$= \dfrac{5(y - x + 2)}{6(y - x + 2)}$

$= \dfrac{5}{6} \cdot \dfrac{y - x + 2}{y - x + 2}$

$= \dfrac{5}{6}$

51. $\dfrac{1}{x^2 - \dfrac{1}{x}} = \dfrac{1}{x^2 \cdot \dfrac{x}{x} - \dfrac{1}{x}} = \dfrac{1}{\dfrac{x^3}{x} - \dfrac{1}{x}} =$

$\dfrac{1}{\dfrac{x^3 - 1}{x}} = 1 \cdot \dfrac{x}{x^3 - 1} = \dfrac{x}{x^3 - 1}$

53. $\dfrac{1}{\dfrac{a^3 + b^3}{a + b}} = 1 \cdot \dfrac{a + b}{a^3 + b^3}$

$= \dfrac{1 \cdot \cancel{(a + b)}}{\cancel{(a + b)}(a^2 - ab + b^2)}$

$= \dfrac{1}{a^2 - ab + b^2}$

Chapter 5 Mid-Chapter Review

---

1. True; see page 431 in the text.

2. False; see page 418 in the text.

3. False; although 4 is not in the domain, 5 is. (It does not make the denominator 0.)

4. $\dfrac{7x - 2}{x - 4} - \dfrac{x + 1}{x + 3} = \dfrac{7x - 2}{x - 4} \cdot \dfrac{x + 3}{x + 3} - \dfrac{x + 1}{x + 3} \cdot \dfrac{x - 4}{x - 4}$

$= \dfrac{7x^2 + 19x - 6}{(x - 4)(x + 3)} - \dfrac{x^2 - 3x - 4}{(x + 3)(x - 4)}$

$= \dfrac{7x^2 + 19x - 6 - x^2 + 3x + 4}{(x - 4)(x + 3)}$

$= \dfrac{6x^2 + 22x - 2}{(x - 4)(x + 3)}$

5. $\dfrac{\dfrac{1}{m}+3}{\dfrac{1}{m}-5} = \dfrac{\dfrac{1}{m}+3}{\dfrac{1}{m}-5} \cdot \dfrac{m}{m} = \dfrac{1+3m}{1-5m}$

6. $f(x) = \dfrac{x+5}{x^2-100}$

We set the denominator equal to 0 and solve.

$$x^2 - 100 = 0$$
$$(x+10)(x-10) = 0$$
$$x+10 = 0 \quad \text{or} \quad x-10 = 0$$
$$x = -10 \quad \text{or} \quad x = 10$$

The domain is $\{x | x$ is a real number $and\, x \neq -10$ and $x \neq 10\}$, or $(-\infty, -10) \cup (-10, 10) \cup (10, \infty)$.

7. $g(x) = \dfrac{-3}{x-7}$

We set the denominator equal to 0 and solve.

$$x - 7 = 0$$
$$x = 7$$

The domain is $\{x | x$ is a real number $and\, x \neq 7\}$, or $(-\infty, 7) \cup (7, \infty)$.

8. $h(x) = \dfrac{x^2-9}{x^2+8x-9}$

We set the denominator equal to 0 and solve.

$$x^2 + 8x - 9 = 0$$
$$(x+9)(x-1) = 0$$
$$x+9 = 0 \quad \text{or} \quad x-1 = 0$$
$$x = -9 \quad \text{or} \quad x = 1$$

The domain is $\{x | x$ is a real number $and\, x \neq -9$ and $x \neq 1\}$, or $(-\infty, -9) \cup (-9, 1) \cup (1, \infty)$.

9. $\dfrac{24p^2}{36p^3} = \dfrac{2 \cdot 12 \cdot p^2}{3 \cdot 12 \cdot p^2 \cdot p} = \dfrac{12p^2}{12p^2} \cdot \dfrac{2}{3p} = \dfrac{2}{3p}$

10. $\dfrac{42y-3}{33} = \dfrac{3(14y-1)}{3 \cdot 11} = \dfrac{3}{3} \cdot \dfrac{14y-1}{11} = \dfrac{14y-1}{11}$

11. $\dfrac{x^2-y^2}{x^3+y^3} = \dfrac{(x+y)(x-y)}{(x+y)(x^2-xy+y^2)} = \dfrac{x+y}{x+y} \cdot \dfrac{x-y}{x^2-xy+y^2} = \dfrac{x-y}{x^2-xy+y^2}$

12. $\dfrac{x^2-x-30}{x^2-4x-12} = \dfrac{(x-6)(x+5)}{(x-6)(x+2)} = \dfrac{x-6}{x-6} \cdot \dfrac{x+5}{x+2} = \dfrac{x+5}{x+2}$

13. $\dfrac{9a-18}{9a+18} = \dfrac{9(a-2)}{9(a+2)} = \dfrac{9}{9} \cdot \dfrac{a-2}{a+2} = \dfrac{a-2}{a+2}$

14. $\dfrac{3-t}{t^2-t-6} = \dfrac{-1(t-3)}{(t-3)(t+2)} = \dfrac{-1}{t+2} \cdot \dfrac{t-3}{t-3} = \dfrac{-1}{t+2}$

15. $x^3 = x \cdot x \cdot x$

$14x^2y = 2 \cdot 7 \cdot x \cdot x \cdot y$

$35x^4y^5 = 5 \cdot 7 \cdot x \cdot x \cdot x \cdot x \cdot y \cdot y \cdot y \cdot y \cdot y$

$\text{LCM} = 2 \cdot 5 \cdot 7 \cdot x \cdot x \cdot x \cdot x \cdot y \cdot y \cdot y \cdot y \cdot y$, or $70x^4y^5$

16. $x^2 - 25 = (x+5)(x-5)$

$x^2 - 10x + 25 = (x-5)(x-5)$

$x^2 + 3x - 40 = (x+8)(x-5)$

$\text{LCM} = (x+5)(x-5)(x-5)(x+8)$, or $(x+5)(x-5)^2(x+8)$

17. $\dfrac{45}{x^2-1} \div \dfrac{x+1}{x-1} = \dfrac{45}{(x+1)(x-1)} \cdot \dfrac{x-1}{x+1}$

$= \dfrac{45(x-1)}{(x+1)(x-1)(x+1)}$

$= \dfrac{45(x-1)}{(x+1)(x-1)(x+1)}$

$= \dfrac{45}{(x+1)^2}$

18. $\dfrac{3x-1}{x+6} + \dfrac{x}{x-2}$, LCM is $(x+6)(x-2)$

$= \dfrac{3x-1}{x+6} \cdot \dfrac{x-2}{x-2} + \dfrac{x}{x-2} \cdot \dfrac{x+6}{x+6}$

$= \dfrac{(3x-1)(x-2) + x(x+6)}{(x+6)(x-2)}$

$= \dfrac{3x^2 - 7x + 2 + x^2 + 6x}{(x+6)(x-2)}$

$= \dfrac{4x^2 - x + 2}{(x+6)(x-2)}$

19. $\dfrac{q}{q+2} - \dfrac{q+1}{q}$, LCM is $q(q+2)$

$= \dfrac{q}{q+2} \cdot \dfrac{q}{q} - \dfrac{q+1}{q} \cdot \dfrac{q+2}{q+2}$

$= \dfrac{q^2 - (q+1)(q+2)}{q(q+2)}$

$= \dfrac{q^2 - (q^2 + 3q + 2)}{q(q+2)}$

$= \dfrac{q^2 - q^2 - 3q - 2}{q(q+2)}$

$= \dfrac{-3q-2}{q(q+2)}$

20. $\dfrac{2y}{y^2+2y-3} - \dfrac{3y+1}{y^2+y-2}$

$= \dfrac{2y}{(y+3)(y-1)} - \dfrac{3y+1}{(y+2)(y-1)}$,

$\qquad\qquad\qquad$ LCM is $(y+3)(y-1)(y+2)$

$= \dfrac{2y}{(y+3)(y-1)} \cdot \dfrac{y+2}{y+2} - \dfrac{3y+1}{(y+2)(y-1)} \cdot \dfrac{y+3}{y+3}$

$= \dfrac{2y(y+2) - (3y+1)(y+3)}{(y+3)(y-1)(y+2)}$

$= \dfrac{2y^2 + 4y - (3y^2 + 10y + 3)}{(y+3)(y-1)(y+2)}$

$= \dfrac{2y^2 + 4y - 3y^2 - 10y - 3}{(y+3)(y-1)(y+2)}$

$= \dfrac{-y^2 - 6y - 3}{(y+3)(y-1)(y+2)}$

21. $\dfrac{\frac{1}{b}-1}{\frac{1}{b^2}-1} = \dfrac{\frac{1}{b}-1}{\frac{1}{b^2}-1} \cdot \dfrac{b^2}{b^2}$

$\quad = \dfrac{\left(\frac{1}{b}-1\right)b^2}{\left(\frac{1}{b^2}-1\right)b^2}$

$\quad = \dfrac{b-b^2}{1-b^2}$

$\quad = \dfrac{b(1-b)}{(1+b)(1-b)}$

$\quad = \dfrac{b(1\!\!\!/-\!\!\!/b)}{(1+b)(1\!\!\!/-\!\!\!/b)}$

$\quad = \dfrac{b}{1+b}$

22. $\dfrac{w^2-z^2}{5w-5z} \cdot \dfrac{w-z}{w+z} = \dfrac{(w^2-z^2)(w-z)}{(5w-5z)(w+z)}$

$\quad = \dfrac{(w+z)(w-z)(w-z)}{5(w-z)(w+z)}$

$\quad = \dfrac{(w\!\!\!/+\!\!\!/z)(w\!\!\!/-\!\!\!/z)(w-z)}{5(w\!\!\!/-\!\!\!/z)(w\!\!\!/+\!\!\!/z)}$

$\quad = \dfrac{w-z}{5}$

23. $\dfrac{t^3-8}{2t+3} \cdot \dfrac{2t^2+t-3}{t-2}$

$\quad = \dfrac{(t^3-8)(2t^2+t-3)}{(2t+3)(t-2)}$

$\quad = \dfrac{(t-2)(t^2+2t+4)(2t+3)(t-1)}{(2t+3)(t-2)}$

$\quad = \dfrac{(t\!\!\!/-\!\!\!/2)(t^2+2t+4)(2t\!\!\!/+\!\!\!/3)(t-1)}{(2t\!\!\!/+\!\!\!/3)(t\!\!\!/-\!\!\!/2)\cdot 1}$

$\quad = (t^2+2t+4)(t-1)$

24. $\dfrac{5c}{3}+\dfrac{2a}{5c}$, LCM is $3\cdot 5c$, or $15c$

$\quad = \dfrac{5c}{3}\cdot\dfrac{5c}{5c}+\dfrac{2a}{5c}\cdot\dfrac{3}{3}$

$\quad = \dfrac{25c^2+6a}{15c}$

25. $\dfrac{x^2-4x}{x^2+2x} \div \dfrac{x^2-8x+16}{x^2+4x+4}$

$\quad = \dfrac{x^2-4x}{x^2+2x} \cdot \dfrac{x^2+4x+4}{x^2-8x+16}$

$\quad = \dfrac{x(x-4)(x+2)(x+2)}{x(x+2)(x-4)(x-4)}$

$\quad = \dfrac{x\!\!\!/(x\!\!\!/-\!\!\!/4)(x\!\!\!/+\!\!\!/2)(x+2)}{x\!\!\!/(x\!\!\!/+\!\!\!/2)(x\!\!\!/-\!\!\!/4)(x-4)}$

$\quad = \dfrac{x+2}{x-4}$

26.

$$\begin{array}{r} 3x + 2 \\ 2x-3\,\overline{\smash{)}\,6x^2 - 5x + 11} \\ \underline{6x^2 - 9x} \\ 4x + 11 \\ \underline{4x - 6} \\ 17 \end{array}$$

The answer is $3x+2$, R 17, or $3x+2+\dfrac{17}{2x-3}$.

27.

$$\begin{array}{r} x^3 - x^2 + x - 1 \\ x+1\,\overline{\smash{)}\,x^4 + 0x^3 + 0x^2 + 0x - 1} \\ \underline{x^4 + x^3} \\ -x^3 + 0x^2 \\ \underline{-x^3 - x^2} \\ x^2 + 0x \\ \underline{x^2 + x} \\ -x - 1 \\ \underline{-x - 1} \\ 0 \end{array}$$

The answer is $x^3 - x^2 + x - 1$.

28. $(2x^3-x^2+5x-4)\div(x+2) = (2x^3-x^2+5x-4)\div[x-(-2)]$

$$\begin{array}{r|rrrr} -2 & 2 & -1 & 5 & -4 \\ & & -4 & 10 & -30 \\ \hline & 2 & -5 & 15 & |-34 \end{array}$$

The answer is $2x^2-5x+15$, R -34, or

$2x^2-5x+15+\dfrac{-34}{x+2}$.

29. $\begin{array}{r|rrr} 6 & 1 & -4 & -12 \\ & & 6 & 12 \\ \hline & 1 & 2 & |\ 0 \end{array}$

The answer is $x+2$.

30. $(x^4-3x^2+2)\div(x+3) = (x^4-3x^2+2)\div[x-(-3)]$

$$\begin{array}{r|rrrrr} -3 & 1 & 0 & -3 & 0 & 2 \\ & & -3 & 9 & -18 & 54 \\ \hline & 1 & -3 & 6 & -18 & |\ 56 \end{array}$$

The answer is $x^3-3x^2+6x-18$ R 56, or

$x^3-3x^2+6x-18+\dfrac{56}{x+3}$.

31.

$$\begin{array}{r} 3x - 1 \\ 5x+1\,\overline{\smash{)}\,15x^2 - 2x + 6} \\ \underline{15x^2 + 3x} \\ -5x + 6 \\ \underline{-5x - 1} \\ 7 \end{array}$$

The answer is $3x-1$, R 7, or $3x-1+\dfrac{7}{5x+1}$.

32. For a, a remainder of 0 indicates that $x-a$ is a factor of the polynomial. The quotient is a polynomial of one degree less than the original polynomial, and it can be factored further, if possible, using synthetic division again or another factoring method.

33. Addition, subtraction, and multiplication of polynomials always result in a polynomial, because these operations always result in a monomial or a sum of monomials. Division of polynomials does not always result in a polynomial, because the quotient is not always a monomial or a sum of monomials. Example 1 in Section 5.3 in the text illustrates this.

34. No; when we simplify a rational expression by removing a factor of 1, we are actually reversing the multiplication process.

35. Janine's answer was correct. It is equivalent to the answer at the back of the book:

$$\frac{3-x}{x-5} = \frac{-x+3}{x-5} = \frac{-1(-x+3)}{-1(x-5)} = \frac{x-3}{-x+5} = \frac{x-3}{5-x}.$$

36. Nancy's misconception is that x is a factor of the numerator. $\left(\dfrac{x+2}{x} = 3 \text{ only for } x = 1.\right)$

37. Most would agree that it is easier to find the LCM of all the denominators, bd, and then to multiply by $bd/(bd)$ than it is to add in the numerator, subtract in the denominator, and then divide the numerator by the denominator.

Exercise Set 5.5

RC1. Rational expression

RC3. Rational expression

RC5. Solutions

RC7. Rational expression

1. $\dfrac{y}{10} = \dfrac{2}{5} + \dfrac{3}{8}$, LCM is 40

$40 \cdot \dfrac{y}{10} = 40 \cdot \left(\dfrac{2}{5} + \dfrac{3}{8}\right)$ Multiplying by the LCM

$4y = 40 \cdot \dfrac{2}{5} + 40 \cdot \dfrac{3}{8}$ Removing parentheses

$4y = 16 + 15$

$4y = 31$

$y = \dfrac{31}{4}$

Check: $\dfrac{y}{10} = \dfrac{2}{5} + \dfrac{3}{8}$

$$\frac{\frac{31}{4}}{10} \; ? \; \frac{2}{5} + \frac{3}{8}$$

$$\frac{31}{4} \cdot \frac{1}{10} \;\Big|\; \frac{16}{40} + \frac{15}{40}$$

$$\frac{31}{40} \;\Big|\; \frac{31}{40} \quad \text{TRUE}$$

The solution is $\dfrac{31}{4}$.

3. $\dfrac{1}{4} - \dfrac{5}{6} = \dfrac{1}{a}$, LCM is $12a$

$12a \cdot \left(\dfrac{1}{4} - \dfrac{5}{6}\right) = 12a \cdot \dfrac{1}{a}$ Multiplying by the LCM

$12a \cdot \dfrac{1}{4} - 12a \cdot \dfrac{5}{6} = 12$

$3a - 10a = 12$

$-7a = 12$

$a = -\dfrac{12}{7}$

Check: $\dfrac{1}{4} - \dfrac{5}{6} = \dfrac{1}{a}$

$$\frac{1}{4} - \frac{5}{6} \; ? \; \frac{1}{-\frac{12}{7}}$$

$$\frac{3}{12} - \frac{10}{12} \;\Big|\; 1 \cdot \left(-\frac{7}{12}\right)$$

$$-\frac{7}{12} \;\Big|\; -\frac{7}{12} \quad \text{TRUE}$$

The solution is $-\dfrac{12}{7}$.

5. $\dfrac{x}{3} - \dfrac{x}{4} = 12$, LCM is 12

$12 \cdot \left(\dfrac{x}{3} - \dfrac{x}{4}\right) = 12 \cdot 12$

$12 \cdot \dfrac{x}{3} - 12 \cdot \dfrac{x}{4} = 144$

$4x - 3x = 144$

$x = 144$

Check: $\dfrac{x}{3} - \dfrac{x}{4} = 12$

$$\frac{144}{3} - \frac{144}{4} \; ? \; 12$$

$$48 - 36 \;\Big|\;$$

$$12 \;\Big|\; \quad \text{TRUE}$$

The solution is 144.

7. $x + \dfrac{8}{x} = -9$, LCM is x

$x\left(x + \dfrac{8}{x}\right) = x(-9)$

$x \cdot x + x \cdot \dfrac{8}{x} = -9x$

$x^2 + 8 = -9x$

$x^2 + 9x + 8 = 0$

$(x+1)(x+8) = 0$

$x + 1 = 0 \quad or \quad x + 8 = 0$ Principle of zero products

$x = -1 \quad or \quad x = -8$

Check:

For -1:   For -8:

$$x + \frac{8}{x} = -9$$ $$x + \frac{8}{x} = -9$$

$$-1 + \frac{8}{-1} \; ? \; -9$$ $$-8 + \frac{8}{-8} \; ? \; -9$$

$$\begin{array}{c|c} -1 - 8 & \\ -9 & \text{TRUE} \end{array}$$ $$\begin{array}{c|c} -8 - 1 & \\ -9 & \text{TRUE} \end{array}$$

The solutions are -1 and -8.

9. $\dfrac{3}{y} + \dfrac{7}{y} = 5$, LCM is y

$$y\left(\frac{3}{y} + \frac{7}{y}\right) = y \cdot 5$$

$$y \cdot \frac{3}{y} + y \cdot \frac{7}{y} = 5y$$

$$3 + 7 = 5y$$

$$10 = 5y$$

$$2 = y$$

Check: $\dfrac{3}{y} + \dfrac{7}{y} = 5$

$$\begin{array}{c|c} \dfrac{3}{2} + \dfrac{7}{2} \; ? \; 5 & \\ \dfrac{10}{2} & \\ 5 & \text{TRUE} \end{array}$$

The solution is 2.

11. $\dfrac{1}{2} = \dfrac{z-5}{z+1}$, LCM is $2(z+1)$

$$2(z+1) \cdot \frac{1}{2} = 2(z+1) \cdot \frac{z-5}{z+1}$$

$$z + 1 = 2(z-5)$$

$$z + 1 = 2z - 10$$

$$11 = z$$

Check: $\dfrac{1}{2} = \dfrac{z-5}{z+1}$

$$\begin{array}{c|c} \dfrac{1}{2} \; ? \; \dfrac{11-5}{11+1} & \\ \dfrac{6}{12} & \\ \dfrac{1}{2} & \text{TRUE} \end{array}$$

The solution is 11.

13. $\dfrac{3}{y+1} = \dfrac{2}{y-3}$, LCM is $(y+1)(y-3)$

$$(y+1)(y-3) \cdot \frac{3}{y+1} = (y+1)(y-3) \cdot \frac{2}{y-3}$$

$$3(y-3) = 2(y+1)$$

$$3y - 9 = 2y + 2$$

$$y = 11$$

Check: $\dfrac{3}{y+1} = \dfrac{2}{y-3}$

$$\begin{array}{c|c} \dfrac{3}{11+1} \; ? \; \dfrac{2}{11-3} & \\ \dfrac{3}{12} & \dfrac{2}{8} \\ \dfrac{1}{4} & \dfrac{1}{4} \quad \text{TRUE} \end{array}$$

The solution is 11.

15. $\dfrac{y-1}{y-3} = \dfrac{2}{y-3}$, LCM is $y-3$

$$(y-3) \cdot \frac{y-1}{y-3} = (y-3) \cdot \frac{2}{y-3}$$

$$y - 1 = 2$$

$$y = 3$$

Check: $\dfrac{y-1}{y-3} = \dfrac{2}{y-3}$

$$\begin{array}{c|c} \dfrac{3-1}{3-3} \; ? \; \dfrac{2}{3-3} & \\ \dfrac{2}{0} & \dfrac{2}{0} \quad \text{UNDEFINED} \end{array}$$

We know that 3 is not a solution of the original equation, because it results in division by 0. The equation has no solution.

17. $\dfrac{x+1}{x} = \dfrac{3}{2}$, LCM is $2x$

$$2x \cdot \frac{x+1}{x} = 2x \cdot \frac{3}{2}$$

$$2(x+1) = x \cdot 3$$

$$2x + 2 = 3x$$

$$2 = x$$

Check: $\dfrac{x+1}{x} = \dfrac{3}{2}$

$$\begin{array}{c|c} \dfrac{2+1}{2} \; ? \; \dfrac{3}{2} & \\ \dfrac{3}{2} & \text{TRUE} \end{array}$$

The solution is 2.

19.
$$\frac{1}{2} - \frac{4}{9x} = \frac{4}{9} - \frac{1}{6x}, \text{ LCM is } 18x$$

$$18x\left(\frac{1}{2} - \frac{4}{9x}\right) = 18x\left(\frac{4}{9} - \frac{1}{6x}\right)$$

$$18x \cdot \frac{1}{2} - 18x \cdot \frac{4}{9x} = 18x \cdot \frac{4}{9} - 18x \cdot \frac{1}{6x}$$

$$9x - 8 = 8x - 3$$

$$x = 5$$

Since 5 checks, it is the solution.

21.
$$\frac{60}{x} - \frac{60}{x-5} = \frac{2}{x}, \text{ LCM is } x(x-5)$$

$$x(x-5)\left(\frac{60}{x} - \frac{60}{x-5}\right) = x(x-5) \cdot \frac{2}{x}$$

$$60(x-5) - 60x = 2(x-5)$$

$$60x - 300 - 60x = 2x - 10$$

$$-300 = 2x - 10$$

$$-290 = 2x$$

$$-145 = x$$

Since -145 checks, it is the solution.

23.
$$\frac{7}{5x-2} = \frac{5}{4x}, \text{ LCM is } 4x(5x-2)$$

$$4x(5x-2)\,\frac{7}{5x-2} = 4x(5x-2) \cdot \frac{5}{4x}$$

$$4x \cdot 7 = 5(5x-2)$$

$$28x = 25x - 10$$

$$3x = -10$$

$$x = -\frac{10}{3}$$

Since $-\dfrac{10}{3}$ checks, it is the solution.

25.
$$\frac{x}{x-2} + \frac{x}{x^2-4} = \frac{x+3}{x+2}$$

$$\frac{x}{x-2} + \frac{x}{(x+2)(x-2)} = \frac{x+3}{x+2}$$

$$\text{LCM is } (x+2)(x-2)$$

$$(x+2)(x-2)\left(\frac{x}{x-2} + \frac{x}{(x+2)(x-2)}\right) =$$

$$(x+2)(x-2) \cdot \frac{x+3}{x+2}$$

$$x(x+2) + x =$$

$$(x-2)(x+3)$$

$$x^2 + 2x + x = x^2 + x - 6$$

$$3x = x - 6$$

$$2x = -6$$

$$x = -3$$

Since -3 checks, it is the solution.

27.
$$\frac{6}{x^2-4x+3} - \frac{1}{x-3} = \frac{1}{4x-4}$$

$$\frac{6}{(x-3)(x-1)} - \frac{1}{x-3} = \frac{1}{4(x-1)}$$

$$\text{LCM is } 4(x-3)(x-1)$$

$$4(x-3)(x-1)\left(\frac{6}{(x-3)(x-1)} - \frac{1}{x-3}\right) =$$

$$4(x-3)(x-1) \cdot \frac{1}{4(x-1)}$$

$$4 \cdot 6 - 4(x-1) = x - 3$$

$$24 - 4x + 4 = x - 3$$

$$-5x = -31$$

$$x = \frac{31}{5}$$

Since $\dfrac{31}{5}$ checks, it is the solution.

29.
$$\frac{5}{y+3} = \frac{1}{4y^2-36} + \frac{2}{y-3}$$

$$\frac{5}{y+3} = \frac{1}{4(y+3)(y-3)} + \frac{2}{y-3}$$

$$\text{LCM is } 4(y+3)(y-3)$$

$$4(y+3)(y-3) \cdot \frac{5}{y+3} =$$

$$4(y+3)(y-3)\left(\frac{1}{4(y+3)(y-3)} + \frac{2}{y-3}\right)$$

$$4 \cdot 5(y-3) = 1 + 4 \cdot 2(y+3)$$

$$20y - 60 = 1 + 8y + 24$$

$$12y = 85$$

$$y = \frac{85}{12}$$

Since $\dfrac{85}{12}$ checks, it is the solution.

31.
$$\frac{a}{2a-6} - \frac{3}{a^2-6a+9} = \frac{a-2}{3a-9}$$

$$\frac{a}{2(a-3)} - \frac{3}{(a-3)(a-3)} = \frac{a-2}{3(a-3)}$$

$$\text{LCM is } 2 \cdot 3(a-3)(a-3)$$

$$6(a-3)(a-3)\left(\frac{a}{2(a-3)} - \frac{3}{(a-3)(a-3)}\right) =$$

$$6(a-3)(a-3) \cdot \frac{a-2}{3(a-3)}$$

$$3a(a-3) - 6 \cdot 3 = 2(a-3)(a-2)$$

$$3a^2 - 9a - 18 = 2(a^2 - 5a + 6)$$

$$3a^2 - 9a - 18 = 2a^2 - 10a + 12$$

$$a^2 + a - 30 = 0$$

$$(a+6)(a-5) = 0$$

$$a + 6 = 0 \quad or \quad a - 5 = 0$$

$$a = -6 \quad or \qquad a = 5$$

Both -6 and 5 check. The solutions are -6 and 5.

33.

$$\frac{2x+3}{x-1} = \frac{10}{x^2-1} + \frac{2x-3}{x+1}$$

$$\frac{2x+3}{x-1} = \frac{10}{(x+1)(x-1)} + \frac{2x-3}{x+1}$$

$$\text{LCM is } (x+1)(x-1)$$

$$(x+1)(x-1) \cdot \frac{2x+3}{x-1} =$$

$$(x+1)(x-1)\left(\frac{10}{(x+1)(x-1)} + \frac{2x-3}{x+1}\right)$$

$$(x+1)(2x+3) = 10 + (x-1)(2x-3)$$

$$2x^2 + 5x + 3 = 10 + 2x^2 - 5x + 3$$

$$5x + 3 = 13 - 5x$$

$$10x = 10$$

$$x = 1$$

We know that 1 is not a solution of the original equation, because it results in division by 0. The equation has no solution.

35.

$$\frac{3x}{x+2} + \frac{72}{x^3+8} = \frac{24}{x^2-2x+4}$$

$$\frac{3x}{x+2} + \frac{72}{(x+2)(x^2-2x+4)} = \frac{24}{x^2-2x+4},$$

$$\text{LCM is } (x+2)(x^2-2x+4)$$

$$(x+2)(x^2-2x+4)\left(\frac{3x}{x+2} + \frac{72}{(x+2)(x^2-2x+4)}\right) =$$

$$(x+2)(x^2-2x+4) \cdot \frac{24}{x^2-2x+4}$$

$$(x^2-2x+4)(3x) + 72 = (x+2)24$$

$$3x^3 - 6x^2 + 12x + 72 = 24x + 48$$

$$3x^3 - 6x^2 - 12x + 24 = 0$$

$$3(x^3 - 2x^2 - 4x + 8) = 0$$

$$3[x^2(x-2) - 4(x-2)] = 0$$

$$3(x^2-4)(x-2) = 0$$

$$3(x+2)(x-2)(x-2) = 0$$

$$x+2 = 0 \quad or \quad x-2 = 0 \quad or \quad x-2 = 0$$

$$x = -2 \quad or \quad\quad x = 2 \quad or \quad\quad x = 2$$

Only 2 checks. (-2 makes a denominator 0.) The solution is 2.

37.

$$\frac{5x}{x-7} - \frac{35}{x+7} = \frac{490}{x^2-49}$$

$$\frac{5x}{x-7} - \frac{35}{x+7} = \frac{490}{(x+7)(x-7)}$$

$$\text{LCM is } (x+7)(x-7)$$

$$(x+7)(x-7)\left(\frac{5x}{x-7} - \frac{35}{x+7}\right) =$$

$$(x+7)(x-7) \cdot \frac{490}{(x+7)(x-7)}$$

$$5x(x+7) - 35(x-7) = 490$$

$$5x^2 + 35x - 35x + 245 = 490$$

$$5x^2 + 245 = 490$$

$$5x^2 - 245 = 0$$

$$5(x^2 - 49) = 0$$

$$5(x+7)(x-7) = 0$$

$$x+7 = 0 \quad or \quad x-7 = 0$$

$$x = -7 \quad or \quad\quad x = 7$$

The numbers -7 and 7 are possible solutions. Each makes a denominator 0, so the equation has no solution.

39.

$$\frac{x^2}{x^2-4} = \frac{x}{x+2} - \frac{2x}{2-x}$$

$$\frac{x^2}{(x+2)(x-2)} = \frac{x}{x+2} - \frac{2x}{2-x} \cdot \frac{-1}{-1}$$

$$\frac{x^2}{(x+2)(x-2)} = \frac{x}{x+2} - \frac{-2x}{x-2}$$

$$\text{LCM is } (x+2)(x-2)$$

$$(x+2)(x-2) \cdot \frac{x^2}{(x+2)(x-2)} =$$

$$(x+2)(x-2)\left(\frac{x}{x+2} - \frac{-2x}{x-2}\right)$$

$$x^2 = (x-2)x - (-2x)(x+2)$$

$$x^2 = x^2 - 2x + 2x^2 + 4x$$

$$0 = 2x^2 + 2x$$

$$0 = 2x(x+1)$$

$$2x = 0 \quad or \quad x+1 = 0$$

$$x = 0 \quad or \quad\quad x = -1$$

Both numbers check. The solutions are 0 and -1.

41. We find all values of x for which $2x - \dfrac{6}{x} = 1$. First note that $x \neq 0$. Then multiply on both sides by the LCD, x.

$$x\left(2x - \frac{6}{x}\right) = x \cdot 1$$

$$x \cdot 2x - x \cdot \frac{6}{x} = x$$

$$2x^2 - 6 = x$$

$$2x^2 - x - 6 = 0$$

$$(2x+3)(x-2) = 0$$

$$x = -\frac{3}{2} \quad or \quad x = 2$$

Both values check. The solutions are $-\dfrac{3}{2}$ and 2.

43. We find all values of x for which $\dfrac{x-3}{x+2} = \dfrac{1}{5}$. First note that $x \neq -2$. Then multiply on both sides by the LCD, $5(x+2)$.

$$5(x+2) \cdot \frac{x-3}{x+2} = 5(x+2) \cdot \frac{1}{5}$$
$$5(x-3) = x+2$$
$$5x - 15 = x + 2$$
$$4x = 17$$
$$x = \frac{17}{4}$$

This value checks. The solution is $\dfrac{17}{4}$.

45. We find all values of x for which $\dfrac{6}{x} - \dfrac{6}{2x} = 5$. First note that $x \neq 0$. Then multiply on both sides by the LCD, $2x$.

$$2x\left(\frac{6}{x} - \frac{6}{2x}\right) = 2x \cdot 5$$
$$2x \cdot \frac{6}{x} - 2x \cdot \frac{6}{2x} = 10x$$
$$12 - 6 = 10x$$
$$6 = 10x$$
$$\frac{3}{5} = x$$

This value checks. The solution is $\dfrac{3}{5}$.

47.
$$1 - t^6$$
$$= 1^2 - (t^3)^2$$
$$= (1 + t^3)(1 - t^3)$$
$$= (1+t)(1-t+t^2)(1-t)(1+t+t^2)$$

49.
$$a^3 - 8b^3$$
$$= a^3 - (2b)^3$$
$$= (a - 2b)(a^2 + 2ab + 4b^2)$$

51. $(x-3)(x+4) = 0$
$$x - 3 = 0 \quad or \quad x + 4 = 0$$
$$x = 3 \quad or \qquad x = -4$$
The solutions are 3 and -4.

53. $12x^2 - 11x + 2 = 0$
$$(4x - 1)(3x - 2) = 0$$
$$4x - 1 = 0 \quad or \quad 3x - 2 = 0$$
$$4x = 1 \quad or \qquad 3x = 2$$
$$x = \frac{1}{4} \quad or \qquad x = \frac{2}{3}$$
The solutions are $\dfrac{1}{4}$ and $\dfrac{2}{3}$.

55. We have the data points $(2001, 15{,}054)$ and $(2012, 4100)$.

$$\text{Rate of change} = \frac{4100 - 15{,}054}{2012 - 2001} = \frac{-10{,}954}{11} \approx -996$$

The number of new housing permits issued decreased at a rate of about 996 permits per year.

57. a) Graph $y_1 = \dfrac{x+3}{x+2} - \dfrac{x+4}{x+3}$ and $y_2 = \dfrac{x+5}{x+4} - \dfrac{x+6}{x+5}$ and use the INTERSECT feature to find the point of intersection of the graphs. It is $(-3.5, 1.\overline{3})$.

b) Set $f(x)$ equal to $g(x)$ and solve for x.

$$\frac{x+3}{x+2} - \frac{x+4}{x+3} = \frac{x+5}{x+4} - \frac{x+6}{x+5}$$

Note that $x \neq -2$ and $x \neq -3$ and $x \neq -4$ and $x \neq -5$.

$$(x+2)(x+3)(x+4)(x+5)\left(\frac{x+3}{x+2} - \frac{x+4}{x+3}\right) =$$
$$(x+2)(x+3)(x+4)(x+5)\left(\frac{x+5}{x+4} - \frac{x+6}{x+5}\right)$$
$$(x+3)(x+4)(x+5)(x+3) - (x+2)(x+4)(x+5)(x+4) =$$
$$(x+2)(x+3)(x+5)(x+5) - (x+2)(x+3)(x+4)(x+6)$$
$$x^4 + 15x^3 + 83x^2 + 201x + 180 -$$
$$(x^4 + 15x^3 + 82x^2 + 192x + 160) =$$
$$x^4 + 15x^3 + 81x^2 + 185x + 150 -$$
$$(x^4 + 15x^3 + 80x^2 + 180x + 144)$$
$$x^2 + 9x + 20 = x^2 + 5x + 6$$
$$4x = -14$$
$$x = -\frac{7}{2}$$

This value checks. When $x = -\dfrac{7}{2}$, $f(x) = g(x)$. This confirms that the graphs of $f(x)$ and $g(x)$ intersect when $x = -\dfrac{7}{2}$, or -3.5.

c) Answers will vary.

Exercise Set 5.6

RC1. If two triangles are similar, then their corresponding angles have the same measures and their corresponding sides are proportional.

RC3. An equality of ratios, $\dfrac{A}{B} = \dfrac{C}{D}$, is called a proportion.

RC5. Rate equals distance divided by time.

RC7. The numbers named in a true proportion are said to be proportional to each other.

1. Familiarize. Let $t =$ the time it will take Jose and Miguel to paint the house, working together, in hours.

Translate. Using the work principle, we get the following equation.

$$\frac{t}{28} + \frac{t}{36} = 1$$

Solve.

$$\frac{t}{28} + \frac{t}{36} = 1, \text{ LCM is } 252$$

$$252\left(\frac{t}{28} + \frac{t}{36}\right) = 252 \cdot 1$$

$$9t + 7t = 252$$

$$16t = 252$$

$$t = \frac{252}{16} = \frac{63}{4}, \text{ or } 15\frac{3}{4}$$

Check. We verify the work principle.

$$\frac{\frac{63}{4}}{28} + \frac{\frac{63}{4}}{36} = \frac{63}{4} \cdot \frac{1}{28} + \frac{63}{4} \cdot \frac{1}{36} = \frac{9}{16} + \frac{7}{16} = 1$$

The answer checks.

State. It will take Jose and Miguel $15\frac{3}{4}$ hr to paint the house, working together.

3. **Familiarize.** Let $t =$ the time it will take the machines to print the order, working together, in hours.

   Translate. Using the work principle, we get the following equation.

   $$\frac{t}{30} + \frac{t}{24} = 1$$

   Solve.

   $$\frac{t}{30} + \frac{t}{24} = 1, \text{ LCM is } 120$$

   $$120\left(\frac{t}{30} + \frac{t}{24}\right) = 120 \cdot 1$$

   $$4t + 5t = 120$$

   $$9t = 120$$

   $$t = \frac{40}{3}, \text{ or } 13\frac{1}{3}$$

   Check. We verify the work principle.

   $$\frac{\frac{40}{3}}{30} + \frac{\frac{40}{3}}{24} = \frac{40}{3} \cdot \frac{1}{30} + \frac{40}{3} \cdot \frac{1}{24} = \frac{4}{9} + \frac{5}{9} = 1$$

   The answer checks.

   State. It will take the machines $13\frac{1}{3}$ hr to print the order, working together.

5. **Familiarize.** Let $b =$ the number of hours it will take machine B to do the job. Then $b - 4 =$ the number of hours the job will take machine A.

   Translate. Using the work principle, we get the following equation.

   $$\frac{1.5}{b} + \frac{1.5}{b - 4} = 1$$

Solve.

$$\frac{1.5}{b} + \frac{1.5}{b - 4} = 1, \text{ LCM is } b(b - 4)$$

$$b(b - 4)\left(\frac{1.5}{b} + \frac{1.5}{b - 4}\right) = b(b - 4)$$

$$(b - 4)1.5 + b(1.5) = b^2 - 4b$$

$$1.5b - 6 + 1.5b = b^2 - 4b$$

$$3b - 6 = b^2 - 4b$$

$$0 = b^2 - 7b + 6$$

$$0 = (b - 6)(b - 1)$$

$$b - 6 = 0 \quad or \quad b - 1 = 0$$

$$b = 6 \quad or \quad \quad b = 1$$

If $b = 6$, then $b - 4 = 6 - 4 = 2$.

If $b = 1$, then $b - 4 = 1 - 4 = -3$. The number of hours cannot be negative, so 1 cannot be a solution.

Check. We verify the work principle.

$$\frac{1.5}{6} + \frac{1.5}{6 - 4} = \frac{1.5}{6} + \frac{1.5}{2} = 0.25 + 0.75 = 1$$

The answer checks.

State. It will take machine A 2 hr to do the job, and it will take machine B 6 hr.

7. **Familiarize.** Let $t =$ the time it will take Ryan and Ethan to clear the land, working together.

   Translate. Using the work principle, we get the following equation.

   $$\frac{t}{7.5} + \frac{t}{10.5} = 1$$

   Solve. First we multiply by $\frac{1}{10}$ on both sides to clear the decimals.

   $$\frac{1}{10}\left(\frac{t}{7.5} + \frac{t}{10.5}\right) = \frac{1}{10} \cdot 1$$

   $$\frac{t}{75} + \frac{t}{105} = \frac{1}{10}, \text{ LCM is } 525$$

   $$525\left(\frac{t}{75} + \frac{t}{105}\right) = 525 \cdot \frac{1}{10}$$

   $$7t + 5t = \frac{105}{2}$$

   $$12t = \frac{105}{2}$$

   $$t = \frac{105}{2} \cdot \frac{1}{12}$$

   $$t = \frac{35}{8}, \text{ or } 4\frac{3}{8}, \text{ or } 4.375$$

   Check. We verify the work principle.

   $$\frac{4.375}{7.5} + \frac{4.375}{10.5} = \frac{7}{12} + \frac{5}{12} = 1$$

   The answer checks.

   State. It will take Ryan and Ethan 4.375 hr, or $4\frac{3}{8}$ hr, to clear the land, working together.

9. *Familiarize*. Let $s =$ the number of minutes it will take Cole to skim the pool. Then $3s =$ the number of minutes it will take Jim to skim the pool.

***Translate*.** Using the work principle, we get the following equation.

$$\frac{6}{s} + \frac{6}{3s} = 1$$

***Solve*.** We solve the equation.

$$\frac{6}{s} + \frac{6}{3s} = 1$$

$$\frac{6}{s} + \frac{2}{s} = 1 \quad \text{Simplifying } \frac{6}{3s}$$

$$\frac{8}{s} = 1$$

$$s \cdot \frac{8}{s} = s \cdot 1$$

$$8 = s$$

If $s = 8$, then $3s = 3 \cdot 8 = 24$.

***Check*.** We verify the work principle.

$$\frac{6}{8} + \frac{6}{24} = \frac{3}{4} + \frac{1}{4} = 1$$

The answer checks.

***State*.** It will take Cole 8 min to skim the pool, and it will take Jim 24 min.

11. *Familiarize*. Let $t =$ the number of three-point field goals that will be scored in the 82-game season.

***Translate*.** We translate to a proportion.

$$\begin{array}{l} \text{Goals scored} \rightarrow \\ \text{Games} \rightarrow \end{array} \frac{9}{15} = \frac{t}{82} \begin{array}{l} \leftarrow \text{Goals scored} \\ \leftarrow \quad \text{Games} \end{array}$$

***Solve*.** We solve the proportion.

$$\frac{9}{15} = \frac{t}{82}$$

$$9 \cdot 82 = 15 \cdot t \quad \text{Equating cross products}$$

$$738 = 15t$$

$$49 \approx t$$

***Check*.** We substitute into the proportion and check cross products.

$$\frac{9}{15} = \frac{49}{82}; \; 9 \cdot 82 = 738; \; 15 \cdot 49 = 735$$

The cross products are approximately the same, so the answer checks. (Remember that we rounded the value of t.)

***State*.** The player would make about 49 three-point field goals in the 82-game season.

13. *Familiarize*. Let $l =$ the length of the model, in inches.

***Translate*.** We translate to a proportion.

$$\begin{array}{l} \text{Actual length} \rightarrow \\ \text{Model length} \rightarrow \end{array} \frac{15}{l} = \frac{7}{3.5} \begin{array}{l} \leftarrow \text{Actual width} \\ \leftarrow \text{Model width} \end{array}$$

***Solve*.** We solve the proportion.

$$\frac{15}{l} = \frac{7}{3.5}$$

$$15 \cdot 3.5 = l \cdot 7 \quad \text{Equating cross products}$$

$$52.5 = 7l$$

$$7.5 = l$$

***Check*.** We substitute into the proportion and check cross products.

$$\frac{15}{7.5} = \frac{7}{3.5}; \; 15 \cdot 3.5 = 52.5; \; 7.5 \cdot 7 = 52.5$$

The cross products are the same so the answer checks.

***State*.** The model is 7.5 in. long.

15. *Familiarize*. Let $w =$ the number of pounds the astronaut will weigh on the moon.

***Translate*.** We translate to a proportion.

$$\begin{array}{l} \text{Weight on moon} \rightarrow \\ \text{Weight on earth} \rightarrow \end{array} \frac{0.16}{1} = \frac{w}{180} \begin{array}{l} \leftarrow \text{Weight on moon} \\ \leftarrow \text{Weight on earth} \end{array}$$

***Solve*.** We solve the proportion.

$$\frac{0.16}{1} = \frac{w}{180}$$

$$0.16 \cdot 180 = 1 \cdot w \quad \text{Equating cross products}$$

$$28.8 = w$$

***Check*.** We substitute into the proportion and check cross products.

$$\frac{0.16}{1} = \frac{28.8}{180}; \; 0.16(180) = 28.8; \; 1(28.8) = 28.8$$

Since the cross products are the same, the answer checks.

***State*.** A 180-lb astronaut will weigh 28.8 lb on the moon.

17. *Familiarize*. Let $t =$ the number of trees required to produce 638 kg of coffee.

***Translate*.** We translate to a proportion.

$$\begin{array}{l} \text{Trees} \rightarrow \\ \text{Coffee} \rightarrow \end{array} \frac{14}{7.7} = \frac{t}{638} \begin{array}{l} \leftarrow \text{Trees} \\ \leftarrow \text{Coffee} \end{array}$$

***Solve*.** We solve the proportion.

$$\frac{14}{7.7} = \frac{t}{638}$$

$$14 \cdot 638 = 7.7 \cdot t \quad \text{Equating cross products}$$

$$8932 = 7.7t$$

$$1160 = t$$

***Check*.** We substitute into the proportion and check cross products.

$$\frac{14}{7.7} = \frac{1160}{638}; \; 14 \cdot 638 = 8932; \; 7.7 \times 1160 = 8932$$

The cross products are the same, so the answer checks.

***State*.** It takes the beans from 1160 trees to produce 638 kg of coffee.

19. *Familiarize*. Let $T =$ the number of tons of grapes required to produce 7850 bottles of wine.

***Translate*.** We translate to a proportion.

$$\begin{array}{l} \text{Grapes} \rightarrow \\ \text{Bottles} \rightarrow \end{array} \frac{8}{4320} = \frac{T}{7850} \begin{array}{l} \leftarrow \text{Grapes} \\ \leftarrow \text{Bottles} \end{array}$$

***Solve*.** We solve the proportion.

$$\frac{8}{4320} = \frac{T}{7850}$$

$$8 \cdot 7850 = 4320 \cdot T \quad \text{Equating cross products}$$

$$62,800 = 4320T$$

$$14.5 \approx T$$

Check. We substitute into the proportion and check cross products.

$$\frac{8}{4320} = \frac{14.5}{7850}; \; 8 \cdot 7850 = 62,800; \; 4320 \times 14.5 = 62,640$$

The cross products are approximately the same, so the answer checks. (Remember that we rounded the value of T.)

State. Producing 7850 bottles of wine will require about 14.5 tons of grapes.

21. **Familiarize**. Let $T =$ the number of trout in the lake.

Translate. We translate to a proportion.

$$
\begin{array}{l}
\text{Trout tagged} \\
\text{originally} \\
\text{Trout in} \\
\text{lake}
\end{array}
\begin{array}{l}
\rightarrow \\
\rightarrow
\end{array}
\frac{112}{T} = \frac{32}{82}
\begin{array}{l}
\leftarrow \;\text{caught later} \\
\leftarrow \;\text{Trout caught} \\
\quad\;\; \text{later}
\end{array}
$$

Solve. We solve the proportion.

$$\frac{112}{T} = \frac{32}{82}$$

$$112 \cdot 82 = T \cdot 32 \quad \text{Equating cross products}$$

$$\frac{112 \cdot 82}{32} = T$$

$$287 = T$$

Check. We substitute into the proportion and check cross products.

$$\frac{112}{287} = \frac{32}{82}; \; 112 \cdot 82 = 9184; \; 287 \cdot 32 = 9184$$

Since the cross products are the same, the answer checks.

State. There are 287 trout in the lake.

23. **Familiarize**. Let $l =$ the length of one piece of rope, in feet. Then $28 - l =$ the length of the other piece.

Translate. The ratio of the lengths is 3 to 5, so we have

$$\frac{l}{28 - l} = \frac{3}{5}.$$

Solve. We begin by equating cross products.

$$\frac{l}{28 - l} = \frac{3}{5}$$

$$l \cdot 5 = (28 - l)3$$

$$5l = 84 - 3l$$

$$8l = 84$$

$$l = \frac{21}{2}, \text{ or } 10\frac{1}{2}$$

If $l = 10\frac{1}{2}$, then $28 - l = 28 - 10\frac{1}{2} = 17\frac{1}{2}$.

Check. We check to see if the ratio of $10\frac{1}{2}$ to $17\frac{1}{2}$ is 3 to 5, or $\frac{3}{5}$.

$$\frac{10\frac{1}{2}}{17\frac{1}{2}} = \frac{\frac{21}{2}}{\frac{35}{2}} = \frac{21}{2} \cdot \frac{2}{35} = \frac{21 \cdot 2}{2 \cdot 35} = \frac{3 \cdot 7 \cdot 2}{2 \cdot 5 \cdot 7} = \frac{3}{5}$$

The answer checks.

State. The rope should be cut so that the length of one piece is $10\frac{1}{2}$ ft. Then the length of the other piece will be $17\frac{1}{2}$ ft.

25. **Familiarize**. Let $s =$ the number of kilograms of stone required for a 65 m^2 retaining wall.

Translate. We translate to a proportion.

$$
\begin{array}{l}
\text{Stone} \rightarrow \\
\text{Area} \rightarrow
\end{array}
\frac{1017}{3.2} = \frac{s}{65}
\begin{array}{l}
\leftarrow \text{Stone} \\
\leftarrow \text{Area}
\end{array}
$$

Solve. We begin by equating cross products.

$$\frac{1017}{3.2} = \frac{s}{65}$$

$$1017 \cdot 65 = 3.2 \cdot s$$

$$\frac{1017 \cdot 65}{3.2} = s$$

$$20,658 \approx s \qquad \text{Rounding}$$

Check. We substitute into the proportion and check cross products.

$$\frac{1017}{3.2} = \frac{20,658}{65}; \; 1017 \cdot 65 = 66,105;$$

$$3.2(20,658) = 66,105.6 \approx 66,105$$

The cross products are about the same, so the answer checks. (Remember that we rounded to get the answer.)

State. About 20,658 kg of stone are needed.

27. **Familiarize**. Using the labels on the drawing in the text, we let $w =$ the speed of the wind, in mph. Then $570 + w =$ the speed of the plane with the wind and $570 - w =$ the speed against the wind. We organize the information in a table.

	Distance	Speed	Time
With wind	2420	$570 + w$	$\dfrac{2420}{570 + w}$
Against wind	2140	$570 - w$	$\dfrac{2140}{570 - w}$

Translate. Since the times are the same, we have

$$\frac{2420}{570 + w} = \frac{2140}{570 - w}.$$

Solve.

$$\frac{2420}{570 + w} = \frac{2140}{570 - w},$$

$$\text{LCM is } (570 + w)(570 - w)$$

$$(570+w)(570-w) \cdot \frac{2420}{570-w} = (570+w)(570-w) \cdot \frac{2140}{570-w}$$

$$2420(570 - w) = 2140(570 + w)$$

$$1,379,400 - 2420w = 1,219,800 + 2140w$$

$$159,600 = 4560w$$

$$35 = w$$

Check. If the speed of the wind is 35 mph, then the plane's speed with the wind is $570 + 35$, or 605 mph, and the time it takes to travel 2420 mi is $\dfrac{2420}{605}$, or 4 hr. The plane's speed against the wind is $570 - 35$, or 535 mph, and the time it takes to travel 2140 mi is $\dfrac{2140}{535}$, or 4 hr. The times are the same, so the answer checks.

State. The speed of the wind is 35 mph.

29. *Familiarize*. We first make a drawing. Let r = the kayak's speed in still water in mph. Then $r - 3$ = the speed upstream and $r + 3$ = the speed downstream.

$$\text{Upstream} \quad 4 \text{ miles} \quad r - 3 \text{ mph}$$

$$10 \text{ miles} \quad r + 3 \text{ mph} \quad \text{Downstream}$$

We organize the information in a table. The time is the same both upstream and downstream so we use t for each time.

	Distance	Speed	Time
Upstream	4	$r - 3$	t
Downstream	10	$r + 3$	t

Translate. Using the formula Time = Distance/Rate in each row of the table and the fact that the times are the same, we can write an equation.

$$\frac{4}{r-3} = \frac{10}{r+3}$$

Solve. We solve the equation.

$$\frac{4}{r-3} = \frac{10}{r+3}, \text{ LCM is } (r-3)(r+3)$$

$$(r-3)(r+3) \cdot \frac{4}{r-3} = (r-3)(r+3) \cdot \frac{10}{r+3}$$

$$4(r+3) = 10(r-3)$$

$$4r + 12 = 10r - 30$$

$$42 = 6r$$

$$7 = r$$

Check. If the kayak's speed in still water is 7 mph, then the speed upstream is $7 - 3$, or 4 mph, and the time it takes to travel 4 mi is $\frac{4}{4}$, or 1 hr. The speed downstream is $7+3$, or 10 mph, and the time it takes to travel 10 mi is $\frac{10}{10}$, or 1 hr. The times are the same, so the answer checks.

State. The speed of the kayak in still water is 7 mph.

31. *Familiarize*. Let r = the speed of the express buses, in mph. Then $r - 15$ = the speed of the trolley. We organize the information in a table.

	Distance	Speed	Time
Express	132	r	t
Trolley	99	$r - 15$	t

Translate. Using the formula Time = Distance/Rate in each row of the table and the fact that the times are the same, we can write an equation.

$$\frac{132}{r} = \frac{99}{r-15}$$

Solve.

$$\frac{132}{r} = \frac{99}{r-15}, \text{ LCM is } r(r-15)$$

$$r(r-15) \cdot \frac{132}{r} = r(r-15) \cdot \frac{99}{r-15}$$

$$132(r-15) = 99r$$

$$132r - 1980 = 99r$$

$$-1980 = -33r$$

$$60 = r$$

Check. At a speed of 60 mph, the time it takes the express buses to travel 132 mi is $\frac{132}{60}$, or 2.2 hr. At a speed of $60 - 15$, or 45 mph, the time it takes the trolley to travel 99 mi is $\frac{99}{45}$, or 2.2 hr. The times are the same, so the answer checks.

State. The express buses' speed is 60 mph, and the trolley's speed is 45 mph.

33. *Familiarize*. We first make a drawing. Let r = Thomas' speed on a nonmoving sidewalk in ft/sec. Then his speed moving forward on the moving sidewalk is $r + 1.8$, and his speed in the opposite direction is $r - 1.8$.

$$\text{Forward} \quad r + 1.8 \quad 105 \text{ ft}$$

$$\text{Opposite}$$
$$51 \text{ ft} \quad r - 1.8 \quad \text{direction}$$

We organize the information in a table. The time is the same both forward and in the opposite direction so we use t for each time.

	Distance	Speed	Time
Forward	105	$r + 1.8$	t
Opposite direction	51	$r - 1.8$	t

Translate. Using the formula Time = Distance/Rate in each row of the table and the fact that the times are the same, we can write an equation.

$$\frac{105}{r+1.8} = \frac{51}{r-1.8}$$

Solve. We solve the equation.

$$\frac{105}{r+1.8} = \frac{51}{r-1.8},$$

$$\text{LCM is } (r+1.8)(r-1.8)$$

$$(r+1.8)(r-1.8) \cdot \frac{105}{r+1.8} = (r+1.8)(r-1.8) \cdot \frac{51}{r-1.8}$$

$$105(r-1.8) = 51(r+1.8)$$

$$105r - 189 = 51r + 91.8$$

$$54r = 280.8$$

$$r = 5.2$$

Check. If Thomas' speed on a nonmoving sidewalk is 5.2 ft/sec, then his speed moving forward on the moving sidewalk is $5.2 + 1.8$, or 7 ft/sec, and his speed moving in the opposite direction on the sidewalk is $5.2 - 1.8$, or

3.4 ft/sec. Moving 105 ft at 7 ft/sec takes $\dfrac{105}{7}$, or 15 sec.
Moving 51 ft at 3.4 ft/sec takes $\dfrac{51}{3.4}$, or 15 sec. Since the times are the same, the answer checks.

State. Thomas would be walking 5.2 ft/sec on a nonmoving sidewalk.

35. The domain is the set of all x-values on the graph, $[-5, 5]$. The range is the set of all y-values on the graph, $[-4, 3]$.

37. The domain is the set of all x-values on the graph, $[-5, 5]$. The range is the set of all y-values on the graph, $[-5, 3]$.

39. Graph: $x - 4y \geq 4$

We first graph the line $x - 4y = 4$. The intercepts are $(0, -1)$ and $(4, 0)$. We draw the line solid since the inequality symbol is $\geq$. To determine which half-plane to shade, we consider a test point not on the line. We try $(0, 0)$:

$$\dfrac{x - 4y \geq 4}{0 - 4 \cdot 0 \;?\; 4}$$
$$0 \;\Big|\; \text{FALSE}$$

Since $0 \geq 4$ is false, we shade the half-plane that does not contain $(0, 0)$.

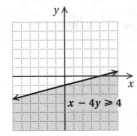

41. Graph: $f(x) = |x + 3|$

We select x-values and find the corresponding y-values. We plot these ordered pairs and draw the graph.

x	y
-5	2
-3	0
-1	2
0	3
2	5

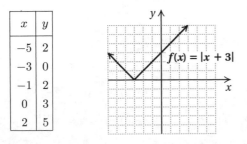

43. **Familiarize**. If the trucks can do the job in t hr working together then $t+1$, $t+6$, and $t+t$, or $2t$, represent the times it takes A, B, and C, respectively, to do the job alone.

Translate. We use the work principle.

$$\frac{t}{t+1} + \frac{t}{t+6} + \frac{t}{2t} = 1, \text{ or}$$
$$\frac{t}{t+1} + \frac{t}{t+6} + \frac{1}{2} = 1$$

Solve.

$$\frac{t}{t+1} + \frac{t}{t+6} + \frac{1}{2} = 1$$
$$\frac{t}{t+1} + \frac{t}{t+6} = \frac{1}{2}, \text{ LCM is } 2(t+1)(t+6)$$
$$2(t+1)(t+6)\left(\frac{t}{t+1} + \frac{t}{t+6}\right) = 2(t+1)(t+6) \cdot \frac{1}{2}$$
$$2t(t+6) + 2t(t+1) = (t+1)(t+6)$$
$$2t^2 + 12t + 2t^2 + 2t = t^2 + 7t + 6$$
$$4t^2 + 14t = t^2 + 7t + 6$$
$$3t^2 + 7t - 6 = 0$$
$$(3t - 2)(t + 3) = 0$$
$$3t - 2 = 0 \quad or \quad t + 3 = 0$$
$$3t = 2 \quad or \qquad t = -3$$
$$t = \frac{2}{3} \quad or \qquad t = -3$$

Check. The time cannot be negative, so we check only $\dfrac{2}{3}$. We verify the work principle.

$$\frac{\frac{2}{3}}{\frac{2}{3}+1} + \frac{\frac{2}{3}}{\frac{2}{3}+6} + \frac{1}{2} = \frac{\frac{2}{3}}{\frac{5}{3}} + \frac{\frac{2}{3}}{\frac{20}{3}} + \frac{1}{2} =$$
$$\frac{2}{3} \cdot \frac{3}{5} + \frac{2}{3} \cdot \frac{3}{20} + \frac{1}{2} = \frac{2}{5} + \frac{1}{10} + \frac{1}{2} = \frac{4}{10} + \frac{1}{10} + \frac{5}{10} =$$
$$\frac{10}{10} = 1$$

The answer checks.

State. We find that t is $\dfrac{2}{3}$ hr.

45. **Familiarize**. Let $x = $ the number of miles driven in the city and $y = $ the number of miles driven on the highway. Then $\dfrac{x}{22.5} = $ the number of gallons of gasoline used in city driving and $\dfrac{y}{30} = $ the number of gallons used in highway driving.

Translate. The total number of miles is 465, so we have one equation:

$$x + y = 465$$

The total amount of gasoline used is 18.4 gal, so we have a second equation.

$$\frac{x}{22.5} + \frac{y}{30} = 18.4$$

After clearing fractions, we have the following system of equations:

$$x + y = 465,$$
$$30x + 22.5y = 12,420$$

Solve.   The solution of the system of equations is $(261, 204)$.

Check. If 261 city miles and 204 highway miles are driven, the total number of miles is $261 + 204$, or 465. In 261 city miles $261/22.5$ or 11.6 gal, of gasoline are used; in 204 highway miles $204/30$, or 6.8 gal, are used. Then the total amount of gasoline used is $11.6 + 6.8$, or 18.4 gal. The answer checks.

State. 261 mi were driven in the city, and 204 mi were driven on the highway.

Exercise Set 5.7

RC1.
$$\frac{x}{y} = \frac{w}{t}$$
$$yt \cdot \frac{x}{y} = yt \cdot \frac{w}{t}$$
$$xt = yw$$
$$\frac{xt}{w} = y, \text{ or } y = \frac{xt}{w}$$
The answer is (c).

RC3.
$$\frac{1}{w} = \frac{1}{x} + \frac{1}{y}$$
$$wxy \cdot \frac{1}{w} = wxy\left(\frac{1}{x} + \frac{1}{y}\right)$$
$$xy = wxy \cdot \frac{1}{x} + wxy \cdot \frac{1}{y}$$
$$xy = wy + wx$$
$$xy - wy = wx$$
$$y(x - w) = wx$$
$$y = \frac{wx}{x - w}$$
The answer is (e).

1.
$$\frac{W_1}{W_2} = \frac{d_1}{d_2}$$
$$W_2 d_2 \cdot \frac{W_1}{W_2} = W_2 d_2 \cdot \frac{d_1}{d_2} \quad \text{Multiplying by the LCM}$$
$$d_2 W_1 = W_2 d_1$$
$$\frac{d_2 W_1}{d_1} = W_2 \quad \text{Dividing by } d_1$$

3.
$$\frac{1}{R} = \frac{1}{r_1} + \frac{1}{r_2}$$
$$R r_1 r_2 \cdot \frac{1}{R} = R r_1 r_2 \left(\frac{1}{r_1} + \frac{1}{r_2}\right) \quad \begin{array}{l}\text{Multiplying by} \\ \text{the LCM}\end{array}$$
$$r_1 r_2 = R r_2 + R r_1$$
$$r_1 r_2 - R r_2 = R r_1 \quad \text{Subtracting } R r_2$$
$$r_2(r_1 - R) = R r_1 \quad \text{Factoring}$$
$$r_2 = \frac{R r_1}{r_1 - R} \quad \text{Dividing by } r_1 - R$$

5.
$$s = \frac{(v_1 + v_2)t}{2}$$
$$2 \cdot s = 2 \cdot \frac{(v_1 + v_2)t}{2} \quad \text{Multiplying by 2}$$
$$2s = (v_1 + v_2)t$$
$$\frac{2s}{v_1 + v_2} = t \quad \text{Dividing by } v_1 + v_2$$

7.
$$R = \frac{gs}{g + s}$$
$$(g + s) \cdot R = (g + s) \cdot \frac{gs}{g + s} \quad \text{Multiplying by } g + s$$
$$R(g + s) = gs$$
$$Rg + Rs = gs \quad \text{Removing parentheses}$$
$$Rg = gs - Rs \quad \text{Subtracting } Rs$$
$$Rg = s(g - R) \quad \text{Factoring}$$
$$\frac{Rg}{g - R} = s \quad \text{Dividing by } g - R$$

9.
$$\frac{1}{p} + \frac{1}{q} = \frac{1}{f}$$
$$pqf\left(\frac{1}{p} + \frac{1}{q}\right) = pqf \cdot \frac{1}{f} \quad \text{Multiplying by } pqf$$
$$pqf \cdot \frac{1}{p} + pqf \cdot \frac{1}{q} = pq$$
$$qf + pf = pq$$
$$qf = pq - pf \quad \text{Subtracting } pf$$
$$qf = p(q - f) \quad \text{Factoring}$$
$$\frac{qf}{q - f} = p \quad \text{Dividing by } q - f$$

11.
$$\frac{t}{a} + \frac{t}{b} = 1$$
$$ab\left(\frac{t}{a} + \frac{t}{b}\right) = ab \cdot 1 \quad \text{Multiplying by } ab$$
$$ab \cdot \frac{t}{a} + ab \cdot \frac{t}{b} = ab \quad \text{Removing parentheses}$$
$$bt + at = ab$$
$$bt = ab - at \quad \text{Subtracting } at$$
$$bt = a(b - t) \quad \text{Factoring}$$
$$\frac{bt}{b - t} = a \quad \text{Dividing by } b - t$$

13.
$$I = \frac{nE}{E + nr}$$
$$(E + nr)I = (E + nr) \cdot \frac{nE}{E + nr} \quad \begin{array}{l}\text{Multiplying by} \\ E + nr\end{array}$$
$$EI + nrI = nE$$
$$nrI = nE - EI \quad \text{Subtracting } EI$$
$$nrI = E(n - I) \quad \text{Factoring}$$
$$\frac{nrI}{n - I} = E \quad \text{Dividing by } n - I$$

15.
$$I = \frac{704.5W}{H^2}$$
$$H^2 I = 704.5W \quad \text{Multiplying by } H^2$$
$$H^2 = \frac{704.5W}{I} \quad \text{Dividing by } I$$

17.
$$\frac{E}{e} = \frac{R+r}{r}$$

$$er \cdot \frac{E}{e} = er \cdot \frac{R+r}{r} \quad \text{Multiplying by } er$$

$$Er = eR + er$$

$$Er - er = eR \qquad \text{Subtracting } er$$

$$r(E - e) = eR \qquad \text{Factoring}$$

$$r = \frac{eR}{E-e} \qquad \text{Dividing by } E - e$$

19.
$$V = \frac{1}{3}\pi h^2(3R - h)$$

$$\frac{3}{\pi h^2} \cdot V = \frac{3}{\pi h^2} \cdot \frac{1}{3}\pi h^2(3R - h) \quad \text{Multiplying by } \frac{3}{\pi h^2}$$

$$\frac{3V}{\pi h^2} = 3R - h$$

$$\frac{3V}{\pi h^2} + h = 3R \qquad \text{Adding } h$$

$$\frac{1}{3}\left(\frac{3V}{\pi h^2} + h\right) = \frac{1}{3} \cdot 3R \qquad \text{Multiplying by } \frac{1}{3}$$

$$\frac{V}{\pi h^2} + \frac{h}{3} = R, \text{ or}$$

$$\frac{3V + \pi h^3}{3\pi h^2} = R$$

21.
$$S = 2\pi rh + 2\pi r^2$$

$$S - 2\pi r^2 = 2\pi rh \qquad \text{Subtracting } 2\pi r^2$$

$$\frac{S - 2\pi r^2}{2\pi r} = h \qquad \text{Dividing by } 2\pi r$$

23.
$$v = \frac{d_2 - d_1}{t_2 - t_1}$$

$$(t_2 - t_1)v = (t_2 - t_1) \cdot \frac{d_2 - d_1}{t_2 - t_1} \quad \text{Multiplying by } t_2 - t_1$$

$$(t_2 - t_1)v = d_2 - d_1$$

$$t_2 - t_1 = \frac{d_2 - d_1}{v} \qquad \text{Dividing by } v$$

$$t_2 = \frac{d_2 - d_1}{v} + t_1, \text{ or } \frac{d_2 - d_1 + t_1 v}{v}$$

25. Familiarize. Let d, n, and q represent the number of rolls of dimes, nickels and quarters, respectively. The value of a roll of dimes is 50($0.10), or $5; the value of a roll of nickels is 40($0.05), or $2; and the value of a roll of quarters is 40($0.25), or $10. Then the values of d rolls of dimes, n rolls of nickels and q rolls of quarters are $5d$, $2n$, and $10q$, respectively.

Translate. There are 12 rolls of coins, so we have

$$d + n + q = 12.$$

The total value of the coins is $70, so we have

$$5d + 2n + 10q = 70.$$

There are 3 more rolls of nickels than dimes, so we have a third equation

$$n = d + 3.$$

We have a system of equations:
$$d + n + q = 12,$$
$$5d + 2n + 10q = 70,$$
$$n = d + 3.$$

Solve. Solving the system of equations, we obtain $(2, 5, 5)$.

Check. The number of rolls of coins is $2 + 5 + 5$, or 12. The total value of the coins is $2 \cdot \$5 \cdot 5 \cdot \$2 + 5 \cdot \$10$, or $\$10 + \$10 + \$50$, or $70. Also, the number of rolls of nickels, 5, is 3 more than the number of rolls of dimes, 2. The answer checks.

State. There were 2 rolls of dimes, 5 rolls of nickels, and 5 rolls of quarters.

27. $f(x) = x^3 - x$
$$f(2) = 2^3 - 2 = 8 - 2 = 6$$

29. $f(x) = x^3 - x$
$$f(2a) = (2a)^3 - 2a = 8a^3 - 2a$$

31. First we find the slope.
$$m = \frac{5 - (-3)}{-2 - 8} = \frac{8}{-10} = -\frac{4}{5}$$

Now we use the slope and one of the given points to find b. We will use $(-2, 5)$. Substitute -2 for x, 5 for y, and $-\frac{4}{5}$ for m in $y = mx + b$ and solve for b.

$$y = mx + b$$
$$5 = -\frac{4}{5}(-2) + b$$
$$5 = \frac{8}{5} + b$$
$$\frac{17}{5} = b$$

Use $y = mx + b$ again, substituting $-\frac{4}{5}$ for m and $\frac{17}{5}$ for b. The equation is

$$y = -\frac{4}{5}x + \frac{17}{5}.$$

Exercise Set 5.8

RC1. (f)

RC3. (h)

RC5. (c)

RC7. (g)

1. $y = kx$

$40 = k \cdot 8$ Substituting

$5 = k$ Solving for k

The variation constant is 5.

The equation of variation is $y = 5x$.

3. $y = kx$

$4 = k \cdot 30$ Substituting

$\dfrac{4}{30} = k,$ or   Solving for k

$\dfrac{2}{15} = k$ Simplifying

The variation constant is $\dfrac{2}{15}$.

The equation of variation is $y = \dfrac{2}{15}x$.

5. $y = kx$

$0.9 = k \cdot 0.4$ Substituting

$\dfrac{0.9}{0.4} = k,$ or

$\dfrac{9}{4} = k$

The variation constant is $\dfrac{9}{4}$.

The equation of variation is $y = \dfrac{9}{4}x$.

7. First we find k.

$T = kW$ T varies directly as W

$75 = k \cdot 1500$ Substituting

$0.05 = k$ Variation constant

$T = 0.05W$ Equation of variation

$T = 0.05(3500)$ Substituting

$T = 175$

To ship 3500 tons of metal, 175 trucks are needed.

9. Let $F =$ the number of grams of fat and $w =$ the weight.

$F = kw$ F varies directly as w.

$60 = k \cdot 120$ Substituting

$\dfrac{60}{120} = k,$ or   Solving for k

$\dfrac{1}{2} = k$ Variation constant

$F = \dfrac{1}{2}w$ Equation of variation

$F = \dfrac{1}{2} \cdot 180$ Substituting

$F = 90$

The maximum daily fat intake for a person weighing 180 lb is 90 g.

11. Let $m =$ the mass of the body.

$W = km$ W varies directly as m.

$64 = k \cdot 96$ Substituting

$\dfrac{64}{96} = k$ Solving for k

$\dfrac{2}{3} = k$ Variation constant

$W = \dfrac{2}{3}m$ Equation of variation

$W = \dfrac{2}{3} \cdot 60$ Substituting

$W = 40$

There are 40 kg of water in a 60-kg person.

13. Let $p =$ the number of people using the cans.

$N = kp$ N varies directly as p.

$60,000 = k \cdot 250$ Substituting

$\dfrac{60,000}{250} = k$ Solving for k

$240 = k$ Variation constant

$N = 240p$ Equation of variation

$N = 240(318,172)$ Substituting

$N = 76,361,280$

In St. Louis 76,361,280 cans are used each year.

15. $y = \dfrac{k}{x}$

$14 = \dfrac{k}{7}$ Substituting

$7 \cdot 14 = k$ Solving for k

$98 = k$

The variation constant is 98.

The equation of variation is $y = \dfrac{98}{x}$.

17. $y = \dfrac{k}{x}$

$3 = \dfrac{k}{12}$ Substituting

$12 \cdot 3 = k$ Solving for k

$36 = k$

The variation constant is 36.

The equation of variation is $y = \dfrac{36}{x}$.

19. $y = \dfrac{k}{x}$

$0.1 = \dfrac{k}{0.5}$ Substituting

$0.5(0.1) = k$ Solving for k

$0.05 = k$

The variation constant is 0.05.

The equation of variation is $y = \dfrac{0.05}{x}$.

21. $T = \dfrac{k}{P}$ T varies inversely as P.

$5 = \dfrac{k}{7}$ Substituting

$35 = k$ Variation constant

$T = \dfrac{35}{P}$ Equation of variation

$T = \dfrac{35}{10}$ Substituting

$T = 3.5$

It will take 10 bricklayers 3.5 hr to complete the job.

23. $I = \dfrac{k}{R}$ I varies inversely as R.

$\dfrac{1}{2} = \dfrac{k}{240}$ Substituting

$240 \cdot \dfrac{1}{2} = k$

$120 = k$ Variation constant

$I = \dfrac{120}{R}$ Equation of variation

$I = \dfrac{120}{540}$ Substituting

$I = \dfrac{2}{9}$

When the resistance is 540 ohms, the current is $\dfrac{2}{9}$ ampere.

25. $W = \dfrac{k}{L}$ W varies inversely as L

$1200 = \dfrac{k}{12}$ Substituting

$1200 \cdot 12 = k$

$14,400 = k$ Variation constant

$W = \dfrac{14,400}{L}$ Equation of variation

$W = \dfrac{14,400}{15}$ Substituting

$W = 960$

A 15-ft beam can support 960 kg.

27. $t = \dfrac{k}{r}$ t varies inversely as r

$5 = \dfrac{k}{80}$ Substituting

$5 \cdot 80 = k$

$400 = k$ Variation constant

$t = \dfrac{400}{r}$ Equation of variation

$t = \dfrac{400}{70}$ Substituting

$t = \dfrac{40}{7}$, or $5\dfrac{5}{7}$

It takes $5\dfrac{5}{7}$ hr to drive the distance at 70 km/h.

29. $y = kx^2$

$0.15 = k(0.1)^2$ Substituting

$0.15 = 0.01k$

$\dfrac{0.15}{0.01} = k$

$15 = k$

The equation of variation is $y = 15x^2$.

31. $y = \dfrac{k}{x^2}$

$0.15 = \dfrac{k}{(0.1)^2}$ Substituting

$0.15 = \dfrac{k}{0.01}$

$0.15(0.01) = k$

$0.0015 = k$

The equation of variation is $y = \dfrac{0.0015}{x^2}$.

33. $y = kxz$

$56 = k \cdot 7 \cdot 8$ Substituting

$56 = 56k$

$1 = k$

The equation of variation is $y = xz$.

35. $y = kxz^2$

$105 = k \cdot 14 \cdot 5^2$ Substituting

$105 = 350k$

$\dfrac{105}{350} = k$

$\dfrac{3}{10} = k$

The equation of variation is $y = \dfrac{3}{10}xz^2$.

37. $y = k\dfrac{xz}{wp}$

$\dfrac{3}{28} = k\dfrac{3 \cdot 10}{7 \cdot 8}$ Substituting

$\dfrac{3}{28} = k \cdot \dfrac{30}{56}$

$\dfrac{3}{28} \cdot \dfrac{56}{30} = k$

$\dfrac{1}{5} = k$

The equation of variation is $y = \dfrac{xz}{5wp}$.

39. $I = \dfrac{k}{d^2}$

$90 = \dfrac{k}{5^2}$ Substituting

$90 = \dfrac{k}{25}$

$2250 = k$

The equation of variation is $I = \dfrac{2250}{d^2}$.

Substitute 40 for I and find d.

$40 = \dfrac{2250}{d^2}$

$40d^2 = 2250$

$d^2 = 56.25$

$d = 7.5$

The distance from 5 m to 7.5 m is $7.5 - 5$, or 2.5 m, so it is 2.5 m further to a point where the intensity is 40 W/m^2.

41.

$$W = \frac{k}{d^2}$$

$$220 = \frac{k}{(3978)^2} \qquad \text{Substituting}$$

$$220 = \frac{k}{15,824,484}$$

$$3,481,386,480 = k$$

$$W = \frac{3,481,386,480}{d^2}$$

When the astronaut is 200 mi above the surface of the earth, he is $3978 + 200$, or 4178 mi, from the center of the earth. Substitute 4178 for d and find W.

$$W = \frac{3,481,386,480}{(4178)^2}$$

$$W = \frac{3,481,386,480}{17,455,684}$$

$$W \approx 199.4$$

The astronaut's weight is about 199.4 lb when he is 200 mi above the surface of the earth.

43. $E = \frac{kR}{I}$

We first find k.

$$3.75 = \frac{k \cdot 89}{213.1}$$

$$\frac{213.1(3.75)}{89} = k$$

$$8.98 \approx k$$

The equation of variation is $E = \frac{8.98R}{I}$.

Substitute 3.75 for E and 235 for I and solve for R.

$$3.75 = \frac{8.98R}{235}$$

$$\frac{235(3.75)}{8.98} = R$$

$$98 \approx R$$

Jon Lester would have given up about 98 earned runs if he had pitched 235 innings.

45. $Q = kd^2$

We first find k.

$$225 = k \cdot 5^2$$

$$225 = 25k$$

$$9 = k$$

The equation of variation is $Q = 9d^2$.

Substitute 9 for d and compute Q.

$$Q = 9 \cdot 9^2$$

$$Q = 9 \cdot 81$$

$$Q = 729$$

729 gallons of water are emptied by a pipe that is 9 in. in diameter.

47. Interval notation for $\{y | y \geq 8\}$ is $[8, \infty)$.

49. Interval notation for $\{t | -5 \leq t < 15\}$ is $[-5, 15)$.

51. Interval notation for $\left\{q | q < -\frac{1}{2}\right\}$ is $\left(-\infty, -\frac{1}{2}\right)$.

53. $3x - y = 7$, (1)

$y = 1 - x$ (2)

Substitute $1 - x$ for y in Equation (1) and solve for x.

$$3x - y = 7$$

$$3x - (1 - x) = 7$$

$$3x - 1 + x = 7$$

$$4x - 1 = 7$$

$$4x = 8$$

$$x = 2$$

Now substitute 2 for x in Equation (2) and find y.

$$y = 1 - x = 1 - 2 = -1$$

The solution is $(2, -1)$.

55. a) $7xy = 14$

$$y = \frac{2}{x}$$

Inversely

b) $x - 2y = 12$

$$y = \frac{x}{2} - 6$$

Neither

c) $-2x + 3y = 0$

$$3y = 2x$$

$$y = \frac{2}{3}x$$

Directly

d) $x = \frac{3}{4}y$

$$y = \frac{4}{3}x$$

Directly

Chapter 5 Vocabulary Reinforcement

1. An equality of ratios, $A/B = C/D$, is called a <u>proportion</u>.

2. An expression that consists of the quotient of two polynomials, where the polynomial in the denominator is nonzero, is called a <u>rational</u> expression.

3. A <u>rational</u> equation is an equation containing one or more rational expressions.

4. If a situation gives rise to a function $f(x) = k/x$, or $y = k/x$, where k is a <u>positive</u> constant, we say that we have <u>inverse</u> variation.

5. A <u>complex</u> rational expression is a rational expression that contains rational expressions within its numerator and/or its denominator.

6. If a situation gives rise to a linear function $f(x) = kx$, or $y = kx$, where k is a <u>positive</u> constant, we say that we have <u>direct</u> variation. The <u>number</u> k is called the variation <u>constant</u>.

Chapter 5 Concept Reinforcement

1. True; see page 472 in the text.

2. True; see page 445 in the text.

Chapter 5 Study Guide

1. $f(x) = \dfrac{x^2 + 3x - 28}{x^2 + 3x - 54}$

We set the denominator equal to 0 and solve.

$$x^2 + 3x - 54 = 0$$
$$(x + 9)(x - 6) = 0$$
$$x + 9 = 0 \quad or \quad x - 6 = 0$$
$$x = -9 \quad or \qquad x = 6$$

The domain is $\{x | x \text{ is a real number } and\ x \neq -9\ and\ x \neq 6\}$, or $(-\infty, -9) \cup (-9, 6) \cup (6, \infty)$.

2. $\dfrac{b^2 - 9}{b^2 - 5b - 24} = \dfrac{(b+3)(b-3)}{(b-8)(b+3)} = \dfrac{b+3}{b+3} \cdot \dfrac{b-3}{b-8} = \dfrac{b-3}{b-8}$

3. $\dfrac{w^3 - 125}{w^3 + 8w^2 + 15w} \div \dfrac{w - 5}{w^3 - 25w}$

$= \dfrac{w^3 - 125}{w^3 + 8w^2 + 15w} \cdot \dfrac{w^3 - 25w}{w - 5}$

$= \dfrac{(w \cancel{-5})(w^2 + 5w + 25)(\cancel{w})(w \cancel{+5})(w - 5)}{\cancel{w}(w \cancel{+5})(w + 3)(w \cancel{-5})}$

$= \dfrac{(w^2 + 5w + 25)(w - 5)}{w + 3}$

4. $x^4 = x \cdot x \cdot x \cdot x$

$x^5 - 9x^3 = x^3(x^2 - 9) = x \cdot x \cdot x(x + 3)(x - 3)$

$2x^2 + 11x + 15 = (2x + 5)(x + 3)$

$\text{LCM} = x^4(x + 3)(x - 3)(2x + 5)$

5. $\dfrac{r + s}{r^2 + rs - 2s^2} - \dfrac{5s}{r^2 - s^2}$

$= \dfrac{r + s}{(r + 2s)(r - s)} - \dfrac{5s}{(r + s)(r - s)}$

$\qquad\qquad\qquad \text{LCM is } (r + 2s)(r - s)(r + s)$

$= \dfrac{r + s}{(r + 2s)(r - s)} \cdot \dfrac{r + s}{r + s} - \dfrac{5s}{(r + s)(r - s)} \cdot \dfrac{r + 2s}{r + 2s}$

$= \dfrac{(r + s)(r + s) - 5s(r + 2s)}{(r + 2s)(r - s)(r + s)}$

$= \dfrac{r^2 + 2rs + s^2 - 5rs - 10s^2}{(r + 2s)(r - s)(r + s)}$

$= \dfrac{r^2 - 3rs - 9s^2}{(r + 2s)(r - s)(r + s)}$

6.
$$
\begin{array}{r}
y - 4 \\
y - 1 \overline{\smash{)}\, y^2 - 5y + 9} \\
\underline{y^2 - y} \\
-4y + 9 \\
\underline{-4y + 4} \\
5
\end{array}
$$

The answer is $y - 4$, R 5, or $y - 4 + \dfrac{5}{y - 1}$.

7. $(x^3 - 5x^2 - 1) \div x + 3 = (x^3 - 5x^2 - 1) \div [x - (-3)]$

$$
\begin{array}{r|rrrr}
-3 & 1 & -5 & 0 & -1 \\
   &   & -3 & 24 & -72 \\
\hline
   & 1 & -8 & 24 & -73
\end{array}
$$

The answer is $x^2 - 8x + 24$, R -73, or $x^2 - 8x + 24 + \dfrac{-73}{x + 3}$.

8. $\dfrac{\dfrac{2}{a} + \dfrac{8}{b}}{\dfrac{8}{a} - \dfrac{2}{b}} = \dfrac{\dfrac{2}{a} + \dfrac{8}{b}}{\dfrac{8}{a} - \dfrac{2}{b}} \cdot \dfrac{ab}{ab}$

$= \dfrac{\dfrac{2}{a} \cdot ab + \dfrac{8}{b} \cdot ab}{\dfrac{8}{a} \cdot ab - \dfrac{2}{b} \cdot ab}$

$= \dfrac{2b + 8a}{8b - 2a}$

$= \dfrac{2(b + 4a)}{2(4b - a)}$

$= \dfrac{b + 4a}{4b - a}$

9. $\dfrac{5}{x - 4} - \dfrac{3}{x + 5} = \dfrac{4}{x^2 + x - 20}$

$\dfrac{5}{x - 4} - \dfrac{3}{x + 5} = \dfrac{4}{(x + 5)(x - 4)}$

$\qquad\qquad\qquad \text{LCM is } (x + 5)(x - 4)$

$(x+5)(x-4)\left(\dfrac{5}{x-4} - \dfrac{3}{x+5}\right) = (x+5)(x-4) \cdot \dfrac{4}{(x+5)(x-4)}$

$5(x + 5) - 3(x - 4) = 4$

$5x + 25 - 3x + 12 = 4$

$2x + 37 = 4$

$2x = -33$

$x = -\dfrac{33}{2}$

The number $-\dfrac{33}{2}$ checks. It is the solution.

10. $\qquad y = kx$

$\qquad 62 = k \cdot \dfrac{2}{3}$

$\dfrac{3}{2} \cdot 62 = k$

$\qquad 93 = k \qquad$ Variation constant

$\qquad y = 93x \qquad$ Equation of variation

11.
$$y = \frac{k}{x}$$

$$\frac{3}{10} = \frac{k}{15}$$

$$15 \cdot \frac{3}{10} = k$$

$$\frac{9}{2} = k \qquad \text{Variation constant}$$

$$y = \frac{\frac{9}{2}}{x}, \text{ or } y = \frac{9}{2x} \qquad \text{Equation of variation}$$

Chapter 5 Review Exercises

1. $\dfrac{x^2 - 3x + 2}{x^2 - 9}$

We set the denominator equal to 0 and solve.
$$x^2 - 9 = 0$$
$$(x + 3)(x - 3) = 0$$
$$x + 3 = 0 \quad or \quad x - 3 = 0$$
$$x = -3 \quad or \qquad x = 3$$

The expression is not defined for the numbers -3 and 3.

2. In Exercise 1 we found that the replacements for which $\dfrac{x^2 - 3x + 2}{x^2 - 9}$ is not defined are -3 and 3. Then the domain of $f(x)$ is $\{x | x$ is a real number $and \ x \neq -3 \ and \ x \neq 3\}$, or $(-\infty, -3) \cup (-3, 3) \cup (3, \infty)$.

3. $\dfrac{4x^2 - 7x - 2}{12x^2 + 11x + 2}$

$$= \frac{(4x + 1)(x - 2)}{(4x + 1)(3x + 2)}$$

$$= \frac{(4x + 1)(x - 2)}{(4x + 1)(3x + 2)}$$

$$= \frac{x - 2}{3x + 2}$$

4. $\dfrac{a^2 + 2a + 4}{a^3 - 8} = \dfrac{a^2 + 2a + 4}{(a - 2)(a^2 + 2a + 4)}$

$$= \frac{a^2 + 2a + 4}{a^2 + 2a + 4} \cdot \frac{1}{a - 2}$$

$$= \frac{1}{a - 2}$$

5. $6x^3 = 2 \cdot 3 \cdot x \cdot x \cdot x$

$16x^2 = 2 \cdot 2 \cdot 2 \cdot 2 \cdot x \cdot x$

$\text{LCM} = 2 \cdot 2 \cdot 2 \cdot 2 \cdot 3 \cdot x \cdot x \cdot x, \text{ or } 48x^3$

6. $x^2 - 49 = (x + 7)(x - 7)$

$3x + 1 = 3x + 1$

$\text{LCM} = (x + 7)(x - 7)(3x + 1)$

7. $x^2 + x - 20 = (x + 5)(x - 4)$

$x^2 + 3x - 10 = (x + 5)(x - 2)$

$\text{LCM} = (x + 5)(x - 4)(x - 2)$

8. $\dfrac{y^2 - 64}{2y + 10} \cdot \dfrac{y + 5}{y + 8} = \dfrac{(y^2 - 64)(y + 5)}{(2y + 10)(y + 8)}$

$$= \frac{(y + 8)(y - 8)(y + 5)}{2(y + 5)(y + 8)}$$

$$= \frac{(y + 8)(y - 8)(y + 5)}{2(y + 5)(y + 8)}$$

$$= \frac{y - 8}{2}$$

9. $\dfrac{x^3 - 8}{x^2 - 25} \cdot \dfrac{x^2 + 10x + 25}{x^2 + 2x + 4}$

$$= \frac{(x^3 - 8)(x^2 + 10x + 25)}{(x^2 - 25)(x^2 + 2x + 4)}$$

$$= \frac{(x - 2)(x^2 + 2x + 4)(x + 5)(x + 5)}{(x + 5)(x - 5)(x^2 + 2x + 4)}$$

$$= \frac{(x^2 + 2x + 4)(x + 5)}{(x^2 + 2x + 4)(x + 5)} \cdot \frac{(x - 2)(x + 5)}{x - 5}$$

$$= \frac{(x - 2)(x + 5)}{x - 5}$$

10. $\dfrac{9a^2 - 1}{a^2 - 9} \div \dfrac{3a + 1}{a + 3} = \dfrac{9a^2 - 1}{a^2 - 9} \cdot \dfrac{a + 3}{3a + 1}$

$$= \frac{(9a^2 - 1)(a + 3)}{(a^2 - 9)(3a + 1)}$$

$$= \frac{(3a + 1)(3a - 1)(a + 3)}{(a + 3)(a - 3)(3a + 1)}$$

$$= \frac{(3a + 1)(3a - 1)(a + 3)}{(a + 3)(a - 3)(3a + 1)}$$

$$= \frac{3a - 1}{a - 3}$$

11. $\dfrac{x^3 - 64}{x^2 - 16} \div \dfrac{x^2 + 5x + 6}{x^2 - 3x - 18}$

$$= \frac{x^3 - 64}{x^2 - 16} \cdot \frac{x^2 - 3x - 18}{x^2 + 5x + 6}$$

$$= \frac{(x^3 - 64)(x^2 - 3x - 18)}{(x^2 - 16)(x^2 + 5x + 6)}$$

$$= \frac{(x - 4)(x^2 + 4x + 16)(x - 6)(x + 3)}{(x + 4)(x - 4)(x + 2)(x + 3)}$$

$$= \frac{(x - 4)(x^2 + 4x + 16)(x - 6)(x + 3)}{(x + 4)(x - 4)(x + 2)(x + 3)}$$

$$= \frac{(x^2 + 4x + 16)(x - 6)}{(x + 4)(x + 2)}$$

12.
$$\frac{x}{x^2+5x+6} - \frac{2}{x^2+3x+2}$$
$$= \frac{x}{(x+2)(x+3)} - \frac{2}{(x+1)(x+2)}$$
$$\text{LCD} = (x+2)(x+3)(x+1)$$
$$= \frac{x}{(x+2)(x+3)} \cdot \frac{x+1}{x+1} - \frac{2}{(x+1)(x+2)} \cdot \frac{x+3}{x+3}$$
$$= \frac{x^2+x-(2x+6)}{(x+2)(x+3)(x+1)}$$
$$= \frac{x^2+x-2x-6}{(x+2)(x+3)(x+1)}$$
$$= \frac{x^2-x-6}{(x+2)(x+3)(x+1)}$$
$$= \frac{(x+2)(x-3)}{(x+2)(x+3)(x+1)}$$
$$= \frac{(x\!\!\not{+}2)(x-3)}{(x\!\!\not{+}2)(x+3)(x+1)}$$
$$= \frac{x-3}{(x+3)(x+1)}$$

13.
$$\frac{2x^2}{x-y} + \frac{2y^2}{x+y} \quad \text{LCD} = (x-y)(x+y)$$
$$= \frac{2x^2}{x-y} \cdot \frac{x+y}{x+y} + \frac{2y^2}{x+y} \cdot \frac{x-y}{x-y}$$
$$= \frac{2x^3+2x^2y+2xy^2-2y^3}{(x-y)(x+y)}$$

14.
$$\frac{3}{y+4} - \frac{y}{y-1} + \frac{y^2+3}{y^2+3y-4}$$
$$= \frac{3}{y+4} - \frac{y}{y-1} + \frac{y^2+3}{(y-1)(y+4)}$$
$$\text{LCD} = (y+4)(y-1)$$
$$= \frac{3}{y+4} \cdot \frac{y-1}{y-1} - \frac{y}{y-1} \cdot \frac{y+4}{y+4} + \frac{y^2+3}{(y-1)(y+4)}$$
$$= \frac{3y-3-(y^2+4y)+y^2+3}{(y+4)(y-1)}$$
$$= \frac{3y-3-y^2-4y+y^2+3}{(y+4)(y-1)}$$
$$= \frac{-y}{(y+4)(y-1)}$$

15.
$$(16ab^3c - 10ab^2c^2 + 12a^2b^2c) \div (4ab)$$
$$= \frac{16ab^3c - 10ab^2c^2 + 12a^2b^2c}{4ab}$$
$$= \frac{16ab^3c}{4ab} - \frac{10ab^2c^2}{4ab} + \frac{12a^2b^2c}{4ab}$$
$$= 4b^2c - \frac{5}{2}bc^2 + 3abc$$

16.
$$\begin{array}{r} y - 14 \\ y-6 \overline{)\ y^2 - 20y + 64\ } \\ \underline{y^2 - 6y} \\ -14y + 64 \\ \underline{-14y + 84} \\ -20 \end{array}$$

The answer is $y-14$, R -20, or $y-14+\dfrac{-20}{y-6}$.

17.
$$\begin{array}{r} 6x^2 \qquad\qquad - 9 \\ x^2+2 \overline{)\ 6x^4 + 0x^3 + 3x^2 + 5x + 4\ } \\ \underline{6x^4 \qquad + 12x^2} \\ -9x^2 \\ \underline{-9x^2 \qquad - 18} \\ 5x + 22 \end{array}$$

The answer is $6x^2-9$, R $(5x+22)$, or $6x^2-9+\dfrac{5x+22}{x^2+2}$.

18.
$$\begin{array}{r|rrrr} 4 & 1 & 5 & 4 & -7 \\ & & 4 & 36 & 160 \\ \hline & 1 & 9 & 40 & |\ 153 \end{array}$$

The answer is $x^2+9x+40$, R 153, or $x^2+9x+40+\dfrac{153}{x-4}$.

19. $(3x^4 - 5x^3 + 2x - 7) \div (x+1) =$
$$(3x^4 - 5x^3 + 0x^2 + 2x - 7) \div [x - (-1)]$$
$$\begin{array}{r|rrrrr} -1 & 3 & -5 & 0 & 2 & -7 \\ & & -3 & 8 & -8 & 6 \\ \hline & 3 & -8 & 8 & -6 & |\ -1 \end{array}$$

The answer is $3x^3 - 8x^2 + 8x - 6$, R -1, or
$3x^3 - 8x^2 + 8x - 6 + \dfrac{-1}{x+1}$.

20.
$$\frac{3+\dfrac{3}{y}}{4+\dfrac{4}{y}} = \frac{3+\dfrac{3}{y}}{4+\dfrac{4}{y}} \cdot \frac{y}{y}$$
$$= \frac{\left(3+\dfrac{3}{y}\right)\cdot y}{\left(4+\dfrac{4}{y}\right)\cdot y}$$
$$= \frac{3\cdot y + \dfrac{3}{y}\cdot y}{4\cdot y + \dfrac{4}{y}\cdot y}$$
$$= \frac{3y+3}{4y+4}$$
$$= \frac{3(y+1)}{4(y+1)}$$
$$= \frac{3(y\!\!\not{+}1)}{4(y\!\!\not{+}1)}$$
$$= \frac{3}{4}$$

21. $\dfrac{\dfrac{2}{a}+\dfrac{2}{b}}{\dfrac{4}{a^3}+\dfrac{4}{b^3}} = \dfrac{\dfrac{2}{a}\cdot\dfrac{b}{b}+\dfrac{2}{b}\cdot\dfrac{a}{a}}{\dfrac{4}{a^3}\cdot\dfrac{b^3}{b^3}+\dfrac{4}{b^3}\cdot\dfrac{a^3}{a^3}}$

$= \dfrac{\dfrac{2b}{ab}+\dfrac{2a}{ab}}{\dfrac{4b^3}{a^3b^3}+\dfrac{4a^3}{a^3b^3}}$

$= \dfrac{\dfrac{2b+2a}{ab}}{\dfrac{4b^3+4a^3}{a^3b^3}}$

$= \dfrac{2b+2a}{ab}\cdot\dfrac{a^3b^3}{4b^3+4a^3}$

$= \dfrac{(2b+2a)\cdot a^3b^3}{ab(4b^3+4a^3)}$

$= \dfrac{2(b+a)\cdot ab\cdot a^2b^2}{ab\cdot 2\cdot 2(b+a)(b^2-ab+a^2)}$

$= \dfrac{2(b+a)\cdot ab}{2(b+a)\cdot ab}\cdot\dfrac{a^2b^2}{2(b^2-ab+a^2)}$

$= \dfrac{a^2b^2}{2(b^2-ab+a^2)}$

22. $\dfrac{\dfrac{x^2-5x-36}{x^2-36}}{\dfrac{x^2+x-12}{x^2-12x+36}}$

$= \dfrac{x^2-5x-36}{x^2-36}\cdot\dfrac{x^2-12x+36}{x^2+x-12}$

$= \dfrac{(x^2-5x-36)(x^2-12x+36)}{(x^2-36)(x^2+x-12)}$

$= \dfrac{(x-9)(x+4)(x-6)(x-6)}{(x+6)(x-6)(x+4)(x-3)}$

$= \dfrac{(x-9)(x+4)(x-6)(x-6)}{(x+6)(x-6)(x+4)(x-3)}$

$= \dfrac{(x-9)(x-6)}{(x+6)(x-3)}$

23. $\dfrac{\dfrac{4}{x+3}-\dfrac{2}{x^2-3x+2}}{\dfrac{3}{x-2}+\dfrac{1}{x^2+2x-3}}$

$= \dfrac{\dfrac{4}{x+3}-\dfrac{2}{(x-1)(x-2)}}{\dfrac{3}{x-2}+\dfrac{1}{(x+3)(x-1)}}$

$= \dfrac{\dfrac{4}{x+3}-\dfrac{2}{(x-1)(x-2)}}{\dfrac{3}{x-2}+\dfrac{1}{(x+3)(x-1)}}\cdot\dfrac{(x+3)(x-1)(x-2)}{(x+3)(x-1)(x-2)}$

$= \dfrac{\dfrac{4}{x+3}\cdot(x+3)(x-1)(x-2)-\dfrac{2}{(x-1)(x-2)}\cdot(x+3)(x-1)(x-2)}{\dfrac{3}{x-2}\cdot(x+3)(x-1)(x-2)+\dfrac{1}{(x+3)(x-1)}\cdot(x+3)(x-1)(x-2)}$

$= \dfrac{4(x-1)(x-2)-2(x+3)}{3(x+3)(x-1)+(x-2)}$

$= \dfrac{4(x^2-3x+2)-2x-6}{3(x^2+2x-3)+x-2}$

$= \dfrac{4x^2-12x+8-2x-6}{3x^2+6x-9+x-2}$

$= \dfrac{4x^2-14x+2}{3x^2+7x-11}$

$= \dfrac{2(2x^2-7x+1)}{3x^2+7x-11}$

24. $\dfrac{x}{4}+\dfrac{x}{7}=1,\ \text{LCM is }28$

$28\left(\dfrac{x}{4}+\dfrac{x}{7}\right)=28\cdot 1$

$28\cdot\dfrac{x}{4}+28\cdot\dfrac{x}{7}=28$

$7x+4x=28$

$11x=28$

$x=\dfrac{28}{11}$

Since $\dfrac{28}{11}$ checks, it is the solution.

25. $\dfrac{5}{3x+2}=\dfrac{3}{2x},\ \text{LCM is }2x(3x+2)$

$2x(3x+2)\cdot\dfrac{5}{3x+2}=2x(3x+2)\cdot\dfrac{3}{2x}$

$2x\cdot 5=3(3x+2)$

$10x=9x+6$

$x=6$

Since 6 checks, it is the solution.

26.
$$\frac{4x}{x+1} + \frac{4}{x} + 9 = \frac{4}{x^2 + x}$$

$$\frac{4x}{x+1} + \frac{4}{x} + 9 = \frac{4}{x(x+1)},$$
$$\text{LCM is } x(x+1)$$

$$x(x+1)\left(\frac{4x}{x+1} + \frac{4}{x} + 9\right) = x(x+1) \cdot \frac{4}{x(x+1)}$$

$$x \cdot 4x + 4(x+1) + 9x(x+1) = 4$$

$$4x^2 + 4x + 4 + 9x^2 + 9x = 4$$

$$13x^2 + 13x + 4 = 4$$

$$13x^2 + 13x = 0$$

$$13x(x+1) = 0$$

$$13x = 0 \ \ or \ \ x + 1 = 0$$

$$x = 0 \ \ or \ \ \ \ \ x = -1$$

The numbers 0 and -1 each make a denominator 0, so there are no solutions.

27.
$$\frac{90}{x^2 - 3x + 9} - \frac{5x}{x+3} = \frac{405}{x^3 + 27}$$

$$\frac{90}{x^2 - 3x + 9} - \frac{5x}{x+3} = \frac{405}{(x+3)(x^2 - 3x + 9)}$$
$$\text{LCM is } (x+3)(x^2 - 3x + 9)$$

$$(x+3)(x^2 - 3x + 9) \cdot \left(\frac{90}{x^2 - 3x + 9} - \frac{5x}{x+3}\right) =$$
$$(x+3)(x^2 - 3x + 9) \cdot \frac{405}{(x+3)(x^2 + 3x + 9)}$$

$$90(x+3) - 5x(x^2 - 3x + 9) = 405$$

$$90x + 270 - 5x^3 + 15x^2 - 45x = 405$$

$$-5x^3 + 15x^2 + 45x + 270 = 405$$

$$-5x^3 + 15x^2 + 45x - 135 = 0$$

$$5x^3 - 15x^2 - 45x + 135 = 0 \ \ \ \ \ \text{Multiplying by } -1$$

$$5(x^3 - 3x^2 - 9x + 27) = 0$$

$$5[x^2(x - 3) - 9(x - 3)] = 0$$

$$5(x^2 - 9)(x - 3) = 0$$

$$5(x+3)(x-3)(x-3) = 0$$

$$x + 3 = 0 \ \ \ or \ \ x - 3 = 0 \ \ or \ \ x - 3 = 0$$

$$x = -3 \ \ or \ \ \ \ \ \ x = 3 \ \ or \ \ \ \ \ \ x = 3$$

We know that -3 is not a solution of the original equation, because it results in division by 0. Since 3 checks, it is the solution.

28.
$$\frac{2}{x-3} + \frac{1}{4x + 20} = \frac{1}{x^2 + 2x - 15}$$

$$\frac{2}{x-3} + \frac{1}{4(x+5)} = \frac{1}{(x-3)(x+5)}$$
$$\text{LCM is } 4(x-3)(x+5)$$

$$4(x-3)(x+5)\left(\frac{2}{x-3} + \frac{1}{4(x+5)}\right) = 4(x-3)(x+5) \cdot \frac{1}{(x-3)(x+5)}$$

$$4(x+5) \cdot 2 + (x-3) = 4$$

$$8x + 40 + x - 3 = 4$$

$$9x + 37 = 4$$

$$9x = -33$$

$$x = -\frac{11}{3}$$

Since $-\frac{11}{3}$ checks, it is the solution.

29. $\dfrac{6}{x} + \dfrac{4}{x} = 5$

$$\frac{10}{x} = 5 \ \ \ \text{Adding}$$

$$x \cdot \frac{10}{x} = x \cdot 5$$

$$10 = 5x$$

$$2 = x$$

Since 2 checks, the value of x for which $f(x) = 5$ is 2.

30. *Familiarize*. Let $t = $ the time it takes them to paint the house, working together.

***Translate*.** Using the work principle, we get the following equation:

$$\frac{t}{12} + \frac{t}{9} = 1$$

***Solve*.** We solve the equation.

$$\frac{t}{12} + \frac{t}{9} = 1, \ \text{LCM is } 36$$

$$36\left(\frac{t}{12} + \frac{t}{9}\right) = 36 \cdot 1$$

$$36 \cdot \frac{t}{12} + 36 \cdot \frac{t}{9} = 36$$

$$3t + 4t = 36$$

$$7t = 36$$

$$t = \frac{36}{7}, \ \text{or } 5\frac{1}{7}$$

***Check*.** We verify the work principle.

$$\frac{\frac{36}{7}}{12} + \frac{\frac{36}{7}}{9} = \frac{36}{7} \cdot \frac{1}{12} + \frac{36}{7} \cdot \frac{1}{9} = \frac{3}{7} + \frac{4}{7} = 1$$

***State*.** It will take them $5\frac{1}{7}$ hr to paint the house, working together.

31. *Familiarize*. Let $r = $ the speed of the boat in still water. Then the boat's speed traveling downstream is $r + 6$ and the speed upstream is $r - 6$. Since the time is the same downstream and upstream, we let t represent each time.

	Distance	Speed	Time
Downstream	50	$r+6$	t
Upstream	30	$r-6$	t

Translate. Using the formula Time = Distance/Rate in each row of the table and the fact that the times are the same, we can write an equation.

$$\frac{50}{r+6} = \frac{30}{r-6}$$

Solve.

$$\frac{50}{r+6} = \frac{30}{r-6},$$

LCM is $(r+6)(r-6)$

$$(r+6)(r-6) \cdot \frac{50}{r+6} = (r+6)(r-6) \cdot \frac{30}{r-6}$$

$$50(r-6) = 30(r+6)$$

$$50r - 300 = 30r + 180$$

$$20r = 480$$

$$r = 24$$

Check. If the speed of the boat in still water is 24 mph, then the speed downstream is $24+6$, or 30 mph. It would take 50/30, or 5/3 hr, to travel 50 mi downstream. The speed upstream is $24-6$, or 18 mph. It would take 30/18, or 5/3 hr, to travel 30 mi upstream. The times are the same, so the answer checks.

State. The speed of the boat in still water is 24 mph.

32. Familiarize. Let $d =$ the number of miles Fred will travel in 15 days.

Translate. We translate to a proportion.

$$\begin{array}{c} \text{Miles} \rightarrow \\ \text{Days} \rightarrow \end{array} \frac{800}{3} = \frac{d}{15} \begin{array}{c} \leftarrow \text{Miles} \\ \leftarrow \text{Days} \end{array}$$

Solve.

$$\frac{800}{3} = \frac{d}{15}$$

$$800 \cdot 15 = 3 \cdot d$$

$$\frac{800 \cdot 15}{3} = d$$

$$4000 = d$$

Check. We substitute in the proportion and check cross products.

$$\frac{800}{3} = \frac{4000}{15}; \ 800 \cdot 15 = 12,000; \ 3 \cdot 4000 = 12,000$$

Since the cross products are the same the answer checks.

State. Fred will travel 4000 mi in 15 days.

33. First we solve for d.

$$W = \frac{cd}{c+d}$$

$$W(c+d) = cd \qquad \text{Multiplying by } c+d$$

$$Wc + Wd = cd$$

$$Wc = cd - Wd$$

$$Wc = d(c - W)$$

$$\frac{Wc}{c - W} = d$$

Now we solve for c.

$$W = \frac{cd}{c+d}$$

$$W(c+d) = cd \qquad \text{Multiplying by } c+d$$

$$Wc + Wd = cd$$

$$Wd = cd - Wc$$

$$Wd = c(d - W)$$

$$\frac{Wd}{d - W} = c$$

34. First we solve for b.

$$S = \frac{p}{a} + \frac{t}{b}$$

$$ab \cdot S = ab\left(\frac{p}{a} + \frac{t}{b}\right)$$

$$abS = ab \cdot \frac{p}{a} + ab \cdot \frac{t}{b}$$

$$abS = bp + at$$

$$abS - bp = at$$

$$b(aS - p) = at$$

$$b = \frac{at}{aS - p}$$

Now we solve for t.

$$S = \frac{p}{a} + \frac{t}{b}$$

$$ab \cdot S = ab\left(\frac{p}{a} + \frac{t}{b}\right)$$

$$abS = ab \ \frac{p}{a} \ | \ ab \ \frac{t}{b}$$

$$abS = bp + at$$

$$abS - bp = at$$

$$\frac{abS - bp}{a} = t$$

35.

$$y = kx$$

$$100 = k \cdot 25$$

$$4 = k$$

$$y = 4x \qquad \text{Equation of variation}$$

36.

$$y = \frac{k}{x}$$

$$100 = \frac{k}{25}$$

$$2500 = k$$

$$y = \frac{2500}{x} \qquad \text{Equation of variation}$$

37.

$$t = \frac{k}{r} \qquad t \text{ varies inversely as } r.$$

$$35 = \frac{k}{800} \qquad \text{Substituting}$$

$$28,000 = k$$

$$t = \frac{28,000}{r} \qquad \text{Equation of variation}$$

$$t = \frac{28,000}{1400}$$

$$t = 20$$

It will take the pump 20 min to empty the tank at the rate of 1400 kL per minute.

38. $N = ka$ N varies directly as a.

$87 = k \cdot 28$ Substituting

$\dfrac{87}{28} = k$

$N = \dfrac{87}{28}a$

$N = \dfrac{87}{28} \cdot 25$

$N \approx 77.7$

Ellen's score would have been about 77.7 if she had answered 25 questions correctly.

39. $P = kC^2$ P varies directly as the square of C.

$180 = k \cdot 6^2$ Substituting

$180 = 36k$

$5 = k$

$P = 5C^2$ Equation of variation

$P = 5 \cdot 10^2$

$P = 5 \cdot 100$

$P = 500$

The circuit expends 500 watts of heat when the current is 10 amperes.

40. $f(x) = \dfrac{x^2 - x}{x^2 - 2x - 35}$

We set the denominator equal to 0 and solve.

$x^2 - 2x - 35 = 0$

$(x - 7)(x + 5) = 0$

$x - 7 = 0 \ \ or \ \ x + 5 = 0$

$x = 7 \ \ or \ \ \ \ \ \ x = -5$

We see that the domain is all real numbers except 7 and -5, so answer B is correct.

41. $x^5 = x \cdot x \cdot x \cdot x \cdot x$

$x - 4$

$x^2 - 4 = (x + 2)(x - 2)$

$x^2 - 4x = x(x - 4)$

LCM $= x \cdot x \cdot x \cdot x \cdot x(x - 4)(x + 2)(x - 2)$, or

$x^5(x - 4)(x + 2)(x - 2)$

Answer C is correct.

42. The reciprocal of $\dfrac{a - b}{a^3 - b^3}$ is $\dfrac{a^3 - b^3}{a - b}$.

$\dfrac{a^3 - b^3}{a - b} = \dfrac{(a - b)(a^2 + ab + b^2)}{(a - b) \cdot 1}$

$= \dfrac{(a\!\!\!\diagup\!\!\!-b)(a^2 + ab + b^2)}{(a\!\!\!\diagup\!\!\!-b) \cdot 1}$

$= a^2 + ab + b^2$

43.

$\dfrac{5}{x - 13} - \dfrac{5}{x} = \dfrac{65}{x^2 - 13x}$

$\dfrac{5}{x - 13} - \dfrac{5}{x} = \dfrac{65}{x(x - 13)}$

LCM is $x(x - 13)$

$x(x - 13)\left(\dfrac{5}{x - 13} - \dfrac{5}{x}\right) = x(x - 13) \cdot \dfrac{65}{x(x - 13)}$

$5x - 5(x - 13) = 65$

$5x - 5x + 65 = 65$

$65 = 65$

We get a true equation. Thus, all real numbers except those that make a denominator 0 are solutions of the equation. The numbers that make a denominator 0 are 13 and 0, so all real numbers except 13 and 0 are solutions.

Chapter 5 Discussion and Writing Exercises

1. When adding or subtracting rational expressions, we use the LCM of the denominators (the LCD). When solving a rational equation or when solving a formula for a given letter, we multiply by the LCM of all the denominators to clear fractions. When simplifying a complex rational expression, we can use the LCM in either of two ways. We can multiply by a/a, where a is the LCM of all the denominators occurring in the expression. Or we can use the LCM to add or subtract as necessary in the numerator and in the denominator.

2. Rational equations differ from those previously studied because they contain variables in denominators. Because of this, possible solutions must be checked in the original equation to avoid division by 0.

3. Assuming all algebraic procedures have been performed correctly, a possible solution of a rational equation would fail to be an actual solution only if it were not in the domain of one of the rational expressions in the equation. This occurs when the number in question makes a denominator 0.

4. Let $y = k_1 x$ and $x = \dfrac{k_2}{z}$. Then $y = k_1 \cdot \dfrac{k_2}{z}$, or $y = \dfrac{k_1 k_2}{z}$, so y varies inversely as z.

5. Answers may vary. From Example 4 in Section 5.5 we see that one form of such an equation is $\dfrac{x^2}{x - a} = \dfrac{a^2}{x - a}$.

6. Answers may vary. Many would probably argue that it is easier to solve $\dfrac{1}{a} + \dfrac{1}{b} = \dfrac{1}{x}$ since it is easier for them to multiply a and b than 38 and 47. Others might argue that it is easier to solve $\dfrac{1}{38} + \dfrac{1}{47} = \dfrac{1}{x}$ since it is easier for them to work with constants than variables.

Chapter 5 Test

1. $\dfrac{x^2 - 16}{x^2 - 3x + 2}$

We set the denominator equal to 0 and solve.

$$x^2 - 3x + 2 = 0$$
$$(x - 1)(x - 2) = 0$$
$$x - 1 = 0 \ \ or \ \ x - 2 = 0$$
$$x = 1 \ \ or \ \ \ \ \ \ x = 2$$

The expression is not defined for the numbers 1 and 2.

2. In Exercise 1 we found that the replacements for which $\dfrac{x^2 - 16}{x^2 - 3x + 2}$ is not defined are 1 and 2. Then the domain of $f(x) = \dfrac{x^2 - 16}{x^2 - 3x + 2}$ is $(-\infty, 1) \cup (1, 2) \cup (2, \infty)$.

3. $\dfrac{12x^2 + 11x + 2}{4x^2 - 7x - 2} = \dfrac{(4x + 1)(3x + 2)}{(4x + 1)(x - 2)}$

$$= \dfrac{(4x+1)(3x+2)}{(4x+1)(x - 2)}$$

$$= \dfrac{3x + 2}{x - 2}$$

4. $\dfrac{p^3 + 1}{p^2 - p - 2} = \dfrac{(p + 1)(p^2 - p + 1)}{(p + 1)(p - 2)}$

$$= \dfrac{(p+1)(p^2 - p + 1)}{(p+1)(p - 2)}$$

$$= \dfrac{p^2 - p + 1}{p - 2}$$

5. $x^2 + x - 6 = (x + 3)(x - 2)$

$x^2 + 8x + 15 = (x + 3)(x + 5)$

LCM $= (x + 3)(x - 2)(x + 5)$

6. $\dfrac{2x^2 + 20x + 50}{x^2 - 4} \cdot \dfrac{x + 2}{x + 5}$

$$= \dfrac{(2x^2 + 20x + 50)(x + 2)}{(x^2 - 4)(x + 5)}$$

$$= \dfrac{2(x^2 + 10x + 25)(x + 2)}{(x + 2)(x - 2)(x + 5)}$$

$$= \dfrac{2(x + 5)(x + 5)(x + 2)}{(x + 2)(x - 2)(x + 5)}$$

$$= \dfrac{2(x+5)(x + 5)(x+2)}{(x+2)(x - 2)(x+5)}$$

$$= \dfrac{2(x + 5)}{x - 2}$$

7. $\dfrac{x}{x^2 + 11x + 30} - \dfrac{5}{x^2 + 9x + 20}$

$$= \dfrac{x}{(x + 5)(x + 6)} - \dfrac{5}{(x + 4)(x + 5)}$$

$$\text{LCM is} (x + 5)(x + 6)(x + 4)$$

$$= \dfrac{x}{(x + 5)(x + 6)} \cdot \dfrac{x + 4}{x + 4} - \dfrac{5}{(x + 4)(x + 5)} \cdot \dfrac{x + 6}{x + 6}$$

$$= \dfrac{x^2 + 4x - (5x + 30)}{(x + 5)(x + 6)(x + 4)}$$

$$= \dfrac{x^2 + 4x - 5x - 30}{(x + 5)(x + 6)(x + 4)}$$

$$= \dfrac{x^2 - x - 30}{(x + 5)(x + 6)(x + 4)}$$

$$= \dfrac{(x + 5)(x - 6)}{(x + 5)(x + 6)(x + 4)}$$

$$= \dfrac{(x+5)(x - 6)}{(x+5)(x + 6)(x + 4)}$$

$$= \dfrac{x - 6}{(x + 6)(x + 4)}$$

8. $\dfrac{y^2 - 16}{2y + 6} \div \dfrac{y - 4}{y + 3} = \dfrac{y^2 - 16}{2y + 6} \cdot \dfrac{y + 3}{y - 4}$

$$= \dfrac{(y^2 - 16)(y + 3)}{(2y + 6)(y - 4)}$$

$$= \dfrac{(y + 4)(y - 4)(y + 3)}{2(y + 3)(y - 4)}$$

$$= \dfrac{(y + 4)(y - 4)(y + 3)}{2(y + 3)(y - 4)}$$

$$= \dfrac{y + 4}{2}$$

9. $\dfrac{x^2}{x - y} + \dfrac{y^2}{y - x} = \dfrac{x^2}{x - y} + \dfrac{y^2}{y - x} \cdot \dfrac{-1}{-1}$

$$= \dfrac{x^2}{x - y} + \dfrac{-y^2}{x - y}$$

$$= \dfrac{x^2 - y^2}{x - y}$$

$$= \dfrac{(x + y)(x - y)}{(x - y)}$$

$$= \dfrac{(x + y)(x - y)}{(x - y) \cdot 1}$$

$$= x + y$$

10. $\dfrac{1}{x + 1} - \dfrac{x + 2}{x^2 - 1} + \dfrac{3}{x - 1}$

$$= \dfrac{1}{x + 1} - \dfrac{x + 2}{(x + 1)(x - 1)} + \dfrac{3}{x - 1} \quad \text{LCM is } (x + 1)(x - 1)$$

$$= \dfrac{1}{x + 1} \cdot \dfrac{x - 1}{x - 1} - \dfrac{x + 2}{(x + 1)(x - 1)} + \dfrac{3}{x - 1} \cdot \dfrac{x + 1}{x + 1}$$

$$= \dfrac{x - 1 - (x + 2) + 3x + 3}{(x + 1)(x - 1)}$$

$$= \dfrac{x - 1 - x - 2 + 3x + 3}{(x + 1)(x - 1)}$$

$$= \dfrac{3x}{(x + 1)(x - 1)}$$

11.
$$\frac{a}{a-b} + \frac{b}{a^2+ab+b^2} - \frac{2}{a^3-b^3}$$

$$= \frac{a}{a-b} + \frac{b}{a^2+ab+b^2} - \frac{2}{(a-b)(a^2+ab+b^2)}$$

$$\text{LCM is } (a-b)(a^2+ab+b^2)$$

$$= \frac{a}{a-b} \cdot \frac{a^2+ab+b^2}{a^2+ab+b^2} + \frac{b}{a^2+ab+b^2} \cdot \frac{a-b}{a-b} -$$

$$\frac{2}{(a-b)(a^2+ab+b^2)}$$

$$= \frac{a^3 + a^2b + ab^2 + ab - b^2 - 2}{(a-b)(a^2+ab+b^2)}$$

12.
$$(20r^2s^3 + 15r^2s^2 - 10r^3s^3) \div (5r^2s)$$

$$= \frac{20r^2s^3 + 15r^2s^2 - 10r^3s^3}{5r^2s}$$

$$= \frac{20r^2s^3}{5r^2s} + \frac{15r^2s^2}{5r^2s} - \frac{10r^3s^3}{5r^2s}$$

$$= 4s^2 + 3s - 2rs^2$$

13.
$$\begin{array}{r} y^2 - 5y + 25 \\ y+5 \overline{)\, y^3 + 0y^2 + 0y + 125} \\ \underline{y^3 + 5y^2} \\ -5y^2 + 0y \\ \underline{-5y^2 - 25y} \\ 25y + 125 \\ \underline{25y + 125} \\ 0 \end{array}$$

The answer is $y^2 - 5y + 25$.

14.
$$\begin{array}{r} 4x^2 + 3x - 4 \\ x^2+1 \overline{)\, 4x^4 + 3x^3 + 0x^2 - 5x - 2} \\ \underline{4x^4 \qquad + 4x^2} \\ 3x^3 - 4x^2 \\ \underline{3x^3 \qquad + 3x} \\ -4x^2 - 8x \\ \underline{-4x^2 \qquad -4} \\ -8x + 2 \end{array}$$

The answer is $4x^2 + 3x - 4$, R $(-8x + 2)$, or
$4x^2 + 3x - 4 + \dfrac{-8x+2}{x^2+1}$.

15. $(x^3 + 3x^2 + 2x - 6) \div (x - 3)$

$$\begin{array}{r|rrrr} 3 & 1 & 3 & 2 & -6 \\ & & 3 & 18 & 60 \\ \hline & 1 & 6 & 20 & | \, 54 \end{array}$$

The answer is $x^2 + 6x + 20$, R 54, or $x^2 + 6x + 20 + \dfrac{54}{x-3}$.

16. $(3x^3 + 22x^2 - 160) \div (x+4) = (3x^3 + 22x^2 + 0x - 160) \div [x - (-4)]$

$$\begin{array}{r|rrrr} -4 & 3 & 22 & 0 & -160 \\ & & -12 & -40 & 160 \\ \hline & 3 & 10 & -40 & | \, 0 \end{array}$$

The answer is $3x^2 + 10x - 40$.

17. $\dfrac{1 - \dfrac{1}{x^2}}{1 - \dfrac{1}{x}}$

The LCM of the denominators is x^2. We multiply by 1 using x^2/x^2.

$$\frac{1 - \dfrac{1}{x^2}}{1 - \dfrac{1}{x}} = \frac{1 - \dfrac{1}{x^2}}{1 - \dfrac{1}{x}} \cdot \frac{x^2}{x^2}$$

$$= \frac{\left(1 - \dfrac{1}{x^2}\right) \cdot x^2}{\left(1 - \dfrac{1}{x}\right) \cdot x^2}$$

$$= \frac{1 \cdot x^2 - \dfrac{1}{x^2} \cdot x^2}{1 \cdot x^2 - \dfrac{1}{x} \cdot x^2}$$

$$= \frac{x^2 - 1}{x^2 - x}$$

$$= \frac{(x+1)(x-1)}{x(x-1)}$$

$$= \frac{(x+1)(x-1)}{x(x-1)}$$

$$= \frac{x+1}{x}$$

18. $\dfrac{\dfrac{1}{a^3} + \dfrac{1}{b^3}}{\dfrac{1}{a} + \dfrac{1}{b}}$

The LCM of the denominators is a^3b^3. We multiply by 1 using $(a^3b^3)/(a^3b^3)$.

$$\frac{\dfrac{1}{a^3} + \dfrac{1}{b^3}}{\dfrac{1}{a} + \dfrac{1}{b}} = \frac{\dfrac{1}{a^3} + \dfrac{1}{b^3}}{\dfrac{1}{a} + \dfrac{1}{b}} \cdot \frac{a^3b^3}{a^3b^3}$$

$$= \frac{\left(\dfrac{1}{a^3} + \dfrac{1}{b^3}\right) \cdot a^3b^3}{\left(\dfrac{1}{a} + \dfrac{1}{b}\right) \cdot a^3b^3}$$

$$= \frac{\dfrac{1}{a^3} \cdot a^3b^3 + \dfrac{1}{b^3} \cdot a^3b^3}{\dfrac{1}{a} \cdot a^3b^3 + \dfrac{1}{b} \cdot a^3b^3}$$

$$= \frac{b^3 + a^3}{a^2b^3 + a^3b^2}$$

$$= \frac{(b+a)(b^2 - ba + a^2)}{a^2b^2(b+a)}$$

$$= \frac{(b+a)(b^2 - ba + a^2)}{a^2b^2(b+a)}$$

$$= \frac{b^2 - ba + a^2}{a^2b^2}$$

19.
$$\frac{2}{x-1} + \frac{2}{x+2} = 1 \quad \text{LCM is } (x-1)(x+2)$$

$$(x-1)(x+2)\left(\frac{2}{x-1} + \frac{2}{x+2}\right) = (x-1)(x+2) \cdot 1$$

$$2(x+2) + 2(x-1) = x^2 + x - 2$$

$$2x + 4 + 2x - 2 = x^2 + x - 2$$

$$4x + 2 = x^2 + x - 2$$

$$0 = x^2 - 3x - 4$$

$$0 = (x-4)(x+1)$$

$$x - 4 = 0 \quad or \quad x + 1 = 0$$

$$x = 4 \quad or \quad x = -1$$

Both values check. The values of x for which $f(x) = 1$ are 4 and −1.

20.
$$\frac{2}{x-1} = \frac{3}{x+3} \quad \text{LCM is } (x-1)(x+3)$$

$$(x-1)(x+3) \cdot \frac{2}{x-1} = (x-1)(x+3) \cdot \frac{3}{x+3}$$

$$2(x+3) = 3(x-1)$$

$$2x + 6 = 3x - 3$$

$$9 = x$$

This value checks. The solution is 9.

21.
$$\frac{7x}{x+3} + \frac{21}{x-3} = \frac{126}{x^2 - 9}$$

$$\frac{7x}{x+3} + \frac{21}{x-3} = \frac{126}{(x+3)(x-3)}$$

$$\text{LCM is } (x+3)(x-3)$$

$$(x+3)(x-3)\left(\frac{7x}{x+3} + \frac{21}{x-3}\right) =$$

$$(x+3)(x-3) \cdot \frac{126}{(x+3)(x-3)}$$

$$7x(x-3) + 21(x+3) = 126$$

$$7x^2 - 21x + 21x + 63 = 126$$

$$7x^2 + 63 = 126$$

$$7x^2 - 63 = 0$$

$$7(x^2 - 9) = 0$$

$$7(x+3)(x-3) = 0$$

$$x + 3 = 0 \quad or \quad x - 3 = 0$$

$$x = -3 \quad or \quad x = 3$$

We know that neither number can be a solution of the original equation because each one results in division by 0. Thus, the equation has no solution.

22.
$$\frac{2x}{x+7} = \frac{5}{x+1} \quad \text{LCM is } (x+7)(x+1)$$

$$(x+7)(x+1) \cdot \frac{2x}{x+7} = (x+7)(x+1) \cdot \frac{5}{x+1}$$

$$2x(x+1) = 5(x+7)$$

$$2x^2 + 2x = 5x + 35$$

$$2x^2 - 3x - 35 = 0$$

$$(2x+7)(x-5) = 0$$

$$2x + 7 = 0 \quad or \quad x - 5 = 0$$

$$2x = -7 \quad or \quad x = 5$$

$$x = -\frac{7}{2} \quad or \quad x = 5$$

Both values check. The solutions are $-\frac{7}{2}$ and 5.

23.
$$\frac{1}{3x - 6} - \frac{1}{x^2 - 4} = \frac{3}{x+2}$$

$$\frac{1}{3(x-2)} - \frac{1}{(x+2)(x-2)} = \frac{3}{x+2}$$

$$\text{LCM is } 3(x-2)(x+2)$$

$$3(x-2)(x+2)\left(\frac{1}{3(x-2)} - \frac{1}{(x+2)(x-2)}\right) =$$

$$3(x-2)(x+2) \cdot \frac{3}{x+2}$$

$$x + 2 - 3 = 9(x-2)$$

$$x - 1 = 9x - 18$$

$$-8x = -17$$

$$x = \frac{17}{8}$$

This value checks. The solution is $\frac{17}{8}$.

24. *Familiarize*. Let $s -$ the number of hours it takes Jessie to complete the puzzle working alone. Then $s + 4 =$ Rachel's time.

***Translate*.** We use the work principle.

$$\frac{t}{a} + \frac{t}{b} = 1$$

$$\frac{1.5}{s} + \frac{1.5}{s+4} = 1$$

***Solve*.** We solve the equation.

$$\frac{1.5}{s} + \frac{1.5}{s+4} = 1 \quad \text{LCM is } s(s+4)$$

$$s(s+4)\left(\frac{1.5}{s} + \frac{1.5}{s+4}\right) = s(s+4) \cdot 1$$

$$1.5(s+4) + 1.5s = s^2 + 4s$$

$$1.5s + 6 + 1.5s = s^2 + 4s$$

$$3s + 6 = s^2 + 4s$$

$$0 = s^2 + s - 6$$

$$0 = (s+3)(s-2)$$

$$s + 3 = 0 \quad or \quad s - 2 = 0$$

$$s = -3 \quad or \quad s = 2$$

***Check*.** The time cannot be negative, so we check only 2. If $s = 2$, then $s + 4 = 2 + 4 = 6$. In 1.5 hr Jessie does $\frac{1.5}{2}$, or 0.75, of the job and Rachel does $\frac{1.5}{6}$, or 0.25, of the job. Together they do $0.75 + 0.25$, or 1 entire job. The answer checks.

***State*.** It would take Jessie 2 hr to complete the puzzle working alone.

25. *Familiarize*. Let $w =$ the speed of the wind. Then David's speed against the wind is $12 - w$, and the speed with the wind is $12 + w$. We organize the information in a table.

	Distance	Speed	Time
Against wind	8	$12 - w$	t
With wind	14	$12 + w$	t

Translate. Using the formula Time = Distance/Rate in each row of the table and the fact that the times are the same, we can write an equation.

$$\frac{8}{12 - w} = \frac{14}{12 + w}$$

Solve. We solve the equation.

$$\frac{8}{12 - w} = \frac{14}{12 + w}$$

$$\text{LCM is } (12 - w)(12 + w)$$

$$(12-w)(12+w) \cdot \frac{8}{12-w} = (12-w)(12+w) \cdot \frac{14}{12+w}$$

$$8(12 + w) = 14(12 - w)$$

$$96 + 8w = 168 - 14w$$

$$22w = 72$$

$$w = \frac{36}{11}, \text{ or } 3\frac{3}{11}$$

Check. David's speed against a $3\frac{3}{11}$ mph wind is $12 - 3\frac{3}{11}$, or $8\frac{8}{11}$ mph. At this speed David travels 8 mi in $8 \div \left(8\frac{8}{11}\right)$, or $\frac{11}{12}$ hr. The speed with the wind is $12 + 3\frac{3}{11}$, or $15\frac{3}{11}$ mph. At this speed David travels 14 mi in $14 \div \left(15\frac{3}{11}\right)$, or $\frac{11}{12}$ hr. The times are the same, so the answer checks.

State. The speed of the wind is $3\frac{3}{11}$ mph.

26. **Familiarize.** Let $p =$ the number of gallons of paint needed to paint 6000 ft^2 of clapboard.

   Translate. We translate to a proportion.

$$\text{Paint} \rightarrow \frac{4}{1700} = \frac{p}{6000} \leftarrow \text{Paint}$$
$$\text{Area} \rightarrow \qquad \qquad \leftarrow \text{Area}$$

   Solve. We solve the proportion.

$$\frac{4}{1700} = \frac{p}{6000}$$

$$4 \cdot 6000 = 1700 \cdot p \quad \text{Equating cross products}$$

$$\frac{4 \cdot 6000}{1700} = p$$

$$\frac{240}{17} = p, \text{ or}$$

$$14\frac{2}{17} = p$$

   Check. We substitute into the proportion and check cross products.

$$\frac{4}{1700} = \frac{240/17}{6000}; \; 4 \cdot 6000 = 24,000; \; 1700 \cdot \frac{240}{17} = 24,000$$

Since the cross products are the same, the answer checks.

State. $14\frac{2}{17}$ gal of paint would be needed.

27. Solve for a:

$$T = \frac{ab}{a - b}$$

$$(a - b)T = (a - b) \cdot \frac{ab}{a - b}$$

$$aT - bT = ab$$

$$aT - ab = bT$$

$$a(T - b) = bT$$

$$a = \frac{bT}{T - b}$$

Solve for b:

$$T = \frac{ab}{a - b}$$

$$(a - b)T = (a - b) \cdot \frac{ab}{a - b}$$

$$aT - bT = ab$$

$$aT = ab + bT$$

$$aT = b(a + T)$$

$$\frac{aT}{a + T} = b$$

28.
$$Q = \frac{2}{a} - \frac{t}{b}$$

$$ab \cdot Q = ab\left(\frac{2}{a} - \frac{t}{b}\right)$$

$$abQ = ab \cdot \frac{2}{a} - ab \cdot \frac{t}{b}$$

$$abQ = 2b - at$$

$$abQ + at = 2b$$

$$a(bQ + t) = 2b$$

$$a = \frac{2b}{bQ + t}$$

29. $Q = kxy$

$$25 = k \cdot 2 \cdot 5$$

$$25 = 10k$$

$$\frac{5}{2} = k$$

$$Q = \frac{5}{2}xy \quad \text{Equation of variation}$$

30. $y = \dfrac{k}{x}$

$$10 = \frac{k}{25}$$

$$250 = k$$

$$y = \frac{250}{x} \quad \text{Equation of variation}$$

31. We first find an equation of variation.

$$I = kt$$

$$550 = k \cdot 40$$

$$13.75 = k$$

$$I = 13.75t \quad \text{Equation of variation}$$

Now we use the equation to find the pay for working 72 hr.

$$I = 13.75t$$
$$I = 13.75(72)$$
$$I = \$990$$

Kaylee is paid \$990 for working 72 hr.

32. We first find an equation of variation.

$$t = \frac{k}{r}$$
$$5 = \frac{k}{60}$$
$$300 = k$$
$$t = \frac{300}{r} \quad \text{Equation of variation}$$

Now we use the equation to find how long it would take to drive the same distance at 40 km/h.

$$t = \frac{300}{r}$$
$$t = \frac{300}{40}$$
$$t = \frac{15}{2}, \text{ or } 7\frac{1}{2}$$

It would take $7\frac{1}{2}$ hr at a speed of 40 km/h.

33. First we find an equation of variation.

$$A = kr^2$$
$$314 = k \cdot 5^2$$
$$314 = 25k$$
$$12.56 = k$$
$$A = 12.56r^2 \quad \text{Equation of variation}$$

Now we use the equation to find the area when the radius is 7 cm.

$$A = 12.56r^2$$
$$A = 12.56 \cdot 7^2$$
$$A = 12.56(49)$$
$$A = 615.44$$

The area is 615.44 cm^2.

34. $6x^2 = 2 \cdot 3 \cdot x \cdot x$

$3x^2 - 3y^2 = 3(x^2 - y^2) = 3(x + y)(x - y)$

$x^2 - 2xy - 3y^2 = (x - 3y)(x + y)$

LCM $= 2 \cdot 3 \cdot x \cdot x(x + y)(x - y)(x - 3y)$

$\quad = 6x^2(x + y)(x - y)(x - 3y)$

Answer D is correct.

35.

$$\frac{6}{x - 15} - \frac{6}{x} = \frac{90}{x^2 - 15x}$$

$$\frac{6}{x - 15} - \frac{6}{x} = \frac{90}{x(x - 15)} \quad \text{LCM is } x(x - 15)$$

$$x(x - 15)\left(\frac{6}{x - 15} - \frac{6}{x}\right) = x(x - 15) \cdot \frac{90}{x(x - 15)}$$

$$6x - 6(x - 15) = 90$$
$$6x - 6x + 90 = 90$$
$$90 = 90$$

We get an equation that is true for all values of x. Thus, all real numbers except those that result in division by 0 in the original equation are solutions. We see that division by 0 results when $x = 0$ or $x = 15$, so all real numbers except 0 and 15 are solutions.

36. To find the x-intercept we set $f(x)$ equal to 0 and solve for x.

$$\frac{\frac{5}{x + 4} - \frac{3}{x - 2}}{\frac{2}{x - 3} + \frac{1}{x + 4}} = 0$$

$$\frac{5}{x + 4} - \frac{3}{x - 2} = 0 \quad \text{Multiplying by}$$
$$\frac{2}{x - 3} + \frac{1}{x + 4}$$

$$(x+4)(x-2)\left(\frac{5}{x+4} - \frac{3}{x-2}\right) = (x+4)(x-2) \cdot 0$$

$$5(x - 2) - 3(x + 4) = 0$$
$$5x - 10 - 3x - 12 = 0$$
$$2x - 22 = 0$$
$$2x = 22$$
$$x = 11$$

The x-intercept is $(11, 0)$.

To find the y-intercept we find $f(0)$.

$$\frac{\frac{5}{0 + 4} - \frac{3}{0 - 2}}{\frac{2}{0 - 3} + \frac{1}{0 + 4}} = \frac{\frac{5}{4} + \frac{3}{2}}{-\frac{2}{3} + \frac{1}{4}}$$

$$= \frac{\frac{5}{4} + \frac{6}{4}}{-\frac{8}{12} + \frac{3}{12}}$$

$$= \frac{\frac{11}{4}}{-\frac{5}{12}}$$

$$= \frac{11}{4} \cdot \left(-\frac{12}{5}\right)$$

$$= -\frac{11 \cdot 12}{4 \cdot 5}$$

$$= -\frac{11 \cdot 3 \cdot 4}{4 \cdot 5}$$

$$= -\frac{11 \cdot 3 \cdot \cancel{4}}{\cancel{4} \cdot 5}$$

$$= -\frac{33}{5}$$

The y-intercept is $\left(0, -\frac{33}{5}\right)$.

Cumulative Review Chapters 1 - 5

1. Graph $y = -5x + 4$.

We find three ordered pairs that are solutions, plot these points, and draw the line through them.

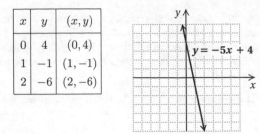

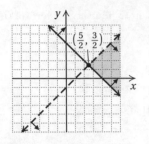

2. Graph $3x - 18 = 0$.

$$3x - 18 = 0$$
$$3x = 18$$
$$x = 6$$

This is the equation of a vertical line with x-intercept $(6, 0)$.

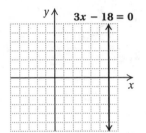

3. Graph $x + 3y < 4$.

First we graph the equation $x + 3y = 4$. We use a dashed line because the inequality symbol is $<$. Next we test a point not on the line. We will use $(0, 0)$.

$$\frac{x + 3y < 4}{0 + 3 \cdot 0 \; ? \; 4}$$
$$\quad 0 \; \bigg| \quad \text{TRUE}$$

Since $0 < 4$ is true, we shade the half-plane that contains $(0, 0)$.

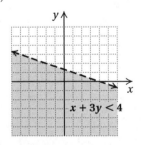

4. $x + y \geq 4$,

   $x - y > 1$

First we graph $x + y = 4$ using a solid line because the inequality symbol in the first inequality is $\geq$. Then we graph $x - y = 1$ using a dashed line because the inequality symbol in the second inequality is $>$. We indicate the region for each inequality and shade the region where they overlap.

To find the coordinates of the vertex, we solve the system of equations

$$x + y = 4,$$
$$x - y = 1.$$

The coordinates are $\left(\dfrac{5}{2}, \dfrac{3}{2} \right)$.

5. $g(x) = |x - 4| + 5$

$$g(-2) = |-2 - 4| + 5 = |-6| + 5 = 6 + 5 = 11$$

6. $f(x) = \dfrac{x - 2}{x^2 - 25}$

We set the denominator equal to zero and solve.

$$x^2 - 25 = 0$$
$$(x + 5)(x - 5) = 0$$
$$x + 5 = 0 \quad or \quad x - 5 = 0$$
$$x = -5 \quad or \qquad x = 5$$

The domain $\{x | x \text{ is a real number } and \; x \neq -5 \; and \; x \neq 5\}$, or $(-\infty, -5) \cup (-5, 5) \cup (5, \infty)$.

7. From the graph we see that the inputs extend from -5 to 5, not including -5 and 5. The domain is $(-5, 5)$. From the graph we see that the outputs extend from -2 to 4, including -2 and 4. The range is $[-2, 4]$.

8. $(6m - n)^2 = (6m)^2 - 2 \cdot 6m \cdot n + n^2 = 36m^2 - 12mn + n^2$

9. $(3a - 4b)(5a + 2b) = 15a^2 + 6ab - 20ab - 8b^2 = 15a^2 - 14ab - 8b^2$

10. $\dfrac{y^2 - 4}{3y + 33} \cdot \dfrac{y + 11}{y + 2} = \dfrac{(y^2 - 4)(y + 11)}{(3y + 33)(y + 2)} = \dfrac{(y+2)(y-2)(y+11)}{3(y+11)(y+2)} = \dfrac{y - 2}{3}$

11. $\dfrac{9x^2 - 25}{x^2 - 16} \div \dfrac{3x + 5}{x - 4} = \dfrac{9x^2 - 25}{x^2 - 16} \cdot \dfrac{x - 4}{3x + 5}$

$= \dfrac{(3x+5)(3x - 5)(x-4)}{(x + 4)(x-4)(3x+5)}$

$= \dfrac{3x - 5}{x + 4}$

12. $\dfrac{2x+1}{4x-12} - \dfrac{x-2}{5x-15}$

$= \dfrac{2x+1}{4(x-3)} - \dfrac{x-2}{5(x-3)}$

$\qquad\qquad$ LCM is $4 \cdot 5 \cdot (x-3)$, or $20(x-3)$

$= \dfrac{2x+1}{4(x-3)} \cdot \dfrac{5}{5} - \dfrac{x-2}{5(x-3)} \cdot \dfrac{4}{4}$

$= \dfrac{5(2x+1) - 4(x-2)}{20(x-3)}$

$= \dfrac{10x+5 - 4x+8}{20(x-3)}$

$= \dfrac{6x+13}{20(x-3)}$

13. $\dfrac{1 - \dfrac{2}{y^2}}{1 - \dfrac{1}{y^3}} = \dfrac{1 - \dfrac{2}{y^2}}{1 - \dfrac{1}{y^3}} \cdot \dfrac{y^3}{y^3}$

$\qquad = \dfrac{1 \cdot y^3 - \dfrac{2}{y^2} \cdot y^3}{1 \cdot y^3 - \dfrac{1}{y^3} \cdot y^3}$

$\qquad = \dfrac{y^3 - 2y}{y^3 - 1}$

$\qquad = \dfrac{y(y^2 - 2)}{(y-1)(y^2+y+1)}$

14. $(6p^2 - 2p + 5) - (-10p^2 + 6p + 5) =$

$6p^2 - 2p + 5 + 10p^2 - 6p - 5 = 16p^2 - 8p$

15. $\dfrac{2}{x+2} + \dfrac{3}{x-2} - \dfrac{x+1}{x^2-4}$

$= \dfrac{2}{x+2} + \dfrac{3}{x-2} - \dfrac{x+1}{(x+2)(x-2)}$

$\qquad\qquad$ LCM is $(x+2)(x-2)$

$= \dfrac{2}{x+2} \cdot \dfrac{x-2}{x-2} + \dfrac{3}{x-2} \cdot \dfrac{x+2}{x+2} - \dfrac{x+1}{(x+2)(x-2)}$

$= \dfrac{2(x-2) + 3(x+2) - (x+1)}{(x+2)(x-2)}$

$= \dfrac{2x-4 + 3x+6 - x-1}{(x+2)(x-2)}$

$= \dfrac{4x+1}{(x+2)(x-2)}$

16. $(2x^3 - 7x^2 + x - 3) \div (x+2) = (2x^3 - 7x^2 + x - 3) \div [x - (-2)]$

$$
\begin{array}{r|rrrr}
-2 & 2 & -7 & 1 & -3 \\
   &   & -4 & 22 & -46 \\
\hline
   & 2 & -11 & 23 & -49
\end{array}
$$

The answer is $2x^2 - 11x + 23$, R -49, or

$2x^2 - 11x + 23 + \dfrac{-49}{x+2}$.

17. $9y - (5y - 3) = 33$

$9y - 5y + 3 = 33$

$4y + 3 = 33$

$4y = 30$

$y = \dfrac{30}{4} = \dfrac{15}{2}$

The solution is $\dfrac{15}{2}$.

18. $-3 < -2x - 6 < 0$

$3 < -2x < 6$

$-\dfrac{3}{2} > x > -3$

The solution set is $\left\{x \mid -3 < x < -\dfrac{3}{2}\right\}$, or $\left(-3, -\dfrac{3}{2}\right)$.

19. $\dfrac{3x}{x-2} - \dfrac{6}{x+2} = \dfrac{24}{x^2-4}$

$\dfrac{3x}{x-2} - \dfrac{6}{x+2} = \dfrac{24}{(x+2)(x-2)}$

$\qquad\qquad$ LCM is $(x+2)(x-2)$

$(x+2)(x-2)\left(\dfrac{3x}{x-2} - \dfrac{6}{x+2}\right) = (x+2)(x-2) \cdot \dfrac{24}{(x+2)(x-2)}$

$3x(x+2) - 6(x-2) = 24$

$3x^2 + 6x - 6x + 12 = 24$

$3x^2 + 12 = 24$

$3x^2 - 12 = 0$

$3(x^2 - 4) = 0$

$3(x+2)(x-2) = 0$

$x+2 = 0 \quad or \quad x-2 = 0$

$x = -2 \quad or \qquad x = 2$

Both possible solutions make a denominator 0. The equation has no solution.

20. $P = \dfrac{3a}{a+b}$

$P(a+b) = 3a$

$Pa + Pb = 3a$

$Pb = 3a - Pa$

$Pb = a(3 - P)$

$\dfrac{Pb}{3-P} = a$

21. $F = \dfrac{9}{5}C + 32$

$F - 32 = \dfrac{9}{5}C$

$\dfrac{5}{9}(F - 32) = C$

22. $|x| \geq 2.1$

$x \leq -2.1 \quad or \quad x \geq 2.1$

The solution set is $\{x \mid x \leq -2.1 \ or \ x \geq 2.1\}$, or $(-\infty, -2.1] \cup [2.1, \infty)$.

23.
$$\frac{6}{x-5} = \frac{2}{2x}$$

$$\frac{6}{x-5} = \frac{1}{x} \qquad \left(\frac{2}{2x} = \frac{2 \cdot 1}{2 \cdot x} = \frac{2}{2} \cdot \frac{1}{x} = \frac{1}{x}\right)$$

LCM is $x(x-5)$

$$x(x-5) \cdot \frac{6}{x-5} = x(x-5) \cdot \frac{1}{x}$$

$$6x = x - 5$$

$$5x = -5$$

$$x = -1$$

The solution is -1.

24. $8x = 1 + 16x^2$

$$0 = 16x^2 - 8x + 1$$

$$0 = (4x - 1)(4x - 1)$$

$$4x - 1 = 0 \quad or \quad 4x - 1 = 0$$

$$4x = 1 \quad or \qquad 4x = 1$$

$$x = \frac{1}{4} \quad or \qquad x = \frac{1}{4}$$

The solution is $\frac{1}{4}$.

25. $14 + 3x = 2x^2$

$$0 = 2x^2 - 3x - 14$$

$$0 = (2x - 7)(x + 2)$$

$$2x - 7 = 0 \quad or \quad x + 2 = 0$$

$$2x = 7 \quad or \qquad x = -2$$

$$x = \frac{7}{2} \quad or \qquad x = -2$$

The solutions are $\frac{7}{2}$ and -2.

26. $4x - 2y = 6,$ (1)

$6x - 3y = 9$ (2)

Multiply Equation (1) by 3 and Equation (2) by -2 and then add.

$$\begin{array}{r} 12x - 6y = 18 \\ -12x + 6y = -18 \\ \hline 0 = 0 \end{array}$$

We get an equation that is true for all values of x and y. The system of equations has an infinite number of solutions.

27. $4x + 5y = -3,$ (1)

$x = 1 - 3y$ (2)

Substitute $1 - 3y$ for x in Equation (1) and solve for y.

$$4(1 - 3y) + 5y = -3$$

$$4 - 12y + 5y = -3$$

$$4 - 7y = -3$$

$$-7y = -7$$

$$y = 1$$

Substitute 1 for y in Equation (2) and find x.

$$x = 1 - 3 \cdot 1 = 1 - 3 = -2$$

The solution is $(-2, 1)$.

28. $x + 2y - 2z = 9,$ (1)

$2x - 3y + 4z = -4$ (2)

$5x - 4y + 2z = 5$ (3)

First add Equations (1) and (3) to eliminate z.

$$\begin{array}{r} x + 2y - 2z = 9 \\ 5x - 4y + 2z = 5 \\ \hline 6x - 2y = 14 \quad (4) \end{array}$$

Now multiply Equation (1) by 2 and then add it and Equation (2).

$$\begin{array}{r} 2x + 4y - 4z = 18 \\ 2x - 3y + 4z = -4 \\ \hline 4x + y = 14 \quad (5) \end{array}$$

Now solve the system of Equations (4) and (5). We multiply Equation (5) by 2 and then add.

$$\begin{array}{r} 6x - 2y = 14 \\ 8x + 2y = 28 \\ \hline 14x = 42 \\ x = 3 \end{array}$$

Substitute 3 for x in Equation (5) and solve for y.

$$4 \cdot 3 + y = 14$$

$$12 + y = 14$$

$$y = 2$$

Finally, substitute 3 for x and 2 for y in Equation (3) and solve for z.

$$5 \cdot 3 - 4 \cdot 2 + 2z = 5$$

$$15 - 8 + 2z = 5$$

$$7 + 2z = 5$$

$$2z = -2$$

$$z = -1$$

The solution is $(3, 2, -1)$.

29. $x + 6y + 4z = -2,$ (1)

$4x + 4y + z = 2,$ (2)

$3x + 2y - 4z = 5$ (3)

First add Equations (1) and (3) to eliminate z.

$$\begin{array}{r} x + 6y + 4z = -2 \\ 3x + 2y - 4z = 5 \\ \hline 4x + 8y = 3 \quad (4) \end{array}$$

Now multiply Equation (2) by 4 and then add it and Equation (3).

$$\begin{array}{r} 16x + 16y + 4z = 8 \\ 3x + 2y - 4z = 5 \\ \hline 19x + 18y = 13 \quad (5) \end{array}$$

Now solve the system of Equations (4) and (5). Multiply Equation (4) by 19 and Equation (5) by -4 and then add.

$$\begin{array}{r} 76x + 152y = 57 \\ -76x - 72y = -52 \\ \hline 80y = 5 \\ y = \frac{5}{80} = \frac{1}{16} \end{array}$$

Substitute $\dfrac{1}{16}$ for y in Equation (4) and solve for x.

$$4x + 8 \cdot \frac{1}{16} = 3$$

$$4x + \frac{1}{2} = 3$$

$$4x = \frac{5}{2}$$

$$x = \frac{1}{4} \cdot \frac{5}{2} = \frac{5}{8}$$

Finally, substitute $\dfrac{5}{8}$ for x and $\dfrac{1}{16}$ for y in Equation (2) and solve for z.

$$4 \cdot \frac{5}{8} + 4 \cdot \frac{1}{16} + z = 2$$

$$\frac{5}{2} + \frac{1}{4} + z = 2$$

$$\frac{11}{4} + z = 2$$

$$z = -\frac{3}{4}$$

The solution is $\left(\dfrac{5}{8}, \dfrac{1}{16}, -\dfrac{3}{4} \right)$.

30. $4x^3 + 18x^2 = 2x^2 \cdot 2x + 2x^2 \cdot 9 = 2x^2(2x + 9)$

31. $\quad 8a^3 - 4a^2 - 6a + 3$

$= 4a^2(2a - 1) - 3(2a - 1)$

$= (2a - 1)(4a^2 - 3)$

32. $x^2 + 8x - 84$

We look for two numbers whose product is -84 and whose sum is 8. The numbers we need are 14 and -6.

$x^2 + 8x - 84 = (x + 14)(x - 6)$

33. $6x^2 + 11x - 10$

Using trial and error or the ac-method, we have

$6x^2 + 11x - 10 = (2x + 5)(3x - 2)$.

34. $16y^2 - 81 = (4y)^2 - 9^2 = (4y + 9)(4y - 9)$

35. $t^2 - 16t + 64 = t^2 - 2 \cdot t \cdot 8 + 8^2 = (t - 8)^2$

36. $\quad 64x^3 + 8 = 8(8x^3 + 1) = 8[(2x)^3 + 1^3] =$

$8(2x + 1)(4x^2 - 2x + 1)$

37. $\quad 0.027b^3 - 0.008c^3 = (0.3b)^3 - (0.2c)^3 =$

$(0.3b - 0.2c)(0.09b^2 + 0.06bc + 0.04c^2)$

38. $\quad x^6 - x^2 = x^2(x^4 - 1)$

$= x^2(x^2 + 1)(x^2 - 1)$

$= x^2(x^2 + 1)(x + 1)(x - 1)$

39. $20x^2 + 7x - 3$

Using trial and error or the ac-method, we have

$20x^2 + 7x - 3 = (4x - 1)(5x + 3)$.

40. We will use the point-slope equation.

$$y - y_1 = m(x - x_1)$$

$$y - (-2) = -\frac{1}{2}(x - 2)$$

$$y + 2 = -\frac{1}{2}x + 1$$

$$y = -\frac{1}{2}x - 1$$

41. First we find the slope of the given line.

$$2x + y = 5$$

$$y = -2x + 5$$

The slope is -2. The slope of a line perpendicular to this line is the opposite of the reciprocal of -2, or $\dfrac{1}{2}$. We use the slope-intercept equation to find the equation of the desired line.

$$y = mx + b$$

$$-1 = \frac{1}{2} \cdot 3 + b$$

$$-1 = \frac{3}{2} + b$$

$$-\frac{5}{2} = b$$

The equation is $y = \dfrac{1}{2}x - \dfrac{5}{2}$.

42. Familiarize. Let x, y, and z represent the number of wins, losses, and ties, respectively.

Translate. The team played 81 games, so we have

$$x + y + z = 81.$$

Wins are 3 times ties less 1.

$$\begin{array}{ccccccc} \downarrow & \downarrow & \downarrow & \downarrow & \downarrow & \downarrow & \downarrow \\ x & = & 3 & \cdot & z & - & 1 \end{array}$$

Losses are wins less 8.

$$\begin{array}{ccccc} \downarrow & \downarrow & \downarrow & \downarrow & \downarrow \\ y & = & x & - & 8 \end{array}$$

We have a system of equations.

$$x + y + z = 81,$$

$$x = 3z - 1,$$

$$y = x - 8$$

Solve. Solving the system of equations, we get $(38, 30, 13)$.

Check. The number of games is $38 + 30 + 13$, or 81. The number of wins, 38, is 1 less than three times the number of ties, 13. That is, $38 = 3 \cdot 13 - 1$. Also, the number of losses, 30, is 8 less than 38, the number of wins. The answer checks.

State. The team had 38 wins, 30 losses, and 13 ties.

43. Let $W =$ the amount of waste generated, in pounds, and $c =$ the number of customers. First we find the variation constant.

$$W = kc$$
$$238 = k \cdot 2000$$
$$0.119 = k \qquad \text{Variation constant}$$
$$W = 0.119c \qquad \text{Equation of variation}$$
$$W = 0.119(1700)$$
$$W = 202.3$$

A fast-food restaurant that serves 1700 customers per day generates 202.3 lb of waste daily.

44.
$$\frac{x}{x-4} - \frac{4}{x+3} = \frac{28}{x^2 - x - 12}$$
$$\frac{x}{x-4} - \frac{4}{x+3} = \frac{28}{(x-4)(x+3)}$$

$$\text{LCM is } (x-4)(x+3)$$

$$(x-4)(x+3)\left(\frac{x}{x-4} - \frac{4}{x+3}\right) = (x-4)(x+3) \cdot \frac{28}{(x-4)(x+3)}$$
$$x(x+3) - 4(x-4) = 28$$
$$x^2 + 3x - 4x + 16 = 28$$
$$x^2 - x + 16 = 28$$
$$x^2 - x - 12 = 0$$
$$(x-4)(x+3) = 0$$
$$x - 4 = 0 \quad or \quad x + 3 = 0$$
$$x = 4 \quad or \qquad x = -3$$

Both numbers make a denominator 0, so the equation has no solution. Answer A is correct.

45.
$$x^2 - x - 6 = 6$$
$$x^2 - x - 12 = 0$$
$$(x-4)(x+3) = 0$$
$$x - 4 = 0 \quad or \quad x + 3 = 0$$
$$x = 4 \quad or \qquad x = -3$$

The solutions are 4 and -3. Answer C is correct.

46. Familiarize. Let $t =$ the number of hours it would take the ships to fill the tank, working together.

Translate. We use the work principle.

$$\frac{t}{10} + \frac{t}{15} = 1$$

Solve. We begin by multiplying by the LCM of the denominators, 30, on both sides of the equation.

$$30\left(\frac{t}{10} + \frac{t}{15}\right) = 30 \cdot 1$$
$$3t + 2t = 30$$
$$5t = 30$$
$$t = 6$$

Check. We verify the work principle.

$$\frac{6}{10} + \frac{6}{15} = \frac{18}{30} + \frac{12}{30} = \frac{30}{30} = 1$$

The answer checks.

State. It would take the ships 6 hr to fill the tank, working together. Answer B is correct.

47. We substitute for x and y, using the three given points.

$$2 = a \cdot 4^2 + b \cdot 4 + c, \qquad 2 = 16a + 4b + c,$$
$$0 = a \cdot 2^2 + b \cdot 2 + c, \quad or \quad 0 = 4a + 2b + c,$$
$$2 = a \cdot 1^2 + b \cdot 1 + c \qquad 2 = a + b + c$$

Solving the system of equations, we get $(1, -5, 6)$, so $a = 1$, $b = -5$, and $c = 6$.

48.
$$16x^3 = x$$
$$16x^3 - x = 0$$
$$x(16x^2 - 1) =$$
$$x(4x + 1)(4x - 1) = 0$$
$$x = 0 \quad or \quad 4x + 1 = 0 \quad or \quad 4x - 1 = 0$$
$$x = 0 \quad or \qquad 4x = -1 \quad or \qquad 4x = 1$$
$$x = 0 \quad or \qquad x = -\frac{1}{4} \quad or \qquad x = \frac{1}{4}$$

The solutions are 0, $-\dfrac{1}{4}$, and $\dfrac{1}{4}$.

49.
$$\frac{18}{x-9} + \frac{10}{x+5} = \frac{28x}{x^2 - 4x - 45}$$
$$\frac{18}{x-9} + \frac{10}{x+5} = \frac{28x}{(x-9)(x+5)}$$

$$\text{LCM is } (x-9)(x+5)$$

$$(x-9)(x+5)\left(\frac{18}{x-9} + \frac{10}{x+5}\right) = (x-9)(x+5) \cdot \frac{28x}{(x-9)(x+5)}$$
$$18(x+5) + 10(x-9) = 28x$$
$$18x + 90 + 10x - 90 = 28x$$
$$28x = 28x$$
$$0 = 0 \qquad \text{Subtracting 28x}$$

We get an equation that is true for all values of x, so the solutions of the equation are all real numbers that do not make a denominator 0. Since $x - 9 = 0$ when $x = 9$ and $x + 5 = 0$ when $x = -5$, all real numbers except 9 and -5 are solutions.

Chapter 6

Radical Expressions, Equations, and Functions

RC1. The domain of $f(x) = \sqrt{9-x}$ is the set of all x-values for which $9 - x \geq 0$.

$$9 - x \geq 0$$
$$9 \geq x$$

The domain is $(-\infty, 9]$. The answer is (j).

RC3. The domain of $g(x) = \sqrt{x-3}$ is the set of all x-values for which $x - 3 \geq 0$.

$$x - 3 \geq 0$$
$$x \geq 3$$

The domain is $[3, \infty)$. The answer is (h).

RC5. $f(x) = 3 - x$ is defined for all real numbers. Thus the domain $(-\infty, \infty)$. The answer is (i).

1. The square roots of 16 are 4 and -4, because $4^2 = 16$ and $(-4)^2 = 16$.

3. The square roots of 144 are 12 and -12, because $12^2 = 144$ and $(-12)^2 = 144$.

5. The square roots of 400 are 20 and -20, because $20^2 = 400$ and $(-20)^2 = 400$.

7. $-\sqrt{\dfrac{49}{36}} = -\dfrac{7}{6}$ Since $\sqrt{\dfrac{49}{36}} = \dfrac{7}{6}$, $-\sqrt{\dfrac{49}{36}} = -\dfrac{7}{6}$.

9. $\sqrt{196} = 14$ Remember, $\sqrt{}$ indicates the principle square root.

11. $\sqrt{0.0036} = 0.06$

13. $\sqrt{-225}$ does not exist as a real number because negative numbers do not have real-number square roots.

15. $\sqrt{347} \approx 18.628$

17. $\sqrt{\dfrac{285}{74}} \approx 1.962$

19. $9\sqrt{y^2 + 16}$

The radicand is the expression written under the radical sign, $y^2 + 16$.

21. $x^4 y^5 \sqrt{\dfrac{x}{y-1}}$

The radicand is the expression written under the radical sign, $\dfrac{x}{y-1}$.

23. $f(x) = \sqrt{5x - 10}$

$f(6) = \sqrt{5 \cdot 6 - 10} = \sqrt{20} \approx 4.472$
$f(2) = \sqrt{5 \cdot 2 - 10} = \sqrt{0} = 0$
$f(1) = \sqrt{5 \cdot 1 - 10} = \sqrt{-5}$

Since negative numbers do not have real-number square roots, $f(1)$ does not exist as a real number.

$f(-1) = \sqrt{5(-1) - 10} = \sqrt{-15}$

Since negative numbers do not have real-number square roots, $f(-1)$ does not exist as a real number.

25. $g(x) = \sqrt{x^2 - 25}$

$g(-6) = \sqrt{(-6)^2 - 25} = \sqrt{11} \approx 3.317$
$g(3) = \sqrt{3^2 - 25} = \sqrt{-16}$

Since negative numbers do not have real-number square roots, $g(3)$ does not exist as a real number.

$g(6) = \sqrt{6^2 - 25} = \sqrt{11} \approx 3.317$
$g(13) = \sqrt{13^2 - 25} = \sqrt{144} = 12$

27. The domain of $f(x) = \sqrt{5x - 10}$ is the set of all x-values for which $5x - 10 \geq 0$.

$$5x - 10 \geq 0$$
$$5x \geq 10$$
$$x \geq 2$$

The domain is $\{x | x \geq 2\}$, or $[2, \infty)$.

29. $N(a) = 2.5\sqrt{a}$

$N(66) = 2.5\sqrt{66} \approx 20.3$

About 21 spaces will be needed.

$N(10) = 2.5\sqrt{100} = 2.5(10) = 25$ spaces

31. Graph: $f(x) = 2\sqrt{x}$.

We find some ordered pairs, plot points, and draw the curve.

x	$f(x)$	$(x, f(x))$
0	0	$(0, 0)$
1	2	$(1, 2)$
2	2.8	$(2, 2.8)$
3	3.5	$(3, 3.5)$
4	4	$(4, 4)$
5	4.5	$(5, 4.5)$

33. Graph: $F(x) = -3\sqrt{x}$.

We find some ordered pairs, plot points, and draw the curve.

x	$f(x)$	$(x, f(x))$
0	0	$(0,0)$
1	-3	$(1,-3)$
2	-4.2	$(2,-4.2)$
3	-5.2	$(3,-5.2)$
4	-6	$(4,-6)$
5	-6.7	$(5,-6.7)$

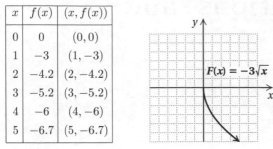

35. Graph: $f(x) = \sqrt{x}$.

We find some ordered pairs, plot points, and draw the curve.

x	$f(x)$	$(x, f(x))$
0	0	$(0,0)$
1	1	$(1,1)$
2	1.4	$(2,1.4)$
3	1.7	$(3,1.7)$
4	2	$(4,2)$
5	2.2	$(5,2.2)$

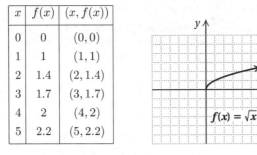

37. Graph: $f(x) = \sqrt{x-2}$.

We find some ordered pairs, plot points, and draw the curve.

x	$f(x)$	$(x, f(x))$
2	0	$(2,0)$
3	1	$(3,1)$
4	1.4	$(4,1.4)$
5	1.7	$(5,1.7)$
7	2.2	$(7,2.2)$
9	2.6	$(9,2.6)$

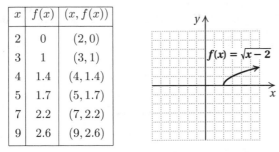

39. Graph: $f(x) = \sqrt{12-3x}$.

We find some ordered pairs, plot points, and draw the curve.

x	$f(x)$	$(x, f(x))$
-5	5.2	$(-5,5.2)$
-3	4.6	$(-3,4.6)$
-1	3.9	$(-1,3.9)$
0	3.5	$(0,3.5)$
2	2.4	$(2,2.4)$
4	0	$(4,0)$

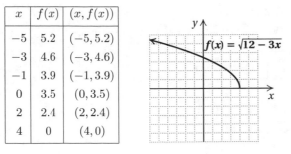

41. Graph: $g(x) = \sqrt{3x+9}$.

We find some ordered pairs, plot points, and draw the curve.

x	$f(x)$	$(x, f(x))$
-3	0	$(-3,0)$
-1	2.4	$(-1,2.4)$
0	3	$(0,3)$
1	3.5	$(1,3.5)$
3	4.2	$(3,4.2)$
5	4.9	$(5,4.9)$

43. $\sqrt{16x^2} = \sqrt{(4x)^2} = |4x| = 4|x|$

(The absolute value is used to ensure that the principal square root is nonnegative.)

45. $\sqrt{(-12c)^2} = |-12c| = |-12| \cdot |c| = 12|c|$

(The absolute value is used to ensure that the principal square root is nonnegative.)

47. $\sqrt{(p+3)^2} = |p+3|$

(The absolute value is used to ensure that the principal square root is nonnegative.)

49. $\sqrt{x^2 - 4x + 4} = \sqrt{(x-2)^2} = |x-2|$

(The absolute value is used to ensure that the principal square root is nonnegative.)

51. $\sqrt[3]{27} = 3 \quad [3^3 = 27]$

53. $\sqrt[3]{-64x^3} = -4x \quad [(-4x)^3 = -64x^3]$

55. $\sqrt[3]{-216} = -6 \quad [(-6)^3 = -216]$

57. $\sqrt[3]{0.343(x+1)^3} = 0.7(x+1)$

$$[(0.7(x+1))^3 = 0.343(x+1)^3]$$

59.
$$f(x) = \sqrt[3]{x+1}$$
$$f(7) = \sqrt[3]{7+1} = \sqrt[3]{8} = 2$$
$$f(26) = \sqrt[3]{26+1} = \sqrt[3]{27} = 3$$
$$f(-9) = \sqrt[3]{-9+1} = \sqrt[3]{-8} = -2$$
$$f(-65) = \sqrt[3]{-65+1} = \sqrt[3]{-64} = -4$$

61.
$$f(x) = -\sqrt[3]{3x+1}$$
$$f(0) = -\sqrt[3]{3 \cdot 0 + 1} = -\sqrt[3]{1} = -1$$
$$f(-7) = -\sqrt[3]{3(-7)+1} = -\sqrt[3]{-20}, or \sqrt[3]{20} \approx 2.7144$$
$$f(21) = -\sqrt[3]{3 \cdot 21 + 1} = -\sqrt[3]{64} = -4$$
$$f(333) = -\sqrt[3]{3 \cdot 333 + 1} = -\sqrt[3]{1000} = -10$$

63. $-\sqrt[4]{625} = -5 \quad$ Since $5^4 = 625$, then $\sqrt[4]{625} = 5$ and $-\sqrt{625} = -5$.

65. $\sqrt[5]{-1} = -1 \quad$ Since $(-1)^5 = -1$

67. $\sqrt[5]{-\dfrac{32}{243}} = -\dfrac{2}{3} \quad$ Since $\left(-\dfrac{2}{3}\right)^5 = -\dfrac{32}{243}$

69. $\sqrt[6]{x^6} = |x|$

The index is even so we use absolute-value notation.

71. $\sqrt[4]{(5a)^4} = |5a| = 5|a|$

The index is even so we use absolute-value notation.

73. $\sqrt[10]{(-6)^{10}} = |-6| = 6$

75. $\sqrt[414]{(a+b)^{414}} = |a+b|$

The index is even so we use absolute-value notation.

77. $\sqrt[7]{y^7} = y$

We do not use absolute-value notation when the index is odd.

79. $\sqrt[5]{(x-2)^5} = x - 2$

We do not use absolute-value notation when the index is odd.

81. $\quad x^2 + x - 2 = 0$

$(x+2)(x-1) = 0$ Factoring

$x + 2 = 0 \quad or \quad x - 1 = 0$ Principle of zero products

$\qquad x = -2 \quad or \qquad x = 1$

The solutions are -2 and 1.

83. $\qquad 4x^2 - 49 = 0$

$(2x+7)(2x-7) = 0$ Factoring

$2x + 7 = 0 \quad or \quad 2x - 7 = 0$ Principle of zero products

$2x = -7 \quad or \qquad 2x = 7$

$x = -\dfrac{7}{2} \quad or \qquad x = \dfrac{7}{2}$

The solutions are $-\dfrac{7}{2}$ and $\dfrac{7}{2}$.

85. $\qquad 3x^2 + x = 10$

$3x^2 + x - 10 = 0$

$(3x-5)(x+2) = 0$

$3x - 5 = 0 \quad or \quad x + 2 = 0$

$3x = 5 \quad or \qquad x = -2$

$x = \dfrac{5}{3} \quad or \qquad x = -2$

The solutions are $\dfrac{5}{3}$ and -2.

87. $4x^3 - 20x^2 + 25x = 0$

$x(4x^2 - 20x + 25) = 0$

$x(2x-5)(2x-5) = 0$

$x = 0 \quad or \quad 2x - 5 = 0 \quad or \quad 2x - 5 = 0$

$x = 0 \quad or \qquad 2x = 5 \quad or \qquad 2x = 5$

$x = 0 \quad or \qquad x = \dfrac{5}{2} \quad or \qquad x = \dfrac{5}{2}$

The solutions are 0 and $\dfrac{5}{2}$.

89. $(a^3b^2c^5)^3 = a^{3\cdot3}b^{2\cdot3}c^{5\cdot3} = a^9b^6c^{15}$

91. $f(x) = \dfrac{\sqrt{x+3}}{\sqrt{2-x}}$

In the numerator we must have $x + 3 \geq 0$, or $x \geq -3$, and in the denominator we must have $2 - x > 0$, or $x < 2$. Thus, we have $x \geq -3$ *and* $x < 2$, so

Domain of $f = \{x | -3 \leq x < 2\}$, or $[-3, 2)$.

93. From 3 on the x-axis, go up to the graph and across to the y-axis to find $f(3) = \sqrt{3} \approx 1.7$.

From 5 on the x-axis, go up to the graph and across to the y-axis to find $f(5) = \sqrt{5} \approx 2.2$.

From 10 on the x-axis, go up to the graph and across to the y-axis to find $f(10) = \sqrt{10} \approx 3.2$.

95. a) $f(x) = \sqrt[3]{x}$

Domain $= (-\infty, \infty)$; range $= (-\infty, \infty)$

b) $g(x) = \sqrt[3]{4x - 5}$

Domain $= (-\infty, \infty)$; range $= (-\infty, \infty)$

c) $q(x) = 2 - \sqrt{x+3}$

Domain $= [-3, \infty)$; range $= (-\infty, 2]$

d) $h(x) = \sqrt[4]{x}$

Domain $= [0, \infty)$; range $= [0, \infty)$

e) $t(x) = \sqrt[4]{x - 3}$

Domain $= [3, \infty)$; range $= [0, \infty)$

Exercise Set 6.2

RC1. $\dfrac{c^2}{c^5} = c^{2-5}$; the answer is (h).

RC3. $\sqrt{c} = c^{1/2}$; the answer is (c).

RC5. $\sqrt{c^5} = (c^5)^{1/2} = c^{5/2}$; the answer is (e).

RC7. $-c^5 \cdot c^2 = -c^{5+2}$; the answer is (f).

1. $y^{1/7} = \sqrt[7]{y}$

3. $(8)^{1/3} = \sqrt[3]{8} = 2$

5. $(a^3b^3)^{1/5} = \sqrt[5]{a^3b^3}$

7. $16^{3/4} = \sqrt[4]{16^3} = (\sqrt[4]{16})^3 = 2^3 = 8$

9. $49^{3/2} = \sqrt{49^3} = (\sqrt{49})^3 = 7^3 = 343$

11. $\sqrt{17} = 17^{1/2}$

13. $\sqrt[3]{18} = 18^{1/3}$

15. $\sqrt[5]{xy^2z} = (xy^2z)^{1/5}$

17. $(\sqrt{3mn})^3 = (3mn)^{3/2}$

19. $(\sqrt[7]{8x^2y})^5 = (8x^2y)^{5/7}$

21. $27^{-1/3} = \dfrac{1}{27^{1/3}} = \dfrac{1}{\sqrt[3]{27}} = \dfrac{1}{3}$

23. $100^{-3/2} = \dfrac{1}{100^{3/2}} = \dfrac{1}{(\sqrt{100})^3} = \dfrac{1}{10^3} = \dfrac{1}{1000}$

25. $3x^{-1/4} = 3 \cdot \dfrac{1}{x^{1/4}} = \dfrac{3}{x^{1/4}}$

27. $(2rs)^{-3/4} = \dfrac{1}{(2rs)^{3/4}}$

29. $2a^{3/4}b^{-1/2}c^{2/3} = 2 \cdot a^{3/4} \cdot \dfrac{1}{b^{1/2}} \cdot c^{2/3} = \dfrac{2a^{3/4}c^{2/3}}{b^{1/2}}$

31. $\left(\dfrac{7x}{8yz}\right)^{-3/5} = \left(\dfrac{8yz}{7x}\right)^{3/5}$　$\left(\text{Since } \left(\dfrac{a}{b}\right)^{-n} = \left(\dfrac{b}{a}\right)^n\right)$

33. $\dfrac{1}{x^{-2/3}} = x^{2/3}$

35. $2^{-1/3}x^4y^{-2/7} = \dfrac{1}{2^{1/3}} \cdot x^4 \cdot \dfrac{1}{y^{2/7}} = \dfrac{x^4}{2^{1/3}y^{2/7}}$

37. $\dfrac{7x}{\sqrt[3]{z}} = \dfrac{7x}{z^{1/3}}$

39. $\dfrac{5a}{3c^{-1/2}} = \dfrac{5a}{3} \cdot c^{1/2} = \dfrac{5ac^{1/2}}{3}$

41. $5^{3/4} \cdot 5^{1/8} = 5^{3/4+1/8} = 5^{6/8+1/8} = 5^{7/8}$

43. $\dfrac{7^{5/8}}{7^{3/8}} = 7^{5/8-3/8} = 7^{2/8} = 7^{1/4}$

45. $\dfrac{4.9^{-1/6}}{4.9^{-2/3}} = 4.9^{-1/6-(-2/3)} = 4.9^{-1/6+4/6} = 4.9^{3/6} = 4.9^{1/2}$

47. $(6^{3/8})^{2/7} = 6^{3/8 \cdot 2/7} = 6^{6/56} = 6^{3/28}$

49. $a^{2/3} \cdot a^{5/4} = a^{2/3+5/4} = a^{8/12+15/12} = a^{23/12}$

51. $(a^{2/3} \cdot b^{5/8})^4 = (a^{2/3})^4(b^{5/8})^4 = a^{8/3}b^{20/8} = a^{8/3}b^{5/2}$

53. $(x^{2/3})^{-3/7} = x^{2/3(-3/7)} = x^{-2/7} = \dfrac{1}{x^{2/7}}$

55. $\left(\dfrac{x^{3/4}}{y^{1/2}}\right)^{-2/3} = \left(\dfrac{y^{1/2}}{x^{3/4}}\right)^{2/3} = \dfrac{y^{1/2 \cdot 2/3}}{x^{3/4 \cdot 2/3}} = \dfrac{y^{1/3}}{x^{1/2}}$

57. $(m^{-1/4} \cdot n^{-5/6})^{-12/5} = m^{-1/4 \cdot (-12/5)}n^{-5/6 \cdot (-12/5)} = m^{3/5}n^2$

59. $\sqrt[6]{a^2} = a^{2/6}$　Converting to exponential notation
$\qquad = a^{1/3}$　Simplifying the exponent
$\qquad = \sqrt[3]{a}$　Returning to radical notation

61. $\sqrt[3]{x^{15}} = x^{15/3}$　Converting to exponential notation
$\qquad = x^5$　Simplifying

63. $\sqrt[6]{x^{-18}} = x^{-18/6}$　Converting to exponential notation
$\qquad = x^{-3}$　Simplifying
$\qquad = \dfrac{1}{x^3}$

65. $(\sqrt[3]{ab})^{15} = (ab)^{15/3}$　Converting to exponential notation
$\qquad = (ab)^5$　Simplifying the exponent
$\qquad = a^5b^5$　Using the law of exponents

67. $\sqrt[14]{128} = \sqrt[14]{2^7} = 2^{7/14} = 2^{1/2} = \sqrt{2}$

69. $\sqrt[6]{4x^2} = (2^2x^2)^{1/6} = 2^{2/6}x^{2/6}$
$\qquad = 2^{1/3}x^{1/3} = (2x)^{1/3} = \sqrt[3]{2x}$

71. $\sqrt{x^4y^6} = (x^4y^6)^{1/2} = x^{4/2}y^{6/2} = x^2y^3$

73. $\sqrt[5]{32c^{10}d^{15}} = (2^5c^{10}d^{15})^{1/5} = 2^{5/5}c^{10/5}d^{15/5}$
$\qquad = 2c^2d^3$

75. $\sqrt[3]{7} \cdot \sqrt[4]{5} = 7^{1/3} \cdot 5^{1/4} = 7^{4/12} \cdot 5^{3/12} = (7^4 \cdot 5^3)^{1/12} = \sqrt[12]{7^4 \cdot 5^3}$

77. $\sqrt[4]{5} \cdot \sqrt[5]{7} = 5^{1/4} \cdot 7^{1/5} = 5^{5/20} \cdot 7^{4/20} = (5^5 \cdot 7^4)^{1/20} = \sqrt[20]{5^5 \cdot 7^4}$

79. $\sqrt{x}\sqrt[3]{2x} = x^{1/2} \cdot (2x)^{1/3} = x^{3/6} \cdot (2x)^{2/6} = [x^3(2x)^2]^{1/6} = (x^3 \cdot 4x^2)^{1/6} = (4x^5)^{1/6} = \sqrt[6]{4x^5}$

81. $(\sqrt[5]{a^2b^4})^{15} = (a^2b^4)^{15/5} = (a^2b^4)^3 = a^6b^{12}$

83. $\sqrt[3]{\sqrt[6]{m}} = \sqrt[3]{m^{1/6}} = (m^{1/6})^{1/3} = m^{1/18} = \sqrt[18]{m}$

85. $x^{1/3} \cdot y^{1/4} \cdot z^{1/6} = x^{4/12} \cdot y^{3/12} \cdot z^{2/12} = (x^4y^3z^2)^{1/12} = \sqrt[12]{x^4y^3z^2}$

87. $\left(\dfrac{c^{-4/5}d^{5/9}}{c^{3/10}d^{1/6}}\right)^3 = (c^{-4/5-3/10}d^{5/9-1/6})^3 =$
$(c^{-8/10-3/10}d^{10/18-3/18})^3 = (c^{-11/10}d^{7/18})^3 =$
$c^{-33/10}d^{7/6} = c^{-99/30}d^{35/30} = (c^{-99}d^{35})^{1/30} =$
$\left(\dfrac{d^{35}}{c^{99}}\right)^{1/30} = \sqrt[30]{\dfrac{d^{35}}{c^{99}}}$

89. $|7x - 5| = 9$
$7x - 5 = -9 \quad or \quad 7x - 5 = 9$
$\qquad 7x = -4 \quad or \qquad 7x = 14$
$\qquad x = -\dfrac{4}{7} \quad or \qquad x = 2$
The solution set is $\left\{-\dfrac{4}{7}, 2\right\}$.

91. $8 - |2x + 5| = -2$
$\qquad -|2x + 5| = -10$
$\qquad |2x + 5| = 10$
$2x + 5 = -10 \quad or \quad 2x + 5 = 10$
$\qquad 2x = -15 \quad or \qquad 2x = 5$
$\qquad x = -\dfrac{15}{2} \quad or \qquad x = \dfrac{5}{2}$
The solution set is $\left\{-\dfrac{15}{2}, \dfrac{5}{2}\right\}$.

93.

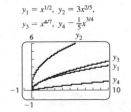

Exercise Set 6.3

RC1. True; see page 510 in the text.

RC3. True; see page 511 in the text.

1. $\sqrt{24} = \sqrt{4 \cdot 6} = \sqrt{4}\,\sqrt{6} = 2\sqrt{6}$

3. $\sqrt{90} = \sqrt{9 \cdot 10} = \sqrt{9}\,\sqrt{10} = 3\sqrt{10}$

5. $\sqrt[3]{250} = \sqrt[3]{125 \cdot 2} = \sqrt[3]{125}\,\sqrt[3]{2} = 5\sqrt[3]{2}$

7. $\sqrt{180x^4} = \sqrt{36 \cdot 5 \cdot x^4} = \sqrt{36x^4}\,\sqrt{5} = 6x^2\sqrt{5}$

9. $\sqrt[3]{54x^8} = \sqrt[3]{27 \cdot 2 \cdot x^6 \cdot x^2} = \sqrt[3]{27x^6}\,\sqrt[3]{2x^2} = 3x^2\sqrt[3]{2x^2}$

11. $\sqrt[3]{80t^8} = \sqrt[3]{8 \cdot 10 \cdot t^6 \cdot t^2} = \sqrt[3]{8t^6}\,\sqrt[3]{10t^2} = 2t^2\sqrt[3]{10t^2}$

13. $\sqrt[4]{80} = \sqrt[4]{16 \cdot 5} = \sqrt[4]{16}\,\sqrt[4]{5} = 2\sqrt[4]{5}$

15. $\sqrt{32a^2b} = \sqrt{16a^2 \cdot 2b} = \sqrt{16a^2} \cdot \sqrt{2b} = 4a\sqrt{2b}$

17. $\sqrt[4]{243x^8y^{10}} = \sqrt[4]{81x^8y^8 \cdot 3y^2} = \sqrt[4]{81x^8y^8}\,\sqrt[4]{3y^2} = 3x^2y^2\sqrt[4]{3y^2}$

19. $\sqrt[5]{96x^7y^{15}} = \sqrt[5]{32x^5y^{15} \cdot 3x^2} = \sqrt[5]{32x^5y^{15}}\,\sqrt[5]{3x^2} = 2xy^3\sqrt[5]{3x^2}$

21. $\sqrt{10}\sqrt{5} = \sqrt{10 \cdot 5} = \sqrt{50} = \sqrt{25 \cdot 2} = 5\sqrt{2}$

23. $\sqrt{15}\,\sqrt{6} = \sqrt{15 \cdot 6} = \sqrt{90}$
$= \sqrt{9 \cdot 10} = \sqrt{9}\,\sqrt{10} = 3\sqrt{10}$

25. $\sqrt[3]{2}\sqrt[3]{4} = \sqrt[3]{2 \cdot 4} = \sqrt[3]{8} = 2$

27. $\sqrt{45}\,\sqrt{60} = \sqrt{45 \cdot 60} = \sqrt{2700}$
$= \sqrt{900 \cdot 3} = \sqrt{900}\,\sqrt{3} = 30\sqrt{3}$

29. $\sqrt{3x^3}\sqrt{6x^5} = \sqrt{18x^8} = \sqrt{9x^8 \cdot 2} = 3x^4\sqrt{2}$

31. $\sqrt{5b^3}\,\sqrt{10c^4} - \sqrt{5b^3 \cdot 10c^4}$
$= \sqrt{50b^3c^4}$
$= \sqrt{25 \cdot 2 \cdot b^2 \cdot b \cdot c^4}$
$= \sqrt{25b^2c^4}\,\sqrt{2b}$
$= 5bc^2\,\sqrt{2b}$

33. $\sqrt[3]{5a^2}\,\sqrt[3]{2a} = \sqrt[3]{5a^2 \cdot 2a} = \sqrt[3]{10a^3} = \sqrt[3]{a^3 \cdot 10} = a\sqrt[3]{10}$

35. $\sqrt[3]{y^4}\,\sqrt[3]{16y^5} = \sqrt[3]{y^4 \cdot 16y^5}$
$= \sqrt[3]{16y^9}$
$= \sqrt[3]{8 \cdot 2 \cdot y^9}$
$= \sqrt[3]{8y^9}\,\sqrt[3]{2}$
$= 2y^3\,\sqrt[3]{2}$

37. $\sqrt[4]{16}\,\sqrt[4]{64} = \sqrt[4]{16 \cdot 64} = \sqrt[4]{1024} = \sqrt[4]{256 \cdot 4} = \sqrt[4]{256}\,\sqrt[4]{4} = 4\sqrt[4]{4}$

39. $\sqrt{12a^3b}\,\sqrt{8a^4b^2} = \sqrt{12a^3b \cdot 8a^4b^2} = $
$\sqrt{96a^7b^3} = \sqrt{16a^6b^2 \cdot 6ab} = \sqrt{16a^6b^2}\,\sqrt{6ab} = $
$4a^3b\sqrt{6ab}$

41. $\sqrt{2}\,\sqrt[3]{5}$
$= 2^{1/2} \cdot 5^{1/3}$ Converting to exponential notation
$= 2^{3/6} \cdot 5^{2/6}$ Rewriting so that exponents have a common denominator
$= (2^3 \cdot 5^2)^{1/6}$ Using $a^nb^n = (ab)^n$
$= \sqrt[6]{2^3 \cdot 5^2}$ Converting to radical notation
$= \sqrt[6]{8 \cdot 25}$ Simplifying
$= \sqrt[6]{200}$ Multiplying

43. $\sqrt[4]{3}\,\sqrt{2}$
$= 3^{1/4} \cdot 2^{1/2}$ Converting to exponential notation
$= 3^{1/4} \cdot 2^{2/4}$ Rewriting so that exponents have a common denominator
$= (3 \cdot 2^2)^{1/4}$ Using $a^nb^n = (ab)^n$
$= \sqrt[4]{3 \cdot 2^2}$ Converting to radical notation
$= \sqrt[4]{3 \cdot 4}$ Squaring 2
$= \sqrt[4]{12}$ Multiplying

45. $\sqrt{a}\,\sqrt[4]{a^3}$
$= a^{1/2} \cdot a^{3/4}$ Converting to exponential notation
$= a^{5/4}$ Adding exponents
$= a^{1+1/4}$ Writing 5/4 as a mixed number
$= a \cdot a^{1/4}$ Factoring
$= a\sqrt[4]{a}$ Returning to radical notation

47. $\sqrt[5]{b^2}\,\sqrt{b^3}$
$= b^{2/5} \cdot b^{3/2}$ Converting to exponential notation
$= b^{19/10}$ Adding exponents
$= b^{1+9/10}$ Writing 19/10 as a mixed number
$= b \cdot b^{9/10}$ Factoring
$= b\,\sqrt[10]{b^9}$ Returning to radical notation

49. $\sqrt{xy^3}\,\sqrt[3]{x^2y} = (xy^3)^{1/2}(x^2y)^{1/3}$
$= (xy^3)^{3/6}(x^2y)^{2/6}$
$= [(xy^3)^3(x^2y)^2]^{1/6}$
$= \sqrt[6]{x^3y^9 \cdot x^4y^2}$
$= \sqrt[6]{x^7y^{11}}$
$= \sqrt[6]{x^6y^6 \cdot xy^5}$
$= xy\,\sqrt[6]{xy^5}$

51. $\sqrt{2a^3b}\,\sqrt[4]{8ab^2} = (2a^3b)^{1/2}(8ab^2)^{1/4}$
$= (2a^3b)^{2/4}(8ab^2)^{1/4}$
$= [(2a^3b)^2(8ab^2)]^{1/4}$
$= \sqrt[4]{4a^6b^2 \cdot 8ab^2}$
$= \sqrt[4]{32a^7b^4}$
$= \sqrt[4]{16a^4b^4 \cdot 2a^3}$
$= 2ab\,\sqrt[4]{2a^3}$

53. $\dfrac{\sqrt{90}}{\sqrt{5}} = \sqrt{\dfrac{90}{5}} = \sqrt{18} = \sqrt{9 \cdot 2} = \sqrt{9}\,\sqrt{2} = 3\sqrt{2}$

55. $\dfrac{\sqrt{35q}}{\sqrt{7q}} = \sqrt{\dfrac{35q}{7q}} = \sqrt{5}$

57. $\dfrac{\sqrt[3]{54}}{\sqrt[3]{2}} = \sqrt[3]{\dfrac{54}{2}} = \sqrt[3]{27} = 3$

59. $\dfrac{\sqrt{56xy^3}}{\sqrt{8x}} = \sqrt{\dfrac{56xy^3}{8x}} = \sqrt{7y^3} = \sqrt{y^2 \cdot 7y} =$
$\sqrt{y^2}\,\sqrt{7y} = y\sqrt{7y}$

61. $\dfrac{\sqrt[3]{96a^4b^2}}{\sqrt[3]{12a^2b}} = \sqrt[3]{\dfrac{96a^4b^2}{12a^2b}} = \sqrt[3]{8a^2b} = \sqrt[3]{8}\,\sqrt[3]{a^2b} = 2\sqrt[3]{a^2b}$

63. $\dfrac{\sqrt{128xy}}{2\sqrt{2}} = \dfrac{1}{2}\dfrac{\sqrt{128xy}}{\sqrt{2}} = \dfrac{1}{2}\sqrt{\dfrac{128xy}{2}} = \dfrac{1}{2}\sqrt{64xy} =$
$\dfrac{1}{2}\sqrt{64}\,\sqrt{xy} = \dfrac{1}{2} \cdot 8\sqrt{xy} = 4\sqrt{xy}$

65. $\dfrac{\sqrt[4]{48x^9y^{13}}}{\sqrt[4]{3xy^5}} = \sqrt[4]{\dfrac{48x^9y^{13}}{3xy^5}} = \sqrt[4]{16x^8y^8} = 2x^2y^2$

67. $\dfrac{\sqrt[3]{a}}{\sqrt{a}}$

$= \dfrac{a^{1/3}}{a^{1/2}}$ Converting to exponential notation

$= a^{1/3-1/2}$ Subtracting exponents

$= a^{2/6-3/6}$

$= a^{-1/6}$

$= \dfrac{1}{a^{1/6}}$

$= \dfrac{1}{\sqrt[6]{a}}$ Converting to radical notation

69. $\dfrac{\sqrt[3]{a^2}}{\sqrt[4]{a}}$

$= \dfrac{a^{2/3}}{a^{1/4}}$ Converting to exponential notation

$= a^{2/3-1/4}$ Subtracting exponents

$= a^{5/12}$ Converting back

$= \sqrt[12]{a^5}$ to radical notation

71. $\dfrac{\sqrt[4]{x^2y^3}}{\sqrt[3]{xy}}$

$= \dfrac{(x^2y^3)^{1/4}}{(xy)^{1/3}}$ Converting to exponential notation

$= \dfrac{x^{2/4}y^{3/4}}{x^{1/3}y^{1/3}}$ Using the power and product rules

$= x^{2/4-1/3}y^{3/4-1/3}$ Subtracting exponents

$= x^{2/12}y^{5/12}$

$= (x^2y^5)^{1/12}$ Converting back to

$= \sqrt[12]{x^2y^5}$ radical notation

73. $\sqrt{\dfrac{25}{36}} = \dfrac{\sqrt{25}}{\sqrt{36}} = \dfrac{5}{6}$

75. $\sqrt{\dfrac{16}{49}} = \dfrac{\sqrt{16}}{\sqrt{49}} = \dfrac{4}{7}$

77. $\sqrt[3]{\dfrac{125}{27}} = \dfrac{\sqrt[3]{125}}{\sqrt[3]{27}} = \dfrac{5}{3}$

79. $\sqrt{\dfrac{49}{y^2}} = \dfrac{\sqrt{49}}{\sqrt{y^2}} = \dfrac{7}{y}$

81. $\sqrt{\dfrac{25y^3}{x^4}} = \dfrac{\sqrt{25y^3}}{\sqrt{x^4}} = \dfrac{\sqrt{25y^2 \cdot y}}{\sqrt{x^4}} = \dfrac{\sqrt{25y^2}\,\sqrt{y}}{\sqrt{x^4}} =$
$\dfrac{5y\sqrt{y}}{x^2}$

83. $\sqrt[3]{\dfrac{81y^5}{64}} = \dfrac{\sqrt[3]{81y^5}}{\sqrt[3]{64}} = \dfrac{\sqrt[3]{27y^3 \cdot 3y^2}}{\sqrt[3]{64}} = \dfrac{\sqrt[3]{27y^3}\,\sqrt[3]{3y^2}}{\sqrt[3]{64}} =$
$\dfrac{3y\sqrt[3]{3y^2}}{4}$

85. $\sqrt[3]{\dfrac{27a^4}{8b^3}} = \dfrac{\sqrt[3]{27a^4}}{\sqrt[3]{8b^3}} = \dfrac{\sqrt[3]{27a^3 \cdot a}}{\sqrt[3]{8b^3}} = \dfrac{\sqrt[3]{27a^3}\,\sqrt[3]{a}}{\sqrt[3]{8b^3}} =$
$\dfrac{3a\sqrt[3]{a}}{2b}$

87. $\sqrt[4]{\dfrac{81x^4}{16}} = \dfrac{\sqrt[4]{81x^4}}{\sqrt[4]{16}} = \dfrac{3x}{2}$

89. $\sqrt[4]{\dfrac{16a^{12}}{b^4c^{16}}} = \dfrac{\sqrt[4]{16a^{12}}}{\sqrt[4]{b^4c^{16}}} = \dfrac{2a^3}{bc^4}$

91. $\sqrt[5]{\dfrac{32x^8}{y^{10}}} = \dfrac{\sqrt[5]{32x^8}}{\sqrt[5]{y^{10}}} = \dfrac{\sqrt[5]{32 \cdot x^5 \cdot x^3}}{\sqrt[5]{y^{10}}} = \dfrac{\sqrt[5]{32x^5}\,\sqrt[5]{x^3}}{\sqrt[5]{y^{10}}} =$
$\dfrac{2x\sqrt[5]{x^3}}{y^2}$

93. $\sqrt[5]{\dfrac{w^7}{z^{10}}} = \dfrac{\sqrt[5]{w^7}}{\sqrt[5]{z^{10}}} = \dfrac{\sqrt[5]{w^5 \cdot w^2}}{\sqrt[5]{z^{10}}} = \dfrac{\sqrt[5]{w^5}\,\sqrt[5]{w^2}}{\sqrt[5]{z^{10}}} =$
$\dfrac{w\sqrt[5]{w^2}}{z^2}$

95. $\sqrt[6]{\dfrac{x^{13}}{y^6z^{12}}} = \dfrac{\sqrt[6]{x^{13}}}{\sqrt[6]{y^6z^{12}}} = \dfrac{\sqrt[6]{x^{12} \cdot x}}{\sqrt[6]{y^6z^{12}}} = \dfrac{\sqrt[6]{x^{12}}\,\sqrt[6]{x}}{\sqrt[6]{y^6z^{12}}} = \dfrac{x^2\sqrt[6]{x}}{yz^2}$

97. *Familiarize.* Let $n =$ the number.

Translate.

$$\underbrace{\text{A number}}_{\downarrow \atop n} \;\; \underbrace{\text{plus}}_{\downarrow \atop +} \;\; \underbrace{\text{a number squared}}_{\downarrow \atop n^2} \;\; \underbrace{\text{is}}_{\downarrow \atop =} \;\; \underbrace{90.}_{\downarrow \atop 90}$$

Solve.
$$n + n^2 = 90$$
$$n^2 + n - 90 = 0$$
$$(n+10)(n-9) = 0$$
$$n + 10 = 0 \quad or \quad n - 9 = 0$$
$$n = -10 \quad or \quad n = 9$$

Check. $-10 + (-10)^2 = -10 + 100 = 90$, and $9 + 9^2 = 9 + 81 = 90$. The answer checks.

State. The number is -10 or 9.

99.
$$\frac{12x}{x-4} - \frac{3x^2}{x+4} = \frac{384}{x^2-16}$$

$$\frac{12x}{x-4} - \frac{3x^2}{x+4} = \frac{384}{(x+4)(x-4)},$$
$$\text{LCM is } (x+4)(x-4).$$

$$(x+4)(x-4)\left[\frac{12x}{x-4} - \frac{3x^2}{x+4}\right] = (x+4)(x-4)\cdot\frac{384}{(x+4)(x-4)}$$

$$12x(x+4) - 3x^2(x-4) = 384$$

$$12x^2 + 48x - 3x^3 + 12x^2 = 384$$

$$-3x^3 + 24x^2 + 48x - 384 = 0$$

$$-3(x^3 - 8x^2 - 16x + 128) = 0$$

$$-3[x^2(x-8) - 16(x-8)] = 0$$

$$-3(x-8)(x^2-16) = 0$$

$$-3(x-8)(x+4)(x-4) = 0$$

$$x - 8 = 0 \;\; or \;\; x + 4 = 0 \;\;\; or \;\; x - 4 = 0$$

$$x = 8 \;\; or \;\;\;\;\; x = -4 \;\; or \;\;\;\;\; x = 4$$

Check: For 8:

$$\frac{12x}{x-4} - \frac{3x^2}{x+4} = \frac{384}{x^2-16}$$

$$\frac{12\cdot 8}{8-4} - \frac{3\cdot 8^2}{8+4} \;?\; \frac{384}{8^2-16}$$

$$\frac{96}{4} - \frac{192}{12} \;\bigg|\; \frac{384}{48}$$

$$24 - 16 \;\bigg|\; 8$$

$$8 \;\bigg|\;\;\;\;\; \text{TRUE}$$

8 is a solution.

For -4:

$$\frac{12x}{x-4} - \frac{3x^2}{x+4} = \frac{384}{x^2-16}$$

$$\frac{12(-4)}{-4-4} - \frac{3(-4)^2}{-4+4} \;?\; \frac{384}{(-4)^2-16}$$

$$\frac{-48}{-8} - \frac{48}{0} \;\bigg|\; \frac{384}{16-16} \;\;\; \text{UNDEFINED}$$

-4 is not a solution.

For 4:

$$\frac{12x}{x-4} - \frac{3x^2}{x+4} = \frac{384}{x^2-16}$$

$$\frac{12\cdot 4}{4-4} - \frac{3\cdot 4^2}{4+4} \;?\; \frac{384}{4^2-16}$$

$$\frac{48}{0} - \frac{48}{8} \;\bigg|\; \frac{384}{16-16} \;\;\; \text{UNDEFINED}$$

4 is not a solution.

The solution is 8.

101. a) $T = 2\pi\sqrt{\dfrac{65}{980}} \approx 1.62$ sec

b) $T = 2\pi\sqrt{\dfrac{98}{980}} \approx 1.99$ sec

c) $T = 2\pi\sqrt{\dfrac{120}{980}} \approx 2.20$ sec

103. $\dfrac{\sqrt{44x^2y^9z}\sqrt{22y^9z^6}}{(\sqrt{11xy^8z^2})^2} = \dfrac{\sqrt{44\cdot 22x^2y^{18}z^7}}{\sqrt{11\cdot 11x^2y^{16}z^4}} =$

$$\sqrt{\frac{44\cdot 22x^2y^{18}z^7}{11\cdot 11x^2y^{16}z^4}} = \sqrt{4\cdot 2y^2z^3} = \sqrt{4y^2z^2\cdot 2z} = 2yz\sqrt{2z}$$

Exercise Set 6.4

RC1. Yes

RC3. No; the radicands are not the same.

RC5. Yes

RC7. No; the indexes are not the same.

1. $7\sqrt{5} + 4\sqrt{5} = (7+4)\sqrt{5}$ Factoring out $\sqrt{5}$
$$= 11\sqrt{5}$$

3. $6\sqrt[3]{7} - 5\sqrt[3]{7} = (6-5)\sqrt[3]{7}$ Factoring out $\sqrt[3]{7}$
$$= \sqrt[3]{7}$$

5. $4\sqrt[3]{y} + 9\sqrt[3]{y} = (4+9)\sqrt[3]{y} = 13\sqrt[3]{y}$

7. $5\sqrt{6} - 9\sqrt{6} - 4\sqrt{6} = (5-9-4)\sqrt{6} = -8\sqrt{6}$

9. $4\sqrt[3]{3} - \sqrt{5} + 2\sqrt[3]{3} + \sqrt{5} =$
$$(4+2)\sqrt[3]{3} + (-1+1)\sqrt{5} = 6\sqrt[3]{3}$$

11. $8\sqrt{27} - 3\sqrt{3} = 8\sqrt{9\cdot 3} - 3\sqrt{3}$ Factoring the
$$= 8\sqrt{9}\cdot\sqrt{3} - 3\sqrt{3} \quad \text{first radical}$$
$$= 8\cdot 3\sqrt{3} - 3\sqrt{3} \quad \text{Taking the square root}$$
$$= 24\sqrt{3} - 3\sqrt{3}$$
$$= (24-3)\sqrt{3} \quad \text{Factoring out } \sqrt{3}$$
$$= 21\sqrt{3}$$

13. $8\sqrt{45} + 7\sqrt{20} = 8\sqrt{9\cdot 5} + 7\sqrt{4\cdot 5}$ Factoring the radicals
$$= 8\sqrt{9}\cdot\sqrt{5} + 7\sqrt{4}\cdot\sqrt{5}$$
$$= 8\cdot 3\sqrt{5} + 7\cdot 2\sqrt{5} \quad \text{Taking the square roots}$$
$$= 24\sqrt{5} + 14\sqrt{5}$$
$$= (24+14)\sqrt{5} \quad \text{Factoring out } \sqrt{5}$$
$$= 38\sqrt{5}$$

15. $18\sqrt{72} + 2\sqrt{98} = 18\sqrt{36\cdot 2} + 2\sqrt{49\cdot 2} =$
$$18\sqrt{36}\cdot\sqrt{2} + 2\sqrt{49}\cdot\sqrt{2} = 18\cdot 6\sqrt{2} + 2\cdot 7\sqrt{2} =$$
$$108\sqrt{2} + 14\sqrt{2} = (108+14)\sqrt{2} = 122\sqrt{2}$$

17. $3\sqrt[3]{16} + \sqrt[3]{54} = 3\sqrt[3]{8\cdot 2} + \sqrt[3]{27\cdot 2} =$
$$3\sqrt[3]{8}\cdot\sqrt[3]{2} + \sqrt[3]{27}\cdot\sqrt[3]{2} = 3\cdot 2\sqrt[3]{2} + 3\sqrt[3]{2} =$$
$$6\sqrt[3]{2} + 3\sqrt[3]{2} = (6+3)\sqrt[3]{2} = 9\sqrt[3]{2}$$

19. $2\sqrt{128} - \sqrt{18} + 4\sqrt{32} =$
$$2\sqrt{64\cdot 2} - \sqrt{9\cdot 2} + 4\sqrt{16\cdot 2} =$$
$$2\sqrt{64}\cdot\sqrt{2} - \sqrt{9}\cdot\sqrt{2} + 4\sqrt{16}\cdot\sqrt{2} =$$
$$2\cdot 8\sqrt{2} - 3\sqrt{2} + 4\cdot 4\sqrt{2} = 16\sqrt{2} - 3\sqrt{2} + 16\sqrt{2} =$$
$$(16-3+16)\sqrt{2} = 29\sqrt{2}$$

21. $\sqrt{5a} + 2\sqrt{45a^3} = \sqrt{5a} + 2\sqrt{9a^2 \cdot 5a} =$
$\sqrt{5a} + 2\sqrt{9a^2} \cdot \sqrt{5a} = \sqrt{5a} + 2 \cdot 3a\sqrt{5a} =$
$\sqrt{5a} + 6a\sqrt{5a} = (1 + 6a)\sqrt{5a}$

23. $\sqrt[3]{24x} - \sqrt[3]{3x^4} = \sqrt[3]{8 \cdot 3x} - \sqrt[3]{x^3 \cdot 3x} =$
$\sqrt[3]{8} \cdot \sqrt[3]{3x} - \sqrt[3]{x^3} \cdot \sqrt[3]{3x} = 2\sqrt[3]{3x} - x\sqrt[3]{3x} =$
$(2 - x)\sqrt[3]{3x}$

25. $7\sqrt{27x^3} + \sqrt{3x} = 7\sqrt{9x^2 \cdot 3x} + \sqrt{3x} =$
$7 \cdot \sqrt{9x^2} \cdot \sqrt{3x} + \sqrt{3x} = 7 \cdot 3x \cdot \sqrt{3x} + \sqrt{3x} =$
$21x\sqrt{3x} + \sqrt{3x} = (21x + 1)\sqrt{3x}$

27. $\sqrt{4} + \sqrt{18} = 2 + \sqrt{9 \cdot 2} = 2 + \sqrt{9} \cdot \sqrt{2} = 2 + 3\sqrt{2}$

29. $5\sqrt[3]{32} - \sqrt[3]{108} + 2\sqrt[3]{256} =$
$5\sqrt[3]{8 \cdot 4} - \sqrt[3]{27 \cdot 4} + 2\sqrt[3]{64 \cdot 4} =$
$5\sqrt[3]{8} \cdot \sqrt[3]{4} - \sqrt[3]{27} \cdot \sqrt[3]{4} + 2\sqrt[3]{64} \cdot \sqrt[3]{4} =$
$5 \cdot 2\sqrt[3]{4} - 3\sqrt[3]{4} + 2 \cdot 4\sqrt[3]{4} = 10\sqrt[3]{4} - 3\sqrt[3]{4} + 8\sqrt[3]{4} =$
$(10 - 3 + 8)\sqrt[3]{4} = 15\sqrt[3]{4}$

31. $\sqrt[3]{6x^4} + \sqrt[3]{48x} - \sqrt[3]{6x}$
$= \sqrt[3]{x^3 \cdot 6x} + \sqrt[3]{8 \cdot 6x} - \sqrt[3]{6x}$
$= \sqrt[3]{x^3} \cdot \sqrt[3]{6x} + \sqrt[3]{8} \cdot \sqrt[3]{6x} - \sqrt[3]{6x}$
$= x\sqrt[3]{6x} + 2\sqrt[3]{6x} - \sqrt[3]{6x}$
$= (x + 2 - 1)\sqrt[3]{6x}$
$= (x + 1)\sqrt[3]{6x}$

33. $\sqrt{4a - 4} + \sqrt{a - 1} = \sqrt{4(a - 1)} + \sqrt{a - 1}$
$= \sqrt{4} \cdot \sqrt{a - 1} + \sqrt{a - 1}$
$= 2\sqrt{a - 1} + \sqrt{a - 1}$
$= (2 + 1)\sqrt{a - 1}$
$= 3\sqrt{a - 1}$

35. $\sqrt{x^3 - x^2} + \sqrt{9x - 9} = \sqrt{x^2(x - 1)} + \sqrt{9(x - 1)}$
$= \sqrt{x^2}\sqrt{x - 1} + \sqrt{9}\sqrt{x - 1}$
$= x\sqrt{x - 1} + 3\sqrt{x - 1}$
$= (x + 3)\sqrt{x - 1}$

37. $\sqrt{5}(4 - 2\sqrt{5}) = \sqrt{5} \cdot 4 - 2(\sqrt{5})^2$ Distributive law
$= 4\sqrt{5} - 2 \cdot 5$
$= 4\sqrt{5} - 10$

39. $\sqrt{3}(\sqrt{2} - \sqrt{7}) = \sqrt{3}\,\sqrt{2} - \sqrt{3}\,\sqrt{7}$ Distributive law
$= \sqrt{6} - \sqrt{21}$

41. $\sqrt{3}(-4\sqrt{3} + 6) = \sqrt{3} \cdot (-4\sqrt{3}) + \sqrt{3} \cdot 6 =$
$-4 \cdot 3 + 6\sqrt{3} = -12 + 6\sqrt{3}$

43. $\sqrt{3}(2\sqrt{5} - 3\sqrt{4}) = \sqrt{3}(2\sqrt{5} - 3 \cdot 2) =$
$\sqrt{3} \cdot 2\sqrt{5} - \sqrt{3} \cdot 6 = 2\sqrt{15} - 6\sqrt{3}$

45. $\sqrt[3]{2}(\sqrt[3]{4} - 2\sqrt[3]{32}) = \sqrt[3]{2} \cdot \sqrt[3]{4} - \sqrt[3]{2} \cdot 2\sqrt[3]{32} =$
$\sqrt[3]{8} - 2\sqrt[3]{64} = 2 - 2 \cdot 4 = 2 - 8 = -6$

47. $3\sqrt[3]{y}(2\sqrt[3]{y^2} - 4\sqrt[3]{y}) = 3\sqrt[3]{y} \cdot 2\sqrt[3]{y^2} - 3\sqrt[3]{y} \cdot 4\sqrt[3]{y} =$
$6\sqrt[3]{y^3} - 12\sqrt[3]{y^2} = 6y - 12\sqrt[3]{y^2}$

49. $\sqrt[3]{a}(\sqrt[3]{2a^2} + \sqrt[3]{16a^2}) = \sqrt[3]{a} \cdot \sqrt[3]{2a^2} + \sqrt[3]{a} \cdot \sqrt[3]{16a^2} =$
$\sqrt[3]{2a^3} + \sqrt[3]{16a^3} = \sqrt[3]{a^3 \cdot 2} + \sqrt[3]{8a^3 \cdot 2} = a\sqrt[3]{2} + 2a\sqrt[3]{2} =$
$3a\sqrt[3]{2}$

51. $(\sqrt{3} - \sqrt{2})(\sqrt{3} + \sqrt{2}) = (\sqrt{3})^2 - (\sqrt{2})^2 = 3 - 2 = 1$

53. $(\sqrt{8} + 2\sqrt{5})(\sqrt{8} - 2\sqrt{5}) = (\sqrt{8})^2 - (2\sqrt{5})^2 =$
$8 - 4 \cdot 5 = 8 - 20 = -12$

55. $(7 + \sqrt{5})(7 - \sqrt{5}) = 7^2 - (\sqrt{5})^2 = 49 - 5 = 44$

57. $(2 - \sqrt{3})(2 + \sqrt{3}) = 2^2 - (\sqrt{3})^2 = 4 - 3 = 1$

59. $(\sqrt{8} + \sqrt{5})(\sqrt{8} - \sqrt{5}) = (\sqrt{8})^2 - (\sqrt{5})^2 = 8 - 5 = 3$

61. $(3 + 2\sqrt{7})(3 - 2\sqrt{7}) = 3^2 - (2\sqrt{7})^2 =$
$9 - 4 \cdot 7 = 9 - 28 = -19$

63. $(\sqrt{a} + \sqrt{b})(\sqrt{a} - \sqrt{b}) = (\sqrt{a})^2 - (\sqrt{b})^2 = a - b$

65. $(3 - \sqrt{5})(2 + \sqrt{5})$
$= 3 \cdot 2 + 3\sqrt{5} - 2\sqrt{5} - (\sqrt{5})^2$ Using FOIL
$= 6 + 3\sqrt{5} - 2\sqrt{5} - 5$
$= 1 + \sqrt{5}$ Simplifying

67. $(\sqrt{3} + 1)(2\sqrt{3} + 1)$
$= \sqrt{3} \cdot 2\sqrt{3} + \sqrt{3} \cdot 1 + 1 \cdot 2\sqrt{3} + 1^2$ Using FOIL
$= 2 \cdot 3 + \sqrt{3} + 2\sqrt{3} + 1$
$= 7 + 3\sqrt{3}$ Simplifying

69. $(2\sqrt{7} - 4\sqrt{2})(3\sqrt{7} + 6\sqrt{2}) =$
$2\sqrt{7} \cdot 3\sqrt{7} + 2\sqrt{7} \cdot 6\sqrt{2} - 4\sqrt{2} \cdot 3\sqrt{7} - 4\sqrt{2} \cdot 6\sqrt{2} =$
$6 \cdot 7 + 12\sqrt{14} - 12\sqrt{14} - 24 \cdot 2 =$
$42 + 12\sqrt{14} - 12\sqrt{14} - 48 = -6$

71. $(\sqrt{a} + \sqrt{2})(\sqrt{a} + \sqrt{3}) =$
$(\sqrt{a})^2 + \sqrt{a} \cdot \sqrt{3} + \sqrt{2} \cdot \sqrt{a} + \sqrt{2} \cdot \sqrt{3} =$
$a + \sqrt{3a} + \sqrt{2a} + \sqrt{6}$

73. $(2\sqrt[3]{3} + \sqrt[3]{2})(\sqrt[3]{3} - 2\sqrt[3]{2}) =$
$2\sqrt[3]{3} \cdot \sqrt[3]{3} - 2\sqrt[3]{3} \cdot 2\sqrt[3]{2} + \sqrt[3]{2} \cdot \sqrt[3]{3} - \sqrt[3]{2} \cdot 2\sqrt[3]{2} =$
$2\sqrt[3]{9} - 4\sqrt[3]{6} + \sqrt[3]{6} - 2\sqrt[3]{4} = 2\sqrt[3]{9} - 3\sqrt[3]{6} - 2\sqrt[3]{4}$

75. $(2 + \sqrt{3})^2 = 2^2 + 4\sqrt{3} + (\sqrt{3})^2$ Squaring a binomial
$= 4 + 4\sqrt{3} + 3$
$= 7 + 4\sqrt{3}$

77. $(\sqrt[5]{9} - \sqrt[5]{3})(\sqrt[5]{8} + \sqrt[5]{27})$
$= \sqrt[5]{9} \cdot \sqrt[5]{8} + \sqrt[5]{9} \cdot \sqrt[5]{27} - \sqrt[5]{3} \cdot \sqrt[5]{8} - \sqrt[5]{3} \cdot \sqrt[5]{27}$
                        Using FOIL
$= \sqrt[5]{72} + \sqrt[5]{243} - \sqrt[5]{24} - \sqrt[5]{81}$
$= \sqrt[5]{72} + 3 - \sqrt[5]{24} - \sqrt[5]{81}$

79. $\dfrac{x^3 + 4x}{x^2 - 16} \div \dfrac{x^2 + 8x + 15}{x^2 + x - 20}$

$= \dfrac{x^3 + 4x}{x^2 - 16} \cdot \dfrac{x^2 + x - 20}{x^2 + 8x + 15}$

$= \dfrac{(x^3 + 4x)(x^2 + x - 20)}{(x^2 - 16)(x^2 + 8x + 15)}$

$= \dfrac{x(x^2 + 4)(x + 5)(x - 4)}{(x + 4)(x - 4)(x + 3)(x + 5)}$

$= \dfrac{x(x^2 + 4)\cancel{(x + 5)}\cancel{(x - 4)}}{(x + 4)\cancel{(x - 4)}(x + 3)\cancel{(x + 5)}}$

$= \dfrac{x(x^2 + 4)}{(x + 4)(x + 3)}$

81. $\dfrac{a^3 + 8}{a^2 - 4} \cdot \dfrac{a^2 - 4a + 4}{a^2 - 2a + 4}$

$= \dfrac{(a^3 + 8)(a^2 - 4a + 4)}{(a^2 - 4)(a^2 - 2a + 4)}$

$= \dfrac{(a + 2)(a^2 - 2a + 4)(a - 2)(a - 2)}{(a + 2)(a - 2)(a^2 - 2a + 4)(1)}$

$= \dfrac{(a + 2)(a^2 - 2a + 4)(a - 2)}{(a + 2)(a^2 - 2a + 4)(a - 2)} \cdot \dfrac{a - 2}{1}$

$= a - 2$

83. $\dfrac{x - \dfrac{1}{3}}{x + \dfrac{1}{4}} = \dfrac{x - \dfrac{1}{3}}{x + \dfrac{1}{4}} \cdot \dfrac{12}{12}$

$= \dfrac{\left(x - \dfrac{1}{3}\right)(12)}{\left(x + \dfrac{1}{4}\right)(12)}$

$= \dfrac{12x - 4}{12x + 3}, \text{ or } \dfrac{4(3x - 1)}{3(4x + 1)}$

85. $\dfrac{\dfrac{1}{p} - \dfrac{1}{q}}{\dfrac{1}{p^2} - \dfrac{1}{q^2}} = \dfrac{\dfrac{1}{p} - \dfrac{1}{q}}{\dfrac{1}{p^2} - \dfrac{1}{q^2}} \cdot \dfrac{p^2 q^2}{p^2 q^2}$

$= \dfrac{\left(\dfrac{1}{p} - \dfrac{1}{q}\right)(p^2 q^2)}{\left(\dfrac{1}{p^2} - \dfrac{1}{q^2}\right)(p^2 q^2)}$

$= \dfrac{p^2 q^2 \cdot \dfrac{1}{p} - p^2 q^2 \cdot \dfrac{1}{q}}{p^2 q^2 \cdot \dfrac{1}{p^2} - p^2 q^2 \cdot \dfrac{1}{q^2}}$

$= \dfrac{pq^2 - p^2 q}{q^2 - p^2}$

$= \dfrac{pq(q - p)}{(q + p)(q - p)}$

$= \dfrac{pq\cancel{(q - p)}}{(q + p)\cancel{(q - p)}}$

$= \dfrac{pq}{q + p}$

87. $|3x + 7| = 22$

$3x + 7 = -22 \quad or \quad 3x + 7 = 22$

$3x = -29 \quad or \qquad 3x = 15$

$x = -\dfrac{29}{3} \quad or \qquad x = 5$

The solutions are $-\dfrac{29}{3}$ and 5.

89. $|3x + 7| \geq 22$

$3x + 7 \leq -22 \quad or \quad 3x + 7 \geq 22$

$3x \leq -29 \quad or \qquad 3x \geq 15$

$x \leq -\dfrac{29}{3} \quad or \qquad x \geq 5$

The solution set is $\left\{x \middle| x \leq -\dfrac{29}{3} \text{ or } x \geq 5\right\}$, or

$\left(-\infty, -\dfrac{29}{3}\right] \cup [5, \infty)$.

91.

$f(x) = \sqrt{(x - 2)^2}$

Since $(x - 2)^2$ is nonnegative for all values of x, the domain of f is $\{x | x \text{ is a real number}\}$, or $(-\infty, \infty)$.

93. $\sqrt{9 + 3\sqrt{5}}\sqrt{9 - 3\sqrt{5}} = \sqrt{(9 + 3\sqrt{5})(9 - 3\sqrt{5})} =$

$\sqrt{9^2 - (3\sqrt{5})^2} = \sqrt{81 - 9 \cdot 5} = \sqrt{81 - 45} = \sqrt{36} = 6$

95. $(\sqrt{3} + \sqrt{5} - \sqrt{6})^2 = [(\sqrt{3} + \sqrt{5}) - \sqrt{6}]^2 =$

$(\sqrt{3} + \sqrt{5})^2 - 2(\sqrt{3} + \sqrt{5})(\sqrt{6}) + (\sqrt{6})^2 =$

$3 + 2\sqrt{15} + 5 - 2\sqrt{18} - 2\sqrt{30} + 6 =$

$14 + 2\sqrt{15} - 2\sqrt{9 \cdot 2} - 2\sqrt{30} =$

$14 + 2\sqrt{15} - 6\sqrt{2} - 2\sqrt{30}$

97. $(\sqrt[3]{9} - 2)(\sqrt[3]{9} + 4)$

$= \sqrt[3]{9}\sqrt[3]{9} + 4\sqrt[3]{9} - 2\sqrt[3]{9} - 2 \cdot 4$

$= \sqrt[3]{81} + 2\sqrt[3]{9} - 8$

$= \sqrt[3]{27 \cdot 3} + 2\sqrt[3]{9} - 8$

$= 3\sqrt[3]{3} + 2\sqrt[3]{9} - 8$

Chapter 6 Mid-Chapter Review

1. False; negative numbers do not have real-number square roots, and 0 has only one square root, 0.

2. True; see page 496 in the text.

3. False; $(a^{m/n})^{-1} = \dfrac{1}{a^{m/n}} \neq a^{n/m}$.

4. True; see page 510 in the text.

5. $\sqrt{6}\sqrt{10} = \sqrt{6 \cdot 10} = \sqrt{2 \cdot 3 \cdot 2 \cdot 5} = 2\sqrt{15}$

6. $5\sqrt{32} - 3\sqrt{18} = 5\sqrt{16 \cdot 2} - 3\sqrt{9 \cdot 2}$
$$= 5 \cdot 4\sqrt{2} - 3 \cdot 3\sqrt{2}$$
$$= 20\sqrt{2} - 9\sqrt{2}$$
$$= 11\sqrt{2}$$

7. $\sqrt{81} = 9$ Remember, $\sqrt{}$ indicates the positive square root.

8. $-\sqrt{144} = -12$ Since $\sqrt{144} = 12$, $-\sqrt{144} = -12$.

9. $\sqrt{\dfrac{16}{25}} = \sqrt{\left(\dfrac{4}{5}\right)^2} = \dfrac{4}{5}$

10. $\sqrt{-9}$ does not exist as a real number because negative numbers do not have real-number square roots.

11. $f(x) = \sqrt{2x+3}$
$$f(3) = \sqrt{2 \cdot 3 + 3} = \sqrt{6+3} = \sqrt{9} = 3$$
$$f(-2) = \sqrt{2(-2)+3} = \sqrt{-4+3} = \sqrt{-1}$$

         Since negative numbers do not have real-number square roots, $f(-2)$ does not exist as a real number.

12. The domain of $f(x) = \sqrt{4-x}$ is the set of all x-values for which $4 - x \geq 0$.
$$4 - x \geq 0$$
$$4 \geq x$$

   The domain is $\{x | x \leq 4\}$, or $(-\infty, 4]$.

13. Graph $f(x) = -2\sqrt{x}$.

   We find some ordered pairs, plot points, and draw the curve.

x	$f(x)$	$(x, f(x))$
0	0	$(0,0)$
1	-2	$(1,-2)$
4	-4	$(4,-4)$
9	-6	$(9,-6)$

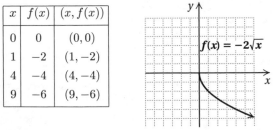

14. Graph $g(x) = \sqrt{x+1}$.

   We find some ordered pairs, plot points, and draw the curve.

x	$g(x)$	$(x, g(x))$
-1	0	$(-1,0)$
0	1	$(0,1)$
3	2	$(3,2)$
8	3	$(8,3)$

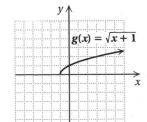

15. $\sqrt{36z^2} = \sqrt{(6z)^2} = |6z| = 6|z|$

16. $\sqrt{x^2 - 8x + 16} = \sqrt{(x-4)^2} = |x-4|$

17. $\sqrt[3]{-64} = -4$ $[(-4)^3 = -64]$

18. $-\sqrt[3]{27a^3} = -3a$ Since $(3a)^3 = 27a^3$, then $\sqrt[3]{27a^3} = 3a$ and $-\sqrt[3]{27a^3} = -3a$.

19. $\sqrt[5]{32} = 2$ $(2^5 = 32)$

20. $\sqrt[10]{y^{10}} = |y|$

21. $125^{1/3} = \sqrt[3]{125} = 5$

22. $(a^3b)^{1/4} = \sqrt[4]{a^3b}$

23. $\sqrt[5]{16} = 16^{1/5}$

24. $\sqrt[3]{6m^2n} = (6m^2n)^{1/3}$

25. $3^{1/4} \cdot 3^{-5/8} = 3^{1/4+(-5/8)} = 3^{-3/8} = \dfrac{1}{3^{3/8}}$

26. $\dfrac{7^{6/5}}{7^{2/5}} = 7^{6/5-2/5} = 7^{4/5}$

27. $(x^{3/4}y^{-2/3})^2 = (x^{3/4})^2(y^{-2/3})^2 = x^{(3/4)\cdot2}y^{(-2/3)\cdot2} =$
$$x^{3/2}y^{-4/3} = \dfrac{x^{3/2}}{y^{4/3}}$$

28. $(n^{-3/5})^{5/4} = n^{-3/5\cdot5/4} = n^{-3/4} = \dfrac{1}{n^{3/4}}$

29. $\sqrt[6]{16} = \sqrt[6]{2^4} = 2^{4/6} = 2^{2/3} = \sqrt[3]{2^2} = \sqrt[3]{4}$

30. $(\sqrt[10]{ab})^5 = [(ab)^{1/10}]^5 = (ab)^{5/10} = (ab)^{1/2} = \sqrt{ab}$

31. $\sqrt{y}\,\sqrt[3]{y} = y^{1/2} \cdot y^{1/3} = y^{5/6} = \sqrt[6]{y^5}$

32. $a^{2/3}b^{3/5} = a^{10/15}b^{9/15} = (a^{10}b^9)^{1/15} = \sqrt[15]{a^{10}b^9}$

33. $\sqrt{5}\sqrt{15} = \sqrt{5 \cdot 15} = \sqrt{5 \cdot 3 \cdot 5} = \sqrt{5 \cdot 5}\,\sqrt{3} = 5\sqrt{3}$

34. $\sqrt[3]{4x^2y}\,\sqrt[3]{6xy^4} = \sqrt[3]{4x^2y \cdot 6xy^4} = \sqrt[3]{24x^3y^5} =$
$$\sqrt[3]{8 \cdot 3 \cdot x^3 \cdot y^3 \cdot y^2} = \sqrt[3]{8x^3y^3}\,\sqrt[3]{3y^2} = 2xy\sqrt[3]{3y^2}$$

35. $\dfrac{\sqrt[3]{80}}{\sqrt[3]{2}} = \sqrt[3]{\dfrac{80}{2}} = \sqrt[3]{40} = \sqrt[3]{8 \cdot 5} = \sqrt[3]{8}\,\sqrt[3]{5} = 2\sqrt[3]{5}$

36. $\sqrt{\dfrac{49a^5}{b^8}} = \dfrac{\sqrt{49a^5}}{\sqrt{b^8}} = \dfrac{\sqrt{49 \cdot a^4 \cdot a}}{\sqrt{b^8}} = \dfrac{\sqrt{49a^4}\sqrt{a}}{\sqrt{b^8}} = \dfrac{7a^2\sqrt{a}}{b^4}$

37. $5\sqrt{7} + 6\sqrt{7} = (5+6)\sqrt{7} = 11\sqrt{7}$

38. $3\sqrt{18x^3} - 6\sqrt{32x} = 3\sqrt{9x^2 \cdot 2x} - 6\sqrt{16 \cdot 2x} =$
$$3 \cdot 3x\sqrt{2x} - 6 \cdot 4\sqrt{2x} = 9x\sqrt{2x} - 24\sqrt{2x} = (9x-24)\sqrt{2x}$$

39. $\sqrt{3}(2 - 5\sqrt{3}) = 2\sqrt{3} - 5\sqrt{9} = 2\sqrt{3} - 5 \cdot 3 = 2\sqrt{3} - 15$

40. $(1-\sqrt{x})(3 - \sqrt{x}) = 3 - \sqrt{x} - 3\sqrt{x} + \sqrt{x^2} = 3 - 4\sqrt{x} + x$

41. $(\sqrt{m} - \sqrt{n})(\sqrt{m} + \sqrt{n}) = (\sqrt{m})^2 - (\sqrt{n})^2 = m - n$

42. $(\sqrt{7}+2)^2 = (\sqrt{7})^2 + 2 \cdot \sqrt{7} \cdot 2 + 2^2 = 7 + 4\sqrt{7} + 4 = 11 + 4\sqrt{7}$

43. $(2\sqrt{3}+3\sqrt{5})(3\sqrt{3}-4\sqrt{5})=6\sqrt{9}-8\sqrt{15}+9\sqrt{15}-12\sqrt{25}=$

$6\cdot3+\sqrt{15}-12\cdot5=18+\sqrt{15}-60=-42+\sqrt{15}$

44. Yes; since x^2 is nonnegative for any value of x, the nth root of x^2 exists regardless of whether n is even or odd. Thus, the nth root of x^2 always exists.

45. Formulate an expression containing a radical term with an even index and a radicand R such that the solution of the inequality $R\geq0$ is $\{x|x\leq5\}$. One expression is $\sqrt{5-x}$. Other expressions could be formulated as $a\sqrt[k]{b(5-x)}+c$, where $a\neq0$, $b>0$ and k is an even integer.

46. Since $x^6\geq0$ and $x^2\geq0$ for any value of x, $\sqrt[3]{x^6}=x^2$. However, $x^3\geq0$ only for $x\geq0$, so $\sqrt{x^6}=x^3$ only when $x\geq0$.

47. No; for example, $\dfrac{\sqrt{8}}{\sqrt{2}}=\sqrt{\dfrac{8}{2}}=\sqrt{4}=2$.

Exercise Set 6.5

RC1. (g)

RC3. (e)

RC5. (c)

1. $\sqrt{\dfrac{5}{3}}=\sqrt{\dfrac{5}{3}\cdot\dfrac{3}{3}}=\sqrt{\dfrac{15}{9}}=\dfrac{\sqrt{15}}{\sqrt{9}}=\dfrac{\sqrt{15}}{3}$

3. $\sqrt{\dfrac{11}{2}}=\sqrt{\dfrac{11}{2}\cdot\dfrac{2}{2}}=\sqrt{\dfrac{22}{4}}=\dfrac{\sqrt{22}}{\sqrt{4}}=\dfrac{\sqrt{22}}{2}$

5. $\dfrac{2\sqrt{3}}{7\sqrt{5}}=\dfrac{2\sqrt{3}}{7\sqrt{5}}\cdot\dfrac{\sqrt{5}}{\sqrt{5}}=\dfrac{2\sqrt{15}}{7\sqrt{5^2}}=\dfrac{2\sqrt{15}}{7\cdot5}=\dfrac{2\sqrt{15}}{35}$

7. $\sqrt[3]{\dfrac{16}{9}}=\sqrt[3]{\dfrac{16}{9}\cdot\dfrac{3}{3}}=\sqrt[3]{\dfrac{48}{27}}=\dfrac{\sqrt[3]{8\cdot6}}{\sqrt[3]{27}}=\dfrac{2\sqrt[3]{6}}{3}$

9. $\dfrac{\sqrt[3]{3a}}{\sqrt[3]{5c}}=\dfrac{\sqrt[3]{3a}}{\sqrt[3]{5c}}\cdot\dfrac{\sqrt[3]{5^2c^2}}{\sqrt[3]{5^2c^2}}=\dfrac{\sqrt[3]{75ac^2}}{\sqrt[3]{5^3c^3}}=\dfrac{\sqrt[3]{75ac^2}}{5c}$

11. $\dfrac{\sqrt[3]{2y^4}}{\sqrt[3]{6x^4}}=\dfrac{\sqrt[3]{2y^4}}{\sqrt[3]{6x^4}}\cdot\dfrac{\sqrt[3]{6^2x^2}}{\sqrt[3]{6^2x^2}}=\dfrac{\sqrt[3]{72x^2y^4}}{\sqrt[3]{6^3x^6}}=\dfrac{\sqrt[3]{8y^3\cdot9x^2y}}{6x^2}=$

$\dfrac{2y\sqrt[3]{9x^2y}}{6x^2}=\dfrac{y\sqrt[3]{9x^2y}}{3x^2}$

13. $\dfrac{1}{\sqrt[4]{st}}=\dfrac{1}{\sqrt[4]{st}}\cdot\dfrac{\sqrt[4]{s^3t^3}}{\sqrt[4]{s^3t^3}}=\dfrac{\sqrt[4]{s^3t^3}}{\sqrt[4]{s^4t^4}}=\dfrac{\sqrt[4]{s^3t^3}}{st}$

15. $\sqrt{\dfrac{3x}{20}}=\sqrt{\dfrac{3x}{20}\cdot\dfrac{5}{5}}=\sqrt{\dfrac{15x}{100}}=\dfrac{\sqrt{15x}}{\sqrt{100}}=\dfrac{\sqrt{15x}}{10}$

17. $\sqrt[3]{\dfrac{4}{5x^5y^2}}=\sqrt[3]{\dfrac{4}{5x^5y^2}\cdot\dfrac{25xy}{5^2xy}}=\sqrt[3]{\dfrac{100xy}{5^3x^6y^3}}=$

$\dfrac{\sqrt[3]{100xy}}{\sqrt[3]{5^3x^6y^3}}=\dfrac{\sqrt[3]{100xy}}{5x^2y}$

19. $\sqrt[4]{\dfrac{1}{8x^7y^3}}=\sqrt[4]{\dfrac{1}{2^3x^7y^3}\cdot\dfrac{2xy}{2xy}}=\sqrt[4]{\dfrac{2xy}{2^4x^8y^4}}=\dfrac{\sqrt[4]{2xy}}{\sqrt[4]{2^4x^8y^4}}=$

$\dfrac{\sqrt[4]{2xy}}{2x^2y}$

21. $\dfrac{9}{6-\sqrt{10}}=\dfrac{9}{6-\sqrt{10}}\cdot\dfrac{6+\sqrt{10}}{6+\sqrt{10}}=\dfrac{9(6+\sqrt{10})}{6^2-(\sqrt{10})^2}=$

$\dfrac{9(6+\sqrt{10})}{36-10}=\dfrac{54+9\sqrt{10}}{26}$

23. $\dfrac{-4\sqrt{7}}{\sqrt{5}+\sqrt{3}}=\dfrac{-4\sqrt{7}}{\sqrt{5}+\sqrt{3}}\cdot\dfrac{\sqrt{5}-\sqrt{3}}{\sqrt{5}-\sqrt{3}}=$

$\dfrac{-4\sqrt{7}(\sqrt{5}-\sqrt{3})}{(\sqrt{5})^2-(\sqrt{3})^2}=\dfrac{-4\sqrt{7}(\sqrt{5}-\sqrt{3})}{5-3}=$

$\dfrac{-4\sqrt{7}(\sqrt{5}-\sqrt{3})}{2}=-2\sqrt{7}(\sqrt{5}-\sqrt{3})=-2\sqrt{35}+2\sqrt{21}$

25. $\dfrac{6\sqrt{3}}{3\sqrt{2}-\sqrt{5}}=\dfrac{6\sqrt{3}}{3\sqrt{2}-\sqrt{5}}\cdot\dfrac{3\sqrt{2}+\sqrt{5}}{3\sqrt{2}+\sqrt{5}}=\dfrac{6\sqrt{3}(3\sqrt{2}+\sqrt{5})}{(3\sqrt{2})^2-(\sqrt{5})^2}=$

$\dfrac{18\sqrt{6}+6\sqrt{15}}{9\cdot2-5}=\dfrac{18\sqrt{6}+6\sqrt{15}}{18-5}=\dfrac{18\sqrt{6}+6\sqrt{15}}{13}$

27. $\dfrac{3+\sqrt{5}}{\sqrt{2}+\sqrt{5}}=\dfrac{3+\sqrt{5}}{\sqrt{2}+\sqrt{5}}\cdot\dfrac{\sqrt{2}-\sqrt{5}}{\sqrt{2}-\sqrt{5}}=$

$\dfrac{3\sqrt{2}-3\sqrt{5}+\sqrt{10}-(\sqrt{5})^2}{(\sqrt{2})^2-(\sqrt{5})^2}=\dfrac{3\sqrt{2}-3\sqrt{5}+\sqrt{10}-5}{2-5}=$

$\dfrac{3\sqrt{2}-3\sqrt{5}+\sqrt{10}-5}{-3}$

29. $\dfrac{\sqrt{3}-\sqrt{2}}{\sqrt{3}-\sqrt{7}}=\dfrac{\sqrt{3}-\sqrt{2}}{\sqrt{3}-\sqrt{7}}\cdot\dfrac{\sqrt{3}+\sqrt{7}}{\sqrt{3}+\sqrt{7}}=$

$\dfrac{(\sqrt{3})^2+\sqrt{21}-\sqrt{6}-\sqrt{14}}{(\sqrt{3})^2-(\sqrt{7})^2}=\dfrac{3+\sqrt{21}-\sqrt{6}-\sqrt{14}}{3-7}=$

$\dfrac{3+\sqrt{21}-\sqrt{6}-\sqrt{14}}{-4}$

31. $\dfrac{\sqrt{5}-2\sqrt{6}}{\sqrt{3}-4\sqrt{5}}=\dfrac{\sqrt{5}-2\sqrt{6}}{\sqrt{3}-4\sqrt{5}}\cdot\dfrac{\sqrt{3}+4\sqrt{5}}{\sqrt{3}+4\sqrt{5}}=$

$\dfrac{\sqrt{15}+4\cdot5-2\sqrt{18}-8\sqrt{30}}{(\sqrt{3})^2-(4\sqrt{5})^2}=$

$\dfrac{\sqrt{15}+20-2\sqrt{9\cdot2}-8\sqrt{30}}{3-16\cdot5}=$

$\dfrac{\sqrt{15}+20-6\sqrt{2}-8\sqrt{30}}{-77}$

33. $\dfrac{2-\sqrt{a}}{3+\sqrt{a}}=\dfrac{2-\sqrt{a}}{3+\sqrt{a}}\cdot\dfrac{3-\sqrt{a}}{3-\sqrt{a}}=\dfrac{6-2\sqrt{a}-3\sqrt{a}+a}{9-a}=$

$\dfrac{6-5\sqrt{a}+a}{9-a}$

35. $\dfrac{2+3\sqrt{x}}{3+2\sqrt{x}}=\dfrac{2+3\sqrt{x}}{3+2\sqrt{x}}\cdot\dfrac{3-2\sqrt{x}}{3-2\sqrt{x}}=$

$\dfrac{6-4\sqrt{x}+9\sqrt{x}-6\cdot x}{9-4\cdot x}=\dfrac{6+5\sqrt{x}-6x}{9-4x}$

37. $\dfrac{5\sqrt{3}-3\sqrt{2}}{3\sqrt{2}-2\sqrt{3}} = \dfrac{5\sqrt{3}-3\sqrt{2}}{3\sqrt{2}-2\sqrt{3}} \cdot \dfrac{3\sqrt{2}+2\sqrt{3}}{3\sqrt{2}+2\sqrt{3}} =$

$\dfrac{15\sqrt{6}+10\cdot 3-9\cdot 2-6\sqrt{6}}{9\cdot 2-4\cdot 3} = \dfrac{12+9\sqrt{6}}{6} =$

$\dfrac{3(4+3\sqrt{6})}{3\cdot 2} = \dfrac{4+3\sqrt{6}}{2}$

39. $\dfrac{\sqrt{x}-\sqrt{y}}{\sqrt{x}+\sqrt{y}} = \dfrac{\sqrt{x}-\sqrt{y}}{\sqrt{x}+\sqrt{y}} \cdot \dfrac{\sqrt{x}-\sqrt{y}}{\sqrt{x}-\sqrt{y}} =$

$\dfrac{x-\sqrt{xy}-\sqrt{xy}+y}{x-y} = \dfrac{x-2\sqrt{xy}+y}{x-y}$

41. $\dfrac{1}{2} - \dfrac{1}{3} = \dfrac{5}{t}$, LCM is $6t$

$6t\left(\dfrac{1}{2} - \dfrac{1}{3}\right) = 6t\left(\dfrac{5}{t}\right)$

$3t - 2t = 30$

$t = 30$

Check:

$\dfrac{1}{2} - \dfrac{1}{3} = \dfrac{5}{t}$

$\dfrac{1}{2} - \dfrac{1}{3} \; ? \; \dfrac{5}{30}$

$\dfrac{3}{6} - \dfrac{2}{6} \;\Big|\; \dfrac{1}{6}$

$\dfrac{1}{6} \;\Big|\;$ TRUE

The solution is 30.

43. $\dfrac{1}{x^3-y^2} \div \dfrac{1}{(x-y)(x^2+xy+y^2)}$

$= \dfrac{1}{(x-y)(x^2+xy+y^2)} \cdot \dfrac{(x-y)(x^2+xy+y^2)}{1}$

$= \dfrac{(x-y)(x^2+xy+y^2)}{(x-y)(x^2+xy+y^2)}$

$= 1$

45. Left to the student

47. $\sqrt{a^2-3} - \dfrac{a^2}{\sqrt{a^2-3}}$

$= \sqrt{a^2-3} - \dfrac{a^2}{\sqrt{a^2-3}} \cdot \dfrac{\sqrt{a^2-3}}{\sqrt{a^2-3}}$

$= \sqrt{a^2-3} - \dfrac{a^2\sqrt{a^2-3}}{a^2-3}$

$= \sqrt{a^2-3} \cdot \dfrac{a^2-3}{a^2-3} - \dfrac{a^2\sqrt{a^2-3}}{a^2-3}$

$= \dfrac{a^2\sqrt{a^2-3}-3\sqrt{a^2-3}-a^2\sqrt{a^2-3}}{a^2-3}$

$= \dfrac{-3\sqrt{a^2-3}}{a^2-3}$, or $-\dfrac{3\sqrt{a^2-3}}{a^2-3}$

Exercise Set 6.6

RC1. The equation $\sqrt{4-11x} = 3$ is a <u>radical</u> equation.

RC3. To solve an equation with a radical term, we first <u>isolate</u> the radical term on one side of the equation.

RC5. A check is essential when we raise both sides of an equation to an <u>even</u> power.

1. $\sqrt{2x-3} = 4$

$(\sqrt{2x-3})^2 = 4^2$　Principle of powers

$2x - 3 = 16$

$2x = 19$

$x = \dfrac{19}{2}$

Check:　$\sqrt{2x-3} = 4$

$\sqrt{2\cdot\dfrac{19}{2}-3} \; ? \; 4$

$\sqrt{19-3}$

$\sqrt{16}$

$4 \;\Big|\;$ TRUE

The solution is $\dfrac{19}{2}$.

3. $\sqrt{6x}+1 = 8$

$\sqrt{6x} = 7$　Subtracting to isolate the radical

$(\sqrt{6x})^2 = 7^2$　Principle of powers

$6x = 49$

$x = \dfrac{49}{6}$

Check:　$\sqrt{6x}+1 = 8$

$\sqrt{6\cdot\dfrac{49}{6}}+1 \; ? \; 8$

$\sqrt{49}+1$

$7+1$

$8 \;\Big|\;$ TRUE

The solution is $\dfrac{49}{6}$.

5. $\sqrt{y+7}-4 = 4$

$\sqrt{y+7} = 8$　Adding to isolate the radical

$(\sqrt{y+7})^2 = 8^2$　Principle of powers

$y+7 = 64$

$y = 57$

Check: $\dfrac{\sqrt{y+7}-4=4}{\sqrt{57+7}-4 \ ? \ 4}$
$\sqrt{64}-4$
$8-4$
$4 \ | \ $ TRUE

The solution is 57.

7. $\sqrt{5y+8}=10$

$(\sqrt{5y+8})^2=10^2$ Principle of powers

$5y+8=100$

$5y=92$

$y=\dfrac{92}{5}$

Check: $\dfrac{\sqrt{5y+8}=10}{\sqrt{5\cdot\dfrac{92}{5}+8} \ ? \ 10}$
$\sqrt{92+8}$
$\sqrt{100}$
$10 \ | \ $ TRUE

The solution is $\dfrac{92}{5}$.

9. $\sqrt[3]{x}=-1$

$(\sqrt[3]{x})^3=(-1)^3$ Principle of powers

$x=-1$

Check: $\dfrac{\sqrt[3]{x}=-1}{\sqrt[3]{-1} \ ? \ -1}$
$-1 \ | \ $ TRUE

The solution is -1.

11. $\sqrt{x+2}=-4$

$(\sqrt{x+2})^2=(-4)^2$

$x+2=16$

$x=14$

Check: $\dfrac{\sqrt{x+2}=-4}{\sqrt{14+2} \ ? \ -4}$
$\sqrt{16}$
$4 \ | \ $ FALSE

The number 14 does not check. The equation has no solution. We might have observed at the outset that this equation has no solution because the principle square root of a number is never negative.

13. $\sqrt[3]{x+5}=2$

$(\sqrt[3]{x+5})^3=2^3$

$x+5=8$

$x=3$

Check: $\dfrac{\sqrt[3]{x+5}=2}{\sqrt[3]{3+5} \ ? \ 2}$
$\sqrt[3]{8}$
$2 \ | \ $ TRUE

The solution is 3.

15. $\sqrt[4]{y-3}=2$

$(\sqrt[4]{y-3})^4=2^4$

$y-3=16$

$y=19$

Check: $\dfrac{\sqrt[4]{y-3}=2}{\sqrt[4]{19-3} \ ? \ 2}$
$\sqrt[4]{16}$
$2 \ | \ $ TRUE

The solution is 19.

17. $\sqrt[3]{6x+9}+8=5$

$\sqrt[3]{6x+9}=-3$

$(\sqrt[3]{6x+9})^3=(-3)^3$

$6x+9=-27$

$6x=-36$

$x=-6$

Check: $\dfrac{\sqrt[3]{6x+9}+8=5}{\sqrt[3]{6(-6)+9}+8 \ ? \ 5}$
$\sqrt[3]{-27}+8$
$-3+8$
$5 \ | \ $ TRUE

The solution is -6.

19. $8=\dfrac{1}{\sqrt{x}}$

$8\cdot\sqrt{x}=\dfrac{1}{\sqrt{x}}\cdot\sqrt{x}$

$8\sqrt{x}=1$

$(8\sqrt{x})^2=1^2$

$64x=1$

$x=\dfrac{1}{64}$

Check:

$$8 = \frac{1}{\sqrt{x}}$$

$$\begin{array}{c|c} 8 \ ? \ \dfrac{1}{\sqrt{\dfrac{1}{64}}} \\[2ex] \dfrac{1}{\dfrac{1}{8}} \\[2ex] 8 & \text{TRUE} \end{array}$$

The solution is $\dfrac{1}{64}$.

21.
$$x - 7 = \sqrt{x - 5}$$
$$(x - 7)^2 = (\sqrt{x - 5})^2$$
$$x^2 - 14x + 49 = x - 5$$
$$x^2 - 15x + 54 = 0$$
$$(x - 6)(x - 9) = 0$$
$$x - 6 = 0 \ \ or \ \ x - 9 = 0$$
$$x = 6 \ \ or \ \ \ \ \ \ x = 9$$

Check: For 6:
$$\begin{array}{c|c} \multicolumn{2}{c}{x - 7 = \sqrt{x - 5}} \\ \hline 6 - 7 \ ? \ \sqrt{6 - 5} \\ -1 & \sqrt{1} \\ -1 & 1 \ \text{FALSE} \end{array}$$

Check: For 9:
$$\begin{array}{c|c} \multicolumn{2}{c}{x - 7 = \sqrt{x - 5}} \\ \hline 9 - 7 \ ? \ \sqrt{9 - 5} \\ 2 & \sqrt{4} \\ 2 & 2 \ \text{TRUE} \end{array}$$

The number 6 does not check, but 9 does. The solution is 9.

23.
$$2\sqrt{x + 1} + 7 = x$$
$$2\sqrt{x + 1} = x - 7$$
$$(2\sqrt{x + 1})^2 = (x - 7)^2$$
$$4(x + 1) = x^2 - 14x + 49$$
$$4x + 4 = x^2 - 14x + 49$$
$$0 = x^2 - 18x + 45$$
$$0 = (x - 3)(x - 15)$$
$$x - 3 = 0 \ \ or \ \ x - 15 = 0$$
$$x = 3 \ \ or \ \ \ \ \ \ x = 15$$

Check: For 3:
$$\begin{array}{c|c} \multicolumn{2}{c}{2\sqrt{x + 1} + 7 = x} \\ \hline 2\sqrt{3 + 1} + 7 \ ? \ 3 \\ 2\sqrt{4} + 7 \\ 2 \cdot 2 + 7 \\ 4 + 7 \\ 11 & 3 \ \text{FALSE} \end{array}$$

Check: For 15:
$$\begin{array}{c|c} \multicolumn{2}{c}{2\sqrt{x + 1} + 7 = x} \\ \hline 2\sqrt{15 + 1} + 7 \ ? \ 15 \\ 2\sqrt{16} + 7 \\ 2 \cdot 4 + 7 \\ 8 + 7 \\ 15 & 15 \ \text{TRUE} \end{array}$$

The number 3 does not check, but 15 does. The solution is 15.

25.
$$3\sqrt{x - 1} - 1 = x$$
$$3\sqrt{x - 1} = x + 1$$
$$(3\sqrt{x - 1})^2 = (x + 1)^2$$
$$9(x - 1) = x^2 + 2x + 1$$
$$9x - 9 = x^2 + 2x + 1$$
$$0 = x^2 - 7x + 10$$
$$0 = (x - 2)(x - 5)$$
$$x - 2 = 0 \ \ or \ \ x - 5 = 0$$
$$x = 2 \ \ or \ \ \ \ \ \ x = 5$$

Check: For 2:
$$\begin{array}{c|c} \multicolumn{2}{c}{3\sqrt{x - 1} - 1 = x} \\ \hline 3\sqrt{2 - 1} - 1 \ ? \ 2 \\ 3\sqrt{1} - 1 \\ 3 \cdot 1 - 1 \\ 3 - 1 \\ 2 & 2 \ \text{TRUE} \end{array}$$

Check: For 5:
$$\begin{array}{c|c} \multicolumn{2}{c}{3\sqrt{x - 1} - 1 = x} \\ \hline 3\sqrt{5 - 1} - 1 \ ? \ 5 \\ 3\sqrt{4} - 1 \\ 3 \cdot 2 - 1 \\ 6 - 1 \\ 5 & 5 \ \text{TRUE} \end{array}$$

Both numbers check. The solutions are 2 and 5.

27.
$$x - 3 = \sqrt{27 - 3x}$$
$$(x - 3)^2 = (\sqrt{27 - 3x})^2$$
$$x^2 - 6x + 9 = 27 - 3x$$
$$x^2 - 3x - 18 = 0$$
$$(x - 6)(x + 3) = 0$$
$$x - 6 = 0 \ \ or \ \ x + 3 = 0$$
$$x = 6 \ \ or \ \ \ \ \ \ x = -3$$

Check: For 6:
$$\begin{array}{c|c} \multicolumn{2}{c}{x - 3 = \sqrt{27 - 3x}} \\ \hline 6 - 3 \ ? \ \sqrt{27 - 3 \cdot 6} \\ 3 & \sqrt{27 - 18} \\ & \sqrt{9} \\ 3 & 3 \ \text{TRUE} \end{array}$$

Check: For -3:

$$x - 3 = \sqrt{27 - 3x}$$

$$\frac{-3 - 3 \ ? \ \sqrt{27 - 3(-3)}}{}$$

$$-6 \ \bigg| \ \sqrt{27 + 9}$$

$$\sqrt{36}$$

$$-6 \ \bigg| \ 6 \ \text{FALSE}$$

The number 6 checks but -3 does not. The solution is 6.

29. $\sqrt{3y + 1} = \sqrt{2y + 6}$

$(\sqrt{3y + 1})^2 = (\sqrt{2y + 6})^2$

$3y + 1 = 2y + 6$

$y = 5$

Check: $\dfrac{\sqrt{3y + 1} = \ \sqrt{2y + 6}}{}$

$\sqrt{3 \cdot 5 + 1} \ ? \ \sqrt{2 \cdot 5 + 6}$

$\sqrt{16} \ \bigg| \ \sqrt{16} \ \text{TRUE}$

The solution is 5.

31. $\sqrt{y - 5} + \sqrt{y} = 5$

$\sqrt{y - 5} = 5 - \sqrt{y}$ Isolating one radical

$(\sqrt{y - 5})^2 = (5 - \sqrt{y})^2$

$y - 5 = 25 - 10\sqrt{y} + y$

$10\sqrt{y} = 30$ Isolating the remaining radical

$\sqrt{y} = 3$ Dividing by 10

$(\sqrt{y})^2 = 3^2$

$y = 9$

The number 9 checks, so it is the solution.

33. $3 + \sqrt{z - 6} = \sqrt{z + 9}$

$(3 + \sqrt{z - 6})^2 = (\sqrt{z + 9})^2$

$9 + 6\sqrt{z - 6} + z - 6 = z + 9$

$6\sqrt{z - 6} = 6$

$\sqrt{z - 6} = 1$ Dividing by 6

$(\sqrt{z - 6})^2 = 1^2$

$z - 6 = 1$

$z = 7$

The number 7 checks, so it is the solution.

35. $\sqrt{20 - x} + 8 = \sqrt{9 - x} + 11$

$\sqrt{20 - x} = \sqrt{9 - x} + 3$ Isolating one radical

$(\sqrt{20 - x})^2 = (\sqrt{9 - x} + 3)^2$

$20 - x = 9 - x + 6\sqrt{9 - x} + 9$

$2 = 6\sqrt{9 - x}$ Isolating the remaining radical

$1 = 3\sqrt{9 - x}$ Dividing by 2

$1^2 = (3\sqrt{9 - x})^2$

$1 = 9(9 - x)$

$1 = 81 - 9x$

$9x = 80$

$x = \dfrac{80}{9}$

The number $\dfrac{80}{9}$ checks, so it is the solution.

37. $\sqrt{4y + 1} - \sqrt{y - 2} = 3$

$\sqrt{4y + 1} = 3 + \sqrt{y - 2}$ Isolating one radical

$(\sqrt{4y + 1})^2 = (3 + \sqrt{y - 2})^2$

$4y + 1 = 9 + 6\sqrt{y - 2} + y - 2$

$3y - 6 = 6\sqrt{y - 2}$ Isolating the remaining radical

$y - 2 = 2\sqrt{y - 2}$ Multiplying by $\dfrac{1}{3}$

$(y - 2)^2 = (2\sqrt{y - 2})^2$

$y^2 - 4y + 4 = 4(y - 2)$

$y^2 - 4y + 4 = 4y - 8$

$y^2 - 8y + 12 = 0$

$(y - 6)(y - 2) = 0$

$y - 6 = 0 \ \text{ or } \ y - 2 = 0$

$y = 6 \ \text{ or } \ \ \ \ \ \ \ y = 2$

The numbers 6 and 2 check, so they are the solutions.

39. $\sqrt{x + 2} + \sqrt{3x + 4} = 2$

$\sqrt{x + 2} = 2 - \sqrt{3x + 4}$ Isolating one radical

$(\sqrt{x + 2})^2 = (2 - \sqrt{3x - 4})^2$

$x + 2 = 4 - 4\sqrt{3x + 4} + 3x + 4$

$-2x - 6 = -4\sqrt{3x + 4}$ Isolating the remaining radical

$x + 3 = 2\sqrt{3x + 4}$ Dividing by -2

$(x + 3)^2 = (2\sqrt{3x + 4})^2$

$x^2 + 6x + 9 = 4(3x + 4)$

$x^2 + 6x + 9 = 12x + 16$

$x^2 - 6x - 7 = 0$

$(x - 7)(x + 1) = 0$

$x - 7 = 0 \ \text{ or } \ x + 1 = 0$

$x = 7 \ \text{ or } \ \ \ \ \ \ x = -1$

Check: For 7: $\dfrac{\sqrt{x+2}+\sqrt{3x+4}=2}{\sqrt{7+2}+\sqrt{3\cdot 7+4}\ ?\ 2}$

$$\sqrt{9}+\sqrt{25}\ \Big|$$

$$8\ \Big|\ \text{FALSE}$$

Check: For -1: $\dfrac{\sqrt{x+2}+\sqrt{3x+4}=2}{\sqrt{-1+2}+\sqrt{3(-1)+4}\ ?\ 2}$

$$\sqrt{1}+\sqrt{1}\ \Big|$$

$$2\ \Big|\ \text{TRUE}$$

Since -1 checks but 7 does not, the solution is -1.

41. $\sqrt{3x-5}+\sqrt{2x+3}+1=0$

$$\sqrt{3x-5}+1=-\sqrt{2x+3}$$
$$(\sqrt{3x-5}+1)^2=(-\sqrt{2x+3})^2$$
$$3x-5+2\sqrt{3x-5}+1=2x+3$$
$$2\sqrt{3x-5}=-x+7$$
$$(2\sqrt{3x-5})^2=(-x+7)^2$$
$$4(3x-5)=x^2-14x+49$$
$$12x-20=x^2-14x+49$$
$$0=x^2-26x+69$$
$$0=(x-23)(x-3)$$

$$\text{-}x-23=0\quad or\quad x-3=0$$
$$x=23\quad or\qquad x=3$$

Neither number checks. There is no solution. (At the outset we might have observed that there is no solution since the sum on the left side of the equation must be at least 1.)

43. $2\sqrt{t-1}-\sqrt{3t-1}=0$

$$2\sqrt{t-1}=\sqrt{3t-1}$$
$$(2\sqrt{t-1})^2=(\sqrt{3t-1})^2$$
$$4(t-1)=3t-1$$
$$4t-4=3t-1$$
$$t=3$$

Since 3 checks, it is the solution.

45. $D=1.2\sqrt{h}$

$$D=1.2\sqrt{1353}$$
$$D\approx 44.1$$

A tourist can see about 44.1 mi to the horizon.

47. $\qquad D=1.2\sqrt{h}$

$$31.3=1.2\sqrt{h}$$
$$(31.3)^2=(1.2\sqrt{h})^2$$
$$979.69=1.44h$$
$$680\approx h$$

The height of Sarah's eyes is about 680 ft.

49. $\qquad D=1.2\sqrt{h}$

$$30.4=1.2\sqrt{h}$$
$$(30.4)^2=(1.2\sqrt{h})^2$$
$$924.16=1.44h$$
$$642\approx h$$

The height of the tower is about 642 ft.

51. At 55 mph: $r=2\sqrt{5L}$

$$55=2\sqrt{5L}$$
$$27.5=\sqrt{5L}$$
$$(27.5)^2=(\sqrt{5L})^2$$
$$756.25=5L$$
$$151.25=L$$

At 55 mph, a car will skid 151.25 ft.

At 75 mph: $r=2\sqrt{5L}$

$$75=2\sqrt{5L}$$
$$37.5=\sqrt{5L}$$
$$(37.5)^2=(\sqrt{5L})^2$$
$$1406.25=5L$$
$$281.25=L$$

At 75 mph, a car will skid 281.25 ft.

53. $\qquad S=21.9\sqrt{5t+2457}$

$$1176=21.9\sqrt{5t+2457}$$
$$\frac{1176}{21.9}=\sqrt{5t+2457}$$
$$\left(\frac{1176}{21.9}\right)^2=(\sqrt{5t+2457})^2$$
$$2883.5\approx 5t+2457$$
$$426.5\approx 5t$$
$$85\approx t$$

The temperature was approximately 85°F.

55. $\qquad T=2\pi\sqrt{\dfrac{L}{32}}$

$$1=2(3.14)\sqrt{\dfrac{L}{32}}$$
$$1=6.28\sqrt{\dfrac{L}{32}}$$
$$1^2=\left(6.28\sqrt{\dfrac{L}{32}}\right)^2$$
$$1=39.4384\cdot\dfrac{L}{32}$$
$$\frac{32}{39.4384}=L$$
$$0.81\approx L$$

The pendulum is about 0.81 ft long.

57.
$$T = 2\pi\sqrt{\frac{L}{32}}$$

$$2.2 = 2(3.14)\sqrt{\frac{L}{32}}$$

$$\frac{1.1}{3.14} = \sqrt{\frac{L}{32}} \quad \text{Dividing by } 2(3.14) \text{ and simplifying}$$

$$\left(\frac{1.1}{3.14}\right)^2 = \left(\sqrt{\frac{L}{32}}\right)^2$$

$$0.128 \approx \frac{L}{32}$$

$$3.9 \approx L$$

The length of the pendulum is about 3.9 ft.

59. *Familiarize*. Let $t = $ the time it will take Julia and George to paint the room working together.

Translate. We use the work principle.
$$\frac{t}{a} + \frac{t}{b} = 1$$
$$\frac{t}{8} + \frac{t}{10} = 1$$

Solve. We first multiply by 40 to clear fractions.
$$40\left(\frac{t}{8} + \frac{t}{10}\right) = 40 \cdot 1$$
$$40 \cdot \frac{t}{8} + 40 \cdot \frac{t}{10} = 40$$
$$5t + 4t = 40$$
$$9t = 40$$
$$t = \frac{40}{9}, \text{ or } 4\frac{4}{9}$$

Check. In $\frac{40}{9}$ hr, Julia does $\frac{40}{9}\left(\frac{1}{8}\right)$, or $\frac{5}{9}$, of the job and George does $\frac{40}{9}\left(\frac{1}{10}\right)$, or $\frac{4}{9}$, of the job. Together they do $\frac{5}{9} + \frac{4}{9}$, or 1 entire job. The answer checks.

State. It will take them $4\frac{4}{9}$ hr to paint the room, working together.

61. *Familiarize*. Let $d = $ the distance the cyclist would travel in 56 days at the same rate.

Translate. We translate to a proportion.
$$\begin{array}{c} \text{Distance} \rightarrow \\ \text{Days} \rightarrow \end{array} \frac{702}{14} = \frac{d}{56} \begin{array}{c} \leftarrow \text{Distance} \\ \leftarrow \text{Days} \end{array}$$

Solve. We equate cross products.
$$\frac{702}{14} = \frac{d}{56}$$
$$702 \cdot 56 = 14 \cdot d$$
$$\frac{702 \cdot 56}{14} = d$$
$$2808 = d$$

Check. We substitute in the proportion and check cross products.
$$\frac{702}{14} = \frac{2808}{56}; \ 702 \cdot 56 = 39,312; \ 14 \cdot 2808 = 39,312$$
The cross products are the same, so the answer checks.

State. The cyclist would have traveled 2808 mi in 56 days.

63.
$$x^2 + 2.8x = 0$$
$$x(x + 2.8) = 0$$
$$x = 0 \ \ or \ \ x + 2.8 = 0$$
$$x = 0 \ \ or \ \ \ \ \ \ \ \ \ x = -2.8$$
The solutions are 0 and -2.8.

65.
$$x^2 - 64 = 0$$
$$(x + 8)(x - 8) = 0$$
$$x + 8 = 0 \ \ \ or \ \ x - 8 = 0$$
$$x = -8 \ \ or \ \ \ \ \ \ \ x = 8$$
The solutions are -8 and 8.

67. $f(x) = x^2$
$$f(a + h) - f(a) = (a + h)^2 - a^2$$
$$= a^2 + 2ah + h^2 - a^2$$
$$= 2ah + h^2$$

69. $f(x) = 2x^2 - 3x$
$$f(a + h) - f(a)$$
$$= 2(a + h)^2 - 3(a + h) - (2a^2 - 3a)$$
$$= 2(a^2 + 2ah + h^2) - 3a - 3h - 2a^2 + 3a$$
$$= 2a^2 + 4ah + 2h^2 - 3a - 3h - 2a^2 + 3a$$
$$= 4ah + 2h^2 - 3h$$

71. Left to the student

73.
$$\sqrt{\sqrt{y + 49} - \sqrt{y}} = \sqrt{7}$$
$$\left(\sqrt{\sqrt{y + 49} - \sqrt{y}}\right)^2 = (\sqrt{7})^2$$
$$\sqrt{y + 49} - \sqrt{y} = 7$$
$$\sqrt{y + 49} = 7 + \sqrt{y}$$
$$(\sqrt{y + 49})^2 = (7 + \sqrt{y})^2$$
$$y + 49 = 49 + 14\sqrt{y} + y$$
$$0 = 14\sqrt{y}$$
$$0 = \sqrt{y}$$
$$0^2 = (\sqrt{y})^2$$
$$0 = y$$
The number 0 checks and is the solution.

75.

$$\sqrt{\sqrt{x^2+9x+34}} = 2$$

$$\left(\sqrt{\sqrt{x^2+9x+34}}\right)^2 = 2^2$$

$$\sqrt{x^2+9x+34} = 4$$

$$(\sqrt{x^2+9x+34})^2 = 4^2$$

$$x^2+9x+34 = 16$$

$$x^2+9x+18 = 0$$

$$(x+6)(x+3) = 0$$

$$x+6 = 0 \quad \text{or} \quad x+3 = 0$$

$$x = -6 \quad \text{or} \quad x = -3$$

Both values check. The solutions are -6 and -3.

77.

$$\sqrt{x-2} - \sqrt{x+2} + 2 = 0$$

$$\sqrt{x-2} + 2 = \sqrt{x+2}$$

$$(\sqrt{x-2}+2)^2 = (\sqrt{x+2})^2$$

$$(x-2) + 4\sqrt{x-2} + 4 = x+2$$

$$4\sqrt{x-2} = 0$$

$$\sqrt{x-2} = 0$$

$$(\sqrt{x-2})^2 = 0^2$$

$$x-2 = 0$$

$$x = 2$$

The number 2 checks, so it is the solution.

79.

$$\sqrt{a^2+30a} = a + \sqrt{5a}$$

$$(\sqrt{a^2+30a})^2 = (a+\sqrt{5a})^2$$

$$a^2+30a = a^2 + 2a\sqrt{5a} + 5a$$

$$25a = 2a\sqrt{5a}$$

$$(25a)^2 = (2a\sqrt{5a})^2$$

$$625a^2 = 4a^2 \cdot 5a$$

$$625a^2 = 20a^3$$

$$0 = 20a^3 - 625a^2$$

$$0 = 5a^2(4a - 125)$$

$$5a^2 = 0 \quad \text{or} \quad 4a - 125 = 0$$

$$a^2 = 0 \quad \text{or} \quad 4a = 125$$

$$a = 0 \quad \text{or} \quad a = \frac{125}{4}$$

Both values check. The solutions are 0 and $\frac{125}{4}$.

81.

$$\frac{x-1}{\sqrt{x^2+3x+6}} = \frac{1}{4},$$

$$\text{LCM} = 4\sqrt{x^2+3x+6}$$

$$4\sqrt{x^2+3x+6} \cdot \frac{x-1}{\sqrt{x^2+3x+6}} = 4\sqrt{x^2+3x+6} \cdot \frac{1}{4}$$

$$4x - 4 = \sqrt{x^2+3x+6}$$

$$16x^2 - 32x + 16 = x^2 + 3x + 6$$

$$\text{Squaring both sides}$$

$$15x^2 - 35x + 10 = 0$$

$$3x^2 - 7x + 2 = 0 \qquad \text{Dividing by 5}$$

$$(3x-1)(x-2) = 0$$

$$3x - 1 = 0 \quad \text{or} \quad x - 2 = 0$$

$$3x = 1 \quad \text{or} \quad x = 2$$

$$x = \frac{1}{3} \quad \text{or} \quad x = 2$$

The number 2 checks but $\frac{1}{3}$ does not. The solution is 2.

83.

$$\sqrt{y+1} - \sqrt{2y-5} = \sqrt{y-2}$$

$$(\sqrt{y+1} - \sqrt{2y-5})^2 = (\sqrt{y-2})^2$$

$$y+1 - 2\sqrt{(y+1)(2y-5)} + 2y - 5 = y - 2$$

$$-2\sqrt{2y^2-3y-5} = -2y+2$$

$$\sqrt{2y^2-3y-5} = y-1$$

$$\text{Dividing by } -2$$

$$(\sqrt{2y^2-3y-5})^2 = (y-1)^2$$

$$2y^2 - 3y - 5 = y^2 - 2y + 1$$

$$y^2 - y - 6 = 0$$

$$(y-3)(y+2) = 0$$

$$y - 3 = 0 \quad \text{or} \quad y + 2 = 0$$

$$y = 3 \quad \text{or} \quad y = -2$$

The number 3 checks but -2 does not. The solution is 3.

Exercise Set 6.7

RC1. (e)

RC3. (g)

1. $a = 3, \quad b = 5$

Find c.

$$c^2 = a^2 + b^2 \qquad \text{Pythagorean theorem}$$

$$c^2 = 3^2 + 5^2 \qquad \text{Substituting}$$

$$c^2 = 9 + 25$$

$$c^2 = 34$$

$$c = \sqrt{34} \qquad \text{Exact answer}$$

$$c \approx 5.831 \qquad \text{Approximation}$$

3. $a = 15, \quad b = 15$

Find c.

$c^2 = a^2 + b^2$ Pythagorean theorem

$c^2 = 15^2 + 15^2$ Substituting

$c^2 = 225 + 225$

$c^2 = 450$

$c = \sqrt{450}$ Exact answer

$c \approx 21.213$ Approximation

5. $b = 12, \quad c = 13$

Find a.

$a^2 + b^2 = c^2$ Pythagorean theorem

$a^2 + 12^2 = 13^2$ Substituting

$a^2 + 144 = 169$

$a^2 = 25$

$a = 5$

7. $c = 7, \quad a = \sqrt{6}$

Find b.

$c^2 = a^2 + b^2$ Pythagorean theorem

$7^2 = (\sqrt{6})^2 + b^2$ Substituting

$49 = 6 + b^2$

$43 = b^2$

$\sqrt{43} = b$ Exact answer

$6.557 \approx b$ Approximation

9. $b = 1, \quad c = \sqrt{13}$

Find a.

$a^2 + b^2 = c^2$ Pythagorean theorem

$a^2 + 1^2 = (\sqrt{13})^2$ Substituting

$a^2 + 1 = 13$

$a^2 = 12$

$a = \sqrt{12}$ Exact answer

$a \approx 3.464$ Approximation

11. $a = 1, \quad c = \sqrt{n}$

Find b.

$a^2 + b^2 = c^2$

$1^2 + b^2 = (\sqrt{n})^2$

$1 + b^2 = n$

$b^2 = n - 1$

$b = \sqrt{n-1}$

13. $L = \dfrac{0.000169 d^{2.27}}{h}$

$L = \dfrac{0.000169(200)^{2.27}}{4}$

≈ 7.1

The length of the letters should be about 7.1 ft.

15. We make a drawing and let $d = $ the length of the guy wire.

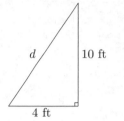

We use the Pythagorean theorem to find d.

$d^2 = 4^2 + 10^2$

$d^2 = 16 + 100$

$d^2 = 116$

$d = \sqrt{116}$

$d \approx 10.770$

The wire is $\sqrt{116}$ ft, or about 10.770 ft long.

17. Let $d = $ the distance from the center of the base to a corner of the base, in feet. We have a right triangle with legs of length 71 and d and hypotenuse of length 100.

We use the Pythagorean equation to find d.

$71^2 + d^2 = 100^2$

$5041 + d^2 = 10,000$

$d^2 = 4959$

$d = \sqrt{4959} \approx 70.4$

The distance from the center of the base to a corner of the base is $\sqrt{4959}$ ft, or about 70.4 ft.

19. We add labels to the drawing in the text. We let $h = $ the height of the bulge.

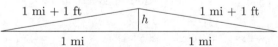

Note that 1 mi = 5280 ft, so 1 mi + 1 ft = 5280 + 1, or 5281 ft.

We use the Pythagorean equation to find h.

$5281^2 = 5280^2 + h^2$

$27,888,961 = 27,878,400 + h^2$

$10,561 = h^2$

$\sqrt{10,561} = h$

$102.767 \approx h$

The bulge is $\sqrt{10,561}$ ft, or about 102.767 ft high.

21. We first make a drawing. A point on the x-axis has coordinates $(x, 0)$ and is $|x|$ units from the origin.

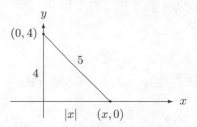

We use the Pythagorean equation to find x.

$$4^2 + |x|^2 = 5^2$$
$$16 + x^2 = 25 \quad |x|^2 = x^2$$
$$x^2 - 9 = 0 \quad \text{Subtracting 25}$$
$$(x + 3)(x - 3) = 0$$
$$x - 3 = 0 \quad \text{or} \quad x + 3 = 0$$
$$x = 3 \quad \text{or} \quad x = -3$$

The points are $(3, 0)$ and $(-3, 0)$.

23. We make a drawing, letting $d =$ the distance the wire will run diagonally, disregarding the slack.

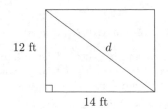

We use the Pythagorean equation to find d.

$$d^2 = 12^2 + 14^2$$
$$d^2 = 144 + 196$$
$$d^2 = 340$$
$$d = \sqrt{340}$$

Adding 4 ft of slack on each end, we find that a wire of length $\sqrt{340} + 4 + 4$, or $\sqrt{340} + 8$ ft should be purchased. This is approximately 26.439 ft of wire.

25. Referring to the drawing in the text, we let $t =$ the travel. Then we use the Pythagorean equation.

$$t^2 = (17.75)^2 + (10.25)^2$$
$$t^2 = 315.0625 + 105.0625$$
$$t^2 = 420.125$$
$$t = \sqrt{420.125}$$
$$t \approx 20.497$$

The travel is $\sqrt{420.125}$ in., or about 20.497 in.

27. We make a drawing. Let $x =$ the width of the rectangle. Then $x + 1 =$ the length.

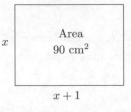

We first find the length and width of the rectangle. Recall the formula for the area of a rectangle, $A = lw$. We substitute 90 for A, $x + 1$ for l, and x for w in this formula and solve for x.

$$90 = (x + 1)x$$
$$90 = x^2 + x$$
$$0 = x^2 + x - 90$$
$$0 = (x + 10)(x - 9)$$
$$x + 10 = 0 \quad \text{or} \quad x - 9 = 0$$
$$x = -10 \quad \text{or} \quad x = 9$$

Since the width cannot be negative, we know that the width is 9 cm. Thus the length is 10 cm. (These numbers check since 9 and 10 are consecutive integers and the area of a rectangle with width 9 cm and length 10 cm is $10 \cdot 9$, or 90 cm^2.)

Now we find the length of the diagonal of the rectangle. We make another drawing, letting $d =$ the length of the diagonal.

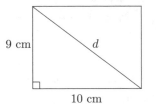

We use the Pythagorean equation to find d.

$$d^2 = 9^2 + 10^2$$
$$d^2 = 81 + 100$$
$$d^2 = 181$$
$$d = \sqrt{181}$$
$$d \approx 13.454$$

The length of the diagonal is $\sqrt{181}$ cm, or about 13.454 cm.

29. We add some labels to the drawing in the text.

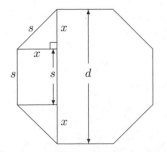

Note that $d = s + 2x$. We use the Pythagorean equation to find x.

$$x^2 + x^2 = s^2$$
$$2x^2 = s^2$$
$$x^2 = \frac{s^2}{2}$$
$$x = \sqrt{\frac{s^2}{2}}$$
$$x = \frac{s}{\sqrt{2}}$$
$$x = \frac{s\sqrt{2}}{2} \quad \text{Rationalizing the denominator}$$

Then $d = s + 2x = s + 2\left(\dfrac{s\sqrt{2}}{2}\right) = s + s\sqrt{2}$.

31. *Familiarize.* Let r = the speed of the Carmel Crawler. Then $r + 14$ = the speed of the Zionsville Flash. We organize the information in a table.

	Distance	Speed	Time
Crawler	230	r	t
Flash	290	$r + 14$	t

Translate. Using the formula $t = d/r$ and noting that the times are the same, we have
$$\frac{230}{r} = \frac{290}{r + 14}.$$

Solve. We first clear fractions by multiplying by the LCM of the denominators, $r(r + 14)$.
$$r(r + 14) \cdot \frac{230}{r} = r(r + 14) \cdot \frac{290}{r + 14}$$
$$230(r + 14) = 290r$$
$$230r + 3220 = 290r$$
$$3220 = 60r$$
$$\frac{161}{3} = r, \text{ or}$$
$$53\frac{2}{3} = r$$

If $r = 53\dfrac{2}{3}$, then $r + 14 = 67\dfrac{2}{3}$.

Check. At $53\dfrac{2}{3}$, or $\dfrac{161}{3}$ mph, the Crawler travels 230 mi in $\dfrac{230}{161/3}$, or about 4.3 hr. At $67\dfrac{2}{3}$, or $\dfrac{203}{3}$ mph, the Flash travels 290 mi in $\dfrac{290}{203/3}$, or about 4.3 hr. Since the times are the same, the answer checks.

State. The Carmel Crawler's speed is $53\dfrac{2}{3}$ mph, and the Zionsville Flash's speed is $67\dfrac{2}{3}$ mph.

33.
$$2x^2 + 11x - 21 = 0$$
$$(2x - 3)(x + 7) = 0$$
$$2x - 3 = 0 \quad or \quad x + 7 = 0$$
$$x = \frac{3}{2} \quad or \quad x = -7$$

The solutions are $\dfrac{3}{2}$ and -7.

35.
$$\frac{x + 2}{x + 3} = \frac{x - 4}{x - 5},$$
$$\text{LCM is } (x + 3)(x - 5)$$
$$(x + 3)(x - 5) \cdot \frac{x + 2}{x + 3} = (x + 3)(x - 5) \cdot \frac{x - 4}{x - 5}$$
$$(x - 5)(x + 2) = (x + 3)(x - 4)$$
$$x^2 - 3x - 10 = x^2 - x - 12$$
$$-2x = -2$$
$$x = 1$$

The number 1 checks, so it is the solution.

37.
$$\frac{x - 5}{x - 7} = \frac{4}{3}, \text{ LCM is } 3(x - 7)$$
$$3(x - 7) \cdot \frac{x - 5}{x - 7} = 3(x - 7) \cdot \frac{4}{3}$$
$$3(x - 5) = 4(x - 7)$$
$$3x - 15 = 4x - 28$$
$$13 = x$$

The number 13 checks and is the solution.

39.

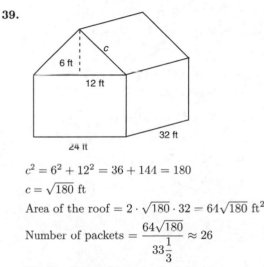

$$c^2 = 6^2 + 12^2 = 36 + 144 = 180$$
$$c = \sqrt{180} \text{ ft}$$
Area of the roof $= 2 \cdot \sqrt{180} \cdot 32 = 64\sqrt{180} \text{ ft}^2$
Number of packets $= \dfrac{64\sqrt{180}}{33\dfrac{1}{3}} \approx 26$

Kit should buy 26 packets of shingles.

Exercise Set 6.8

RC1. True; see page 551 in the text.

RC3. False; see page 551 in the text.

RC5. True; see page 554 in the text.

1. $\sqrt{-35} = \sqrt{-1 \cdot 35} = \sqrt{-1} \cdot \sqrt{35} = i\sqrt{35}$, or $\sqrt{35}i$

3. $\sqrt{-16} = \sqrt{-1 \cdot 16} = \sqrt{-1} \cdot \sqrt{16} = i \cdot 4 = 4i$

5. $-\sqrt{-12} = -\sqrt{-1 \cdot 12} = -\sqrt{-1} \cdot \sqrt{12} = -i \cdot 2\sqrt{3} = -2\sqrt{3}i$, or $-2i\sqrt{3}$

7. $\sqrt{-3} = \sqrt{-1 \cdot 3} = \sqrt{-1} \cdot \sqrt{3} = i\sqrt{3}$, or $\sqrt{3}i$

9. $\sqrt{-81} = \sqrt{-1 \cdot 81} = \sqrt{-1} \cdot \sqrt{81} = i \cdot 9 = 9i$

11. $\sqrt{-98} = \sqrt{-1 \cdot 98} = \sqrt{-1} \cdot \sqrt{98} = i \cdot 7\sqrt{2} = 7\sqrt{2}i$, or $7i\sqrt{2}$

13. $-\sqrt{-49} = -\sqrt{-1 \cdot 49} = -\sqrt{-1} \cdot \sqrt{49} = -i \cdot 7 = -7i$

15. $4 - \sqrt{-60} = 4 - \sqrt{-1 \cdot 60} = 4 - \sqrt{-1} \cdot \sqrt{60} = 4 - i \cdot 2\sqrt{15} = 4 - 2\sqrt{15}i$, or $4 - 2i\sqrt{15}$

17. $\sqrt{-4} + \sqrt{-12} = \sqrt{-1 \cdot 4} + \sqrt{-1 \cdot 12} = \sqrt{-1} \cdot \sqrt{4} + \sqrt{-1} \cdot \sqrt{12} = i \cdot 2 + i \cdot 2\sqrt{3} = (2 + 2\sqrt{3})i$

19. $\quad (7 + 2i) + (5 - 6i)$
$= (7 + 5) + (2 - 6)i \quad$ Collecting like terms
$= 12 - 4i$

21. $\quad (4 - 3i) + (5 - 2i)$
$= (4 + 5) + (-3 - 2)i \quad$ Collecting like terms
$= 9 - 5i$

23. $(9 - i) + (-2 + 5i) = (9 - 2) + (-1 + 5)i$
$= 7 + 4i$

25. $(6 - i) - (10 + 3i) = (6 - 10) + (-1 - 3)i$
$= -4 - 4i$

27. $(4 - 2i) - (5 - 3i) = (4 - 5) + [-2 - (-3)]i$
$= -1 + i$

29. $(9 + 5i) - (-2 - i) = [9 - (-2)] + [5 - (-1)]i$
$= 11 + 6i$

31. $\sqrt{-36} \cdot \sqrt{-9} = \sqrt{-1} \cdot \sqrt{36} \cdot \sqrt{-1} \cdot \sqrt{9}$
$= i \cdot 6 \cdot i \cdot 3$
$= i^2 \cdot 18$
$= -1 \cdot 18 \qquad i^2 = -1$
$= -18$

33. $\sqrt{-7} \cdot \sqrt{-2} = \sqrt{-1} \cdot \sqrt{7} \cdot \sqrt{-1} \cdot \sqrt{2}$
$= i \cdot \sqrt{7} \cdot i \cdot \sqrt{2}$
$= i^2(\sqrt{14})$
$= -1(\sqrt{14}) \qquad i^2 = -1$
$= -\sqrt{14}$

35. $-3i \cdot 7i = -21 \cdot i^2$
$= -21(-1) \qquad i^2 = -1$
$= 21$

37. $-3i(-8 - 2i) = -3i(-8) - 3i(-2i)$
$= 24i + 6i^2$
$= 24i + 6(-1) \qquad i^2 = -1$
$= 24i - 6$
$= -6 + 24i$

39. $\quad (3 + 2i)(1 + i)$
$= 3 + 3i + 2i + 2i^2 \quad$ Using FOIL
$= 3 + 3i + 2i - 2 \qquad i^2 = -1$
$= 1 + 5i$

41. $\quad (2 + 3i)(6 - 2i)$
$= 12 - 4i + 18i - 6i^2 \quad$ Using FOIL
$= 12 - 4i + 18i + 6 \qquad i^2 = -1$
$= 18 + 14i$

43. $(6 - 5i)(3 + 4i) = 18 + 24i - 15i - 20i^2$
$= 18 + 24i - 15i + 20$
$= 38 + 9i$

45. $(7 - 2i)(2 - 6i) = 14 - 42i - 4i + 12i^2$
$= 14 - 42i - 4i - 12$
$= 2 - 46i$

47. $(3 - 2i)^2 = 3^2 - 2 \cdot 3 \cdot 2i + (2i)^2 \quad$ Squaring a binomial
$= 9 - 12i + 4i^2$
$= 9 - 12i - 4 \qquad i^2 = -1$
$= 5 - 12i$

49. $\quad (1 + 5i)^2$
$= 1^2 + 2 \cdot 1 \cdot 5i + (5i)^2 \quad$ Squaring a binomial
$= 1 + 10i + 25i^2$
$= 1 + 10i - 25 \qquad i^2 = -1$
$= -24 + 10i$

51. $(-2 + 3i)^2 = 4 - 12i + 9i^2 = 4 - 12i - 9 = -5 - 12i$

53. $i^7 = i^6 \cdot i = (i^2)^3 \cdot i = (-1)^3 \cdot i = -1 \cdot i = -i$

55. $i^{24} = (i^2)^{12} = (-1)^{12} = 1$

57. $i^{42} = (i^2)^{21} = (-1)^{21} = -1$

59. $i^9 = (i^2)^4 \cdot i = (-1)^4 \cdot i = 1 \cdot i = i$

61. $i^6 = (i^2)^3 = (-1)^3 = -1$

63. $(5i)^3 = 5^3 \cdot i^3 = 125 \cdot i^2 \cdot i = 125(-1)(i) = -125i$

65. $7 + i^4 = 7 + (i^2)^2 = 7 + (-1)^2 = 7 + 1 = 8$

67. $i^{28} - 23i = (i^2)^{14} - 23i = (-1)^{14} - 23i = 1 - 23i$

69. $i^2 + i^4 = -1 + (i^2)^2 = -1 + (-1)^2 = -1 + 1 = 0$

71. $i^5 + i^7 = i^4 \cdot i + i^6 \cdot i = (i^2)^2 \cdot i + (i^2)^3 \cdot i = (-1)^2 \cdot i + (-1)^3 \cdot i = 1 \cdot i + (-1)i = i - i = 0$

73. $1 + i + i^2 + i^3 + i^4 = 1 + i + i^2 + i^2 \cdot i + (i^2)^2$
$= 1 + i + (-1) + (-1) \cdot i + (-1)^2$
$= 1 + i - 1 - i + 1$
$= 1$

75. $5 - \sqrt{-64} = 5 - \sqrt{-1} \cdot \sqrt{64} = 5 - i \cdot 8 = 5 - 8i$

77. $\dfrac{8-\sqrt{-24}}{4} = \dfrac{8-\sqrt{-1}\cdot\sqrt{24}}{4} = \dfrac{8-i\cdot2\sqrt{6}}{4} =$

$\dfrac{2(4-i\sqrt{6})}{2\cdot2} = \dfrac{\cancel{2}(4-i\sqrt{6})}{\cancel{2}\cdot2} = \dfrac{4-i\sqrt{6}}{2} = 2-\dfrac{\sqrt{6}}{2}i$

79. $\dfrac{4+3i}{3-i} = \dfrac{4+3i}{3-i}\cdot\dfrac{3+i}{3+i}$

$= \dfrac{(4+3i)(3+i)}{(3-i)(3+i)}$

$= \dfrac{12+4i+9i+3i^2}{9-i^2}$

$= \dfrac{12+13i-3}{9-(-1)}$

$= \dfrac{9+13i}{10}$

$= \dfrac{9}{10}+\dfrac{13}{10}i$

81. $\dfrac{3-2i}{2+3i} = \dfrac{3-2i}{2+3i}\cdot\dfrac{2-3i}{2-3i}$

$= \dfrac{(3-2i)(2-3i)}{(2+3i)(2-3i)}$

$= \dfrac{6-9i-4i+6i^2}{4-9i^2}$

$= \dfrac{6-13i-6}{4-9(-1)}$

$= \dfrac{-13i}{13}$

$= -i$

83. $\dfrac{8-3i}{7i} = \dfrac{8-3i}{7i}\cdot\dfrac{-7i}{-7i}$

$= \dfrac{-56i+21i^2}{-49i^2}$

$= \dfrac{-21-56i}{49}$

$= -\dfrac{21}{49}-\dfrac{56}{49}i$

$= -\dfrac{3}{7}-\dfrac{8}{7}i$

85. $\dfrac{4}{3+i} = \dfrac{4}{3+i}\cdot\dfrac{3-i}{3-i}$

$= \dfrac{12-4i}{9-i^2}$

$= \dfrac{12-4i}{9-(-1)}$

$= \dfrac{12-4i}{10}$

$= \dfrac{12}{10}-\dfrac{4}{10}i$

$= \dfrac{6}{5}-\dfrac{2}{5}i$

87. $\dfrac{2i}{5-4i} = \dfrac{2i}{5-4i}\cdot\dfrac{5+4i}{5+4i}$

$= \dfrac{10i+8i^2}{25-16i^2}$

$= \dfrac{10i+8(-1)}{25-16(-1)}$

$= \dfrac{-8+10i}{41}$

$= -\dfrac{8}{41}+\dfrac{10}{41}i$

89. $\dfrac{4}{3i} = \dfrac{4}{3i}\cdot\dfrac{-3i}{-3i}$

$= \dfrac{-12i}{-9i^2}$

$= \dfrac{-12i}{-9(-1)}$

$= \dfrac{-12i}{9}$

$= -\dfrac{4}{3}i$

91. $\dfrac{9-4i}{8i} = \dfrac{2-4i}{8i}\cdot\dfrac{-8i}{-8i}$

$= \dfrac{-16i+32i^2}{-64i^2}$

$= \dfrac{-16i+32(-1)}{-64(-1)}$

$= \dfrac{-32-16i}{64}$

$= -\dfrac{32}{64}-\dfrac{16}{64}i$

$= -\dfrac{1}{2}-\dfrac{1}{4}i$

93. $\dfrac{6+3i}{6-3i} = \dfrac{6+3i}{6-3i}\cdot\dfrac{6+3i}{6+3i}$

$= \dfrac{36+18i+18i+9i^2}{36-9i^2}$

$= \dfrac{36+36i-9}{36-9(-1)}$

$= \dfrac{27+36i}{45}$

$= \dfrac{27}{45}+\dfrac{36}{45}i$

$= \dfrac{3}{5}+\dfrac{4}{5}i$

95. Substitute $1-2i$ for x in the equation.

$$x^2-2x+5=0$$

$\dfrac{(1-2i)^2-2(1-2i)+5 \;\;?\;\; 0}{}$

$1-4i+4i^2-2+4i+5$

$1-4i-4-2+4i+5$

$\qquad\qquad 0 \;\Big|\; \text{TRUE}$

$1-2i$ is a solution.

97. Substitute $2 + i$ for x in the equation.

$$\frac{x^2 - 4x - 5 = 0}{(2+i)^2 - 4(2+i) - 5 \ ? \ 0}$$

$$4 + 4i + i^2 - 8 - 4i - 5$$

$$4 + 4i - 1 - 8 - 4i - 5$$

$$-10 \ \Big| \ \text{FALSE}$$

$2 + i$ is not a solution.

99. x-intercept: $2x - 0 = -30$

$$2x = -30$$

$$x = -15$$

The x-intercept is $(-15, 0)$.

y-intercept: $2 \cdot 0 - y = -30$

$$-y = -30$$

$$y = 30$$

The y-intercept is $(0, 30)$.

101. x-intercept: $5x = 10 - 2 \cdot 0$

$$5x = 10$$

$$x = 2$$

The x-intercept is $(2, 0)$.

y-intercept: $5 \cdot 0 = 10 - 2y$

$$0 = 10 - 2y$$

$$2y = 10$$

$$y = 5$$

The y-intercept is $(0, 5)$.

103. $x - 3y = 15$

$$-3y = -x + 15$$

$$y = \frac{1}{3}x - 5 \quad \text{Dividing by } -3$$

The equation is now in the form $y = mx + b$, where m is the slope. Thus, the slope is $\frac{1}{3}$.

105. $\dfrac{3.6 \times 10^{-5}}{1.2 \times 10^{-8}} = \dfrac{3.6}{1.2} \times \dfrac{10^{-5}}{10^{-8}}$

$$= 3 \times 10^3$$

107. $g(2i) = \dfrac{(2i)^4 - (2i)^2}{2i - 1} = \dfrac{16i^4 - 4i^2}{-1 + 2i} = \dfrac{20}{-1 + 2i} =$

$$\dfrac{20}{-1 + 2i} \cdot \dfrac{-1 - 2i}{-1 - 2i} = \dfrac{-20 - 40i}{5} = -4 - 8i;$$

$$g(i + 1) = \dfrac{(i+1)^4 - (i+1)^2}{(i+1) - 1} =$$

$$\dfrac{(i+1)^2[(i+1)^2 - 1]}{i} = \dfrac{2i(2i - 1)}{i} = 2(2i - 1) =$$

$$-2 + 4i;$$

$$g(2i - 1) = \dfrac{(2i-1)^4 - (2i-1)^2}{(2i-1) - 1} =$$

$$\dfrac{(2i-1)^2[(2i-1)^2 - 1]}{2i - 2} = \dfrac{(-3-4i)(-4-4i)}{-2 + 2i} =$$

$$\dfrac{(-3-4i)(-2-2i)}{-1+i} = \dfrac{-2 + 14i}{-1+i} =$$

$$\dfrac{-2 + 14i}{-1+i} \cdot \dfrac{-1-i}{-1-i} = \dfrac{16 - 12i}{2} = 8 - 6i$$

109. $\dfrac{1}{8}\left(-24 - \sqrt{-1024}\right) = \dfrac{1}{8}(-24 - 32i) = -3 - 4i$

111. $7\sqrt{-64} - 9\sqrt{-256} = 7 \cdot 8i - 9 \cdot 16i = 56i - 144i = -88i$

113. $(1 - i)^3(1 + i)^3 =$

$$(1-i)(1+i) \cdot (1-i)(1+i) \cdot (1-i)(1+i) =$$

$$(1 - i^2)(1 - i^2)(1 - i^2) = (1+1)(1+1)(1+1) =$$

$$2 \cdot 2 \cdot 2 = 8$$

115. $\dfrac{6}{1 + \dfrac{3}{i}} = \dfrac{6}{\dfrac{i+3}{i}} = \dfrac{6i}{i+3} = \dfrac{6i}{i+3} \cdot \dfrac{-i+3}{-i+3} =$

$$\dfrac{-6i^2 + 18i}{-i^2 + 9} = \dfrac{6 + 18i}{10} = \dfrac{6}{10} + \dfrac{18}{10}i = \dfrac{3}{5} + \dfrac{9}{5}i$$

117. $\dfrac{i - i^{38}}{1 + i} = \dfrac{i - (i^2)^{19}}{1 + i} = \dfrac{i - (-1)^{19}}{1 + i} = \dfrac{i - (-1)}{1 + i} =$

$$\dfrac{i + 1}{1 + i} = 1$$

Chapter 6 Vocabulary Reinforcement

1. The number c is the <u>cube</u> root of a, written $\sqrt[3]{a}$, if the third power of c is a — that is, if $c^3 = a$, then $\sqrt[3]{a} = c$.

2. A <u>complex</u> number is any number that can be named $a + bi$, where a and b are any real numbers.

3. For any real number a, $\sqrt{a^2} = |a|$. The <u>principal</u> (nonnegative) square root of a^2 is the absolute value of a.

4. To find an equivalent expression without a radical in the denominator is called <u>rationalizing</u> the denominator.

5. The symbol $\sqrt{}$ is called a <u>radical</u>.

6. An <u>imaginary</u> number is a number that can be named bi, where b is some real number and $b \neq 0$.

7. The number c is a <u>square</u> root of a if $c^2 = a$.

8. The expression written under the radical is called the <u>radicand</u>.

9. The <u>conjugate</u> of a complex number $a + bi$ is $a - bi$.

10. In the expression $\sqrt[k]{a}$, we call k the <u>index</u>.

Chapter 6 Concept Reinforcement

1. True; see page 495 in the text.

2. False; $\sqrt[m]{a} \cdot \sqrt[n]{b} = a^{1/m} \cdot b^{1/n} = a^{n/(mn)} \cdot b^{m/(mn)} = (a^n b^m)^{1/mn} = \sqrt[mn]{a^n b^m} \neq \sqrt[mn]{ab}$

3. False; radicals cannot be added unless they have the same index *and* the same radicand.

4. False; if $x^2 = 4$, $x = 2$ *or* $x = -2$.

5. True; see page 551 in the text.

6. True; see page 554 in the text.

Chapter 6 Study Guide

1. $\sqrt{36y^2} = \sqrt{(6y)^2} = |6y| = 6|y|$

2. $\sqrt{a^2 + 4a + 4} = \sqrt{(a+2)^2} = |a+2|$

3. $z^{3/5} = \sqrt[5]{z^3}$

4. $(\sqrt{6ab})^5 = (6ab)^{5/2}$

5. $9^{-3/2} = \dfrac{1}{9^{3/2}} = \dfrac{1}{(\sqrt{9})^3} = \dfrac{1}{3^3} = \dfrac{1}{27}$

6. $\sqrt[8]{a^6 b^2} = (a^6 b^2)^{1/8} = a^{6/8} b^{2/8} = a^{3/4} b^{1/4} = (a^3 b)^{1/4} = \sqrt[4]{a^3 b}$

7. $\sqrt{5y}\sqrt{30y} = \sqrt{5y \cdot 30y} - \sqrt{150y^2} = \sqrt{25y^2 \cdot 6} = \sqrt{25y^2}\sqrt{6} = 5y\sqrt{6}$

8. $\dfrac{\sqrt{20a}}{\sqrt{5}} = \sqrt{\dfrac{20a}{5}} = \sqrt{4a} = 2\sqrt{a}$

9. $\sqrt{48} - 2\sqrt{3} = \sqrt{16 \cdot 3} - 2\sqrt{3}$
$= 4\sqrt{3} - 2\sqrt{3}$
$= (4-2)\sqrt{3}$
$= 2\sqrt{3}$

10. $(5 - \sqrt{x})^2 = 25 - 10\sqrt{x} + (\sqrt{x})^2 = 25 - 10\sqrt{x} + x$

11. $3 + \sqrt{x-1} = x$
$\sqrt{x-1} = x - 3$
$(\sqrt{x-1})^2 = (x-3)^2$
$x - 1 = x^2 - 6x + 9$
$0 = x^2 - 7x + 10$
$0 = (x-2)(x-5)$
$x - 2 = 0 \ \ or \ \ x - 5 = 0$
$x = 2 \ \ or \ \ \ \ \ \ x = 5$

The number 2 does not check, but 5 does. The solution is 5.

12. $\sqrt{x+3} - \sqrt{x-2} = 1$
$\sqrt{x+3} = \sqrt{x-2} + 1$
$(\sqrt{x+3})^2 = (\sqrt{x-2}+1)^2$
$x + 3 = x - 2 + 2\sqrt{x-2} + 1$
$4 = 2\sqrt{x-2}$
$2 = \sqrt{x-2} \qquad \text{Dividing by 2}$
$2^2 = (\sqrt{x-2})^2$
$4 = x - 2$
$6 = x$

The number 6 checks. It is the solution.

13. $(2 - 5i)^2 = 4 - 20i + 25i^2 = 4 - 20i + 25(-1) = 4 - 20i - 25 = -21 - 20i$

14. $\dfrac{3 - 2i}{2 + i} = \dfrac{3 - 2i}{2 + i} \cdot \dfrac{2 - i}{2 - i} = \dfrac{6 - 3i - 4i + 2i^2}{4 - i^2}$
$= \dfrac{6 - 7i + 2(-1)}{4 - (-1)}$
$= \dfrac{6 - 7i - 2}{4 + 1}$
$= \dfrac{4 - 7i}{5}$
$= \dfrac{4}{5} - \dfrac{7}{5}i$

Chapter 6 Review Exercises

1. $\sqrt{778} \approx 27.893$

2. $\sqrt{\dfrac{963.2}{23.68}} \approx 6.378$

3. $f(x) = \sqrt{3x - 16}$
$f(0) = \sqrt{3 \cdot 0 - 16} = \sqrt{-16}$

    Since negative numbers do not have real-number square roots, $f(0)$ does not exist as a real number.

$f(-1) = \sqrt{3(-1) - 16} = \sqrt{-3 - 16} = \sqrt{-19}$

    Since negative numbers do not have real-number square roots, $f(-1)$ does not exist as a real number.

$f(1) = \sqrt{3 \cdot 1 - 16} = \sqrt{3 - 16} = \sqrt{-13}$

    Since negative numbers do not have real-number square roots, $f(1)$ does not exist as a real number.

$f\left(\dfrac{41}{3}\right) = \sqrt{3 \cdot \dfrac{41}{3} - 16} = \sqrt{41 - 16} = \sqrt{25} = 5$

4. The domain of $f(x) = \sqrt{3x - 16}$ is the set of all x-values for which $3x - 16 \geq 0$.
$3x - 16 \geq 0$
$3x \geq 16$
$x \geq \dfrac{16}{3}$

The domain is $\left\{x \middle| x \geq \dfrac{16}{3}\right\}$, or $\left[\dfrac{16}{3}, \infty\right)$.

5. $\sqrt{81a^2} = \sqrt{(9a)^2} = |9a| = |9| \cdot |a| = 9|a|$

6. $\sqrt{(-7z)^2} = |-7z| = |-7| \cdot |z| = 7|z|$

7. $\sqrt{(6-b)^2} = |6-b|$

8. $\sqrt{x^2 + 6x + 9} = \sqrt{(x+3)^2} = |x+3|$

9. $\sqrt[3]{-1000} = -10 \qquad \text{Since } (-10)^3 = -1000$

10. $\sqrt[3]{-\dfrac{1}{27}} = -\dfrac{1}{3}$ Since $\left(-\dfrac{1}{3}\right)^3 = -\dfrac{1}{27}$

11. $f(x) = \sqrt[3]{x+2}$
$f(6) = \sqrt[3]{6+2} = \sqrt[3]{8} = 2$
$f(-10) = \sqrt[3]{-10+2} = \sqrt[3]{-8} = -2$
$f(25) = \sqrt[3]{25+2} = \sqrt[3]{27} = 3$

12. $\sqrt[10]{x^{10}} = |x|$

13. $-\sqrt[13]{(-3)^{13}} = -(-3) = 3$

14. $a^{1/5} = \sqrt[5]{a}$

15. $64^{3/2} = (\sqrt{64})^3 = 8^3 = 512$

16. $\sqrt{31} = 31^{1/2}$

17. $\sqrt[5]{a^2 b^3} = (a^2 b^3)^{1/5}$

18. $49^{-1/2} = \dfrac{1}{49^{1/2}} = \dfrac{1}{\sqrt{49}} = \dfrac{1}{7}$

19. $(8xy)^{-2/3} = \dfrac{1}{(8xy)^{2/3}} = \dfrac{1}{8^{2/3}x^{2/3}y^{2/3}} =$
$\dfrac{1}{(2^3)^{2/3}x^{2/3}y^{2/3}} = \dfrac{1}{2^2 x^{2/3}y^{2/3}} = \dfrac{1}{4x^{2/3}y^{2/3}}$

20. $5a^{-3/4}b^{1/2}c^{-2/3} = 5 \cdot \dfrac{1}{a^{3/4}} \cdot b^{1/2} \cdot \dfrac{1}{c^{2/3}} =$
$\dfrac{5b^{1/2}}{a^{3/4}c^{2/3}}$

21. $\dfrac{3a}{\sqrt[4]{t}} = \dfrac{3a}{t^{1/4}}$

22. $(x^{-2/3})^{3/5} = x^{(-2/3)(3/5)} = x^{-2/5} = \dfrac{1}{x^{2/5}}$

23. $\dfrac{7^{-1/3}}{7^{-1/2}} = 7^{-1/3-(-1/2)} = 7^{-1/3+1/2} = 7^{-2/6+3/6} = 7^{1/6}$

24. $\sqrt[3]{x^{21}} = (x^{21})^{1/3} = x^7$

25. $\sqrt[3]{27x^6} = \sqrt[3]{(3x^2)^3} = [(3x^2)^3]^{1/3} = 3x^2$

26. $x^{1/3}y^{1/4} = x^{4/12}y^{3/12} = (x^4 y^3)^{1/12} = \sqrt[12]{x^4 y^3}$

27. $\sqrt[4]{x}\sqrt[3]{x} = x^{1/4} \cdot x^{1/3} = x^{1/4+1/3} = x^{3/12+4/12} = x^{7/12} =$
$\sqrt[12]{x^7}$

28. $\sqrt{245} = \sqrt{49 \cdot 5} = \sqrt{49}\sqrt{5} = 7\sqrt{5}$

29. $\sqrt[3]{-108} = \sqrt[3]{-27 \cdot 4} = \sqrt[3]{-27}\sqrt[3]{4} = -3\sqrt[3]{4}$

30. $\sqrt[3]{250a^2 b^6} = \sqrt[3]{125b^6 \cdot 2a^2} = \sqrt[3]{125b^6}\sqrt[3]{2a^2}$
$5b^2\sqrt[3]{2a^2}$

31. $\sqrt{\dfrac{49}{36}} = \dfrac{\sqrt{49}}{\sqrt{36}} = \dfrac{7}{6}$

32. $\sqrt[3]{\dfrac{64x^6}{27}} = \dfrac{\sqrt[3]{64x^6}}{\sqrt[3]{27}} = \dfrac{4x^2}{3}$, or $\dfrac{4}{3}x^2$

33. $\sqrt[4]{\dfrac{16x^8}{81y^{12}}} = \dfrac{\sqrt[4]{16x^8}}{\sqrt[4]{81y^{12}}} = \dfrac{2x^2}{3y^3}$

34. $\sqrt{5x}\sqrt{3y} = \sqrt{5x \cdot 3y} = \sqrt{15xy}$

35. $\sqrt[3]{a^5 b}\sqrt[3]{27b} = \sqrt[3]{27a^5 b^2} = \sqrt[3]{27a^3 \cdot a^2 b^2} =$
$3a\sqrt[3]{a^2 b^2}$

36. $\sqrt[3]{a}\sqrt[5]{b^3} = a^{1/3}b^{3/5} = a^{5/15}b^{9/15} = (a^5 b^9)^{1/15} =$
$\sqrt[15]{a^5 b^9}$

37. $\dfrac{\sqrt[3]{60xy^3}}{\sqrt[3]{10x}} = \sqrt[3]{\dfrac{60xy^3}{10x}} = \sqrt[3]{6y^3} = y\sqrt[3]{6}$

38. $\dfrac{\sqrt{75x}}{2\sqrt{3}} = \dfrac{1}{2}\sqrt{\dfrac{75x}{3}} = \dfrac{1}{2}\sqrt{25x} = \dfrac{1}{2} \cdot 5\sqrt{x} = \dfrac{5}{2}\sqrt{x}$

39. $\dfrac{\sqrt[3]{x^2}}{\sqrt[4]{x}} = \dfrac{x^{2/3}}{x^{1/4}} = x^{2/3-1/4} = x^{8/12-3/12} = x^{5/12} =$
$\sqrt[12]{x^5}$

40. $5\sqrt[3]{x} + 2\sqrt[3]{x} = (5+2)\sqrt[3]{x} = 7\sqrt[3]{x}$

41. $2\sqrt{75} - 7\sqrt{3} = 2\sqrt{25 \cdot 3} - 7\sqrt{3} = 2\sqrt{25} \cdot \sqrt{3} - 7\sqrt{3} =$
$2 \cdot 5\sqrt{3} - 7\sqrt{3} = 10\sqrt{3} - 7\sqrt{3} = 3\sqrt{3}$

42. $\sqrt{50} + 2\sqrt{18} + \sqrt{32} = \sqrt{25 \cdot 2} + 2\sqrt{9 \cdot 2} + \sqrt{16 \cdot 2} =$
$\sqrt{25}\sqrt{2} + 2\sqrt{9}\sqrt{2} + \sqrt{16}\sqrt{2} = 5\sqrt{2} + 2 \cdot 3\sqrt{2} + 4\sqrt{2} =$
$5\sqrt{2} + 6\sqrt{2} + 4\sqrt{2} = 15\sqrt{2}$

43. $\sqrt[3]{8x^4} + \sqrt[3]{xy^6} = \sqrt[3]{8x^3 \cdot x} + \sqrt[3]{y^6 \cdot x} =$
$\sqrt[3]{8x^3}\sqrt[3]{x} + \sqrt[3]{y^6}\sqrt[3]{x} = 2x\sqrt[3]{x} + y^2\sqrt[3]{x} =$
$(2x + y^2)\sqrt[3]{x}$

44. $(\sqrt{5} - 3\sqrt{8})(\sqrt{5} + 2\sqrt{8})$
$= (\sqrt{5})^2 + \sqrt{5} \cdot 2\sqrt{8} - 3\sqrt{8} \cdot \sqrt{5} - 3\sqrt{8} \cdot 2\sqrt{8}$
$= 5 + 2\sqrt{40} - 3\sqrt{40} - 6 \cdot 8$
$= 5 - \sqrt{40} - 48$
$= -43 - \sqrt{4 \cdot 10}$
$= -43 - 2\sqrt{10}$

45. $(1 - \sqrt{7})^2 = 1^2 - 2 \cdot 1 \cdot \sqrt{7} + (\sqrt{7})^2$
$= 1 - 2\sqrt{7} + 7$
$= 8 - 2\sqrt{7}$

46. $(\sqrt[3]{27} - \sqrt[3]{2})(\sqrt[3]{27} + \sqrt[3]{2})$
$= (3 - \sqrt[3]{2})(3 + \sqrt[3]{2})$
$= 9 - (\sqrt[3]{2})^2$
$= 9 - \sqrt[3]{2^2}$
$= 9 - \sqrt[3]{4}$

47. $\sqrt{\dfrac{8}{3}} = \sqrt{\dfrac{8}{3} \cdot \dfrac{3}{3}} = \sqrt{\dfrac{24}{9}} = \dfrac{\sqrt{24}}{\sqrt{9}} = \dfrac{\sqrt{4 \cdot 6}}{3} = \dfrac{2\sqrt{6}}{3}$

48. $\dfrac{2}{\sqrt{a} + \sqrt{b}} = \dfrac{2}{\sqrt{a} + \sqrt{b}} \cdot \dfrac{\sqrt{a} - \sqrt{b}}{\sqrt{a} - \sqrt{b}} = \dfrac{2(\sqrt{a} - \sqrt{b})}{(\sqrt{a})^2 - (\sqrt{b})^2} =$
$\dfrac{2\sqrt{a} - 2\sqrt{b}}{a - b}$

49.
$$x - 3 = \sqrt{5 - x}$$
$$(x - 3)^2 = (\sqrt{5 - x})^2$$
$$x^2 - 6x + 9 = 5 - x$$
$$x^2 - 5x + 4 = 0$$
$$(x - 1)(x - 4) = 0$$
$$x - 1 = 0 \ \ or \ \ x - 4 = 0$$
$$x = 1 \ \ or \ \ \ \ x = 4$$

Since 4 checks but 1 does not, the solution is 4.

50.
$$\sqrt[4]{x + 3} = 2$$
$$(\sqrt[4]{x + 3})^4 = 2^4$$
$$x + 3 = 16$$
$$x = 13$$

The number 13 checks, so it is the solution.

51. $\sqrt{x + 8} - \sqrt{3x + 1} = 1$
$$\sqrt{x + 8} = \sqrt{3x + 1} + 1$$
$$(\sqrt{x + 8})^2 = (\sqrt{3x + 1} + 1)^2$$
$$x + 8 = 3x + 1 + 2\sqrt{3x + 1} + 1$$
$$-2x + 6 = 2\sqrt{3x + 1}$$
$$-x + 3 = \sqrt{3x + 1} \qquad \text{Dividing by 2}$$
$$(-x + 3)^2 = (\sqrt{3x + 1})^2$$
$$x^2 - 6x + 9 = 3x + 1$$
$$x^2 - 9x + 8 = 0$$
$$(x - 1)(x - 8) = 0$$
$$x - 1 = 0 \ \ or \ \ x - 8 = 0$$
$$x = 1 \ \ or \ \ \ \ x = 8$$

The number 1 checks, but 8 does not. The solution is 1.

52.
$$d(n) = 0.75\sqrt{2.8n}$$
$$81 = 0.75\sqrt{2.8n}$$
$$108 = \sqrt{2.8n} \qquad \text{Dividing by 0.75}$$
$$(108)^2 = (\sqrt{2.8n})^2$$
$$11,664 = 2.8n$$
$$4166 \approx n$$

The engine produces peak power at about 4166 rpm.

53.
$$d(n) = 0.75\sqrt{2.8n}$$
$$84 = 0.75\sqrt{2.8n}$$
$$112 = \sqrt{2.8n} \qquad \text{Dividing by 0.75}$$
$$(112)^2 = (\sqrt{2.8n})^2$$
$$12,544 = 2.8n$$
$$4480 = n$$

The engine produces peak power at 4480 rpm.

54. Let $s =$ the length of a side of the square, in cm. We use the Pythagorean theorem.
$$s^2 + s^2 = (9\sqrt{2})^2$$
$$2s^2 = 81 \cdot 2$$
$$s^2 = 81$$
$$s = 9$$

The length of a side of the square is 9 cm.

55. Let $w =$ the width of the bookcase, in ft. Then we can refer to the drawing in the text, replacing "?" with w. We use the Pythagorean theorem.
$$5^2 + w^2 = 7^2$$
$$25 + w^2 = 49$$
$$w^2 = 24$$
$$w = \sqrt{24}$$
$$w \approx 4.899$$

The width of the bookcase is $\sqrt{24}$ ft, or approximately 4.899 ft.

56. $a = 7, b = 24$

Find c.
$$a^2 + b^2 = c^2$$
$$7^2 + 24^2 = c^2$$
$$49 + 576 = c^2$$
$$625 = c^2$$
$$25 = c$$

57. $a = 2, c = 5\sqrt{2}$

Find b.
$$a^2 + b^2 = c^2$$
$$2^2 + b^2 = (5\sqrt{2})^2$$
$$4 + b^2 = 25 \cdot 2$$
$$4 + b^2 = 50$$
$$b^2 = 46$$
$$b = \sqrt{46} \qquad \text{Exact answer}$$
$$b \approx 6.782 \qquad \text{Approximation}$$

58. $\sqrt{-25} + \sqrt{-8} = \sqrt{-1 \cdot 25} + \sqrt{-1 \cdot 4 \cdot 2} =$
$$\sqrt{-1} \cdot \sqrt{25} + \sqrt{-1} \cdot \sqrt{4} \cdot \sqrt{2} = i \cdot 5 + i \cdot 2 \cdot \sqrt{2} =$$
$$5i + 2\sqrt{2}i = (5 + 2\sqrt{2})i$$

59. $(-4 + 3i) + (2 - 12i) = (-4 + 2) + (3 - 12)i =$
$$-2 - 9i$$

60. $(4 - 7i) - (3 - 8i) = (4 - 3) + [-7 - (-8)]i = 1 + i$

61. $(2 + 5i)(2 - 5i) = 2^2 - (5i)^2$
$$= 4 - 25i^2$$
$$= 4 + 25 \qquad i^2 = -1$$
$$= 29$$

62. $i^{13} = i^{12} \cdot i = (i^2)^6 \cdot i = (-1)^6 \cdot i = 1 \cdot i = i$

63.　　$(6 - 3i)(2 - i)$

$= 12 - 6i - 6i + 3i^2$　　Using FOIL

$= 12 - 12i - 3$　　　　$i^2 = -1$

$= 9 - 12i$

64.　$\dfrac{-3 + 2i}{5i} = \dfrac{-3 + 2i}{5i} \cdot \dfrac{-5i}{-5i}$

$= \dfrac{15i - 10i^2}{-25i^2}$

$= \dfrac{15i + 10}{25}$

$= \dfrac{10}{25} + \dfrac{15}{25}i$

$= \dfrac{2}{5} + \dfrac{3}{5}i$

65.　$\dfrac{1 - 2i}{3 + i} = \dfrac{1 - 2i}{3 + i} \cdot \dfrac{3 - i}{3 - i}$

$= \dfrac{3 - i - 6i + 2i^2}{9 - i^2}$

$= \dfrac{3 - 7i + 2(-1)}{9 - (-1)}$

$= \dfrac{3 - 7i - 2}{9 + 1}$

$= \dfrac{1 - 7i}{10}$

$= \dfrac{1}{10} - \dfrac{7}{10}i$

66. Graph: $f(x) = \sqrt{x}$.

We find some ordered pairs, plot points, and draw the curve.

x	$f(x)$	$(x, f(x))$
0	0	$(0, 0)$
1	1	$(1, 1)$
2	1.4	$(2, 1.4)$
3	1.7	$(3, 1.7)$
4	2	$(4, 2)$
5	2.2	$(5, 2.2)$

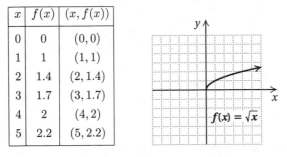

67. We substitute each complex number in the equation to determine if it is a solution.

For $1 - i$:

$$x^2 + 4x + 5 = 0$$

$\overline{(1 - i)^2 + 4(1 - i) + 5 \ ? \ 0}$

$1 - 2i + i^2 + 4 - 4i + 5 \ \Big|$

$1 - 2i - 1 + 4 - 4i + 5 \ \Big|$

$\qquad\qquad\qquad 9 - 6i \ \Big| \ $ FALSE

Choice A is not correct.

For $1 + i$:

$$x^2 + 4x + 5 = 0$$

$\overline{(1 + i)^2 + 4(1 + i) + 5 \ ? \ 0}$

$1 + 2i + i^2 + 4 + 4i + 5 \ \Big|$

$1 + 2i - 1 + 4 + 4i + 5 \ \Big|$

$\qquad\qquad\qquad 9 + 6i \ \Big| \ $ FALSE

Choice B is not correct.

For $2 + i$:

$$x^2 + 4x + 5 = 0$$

$\overline{(2 + i)^2 + 4(2 + i) + 5 \ ? \ 0}$

$4 + 4i + i^2 + 8 + 4i + 5 \ \Big|$

$4 + 4i - 1 + 8 + 4i + 5 \ \Big|$

$\qquad\qquad\qquad 16 + 8i \ \Big| \ $ FALSE

Choice C is not correct.

For $-2 + i$:

$$x^2 + 4x + 5 = 0$$

$\overline{(-2 + i)^2 + 4(-2 + i) + 5 \ ? \ 0}$

$4 - 4i + i^2 - 8 + 4i + 5 \ \Big|$

$4 - 4i - 1 - 8 + 4i + 5 \ \Big|$

$\qquad\qquad\qquad 0 \ \Big| \ $ TRUE

Choice D is correct.

68. $i \cdot i^2 \cdot i^3 \ldots i^{99} \cdot i^{100}$

Pairing i and i^{100}, i^2 and i^{99}, i^3 and i^{98}, and so on, we have 50 pairs of factors equivalent to i^{101}. Thus we have $(i^{101})^{50} = i^{5050} = (i^2)^{2525} = (-1)^{2525} = -1$.

69.　　$\sqrt{11x + \sqrt{6 + x}} = 6$

$\left(\sqrt{11x + \sqrt{6 + x}}\right)^2 = 6^2$

$11x + \sqrt{6 + x} = 36$

$\sqrt{6 + x} = 36 - 11x$

$\left(\sqrt{6 + x}\right)^2 = (36 - 11x)^2$

$6 + x = 1296 - 792x + 121x^2$

$0 = 121x^2 - 793x + 1290$

$0 = (121x - 430)(x - 3)$

$121x - 430 = 0 \quad or \quad x - 3 = 0$

$121x = 430 \quad or \qquad x = 3$

$x = \dfrac{430}{121} \quad or \qquad x = 3$

Since 3 checks but $\dfrac{430}{121}$ does not, the solution is 3.

Chapter 6 Discussion and Writing Exercises

1. $f(x) = (x + 5)^{1/2}(x + 7)^{-1/2}$

Consider $(x + 5)^{1/2}$. Since the exponent is $\dfrac{1}{2}$, $x + 5$

must be nonnegative. Then $x + 5 \geq 0$, or $x \geq -5$.

Consider $(x + 7)^{-1/2}$. Since the exponent is $-\frac{1}{2}$, $x + 7$ must be positive. Then $x + 7 > 0$, or $x > -7$.

Then the domain of $f = \{x | x \geq -5 \text{ and } x > -7\}$, or $\{x | x \geq -5\}$.

2. Since $\sqrt{x}$ exists only for $\{x | x \geq 0\}$, this is the domain of $y = \sqrt{x} \cdot \sqrt{x}$.

3. The distributive law is used to collect radical expressions with the same indices and radicands just as it is used to collect monomials with the same variables and exponents.

4. No; when n is odd, it is true that if $a^n = b^n$, then $a = b$.

5. Use a calculator to show that $\dfrac{5 + \sqrt{2}}{\sqrt{18}} \neq 2$. Explain that we multiply by 1 to rationalize a denominator. In this case we would write 1 as $\sqrt{2}/\sqrt{2}$.

6. When two radical expressions are conjugates, their product contains no radicals. Similarly, the product of a complex number and its conjugate does not contain i.

Chapter 6 Test

1. $\sqrt{148} \approx 12.166$

2. $f(x) = \sqrt{8 - 4x}$

$f(1) = \sqrt{8 - 4 \cdot 1} = \sqrt{8 - 4} = \sqrt{4} = 2$

$f(3) = \sqrt{8 - 4 \cdot 3} = \sqrt{8 - 12} = \sqrt{-4}$

Since negative numbers do not have real-number square roots, $f(3)$ does not exist as a real number.

3. The domain of $f(x) = \sqrt{8 - 4x}$ is the set of all x-values for which $8 - 4x \geq 0$.

$8 - 4x \geq 0$

$8 \geq 4x$

$2 \geq x$

The domain is $\{x | x \leq 2\}$, or $(-\infty, 2]$.

4. $\sqrt{(-3q)^2} = |-3q| = |-3| \cdot |q| = 3|q|$

5. $\sqrt{x^2 + 10x + 25} = \sqrt{(x + 5)^2} = |x + 5|$

6. $\sqrt[3]{-\dfrac{1}{1000}} = -\dfrac{1}{10}$ Since $\left(-\dfrac{1}{10}\right)^3 = -\dfrac{1}{1000}$

7. $\sqrt[5]{x^5} = x$

We do not use absolute-value notation when the index is odd.

8. $\sqrt[10]{(-4)^{10}} = |-4| = 4$

9. $a^{2/3} = \sqrt[3]{a^2}$

10. $32^{3/5} = \sqrt[5]{32^3} = \sqrt[5]{(2^5)^3} = \sqrt[5]{2^{15}} = 2^3 = 8$

11. $\sqrt{37} = 37^{1/2}$

12. $(\sqrt{5xy^2})^5 = [(5xy^2)^{1/2}]^5 = (5xy^2)^{5/2}$

13. $1000^{-1/3} = \dfrac{1}{1000^{1/3}} = \dfrac{1}{\sqrt[3]{1000}} = \dfrac{1}{10}$

14. $8a^{3/4}b^{-3/2}c^{-2/5} = 8 \cdot a^{3/4} \cdot \dfrac{1}{b^{3/2}} \cdot \dfrac{1}{c^{2/5}} = \dfrac{8a^{3/4}}{b^{3/2}c^{2/5}}$

15. $(x^{2/3}y^{-3/4})^{12/5} = (x^{2/3})^{12/5}(y^{-3/4})^{12/5} =$

$x^{24/15}y^{-36/20} = x^{8/5}y^{-9/5} = \dfrac{x^{8/5}}{y^{9/5}}$

16. $\dfrac{2.9^{-5/8}}{2.9^{2/3}} = 2.9^{-5/8 - 2/3} = 2.9^{-15/24 - 16/24} =$

$2.9^{-31/24} = \dfrac{1}{2.9^{31/24}}$

17. $\sqrt[8]{x^2} = x^{2/8} = x^{1/4} = \sqrt[4]{x}$

18. $\sqrt[4]{16x^6} = (16x^6)^{1/4} = (2^4x^6)^{1/4} = (2^4)^{1/4}(x^6)^{1/4} =$

$2x^{3/2} = 2\sqrt{x^3} = 2x\sqrt{x}$

19. $a^{2/5}b^{1/3} = a^{6/15}b^{5/15} = (a^6b^5)^{1/15} = \sqrt[15]{a^6b^5}$

20. $\sqrt[4]{2y}\,\sqrt[3]{y} = (2y)^{1/4}(y^{1/3}) = (2y)^{3/12}(y^{4/12}) =$

$[(2y)^3(y^4)]^{1/12} = (8y^3 \cdot y^4)^{1/12} = (8y^7)^{1/12} =$

$\sqrt[12]{8y^7}$

21. $\sqrt{148} = \sqrt{4 \cdot 37} = \sqrt{4}\sqrt{37} = 2\sqrt{37}$

22. $\sqrt[4]{80} = \sqrt[4]{16 \cdot 5} = \sqrt[4]{16}\sqrt[4]{5} = 2\sqrt[4]{5}$

23. $\sqrt[3]{24a^{11}b^{13}} = \sqrt[3]{8a^9b^{12} \cdot 3a^2b} =$

$\sqrt[3]{8a^9b^{12}}\sqrt[3]{3a^2b} = 2a^3b^4\sqrt[3]{3a^2b}$

24. $\sqrt[3]{\dfrac{16x^5}{y^6}} = \dfrac{\sqrt[3]{16x^5}}{\sqrt[3]{y^6}} = \dfrac{\sqrt[3]{8x^3 \cdot 2x^2}}{\sqrt[3]{y^6}} =$

$\dfrac{\sqrt[3]{8x^3}\sqrt[3]{2x^2}}{\sqrt[3]{y^6}} = \dfrac{2x\sqrt[3]{2x^2}}{y^2}$

25. $\sqrt{\dfrac{25x^2}{36y^4}} = \dfrac{\sqrt{25x^2}}{\sqrt{36y^4}} = \dfrac{5x}{6y^2}$

26. $\sqrt[3]{2x}\,\sqrt[3]{5y^2} = \sqrt[3]{2x \cdot 5y^2} = \sqrt[3]{10xy^2}$

27. $\sqrt[4]{x^3y^2}\,\sqrt{xy} = (x^3y^2)^{1/4}(xy)^{1/2} = x^{3/4}y^{1/2}x^{1/2}y^{1/2} =$

$x^{3/4 + 1/2}y^{1/2 + 1/2} = x^{5/4}y = y\sqrt[4]{x^5} = xy\sqrt[4]{x}$

28. $\dfrac{\sqrt[5]{x^3y^4}}{\sqrt[5]{xy^2}} = \sqrt[5]{\dfrac{x^3y^4}{xy^2}} = \sqrt[5]{x^2y^2}$

29. $\dfrac{\sqrt{300a}}{5\sqrt{3}} = \dfrac{1}{5}\sqrt{\dfrac{300a}{3}} = \dfrac{1}{5}\sqrt{100a} = \dfrac{1}{5} \cdot 10\sqrt{a} = 2\sqrt{a}$

30. $3\sqrt{128} + 2\sqrt{18} + 2\sqrt{32}$

$= 3\sqrt{64 \cdot 2} + 2\sqrt{9 \cdot 2} + 2\sqrt{16 \cdot 2}$

$= 3\sqrt{64}\sqrt{2} + 2\sqrt{9}\sqrt{2} + 2\sqrt{16}\sqrt{2}$

$= 3 \cdot 8\sqrt{2} + 2 \cdot 3\sqrt{2} + 2 \cdot 4\sqrt{2}$

$= 24\sqrt{2} + 6\sqrt{2} + 8\sqrt{2}$

$= 38\sqrt{2}$

31. $(\sqrt{20} + 2\sqrt{5})(\sqrt{20} - 3\sqrt{5})$

$= \sqrt{20}\sqrt{20} - \sqrt{20} \cdot 3\sqrt{5} + 2\sqrt{5}\sqrt{20} - 2\sqrt{5} \cdot 3\sqrt{5}$

$\qquad\qquad\qquad\qquad$ Using FOIL

$= 20 - 3\sqrt{100} + 2\sqrt{100} - 6 \cdot 5$

$= 20 - 3 \cdot 10 + 2 \cdot 10 - 30$

$= 20 - 30 + 20 - 30$

$= -20$

32. $(3 + \sqrt{x})^2 = 3^2 + 2 \cdot 3 \cdot \sqrt{x} + (\sqrt{x})^2 = 9 + 6\sqrt{x} + x$

33. $\dfrac{1 + \sqrt{2}}{3 - 5\sqrt{2}} = \dfrac{1 + \sqrt{2}}{3 - 5\sqrt{2}} \cdot \dfrac{3 + 5\sqrt{2}}{3 + 5\sqrt{2}}$

$\qquad = \dfrac{3 + 5\sqrt{2} + 3\sqrt{2} + 5 \cdot 2}{9 - 25 \cdot 2}$

$\qquad = \dfrac{3 + 8\sqrt{2} + 10}{9 - 50}$

$\qquad = \dfrac{13 + 8\sqrt{2}}{-41}$

34. $\sqrt[5]{x - 3} = 2$

$(\sqrt[5]{x - 3})^5 = 2^5$

$x - 3 = 32$

$x = 35$

The number 35 checks, so it is the solution.

35. $\sqrt{x - 6} = \sqrt{x + 9} - 3$

$(\sqrt{x - 6})^2 = (\sqrt{x + 9} - 3)^2$

$x - 6 = x + 9 - 6\sqrt{x + 9} + 9$

$-24 = -6\sqrt{x + 9}$

$4 = \sqrt{x + 9} \qquad$ Dividing by -6

$4^2 = (\sqrt{x + 9})^2$

$16 = x + 9$

$7 = x$

The number 7 checks, so it is the solution.

36. $\sqrt{x - 1} + 3 = x$

$\sqrt{x - 1} = x - 3$

$(\sqrt{x - 1})^2 = (x - 3)^2$

$x - 1 = x^2 - 6x + 9$

$0 = x^2 - 7x + 10$

$0 = (x - 2)(x - 5)$

$x - 2 = 0 \quad or \quad x - 5 = 0$

$x = 2 \quad or \qquad x = 5$

The number 2 does not check, but 5 does. The solution is 5.

37. We make a drawing, letting s = the length of a side of the square, in feet.

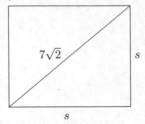

We use the Pythagorean theorem.

$s^2 + s^2 = (7\sqrt{2})^2$

$2s^2 = 49 \cdot 2$

$2s^2 = 98$

$s^2 = 49$

$s = \sqrt{49} = 7$

The length of a side of the square is 7 ft.

38. $D = 1.2\sqrt{h}$

$72 = 1.2\sqrt{h}$

$60 = \sqrt{h}$

$60^2 = (\sqrt{h})^2$

$3600 = h$

The airplane is 3600 ft high.

39. $a = 7$, $b = 7$; find c.

$a^2 + b^2 = c^2 \quad$ Pythagorean equation

$7^2 + 7^2 = c^2 \quad$ Substituting

$49 + 49 = c^2$

$98 = c^2$

$\sqrt{98} = c \quad$ Exact answer

$9.899 \approx c \quad$ Approximation

40. $a = 1$, $c = \sqrt{5}$; find b.

$a^2 + b^2 = c^2 \qquad$ Pythagorean equation

$1^2 + b^2 = (\sqrt{5})^2 \quad$ Substituting

$1 + b^2 = 5$

$b^2 = 4$

$b = 2$

41. $\sqrt{-9} + \sqrt{-64} = \sqrt{-1 \cdot 9} + \sqrt{-1 \cdot 64} = i \cdot 3 + i \cdot 8 =$

$3i + 8i = 11i$

42. $(5 + 8i) - (-2 + 3i) = [5 - (-2)] + (8i - 3i) = 7 + 5i$

43. $(3 - 4i)(3 + 7i) = 9 + 21i - 12i - 28i^2 \quad$ FOIL

$= 9 + 21i - 12i + 28 \qquad i^2 = -1$

$= 37 + 9i$

44. $i^{95} = i^{94} \cdot i = (i^2)^{47} \cdot i = (-1)^{47} \cdot i = -i$

45. $\dfrac{-7+14i}{6-8i} = \dfrac{-7+14i}{6-8i} \cdot \dfrac{6+8i}{6+8i}$

$\qquad = \dfrac{-42-56i+84i+112i^2}{36-64i^2}$

$\qquad = \dfrac{-42-56i+84i-112}{36+64}$

$\qquad = \dfrac{-154+28i}{100}$

$\qquad = -\dfrac{154}{100} + \dfrac{28}{100}i$

$\qquad = -\dfrac{77}{50} + \dfrac{7}{25}i$

46. Substitute $1+2i$ for x in the equation.

$$x^2 + 2x + 5 = 0$$

$$\begin{array}{c|c} (1+2i)^2 + 2(1+2i) + 5 \ ? \ 0 & \\ 1 + 4i + 4i^2 + 2 + 4i + 5 & \\ 1 + 4i - 4 + 2 + 4i + 5 & \\ 4 + 8i & \text{FALSE} \end{array}$$

$1+2i$ is not a solution.

47. $\qquad x - 4 = \sqrt{x-2}$

$\qquad (x-4)^2 = (\sqrt{x-2})^2$

$\qquad x^2 - 8x + 16 = x - 2$

$\qquad x^2 - 9x + 18 = 0$

$\qquad (x-3)(x-6) = 0$

$\qquad x - 3 = 0 \ \ or \ \ x - 6 = 0$

$\qquad x = 3 \ \ or \qquad x = 6$

The number 3 does not check, but 6 does. The solution is 6.

We see that there is exactly one solution, and it is positive. Thus, answer A is correct.

48. $\dfrac{1-4i}{4i(1+4i)^{-1}} = \dfrac{(1-4i)(1+4i)}{4i}$

$\qquad = \dfrac{1 - 16i^2}{4i}$

$\qquad = \dfrac{1 + 16}{4i}$

$\qquad = \dfrac{17}{4i}$

$\qquad = \dfrac{17}{4i} \cdot \dfrac{-4i}{-4i}$

$\qquad = \dfrac{-68i}{-16i^2}$

$\qquad = \dfrac{-68i}{16}$

$\qquad = -\dfrac{17i}{4}, \ or \ -\dfrac{17}{4}i$

49.

$$\sqrt{2x-2} + \sqrt{7x+4} = \sqrt{13x+10}$$

$$(\sqrt{2x-2} + \sqrt{7x+4})^2 = (\sqrt{13x+10})^2$$

$$2x-2+2\sqrt{(2x-2)(7x+4)}+7x+4 = 13x+10$$

$$9x + 2 + 2\sqrt{14x^2 - 6x - 8} = 13x + 10$$

$$2\sqrt{14x^2 - 6x - 8} = 4x + 8$$

$$\sqrt{14x^2 - 6x - 8} = 2x+4 \ \ \text{Dividing by 2}$$

$$(\sqrt{14x^2 - 6x - 8})^2 = (2x+4)^2$$

$$14x^2 - 6x - 8 = 4x^2 + 16x + 16$$

$$10x^2 - 22x - 24 = 0$$

$$5x^2 - 11x - 12 = 0 \ \ \ \text{Dividing by 2}$$

$$(5x+4)(x-3) = 0$$

$$5x + 4 = 0 \qquad or \ \ x - 3 = 0$$

$$5x = -4 \ \ or \qquad x = 3$$

$$x = -\dfrac{4}{5} \ \ or \qquad x = 3$$

The number $-\dfrac{4}{5}$ does not check, but 3 does. The solution is 3.

Cumulative Review Chapters 1 - 6

1. $(2x^2 - 3x + 1) + (6x - 3x^3 + 7x^2 - 4) =$

$-3x^3 + 9x^2 + 3x - 3$

2. $(2x^2 - y)^2 = (2x^2)^2 - 2 \cdot 2x^2 \cdot y + y^2$

$\qquad\qquad = 4x^4 - 4x^2 y + y^2$

3.

$$\begin{array}{r} 5x^2 \ \ - 2x + 1 \\ 3x^2 \ + \ x \ - 2 \\ \hline -10x^2 + 4x - 2 \\ 5x^3 - 2x^2 \ + x \\ 15x^4 - 6x^3 + 3x^2 \\ \hline 15x^4 - x^3 - 9x^2 + 5x - 2 \end{array}$$

4. $\dfrac{x^3 + 64}{x^2 - 49} \cdot \dfrac{x^2 - 14x + 49}{x^2 - 4x + 16}$

$\qquad = \dfrac{(x^3 + 64)(x^2 - 14x + 49)}{(x^2 - 49)(x^2 - 4x + 16)}$

$\qquad = \dfrac{(x+4)(x^2 - 4x + 16)(x-7)(x-7)}{(x+7)(x-7)(x^2 - 4x + 16)}$

$\qquad = \dfrac{(x^2 - 4x + 16)(x-7)}{(x^2 - 4x + 16)(x-7)} \cdot \dfrac{(x+4)(x-7)}{x+7}$

$\qquad = \dfrac{(x+4)(x-7)}{x+7}$

5. $\dfrac{\dfrac{y^2 - 5y - 6}{y^2 - 7y - 18}}{\dfrac{y^2 + 3y + 2}{y^2 + 4y + 4}}$

$= \dfrac{y^2 - 5y - 6}{y^2 - 7y - 18} \cdot \dfrac{y^2 + 4y + 4}{y^2 + 3y + 2}$

$= \dfrac{(y^2 - 5y - 6)(y^2 + 4y + 4)}{(y^2 - 7y - 18)(y^2 + 3y + 2)}$

$= \dfrac{(y - 6)(y + 1)(y + 2)(y + 2)}{(y - 9)(y + 2)(y + 1)(y + 2)}$

$= \dfrac{(y - 6)\cancel{(y+1)}\cancel{(y+2)}\cancel{(y+2)}}{(y - 9)\cancel{(y+2)}\cancel{(y+1)}\cancel{(y+2)}}$

$= \dfrac{y - 6}{y - 9}$

6. $\dfrac{x}{x + 2} + \dfrac{1}{x - 3} - \dfrac{x^2 - 2}{x^2 - x - 6}$

$= \dfrac{x}{x + 2} + \dfrac{1}{x - 3} - \dfrac{x^2 - 2}{(x + 2)(x - 3)}$,

$\qquad\qquad\qquad$ LCM is $(x + 2)(x - 3)$

$= \dfrac{x}{x + 2} \cdot \dfrac{x - 3}{x - 3} + \dfrac{1}{x - 3} \cdot \dfrac{x + 2}{x + 2} - \dfrac{x^2 - 2}{(x + 2)(x - 3)}$

$= \dfrac{x^2 - 3x + x + 2 - (x^2 - 2)}{(x + 2)(x - 3)}$

$= \dfrac{x^2 - 3x + x + 2 - x^2 + 2}{(x + 2)(x - 3)}$

$= \dfrac{-2x + 4}{(x + 2)(x - 3)}$, or $\dfrac{-2(x - 2)}{(x + 2)(x - 3)}$

7.
$$\begin{array}{r}
y^2 + y - 2 \\
y + 2 \overline{\smash{\big)}\, y^3 + 3y^2 + 0y - 5} \\
\underline{y^3 + 2y^2} \\
y^2 \\
\underline{y^2 + 2y} \\
-2y - 5 \\
\underline{-2y - 4} \\
-1
\end{array}$$

The answer is $y^2 + y - 2 + \dfrac{-1}{y + 2}$.

8. $\sqrt[3]{-8x^3} = \sqrt[3]{(-2x)^3} = -2x$

9. $\sqrt{16x^2 - 32x + 16} = \sqrt{16(x^2 - 2x + 1)} =$
$\sqrt{16(x - 1)^2} = 4(x - 1)$

10. $9\sqrt{75} + 6\sqrt{12} = 9\sqrt{25 \cdot 3} + 6\sqrt{4 \cdot 3} =$
$9 \cdot 5\sqrt{3} + 6 \cdot 2\sqrt{3} = 45\sqrt{3} + 12\sqrt{3} = 57\sqrt{3}$

11. $\sqrt{2xy^2}\sqrt{8xy^3} = \sqrt{16x^2y^5} = \sqrt{16x^2y^4 \cdot y} =$
$4xy^2\sqrt{y}$

12. $\dfrac{3\sqrt{5}}{\sqrt{6} - \sqrt{3}} = \dfrac{3\sqrt{5}}{\sqrt{6} - \sqrt{3}} \cdot \dfrac{\sqrt{6} + \sqrt{3}}{\sqrt{6} + \sqrt{3}} = \dfrac{3\sqrt{30} + 3\sqrt{15}}{6 - 3} =$
$\dfrac{3\sqrt{30} + 3\sqrt{15}}{3} = \dfrac{3(\sqrt{30} + \sqrt{15})}{3} = \sqrt{30} + \sqrt{15}$

13. $\sqrt[6]{\dfrac{m^{12}n^{24}}{64}} = \left(\dfrac{m^{12}n^{24}}{2^6}\right)^{1/6} = \dfrac{m^2n^4}{2}$

14. $6^{2/9} \cdot 6^{2/3} = 6^{2/9} \cdot 6^{6/9} = 6^{8/9}$

15. $(6 + i) - (3 - 4i) = (6 - 3) + [1 - (-4)]i = 3 + 5i$

16. $\dfrac{2 - i}{6 + 5i} = \dfrac{2 - i}{6 + 5i} \cdot \dfrac{6 - 5i}{6 - 5i}$

$= \dfrac{12 - 10i - 6i + 5i^2}{36 - 25i^2}$

$= \dfrac{12 - 16i - 5}{36 + 25}$

$= \dfrac{7 - 16i}{61}$

$= \dfrac{7}{61} - \dfrac{16}{61}i$

17. $\dfrac{1}{5} + \dfrac{3}{10}x = \dfrac{4}{5}$

$\dfrac{3}{10}x = \dfrac{3}{5}$

$x = \dfrac{10}{3} \cdot \dfrac{3}{5}$

$x = \dfrac{30}{15} = 2$

The solution is 2.

18. $M = \dfrac{1}{8}(c - 3)$

$8M = c - 3 \qquad$ Multiplying by 8

$8M + 3 = c$

19. $3a - 4 < 10 + 5a$

$-2a - 4 < 10$

$-2a < 14$

$a > -7 \quad$ Reversing the inequality symbol

The solution set is $\{a | a > -7\}$, or $(-7, \infty)$.

20. $-8 < x + 2 < 15$

$-10 < x < 13 \qquad$ Subtracting 2

The solution set is $\{x | -10 < x < 13\}$, or $(-10, 13)$.

21. $|3x - 6| = 2$

$3x - 6 = -2 \;\; or \;\; 3x - 6 = 2$

$3x = 4 \quad or \qquad 3x = 8$

$x = \dfrac{4}{3} \quad or \qquad x = \dfrac{8}{3}$

The solutions are $\dfrac{4}{3}$ and $\dfrac{8}{3}$.

22. $625 = 49y^2$

$0 = 49y^2 - 625$

$0 = (7y + 25)(7y - 25)$

$7y + 25 = 0 \qquad or \;\; 7y - 25 = 0$

$7y = -25 \;\; or \qquad 7y = 25$

$y = -\dfrac{25}{7} \;\; or \qquad y = \dfrac{25}{7}$

The solutions are $-\dfrac{25}{7}$ and $\dfrac{25}{7}$.

23. $3x + 5y = 30$, (1)

$5x + 3y = 34$ (2)

We first multiply Equation (1) by 3 and multiply Equation (2) by -5 and then add.

$$9x + 15y = 90$$
$$\underline{-25x - 15y = -170}$$
$$-16x \qquad = -80$$
$$x = 5$$

Now substitute 5 for x in one of the original equations and solve for y.

$$3x + 5y = 30 \quad (1)$$
$$3 \cdot 5 + 5y = 30$$
$$15 + 5y = 30$$
$$5y = 15$$
$$y = 3$$

The solution is $(5, 3)$.

24. $3x + 2y - z = -7$, (1)

$-x + y + 2z = 9$, (2)

$5x + 5y + z = -1$ (3)

First we eliminate z from one pair of equations.

$$6x + 4y - 2z = -14 \quad \text{Multiplying (1) by 2}$$
$$\underline{-x + y + 2z = 9} \quad (2)$$
$$5x + 5y \qquad = -5 \quad (4)$$

Next we add Equations (1) and (3).

$$3x + 2y - z = -7$$
$$\underline{5x + 5y + z = -1}$$
$$8x + 7y \qquad = -8 \quad (5)$$

Now we solve the system of Equations (4) and (5). We multiply Equation (4) by 7 and multiply Equation (5) by -5 and then add.

$$35x + 35y = -35$$
$$\underline{-40x - 35y = 40}$$
$$-5x \qquad = 5$$
$$x = -1$$

Next we use Equation (4) to find y.

$$5(-1) + 5y = -5 \quad \text{Substituting}$$
$$-5 + 5y = -5$$
$$5y = 0$$
$$y = 0$$

Finally we substitute -1 for x and 0 for y in Equation (3) and solve for z.

$$5(-1) + 5 \cdot 0 + z = -1$$
$$-5 + z = -1$$
$$z = 4$$

The solution is $(-1, 0, 4)$.

25.
$$\frac{6x}{x-5} - \frac{300}{x^2+5x+25} = \frac{2250}{x^3-125}$$
$$\frac{6x}{x-5} - \frac{300}{x^2+5x+25} = \frac{2250}{(x-5)(x^2+5x+25)}$$

LCM is $(x-5)(x^2+5x+25)$

$$(x-5)(x^2+5x+25)\left(\frac{6x}{x-5} - \frac{300}{x^2+5x+25}\right) =$$
$$(x-5)(x^2+5x+25) \cdot \frac{2250}{(x-5)(x^2+5x+25)}$$
$$6x(x^2+5x+25) - 300(x-5) = 2250$$
$$6x^3 + 30x^2 + 150x - 300x + 1500 = 2250$$
$$6x^3 + 30x^2 - 150x + 1500 = 2250$$
$$6x^3 + 30x^2 - 150x - 750 = 0$$
$$6(x^3 + 5x^2 - 25x - 125) = 0$$
$$6[x^2(x+5) - 25(x+5)] = 0$$
$$6(x^2 - 25)(x+5) = 0$$
$$6(x+5)(x-5)(x+5) = 0$$

$x + 5 = 0 \quad or \quad x - 5 = 0 \quad or \quad x + 5 = 0$

$x = -5 \quad or \qquad x = 5 \quad or \qquad x = -5$

The number -5 checks but 5 does not, so the solution is -5.

26.
$$\frac{3x^2}{x+2} + \frac{5x-22}{x-2} = \frac{-48}{x^2-4}$$
$$\frac{3x^2}{x+2} + \frac{5x-22}{x-2} = \frac{-48}{(x+2)(x-2)},$$
$$\text{LCM is } (x+2)(x-2)$$

$$(x+2)(x-2)\left(\frac{3x^2}{x+2} + \frac{5x-22}{x-2}\right) =$$
$$(x+2)(x-2) \cdot \frac{-48}{(x+2)(x-2)}$$
$$3x^2(x-2) + (x+2)(5x-22) = -48$$
$$3x^3 - 6x^2 + 5x^2 - 12x - 44 = -48$$
$$3x^3 - x^2 - 12x - 44 = -48$$
$$3x^3 - x^2 - 12x + 4 = 0$$
$$x^2(3x-1) - 4(3x-1) = 0$$
$$(x^2 - 4)(3x-1) = 0$$
$$(x+2)(x-2)(3x-1) = 0$$

$x + 2 = 0 \quad or \quad x - 2 = 0 \quad or \quad 3x - 1 = 0$

$x = -2 \quad or \qquad x = 2 \quad or \qquad 3x = 1$

$x = -2 \quad or \qquad x = 2 \quad or \qquad x = \dfrac{1}{3}$

The numbers -2 and 2 do not check but $\dfrac{1}{3}$ does, so the solution is $\dfrac{1}{3}$.

27.

$$I = \frac{nE}{R + nr}$$

$$(R + nr) \cdot I = nE \qquad \text{Multiplying by } R + nr$$

$$RI + nrI = nE$$

$$RI = nE - nrI$$

$$R = \frac{nE - nrI}{I}$$

28.

$$\sqrt{4x + 1} - 2 = 3$$

$$\sqrt{4x + 1} = 5$$

$$(\sqrt{4x + 1})^2 = 5^2$$

$$4x + 1 = 25$$

$$4x = 24$$

$$x = 6$$

The number 6 checks, so it is the solution.

29.

$$2\sqrt{1 - x} = \sqrt{5}$$

$$(2\sqrt{1 - x})^2 = (\sqrt{5})^2$$

$$4(1 - x) = 5$$

$$4 - 4x = 5$$

$$-4x = 1$$

$$x = -\frac{1}{4}$$

The number $-\dfrac{1}{4}$ checks, so it is the solution.

30.

$$13 - x = 5 + \sqrt{x + 4}$$

$$8 - x = \sqrt{x + 4}$$

$$(8 - x)^2 = (\sqrt{x + 4})^2$$

$$64 - 16x + x^2 = x + 4$$

$$x^2 - 17x + 60 = 0$$

$$(x - 5)(x - 12) = 0$$

$$x - 5 = 0 \quad or \quad x - 12 = 0$$

$$x = 5 \quad or \qquad x = 12$$

The number 5 checks but 12 does not, so the solution is 5.

31. Graph: $f(x) = -\dfrac{2}{3}x + 2$

Make a list of function values in a table.

$$f(-3) = -\frac{2}{3}(-3) + 2 = 2 + 2 = 4$$

$$f(0) = -\frac{2}{3} \cdot 0 + 2 = 2$$

$$f(3) = -\frac{2}{3} \cdot 3 + 2 = -2 + 2 = 0$$

x	$f(x)$
-3	4
0	2
3	0

Plot these points and connect them.

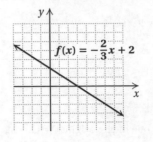

32. Graph: $4x - 2y = 8$

First we will find the intercepts. To find the x-intercept, let $y = 0$ and solve for x.

$$4x - 2 \cdot 0 = 8$$

$$4x = 8$$

$$x = 2$$

The x-intercept is $(2, 0)$.

To find the y-intercept, let $x = 0$ and solve for y.

$$4 \cdot 0 - 2y = 8$$

$$-2y = 8$$

$$y = -4$$

The y-intercept is $(0, -4)$.

We plot these points and draw the line.

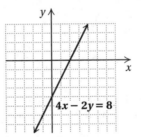

We use a third point as a check. Let $x = 4$.

$$4 \cdot 4 - 2y = 8$$

$$16 - 2y = 8$$

$$-2y = -8$$

$$y = 4$$

We plot $(4, 4)$ and note that it is on the line.

33. Graph: $4x \geq 5y + 20$

First we graph $4x = 5y + 20$. We draw the line solid since the inequality symbol is $\geq$. Test the point $(0, 0)$ to determine if it is a solution.

$$\frac{4x \geq 5y + 20}{4 \cdot 0 \ ? \ 5 \cdot 0 + 20}$$

$$0 \ \bigl| \ 20 \qquad\qquad \text{FALSE}$$

Since $0 \geq 20$ is false, we shade the half-plane that does not contain $(0, 0)$.

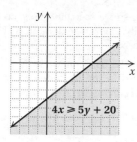

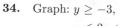

34. Graph: $y \geq -3$,

$\quad\quad y \leq 2x + 3$

Graph the lines $y = -3$ and $y = 2x+3$ using solid lines. Indicate the region for each inequality by arrows, and shade the region where they overlap.

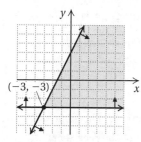

If we solve the system of related equations,

$\quad y = -3$,

$\quad y = 2x + 3$,

we find that the vertex is $(-3, -3)$.

35. Graph: $g(x) = x^2 - x - 2$

Make a list of function values in a table.

$g(-2) = (-2)^2 - (-2) - 2 = 4 + 2 - 2 = 4$

$g(-1) = (-1)^2 - (-1) - 2 = 1 + 1 - 2 = 0$

$y(0) = 0^2 - 0 - 2 = -2$

$g(1) = 1^2 - 1 - 2 = 1 - 1 - 2 = -2$

$g(3) = 3^2 - 3 - 2 = 9 - 3 - 2 = 4$

x	$g(x)$
-2	4
-1	0
0	-2
1	-2
3	4

Plot these points and connect them.

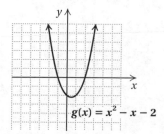

36. $f(x) = |x + 4|$

Make a list of function values in a table.

$f(-5) = |-5 + 4| = |-1| = 1$

$f(-4) = |-4 + 4| = |0| = 0$

$f(-2) = |-2 + 4| = |2| = 2$

$f(0) = |0 + 4| = |4| = 4$

$f(1) = |1 + 4| = |5| = 5$

x	$f(x)$
-5	1
-4	0
-2	2
0	4
1	5

Plot these points and connect them.

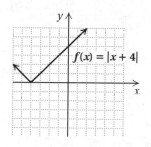

37. Graph: $g(x) = \dfrac{4}{x - 3}$

Make a list of function values in a table. Note that we cannot choose $x = 3$ since it makes the denominator 0. The graph will have two branches.

$g(-5) = \dfrac{4}{-5 - 3} = \dfrac{4}{-8} = -\dfrac{1}{2}$

$g(-1) = \dfrac{4}{-1 - 3} = \dfrac{4}{-4} = -1$

$g(1) = \dfrac{4}{1 - 3} = \dfrac{4}{-2} = -2$

$g(4) = \dfrac{4}{4 - 3} = \dfrac{4}{1} = 4$

$g(5) = \dfrac{4}{5 - 3} = \dfrac{4}{2} = 2$

x	$g(x)$
-5	$-\dfrac{1}{2}$
-1	-1
1	-2
4	4
5	2

Plot these points and connect them.

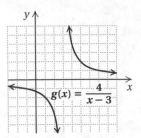

38. Graph: $f(x) = 2 - \sqrt{x}$

We find some ordered pairs, plot points, and draw the curve.

x	$f(x)$	$(x, f(x))$
0	2	$(0, 2)$
1	1	$(1, 1)$
3	0.3	$(3, 0.3)$
4	0	$(4, 0)$
5	−0.2	$(5, -0.2)$

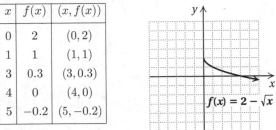

39. $12x^2y^2 - 30xy^3 = 6xy^2 \cdot 2x - 6xy^2 \cdot 5y$
$$= 6xy^2(2x - 5y)$$

40. $3x^2 - 17x - 28$

We will use the *ac*-method.

1) There is no common factor (other than 1 or −1.).

2) Multiply the coefficients of the first and last terms.
$$3(-28) = -84$$

3) Factor −84 so the sum of the factors is −17. The desired factorization is $4(-21)$.

4) Split the middle term as follows:
$$-17x = 4x - 21x$$

5) Factor by grouping:
$$3x^2 - 17x - 28 = 3x^2 + 4x - 21x - 28$$
$$= x(3x + 4) - 7(3x + 4)$$
$$= (x - 7)(3x + 4)$$

41. $y^2 - y - 132$

We look for a pair of factors of −132 whose sum is −1. The numbers we need are 11 and −12.
$$y^2 - y - 132 = (y + 11)(y - 12)$$

42. $27y^3 + 8 = (3y)^3 + 2^3 = (3y + 2)(9y^2 - 6y + 4)$

43. $4x^2 - 625 = (2x)^2 - 25^2 = (2x + 25)(2x - 25)$

44. The set of all x-values in the graph extends from −5 to 5, so the domain is $\{x | -5 \le x \le 5\}$, or $[-5, 5]$.

The set of all y-values in the graph extends from −3 to 4, so the range is $\{y | -3 \le y \le 4\}$, or $[-3, 4]$.

45. No endpoints are indicated, so we see that the graph extends indefinitely horizontally. Thus, the domain is the set of all real numbers, or $(-\infty, \infty)$.

The smallest y-value is −5. No endpoints are indicated, so we see that the graph extends upward indefinitely from $(0, -5)$. Thus, the range is $\{y | y \ge -5\}$, or $[-5, \infty)$.

46. First we find the slope-intercept form of the equation by solving for y.
$$3x - 2y = 8$$
$$-2y = -3x + 8$$
$$y = \frac{-3x + 8}{-2}$$
$$y = \frac{3}{2}x - 4$$

The slope is $\frac{3}{2}$, and the y-intercept is $(0, -4)$.

47. First we find the slope of the given line.
$$3x - y = 5$$
$$-y = -3x + 5$$
$$y = 3x - 5$$

The slope of the given line is 3. The slope of the perpendicular line is the opposite of the reciprocal of 3, or $-\frac{1}{3}$.

We will use the point-slope equation, substituting 1 for x_1, 4 for y_1, and $-\frac{1}{3}$ for m.
$$y - y_1 = m(x - x_1)$$
$$y - 4 = -\frac{1}{3}(x - 1)$$
$$y - 4 = -\frac{1}{3}x + \frac{1}{3}$$
$$y = -\frac{1}{3}x + \frac{13}{3}$$

48.
$$h = \frac{k}{b}$$
$$100 = \frac{k}{20}$$
$$2000 = k$$
$$h = \frac{2000}{b} \quad \text{Equation of variation}$$
$$h = \frac{2000}{16}$$
$$h = 125$$

When the base is 16 ft, the height is 125 ft.

We use the formula for the area of a triangle to find the fixed area.
$$A = \frac{1}{2}bh$$
$$= \frac{1}{2} \cdot 16 \cdot 125$$
$$= 1000 \text{ ft}^2$$

49. *Familiarize*. Let $t =$ the number of hours it will take the combines to harvest the field, working together.

***Translate*.** We use the work principle.

$$\frac{t}{3} + \frac{t}{1.5} = 1$$

***Solve*.** We solve the equation.

$$\frac{t}{3} + \frac{t}{1.5} = 1$$

$$3\left(\frac{t}{3} + \frac{t}{1.5}\right) = 3 \cdot 1 \quad \text{Clearing fractions}$$

$$t + 2t = 3$$

$$3t = 3$$

$$t = 1$$

***Check*.** We verify the work principle.

$$\frac{1}{3} + \frac{1}{1.5} = \frac{1}{3} + \frac{1}{3/2} = \frac{1}{3} + 1 \cdot \frac{2}{3} = \frac{1}{3} + \frac{2}{3} = 1$$

***State*.** It will take the combines 1 hr to harvest the field, working together.

50.
$$V = kd^2$$
$$4 = k \cdot 10^2$$
$$4 = k \cdot 100$$
$$0.04 = k$$
$$V = 0.04d^2 \qquad \text{Equation of variation}$$
$$V = 0.04(40)^2$$
$$V = 0.04(1600)$$
$$V = 64 \text{ L}$$

51. $\sqrt[5]{xy^4} = (xy^4)^{1/5}$

Answer D is correct.

52. Using the work principle, we translate to the equation

$$\frac{t}{3} + \frac{t}{15} = 1,$$

where t is the time required to fill the bin using both spouts together. Now we solve the equation.

$$15 \cdot \left(\frac{t}{3} + \frac{t}{15}\right) = 15 \cdot 1$$

$$5t + t = 15$$

$$6t = 15$$

$$t = \frac{5}{2}$$

This solution checks, so the answer A is correct.

53.

$$\begin{array}{r|rrrr}
3 & 1 & -1 & 2 & 4 \\
  &   & 3 & 6 & 24 \\
\hline
  & 1 & 2 & 8 & 28
\end{array}$$

The answer is $x^2 + 2x + 8$, R 28, so answer A is correct.

54.
$$2x + 6 = 8 + \sqrt{5x + 1}$$
$$2x - 2 = \sqrt{5x + 1}$$
$$(2x - 2)^2 = (\sqrt{5x + 1})^2$$
$$4x^2 - 8x + 4 = 5x + 1$$
$$4x^2 - 13x + 3 = 0$$
$$(4x - 1)(x - 3) = 0$$
$$4x - 1 = 0 \quad or \quad x - 3 = 0$$
$$4x = 1 \quad or \qquad x = 3$$
$$x = \frac{1}{4} \quad or \qquad x = 3$$

The number $\frac{1}{4}$ does not check but 3 does, so answer B is correct.

55.
$$\frac{x + \sqrt{x + 1}}{x - \sqrt{x + 1}} = \frac{5}{11}$$

$$11(x - \sqrt{x + 1}) \cdot \frac{x + \sqrt{x + 1}}{x - \sqrt{x + 1}} = 11(x - \sqrt{x + 1}) \cdot \frac{5}{11}$$

$$11(x + \sqrt{x + 1}) = 5(x - \sqrt{x + 1})$$

$$11x + 11\sqrt{x + 1} = 5x - 5\sqrt{x + 1}$$

$$16\sqrt{x + 1} = -6x$$

$$8\sqrt{x + 1} = -3x \quad \text{Dividing by 2}$$

$$(8\sqrt{x + 1})^2 = (-3x)^2$$

$$64(x + 1) = 9x^2$$

$$64x + 64 = 9x^2$$

$$0 = 9x^2 - 64x - 64$$

$$0 = (9x + 8)(x - 8)$$

$$9x + 8 = 0 \quad or \quad x - 8 = 0$$
$$9x = -8 \quad or \qquad x = 8$$
$$x = -\frac{8}{9} \quad or \qquad x = 8$$

The number $-\frac{8}{9}$ checks but 8 does not, so the solution is $-\frac{8}{9}$.

Chapter 7

Quadratic Equations and Functions

Exercise Set 7.1

RC1. True; see page 573 in the text.

RC3. True; see Example 6 on page 576 in the text.

RC5. False; 0 is a solution of this equation, so when we divide by x on both sides we are dividing by 0.

1. a) $6x^2 = 30$

$\quad x^2 = 5 \qquad$ Dividing by 6

$\quad x = \sqrt{5} \text{ or } x = -\sqrt{5} \quad$ Principle of square roots

Check: $\dfrac{6x^2 = 30}{6(\pm\sqrt{5})^2 \; ? \; 30}$
$\qquad\qquad\quad 6 \cdot 5 \;\Big|$
$\qquad\qquad\qquad 30 \;\Big|\qquad$ TRUE

The solutions are $\sqrt{5}$ and $-\sqrt{5}$, or $\pm\sqrt{5}$.

b) The real-number solutions of the equation $6x^2 = 30$ are the first coordinates of the x-intercepts of the graph of $f(x) = 6x^2 - 30$. Thus, the x-intercepts are $(-\sqrt{5}, 0)$ and $(\sqrt{5}, 0)$.

3. a) $9x^2 + 25 = 0$

$\quad 9x^2 = -25 \qquad$ Subtracting 25

$\quad x^2 = -\dfrac{25}{9} \qquad$ Dividing by 9

$\quad x = \sqrt{-\dfrac{25}{9}} \;\text{ or }\; x = -\sqrt{-\dfrac{25}{9}} \quad$ Principle of square roots

$\quad x = \dfrac{5}{3}i \qquad\text{ or }\; x = -\dfrac{5}{3}i \qquad$ Simplifying

Check: $\dfrac{9x^2 + 25 = 0}{9\left(\pm\dfrac{5}{3}i\right)^2 + 25 \; ? \; 0}$
$\qquad\qquad 9\left(-\dfrac{25}{9}\right) + 25 \;\Big|$
$\qquad\qquad\quad -25 + 25 \;\Big|$
$\qquad\qquad\qquad\qquad 0 \;\Big|\quad$ TRUE

The solutions are $\dfrac{5}{3}i$ and $-\dfrac{5}{3}i$, or $\pm\dfrac{5}{3}i$.

b) Since the equation $9x^2 + 25 = 0$ has no real-number solutions, the graph of $f(x) = 9x^2 + 25$ has no x-intercepts.

5. $2x^2 - 3 = 0$

$\quad 2x^2 = 3$

$\quad x^2 = \dfrac{3}{2}$

$\quad x = \sqrt{\dfrac{3}{2}} \;\text{ or }\; x = -\sqrt{\dfrac{3}{2}} \qquad$ Principle of square roots

$\quad x = \sqrt{\dfrac{3}{2} \cdot \dfrac{2}{2}} \;\text{ or }\; x = -\sqrt{\dfrac{3}{2} \cdot \dfrac{2}{2}} \qquad$ Rationalizing denominators

$\quad x = \dfrac{\sqrt{6}}{2} \;\text{ or }\; x = -\dfrac{\sqrt{6}}{2}$

Check: $\dfrac{2x^2 - 3 = 0}{2\left(\pm\dfrac{\sqrt{6}}{2}\right)^2 - 3 \; ? \; 0}$
$\qquad\qquad 2 \cdot \dfrac{6}{4} - 3 \;\Big|$
$\qquad\qquad\quad 3 - 3 \;\Big|$
$\qquad\qquad\qquad 0 \;\Big|\quad$ TRUE

The solutions are $\dfrac{\sqrt{6}}{2}$ and $-\dfrac{\sqrt{6}}{2}$, or $\pm\dfrac{\sqrt{6}}{2}$. Using a calculator, we find that the solutions are approximately ± 1.225.

7. $(x+2)^2 = 49$

$\quad x + 2 = 7 \;\text{ or }\; x + 2 = -7 \qquad$ Principle of square roots

$\qquad x = 5 \;\text{ or }\qquad x = -9$

The solutions are 5 and -9.

9. $(x-4)^2 = 16$

$\quad x - 4 = 4 \;\text{ or }\; x - 4 = -4 \qquad$ Principle of square roots

$\qquad x = 8 \;\text{ or }\qquad x = 0$

The solutions are 8 and 0.

11. $(x-11)^2 = 7$

$\quad x - 11 = \sqrt{7} \qquad\text{ or }\; x - 11 = -\sqrt{7}$

$\qquad x = 11 + \sqrt{7} \;\text{ or }\qquad x = 11 - \sqrt{7}$

The solutions are $11 + \sqrt{7}$ and $11 - \sqrt{7}$, or $11 \pm \sqrt{7}$. Using a calculator, we find that the solutions are approximately 8.354 and 13.646.

13. $(x-7)^2 = -4$

$\quad x - 7 = \sqrt{-4} \;\text{ or }\; x - 7 = -\sqrt{-4}$

$\quad x - 7 = 2i \qquad\text{ or }\; x - 7 = -2i$

$\qquad x = 7 + 2i \;\text{ or }\qquad x = 7 - 2i$

The solutions are $7 + 2i$ and $7 - 2i$, or $7 \pm 2i$.

15. $(x-9)^2 = 81$

$x - 9 = 9 \quad or \quad x - 9 = -9$

$x = 18 \quad or \qquad x = 0$

The solutions are 18 and 0.

17. $\left(x - \dfrac{3}{2}\right)^2 = \dfrac{7}{2}$

$x - \dfrac{3}{2} = \sqrt{\dfrac{7}{2}} \quad or \quad x - \dfrac{3}{2} = -\sqrt{\dfrac{7}{2}}$

$x - \dfrac{3}{2} = \sqrt{\dfrac{7}{2} \cdot \dfrac{2}{2}} \quad or \quad x - \dfrac{3}{2} = -\sqrt{\dfrac{7}{2} \cdot \dfrac{2}{2}}$

$x - \dfrac{3}{2} = \dfrac{\sqrt{14}}{2} \quad or \quad x - \dfrac{3}{2} = -\dfrac{\sqrt{14}}{2}$

$x = \dfrac{3}{2} + \dfrac{\sqrt{14}}{2} \quad or \qquad x = \dfrac{3}{2} - \dfrac{\sqrt{14}}{2}$

The solutions are $\dfrac{3}{2} + \dfrac{\sqrt{14}}{2}$ and $\dfrac{3}{2} - \dfrac{\sqrt{14}}{2}$, or $\dfrac{3}{2} \pm \dfrac{\sqrt{14}}{2}$. Using a calculator, we find that the solutions are approximately -0.371 and 3.371.

19. $x^2 + 6x + 9 = 64$

$(x+3)^2 = 64$

$x + 3 = 8 \quad or \quad x + 3 = -8$

$x = 5 \quad or \qquad x = -11$

The solutions are 5 and -11.

21. $y^2 - 14y + 49 = 4$

$(y-7)^2 = 4$

$y - 7 = 2 \quad or \quad y - 7 = -2$

$y = 9 \quad or \qquad y = 5$

The solutions are 9 and 5.

23. $x^2 + 4x = 2$ \qquad Original equation

$x^2 + 4x + 4 = 2 + 4$ \quad Adding 4: $\left(\dfrac{4}{2}\right)^2 = 2^2 = 4$

$(x+2)^2 = 6$

$x + 2 = \sqrt{6} \qquad or \quad x + 2 = -\sqrt{6}$ \quad Principle of square roots

$x = -2 + \sqrt{6} \quad or \qquad x = -2 - \sqrt{6}$

The solutions are $-2 \pm \sqrt{6}$.

25. $x^2 - 22x = 11$ \qquad Original equation

$x^2 - 22x + 121 = 11 + 121$ \quad Adding 121: $\left(\dfrac{-22}{2}\right)^2 =$
$(-11)^2 = 121$

$(x-11)^2 = 132$

$x - 11 = \sqrt{132} \qquad or \quad x - 11 = -\sqrt{132}$

$x - 11 = 2\sqrt{33} \qquad or \quad x - 11 = -2\sqrt{33}$

$x = 11 + 2\sqrt{33} \quad or \qquad x = 11 - 2\sqrt{33}$

The solutions are $11 \pm 2\sqrt{33}$.

27. $x^2 + x = 1$

$x^2 + x + \dfrac{1}{4} = 1 + \dfrac{1}{4}$ \quad Adding $\dfrac{1}{4}$: $\left(\dfrac{1}{2}\right)^2 = \dfrac{1}{4}$

$\left(x + \dfrac{1}{2}\right)^2 = \dfrac{5}{4}$

$x + \dfrac{1}{2} = \dfrac{\sqrt{5}}{2} \qquad or \quad x + \dfrac{1}{2} = -\dfrac{\sqrt{5}}{2}$

$x = -\dfrac{1}{2} + \dfrac{\sqrt{5}}{2} \quad or \qquad x = -\dfrac{1}{2} - \dfrac{\sqrt{5}}{2}$

The solutions are $-\dfrac{1}{2} \pm \dfrac{\sqrt{5}}{2}$.

29. $t^2 - 5t = 7$

$t^2 - 5t + \dfrac{25}{4} = 7 + \dfrac{25}{4}$ \quad Adding $\dfrac{25}{4}$: $\left(\dfrac{-5}{2}\right)^2 = \dfrac{25}{4}$

$\left(t - \dfrac{5}{2}\right)^2 = \dfrac{53}{4}$

$t - \dfrac{5}{2} = \dfrac{\sqrt{53}}{2} \quad or \quad t - \dfrac{5}{2} = -\dfrac{\sqrt{53}}{2}$

$t = \dfrac{5}{2} + \dfrac{\sqrt{53}}{2} \quad or \qquad t = \dfrac{5}{2} - \dfrac{\sqrt{53}}{2}$

The solutions are $\dfrac{5}{2} \pm \dfrac{\sqrt{53}}{2}$.

31. $x^2 + \dfrac{3}{2}x = 3$

$x^2 + \dfrac{3}{2}x + \dfrac{9}{16} = 3 + \dfrac{9}{16}$ \quad $\left(\dfrac{1}{2} \cdot \dfrac{3}{2}\right)^2 = \left(\dfrac{3}{4}\right)^2 = \dfrac{9}{16}$

$\left(x + \dfrac{3}{4}\right)^2 = \dfrac{57}{16}$

$x + \dfrac{3}{4} = \dfrac{\sqrt{57}}{4} \quad or \quad x + \dfrac{3}{4} = -\dfrac{\sqrt{57}}{4}$

$x = -\dfrac{3}{4} + \dfrac{\sqrt{57}}{4} \quad or \qquad x = -\dfrac{3}{4} - \dfrac{\sqrt{57}}{4}$

The solutions are $-\dfrac{3}{4} \pm \dfrac{\sqrt{57}}{4}$.

33. $m^2 - \dfrac{9}{2}m = \dfrac{3}{2}$ \qquad Original equation

$m^2 - \dfrac{9}{2}m + \dfrac{81}{16} = \dfrac{3}{2} + \dfrac{81}{16}$ \quad $\left[\dfrac{1}{2}\left(-\dfrac{9}{2}\right)\right]^2 = \left(-\dfrac{9}{4}\right)^2 = \dfrac{81}{16}$

$\left(m - \dfrac{9}{4}\right)^2 = \dfrac{105}{16}$

$m - \dfrac{9}{4} = \dfrac{\sqrt{105}}{4} \quad or \quad m - \dfrac{9}{4} = -\dfrac{\sqrt{105}}{4}$

$m = \dfrac{9}{4} + \dfrac{\sqrt{105}}{4} \quad or \qquad m = \dfrac{9}{4} - \dfrac{\sqrt{105}}{4}$

The solutions are $\dfrac{9}{4} \pm \dfrac{\sqrt{105}}{4}$.

35. $x^2 + 6x - 16 = 0$

$x^2 + 6x \quad\quad = 16$ Adding 16

$x^2 + 6x + \;9 = 16 + 9 \quad \left(\dfrac{6}{2}\right)^2 = 3^2 = 9$

$(x + 3)^2 = 25$

$x + 3 = 5 \;\; or \;\; x + 3 = -5$

$x = 2 \;\; or \;\;\;\;\;\; x = -8$

The solutions are 2 and -8.

37. $x^2 + 22x + 102 = 0$

$x^2 + 22x \quad\quad = -102$ Subtracting 102

$x^2 + 22x + 121 = -102 + 121 \quad \left(\dfrac{22}{2}\right)^2 = 11^2 = 121$

$(x + 11)^2 = 19$

$x + 11 = \sqrt{19} \quad\quad\;\; or \;\; x + 11 = -\sqrt{19}$

$x = -11 + \sqrt{19} \;\; or \quad\quad x = -11 - \sqrt{19}$

The solutions are $-11 \pm \sqrt{19}$.

39. $x^2 - 10x - \;4 = 0$

$x^2 - 10x \quad\quad = 4$ Adding 4

$x^2 - 10x + 25 = 4 + 25 \quad \left(\dfrac{-10}{2}\right)^2 = (-5)^2 = 25$

$(x - 5)^2 = 29$

$x - 5 = \sqrt{29} \quad\quad or \;\; x - 5 = -\sqrt{29}$

$x = 5 + \sqrt{29} \;\; or \quad\quad x = 5 - \sqrt{29}$

The solutions are $5 \pm \sqrt{29}$.

41. a) $x^2 + 7x - \;2 = 0$

$x^2 + 7x \quad\quad = 2$ Adding 2

$x^2 + 7x + \dfrac{49}{4} = 2 + \dfrac{49}{4} \quad \left(\dfrac{7}{2}\right)^2 = \dfrac{49}{4}$

$\left(x + \dfrac{7}{2}\right)^2 = \dfrac{57}{4}$

$x + \dfrac{7}{2} = \dfrac{\sqrt{57}}{2} \quad\quad or \;\; x + \dfrac{7}{2} = -\dfrac{\sqrt{57}}{2}$

$x = -\dfrac{7}{2} + \dfrac{\sqrt{57}}{2} \;\; or \quad\quad x = -\dfrac{7}{2} - \dfrac{\sqrt{57}}{2}$

The solutions are $-\dfrac{7}{2} \pm \dfrac{\sqrt{57}}{2}$.

b) The real-number solutions of the equation $x^2 + 7x - 2 = 0$ are the first coordinates of the x-intercepts of the graph of $f(x) = x^2 + 7x - 2$. Thus, the x-intercepts are $\left(-\dfrac{7}{2} - \dfrac{\sqrt{57}}{2}, 0\right)$ and $\left(-\dfrac{7}{2} + \dfrac{\sqrt{57}}{2}, 0\right)$.

43. a) $2x^2 - 5x + 8 = 0$

$\dfrac{1}{2}(2x^2 - 5x + 8) = \dfrac{1}{2} \cdot 0$ Multiplying by $\dfrac{1}{2}$ to make the x^2-coefficient 1

$x^2 - \dfrac{5}{2}x + \;\;4 = 0$

$x^2 - \dfrac{5}{2}x \quad\quad = -4$ Subtracting 4

$x^2 - \dfrac{5}{2}x + \dfrac{25}{16} = -4 + \dfrac{25}{16}$

$\left[\dfrac{1}{2}\left(-\dfrac{5}{2}\right)\right]^2 = \left(-\dfrac{5}{4}\right)^2 = \dfrac{25}{16}$

$\left(x - \dfrac{5}{4}\right)^2 = -\dfrac{64}{16} + \dfrac{25}{16}$

$\left(x - \dfrac{5}{4}\right)^2 = -\dfrac{39}{16}$

$x - \dfrac{5}{4} = \sqrt{-\dfrac{39}{16}} \quad or \;\; x - \dfrac{5}{4} = -\sqrt{-\dfrac{39}{16}}$

$x - \dfrac{5}{4} = i\sqrt{\dfrac{39}{16}} \quad or \;\; x - \dfrac{5}{4} = -i\sqrt{\dfrac{39}{16}}$

$x = \dfrac{5}{4} + i\dfrac{\sqrt{39}}{4} \quad or \quad\quad x = \dfrac{5}{4} - i\dfrac{\sqrt{39}}{4}$

The solutions are $\dfrac{5}{4} \pm i\dfrac{\sqrt{39}}{4}$.

b) Since the equation $2x^2 - 5x + 8 = 0$ has no real-number solutions, the graph of $f(x) = 2x^2 - 5x + 8$ has no x-intercepts.

45. $x^2 - \dfrac{3}{2}x - \dfrac{1}{2} = 0$

$x^2 - \dfrac{3}{2}x \quad\quad = \dfrac{1}{2}$

$x^2 - \dfrac{3}{2}x + \dfrac{9}{16} = \dfrac{1}{2} + \dfrac{9}{16} \quad \left[\dfrac{1}{2}\left(-\dfrac{3}{2}\right)\right]^2 = \left(-\dfrac{3}{4}\right)^2 = \dfrac{9}{16}$

$\left(x - \dfrac{3}{4}\right)^2 = \dfrac{17}{16}$

$x - \dfrac{3}{4} = \dfrac{\sqrt{17}}{4} \quad\quad or \;\; x - \dfrac{3}{4} = -\dfrac{\sqrt{17}}{4}$

$x = \dfrac{3}{4} + \dfrac{\sqrt{17}}{4} \;\; or \quad\quad x = \dfrac{3}{4} - \dfrac{\sqrt{17}}{4}$

The solutions are $\dfrac{3}{4} \pm \dfrac{\sqrt{17}}{4}$.

47. $2x^2 - 3x - 17 = 0$

$\frac{1}{2}(2x^2 - 3x - 17) = \frac{1}{2} \cdot 0$ Multiplying by $\frac{1}{2}$ to make

$\qquad\qquad\qquad\qquad\qquad$ the x^2-coefficient 1

$x^2 - \frac{3}{2}x - \frac{17}{2} = 0$

$x^2 - \frac{3}{2}x \qquad = \frac{17}{2}$ Adding $\frac{17}{2}$

$x^2 - \frac{3}{2}x + \frac{9}{16} = \frac{17}{2} + \frac{9}{16}$ $\left[\frac{1}{2}\left(-\frac{3}{2}\right)\right]^2 =$

$\qquad\qquad\qquad\qquad\qquad \left(-\frac{3}{4}\right)^2 = \frac{9}{16}$

$\left(x - \frac{3}{4}\right)^2 = \frac{145}{16}$

$x - \frac{3}{4} = \frac{\sqrt{145}}{4} \qquad or \quad x - \frac{3}{4} = -\frac{\sqrt{145}}{4}$

$x = \frac{3}{4} + \frac{\sqrt{145}}{4} \quad or \qquad x = \frac{3}{4} - \frac{\sqrt{145}}{4}$

The solutions are $\frac{3}{4} \pm \frac{\sqrt{145}}{4}$.

49. $3x^2 - 4x - 1 = 0$

$\frac{1}{3}(3x^2 - 4x - 1) = \frac{1}{3} \cdot 0$ Multiplying to make

$\qquad\qquad\qquad\qquad\qquad$ the x^2-coefficient 1

$x^2 - \frac{4}{3}x - \frac{1}{3} = 0$

$x^2 - \frac{4}{3}x \qquad = \frac{1}{3}$ Adding $\frac{1}{3}$

$x^2 - \frac{4}{3}x + \frac{4}{9} = \frac{1}{3} + \frac{4}{9}$ $\left[\frac{1}{2}\left(-\frac{4}{3}\right)\right]^2 =$

$\qquad\qquad\qquad\qquad\qquad \left(-\frac{2}{3}\right)^2 = \frac{4}{9}$

$\left(x - \frac{2}{3}\right)^2 = \frac{7}{9}$

$x - \frac{2}{3} = \frac{\sqrt{7}}{3} \qquad or \quad x - \frac{2}{3} = -\frac{\sqrt{7}}{3}$

$x = \frac{2}{3} + \frac{\sqrt{7}}{3} \quad or \qquad x = \frac{2}{3} - \frac{\sqrt{7}}{3}$

The solutions are $\frac{2}{3} \pm \frac{\sqrt{7}}{3}$.

51. $x^2 + x + 2 = 0$

$x^2 + x \qquad = -2$ Subtracting 2

$x^2 + x + \frac{1}{4} = -2 + \frac{1}{4}$ $\left(\frac{1}{2}\right)^2 = \frac{1}{4}$

$\left(x + \frac{1}{2}\right)^2 = -\frac{7}{4}$

$x + \frac{1}{2} = \sqrt{-\frac{7}{4}} \qquad or \quad x + \frac{1}{2} = -\sqrt{-\frac{7}{4}}$

$x + \frac{1}{2} = i\sqrt{\frac{7}{4}} \qquad or \quad x + \frac{1}{2} = -i\sqrt{\frac{7}{4}}$

$x = -\frac{1}{2} + i\frac{\sqrt{7}}{2} \quad or \qquad x = -\frac{1}{2} - i\frac{\sqrt{7}}{2}$

The solutions are $-\frac{1}{2} \pm i\frac{\sqrt{7}}{2}$.

53. $x^2 - 4x + 13 = 0$

$x^2 - 4x \qquad = -13$ Subtracting 13

$x^2 - 4x + 4 = -13 + 4$ $\left(\frac{-4}{2}\right)^2 = (-2)^2 = 4$

$(x - 2)^2 = -9$

$x - 2 = \sqrt{-9} \quad or \quad x - 2 = -\sqrt{-9}$

$x - 2 = 3i \qquad or \quad x - 2 = -3i$

$x = 2 + 3i \quad or \qquad x = 2 - 3i$

The solutions are $2 \pm 3i$.

55. $V(T) = 48T^2$

$38 = 48T^2$ Substituting 38 for $V(T)$

$\frac{38}{48} = T^2$ Solving for T^2

$\sqrt{\frac{38}{48}} = T$

$0.890 \approx T$

The hang time is about 0.890 sec.

57. $s(t) = 16t^2$

$456 = 16t^2$ Substituting 456 for $s(t)$

$\frac{456}{16} = t^2$ Solving for t^2

$28.5 = t^2$

$\sqrt{28.5} = t$

$5.3 \approx t$

It would take about 5.3 sec for an object to fall from the top of the roller coaster.

59. $s(t) = 16t^2$

$555.427 = 16t^2$

$\frac{555.427}{16} = t^2$

$\sqrt{\frac{555.427}{16}} = t$

$5.9 \approx t$

It would take about 5.9 sec for an object to fall from the top of the monument.

61. $s(t) = 16t^2$

$2080 = 16t^2$

$\frac{2080}{16} = t^2$

$130 = t^2$

$\sqrt{130} = t$

$11.4 \approx t$

It would take about 11.4 sec for an object to fall from the top of the tower.

63. a) First find the slope.

$$m = \frac{124 - 128}{30 - 0} = \frac{-4}{30} = -\frac{2}{15}$$

From the data we see that the y-intercept is $(0, 128)$, so we have $R(t) = -\frac{2}{15}t + 128$, where t is the number of years since 1981.

b) In 2020, $t = 2020 - 1981 = 39$.

$$R(39) = -\frac{2}{15} \cdot 39 + 128 = 122.8$$

We estimate that the record marathon time in 2020 will be 122.8 min.

c) Substitute 122 for $R(t)$ and solve for t.

$$122 = -\frac{2}{15}t + 128$$

$$-6 = -\frac{2}{15}t$$

$$-\frac{15}{2}(-6) = t$$

$$45 = t$$

The marathon record will be 122 min 45 yr after 1981, or in 2026.

65. Graph $f(x) = 5 - 2x$

We find some ordered pairs $(x, f(x))$, plot them, and draw the graph.

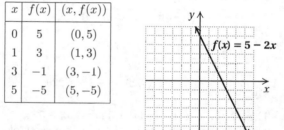

x	$f(x)$	$(x, f(x))$
0	5	$(0, 5)$
1	3	$(1, 3)$
3	-1	$(3, -1)$
5	-5	$(5, -5)$

67. Graph $f(x) = |5 - 2x|$

We find some ordered pairs $(x, f(x))$, plot them, and draw the graph.

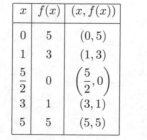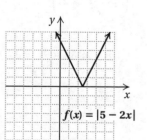

x	$f(x)$	$(x, f(x))$
0	5	$(0, 5)$
1	3	$(1, 3)$
$\frac{5}{2}$	0	$\left(\frac{5}{2}, 0\right)$
3	1	$(3, 1)$
5	5	$(5, 5)$

69. $\sqrt[5]{32x^5} = \sqrt[5]{(2x)^5} = 2x$

71.
$$\frac{4x^3 - 6x^2 - 10x}{3x^3 - 3x} = \frac{2x(2x^2 - 3x - 5)}{3x(x^2 - 1)}$$
$$= \frac{2x(2x - 5)(x + 1)}{3x(x + 1)(x - 1)}$$
$$= \frac{x(x + 1)}{x(x + 1)} \cdot \frac{2(2x - 5)}{3(x - 1)}$$
$$= \frac{2(2x - 5)}{3(x - 1)}$$

73.
$$\frac{\dfrac{t}{t + 1}}{t - \dfrac{1}{t}} = \frac{\dfrac{t}{t + 1}}{t \cdot \dfrac{t}{t} - \dfrac{1}{t}}$$
$$= \frac{\dfrac{t}{t + 1}}{\dfrac{t^2 - 1}{t}}$$
$$= \frac{t}{t + 1} \cdot \frac{t}{t^2 - 1}$$
$$= \frac{t^2}{(t + 1)(t^2 - 1)}$$
$$= \frac{t^2}{(t + 1)(t + 1)(t - 1)}$$
$$= \frac{t^2}{(t + 1)^2(t - 1)}$$

75. Left to the student

77. In order for $x^2 + bx + 64$ to be a trinomial square, the following must be true:

$$\left(\frac{b}{2}\right)^2 = 64$$

$$\frac{b^2}{4} = 64$$

$$b^2 = 256$$

$$b = 16 \quad or \quad b = -16$$

79. $x(2x^2 + 9x - 56)(3x + 10) = 0$

$x(2x - 7)(x + 8)(3x + 10) = 0$

$x = 0 \ or \ 2x - 7 = 0 \ or \ x + 8 = 0 \ or \ 3x + 10 = 0$

$x = 0 \ or \quad x = \frac{7}{2} \ or \quad x = -8 \ or \quad x = -\frac{10}{3}$

The solutions are -8, $-\frac{10}{3}$, 0, and $\frac{7}{2}$.

Exercise Set 7.2

RC1. When we are using the quadratic formula to solve $3x^2 - x - 8 = 0$, the value of a is $\underline{3}$.

RC3. Standard form for the quadratic equation $5x^2 = 9 - x$ is $\underline{5x^2 + x - 9} = 0$.

1. $x^2 + 8x + 2 = 0$

$a = 1,\ b = 8,\ c = 2$

$x = \dfrac{-b \pm \sqrt{b^2 - 4ac}}{2a}$

$x = \dfrac{-8 \pm \sqrt{8^2 - 4 \cdot 1 \cdot 2}}{2 \cdot 1} = \dfrac{-8 \pm \sqrt{64 - 8}}{2}$

$x = \dfrac{-8 \pm \sqrt{56}}{2} = \dfrac{-8 \pm 2\sqrt{14}}{2}$

$x = \dfrac{2(-4 \pm \sqrt{14})}{2} = -4 \pm \sqrt{14}$

The solutions are $-4 + \sqrt{14}$ and $-4 - \sqrt{14}$.

3. $3p^2 = -8p - 1$

$3p^2 + 8p + 1 = 0$ Finding standard form

$a = 3,\ b = 8,\ c = 1$

$p = \dfrac{-b \pm \sqrt{b^2 - 4ac}}{2a}$

$p = \dfrac{-8 \pm \sqrt{8^2 - 4 \cdot 3 \cdot 1}}{2 \cdot 3} = \dfrac{-8 \pm \sqrt{64 - 12}}{6}$

$x = \dfrac{-8 \pm \sqrt{52}}{6} = \dfrac{-8 \pm 2\sqrt{13}}{6}$

$x = \dfrac{2(-4 \pm \sqrt{13})}{2 \cdot 3} = \dfrac{-4 \pm \sqrt{13}}{3}$

The solutions are $\dfrac{-4 + \sqrt{13}}{3}$ and $\dfrac{-4 - \sqrt{13}}{3}$.

5. $x^2 - x + 1 = 0$

$a = 1,\ b = -1,\ c = 1$

$x = \dfrac{-(-1) \pm \sqrt{(-1)^2 - 4 \cdot 1 \cdot 1}}{2 \cdot 1} = \dfrac{1 \pm \sqrt{1 - 4}}{2}$

$x = \dfrac{1 \pm \sqrt{-3}}{2} = \dfrac{1 \pm i\sqrt{3}}{2} = \dfrac{1}{2} \pm i\dfrac{\sqrt{3}}{2}$

The solutions are $\dfrac{1}{2} + i\dfrac{\sqrt{3}}{2}$ and $\dfrac{1}{2} - i\dfrac{\sqrt{3}}{2}$.

7. $x^2 + 13 = 4x$

$x^2 - 4x + 13 = 0$ Finding standard form

$a = 1,\ b = -4,\ c = 13$

$x = \dfrac{-(-4) \pm \sqrt{(-4)^2 - 4 \cdot 1 \cdot 13}}{2 \cdot 1} = \dfrac{4 \pm \sqrt{16 - 52}}{2}$

$x = \dfrac{4 \pm \sqrt{-36}}{2} = \dfrac{4 \pm 6i}{2} = 2 \pm 3i$

The solutions are $2 + 3i$ and $2 - 3i$.

9. $r^2 + 3r = 8$

$r^2 + 3r - 8 = 0$ Finding standard form

$a = 1,\ b = 3,\ c = -8$

$r = \dfrac{-3 \pm \sqrt{3^2 - 4 \cdot 1 \cdot (-8)}}{2 \cdot 1} = \dfrac{-3 \pm \sqrt{9 + 32}}{2}$

$r = \dfrac{-3 \pm \sqrt{41}}{2}$

The solutions are $\dfrac{-3 + \sqrt{41}}{2}$ and $\dfrac{-3 - \sqrt{41}}{2}$.

11. $1 + \dfrac{2}{x} + \dfrac{5}{x^2} = 0$

$x^2 + 2x + 5 = 0$ Multiplying by x^2, the LCM
of the denominators

$a = 1,\ b = 2,\ c = 5$

$x = \dfrac{-2 \pm \sqrt{2^2 - 4 \cdot 1 \cdot 5}}{2 \cdot 1} = \dfrac{-2 \pm \sqrt{4 - 20}}{2}$

$x = \dfrac{-2 \pm \sqrt{-16}}{2} = \dfrac{-2 \pm 4i}{2} = -1 \pm 2i$

The solutions are $-1 + 2i$ and $-1 - 2i$.

13. a) $3x + x(x - 2) = 0$

$3x + x^2 - 2x = 0$

$x^2 + x = 0$

$x(x + 1) = 0$

$x = 0\ \ or\ \ x + 1 = 0$

$x = 0\ \ or\ \ \ \ \ \ \ x = -1$

The solutions are 0 and -1.

b) The solutions of the equation $3x + x(x - 2) = 0$ are the first coordinates of the x-intercepts of the graph of $f(x) = 3x + x(x - 2)$. Thus, the x-intercepts are $(-1, 0)$ and $(0, 0)$.

15. a) $11x^2 - 3x - 5 = 0$

$a = 11,\ b = -3,\ c = -5$

$x = \dfrac{-(-3) \pm \sqrt{(-3)^2 - 4 \cdot 11 \cdot (-5)}}{2 \cdot 11}$

$x = \dfrac{3 \pm \sqrt{9 + 220}}{22} = \dfrac{3 \pm \sqrt{229}}{22}$

The solutions are $\dfrac{3 + \sqrt{229}}{22}$ and $\dfrac{3 - \sqrt{229}}{22}$.

b) The solutions of the equation $11x^2 - 3x - 5 = 0$ are the first coordinates of the x-intercepts of the graph of $f(x) = 11x^2 - 3x - 5$. Thus, the x-intercepts are $\left(\dfrac{3 - \sqrt{229}}{22}, 0\right)$ and $\left(\dfrac{3 + \sqrt{229}}{22}, 0\right)$.

17. a) $25x^2 - 20x + 4 = 0$

$(5x - 2)(5x - 2) = 0$

$5x - 2 = 0\ \ or\ \ 5x - 2 = 0$

$5x = 2\ \ or\ \ \ \ \ \ \ 5x = 2$

$x = \dfrac{2}{5}\ \ or\ \ \ \ \ \ \ x = \dfrac{2}{5}$

The solution is $\dfrac{2}{5}$.

b) The solution of the equation $25x^2 - 20x + 4 = 0$ is the first coordinate of the x-intercept of $f(x) = 25x^2 - 20x + 4$. Thus, the x-intercept is $\left(\dfrac{2}{5}, 0\right)$.

19. $4x(x-2) - 5x(x-1) = 2$

$4x^2 - 8x - 5x^2 + 5x = 2$ Removing parentheses

$-x^2 - 3x = 2$

$-x^2 - 3x - 2 = 0$

$x^2 + 3x + 2 = 0$ Multiplying by -1

$(x+2)(x+1) = 0$

$x + 2 = 0$ or $x + 1 = 0$

$x = -2$ or $x = -1$

The solutions are -2 and -1.

21. $14(x-4) - (x+2) = (x+2)(x-4)$

$14x - 56 - x - 2 = x^2 - 2x - 8$

$13x - 58 = x^2 - 2x - 8$

$0 = x^2 - 15x + 50$

$0 = (x-10)(x-5)$

$x - 10 = 0$ or $x - 5 = 0$

$x = 10$ or $x = 5$

The solutions are 10 and 5.

23. $5x^2 = 17x - 2$

$5x^2 - 17x + 2 = 0$

$a = 5,\ b = -17,\ c = 2$

$x = \dfrac{-(-17) \pm \sqrt{(-17)^2 - 4 \cdot 5 \cdot 2}}{2 \cdot 5}$

$x = \dfrac{17 \pm \sqrt{289 - 40}}{10} = \dfrac{17 \pm \sqrt{249}}{10}$

The solutions are $\dfrac{17 + \sqrt{249}}{10}$ and $\dfrac{17 - \sqrt{249}}{10}$.

25. $x^2 + 5 = 4x$

$x^2 - 4x + 5 = 0$

$a = 1,\ b = -4,\ c = 5$

$x = \dfrac{-(-4) \pm \sqrt{(-4)^2 - 4 \cdot 1 \cdot 5}}{2 \cdot 1} = \dfrac{4 \pm \sqrt{16 - 20}}{2}$

$x = \dfrac{4 \pm \sqrt{-4}}{2} = \dfrac{4 \pm 2i}{2} = 2 \pm i$

The solutions are $2 + i$ and $2 - i$.

27. $x + \dfrac{1}{x} = \dfrac{13}{6}$, LCM is $6x$

$6x\left(x + \dfrac{1}{x}\right) = 6x \cdot \dfrac{13}{6}$

$6x^2 + 6 = 13x$

$6x^2 - 13x + 6 = 0$

$(2x - 3)(3x - 2) = 0$

$2x - 3 = 0$ or $3x - 2 = 0$

$2x = 3$ or $3x = 2$

$x = \dfrac{3}{2}$ or $x = \dfrac{2}{3}$

The solutions are $\dfrac{3}{2}$ and $\dfrac{2}{3}$.

29. $\dfrac{1}{y} + \dfrac{1}{y+2} = \dfrac{1}{3}$, LCM is $3y(y+2)$

$3y(y+2)\left(\dfrac{1}{y} + \dfrac{1}{y+2}\right) = 3y(y+2) \cdot \dfrac{1}{3}$

$3(y+2) + 3y = y(y+2)$

$3y + 6 + 3y = y^2 + 2y$

$6y + 6 = y^2 + 2y$

$0 = y^2 - 4y - 6$

$a = 1,\ b = -4,\ c = -6$

$y = \dfrac{-(-4) \pm \sqrt{(-4)^2 - 4 \cdot 1 \cdot (-6)}}{2 \cdot 1} = \dfrac{4 \pm \sqrt{16 + 24}}{2}$

$y = \dfrac{4 \pm \sqrt{40}}{2} = \dfrac{4 \pm 2\sqrt{10}}{2}$

$y = \dfrac{2(2 \pm \sqrt{10})}{2 \cdot 1} = 2 \pm \sqrt{10}$

The solutions are $2 + \sqrt{10}$ and $2 - \sqrt{10}$.

31. $(2t - 3)^2 + 17t = 15$

$4t^2 - 12t + 9 + 17t = 15$

$4t^2 + 5t - 6 = 0$

$(4t - 3)(t + 2) = 0$

$4t - 3 = 0$ or $t + 2 = 0$

$t = \dfrac{3}{4}$ or $t = -2$

The solutions are $\dfrac{3}{4}$ and -2.

33. $(x - 2)^2 + (x + 1)^2 = 0$

$x^2 - 4x + 4 + x^2 + 2x + 1 = 0$

$2x^2 - 2x + 5 = 0$

$a = 2,\ b = -2,\ c = 5$

$x = \dfrac{-(-2) \pm \sqrt{(-2)^2 - 4 \cdot 2 \cdot 5}}{2 \cdot 2} = \dfrac{2 \pm \sqrt{4 - 40}}{4}$

$x = \dfrac{2 \pm \sqrt{-36}}{4} = \dfrac{2 \pm 6i}{4}$

$x = \dfrac{2(1 \pm 3i)}{2 \cdot 2} = \dfrac{1 \pm 3i}{2} = \dfrac{1}{2} \pm \dfrac{3}{2}i$

The solutions are $\dfrac{1}{2} + \dfrac{3}{2}i$ and $\dfrac{1}{2} - \dfrac{3}{2}i$.

35. $x^3 - 1 = 0$

$(x - 1)(x^2 + x + 1) = 0$

$x - 1 = 0$ or $x^2 + x + 1 = 0$

$x = 1$ or $x = \dfrac{-1 \pm \sqrt{1^2 - 4 \cdot 1 \cdot 1}}{2 \cdot 1}$

$x = 1$ or $x = \dfrac{-1 \pm \sqrt{-3}}{2}$

$x = 1$ or $x = \dfrac{-1 \pm i\sqrt{3}}{2} = -\dfrac{1}{2} \pm i\dfrac{\sqrt{3}}{2}$

The solutions are 1, $-\dfrac{1}{2} + i\dfrac{\sqrt{3}}{2}$, and $-\dfrac{1}{2} - i\dfrac{\sqrt{3}}{2}$.

37. $x^2 + 6x + 4 = 0$

$a = 1,\ b = 6,\ c = 4$

$x = \dfrac{-6 \pm \sqrt{6^2 - 4 \cdot 1 \cdot 4}}{2 \cdot 1} = \dfrac{-6 \pm \sqrt{36 - 16}}{2}$

$x = \dfrac{-6 \pm \sqrt{20}}{2} = \dfrac{-6 \pm \sqrt{4 \cdot 5}}{2}$

$x = \dfrac{-6 \pm 2\sqrt{5}}{2} = \dfrac{2(-3 \pm \sqrt{5})}{2}$

$x = -3 \pm \sqrt{5}$

We can use a calculator to approximate the solutions:

$-3 + \sqrt{5} \approx -0.764;\ -3 - \sqrt{5} \approx -5.236$

The solutions are $-3 + \sqrt{5}$ and $-3 - \sqrt{5}$, or approximately -0.764 and -5.236.

39. $x^2 - 6x + 4 = 0$

$a = 1,\ b = -6,\ c = 4$

$x = \dfrac{-(-6) \pm \sqrt{(-6)^2 - 4 \cdot 1 \cdot 4}}{2 \cdot 1} = \dfrac{6 \pm \sqrt{36 - 16}}{2}$

$x = \dfrac{6 \pm \sqrt{20}}{2} = \dfrac{6 \pm \sqrt{4 \cdot 5}}{2}$

$x = \dfrac{6 \pm 2\sqrt{5}}{2} = \dfrac{2(3 \pm \sqrt{5})}{2}$

$x = 3 \pm \sqrt{5}$

We can use a calculator to approximate the solutions:

$3 + \sqrt{5} \approx 5.236;\ 3 - \sqrt{5} \approx 0.764$

The solutions are $3 + \sqrt{5}$ and $3 - \sqrt{5}$, or approximately 5.236 and 0.764.

41. $2x^2 - 3x - 7 = 0$

$a = 2,\ b = -3,\ c = -7$

$x = \dfrac{-(-3) \pm \sqrt{(-3)^2 - 4 \cdot 2 \cdot (-7)}}{2 \cdot 2} = \dfrac{3 \pm \sqrt{9 + 56}}{4}$

$x = \dfrac{3 \pm \sqrt{65}}{4}$

We can use a calculator to approximate the solutions:

$\dfrac{3 + \sqrt{65}}{4} \approx 2.766;\ \dfrac{3 - \sqrt{65}}{4} \approx -1.266$

The solutions are $\dfrac{3 + \sqrt{65}}{4}$ and $\dfrac{3 - \sqrt{65}}{4}$, or approximately 2.766 and -1.266.

43. $5x^2 = 3 + 8x$

$5x^2 - 8x - 3 = 0$

$a = 5,\ b = -8,\ c = -3$

$x = \dfrac{-(-8) \pm \sqrt{(-8)^2 - 4 \cdot 5 \cdot (-3)}}{2 \cdot 5} = \dfrac{8 \pm \sqrt{64 + 60}}{10}$

$x = \dfrac{8 \pm \sqrt{124}}{10} = \dfrac{8 \pm \sqrt{4 \cdot 31}}{10}$

$x = \dfrac{8 \pm 2\sqrt{31}}{10} = \dfrac{2(4 \pm \sqrt{31})}{2 \cdot 5}$

$x = \dfrac{4 \pm \sqrt{31}}{5}$

We can use a calculator to approximate the solutions:

$\dfrac{4 + \sqrt{31}}{5} \approx 1.914;\ \dfrac{4 - \sqrt{31}}{5} \approx -0.314$

The solutions are $\dfrac{4 + \sqrt{31}}{5}$ and $\dfrac{4 - \sqrt{31}}{5}$, or approximately 1.914 and -0.314.

45.

$\begin{aligned} x &= \sqrt{x + 2} \\ x^2 &= (\sqrt{x + 2})^2 \quad \text{Principle of powers} \\ x^2 &= x + 2 \\ x^2 - x - 2 &= 0 \\ (x - 2)(x + 1) &= 0 \end{aligned}$

$x - 2 = 0\ \ or\ \ x + 1 = 0$

$\quad x = 2\ \ or\ \qquad x = -1$

The number 2 checks but -1 does not, so the solution is 2.

47.

$\begin{aligned} \sqrt{2x - 6} + 11 &= 2 \\ \sqrt{2x - 6} &= -9 \\ (\sqrt{2x - 6})^2 &= (-9)^2 \\ 2x - 6 &= 81 \\ 2x &= 87 \\ x &= \dfrac{87}{2} \end{aligned}$

The number $\dfrac{87}{2}$ does not check, so there is no solution. We might have observed this when we subtracted 11 on both sides to get $\sqrt{2x - 6} = -9$. Since a square root cannot be negative, this equation has no solution.

49.

$\begin{aligned} 2x^2 &= x + 3 \\ 2x^2 - x - 3 &= 0 \\ (2x - 3)(x + 1) &= 0 \end{aligned}$

$2x - 3 = 0\ \ or\ \ x + 1 = 0$

$\quad 2x = 3\ \ or\ \qquad x = -1$

$\quad\ \ x = \dfrac{3}{2}\ \ or\ \qquad x = -1$

The solutions are $\dfrac{3}{2}$ and -1.

51. $\dfrac{3}{x} - \dfrac{1}{4} = \dfrac{7}{2x}$, LCM is $4x$

$4x\left(\dfrac{3}{x} - \dfrac{1}{4}\right) = 4x \cdot \dfrac{7}{2x}$

$4x \cdot \dfrac{3}{x} - 4x \cdot \dfrac{1}{4} = 14$

$12 - x = 14$

$-x = 2$

$x = -2$

The number -2 checks, so it is the solution.

53. The solutions of $2.2x^2 + 0.5x - 1 = 0$ are approximately -0.797 and 0.570.

55. $2x^2 - x - \sqrt{5} = 0$

$a = 2,\ b = -1,\ c = -\sqrt{5}$

$$x = \frac{-(-1) \pm \sqrt{(-1)^2 - 4 \cdot 2 \cdot (-\sqrt{5})}}{2 \cdot 2} = \frac{1 \pm \sqrt{1 + 8\sqrt{5}}}{4}$$

The solutions are $\dfrac{1 + \sqrt{1 + 8\sqrt{5}}}{4}$ and $\dfrac{1 - \sqrt{1 + 8\sqrt{5}}}{4}$.

57. $ix^2 - x - 1 = 0$

$a = i,\ b = -1,\ c = -1$

$$x = \frac{-(-1) \pm \sqrt{(-1)^2 - 4 \cdot i \cdot (-1)}}{2 \cdot i} = \frac{1 \pm \sqrt{1 + 4i}}{2i}$$

$$x = \frac{1 \pm \sqrt{1 + 4i}}{2i} \cdot \frac{i}{i} = \frac{i \pm i\sqrt{1 + 4i}}{2i^2} = \frac{i \pm i\sqrt{1 + 4i}}{-2}$$

$$x = \frac{-i \pm i\sqrt{1 + 4i}}{2}$$

The solutions are $\dfrac{-i + i\sqrt{1 + 4i}}{2}$ and $\dfrac{-i - i\sqrt{1 + 4i}}{2}$.

59.

$$\frac{x}{x+1} = 4 + \frac{1}{3x^2 - 3}$$

$$\frac{x}{x+1} = 4 + \frac{1}{3(x+1)(x-1)},$$

$$\text{LCM is } 3x(x+1)(x-1)$$

$$3(x+1)(x-1) \cdot \frac{x}{x+1} =$$

$$3(x+1)(x-1)\left(4 + \frac{1}{3(x+1)(x-1)}\right)$$

$$3x(x-1) = 12(x+1)(x-1) + 1$$

$$3x^2 - 3x = 12x^2 - 12 + 1$$

$$0 = 9x^2 + 3x - 11$$

$a = 9,\ b = 3,\ c = -11$

$$x = \frac{-3 \pm \sqrt{3^2 - 4 \cdot 9 \cdot (-11)}}{2 \cdot 9} = \frac{-3 \pm \sqrt{9 + 396}}{18}$$

$$x = \frac{-3 \pm \sqrt{405}}{18} = \frac{-3 \pm 9\sqrt{5}}{18}$$

$$x = \frac{3(-1 \pm 3\sqrt{5})}{3 \cdot 6} = \frac{-1 \pm 3\sqrt{5}}{6}$$

The solutions are $\dfrac{-1 + 3\sqrt{5}}{6}$ and $\dfrac{-1 - 3\sqrt{5}}{6}$.

61. Replace $f(x)$ with 13.

$$13 = (x - 3)^2$$

$$\pm\sqrt{13} = x - 3$$

$$3 \pm \sqrt{13} = x$$

The solutions are $3 + \sqrt{13}$ and $3 - \sqrt{13}$.

Exercise Set 7.3

RC1. The answer is (f). See page 598 in the text.

RC3. The answer is (b). See page 594 in the text.

RC5. The answer is (d). See page 596 in the text.

1. *Familiarize.* We make a drawing and label it. We let $x =$ the length of the rectangle. Then $x - 7 =$ the width.

Translate. We use the formula for the area of a rectangle.

$$A = lw$$

$$18 = x(x - 7) \quad \text{Substituting}$$

Solve. We solve the equation.

$$18 = x^2 - 7x$$

$$0 = x^2 - 7x - 18$$

$$0 = (x - 9)(x + 2)$$

$$x - 9 = 0 \quad or \quad x + 2 = 0$$

$$x = 9 \quad or \quad x = -2$$

Check. We only check 9 since the length cannot be negative. If $x = 9$, then $x - 7 = 9 - 7$, or 2, and the area is $9 \cdot 2$, or 18 ft². The value checks.

State. The length of 9 ft, and the width is 2 ft.

3. *Familiarize.* We make a drawing and label it. We let $x =$ the width of the rectangle. Then $2x =$ the length.

Translate.

$$A = lw$$

$$162 = 2x \cdot x \quad \text{Substituting}$$

Solve. We solve the equation.

$$162 = 2x^2$$

$$81 = x^2$$

$$\pm 9 = x$$

Check. We only check 9 since the width cannot be negative. If $x = 9$, then $2x = 2 \cdot 9$, or 18, and the area is $18 \cdot 9$, or 162 yd². The value checks.

State. The length is 18 yd, and the width is 9 yd.

5. *Familiarize.* Let $b =$ the length of the base, in miles. Then $b - 7 =$ the height. Recall that the formula for the area of a triangle is $A = \dfrac{1}{2} \times \text{base} \times \text{height}$.

Translate. The area is 60 mi². We substitute in the formula.

$$\frac{1}{2}b(b - 7) = 60$$

Solve. We solve the equation.

$$\frac{1}{2}b(b - 7) = 60$$

$$b(b - 7) = 120 \quad \text{Multiplying by 2}$$

$$b^2 - 7b = 120$$

$$b^2 - 7b - 120 = 0$$

$$(b - 15)(b + 8) = 0$$

$b - 15 = 0 \quad or \quad b + 8 = 0$

$b = 15 \quad or \qquad b = -8$

Check. We check only 15 since the length of the base cannot be negative. If the base is 15 mi, then the height is $15 - 7$, or 8, mi, and the area is $\frac{1}{2} \cdot 15 \cdot 8$, or 60 mi^2. The answer checks.

State. The length of the base is 15 mi, and the height is 8 mi.

7. **Familiarize**. Let $l =$ the length of the parking lot, in ft. Then $l - 51 =$ the width.

Translate. We will use the Pythagorean theorem.

$$l^2 + (l - 51)^2 = 250^2$$

Solve. We solve the equation.

$$l^2 + (l - 51)^2 = 250^2$$

$$l^2 + l^2 - 102l + 2601 = 62,500$$

$$2l^2 - 102l - 59,899 = 0$$

$$l = \frac{-b \pm \sqrt{b^2 - 4ac}}{2a} = \frac{-(-102) \pm \sqrt{(-102)^2 - 4 \cdot 2 \cdot (-59,899)}}{2 \cdot 2}$$

$$l = \frac{102 \pm \sqrt{10,404 + 479,192}}{4} = \frac{102 \pm \sqrt{489,596}}{4}$$

$$l = \frac{102 \pm \sqrt{4 \cdot 122,399}}{4} = \frac{102 \pm 2\sqrt{122,399}}{4}$$

$$1 = \frac{2(51 \pm \sqrt{122,399})}{2 \cdot 2} = \frac{51 \pm \sqrt{122,399}}{2}$$

Since $\sqrt{122,399} \approx 350$, we see that $\frac{51 - \sqrt{122,399}}{2}$ is a negative number and thus cannot be a solution of the original problem.

If $l = \frac{51 + \sqrt{122,399}}{2}$, then $l - 51 = \frac{51 + \sqrt{122,399}}{2} - 51 =$

$\frac{51 + \sqrt{122,399}}{2} - \frac{102}{2} = \frac{\sqrt{122,399} - 51}{2}$.

Check. As shown above, the width is 51 ft less than the length. We determine whether the length and width satisfy the Pythagorean theorem.

$$\left(\frac{51 + \sqrt{122,399}}{2}\right)^2 + \left(\frac{\sqrt{122,399} - 51}{2}\right)^2 =$$

$$\frac{2601 + 102\sqrt{122,399} + 122,399}{4} +$$

$$\frac{122,399 - 102\sqrt{122,399} + 2601}{4} =$$

$$\frac{250,000}{4} = 62,500 = 250^2.$$

The answer checks.

State. The length of the parking lot is $\frac{51 + \sqrt{122,399}}{2}$ ft, and the width is $\frac{\sqrt{122,399} - 51}{2}$ ft.

9. **Familiarize**. Using the labels on the drawing in the text, we let $x =$ the width of the frame, in inches. Then the dimensions of the part of the mirror that shows are $20 - 2x$ and $14 - 2x$.

Translate. We use the formula for the area of a rectangle.

$$(20 - 2x)(14 - 2x) = 160$$

Solve. We solve the equation.

$$(20 - 2x)(14 - 2x) = 160$$

$$280 - 68x + 4x^2 = 160$$

$$4x^2 - 68x + 120 = 0$$

$$x^2 - 17x + 30 = 0 \quad \text{Dividing by 4}$$

$$(x - 2)(x - 15) = 0$$

$$x - 2 = 0 \quad or \quad x - 5 = 0$$

$$x = 2 \quad or \qquad x = 15$$

If $x = 2$, then $20 - 2x = 20 - 2 \cdot 2 = 20 - 4 = 16$, and $14 - 2x = 14 - 2 \cdot 2 = 14 - 4 = 10$.

If $x = 15$, then $20 - 2x = 20 - 2 \cdot 15 = 20 - 30 = -10$. Since a negative number cannot be a dimension of the mirror, 15 cannot be a solution.

Check. If the frame is 2 in. wide, then the dimensions of the part of the mirror that shows are 16 in. and 10 in. and the area is $16 \cdot 10$, or 160 in^2. The answer checks.

State. The width of the frame is 2 in.

11. **Familiarize**. Using the labels on the drawing in the text, we let x and $x + 2$ represent the lengths of the legs of the right triangle.

Translate. We use the Pythagorean theorem.

$$a^2 + b^2 = c^2$$

$$x^2 + (x + 2)^2 = 10^2 \quad \text{Substituting}$$

Solve. We solve the equation.

$$x^2 + x^2 + 4x + 4 = 100$$

$$2x^2 + 4x + 4 = 100$$

$$2x^2 + 4x - 96 = 0$$

$$x^2 + 2x - 48 = 0 \quad \text{Dividing by 2}$$

$$(x + 8)(x - 6) = 0$$

$$x + 8 = 0 \quad or \quad x - 6 = 0$$

$$x = -8 \quad or \qquad x = 6$$

Check. We only check 6 since the length of a leg cannot be negative. When $x = 6$, then $x + 2 = 8$, and $6^2 + 8^2 = 100 = 10^2$. The number 6 checks.

State. The lengths of the legs are 6 ft and 8 ft.

13. **Familiarize**. Let $x =$ the smaller number. Then $x + 1 =$ the larger number.

Translate.

$$\underbrace{\text{The product of the numbers}}_{x(x+1)} \text{ is } 552.$$

$$x(x+1) \qquad = 552$$

Solve. We solve the equation.

$$x^2 + x = 552$$

$$x^2 + x - 552 = 0$$

$$(x + 24)(x - 23) = 0$$

$$x + 24 = 0 \quad or \quad x - 23 = 0$$

$$x = -24 \quad or \qquad x = 23$$

Check. We only check 23 since a ticket number cannot be negative. If $x = 23$, then $x + 1 = 24$ and $23 \cdot 24 = 552$. The number 23 checks.

State. The numbers are 23 and 24.

15. Familiarize. We make a drawing and label it. We let $x =$ the length and $x - 4 =$ the width.

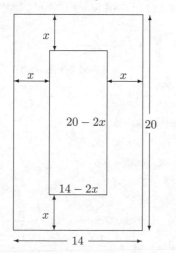

$x - 4$

x

Translate. We use the formula for the area of a rectangle.

$$A = lw$$

$$10 = x(x - 4) \qquad \text{Substituting}$$

Solve. We solve the equation.

$$10 = x^2 - 4x$$

$$0 = x^2 - 4x - 10$$

$$x = \frac{-b \pm \sqrt{b^2 - 4ac}}{2a} = \frac{-(-4) \pm \sqrt{(-4)^2 - 4 \cdot 1 \cdot (-10)}}{2 \cdot 1}$$

$$x = \frac{4 \pm \sqrt{16 + 40}}{2} = \frac{4 \pm \sqrt{56}}{2} = \frac{4 \pm \sqrt{4 \cdot 14}}{2}$$

$$x = \frac{4 \pm 2\sqrt{14}}{2} = 2 \pm \sqrt{14}$$

Check. We only need to check $2 + \sqrt{14}$ since $2 - \sqrt{14}$ is negative and the length cannot be negative. If $x = 2 + \sqrt{14}$, then $x - 4 = (2 + \sqrt{14}) - 4$, or $\sqrt{14} - 2$. Using a calculator we find that the length is $2 + \sqrt{14} \approx 5.742$ ft and the width is $\sqrt{14} - 2 \approx 1.742$ ft, and $(5.742)(1.742) = 10.003 \approx 10$. Our result checks.

State. The length is $2 + \sqrt{14}$ ft ≈ 5.742 ft; the width is $\sqrt{14} - 2$ ft ≈ 1.742 ft.

17. Familiarize. We make a drawing and label it. We let $x =$ the width of the margin.

x

x x

$20 - 2x$ 20

$14 - 2x$

x

14

The length and width of the printed text are represented by $20 - 2x$ and $14 - 2x$. The area of the printed text is 100 in^2.

Translate. We use the formula for the area of a rectangle.

$$A = lw$$

$$100 = (20 - 2x)(14 - 2x)$$

Solve. We solve the equation.

$$100 = 280 - 68x + 4x^2$$

$$0 = 4x^2 - 68x + 180$$

$$0 = x^2 - 17x + 45 \qquad \text{Dividing by 4}$$

$$x = \frac{-b \pm \sqrt{b^2 - 4ac}}{2a} = \frac{-(-17) \pm \sqrt{(-17)^2 - 4 \cdot 1 \cdot 45}}{2 \cdot 1}$$

$$x = \frac{17 \pm \sqrt{289 - 180}}{2} = \frac{17 \pm \sqrt{109}}{2}$$

$$x \approx 13.720 \quad or \quad x \approx 3.280$$

Check. If $x \approx 13.720$, then $20 - 2x \approx -7.440$ and $14 - 2x \approx -13.440$. Since the width of the margin cannot be negative, 13.720 is not a solution. If $x \approx 3.280$, then $20 - 2x \approx 13.440$ and $14 - 2x \approx 7.440$ and $(13.440)(7.440) = 99.99 \approx 100$. The number $\frac{17 - \sqrt{109}}{2} \approx 3.280$ checks.

State. The width of the margin is $\frac{17 - \sqrt{109}}{2}$ in. ≈ 3.280 in.

19. Familiarize. We make a drawing. We let $x =$ the length of the shorter leg and $x + 14 =$ the length of the longer leg.

24

x

$x + 14$

Translate. We use the Pythagorean equation.

$$a^2 + b^2 = c^2$$

$$x^2 + (x + 14)^2 = 24^2 \qquad \text{Substituting}$$

Solve. We solve the equation.

$$x^2 + x^2 + 28x + 196 = 576$$

$$2x^2 + 28x - 380 = 0$$

$$x^2 + 14x - 190 = 0 \qquad \text{Dividing by 2}$$

$$x = \frac{-b \pm \sqrt{b^2 - 4ac}}{2a} = \frac{-14 \pm \sqrt{14^2 - 4 \cdot 1 \cdot (-190)}}{2 \cdot 1}$$

$$x = \frac{-14 \pm \sqrt{196 + 760}}{2} = \frac{-14 \pm \sqrt{956}}{2} = \frac{-14 \pm \sqrt{4 \cdot 239}}{2}$$

$$x = \frac{-14 \pm 2\sqrt{239}}{2} = -7 \pm \sqrt{239}$$

$$x \approx 8.460 \text{ or } x \approx -22.460$$

Check. Since the length of a leg cannot be negative, we only need to check 8.460. If $x = -7 + \sqrt{239} \approx 8.460$, then $x + 14 = -7 + \sqrt{239} + 14 = 7 + \sqrt{239} \approx 22.460$ and $(8.460)^2 + (22.460)^2 = 576.0232 \approx 576 = 24^2$. The number $-7 + \sqrt{239} \approx 8.460$ checks.

State. The lengths of the legs are $-7 + \sqrt{239}$ ft ≈ 8.460 ft and $7 + \sqrt{239}$ ft ≈ 22.460 ft.

21. Familiarize. We first make a drawing, labeling it with the known and unknown information. We can also organize the information in a table. We let r represent the speed and t the time for the first part of the trip.

r mph	t hr	$r - 10$ mph	$4 - t$ hr
	120 mi		100 mi

Trip	Distance	Speed	Time
1st part	120	r	t
2nd part	100	$r - 10$	$4 - t$

Translate. Using $r = \dfrac{d}{t}$, we get two equations from the table, $r = \dfrac{120}{t}$ and $r - 10 = \dfrac{100}{4 - t}$.

Solve. We substitute $\dfrac{120}{t}$ for r in the second equation and solve for t.

$$\frac{120}{t} - 10 = \frac{100}{4 - t}, \text{ LCM is } t(4 - t)$$

$$t(4 - t)\left(\frac{120}{t} - 10\right) = t(4 - t) \cdot \frac{100}{4 - t}$$

$$120(4 - t) - 10t(4 - t) = 100t$$

$$480 - 120t - 40t + 10t^2 = 100t$$

$$10t^2 - 260t + 480 = 0 \quad \text{Standard form}$$

$$t^2 - 26t + 48 = 0 \quad \text{Multiplying by } \frac{1}{10}$$

$$(t - 2)(t - 24) = 0$$

$$t = 2 \ \ or \ \ t = 24$$

Check. Since the time cannot be negative (If $t = 24$, $4 - t = -20$.), we check only 2 hr. If $t = 2$, then $4 - t = 2$. The speed of the first part is $\dfrac{120}{2}$, or 60 mph. The speed of the second part is $\dfrac{100}{2}$, or 50 mph. The speed of the second part is 10 mph slower than the first part. The value checks.

State. The speed of the first part was 60 mph, and speed of the second part was 50 mph.

23. **Familiarize**. We first make a drawing. We also organize the information in a table. We let r = the speed and t = the time of the slower trip.

24 mi	r mph	t hr
24 mi	$r + 2$ mph	$t - 1$ hr

Trip	Distance	Speed	Time
Slower	24	r	t
Faster	24	$r + 2$	$t - 1$

Translate. Using $t = d/r$, we get two equations from the table:

$$t = \frac{24}{r} \text{ and } t - 1 = \frac{24}{r + 2}$$

Solve. We substitute $\dfrac{24}{r}$ for t in the second equation and solve for r.

$$\frac{24}{r} - 1 = \frac{24}{r + 2}, \text{ LCM is } r(r + 2)$$

$$r(r + 2)\left(\frac{24}{r} - 1\right) = r(r + 2) \cdot \frac{24}{r + 2}$$

$$24(r + 2) - r(r + 2) = 24r$$

$$24r + 48 - r^2 - 2r = 24r$$

$$0 = r^2 + 2r - 48$$

$$0 = (r + 8)(r - 6)$$

$$r = -8 \ \ or \ \ r = 6$$

Check. Since negative speed has no meaning in this problem, we check only 6. If $r = 6$, then the time for the slower trip is $\dfrac{24}{6}$, or 4 hours. If $r = 6$, then $r + 2 = 8$ and the time for the faster trip is $\dfrac{24}{8}$, or 3 hours. This is 1 hour less time than the slower trip took, so we have an answer to the problem.

State. The speed is 6 mph.

25. **Familiarize**. We make a drawing and then organize the information in a table. We let r = the speed and t = the time of the Cessna.

600 mi	r mph	t hr
1000 mi	$r + 50$ mph	$t + 1$ hr

Plane	Distance	Speed	Time
Cessna	600	r	t
Beechcraft	1000	$r + 50$	$t + 1$

Translate. Using $t = d/r$, we get two equations from the table:

$$t = \frac{600}{r} \quad \text{and} \quad t + 1 = \frac{1000}{r + 50}$$

Solve. We substitute $\dfrac{600}{r}$ for t in the second equation and solve for r.

$$\frac{600}{r} + 1 = \frac{1000}{r + 50},$$
$$\text{LCM is } r(r + 50)$$

$$r(r + 50)\left(\frac{600}{r} + 1\right) = r(r + 50) \cdot \frac{1000}{r + 50}$$

$$600(r + 50) + r(r + 50) = 1000r$$

$$600r + 30,000 + r^2 + 50r = 1000r$$

$$r^2 - 350r + 30,000 = 0$$

$$(r - 150)(r - 200) = 0$$

$$r = 150 \ \ or \ r = 200$$

Check. If $r = 150$, then the Cessna's time is $\dfrac{600}{150}$, or 4 hr and the Beechcraft's time is $\dfrac{1000}{150 + 50}$, or $\dfrac{1000}{200}$, or 5 hr. If $r = 200$, then the Cessna's time is $\dfrac{600}{200}$, or 3 hr and the Beechcraft's time is $\dfrac{1000}{200 + 50}$, or $\dfrac{1000}{250}$, or 4 hr. Since the Beechcraft's time is 1 hr longer in each case, both values check. There are two solutions.

State. The speed of the Cessna is 150 mph and the speed of the Beechcraft is 200 mph; or the speed of the Cessna is 200 mph and the speed of the Beechcraft is 250 mph.

27. *Familiarize*. We make a drawing and then organize the information in a table. We let r represent the speed and t the time of the trip to Hillsboro.

Hillsboro
40 mi r mph t hr

40 mi $r - 6$ mph $14 - t$ hr

Trip	Distance	Speed	Time
To Hillsboro	40	r	t
Return	40	$r - 6$	$14 - t$

Translate. Using $t = \dfrac{d}{r}$, we get two equations from the table,

$$t = \frac{40}{r} \text{ and } 14 - t = \frac{40}{r - 6}.$$

Solve. We substitute $\dfrac{40}{r}$ for t in the second equation and solve for r.

$$14 - \frac{40}{r} = \frac{40}{r - 6},$$
$$\text{LCM is } r(r - 6)$$
$$r(r - 6)\left(14 - \frac{40}{r}\right) = r(r - 6) \cdot \frac{40}{r - 6}$$
$$14r(r - 6) - 40(r - 6) = 40r$$
$$14r^2 - 84r - 40r + 240 = 40r$$
$$14r^2 - 164r + 240 = 0$$
$$7r^2 - 82r + 120 = 0$$
$$(7r - 12)(r - 10) = 0$$
$$r = \frac{12}{7} \ \text{ or } \ r = 10$$

Check. Since negative speed has no meaning in this problem (If $r = \dfrac{12}{7}$, then $r - 6 = -\dfrac{30}{7}$.), we check only 10 mph.

If $r = 10$, then the time of the trip to Hillsboro is $\dfrac{40}{10}$, or 4 hr. The speed of the return trip is $10 - 6$, or 4 mph, and the time is $\dfrac{40}{4}$, or 10 hr. The total time for the round trip is 4 hr + 10 hr, or 14 hr. The value checks.

State. Naoki's speed on the trip to Hillsboro was 10 mph and it was 4 mph on the return trip.

29. *Familiarize*. We make a drawing and organize the information in a table. Let r represent the speed of the barge in still water, and let t represent the time of the trip upriver.

24 mi $r - 4$ mph t hr
                   Upriver

Downriver   24 mi $r + 4$ mph $5 - t$ hr

Trip	Distance	Speed	Time
Upriver	24	$r - 4$	t
Downriver	24	$r + 4$	$5 - t$

Translate. Using $t = \dfrac{d}{r}$, we get two equations from the table,

$$t = \frac{24}{r - 4} \text{ and } 5 - t = \frac{24}{r + 4}.$$

Solve. We substitute $\dfrac{24}{r - 4}$ for t in the second equation and solve for r.

$$5 - \frac{24}{r - 4} = \frac{24}{r + 4},$$
$$\text{LCM is } (r{-}4)(r+4)$$
$$(r - 4)(r + 4)\left(5 - \frac{24}{r - 4}\right) = (r - 4)(r + 4) \cdot \frac{24}{r + 4}$$
$$5(r - 4)(r + 4) - 24(r + 4) = 24(r - 4)$$
$$5r^2 - 80 - 24r - 96 = 24r - 96$$
$$5r^2 - 48r - 80 = 0$$

We use the quadratic formula.

$$r = \frac{-(-48) \pm \sqrt{(-48)^2 - 4 \cdot 5 \cdot (-80)}}{2 \cdot 5}$$
$$r = \frac{48 \pm \sqrt{3904}}{10}$$
$$r \approx 11 \ \text{ or } \ r \approx -1.5$$

Check. Since negative speed has no meaning in this problem, we check only 11 mph. If $r \approx 11$, then the speed upriver is about $11 - 4$, or 7 mph, and the time is about $\dfrac{24}{7}$, or 3.4 hr. The speed downriver is about $11 + 4$, or 15 mph, and the time is about $\dfrac{24}{15}$, or 1.6 hr. The total time of the round trip is $3.4 + 1.6$, or 5 hr. The value checks.

State. The barge must be able to travel about 11 mph in still water.

31.
$$A = 6s^2$$
$$\frac{A}{6} = s^2 \qquad \text{Dividing by 6}$$
$$\sqrt{\frac{A}{6}} = s \qquad \text{Taking the positive square root}$$

33.
$$F = \frac{Gm_1 m_2}{r^2}$$
$$Fr^2 = Gm_1 m_2 \qquad \text{Multiplying by } r^2$$
$$r^2 = \frac{Gm_1 m_2}{F} \qquad \text{Dividing by } F$$
$$r = \sqrt{\frac{Gm_1 m_2}{F}} \qquad \text{Taking the positive square root}$$

35.
$$E = mc^2$$
$$\frac{E}{m} = c^2 \qquad \text{Dividing by } m$$
$$\sqrt{\frac{E}{m}} = c \qquad \text{Taking the square root}$$

37. $a^2 + b^2 = c^2$

$$b^2 = c^2 - a^2 \quad \text{Subtracting } a^2$$

$$b = \sqrt{c^2 - a^2} \quad \text{Taking the square root}$$

39. $N = \dfrac{k^2 - 3k}{2}$

$$2N = k^2 - 3k$$

$$0 = k^2 - 3k - 2N \quad \text{Standard form}$$

$$a = 1, \ b = -3, \ c = -2N$$

$$k = \frac{-(-3) \pm \sqrt{(-3)^3 - 4 \cdot 1 \cdot (-2N)}}{2 \cdot 1} \quad \begin{array}{l}\text{Using the} \\ \text{quadratic formula}\end{array}$$

$$k = \frac{3 \pm \sqrt{9 + 8N}}{2}$$

Since taking the negative square root would result in a negative answer, we take the positive one.

$$k = \frac{3 + \sqrt{9 + 8N}}{2}$$

41. $A = 2\pi r^2 + 2\pi rh$

$$0 = 2\pi r^2 + 2\pi rh - A \quad \text{Standard form}$$

$$a = 2\pi, \ b = 2\pi h, \ c = -A$$

$$r = \frac{-2\pi h \pm \sqrt{(2\pi h)^2 - 4 \cdot 2\pi \cdot (-A)}}{2 \cdot 2\pi} \quad \begin{array}{l}\text{Using the} \\ \text{quadratic formula}\end{array}$$

$$r = \frac{-2\pi h \pm \sqrt{4\pi^2 h^2 + 8\pi A}}{4\pi}$$

$$r = \frac{-2\pi h \pm 2\sqrt{\pi^2 h^2 + 2\pi A}}{4\pi}$$

$$r = \frac{-\pi h \pm \sqrt{\pi^2 h^2 + 2\pi A}}{2\pi}$$

Since taking the negative square root would result in a negative answer, we take the positive one.

$$r = \frac{-\pi h + \sqrt{\pi^2 h^2 + 2\pi A}}{2\pi}$$

43. $T = 2\pi \sqrt{\dfrac{L}{g}}$

$$\frac{T}{2\pi} = \sqrt{\frac{L}{g}} \quad \text{Dividing by } 2\pi$$

$$\frac{T^2}{4\pi^2} = \frac{L}{g} \quad \text{Squaring}$$

$$gT^2 = 4\pi^2 L \quad \text{Multiplying by } 4\pi^2 g$$

$$g = \frac{4\pi^2 L}{T^2} \quad \text{Dividing by } T^2$$

45. $I = \dfrac{703W}{H^2}$

$$H^2 I = 703W \quad \text{Multiplying by } H^2$$

$$H^2 = \frac{703W}{I} \quad \text{Dividing by } I$$

$$H = \sqrt{\frac{703W}{I}}$$

47.
$$m = \frac{m_0}{\sqrt{1 - \dfrac{v^2}{c^2}}}$$

$$m^2 = \frac{(m_0)^2}{1 - \dfrac{v^2}{c^2}} \quad \text{Principle of powers}$$

$$m^2 \left(1 - \frac{v^2}{c^2}\right) = (m_0)^2$$

$$m^2 - \frac{m^2 v^2}{c^2} = (m_0)^2$$

$$m^2 - (m_0)^2 = \frac{m^2 v^2}{c^2}$$

$$c^2 [m^2 - (m_0)^2] = m^2 v^2$$

$$\frac{c^2 [m^2 - (m_0)^2]}{m^2} = v^2$$

$$\sqrt{\frac{c^2 [m^2 - (m_0)^2]}{m^2}} = v$$

$$\frac{c\sqrt{m^2 - (m_0)^2}}{m} = v$$

49.
$$\frac{1}{x - 1} + \frac{1}{x^2 - 3x + 2}$$

$$= \frac{1}{x - 1} + \frac{1}{(x - 1)(x - 2)}, \quad \text{LCD is } (x - 1)(x - 2)$$

$$= \frac{1}{x - 1} \cdot \frac{x - 2}{x - 2} + \frac{1}{(x - 1)(x - 2)}$$

$$= \frac{x - 2}{(x - 1)(x - 2)} + \frac{1}{(x - 1)(x - 2)}$$

$$= \frac{x - 2 + 1}{(x - 1)(x - 2)}$$

$$= \frac{x - 1}{(x - 1)(x - 2)}$$

$$= \frac{(x - 1) \cdot 1}{(x - 1)(x - 2)}$$

$$= \frac{1}{x - 2}$$

51.
$$\frac{2}{x + 3} - \frac{x}{x - 1} + \frac{x^2 + 2}{x^2 + 2x - 3}$$

$$= \frac{2}{x + 3} - \frac{x}{x - 1} + \frac{x^2 + 2}{(x + 3)(x - 1)},$$
$$\text{LCD is } (x + 3)(x - 1)$$

$$= \frac{2}{x + 3} \cdot \frac{x - 1}{x - 1} - \frac{x}{x - 1} \cdot \frac{x + 3}{x + 3} + \frac{x^2 + 2}{(x + 3)(x - 1)}$$

$$= \frac{2(x - 1)}{(x + 3)(x - 1)} - \frac{x(x + 3)}{(x - 1)(x + 3)} + \frac{x^2 + 2}{(x + 3)(x - 1)}$$

$$= \frac{2(x - 1) - x(x + 3) + x^2 + 2}{(x + 3)(x - 1)}$$

$$= \frac{2x - 2 - x^2 - 3x + x^2 + 2}{(x + 3)(x - 1)}$$

$$= \frac{-x}{(x + 3)(x - 1)}$$

53. $\sqrt{-20} = \sqrt{-1 \cdot 4 \cdot 5} = i \cdot 2 \cdot \sqrt{5} = 2\sqrt{5}i$, or $2i\sqrt{5}$

55.
$$\frac{1}{a-1} = a+1$$
$$\frac{1}{a-1} \cdot a - 1 = (a+1)(a-1)$$
$$1 = a^2 - 1$$
$$2 = a^2$$
$$\pm\sqrt{2} = a$$

57. Let s represent a length of a side of the cube, let S represent the surface area of the cube, and let A represent the surface area of the sphere. Then the diameter of the sphere is s, so the radius r is $s/2$. From Exercise 32, we know, $A = 4\pi r^2$, so when $r = s/2$ we have $A = 4\pi\left(\frac{s}{2}\right)^2 = 4\pi \cdot \frac{s^2}{4} = \pi s^2$. From the formula for the surface area of a cube (See Exercise 31.) we know that $S = 6s^2$, so $\frac{S}{6} = s^2$ and then $A = \pi \cdot \frac{S}{6}$, or $A(S) = \frac{\pi S}{6}$.

59.
$$\frac{w}{l} = \frac{l}{w+l}$$
$$l(w+l) \cdot \frac{w}{l} = l(w+l) \cdot \frac{l}{w+l}$$
$$w(w+l) = l^2$$
$$w^2 + lw = l^2$$
$$0 = l^2 - lw - w^2$$

Use the quadratic formula with $a = 1$, $b = -w$, and $c = -w^2$.
$$l = \frac{-(-w) \pm \sqrt{(-w)^2 - 4 \cdot 1 \cdot (-w^2)}}{2 \cdot 1}$$
$$l = \frac{w \pm \sqrt{w^2 + 4w^2}}{2} = \frac{w \pm \sqrt{5w^2}}{2}$$
$$l = \frac{w \pm w\sqrt{5}}{2}$$

Since $\frac{w - w\sqrt{5}}{2}$ is negative we use the positive square root:
$$l = \frac{w + w\sqrt{5}}{2}$$

Exercise Set 7.4

RC1. Since $b^2 - 4ac > 0$, there are two different real-number solutions. The answer is (b).

RC3. Since $b^2 - 4ac < 0$, there are two different nonreal complex-number solutions. The answer is (c).

RC5. Since $b^2 - 4ac > 0$, there are two different real-number solutions. The answer is (b).

1. $x^2 - 8x + 16 = 0$

$a = 1$, $b = -8$, $c = 16$

We compute the discriminant.

$$b^2 - 4ac = (-8)^2 - 4 \cdot 1 \cdot 16$$
$$= 64 - 64$$
$$= 0$$

Since $b^2 - 4ac = 0$, there is just one solution, and it is a real number.

3. $x^2 + 1 = 0$

$a = 1$, $b = 0$, $c = 1$

We compute the discriminant.

$$b^2 - 4ac = 0^2 - 4 \cdot 1 \cdot 1$$
$$= -4$$

Since $b^2 - 4ac < 0$, there are two nonreal solutions.

5. $x^2 - 6 = 0$

$a = 1$, $b = 0$, $c = -6$

We compute the discriminant.

$$b^2 - 4ac = 0^2 - 4 \cdot 1 \cdot (-6)$$
$$= 24$$

Since $b^2 - 4ac > 0$, there are two real solutions.

7. $4x^2 - 12x + 9 = 0$

$a = 4$, $b = -12$, $c = 9$

We compute the discriminant.

$$b^2 - 4ac = (-12)^2 - 4 \cdot 4 \cdot 9$$
$$= 144 - 144$$
$$= 0$$

Since $b^2 - 4ac = 0$, there is just one solution, and it is a real number.

9. $x^2 - 2x + 4 = 0$

$a = 1$, $b = -2$, $c = 4$

We compute the discriminant.

$$b^2 - 4ac = (-2)^2 - 4 \cdot 1 \cdot 4$$
$$= 4 - 16$$
$$= -12$$

Since $b^2 - 4ac < 0$, there are two nonreal solutions.

11. $9t^2 - 3t = 0$

$a = 9$, $b = -3$, $c = 0$

We compute the discriminant.

$$b^2 - 4ac = (-3)^2 - 4 \cdot 9 \cdot 0$$
$$= 9 - 0$$
$$= 9$$

Since $b^2 - 4ac > 0$, there are two real solutions.

13. $y^2 = \frac{1}{2}y + \frac{3}{5}$

$y^2 - \frac{1}{2}y - \frac{3}{5} = 0$ \qquad Standard form

$a = 1$, $b = -\frac{1}{2}$, $c = -\frac{3}{5}$

We compute the discriminant.

$$b^2 - 4ac = \left(-\frac{1}{2}\right)^2 - 4 \cdot 1 \cdot \left(-\frac{3}{5}\right)$$

$$= \frac{1}{4} + \frac{12}{5}$$

$$= \frac{53}{20}$$

Since $b^2 - 4ac > 0$, there are two real solutions.

15. $4x^2 - 4\sqrt{3}x + 3 = 0$

$a = 4, \; b = -4\sqrt{3}, \; c = 3$

We compute the discriminant.

$$b^2 - 4ac = (-4\sqrt{3})^2 - 4 \cdot 4 \cdot 3$$

$$= 48 - 48$$

$$= 0$$

Since $b^2 - 4ac = 0$, there is just one solution, and it is a real number.

17. The solutions are -4 and 4.

$$x = -4 \quad or \quad x = 4$$

$$x + 4 = 0 \quad or \quad x - 4 = 0$$

$$(x + 4)(x - 4) = 0 \qquad \text{Principle of zero products}$$

$$x^2 - 16 = 0 \qquad (A + B)(A - B) = A^2 - B^2$$

19. The solutions are $-4i$ and $4i$.

$$x = -4i \quad or \quad x = 4i$$

$$x + 4i = 0 \quad or \quad x - 4i = 0$$

$$(x + 4i)(x - 4i) = 0 \qquad \text{Principle of zero products}$$

$$x^2 - 16i^2 = 0 \qquad (A + B)(A - B) = A^2 - B^2$$

$$x^2 - 16(-1) = 0$$

$$x^2 + 16 = 0$$

21. The only solution is 8. It must be a double solution.

$$x = 8 \quad or \quad x = 8$$

$$x - 8 = 0 \quad or \quad x - 8 = 0$$

$$(x - 8)(x - 8) = 0 \qquad \text{Principle of zero products}$$

$$x^2 - 16x + 64 = 0 \qquad (A - B)^2 = A^2 - 2AB + B^2$$

23. The solutions are $-\dfrac{2}{5}$ and $\dfrac{6}{5}$.

$$x = -\frac{2}{5} \quad or \quad x = \frac{6}{5}$$

$$x + \frac{2}{5} = 0 \quad or \quad x - \frac{6}{5} = 0$$

$$5x + 2 = 0 \quad or \quad 5x - 6 = 0 \qquad \text{Clearing fractions}$$

$$(5x + 2)(5x - 6) = 0 \qquad \text{Principle of zero products}$$

$$25x^2 - 20x - 12 = 0 \qquad \text{FOIL}$$

25. The solutions are $\dfrac{k}{3}$ and $\dfrac{m}{4}$.

$$x = \frac{k}{3} \quad or \quad x = \frac{m}{4}$$

$$x - \frac{k}{3} = 0 \quad or \quad x - \frac{m}{4} = 0$$

$$3x - k = 0 \quad or \quad 4x - m = 0 \qquad \text{Clearing fractions}$$

$$(3x - k)(4x - m) = 0 \qquad \text{Principle of zero products}$$

$$12x^2 - 3mx - 4kx + km = 0 \qquad \text{FOIL}$$

$$12x^2 - (3m + 4k)x + km = 0 \qquad \text{Collecting like terms}$$

27. The solutions are $-\sqrt{3}$ and $2\sqrt{3}$.

$$x = -\sqrt{3} \quad or \quad x = 2\sqrt{3}$$

$$x + \sqrt{3} = 0 \quad or \quad x - 2\sqrt{3} = 0$$

$$(x + \sqrt{3})(x - 2\sqrt{3}) = 0 \qquad \text{Principle of zero products}$$

$$x^2 - 2\sqrt{3}x + \sqrt{3}x - 2(\sqrt{3})^2 = 0 \qquad \text{FOIL}$$

$$x^2 - \sqrt{3}x - 6 = 0$$

29. The solutions are $6i$ and $-6i$.

$$x = 6i \quad or \quad x = -6i$$

$$x - 6i = 0 \quad or \quad x + 6i = 0$$

$$(x - 6i)(x + 6i) = 0 \qquad \text{Principle of zero products}$$

$$x^2 - 36i^2 = 0 \qquad (A + B)(A - B) = A^2 - B^2$$

$$x^2 - 36(-1) = 0$$

$$x^2 + 36 = 0$$

31. $x^4 - 6x^2 + 9 = 0$

Let $u = x^2$ and think of x^4 as $(x^2)^2$.

$$u^2 - 6u + 9 = 0 \qquad \text{Substituting } u \text{ for } x^2$$

$$(u - 3)(u - 3) = 0$$

$$u - 3 = 0 \quad or \quad u - 3 = 0$$

$$u = 3 \quad or \quad u = 3$$

Now we substitute x^2 for u and solve the equation:

$$x^2 = 3$$

$$x = \pm\sqrt{3}$$

Both $\sqrt{3}$ and $-\sqrt{3}$ check. They are the solutions.

33. $x - 10\sqrt{x} + 9 = 0$

Let $u = \sqrt{x}$ and think of x as $(\sqrt{x})^2$.

$$u^2 - 10u + 9 = 0 \qquad \text{Substituting } u \text{ for } \sqrt{x}$$

$$(u - 9)(u - 1) = 0$$

$$u - 9 = 0 \quad or \quad u - 1 = 0$$

$$u = 9 \quad or \quad u = 1$$

Now we substitute $\sqrt{x}$ for u and solve these equations:

$$\sqrt{x} = 9 \quad or \quad \sqrt{x} = 1$$

$$x = 81 \quad or \quad x = 1$$

The numbers 81 and 1 both check. They are the solutions.

35. $(x^2 - 6x) - 2(x^2 - 6x) - 35 = 0$

Let $u = x^2 - 6x$.

$$u^2 - 2u - 35 = 0 \qquad \text{Substituting } u \text{ for } x^2 - 6x$$

$$(u - 7)(u + 5) = 0$$

$$u - 7 = 0 \quad or \quad u + 5 = 0$$

$$u = 7 \quad or \quad u = -5$$

Now we substitute $x^2 - 6x$ for u and solve these equations:

$$x^2 - 6x = 7 \quad or \quad x^2 - 6x = -5$$
$$x^2 - 6x - 7 = 0 \quad or \quad x^2 - 6x + 5 = 0$$
$$(x - 7)(x + 1) = 0 \quad or \quad (x - 5)(x - 1) = 0$$
$$x = 7 \ or \ x = -1 \ or \ x = 5 \ or \ x = 1$$

The numbers -1, 1, 5, and 7 check. They are the solutions.

37. $x^{-2} - 5x^{-1} - 36 = 0$

Let $u = x^{-1}$.

$$u^2 - 5u - 36 = 0 \quad \text{Substituting } u \text{ for } x^{-1}$$
$$(u - 9)(u + 4) = 0$$
$$u - 9 = 0 \ or \ u + 4 = 0$$
$$u = 9 \ or \ u = -4$$

Now we substitute x^{-1} for u and solve these equations:

$$x^{-1} = 9 \ or \ x^{-1} = -4$$
$$\frac{1}{x} = 9 \ or \ \frac{1}{x} = -4$$
$$\frac{1}{9} = x \ or \ -\frac{1}{4} = x$$

Both $\dfrac{1}{9}$ and $-\dfrac{1}{4}$ check. They are the solutions.

39. $(1 + \sqrt{x})^2 + (1 + \sqrt{x}) - 6 = 0$

Let $u = 1 + \sqrt{x}$.

$$u^2 + u - 6 = 0 \quad \text{Substituting } u \text{ for } 1 + \sqrt{x}$$
$$(u + 3)(u - 2) = 0$$
$$u + 3 = 0 \quad or \quad u - 2 = 0$$
$$u = -3 \quad or \quad u = 2$$
$$1 + \sqrt{x} = -3 \quad or \quad 1 + \sqrt{x} = 2 \quad \text{Substituting}$$
$$\hspace{7cm} 1 + \sqrt{x} \text{ for } u$$
$$\sqrt{x} = -4 \quad or \quad \sqrt{x} = 1$$
$$\text{No solution} \hspace{2.5cm} x = 1$$

The number 1 checks. It is the solution.

41. $(y^2 - 5y)^2 - 2(y^2 - 5y) - 24 = 0$

Let $u = y^2 - 5y$.

$$u^2 - 2u - 24 = 0 \quad \text{Substituting } u \text{ for } y^2 - 5y$$
$$(u - 6)(u + 4) = 0$$
$$u - 6 = 0 \ or \ u + 4 = 0$$
$$u = 6 \ or \ u = -4$$
$$y^2 - 5y = 6 \ or \ y^2 - 5y = -4$$
$$\hspace{2cm} \text{Substituting } y^2 - 5y \text{ for } u$$
$$y^2 - 5y - 6 = 0 \ or \ y^2 - 5y + 4 = 0$$
$$(y - 6)(y + 1) = 0 \ or \ (y - 4)(y - 1) = 0$$
$$y = 6 \ or \ y = -1 \ or \ y = 4 \ or \ y = 1$$

The numbers -1, 1, 4, and 6 check. They are the solutions.

43. $w^4 - 29w^2 + 100 = 0$

Let $u = w^2$.

$$u^2 - 29u + 100 = 0 \quad \text{Substituting } u \text{ for } w^2$$
$$(u - 4)(u - 25) = 0$$
$$u - 4 = 0 \quad or \quad u - 25 = 0$$
$$u = 4 \quad or \quad u = 25$$
$$w^2 = 4 \quad or \quad w^2 = 25 \quad \text{Substituting } w^2 \text{ for } u$$
$$w = \pm 2 \ or \quad w = \pm 5$$

All four numbers check. They are the solutions.

45. $2x^{-2} + x^{-1} - 1 = 0$

Let $u = x^{-1}$.

$$2u^2 + u - 1 = 0 \quad \text{Substituting } u \text{ for } x^{-1}$$
$$(2u - 1)(u + 1) = 0$$
$$2u - 1 = 0 \quad or \quad u + 1 = 0$$
$$2u = 1 \quad or \quad u = -1$$
$$u = \frac{1}{2} \quad or \quad u = -1$$
$$x^{-1} = \frac{1}{2} \quad or \quad x^{-1} = -1 \quad \text{Substituting } x^{-1}$$
$$\hspace{8cm} \text{for } u$$
$$\frac{1}{x} = \frac{1}{2} \quad or \quad \frac{1}{x} = -1$$
$$x = 2 \quad or \quad x = -1$$

Both 2 and -1 check. They are the solutions.

47. $6x^4 - 19x^2 + 15 = 0$

Let $u = x^2$.

$$6u^2 - 19u + 15 = 0 \quad \text{Substituting } u \text{ for } x^2$$
$$(3u - 5)(2u - 3) = 0$$
$$3u - 5 = 0 \quad or \quad 2u - 3 = 0$$
$$3u = 5 \quad or \quad 2u = 3$$
$$u = \frac{5}{3} \quad or \quad u = \frac{3}{2}$$
$$x^2 = \frac{5}{3} \quad or \quad x^2 = \frac{3}{2} \quad \text{Substituting } x^2$$
$$\hspace{8cm} \text{for } u$$
$$x = \pm\sqrt{\frac{5}{3}} \quad or \quad x = \pm\sqrt{\frac{3}{2}}$$
$$x = \pm\frac{\sqrt{15}}{3} \quad or \quad x = \pm\frac{\sqrt{6}}{2}$$
$$\hspace{4cm} \text{Rationalizing denominators}$$

All four numbers check. They are the solutions.

49. $x^{2/3} - 4x^{1/3} - 5 = 0$

Let $u = x^{1/3}$.

$u^2 - 4u - 5 = 0$ Substituting u for $x^{1/3}$

$(u - 5)(u + 1) = 0$

$u - 5 = 0$ or $u + 1 = 0$

$u = 5$ or $u = -1$

$x^{1/3} = 5$ or $x^{1/3} = -1$ Substituting $x^{1/3}$ for u

$(x^{1/3})^3 = 5^3$ or $(x^{1/3})^3 = (-1)^3$ Principle of powers

$x = 125$ or $x = -1$

Both 125 and -1 check. They are the solutions.

51. $\left(\dfrac{x-4}{x+1}\right)^2 - 2\left(\dfrac{x-4}{x+1}\right) - 35 = 0$

Let $u = \dfrac{x-4}{x+1}$.

$u^2 - 2u - 35 = 0$ Substituting u for $\dfrac{x-4}{x+1}$

$(u - 7)(u + 5) = 0$

$u - 7 = 0$ or $u + 5 = 0$

$u = 7$ or $u = -5$

$\dfrac{x-4}{x+1} = 7$ or $\dfrac{x-4}{x+1} = -5$ Substituting $\dfrac{x-4}{x+1}$ for u

$x - 4 = 7(x+1)$ or $x - 4 = -5(x+1)$

$x - 4 = 7x + 7$ or $x - 4 = -5x - 5$

$-6x = 11$ or $6x = -1$

$x = -\dfrac{11}{6}$ or $x = -\dfrac{1}{6}$

Both $-\dfrac{11}{6}$ and $-\dfrac{1}{6}$ check. They are the solutions.

53. $9\left(\dfrac{x+2}{x+3}\right)^2 - 6\left(\dfrac{x+2}{x+3}\right) + 1 = 0$

Let $u = \dfrac{x+2}{x+3}$.

$9u^2 - 6u + 1 = 0$ Substituting u for $\dfrac{x+2}{x+3}$

$(3u - 1)(3u - 1) = 0$

$3u - 1 = 0$ or $3u - 1 = 0$

$3u = 1$ or $3u = 1$

$u = \dfrac{1}{3}$ or $u = \dfrac{1}{3}$

Now we substitute $\dfrac{x+2}{x+3}$ for u and solve the equation:

$\dfrac{x+2}{x+3} = \dfrac{1}{3}$

$3(x+2) = x+3$ Multiplying by $3(x+3)$

$3x + 6 = x + 3$

$2x = -3$

$x = -\dfrac{3}{2}$

The number $-\dfrac{3}{2}$ checks. It is the solution.

55. $\left(\dfrac{x^2-2}{x}\right)^2 - 7\left(\dfrac{x^2-2}{x}\right) - 18 = 0$

Let $u = \dfrac{x^2-2}{x}$.

$u^2 - 7u - 18 = 0$ Substituting u for $\dfrac{x^2-2}{x}$

$(u - 9)(u + 2) = 0$

$u - 9 = 0$ or $u + 2 = 0$

$u = 9$ or $u = -2$

$\dfrac{x^2-2}{x} = 9$ or $\dfrac{x^2-2}{x} = -2$

Substituting $\dfrac{x^2-2}{x}$ for u

$x^2 - 2 = 9x$ or $x^2 - 2 = -2x$

$x^2 - 9x - 2 = 0$ or $x^2 + 2x - 2 = 0$

$x = \dfrac{-(-9) \pm \sqrt{(-9)^2 - 4 \cdot 1 \cdot (-2)}}{2 \cdot 1}$

$x = \dfrac{9 \pm \sqrt{89}}{2}$

or

$x = \dfrac{-2 \pm \sqrt{2^2 - 4 \cdot 1 \cdot (-2)}}{2 \cdot 1} = \dfrac{-2 \pm \sqrt{12}}{2}$

$x = \dfrac{-2 \pm 2\sqrt{3}}{2} = -1 \pm \sqrt{3}$

All four numbers check. They are the solutions.

57. The x-intercepts occur where $f(x) = 0$. Thus, we must have $5x + 13\sqrt{x} - 6 = 0$.

Let $u = \sqrt{x}$.

$5u^2 + 13u - 6 = 0$ Substituting

$(5u - 2)(u + 3) = 0$

$u = \dfrac{2}{5}$ or $u = -3$

Now replace u with $\sqrt{x}$ and solve these equations:

$\sqrt{x} = \dfrac{2}{5}$ or $\sqrt{x} = -3$

$x = \dfrac{4}{25}$ No solution

The number $\dfrac{4}{25}$ checks. Thus, the x-intercept is $\left(\dfrac{4}{25}, 0\right)$.

59. The x-intercepts occur where $f(x) = 0$. Thus, we must have $(x^2 - 3x)^2 - 10(x^2 - 3x) + 24 = 0$.

Let $u = x^2 - 3x$.

$u^2 - 10u + 24 = 0$ Substituting

$(u - 6)(u - 4) = 0$

$u = 6$ or $u = 4$

Now replace u with $x^2 - 3x$ and solve these equations:

$x^2 - 3x = 6$ or $x^2 - 3x = 4$

$x^2 - 3x - 6 = 0$ or $x^2 - 3x - 4 = 0$

$$x = \frac{-(-3) \pm \sqrt{(-3)^2 - 4(1)(-6)}}{2 \cdot 1} \quad or$$

$$(x - 4)(x + 1) = 0$$

$$x = \frac{3 \pm \sqrt{33}}{2} \quad or \quad x = 4 \ or \ x = -1$$

All four numbers check. Thus, the x-intercepts are $\left(\frac{3 + \sqrt{33}}{2}, 0\right)$, $\left(\frac{3 - \sqrt{33}}{2}, 0\right)$, $(4, 0)$, and $(-1, 0)$.

61. The x-intercepts occur where $f(x) = 0$. Thus, we must have $x^{2/3} + x^{1/3} - 2 = 0$

Let $u = x^{1/3}$.

$$u^2 + u - 2 = 0$$

$$(u + 2)(u - 1) = 0$$

$$u + 2 = 0 \quad or \quad u - 1 = 0$$

$$u = -2 \quad or \quad u = 1$$

$$x^{1/3} = -2 \quad or \quad x^{1/3} = 1$$

$$(x^{1/3})^3 = (-2)^3 \quad or \quad (x^{1/3})^3 = 1^3$$

$$x = -8 \quad or \quad x = 1$$

Both numbers check. Thus, the x-intercepts are $(-8, 0)$ and $(1, 0)$.

63. *Familiarize.* Let $x =$ the number of pounds of Kenyan coffee and $y =$ the number of pounds of Peruvian coffee in the mixture. We organize the information in a table.

Type of Coffee	Kenyan	Peruvian	Mixture
Price per pound	$9.75	$13.25	$11.15
Number of pounds	x	y	50
Total cost	$9.75x	$13.25y	11.15×50, or $557.50

Translate. From the last two rows of the table we get a system of equations.

$$x + y = 50,$$

$$9.75x + 13.25y = 557.50$$

Solve. Solving the system of equations, we get $(30, 20)$.

Check. The total number of pounds in the mixture is $30 + 20$, or 50. The total cost of the mixture is $9.75(30) + 13.25(20) = \$557.50$. The values check.

State. The mixture should consist of 30 lb of Kenyan coffee and 20 lb of Peruvian coffee.

65. $\sqrt{8x} \cdot \sqrt{2x} = \sqrt{8x \cdot 2x} = \sqrt{16x^2} = \sqrt{(4x)^2} = 4x$

67. $\sqrt[4]{9a^2} \cdot \sqrt[4]{18a^3} = \sqrt[4]{9a^2 \cdot 18a^3} =$
$\sqrt[4]{3 \cdot 3 \cdot a^2 \cdot 3 \cdot 3 \cdot 2 \cdot a^2 \cdot a} = \sqrt[4]{3^4 a^4 \cdot 2a} = \sqrt[4]{3^4} \sqrt[4]{a^4} \sqrt[4]{2a} =$
$3a\sqrt[4]{2a}$

69. Graph $f(x) = -\frac{3}{5}x + 4$.

Choose some values for x, find the corresponding values of $f(x)$, plot the points $(x, f(x))$, and draw the graph.

x	$f(x)$	$(x, f(x))$
-5	7	$(-5, 7)$
0	4	$(0, 4)$
5	1	$(5, 1)$

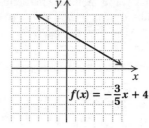

71. Graph $y = 4$.

The graph of $y = 4$ is a horizontal line with y-intercept $(0, 4)$.

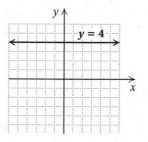

73. Left to the student

75. a) $kx^2 - 2x + k = 0$; one solution is -3

We first find k by substituting -3 for x.

$$k(-3)^2 - 2(-3) + k = 0$$

$$9k + 6 + k = 0$$

$$10k = -6$$

$$k = -\frac{6}{10}$$

$$k = -\frac{3}{5}$$

b) $-\frac{3}{5}x^2 - 2x + \left(-\frac{3}{5}\right) = 0$ Substituting $-\frac{3}{5}$ for k

$$3x^2 + 10x + 3 = 0 \quad \text{Multiplying by } -5$$

$$(3x + 1)(x + 3) = 0$$

$$3x + 1 = 0 \quad or \quad x + 3 = 0$$

$$3x = -1 \quad or \quad x = -3$$

$$x = -\frac{1}{3} \quad or \quad x = -3$$

The other solution is $-\frac{1}{3}$.

77. For $ax^2 + bx + c = 0$, $-\frac{b}{a}$ is the sum of the solutions and $\frac{c}{a}$ is the product of the solutions. Thus $-\frac{b}{a} = \sqrt{3}$ and $\frac{c}{a} = 8$.

$$ax^2 + bx + c = 0$$

$$x^2 + \frac{b}{a}x + \frac{c}{a} = 0 \quad \text{Multiplying by } \frac{1}{a}$$

$$x^2 - \left(-\frac{b}{a}\right)x + \frac{c}{a} = 0$$

$$x^2 - \sqrt{3}x + 8 = 0 \quad \text{Substituting } \sqrt{3} \text{ for } -\frac{b}{a} \text{ and 8 for } \frac{c}{a}$$

79. The graph includes the points $(-3, 0)$, $(0, -3)$, and $(1, 0)$. Substituting in $y = ax^2 + bx + c$, we have three equations.

$$0 = 9a - 3b + c,$$
$$-3 = \qquad\quad c,$$
$$0 = a + b + c$$

The solution of this system of equations is $a = 1$, $b = 2$, $c = -3$.

81. $\dfrac{x}{x-1} - 6\sqrt{\dfrac{x}{x-1}} - 40 = 0$

Let $u = \sqrt{\dfrac{x}{x-1}}$.

$u^2 - 6u - 40 = 0$ Substituting for $\sqrt{\dfrac{x}{x-1}}$

$(u - 10)(u + 4) = 0$

$u = 10 \qquad or \qquad u = -4$

$\sqrt{\dfrac{x}{x-1}} = 10 \qquad or \qquad \sqrt{\dfrac{x}{x-1}} = -4$

$\qquad\qquad\qquad\qquad$ Substituting for u

$\dfrac{x}{x-1} = 100 \qquad or \qquad$ No solution

$x = 100x - 100$ Multiplying by $(x-1)$

$100 = 99x$

$\dfrac{100}{99} = x$

This number checks. It is the solution.

83. $\sqrt{x-3} - \sqrt[4]{x-3} = 12$

$(x-3)^{1/2} - (x-3)^{1/4} - 12 = 0$

Let $u = (x-3)^{1/4}$.

$u^2 - u - 12 = 0$ Substituting for $(x-3)^{1/4}$

$(u - 4)(u + 3) = 0$

$u = 4 \qquad or \qquad u = -3$

$(x-3)^{1/4} = 4 \qquad or \quad (x-3)^{1/4} = -3$

$\qquad\qquad\qquad\qquad$ Substituting for u

$x - 3 = 4^4 \qquad or \qquad$ No solution

$x - 3 = 256$

$x = 259$

This number checks. It is the solution.

85. $x^6 - 28x^3 + 27 = 0$

Let $u = x^3$.

$u^2 - 28u + 27 = 0$ Substituting for x^3

$(u - 27)(u - 1) = 0$

$u = 27 \quad or \quad u = 1$

$x^3 = 27 \quad or \quad x^3 = 1$ Substituting for u

$x = 3 \quad or \quad x = 1$

Both 3 and 1 check. They are the solutions.

Chapter 7 Mid-Chapter Review

1. False; see page 605 in the text.

2. True; see page 588 in the text.

3. True; see page 605 in the text.

4. False; the solutions of a quadratic equation $f(x) = 0$ are the *first* coordinates of the x-intercepts. Thus, for this function, the x-intercepts are $(\sqrt{t}, 0)$ and $(-\sqrt{t}, 0)$.

5.
$$5x^2 + 3x = 4$$
$$\frac{1}{5}(5x^2 + 3x) = \frac{1}{5} \cdot 4$$
$$x^2 + \frac{3}{5}x = \frac{4}{5}$$
$$x^2 + \frac{3}{5}x + \frac{9}{100} = \frac{4}{5} + \frac{9}{100}$$
$$\left(x + \frac{3}{10}\right)^2 = \frac{89}{100}$$

$x + \dfrac{3}{10} = \sqrt{\dfrac{89}{100}} \qquad or \ \ x + \dfrac{3}{10} = -\sqrt{\dfrac{89}{100}}$

$x + \dfrac{3}{10} = \dfrac{\sqrt{89}}{10} \qquad or \ \ x + \dfrac{3}{10} = -\dfrac{\sqrt{89}}{10}$

$x = -\dfrac{3}{10} + \dfrac{\sqrt{89}}{10} \ \ or \qquad x = -\dfrac{3}{10} - \dfrac{\sqrt{89}}{10}$

The solutions are $-\dfrac{3}{10} \pm \dfrac{\sqrt{89}}{10}$.

6.
$$5x^2 + 3x = 4$$
$$5x^2 + 3x - 4 = 0$$
$$5x^2 + 3x + (-4) = 0$$
$$a = 5, \, b = 3, \, c = -4$$
$$x = \frac{-b \pm \sqrt{b^2 - 4ac}}{2a}$$
$$x = \frac{-3 \pm \sqrt{3^2 - 4 \cdot 5 \cdot (-4)}}{2 \cdot 5}$$
$$x = \frac{-3 \pm \sqrt{9 + 80}}{10}$$
$$x = \frac{-3 \pm \sqrt{89}}{10}$$
$$x = -\frac{3}{10} \pm \frac{\sqrt{89}}{10}$$

7.
$$x^2 + 1 = -4x$$
$$x^2 + 4x = -1 \qquad \text{Subtracting 1 and adding } 4x$$
$$x^2 + 4x + 4 = -1 + 4 \quad \left(\frac{1}{2} \cdot 4 = 2; \, 2^2 = 4\right)$$
$$(x + 2)^2 = 3$$

$x + 2 = \sqrt{3} \qquad or \ \ x + 2 = -\sqrt{3}$

$x = -2 + \sqrt{3} \ \ or \qquad x = -2 - \sqrt{3}$

The solutions are $-2 \pm \sqrt{3}$.

8. $2x^2 + 5x - 3 = 0$

$$2x^2 + 5x = 3$$

$$\frac{1}{2}(2x^2 + 5x) = \frac{1}{2} \cdot 3$$

$$x^2 + \frac{5}{2}x = \frac{3}{2}$$

$$x^2 + \frac{5}{2}x + \frac{25}{16} = \frac{3}{2} + \frac{25}{16} \quad \left(\frac{1}{2} \cdot \frac{5}{2} = \frac{5}{4}; \left(\frac{5}{4}\right)^2 = \frac{25}{16}\right)$$

$$\left(x + \frac{5}{4}\right)^2 = \frac{49}{16}$$

$$x + \frac{5}{4} = \sqrt{\frac{49}{16}} \quad or \quad x + \frac{5}{4} = -\sqrt{\frac{49}{16}}$$

$$x + \frac{5}{4} = \frac{7}{4} \quad or \quad x + \frac{5}{4} = -\frac{7}{4}$$

$$x = \frac{2}{4} \quad or \quad x = -\frac{12}{4}$$

$$x = \frac{1}{2} \quad or \quad x = -3$$

The solutions are $\frac{1}{2}$ and -3.

9. $x^2 + 10x - 6 = 0$

$$x^2 + 10x = 6$$

$$x^2 + 10x + 25 = 6 + 25 \quad \left(\frac{1}{2} \cdot 10 = 5; 5^2 = 25\right)$$

$$(x + 5)^2 = 31$$

$$x + 5 = \sqrt{31} \quad or \quad x + 5 = -\sqrt{31}$$

$$x = -5 + \sqrt{31} \quad or \quad x = -5 - \sqrt{31}$$

The solutions are $-5 \pm \sqrt{31}$.

10. $x^2 - x = 5$

$$x^2 - x + \frac{1}{4} = 5 + \frac{1}{4} \quad \left(\frac{1}{2}(-1) = -\frac{1}{2}; \left(-\frac{1}{2}\right)^2 = \frac{1}{4}\right)$$

$$\left(x - \frac{1}{2}\right)^2 = \frac{21}{4}$$

$$x - \frac{1}{2} = \sqrt{\frac{21}{4}} \quad or \quad x - \frac{1}{2} = -\sqrt{\frac{21}{4}}$$

$$x - \frac{1}{2} = \frac{\sqrt{21}}{2} \quad or \quad x - \frac{1}{2} = -\frac{\sqrt{21}}{2}$$

$$x = \frac{1}{2} + \frac{\sqrt{21}}{2} \quad or \quad x = \frac{1}{2} - \frac{\sqrt{21}}{2}$$

The solutions are $\frac{1}{2} \pm \frac{\sqrt{21}}{2}$.

11. $x^2 - 10x + 25 = 0$

$a = 1, b = -10, c = 25$

We compute the discriminant.

$$b^2 - 4ac = (-10)^2 - 4 \cdot 1 \cdot 25$$

$$= 100 - 100$$

$$= 0$$

Since $b^2 - 4ac = 0$, there is only one solution, and it is a real number. Since there is only one solution, there is only one x-intercept.

12. $x^2 - 11 = 0$, or $x^2 + 0x - 11 = 0$

$a = 1, b = 0, c = -11$

We compute the discriminant.

$$b^2 - 4ac = 0^2 - 4 \cdot 1 \cdot (-11)$$

$$= 0 + 44$$

$$= 44$$

Since $b^2 - 4ac > 0$, there are two real solutions and two x-intercepts.

13.
$$y^2 = \frac{1}{3}y - \frac{4}{7}$$

$$y^2 - \frac{1}{3}y + \frac{4}{7} = 0 \qquad \text{Standard form}$$

$$a = 1, b = -\frac{1}{3}, c = \frac{4}{7}$$

We compute the discriminant.

$$b^2 - 4ac = \left(-\frac{1}{3}\right)^2 - 4 \cdot 1 \cdot \frac{4}{7}$$

$$= \frac{1}{9} - \frac{16}{7} = \frac{7}{63} - \frac{144}{63}$$

$$= -\frac{137}{63}$$

Since $b^2 - 4ac < 0$, there are two nonreal solutions. There are no x-intercepts.

14. $x^2 + 5x + 9 = 0$

$a = 1, b = 5, c = 9$

We compute the discriminant.

$$b^2 - 4ac = 5^2 - 4 \cdot 1 \cdot 9$$

$$= 25 - 36$$

$$= -9$$

Since $b^2 - 4ac < 0$, there are two nonreal solutions. There are no x-intercepts.

15.
$$x^2 - 4 = 2x$$

$$x^2 - 2x - 4 = 0$$

$a = 1, b = -2, c = -4$

We compute the discriminant.

$$b^2 - 4ac = (-2)^2 - 4 \cdot 1 \cdot (-4)$$

$$= 4 + 16$$

$$= 20$$

Since $b^2 - 4ac > 0$, there are two real solutions and two x-intercepts.

16. $x^2 - 8x = 0$, or $x^2 - 8x + 0 = 0$

$a = 1, b = -8, c = 0$

We compute the discriminant.

$$b^2 - 4ac = (-8)^2 - 4 \cdot 1 \cdot 0$$

$$= 64 - 0$$

$$= 64$$

Since $b^2 - 4ac > 0$, there are two real solutions and two x-intercepts.

17. The solutions are -1 and 10.

$$x = -1 \ \text{ or } \qquad x = 10$$
$$x + 1 = 0 \quad \text{ or } \ x - 10 = 0$$
$$(x+1)(x-10) = 0$$
$$x^2 - 9x - 10 = 0$$

18. The solutions are -13 and 13.

$$x = -13 \ \text{ or } \qquad x = 13$$
$$x + 13 = 0 \quad \text{ or } \ x - 13 = 0$$
$$(x+13)(x-13) = 0$$
$$x^2 - 169 = 0$$

19. The solutions are $-\sqrt{5}$ and $3\sqrt{5}$.

$$x = -\sqrt{5} \ \text{ or } \qquad x = 3\sqrt{5}$$
$$x + \sqrt{5} = 0 \quad \text{ or } \ x - 3\sqrt{5} = 0$$
$$(x+\sqrt{5})(x-3\sqrt{5}) = 0$$
$$x^2 - 3x\sqrt{5} + x\sqrt{5} - 3 \cdot 5 = 0$$
$$x^2 - 2x\sqrt{5} - 15 = 0$$

20. The solutions are $-4i$ and $4i$.

$$x = -4i \ \text{ or } \qquad x = 4i$$
$$x + 4i = 0 \quad \text{ or } \ x - 4i = 0$$
$$(x+4i)(x-4i) = 0$$
$$x^2 - 16i^2 = 0$$
$$x^2 - 16(-1) = 0$$
$$x^2 + 16 = 0$$

21. There is only one solution, -6. It must be a double solution.

$$x = -6 \ \text{ or } \qquad x = -6$$
$$x + 6 = 0 \quad \text{ or } \ x + 6 = 0$$
$$(x+6)(x+6) = 0$$
$$x^2 + 12x + 36 = 0$$

22. The solutions are $-\dfrac{4}{3}$ and $\dfrac{2}{7}$.

$$x = -\frac{4}{3} \ \text{ or } \qquad x = \frac{2}{7}$$
$$3x = -4 \ \text{ or } \qquad 7x = 2$$
$$3x + 4 = 0 \quad \text{ or } \ 7x - 2 = 0$$
$$(3x+4)(7x-2) = 0$$
$$21x^2 + 22x - 8 = 0$$

23. *Familiarize*. Let $r =$ the speed, in mph, and let $t =$ the time of the trip, in hr, at the speed r. We organize the information in a table.

Trip	Distance	Speed	Time
At actual speed	780	r	t
5 mph faster	780	$r+5$	$t-1$

Translate. Using $t = \dfrac{d}{r}$ in each row of the table, we have

$$t = \frac{780}{r} \text{ and } t - 1 = \frac{780}{r+5}.$$

Solve. We substitute $\dfrac{780}{r}$ for t in the second equation and solve for r.

$$\frac{780}{r} - 1 = \frac{780}{r+5},$$
$$\text{LCM is } r(r+5)$$
$$r(r+5)\left(\frac{780}{r} - 1\right) = r(r+5) \cdot \frac{780}{r+5}$$
$$780(r+5) - r(r+5) = 780r$$
$$780r + 3900 - r^2 - 5r = 780r$$
$$0 = r^2 + 5r - 3900$$
$$0 = (r+65)(r-60)$$
$$r + 65 = 0 \quad \text{ or } \ r - 60 = 0$$
$$r = -65 \ \text{ or } \qquad r = 60$$

Check. Since the speed cannot be negative, we check only 60. If $r = 60$, then the trip takes $\dfrac{780}{60}$, or 13 hr. At $r+5$, or $60+5$, or 65, the trip takes $\dfrac{780}{65}$, or 12 hr. The trip would take 1 hr less at the faster speed, so the answer checks.

State. Jacob's speed was 60 mph.

24.
$$R = as^2$$
$$\frac{R}{a} = s^2$$
$$\pm\sqrt{\frac{R}{a}} = s$$

In many applications, we might want to use only the positive square root, so we would have $s = \sqrt{\dfrac{R}{a}}$.

25.
$$3x^2 + x = 4$$
$$3x^2 + x - 4 = 0$$
$$(3x+4)(x-1) = 0$$
$$3x + 4 = 0 \quad \text{ or } \ x - 1 = 0$$
$$3x = -4 \ \text{ or } \qquad x = 1$$
$$x = -\frac{4}{3} \ \text{ or } \qquad x = 1$$

The solutions are $-\dfrac{4}{3}$ and 1.

26. $x^4 - 8x^2 + 15 = 0$

Let $u = x^2$.

$$u^2 - 8u + 15 = 0$$
$$(u-3)(u-5) = 0$$
$$u - 3 = 0 \qquad \text{ or } \ u - 5 = 0$$
$$u = 3 \qquad \text{ or } \qquad u = 5$$
$$x^2 = 3 \qquad \text{ or } \qquad x^2 = 5$$
$$x = \pm\sqrt{3} \ \text{ or } \qquad x = \pm\sqrt{5}$$

All four numbers check. They are the solutions.

27.
$$4x^2 = 15x - 5$$
$$4x^2 - 15x + 5 = 0$$
$$a = 4, \ b = -15, \ c = 5$$
$$x = \frac{-b \pm \sqrt{b^2 - 4ac}}{2a}$$
$$x = \frac{-(-15) \pm \sqrt{(-15)^2 - 4 \cdot 4 \cdot 5}}{2 \cdot 4} = \frac{15 \pm \sqrt{225 - 80}}{8}$$
$$x = \frac{15 \pm \sqrt{145}}{8}$$
The solutions are $\dfrac{15 \pm \sqrt{145}}{8}$.

28.
$$7x^2 + 2 = -9x$$
$$7x^2 + 9x + 2 = 0$$
$$(7x + 2)(x + 1) = 0$$
$$7x + 2 = 0 \quad or \quad x + 1 = 0$$
$$7x = -2 \quad or \qquad x = -1$$
$$x = -\frac{2}{7} \quad or \qquad x = -1$$
The solutions are $-\dfrac{2}{7}$ and -1.

29.
$$2x + x(x - 1) = 0$$
$$2x + x^2 - x = 0$$
$$x^2 + x - 0$$
$$x(x + 1) = 0$$
$$x = 0 \ or \ x + 1 = 0$$
$$x = 0 \ or \qquad x = -1$$
The solutions are 0 and -1.

30. $(x + 3)^2 = 64$
$$x + 3 = 8 \ or \ x + 3 = -8$$
$$x = 5 \ or \qquad x = -11$$
The solutions are 5 and -11.

31. $49x^2 + 16 = 0$
$$49x^2 = -16$$
$$x^2 = -\frac{16}{49}$$
$$x = \sqrt{-\frac{16}{49}} \ or \ x = -\sqrt{-\frac{16}{49}}$$
$$x = \frac{4}{7}i \qquad or \ x = -\frac{4}{7}i$$
The solutions are $\pm\dfrac{4}{7}i$.

32. $(x^2 - 2)^2 + 2(x^2 - 2) - 24 = 0$
Let $u = x^2 - 2$.
$$u^2 + 2u - 24 = 0$$
$$(u + 6)(u - 4) = 0$$
$$u + 6 = 0 \quad or \quad u - 4 = 0$$
$$u = -6 \ or \qquad u = 4$$
$$x^2 - 2 = -6 \ or \ x^2 - 2 = 4$$
$$x^2 = -4 \ or \qquad x^2 = 6$$

$$x = \sqrt{-4} \ or \ x = -\sqrt{-4} \ or \ x = \sqrt{6} \ or \ x = -\sqrt{6}$$
$$x = 2i \quad or \ x = -2i \quad or \ x = \sqrt{6} \ or \ x = -\sqrt{6}$$
The solutions are $\pm 2i$ and $\pm\sqrt{6}$.

33.
$$r^2 + 5r = 12$$
$$r^2 + 5r - 12 = 0$$
$$a = 1, \ b = 5, \ c = -12$$
$$r = \frac{-b \pm \sqrt{b^2 - 4ac}}{2a} = \frac{-5 \pm \sqrt{5^2 - 4 \cdot 1 \cdot (-12)}}{2 \cdot 1}$$
$$r = \frac{-5 \pm \sqrt{25 + 48}}{2} = \frac{-5 \pm \sqrt{73}}{2}$$
The solutions are $\dfrac{-5 \pm \sqrt{73}}{2}$.

34. $s^2 + 12s + 37 = 0$
$$a = 1, \ b = 12, \ c = 37$$
$$s = \frac{-b \pm \sqrt{b^2 - 4ac}}{2a} = \frac{-12 \pm \sqrt{12^2 - 4 \cdot 1 \cdot 37}}{2 \cdot 1}$$
$$s = \frac{-12 \pm \sqrt{144 - 148}}{2} = \frac{-12 \pm \sqrt{-4}}{2}$$
$$s = \frac{-12 \pm 2i}{2} = \frac{2(-6 \pm i)}{2} = -6 \pm i$$
The solutions are $-6 \pm i$.

35. $\left(x - \dfrac{5}{2}\right)^2 = \dfrac{11}{4}$
$$x - \frac{5}{2} = \sqrt{\frac{11}{4}} \qquad or \ x - \frac{5}{2} = -\sqrt{\frac{11}{4}}$$
$$x - \frac{5}{2} = \frac{\sqrt{11}}{2} \qquad or \ x - \frac{5}{2} = \ \frac{\sqrt{11}}{2}$$
$$x = \frac{5}{2} + \frac{\sqrt{11}}{2} \ or \qquad x = \frac{5}{2} - \frac{\sqrt{11}}{2}$$
The solutions are $\dfrac{5}{2} \pm \dfrac{\sqrt{11}}{2}$, or $\dfrac{5 \pm \sqrt{11}}{2}$.

36.
$$x + \frac{1}{x} = \frac{7}{3}, \text{LCM is 3x}$$
$$3x\left(x + \frac{1}{x}\right) = 3x \cdot \frac{7}{3}$$
$$3x^2 + 3 = 7x$$
$$3x^2 - 7x + 3 = 0$$
$$a = 3, \ b = -7, \ c = 3$$
$$x = \frac{-b \pm \sqrt{b^2 - 4ac}}{2a} = \frac{-(-7) \pm \sqrt{(-7)^2 - 4 \cdot 3 \cdot 3}}{2 \cdot 3}$$
$$x = \frac{7 + \sqrt{49 - 36}}{6} = \frac{7 \pm \sqrt{13}}{6}$$
The solutions are $\dfrac{7 \pm \sqrt{13}}{6}$.

37. $4x + 1 = 4x^2$

$\qquad 0 = 4x^2 - 4x - 1$

$a = 4,\ b = -4,\ c = -1$

$x = \dfrac{-b \pm \sqrt{b^2 - 4ac}}{2a} = \dfrac{-(-4) \pm \sqrt{(-4)^2 - 4 \cdot 4 \cdot (-1)}}{2 \cdot 4}$

$x = \dfrac{4 \pm \sqrt{16 + 16}}{8} = \dfrac{4 \pm \sqrt{32}}{8}$

$x = \dfrac{4 \pm \sqrt{16 \cdot 2}}{8} = \dfrac{4 \pm 4\sqrt{2}}{8} = \dfrac{4(1 \pm \sqrt{2})}{4 \cdot 2}$

$x = \dfrac{1 \pm \sqrt{2}}{2}$

The solutions are $\dfrac{1 \pm \sqrt{2}}{2}$.

38. $\qquad\qquad (x - 3)^2 + (x + 5)^2 = 0$

$x^2 - 6x + 9 + x^2 + 10x + 25 = 0$

$\qquad\qquad\qquad\quad 2x^2 + 4x + 34 = 0$

$\qquad\qquad\qquad\quad\ x^2 + 2x + 17 = 0 \qquad \text{Dividing by 2}$

$a = 1,\ b = 2,\ c = 17$

$x = \dfrac{-b \pm \sqrt{b^2 - 4ac}}{2a} = \dfrac{-2 \pm \sqrt{2^2 - 4 \cdot 1 \cdot 17}}{2 \cdot 1}$

$x = \dfrac{-2 \pm \sqrt{4 - 68}}{2} = \dfrac{-2 \pm \sqrt{-64}}{2}$

$x = \dfrac{-2 \pm 8i}{2} = \dfrac{2(-1 \pm 4i)}{2} = -1 \pm 4i$

The solutions are $-1 \pm 4i$.

39. $\quad b^2 - 16b + 64 = 3$

$\qquad\quad (b - 8)^2 = 3$

$b - 8 = \sqrt{3} \quad or \quad b - 8 = -\sqrt{3}$

$\quad b = 8 + \sqrt{3} \ or \qquad b = 8 - \sqrt{3}$

The solutions are $8 \pm \sqrt{3}$.

40. $(x - 3)^2 = -10$

$x - 3 = \sqrt{-10} \quad or \quad x - 3 = -\sqrt{-10}$

$x - 3 = \sqrt{10}i \quad or \quad x - 3 = -\sqrt{10}i$

$\quad x = 3 + \sqrt{10}i \ or \qquad x = 3 - \sqrt{10}i$

The solutions are $3 \pm \sqrt{10}i$.

41. $\qquad\qquad \dfrac{1}{x} + \dfrac{1}{x + 2} = \dfrac{1}{5}$,

$\qquad\qquad\qquad\qquad\text{LCM is } 5x(x + 2)$

$5x(x + 2)\left(\dfrac{1}{x} + \dfrac{1}{x + 2}\right) = 5x(x + 2) \cdot \dfrac{1}{5}$

$\qquad\qquad 5(x + 2) + 5x = x(x + 2)$

$\qquad\qquad 5x + 10 + 5x = x^2 + 2x$

$\qquad\qquad\quad 10x + 10 = x^2 + 2x$

$\qquad\qquad\qquad\qquad 0 = x^2 - 8x - 10$

$a = 1,\ b = -8,\ c = -10$

$x = \dfrac{-b \pm \sqrt{b^2 - 4ac}}{2a} = \dfrac{-(-8) \pm \sqrt{(-8)^2 - 4 \cdot 1 \cdot (-10)}}{2 \cdot 1}$

$x = \dfrac{8 \pm \sqrt{64 + 40}}{2} = \dfrac{8 \pm \sqrt{104}}{2} = \dfrac{8 \pm \sqrt{4 \cdot 26}}{2}$

$x = \dfrac{8 \pm 2\sqrt{26}}{2} = \dfrac{2(4 \pm \sqrt{26})}{2} = 4 \pm \sqrt{26}$

The solutions are $4 \pm \sqrt{26}$.

42. $x - \sqrt{x} - 6 = 0$

Let $u = \sqrt{x}$.

$\qquad u^2 - u - 6 = 0$

$(u - 3)(u + 2) = 0$

$u - 3 = 0 \quad or \quad u + 2 = 0$

$\quad u = 3 \quad or \qquad u = -2$

$\sqrt{x} = 3 \quad or \quad \sqrt{x} = -2$

$(\sqrt{x})^2 = 3^2 \quad or \quad (\sqrt{x})^2 = (-2)^2$

$\qquad x = 9 \quad or \qquad x = 4$

Only 9 checks. It is the solution.

43. Given the solutions of a quadratic equation, it is possible to find an equation equivalent to the original equation but not necessarily expressed in the same form as the original equation. For example, we can find a quadratic equation with solutions -2 and 4:

$\qquad [x - (-2)](x - 4) = 0$

$\qquad\quad (x + 2)(x - 4) = 0$

$\qquad\qquad x^2 - 2x - 8 = 0$

Now $x^2 - 2x - 8 = 0$ has solutions -2 and 4. However, the original equation might have been in another form, such as $2x(x - 3) - x(x - 4) = 8$.

44. Given the quadratic equation $ax^2 + bx + c = 0$, we find

$x = \dfrac{-b + \sqrt{b^2 - 4ac}}{2a} \ \text{or} \ x = \dfrac{-b - \sqrt{b^2 - 4ac}}{2a}$ using the quadratic formula. Then we have $ax^2 + bx + c =$

$\left(x - \dfrac{-b + \sqrt{b^2 - 4ac}}{2a}\right)\left(x - \dfrac{-b - \sqrt{b^2 - 4ac}}{2a}\right)$.

Consider $5x^2 + 8x - 3$. First use the quadratic formula to solve $5x^2 + 8x - 3 = 0$.

$x = \dfrac{-8 \pm \sqrt{8^2 - 4 \cdot 5 \cdot (-3)}}{2 \cdot 5}$

$x = \dfrac{-8 \pm \sqrt{124}}{10} = \dfrac{-8 \pm 2\sqrt{31}}{10}$

$x = \dfrac{-4 \pm \sqrt{31}}{5}$

Then $5x^2 + 8x - 3 =$

$\left(x - \dfrac{-4 - \sqrt{31}}{5}\right)\left(x - \dfrac{-4 + \sqrt{31}}{5}\right)$.

45. Set the product

$(x - 1)(x - 2)(x - 3)(x - 4)(x - 5)(x - 6)(x - 7)$

equal to 0.

46. Write an equation of the form $a(3x^2+1)^2+b(3x^2+1)+c = 0$ where $a \neq 0$. To ensure that this equation has real-number solutions select a, b, and c so that $b^2 - 4ac \geq 0$ and $3x^2 + 1 \geq 0$.

Exercise Set 7.5

RC1. False; see pages 617, 618, and 619 in the text.

RC3. True; see page 619 in the text.

1. $f(x) = 4x^2$

$f(x) = 4x^2$ is of the form $f(x) = ax^2$. Thus we know that the vertex is $(0,0)$ and $x = 0$ is the line of symmetry.

We know that $f(x) = 0$ when $x = 0$ since the vertex is $(0,0)$.

For $x = 1$, $f(x) = 4x^2 = 4 \cdot 1^2 = 4$.

For $x = -1$, $f(x) = 4x^2 = 4 \cdot (-1)^2 = 4$.

For $x = 2$, $f(x) = 4x^2 = 4 \cdot 2^2 = 16$.

For $x = -2$, $f(x) = 4x^2 = 4 \cdot (-2)^2 = 16$.

We complete the table.

x	$f(x)$	
0	0	← Vertex
1	4	
2	16	
−1	4	
−2	16	

We plot the ordered pairs $(x, f(x))$ from the table and connect them with a smooth curve.

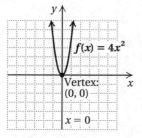

3. $f(x) = \dfrac{1}{3}x^2$ is of the form $f(x) = ax^2$. Thus we know that the vertex is $(0,0)$ and $x = 0$ is the line of symmetry. We choose some numbers for x and find the corresponding values for $f(x)$. Then we plot the ordered pairs $(x, f(x))$ and connect them with a smooth curve.

For $x = 1$, $f(x) = \dfrac{1}{3}x^2 = \dfrac{1}{3} \cdot 1^2 = \dfrac{1}{3}$.

For $x = 3$, $f(x) = \dfrac{1}{3}x^2 = \dfrac{1}{3} \cdot 3^2 = 3$.

For $x = -1$, $f(x) = \dfrac{1}{3}x^2 = \dfrac{1}{3} \cdot (-1)^2 = \dfrac{1}{3}$.

For $x = -3$, $f(x) = \dfrac{1}{3}x^2 = \dfrac{1}{3} \cdot (-3)^2 = 3$.

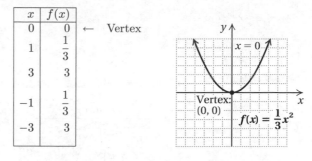

x	$f(x)$	
0	0	← Vertex
1	$\dfrac{1}{3}$	
3	3	
−1	$\dfrac{1}{3}$	
−3	3	

5. $f(x) = (x + 3)^2 = [x - (-3)]^2$ is of the form $f(x) = a(x - h)^2$.

Thus we know that the vertex is $(-3, 0)$ and $x = -3$ is the line of symmetry. We choose some numbers for x and find the corresponding values for $f(x)$. Then we plot the ordered pairs $(x, f(x))$ and connect them with a smooth curve.

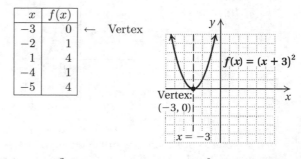

x	$f(x)$	
−3	0	← Vertex
−2	1	
1	4	
−4	1	
−5	4	

7. $f(x) = -4x^2$ is of the form $f(x) = ax^2$. Thus we know that the vertex is $(0,0)$ and $x = 0$ is the line of symmetry. We choose some numbers for x and find the corresponding values for $f(x)$. Then we plot the ordered pairs $(x, f(x))$ and connect them with a smooth curve.

For $x = 1$, $f(x) = -4x^2 = -4 \cdot 1^2 = -4$.

For $x = -1$, $f(x) = -4x^2 = -4 \cdot (-1)^2 = -4$.

For $x = 2$, $f(x) = -4x^2 = -4 \cdot 2^2 = -16$.

For $x = -2$, $f(x) = -4x^2 = -4 \cdot (-2)^2 = -16$.

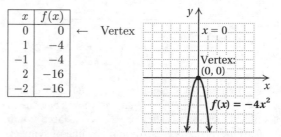

x	$f(x)$	
0	0	← Vertex
1	−4	
−1	−4	
2	−16	
−2	−16	

9. $f(x) = -\dfrac{1}{2}x^2$ is of the form $f(x) = ax^2$. Thus we know that the vertex is $(0,0)$ and $x = 0$ is the line of symmetry. We choose some numbers for x and find the corresponding values for $f(x)$. Then we plot the ordered pairs $(x, f(x))$ and connect them with a smooth curve.

For $x = 2$, $f(x) = -\dfrac{1}{2}x^2 = -\dfrac{1}{2} \cdot 2^2 = -2$.

For $x = -2$, $f(x) = -\dfrac{1}{2}x^2 = -\dfrac{1}{2} \cdot (-2)^2 = -2$.

For $x = 4$, $f(x) = -\dfrac{1}{2}x^2 = -\dfrac{1}{2} \cdot 4^2 = -8$.

For $x = -4$, $f(x) = -\dfrac{1}{2}x^2 = -\dfrac{1}{2} \cdot (-4)^2 = -8$.

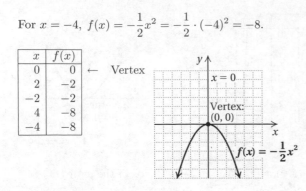

x	$f(x)$	
0	0	← Vertex
2	−2	
−2	−2	
4	−8	
−4	−8	

11. Graph: $f(x) = 2(x - 4)^2$

We choose some values of x and compute $f(x)$. Then we plot these ordered pairs and connect them with a smooth curve.

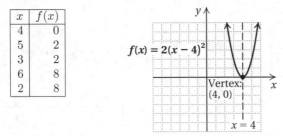

x	$f(x)$
4	0
5	2
3	2
6	8
2	8

The graph of $f(x) = 2(x - 4)^2$ looks like the graph of $f(x) = 2x^2$ except that it is translated 4 units to the right. The vertex is $(4, 0)$, and the line of symmetry is $x = 4$.

13. Graph: $f(x) = -2(x + 2)^2$

We choose some values of x and compute $f(x)$. Then we plot these ordered pairs and connect them with a smooth curve.

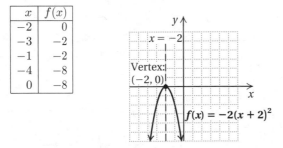

x	$f(x)$
−2	0
−3	−2
−1	−2
−4	−8
0	−8

We can express the equation in the equivalent form $f(x) = -2[x - (-2)]^2$. Then we know that the graph looks like the graph of $f(x) = -2x^2$ translated 2 units to the left. The vertex is $(-2, 0)$, and the line of symmetry is $x = -2$.

15. Graph: $f(x) = 3(x - 1)^2$

We choose some values of x and compute $f(x)$. Then we plot these ordered pairs and connect them with a smooth curve.

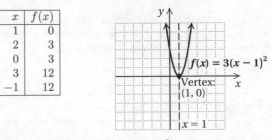

x	$f(x)$
1	0
2	3
0	3
3	12
−1	12

The graph of $f(x) = 3(x - 1)^2$ looks like the graph of $f(x) = 3x^2$ except that it is translated 1 unit to the right. The vertex is $(1, 0)$, and the line of symmetry is $x = 1$.

17. Graph: $f(x) = -\dfrac{3}{2}(x + 2)^2$

We choose some values of x and compute $f(x)$. Then we plot these ordered pairs and connect them with a smooth curve.

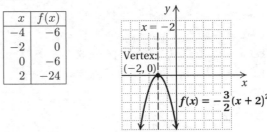

x	$f(x)$
−4	−6
−2	0
0	−6
2	−24

We can express the equation in the equivalent form $f(x) = -\dfrac{3}{2}[x - (-2)]^2$. Then we know that the graph looks like the graph of $f(x) = -\dfrac{3}{2}x^2$ translated 2 units to the left. The vertex is $(-2, 0)$, and the line of symmetry is $x = -2$.

19. Graph: $f(x) = (x - 3)^2 + 1$

We choose some values of x and compute $f(x)$. Then we plot these ordered pairs and connect them with a smooth curve.

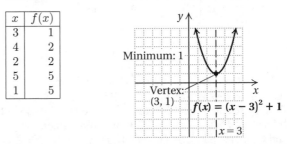

x	$f(x)$
3	1
4	2
2	2
5	5
1	5

The graph of $f(x) = (x - 3)^2 + 1$ looks like the graph of $f(x) = x^2$ except that it is translated 3 units right and 1 unit up. The vertex is $(3, 1)$, and the line of symmetry is $x = 3$. The equation is of the form $f(x) = a(x - h)^2 + k$ with $a = 1$. Since $1 > 0$, we know that 1 is the minimum value.

21. Graph: $f(x) = -3(x + 4)^2 + 1$
$$f(x) = -3[x - (-4)]^2 + 1$$

We choose some values of x and compute $f(x)$. Then we plot these ordered pairs and connect them with a smooth curve.

x	$f(x)$
-4	1
$-3\dfrac{1}{2}$	$\dfrac{1}{4}$
$-4\dfrac{1}{2}$	$\dfrac{1}{4}$
-3	-2
-5	-2
-2	-11
-6	-11

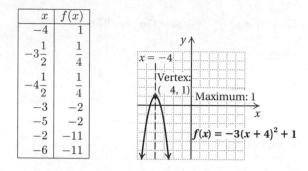

The graph of $f(x) = -3(x+4)^2 + 1$ looks like the graph of $f(x) = 3x^2$ except that it is translated 4 units left and 1 unit up and opens downward. The vertex is $(-4, 1)$, and the line of symmetry is $x = -4$. Since $-3 < 0$, we know that 1 is the maximum value.

23. Graph: $f(x) = \dfrac{1}{2}(x+1)^2 + 4$

$$f(x) = \dfrac{1}{2}[x - (-1)]^2 + 4$$

We choose some values of x and compute $f(x)$. Then we plot these ordered pairs and connect them with a smooth curve.

x	$f(x)$
1	6
2	$8\dfrac{1}{2}$
0	$4\dfrac{1}{2}$
-1	4
-2	$4\dfrac{1}{2}$
-3	6

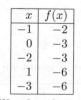

The graph of $f(x) = \dfrac{1}{2}(x+1)^2 + 4$ looks like the graph of $f(x) = \dfrac{1}{2}x^2$ except that it is translated 1 unit left and 4 units up. The vertex is $(-1, 4)$, and the line of symmetry is $x = -1$. Since $\dfrac{1}{2} > 0$, we know that 4 is the minimum value.

25. Graph: $f(x) = -(x+1)^2 - 2$
$$f(x) = -[x - (-1)]^2 + (-2)$$

We choose some values of x and compute $f(x)$.

x	$f(x)$
-1	-2
0	-3
-2	-3
1	-6
-3	-6

We plot these ordered pairs and connect them with a smooth curve.

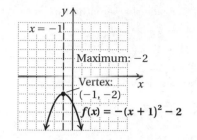

The graph of $f(x) = -(x+1)^2 - 2$ looks like the graph of $f(x) = x^2$ except that it is translated 1 unit left and 2 units down and opens downward. The vertex is $(-1, -2)$, and the line of symmetry is $x = -1$. Since $-1 < 0$, we know that -2 is the maximum value.

27. $m = \dfrac{y_2 - y_1}{x_2 - x_1} = \dfrac{-6 - 0}{5 - (-10)} = \dfrac{-6}{15} = -\dfrac{2}{5}$

Exercise Set 7.6

RC1. The graph of $f(x) = 5x^2 - 10x - 3$ opens <u>upward</u> (because the coefficient of x^2 is positive.)

RC3. $g(x) = (x+3)^2 - 2 = [x - (-3)]^2 + (-2)$, so the graph of $g(x) = (x+3)^2 - 2$ has its <u>vertex</u> at $(-3, -2)$.

1. $f(x) = x^2 - 2x - 3 = (x^2 - 2x) - 3$

We complete the square inside the parentheses. We take half the x-coefficient and square it.

$\dfrac{1}{2} \cdot (-2) = -1$ and $(-1)^2 = 1$

Then we add $1 - 1$ inside the parentheses.

$$
\begin{aligned}
f(x) &= (x^2 - 2x + 1 - 1) - 3 \\
&= (x^2 - 2x + 1) - 1 - 3 \\
&= (x - 1)^2 - 4 \\
&= (x - 1)^2 + (-4)
\end{aligned}
$$

Vertex: $(1, -4)$

Line of symmetry: $x = 1$

The coefficient of x^2 is 1, which is positive, so the graph opens up. This tells us that -4 is a minimum.

We plot a few points and draw the curve.

x	$f(x)$
1	-4
2	-3
0	-3
3	0
-1	0
4	5
-2	5

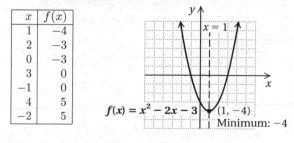

3. $f(x) = -x^2 - 4x - 2 = -(x^2 + 4x) - 2$

We complete the square inside the parentheses. We take half the x-coefficient and square it.

$\dfrac{1}{2} \cdot 4 = 2$ and $2^2 = 4$

Then we add $4 - 4$ inside the parentheses.

$$f(x) = -(x^2 + 4x + 4 - 4) - 2$$
$$= -(x^2 + 4x + 4) + (-1)(-4) - 2$$
$$= -(x+2)^2 + 4 - 2$$
$$= -(x+2)^2 + 2$$
$$= -[x - (-2)]^2 + 2$$

Vertex: $(-2, 2)$

Line of symmetry: $x = -2$

The coefficient of x^2 is -1, which is negative, so the graph opens down. This tells us that 2 is a maximum.

We plot a few points and draw the curve.

x	$f(x)$
-2	2
-4	-2
-3	1
-1	1
0	-2

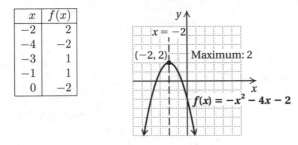

5. $f(x) = 3x^2 - 24x + 50 = 3(x^2 - 8x) + 50$

We complete the square inside the parentheses. We take half the x-coefficient and square it.

$$\frac{1}{2} \cdot (-8) = -4 \text{ and } (-4)^2 = 16$$

Then we add $16 - 16$ inside the parentheses.

$$f(x) = 3(x^2 - 8x + 16 - 16) + 50$$
$$= 3(x^2 - 8x + 16) - 48 + 50$$
$$= 3(x - 4)^2 + 2$$

Vertex: $(4, 2)$

Line of symmetry: $x = 4$

The coefficient of x^2 is 3, which is positive, so the graph opens up. This tells us that 2 is a minimum.

We plot a few points and draw the curve.

x	$f(x)$
4	2
5	5
3	5
6	14
2	14

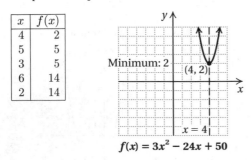

7. $f(x) = -2x^2 - 2x + 3 = -2(x^2 + x) + 3$

We complete the square inside the parentheses. We take half the x-coefficient and square it.

$$\frac{1}{2} \cdot 1 = \frac{1}{2} \text{ and } \left(\frac{1}{2}\right)^2 = \frac{1}{4}$$

Then we add $\frac{1}{4} - \frac{1}{4}$ inside the parentheses.

$$f(x) = -2\left(x^2 + x + \frac{1}{4} - \frac{1}{4}\right) + 3$$
$$= -2\left(x^2 + x + \frac{1}{4}\right) + (-2)\left(-\frac{1}{4}\right) + 3$$
$$= -2\left(x + \frac{1}{2}\right)^2 + \frac{1}{2} + 3$$
$$= -2\left(x + \frac{1}{2}\right)^2 + \frac{7}{2}$$
$$= -2\left[x - \left(-\frac{1}{2}\right)\right]^2 + \frac{7}{2}$$

Vertex: $\left(-\frac{1}{2}, \frac{7}{2}\right)$

Line of symmetry: $x = -\frac{1}{2}$

The coefficient of x^2 is -2, which is negative, so the graph opens down. This tells us that $\frac{7}{2}$ is a maximum.

We plot a few points and draw the curve.

x	$f(x)$
$-\frac{1}{2}$	$\frac{7}{2}$
-2	-1
-1	3
0	3
1	-1

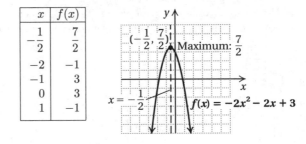

9. $f(x) = 5 - x^2 = -x^2 + 5 = -(x - 0)^2 + 5$

Vertex: $(0, 5)$

Line of symmetry: $x = 0$

The coefficient of x^2 is -1, which is negative, so the graph opens down. This tells us that 5 is a maximum.

We plot a few points and draw the curve.

x	$f(x)$
0	5
1	4
-1	4
2	1
-2	1
3	-4
-3	-4

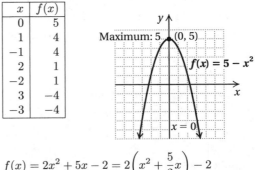

11. $f(x) = 2x^2 + 5x - 2 = 2\left(x^2 + \frac{5}{2}x\right) - 2$

We complete the square inside the parentheses. We take half the x-coefficient and square it.

$$\frac{1}{2} \cdot \frac{5}{2} = \frac{5}{4} \text{ and } \left(\frac{5}{4}\right)^2 = \frac{25}{16}$$

Then we add $\frac{25}{16} - \frac{25}{16}$ inside the parentheses.

$$f(x) = 2\left(x^2 + \frac{5}{2}x + \frac{25}{16} - \frac{25}{16}\right) - 2$$

$$= 2\left(x^2 + \frac{5}{2}x + \frac{25}{16}\right) + 2\left(-\frac{25}{16}\right) - 2$$

$$= 2\left(x + \frac{5}{4}\right)^2 - \frac{25}{8} - 2$$

$$= 2\left(x + \frac{5}{4}\right)^2 - \frac{41}{8}$$

$$= 2\left[x - \left(-\frac{5}{4}\right)\right]^2 + \left(-\frac{41}{8}\right)$$

Vertex: $\left(-\frac{5}{4}, -\frac{41}{8}\right)$

Line of symmetry: $x = -\frac{5}{4}$

The coefficient of x^2 is 2, which is positive, so the graph opens up. This tells us that $-\frac{41}{8}$ is a minimum.

We plot a few points and draw the curve.

x	$f(x)$
$-\frac{5}{4}$	$-\frac{41}{8}$
-3	1
-2	-4
-1	-5
0	-2
1	5

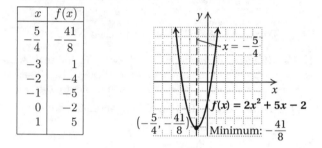

13. $f(x) = x^2 - 6x + 1$

The y-intercept is $(0, f(0))$. Since $f(0) = 0^2 - 6 \cdot 0 + 1 = 1$, the y-intercept is $(0, 1)$.

To find the x-intercepts, we solve $x^2 - 6x + 1 = 0$. Using the quadratic formula gives us $x = 3 \pm 2\sqrt{2}$.

Thus, the x-intercepts are $(3 - 2\sqrt{2}, 0)$ and $(3 + 2\sqrt{2}, 0)$, or approximately $(0.172, 0)$ and $(5.828, 0)$.

15. $f(x) = -x^2 + x + 20$

The y-intercept is $(0, f(0))$. Since $f(0) = -0^2 + 0 + 20 = 20$, the y-intercept is $(0, 20)$.

To find the x-intercepts, we solve $-x^2 + x + 20 = 0$. Factoring and using the principle of zero products gives us $x = -4$ or $x = 5$. Thus, the x-intercepts are $(-4, 0)$ and $(5, 0)$.

17. $f(x) = 4x^2 + 12x + 9$

The y-intercept is $(0, f(0))$. Since $f(0) = 4 \cdot 0^2 + 12 \cdot 0 + 9 = 9$, the y-intercept is $(0, 9)$.

To find the x-intercepts, we solve $4x^2 + 12x + 9 = 0$. Factoring and using the principle of zero products gives us $x = -\frac{3}{2}$. Thus, the x-intercept is $\left(-\frac{3}{2}, 0\right)$.

19. $f(x) = 4x^2 - x + 8$

The y-intercept is $(0, f(0))$. Since $f(0) = 4 \cdot 0^2 - 0 + 8 = 8$, the y-intercept is $(0, 8)$.

To find the x-intercepts, we solve $4x^2 - x + 8 = 0$. Using the quadratic formula gives us $x = \dfrac{1 \pm i\sqrt{127}}{8}$. Since there are no real-number solutions, there are no x-intercepts.

21.
$$D = kw$$
$$420 = k \cdot 28$$
$$\frac{420}{28} = k$$
$$15 = k$$

The equation of variation is $D = 15w$.

23.
$$y = \frac{k}{x}$$
$$125 = \frac{k}{2}$$
$$250 = k \qquad \text{Variation constant}$$
$$y = \frac{250}{x} \qquad \text{Equation of variation}$$

25.
$$y = kx$$
$$125 = k \cdot 2$$
$$\frac{125}{2} = k \qquad \text{Variation constant}$$
$$y = \frac{125}{2}x \qquad \text{Equation of variation}$$

27. a) Minimum: -6.954

b) Maximum: 7.014

29. $f(x) = |x^2 - 1|$

We plot some points and draw the curve. Note that it will lie entirely on or above the x-axis since absolute value is never negative.

x	$f(x)$
-3	8
-2	3
-1	0
0	1
1	0
2	3
3	8

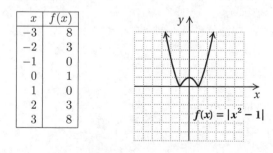

31. $f(x) = |x^2 - 3x - 4|$

We plot some points and draw the curve. Note that it will lie entirely on or above the x-axis since absolute value is never negative.

x	$f(x)$
-4	24
-3	14
-2	6
-1	0
0	4
1	6
2	6
3	4
4	0
5	6
6	14

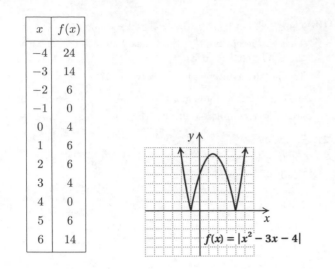

$f(x) = |x^2 - 3x - 4|$

33. The horizontal distance from $(-1, 0)$ to $(3, -5)$ is $|3-(-1)|$, or 4, so by symmetry the other x-intercept is $(3 + 4, 0)$, or $(7, 0)$. Substituting the three ordered pairs $(-1, 0)$, $(3, -5)$, and $(7, 0)$ in the equation $f(x) = ax^2 + bx + c$ yields a system of equations:

$$0 = a(-1)^2 + b(-1) + c,$$
$$-5 = a \cdot 3^2 + b \cdot 3 + c,$$
$$0 = a \cdot 7^2 + b \cdot 7 + c$$

or

$$0 = a - b + c,$$
$$-5 = 9a + 3b + c,$$
$$0 = 49a + 7b + c$$

The solution of this system of equations is $\left(\dfrac{5}{16}, -\dfrac{15}{8}, -\dfrac{35}{16}\right)$, so $f(x) = \dfrac{5}{16}x^2 - \dfrac{15}{8}x - \dfrac{35}{16}$, or $f(x) = \dfrac{5}{16}(x - 3)^2 - 5$.

35. $f(x) = \dfrac{x^2}{8} + \dfrac{x}{4} - \dfrac{3}{8}$

The x-coordinate of the vertex is $-b/2a$:

$$-\frac{b}{2a} = -\frac{\dfrac{1}{4}}{2 \cdot \dfrac{1}{8}} = -\frac{\dfrac{1}{4}}{\dfrac{1}{4}} = -1$$

The second coordinate is $f(-1)$:

$$f(-1) = \frac{(-1)^2}{8} + \frac{-1}{4} - \frac{3}{8}$$
$$= \frac{1}{8} - \frac{1}{4} - \frac{3}{8}$$
$$= -\frac{1}{2}$$

The vertex is $\left(-1, -\dfrac{1}{2}\right)$.

The line of symmetry is $x = -1$.

The coefficient of x^2 is $\dfrac{1}{8}$, which is positive, so the graph opens up. This tells us that $-\dfrac{1}{2}$ is a minimum.

We plot some points and draw the graph.

x	$f(x)$
-5	$\dfrac{3}{2}$
-3	0
-1	$-\dfrac{1}{2}$
0	$-\dfrac{3}{8}$
1	0
3	$\dfrac{3}{2}$
5	4

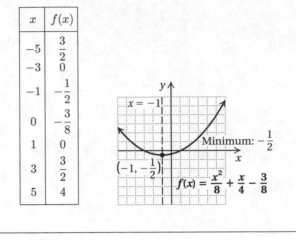

$f(x) = \dfrac{x^2}{8} + \dfrac{x}{4} - \dfrac{3}{8}$

Exercise Set 7.7

RC1. True; we can solve maximum- and minimum-value problems, for example, by finding the coordinates of a vertex.

RC3. True; see pages 637 and 638 in the text.

1. *Familiarize.* Referring to the drawing in the text, we let l = the length of the atrium and w = the width. Then the perimeter of each floor is $2l + 2w$, and the area is $l \cdot w$.

Translate. Using the formula for perimeter we have:

$$2l + 2w = 720$$
$$2l = 720 - 2w$$
$$l = \frac{720 - 2w}{2}$$
$$l = 360 - w$$

Substituting $360 - w$ for l in the formula for area, we get a quadratic function.

$$A = lw = (360 - w)w = 360w - w^2 = -w^2 + 360w$$

Carry out. We complete the square in order to find the vertex of the quadratic function.

$$A = -w^2 + 360w$$
$$= -(w^2 - 360w)$$
$$= -(w^2 - 360w + 32{,}400 - 32{,}400)$$
$$= -(w^2 - 360w + 32{,}400) + (-1)(-32{,}400)$$
$$= -(w - 180)^2 + 32{,}400$$

The vertex is $(180, 32{,}400)$. The coefficient of w^2 is negative, so the graph of the function is a parabola that opens down. This tells us that the function has a maximum value and that value occurs when $w = 180$. When $w = 180$, $l = 360 - w = 360 - 180 = 180$.

Check. We could find the value of the function for some values of w less than 180 and for some values greater than 180, determining that the maximum value we found, 32,400, is larger than these function values. We could also use the graph of the function to check the maximum value. Our answer checks.

State. Floors with dimensions 180 ft by 180 ft will allow an atrium with maximum area.

3. Familiarize. Let x represent the height of the file and y represent the width. We make a drawing.

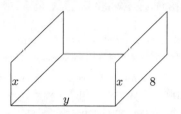

Translate. We have two equations.

$2x + y = 14$

$V = 8xy$

Solve the first equation for y.

$y = 14 - 2x$

Substitute for y in the second equation.

$V = 8x(14 - 2x)$

$V = -16x^2 + 112x$

Carry out. Completing the square, we get

$V = -16\left(x - \frac{7}{2}\right)^2 + 196.$

The maximum function value of 196 occurs when $x = \frac{7}{2}$. When $x = \frac{7}{2}$, $y = 14 - 2 \cdot \frac{7}{2} = 7$.

Check. Check a function value for x less than $\frac{7}{2}$ and for x greater than $\frac{7}{2}$.

$V(3) = -16 \cdot 3^2 + 112 \cdot 3 = 192$

$V(4) = -16 \cdot 4^2 + 112 \cdot 4 = 192$

Since 196 is greater than these numbers, it looks as though we have a maximum.

We could also use the graph of the function to check the maximum value.

State. The file should be $\frac{7}{2}$ in., or 3.5 in., tall.

5. Familiarize and Translate. We want to find the value of x for which $C(x) = 0.1x^2 - 0.7x + 2.425$ is a minimum.

Carry out. We complete the square.

$C(x) = 0.1(x^2 - 7x + 12.25) + 2.425 - 1.225$

$C(x) = 0.1(x - 3.5)^2 + 1.2$

The minimum function value of 1.2 occurs when $x = 3.5$.

Check. Check a function value for x less than 3.5 and for x greater than 3.5.

$C(3) = 0.1(3)^2 - 0.7(3) + 2.425 = 1.225$

$C(4) = 0.1(4)^2 - 0.7(4) + 2.425 = 1.225$

Since 1.2 is less than these numbers, it looks as though we have a minimum.

We could also use the graph of the function to check the minimum value.

State. The shop should build 3.5 hundred, or 350 bicycles.

7. Familiarize. We make a drawing and label it.

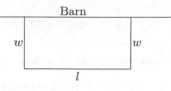

Translate. We have two equations.

$l + 2w = 40$

$A = lw$

Solve the first equation for l.

$l = 40 - 2w$

Substitute for l in the second equation.

$A = (40 - 2w)w = 40w - 2w^2$

$= -2w^2 + 40w$

Carry out. Completing the square, we get

$A = -2(w - 10)^2 + 200$

The maximum function value is 200. It occurs when $w = 10$. When $w - 10$, $l = 40 - 2 \cdot 10 = 20$.

Check. Check a function value for w less than 10 and for w greater than 10.

$A(9) = -2 \cdot 9^2 + 40 \cdot 9 = 198$

$A(11) = -2 \cdot 11^2 + 40 \cdot 11 = 198$

Since 200 is greater than these numbers, it looks as though we have a maximum. We could also use the graph of the function to check the maximum value.

State. The maximum area of 200 ft^2 will occur when the dimensions are 10 ft by 20 ft.

9. Familiarize and Translate. We are given the function $N(x) = -0.4x^2 + 9x + 11$.

Carry out. To find the value of x for which $N(x)$ is a maximum, we first find $-\frac{b}{2a}$:

$-\frac{b}{2a} = -\frac{9}{2(-0.4)} = 11.25$

Now we find the maximum value of the function $N(11.25)$:

$N(11.25) = -0.4(11.25)^2 + 9(11.25) + 11 = 61.625$

Check. We can go over the calculations again. We could also solve the problem again by completing the square. The answer checks.

State. Daily ticket sales will peak 11 days after the concert was announced. About 62 tickets will be sold that day.

11. Find the total profit:

$P(x) = R(x) - C(x)$

$P(x) = (1000x - x^2) - (3000 + 20x)$

$P(x) = -x^2 + 980x - 3000$

To find the maximum value of the total profit and the value of x at which it occurs we complete the square:

$$P(x) = -(x^2 - 980x) - 3000$$
$$= -(x^2 - 980x + 240,100 - 240,100) - 3000$$
$$= -(x^2 - 980x + 240,100) - (-240,100) - 3000$$
$$= -(x - 490)^2 + 237,100$$

The maximum profit of $237,100 occurs at $x = 490$.

13. **Familiarize**. Let x and y represent the numbers.

    Translate. The sum of the numbers is 22, so we have $x + y = 22$. Solving for y, we get $y = 22 - x$. The product of the numbers is xy. Substituting $22 - x$ for y in the product, we get a quadratic function:

    $$P = xy = x(22 - x) = 22x - x^2 = -x^2 + 22x$$

    Carry out. The coefficient of x^2 is negative, so the graph of the function is a parabola that opens down and a maximum exists. We complete the square in order to find the vertex of the quadratic function.

    $$P = -x^2 + 22x$$
    $$= -(x^2 - 22x)$$
    $$= -(x^2 - 22x + 121 - 121)$$
    $$= -(x^2 - 22x + 121) + (-1)(-121)$$
    $$= -(x - 11)^2 + 121$$

    The vertex is $(11, 121)$. This tells us that the maximum product is 121. The maximum occurs when $x = 11$. Note that when $x = 11$, $y = 22 - x = 22 - 11 = 11$, so the numbers that yield the maximum product are 11 and 11.

    Check. We could find the value of the function for some values of x less than 11 and for some greater than 11, determining that the maximum value we found is larger than these function values. We could also use the graph of the function to check the maximum value. Our answer checks.

    State. The maximum product is 121. The numbers 11 and 11 yield this product.

15. **Familiarize**. Let x and y represent the numbers.

    Translate. The difference of the numbers is 4, so we have $x - y = 4$. Solving for x, we get $x = y + 4$. The product of the numbers is xy. Substituting $y + 4$ for x in the product, we get a quadratic function:

    $$P = xy = (y + 4)y = y^2 + 4y$$

    Carry out. The coefficient of y^2 is positive, so the graph of the function opens up and a minimum exists. We complete the square in order to find the vertex of the quadratic function.

    $$P = y^2 + 4y$$
    $$= y^2 + 4y + 4 - 4$$
    $$= (y + 2)^2 - 4$$
    $$= [y - (-2)]^2 + (-4)$$

    The vertex is $(-2, -4)$. This tells us that the minimum product is -4. The minimum occurs when $y = -2$. Note that when $y = -2$, $x = y + 4 = -2 + 4 = 2$, so the numbers that yield the minimum product are 2 and -2.

    Check. We could find the value of the function for some values of y less than -2 and for some greater than -2,

determining that the minimum value we found is smaller than these function values. We could also use the graph of the function to check the minimum value. Our answer checks.

State. The minimum product is -4. The numbers 2 and -2 yield this product.

17. **Familiarize**. We let x and y represent the two numbers, and we let P represent their product.

    Translate. We have two equations.

    $$x + y = -12,$$
    $$P = xy$$

    Solve the first equation for y.

    $$y = -12 - x$$

    Substitute for y in the second equation.

    $$P = x(-12 - x) = -12x - x^2$$
    $$= -x^2 - 12x$$

    Carry out. Completing the square, we get

    $$P = -(x + 6)^2 + 36$$

    The maximum function value is 36. It occurs when $x = -6$. When $x = -6$, $y = -12 - (-6)$, or -6.

    Check. Check a function value for x less than -6 and for x greater than -6.

    $$P(-7) = -(-7)^2 - 12(-7) = 35$$
    $$P(-5) = -(-5)^2 - 12(-5) = 35$$

    Since 36 is greater than these numbers, it looks as though we have a maximum.

    We could also use the graph of the function to check the maximum value.

    State. The maximum product of 36 occurs for the numbers -6 and -6.

19. The data seem to fit a linear function $f(x) = mx + b$.

21. The data fall and then rise in a curved manner fitting a quadratic function $f(x) = ax^2 + bx + c$, $a > 0$.

23. The data fall, then rise, then fall again so they do not fit a linear or a quadratic function but might fit a polynomial function that is neither quadratic nor linear.

25. The data rise and then fall in a curved manner fitting a quadratic function $f(x) = ax^2 + bx + c$, $a < 0$.

27. We look for a function of the form $f(x) = ax^2 + bx + c$. Substituting the data points, we get

    $$4 = a(1)^2 + b(1) + c,$$
    $$-2 = a(-1)^2 + b(-1) + c,$$
    $$13 = a(2)^2 + b(2) + c,$$
    or
    $$4 = a + b + c,$$
    $$-2 = a - b + c,$$
    $$13 = 4a + 2b + c.$$

    Solving this system, we get

    $$a = 2, \ b = 3, \text{ and } c = -1.$$

Therefore the function we are looking for is

$$f(x) = 2x^2 + 3x - 1.$$

29. We look for a function of the form $f(x) = ax^2 + bx + c$. Substituting the data points, we get

$$0 = a(2)^2 + b(2) + c,$$
$$3 = a(4)^2 + b(4) + c,$$
$$-5 = a(12)^2 + b(12) + c,$$

or

$$0 = 4a + 2b + c,$$
$$3 = 16a + 4b + c,$$
$$-5 = 144a + 12b + c.$$

Solving this system, we get

$$a = -\frac{1}{4},\ b = 3,\ c = -5.$$

Therefore the function we are looking for is

$$f(x) = -\frac{1}{4}x^2 + 3x - 5.$$

31. a) We look for a function of the form $A(s) = as^2 + bs + c$, where $A(s)$ represents the number of nighttime accidents (for every 200 million km) and s represents the travel speed (in km/h). We substitute the given values of s and $A(s)$.

$$400 = a(60)^2 + b(60) + c,$$
$$250 = a(80)^2 + b(80) + c,$$
$$250 = a(100)^2 + b(100) + c,$$

or

$$400 = 3600a + 60b + c,$$
$$250 = 6400a + 80b + c,$$
$$250 = 10,000a + 100b + c.$$

Solving the system of equations, we get

$$a = \frac{3}{16},\ b = -\frac{135}{4},\ c = 1750.$$

Thus, the function $A(s) = \frac{3}{16}s^2 - \frac{135}{4}s + 1750$ fits the data.

b) Find $A(50)$.

$$A(50) = \frac{3}{16}(50)^2 - \frac{135}{4}(50) + 1750 = 531.25$$

About 531 accidents per 200,000,000 km driven occur at 50 km/h.

33. We look for a function of the form $D(x) = ax^2 + bx + c$, where $D(x)$ represents the depth of the river, in feet, and x represents the distance from one bank, in feet. We substitute the three given data points.

$$0 = a(0)^2 + b(0) + c,$$
$$20 = a(50)^2 + b(50) + c,$$
$$0 = a(100)^2 + b(100) + c,$$

or

$$0 = c,$$
$$20 = 2500a + 50b + c,$$
$$0 = 10,000a + 100b + c$$

Solving the system of equations, we get $(-0.008, 0.8, 0)$.

Thus, the function $D(x) = -0.008x^2 + 0.8x$ fits the data. Now we find $D(75)$.

$$D(75) = -0.008(75)^2 + 0.8(75) = 15$$

The depth of the river is 15 ft at 75 ft from the bank.

35. a) We look for a function of the form $N(d) = ad^2 + bd + c$. Substituting the data points, we get.

$$12 = a \cdot 6^2 + b \cdot 6 + c$$
$$24 = a \cdot 8^2 + b \cdot 8 + c,$$
$$56 = a \cdot 12^2 + b \cdot 12 + c,$$

or

$$12 = 36a + 6b + c,$$
$$24 = 64a + 8b + c,$$
$$56 = 144a + 12b + c.$$

Solving this system, we get

$$a = \frac{1}{3},\ b = \frac{4}{3},\ \text{and } c = -8.$$

Therefore the function we are looking for is

$$N(d) = \frac{1}{3}d^2 + \frac{4}{3}d - 8.$$

b) $N(9) = \frac{1}{3} \cdot 9^2 + \frac{4}{3} \cdot 9 - 8$

$$= \frac{1}{3} \cdot 81 + \frac{4}{3} \cdot 9 - 8$$
$$= 27 + 12 - 8$$
$$= 31 \text{ servings}$$

37. $(-3x^2 - x - 2) + (x^2 + 3x - 7) = -2x^2 + 2x - 9$

39. $(c^2d + 2y)(c^2d - 2y) = (c^2d)^2 - (2y)^2 = c^4d^2 - 4y^2$

41. $12x^3 - 60x^2 + 75x = 3x(4x^2 - 20x + 25)$
$$= 3x[(2x)^2 - 2(2x)(5) + 5^2]$$
$$= 3x(2x - 5)^2$$

Exercise Set 7.8

RC1. To solve $x^2 - 2 < 0$, we look for intervals for which $f(x) = x^2 - 2$ is <u>negative</u>.

RC3. The inequality $3x < 5 + x^2$ is equivalent to $0 < -3x + 5 + x^2$, or $x^2 - 3x + 5 > 0$. To solve $3x < 5 + x^2$, we look for intervals for which $f(x) = x^2 - 3x + 5$ is <u>positive</u>.

1. $(x - 6)(x + 2) > 0$

The solutions of $(x - 6)(x + 2) = 0$ are 6 and -2. They divide the real-number line into three intervals as shown:

We try test numbers in each interval.

A: Test -3, $y = (-3-6)(-3+2) = 9 > 0$

B: Test 0, $y = (0-6)(0+2) = -12 < 0$

C: Test 7, $y = (7-6)(7+2) = 9 > 0$

The expression is positive for all values of x in intervals A and C. The solution set is $\{x | x < -2 \text{ or } x > 6\}$, or $(-\infty, -2) \cup (6, \infty)$.

From the graph in the text we see that the value of $(x-6)(x+2)$ is positive to the left of -2 and to the right of 6. This verifies the answer we found algebraically.

3. $4 - x^2 \geq 0$

$(2+x)(2-x) \geq 0$

The solutions of $(2+x)(2-x) = 0$ are -2 and 2. They divide the real-number line into three intervals as shown.

We try test numbers in each interval.

A: Test -3, $y = 4 - (-3)^2 = -5 < 0$

B: Test 0, $y = 4 - 0^2 = 4 > 0$

C: Test 3, $y = 4 - 3^2 = -5 < 0$

The expression is positive for values of x in interval B. Since the inequality symbol is $\geq$ we also include the intercepts. The solution set is $\{x | -2 \leq x \leq 2\}$, or $[-2, 2]$.

From the graph we see that $4 - x^2 \geq 0$ at the intercepts and between them. This verifies the answer we found algebraically.

5. $3(x+1)(x-4) \leq 0$

The solutions of $3(x+1)(x-4) = 0$ are -1 and 4. They divide the real-number line into three intervals as shown:

We try test numbers in each interval.

A: Test -2, $y = 3(-2+1)(-2-4) = 18 > 0$

B: Test 0, $y = 3(0+1)(0-4) = -12 < 0$

C: Test 5, $y = 3(5+1)(5-4) = 18 > 0$

The expression is negative for all numbers in interval B. The inequality symbol is $\leq$, so we need to include the intercepts. The solution set is $\{x | -1 \leq x \leq 4\}$, or $[-1, 4]$.

7. $x^2 - x - 2 < 0$

$(x+1)(x-2) < 0$ Factoring

The solutions of $(x+1)(x-2) = 0$ are -1 and 2. They divide the real-number line into three intervals as shown:

We try test numbers in each interval.

A: Test -2, $y = (-2+1)(-2-2) = 4 > 0$

B: Test 0, $y = (0+1)(0-2) = -2 < 0$

C: Test 3, $y = (3+1)(3-2) = 4 > 0$

The expression is negative for all numbers in interval B. The solution set is $\{x | -1 < x < 2\}$, or $(-1, 2)$.

9. $x^2 - 2x + 1 \geq 0$

$(x-1)^2 \geq 0$

The solution of $(x-1)^2 = 0$ is 1. For all real-number values of x except 1, $(x-1)^2$ will be positive. Thus the solution set is $\{x | x \text{ is a real number}\}$, or $(-\infty, \infty)$.

11. $x^2 + 8 < 6x$

$x^2 - 6x + 8 < 0$

$(x-4)(x-2) < 0$

The solutions of $(x-4)(x-2) = 0$ are 4 and 2. They divide the real-number line into three intervals as shown:

We try test numbers in each interval.

A: Test 0, $y = (0-4)(0-2) = 8 > 0$

B: Test 3, $y = (3-4)(3-2) = -1 < 0$

C: Test 5, $y = (5-4)(5-2) = 3 > 0$

The expression is negative for all numbers in interval B. The solution set is $\{x | 2 < x < 4\}$, or $(2, 4)$.

13. $3x(x+2)(x-2) < 0$

The solutions of $3x(x+2)(x-2) = 0$ are 0, -2, and 2. They divide the real-number line into four intervals as shown:

We try test numbers in each interval.

A: Test -3, $y = 3(-3)(-3+2)(-3-2) = -45 < 0$

B: Test -1, $y = 3(-1)(-1+2)(-1-2) = 9 > 0$

C: Test 1, $y = 3(1)(1+2)(1-2) = -9 < 0$

D: Test 3, $y = 3(3)(3+2)(3-2) = 45 > 0$

The expression is negative for all numbers in intervals A and C. The solution set is $\{x | x < -2 \text{ or } 0 < x < 2\}$, or $(-\infty, -2) \cup (0, 2)$.

15. $(x+9)(x-4)(x+1) > 0$

The solutions of $(x+9)(x-4)(x+1) = 0$ are -9, 4, and -1. They divide the real-number line into four intervals as shown:

We try test numbers in each interval.

A: Test -10, $y = (-10+9)(-10-4)(-10+1) = -126 < 0$

B: Test -2, $y = (-2+9)(-2-4)(-2+1) = 42 > 0$

C: Test 0, $y = (0+9)(0-4)(0+1) = -36 < 0$

D: Test 5, $y = (5+9)(5-4)(5+1) = 84 > 0$

The expression is positive for all values of x in intervals B and D. The solution set is $\{x \mid -9 < x < -1 \ or \ x > 4\}$, or $(-9, -1) \cup (4, \infty)$.

17. $(x+3)(x+2)(x-1) < 0$

The solutions of $(x+3)(x+2)(x-1) = 0$ are -3, -2, and 1. They divide the real-number line into four intervals as shown:

We try test numbers in each interval.

A: Test -4, $y = (-4+3)(-4+2)(-4-1) = -10 < 0$

B: Test $-\dfrac{5}{2}$, $y = \left(-\dfrac{5}{2}+3\right)\left(-\dfrac{5}{2}+2\right)\left(-\dfrac{5}{2}-1\right) = \dfrac{7}{8} > 0$

C: Test 0, $y = (0+3)(0+2)(0-1) = -6 < 0$

D: Test 2, $y = (2+3)(2+2)(2-1) = 20 > 0$

The expression is negative for all numbers in intervals A and C. The solution set is $\{x \mid x < -3 \ or \ -2 < x < 1\}$, or $(-\infty, -3) \cup (-2, 1)$.

19. $\dfrac{1}{x-6} < 0$

We write the related equation by changing the $<$ symbol to $=$:

$$\frac{1}{x-6} = 0$$

We solve the related equation.

$$(x-6) \cdot \frac{1}{x-6} = (x-6) \cdot 0$$

$$1 = 0$$

We get a false equation, so the related equation has no solution.

Next we find the numbers for which the rational expression is undefined by setting the denominator equal to 0 and solving:

$$x - 6 = 0$$

$$x = 6$$

We use 6 to divide the number line into two intervals as shown:

We try test numbers in each interval.

A: Test 0,

$$\frac{1}{x-6} < 0$$

$$\frac{1}{0-6} \ ? \ 0$$

$$-\frac{1}{6} \ \Big| \ \text{TRUE}$$

The number 0 is a solution of the inequality, so the interval A is part of the solution set.

B: Test 7,

$$\frac{1}{x-6} < 0$$

$$\frac{1}{7-6} \ ? \ 0$$

$$1 \ \Big| \ \text{FALSE}$$

The number 7 is not a solution of the inequality, so the interval B is not part of the solution set. The solution set is $\{x \mid x < 6\}$, or $(-\infty, 6)$.

21. $\dfrac{x+1}{x-3} > 0$

Solve the related equation.

$$\frac{x+1}{x-3} = 0$$

$$x + 1 = 0$$

$$x = -1$$

Find the numbers for which the rational expression is undefined.

$$x - 3 = 0$$

$$x = 3$$

Use the numbers -1 and 3 to divide the number line into intervals as shown:

Try test numbers in each interval.

A: Test -2,

$$\frac{x+1}{x-3} > 0$$

$$\frac{-2+1}{-2-3} \ ? \ 0$$

$$\frac{-1}{-5}$$

$$\frac{1}{5} \ \Big| \ \text{TRUE}$$

The number -2 is a solution of the inequality, so the interval A is part of the solution set.

B: Test 0,

$$\frac{x+1}{x-3} > 0$$

$$\frac{0+1}{0-3} \ ? \ 0$$

$$-\frac{1}{3} \ \Big| \ \text{FALSE}$$

The number 0 is not a solution of the inequality, so the interval B is not part of the solution set.

C: Test 4,

$$\frac{x+1}{x-3} > 0$$

$$\frac{4+1}{4-3} \ ? \ 0$$

$$\frac{5}{1}$$

$$5 \ \Big| \ \text{TRUE}$$

The number 4 is a solution of the inequality, so the interval C is part of the solution set. The solution set is
$\{x | x < -1 \; or \; x > 3\}$, or $(-\infty, -1) \cup (3, \infty)$.

23. $\dfrac{3x+2}{x-3} \le 0$

Solve the related equation.

$$\frac{3x+2}{x-3} = 0$$
$$3x + 2 = 0$$
$$3x = -2$$
$$x = -\frac{2}{3}$$

Find the numbers for which the rational expression is undefined.

$$x - 3 = 0$$
$$x = 3$$

Use the numbers $-\dfrac{2}{3}$ and 3 to divide the number line into intervals as shown:

Try test numbers in each interval.

A: Test -1, $\dfrac{3x+2}{x-3} \le 0$

$$\frac{3(-1)+2}{-1-3} \; ? \; 0$$
$$\frac{-1}{-4}$$
$$\frac{1}{4} \quad \Big| \quad \text{FALSE}$$

The number -1 is not a solution of the inequality, so the interval A is not part of the solution set.

B: Test 0, $\dfrac{3x+2}{x-3} \le 0$

$$\frac{3 \cdot 0 + 2}{0 - 3} \; ? \; 0$$
$$\frac{2}{-3}$$
$$-\frac{2}{3} \quad \Big| \quad \text{TRUE}$$

The number 0 is a solution of the inequality, so the interval B is part of the solution set.

C: Test 4, $\dfrac{3x+2}{x-3} \le 0$

$$\frac{3 \cdot 4 + 2}{4 - 3} \; ? \; 0$$
$$14 \quad \Big| \quad \text{FALSE}$$

The number 4 is not a solution of the inequality, so the interval C is not part of the solution set. The solution set includes the interval B. The number $-\dfrac{2}{3}$ is also included

since the inequality symbol is $\le$ and $-\dfrac{2}{3}$ is the solution of the related equation. The number 3 is not included because the rational expression is undefined for 3. The solution set is $\left\{x \; \Big| -\dfrac{2}{3} \le x < 3\right\}$, or $\left[-\dfrac{2}{3}, 3\right)$.

25. $\dfrac{x-1}{x-2} > 3$

Solve the related equation.

$$\frac{x-1}{x-2} = 3$$
$$x - 1 = 3(x - 2)$$
$$x - 1 = 3x - 6$$
$$5 = 2x$$
$$\frac{5}{2} = x$$

Find the numbers for which the rational expression is undefined.

$$x - 2 = 0$$
$$x = 2$$

Use the numbers $\dfrac{5}{2}$ and 2 to divide the number line into intervals as shown:

Try test numbers in each interval.

A: Test 0, $\dfrac{x-1}{x-2} > 3$

$$\frac{0-1}{0-2} \; ? \; 3$$
$$\frac{1}{2} \quad \Big| \quad \text{FALSE}$$

The number 0 is not a solution of the inequality, so the interval A is not part of the solution set.

B: Test $\dfrac{9}{4}$, $\dfrac{x-1}{x-2} > 3$

$$\frac{\frac{9}{4} - 1}{\frac{9}{4} - 2} \; ? \; 3$$
$$\frac{\frac{5}{4}}{\frac{1}{4}}$$
$$5 \quad \Big| \quad \text{TRUE}$$

The number $\dfrac{9}{4}$ is a solution of the inequality, so the interval B is part of the solution set.

C: Test 3, $\dfrac{x-1}{x-2} > 3$

$$\frac{3-1}{3-2} \; ? \; 3$$
$$2 \quad \Big| \quad \text{FALSE}$$

The number 3 is not a solution of the inequality, so the interval C is not part of the solution set. The solution set is $\left\{x \middle| 2 < x < \dfrac{5}{2}\right\}$, or $\left(2, \dfrac{5}{2}\right)$.

27. $\dfrac{(x-2)(x+1)}{x-5} < 0$

Solve the related equation.

$$\frac{(x-2)(x+1)}{x-5} = 0$$

$$(x-2)(x+1) = 0$$

$$x = 2 \text{ or } x = -1$$

Find the numbers for which the rational expression is undefined.

$$x - 5 = 0$$

$$x = 5$$

Use the numbers 2, -1, and 5 to divide the number line into intervals as shown:

Try test numbers in each interval.

A: Test -2,

$$\frac{(x-2)(x+1)}{x-5} < 0$$

$$\frac{(-2-2)(-2+1)}{-2-5} \;?\; 0$$

$$\frac{-4(-1)}{-7}$$

$$-\frac{4}{7} \;\middle|\; \text{TRUE}$$

Interval A is part of the solution set.

B: Test 0,

$$\frac{(x-2)(x+1)}{x-5} < 0$$

$$\frac{(0-2)(0+1)}{0-5} \;?\; 0$$

$$\frac{-2 \cdot 1}{-5}$$

$$\frac{2}{5} \;\middle|\; \text{FALSE}$$

Interval B is not part of the solution set.

C: Test 3,

$$\frac{(x-2)(x+1)}{x-5} < 0$$

$$\frac{(3-2)(3+1)}{3-5} \;?\; 0$$

$$\frac{1 \cdot 4}{-2}$$

$$-2 \;\middle|\; \text{TRUE}$$

Interval C is part of the solution set.

D: Test 6,

$$\frac{(x-2)(x+1)}{x-5} < 0$$

$$\frac{(6-2)(6+1)}{6-5} \;?\; 0$$

$$\frac{4 \cdot 7}{1}$$

$$28 \;\middle|\; \text{FALSE}$$

Interval D is not part of the solution set.

The solution set is $\{x | x < -1 \text{ or } 2 < x < 5\}$, or $(-\infty, -1) \cup (2, 5)$.

29. $\dfrac{x+3}{x} \leq 0$

Solve the related equation.

$$\frac{x+3}{x} = 0$$

$$x + 3 = 0$$

$$x = -3$$

Find the numbers for which the rational expression is undefined.

$$x = 0$$

Use the numbers -3 and 0 to divide the number line into intervals as shown:

Try test numbers in each interval.

A: Test -4,

$$\frac{x+3}{x} \leq 0$$

$$\frac{-4+3}{-4} \;?\; 0$$

$$\frac{1}{4} \;\middle|\; \text{FALSE}$$

Interval A is not part of the solution set.

B: Test -1,

$$\frac{x+3}{x} \leq 0$$

$$\frac{-1+3}{-1} \;?\; 0$$

$$-2 \;\middle|\; \text{TRUE}$$

Interval B is part of the solution set.

C: Test 1,

$$\frac{x+3}{x} \leq 0$$

$$\frac{1+3}{1} \;?\; 0$$

$$4 \;\middle|\; \text{FALSE}$$

Interval C is not part of the solution set.

The solution set includes the interval B. The number -3 is also included since the inequality symbol is $\leq$ and -3 is a solution of the related equation. The number 0 is not included because the rational expression is undefined for 0. The solution set is $\{x | -3 \leq x < 0\}$, or $[-3, 0)$.

31. $\dfrac{x}{x-1} > 2$

Solve the related equation.

$$\frac{x}{x-1} = 2$$

$$x = 2x - 2$$

$$2 = x$$

Find the numbers for which the rational expression is undefined.

$$x - 1 = 0$$

$$x = 1$$

Use the numbers 1 and 2 to divide the number line into intervals as shown:

Try test numbers in each interval.

A: Test 0, $\quad \dfrac{x}{x-1} > 2$

$$\frac{0}{0-1} \;?\; 2$$

$$0 \;\big|\; \text{FALSE}$$

Interval A is not part of the solution set.

B: Test $\dfrac{3}{2}$, $\quad \dfrac{x}{x-1} > 2$

$$\frac{\frac{3}{2}}{\frac{3}{2}-1} \;?\; 2$$

$$\frac{\frac{3}{2}}{\frac{1}{2}}$$

$$3 \;\big|\; \text{TRUE}$$

Interval B is part of the solution set.

C: Test 3, $\quad \dfrac{x}{x-1} > 2$

$$\frac{3}{3-1} \;?\; 2$$

$$\frac{3}{2} \;\big|\; \text{FALSE}$$

Interval C is not part of the solution set.

The solution set is $\{x | 1 < x < 2\}$, or $(1,2)$.

33. $\dfrac{x-1}{(x-3)(x+4)} < 0$

Solve the related equation.

$$\frac{x-1}{(x-3)(x+4)} = 0$$

$$x - 1 = 0$$

$$x = 1$$

Find the numbers for which the rational expression is undefined.

$$(x-3)(x+4) = 0$$

$$x = 3 \text{ or } x = -4$$

Use the numbers 1, 3, and -4 to divide the number line into intervals as shown:

Try test numbers in each interval.

A: Test -5, $\quad \dfrac{x-1}{(x-3)(x+4)} < 0$

$$\frac{-5-1}{(-5-3)(-5+4)} \;?\; 0$$

$$\frac{-6}{-8(-1)}$$

$$-\frac{3}{4} \;\big|\; \text{TRUE}$$

Interval A is part of the solution set.

B: Test 0, $\quad \dfrac{x-1}{(x-3)(x+4)} < 0$

$$\frac{0-1}{(0-3)(0+4)} \;?\; 0$$

$$\frac{-1}{-3\cdot 4}$$

$$\frac{1}{12} \;\big|\; \text{FALSE}$$

Interval B is not part of the solution set.

C: Test 2, $\quad \dfrac{x-1}{(x-3)(x+4)} < 0$

$$\frac{2-1}{(2-3)(2+4)} \;?\; 0$$

$$\frac{1}{-1\cdot 6}$$

$$-\frac{1}{6} \;\big|\; \text{TRUE}$$

Interval C is part of the solution set.

D: Test 4, $\quad \dfrac{x-1}{(x-3)(x+4)} < 0$

$$\frac{4-1}{(4-3)(4+4)} \;?\; 0$$

$$\frac{3}{1\cdot 8}$$

$$\frac{3}{8} \;\big|\; \text{FALSE}$$

Interval D is not part of the solution set.

The solution set is $\{x | x < -4 \ or \ 1 < x < 3\}$, or $(-\infty, -4) \cup (1, 3)$.

35. $3 < \dfrac{1}{x}$

Solve the related equation.

$$3 = \dfrac{1}{x}$$

$$x = \dfrac{1}{3}$$

Find the numbers for which the rational expression is undefined.

$$x = 0$$

Use the numbers 0 and $\dfrac{1}{3}$ to divide the number line into intervals as shown:

Try test numbers in each interval.

A: Test -1,

$$3 < \dfrac{1}{x}$$

$$\overline{3 \ ? \ \dfrac{1}{-1}}$$

$$\Big| \ -1 \quad \text{FALSE}$$

Interval A is not part of the solution set.

B: Test $\dfrac{1}{6}$,

$$3 < \dfrac{1}{x}$$

$$\overline{3 \ ? \ \dfrac{1}{\frac{1}{6}}}$$

$$\Big| \ 6 \qquad \text{TRUE}$$

Interval B is part of the solution set.

C: Test 1,

$$3 < \dfrac{1}{x}$$

$$\overline{3 \ ? \ \dfrac{1}{1}}$$

$$\Big| \ 1 \qquad \text{FALSE}$$

Interval C is not part of the solution set.

The solution set is $\left\{x \Big| 0 < x < \dfrac{1}{3}\right\}$, or $\left(0, \dfrac{1}{3}\right)$.

37. $\dfrac{x^2 + x - 2}{x^2 - x - 12} > 0$

$$\dfrac{(x - 1)(x + 2)}{(x + 3)(x - 4)} > 0$$

Solve the related equation.

$$\dfrac{(x - 1)(x + 2)}{(x + 3)(x - 4)} = 0$$

$$(x - 1)(x + 2) = 0$$

$x = 1$ or $x = -2$

Find the numbers for which the rational expression is undefined.

$$(x + 3)(x - 4) = 0$$

$$x = -3 \text{ or } x = 4$$

Use the numbers 1, -2, -3, and 4 to divide the number line into intervals as shown:

Try test numbers in each interval.

A: Test -4, $\qquad \dfrac{(x - 1)(x + 2)}{(x + 3)(x - 4)} > 0$

$$\overline{\dfrac{(-4 - 1)(-4 + 2)}{(-4 + 3)(-4 - 4)} \ ? \ 0}$$

$$\dfrac{-5(-2)}{-1(-8)} \Big|$$

$$\dfrac{5}{4} \ \Big| \quad \text{TRUE}$$

Interval A is part of the solution set.

B: Test $-\dfrac{5}{2}$, $\qquad \dfrac{(x - 1)(x + 2)}{(x + 3)(x - 4)} > 0$

$$\overline{\dfrac{\left(-\frac{5}{2} - 1\right)\left(-\frac{5}{2} + 2\right)}{\left(-\frac{5}{2} + 3\right)\left(-\frac{5}{2} - 4\right)} \ ? \ 0}$$

$$\dfrac{-\frac{7}{2}\left(-\frac{1}{2}\right)}{\frac{1}{2}\left(-\frac{13}{2}\right)} \Big|$$

$$-\dfrac{7}{13} \ \Big| \quad \text{FALSE}$$

Interval B is not part of the solution set.

C: Test 0, $\qquad \dfrac{(x - 1)(x + 2)}{(x + 3)(x - 4)} > 0$

$$\overline{\dfrac{(0 - 1)(0 + 2)}{(0 + 3)(0 - 4)} \ ? \ 0}$$

$$\dfrac{-1 \cdot 2}{3(-4)} \Big|$$

$$\dfrac{1}{6} \ \Big| \quad \text{TRUE}$$

Interval C is part of the solution set.

D: Test 2, $\qquad \dfrac{(x - 1)(x + 2)}{(x + 3)(x - 4)} > 0$

$$\overline{\dfrac{(2 - 1)(2 + 2)}{(2 + 3)(2 - 4)} \ ? \ 0}$$

$$\dfrac{1 \cdot 4}{5(-2)} \Big|$$

$$-\dfrac{2}{5} \ \Big| \quad \text{FALSE}$$

Interval D is not part of the solution set.

E: Test 5,

$$\frac{(x-1)(x+2)}{(x+3)(x-4)} > 0$$

$$\frac{(5-1)(5+2)}{(5+3)(5-4)} \ ? \ 0$$

$$\frac{4 \cdot 7}{8 \cdot 1}$$

$$\frac{7}{2} \quad \Big| \quad \text{TRUE}$$

Interval E is part of the solution set.

The solution set is $\{x | x < -3 \ or \ -2 < x < 1 \ or \ x > 4\}$, or $(-\infty, -3) \cup (-2, 1) \cup (4, \infty)$.

39. $\sqrt[3]{\dfrac{125}{27}} = \dfrac{\sqrt[3]{125}}{\sqrt[3]{27}} = \dfrac{5}{3}$

41. $\sqrt{\dfrac{16a^3}{b^4}} = \dfrac{\sqrt{16a^3}}{\sqrt{b^4}} = \dfrac{\sqrt{16a^2 \cdot a}}{\sqrt{b^4}} = \dfrac{\sqrt{16a^2}\sqrt{a}}{\sqrt{b^4}} = \dfrac{4a}{b^2}\sqrt{a}$

43. $3\sqrt{8} - 5\sqrt{2} = 3\sqrt{4 \cdot 2} - 5\sqrt{2}$
$$= 3\sqrt{4}\sqrt{2} - 5\sqrt{2}$$
$$= 3 \cdot 2\sqrt{2} - 5\sqrt{2}$$
$$= 6\sqrt{2} - 5\sqrt{2}$$
$$= \sqrt{2}$$

45. $5\sqrt[3]{16a^4} + 7\sqrt[3]{2a} = 5\sqrt[3]{8a^3 \cdot 2a} + 7\sqrt[3]{2a}$
$$= 5\sqrt[3]{8a^3}\sqrt[3]{2a} + 7\sqrt[3]{2a}$$
$$= 5 \cdot 2a\sqrt[3]{2a} + 7\sqrt[3]{2a}$$
$$= 10a\sqrt[3]{2a} + 7\sqrt[3]{2a}$$
$$= (10a + 7)\sqrt[3]{2a}$$

47. For Exercise 11, graph $y_1 = x^2 + 8$ and $y_2 = 6x$. Then determine the values of x for which the graph of y_1 lies below the graph of y_2.

For Exercise 22, graph $y_1 = \dfrac{x-2}{x+5}$ and $y_2 = 0$. Then determine the values of x for which the graph of y_1 lies below the graph of y_2. Since the graph of $y_2 = 0$ is the x-axis, this could also be done by graphing $y_1 = \dfrac{x-2}{x+5}$ and determining the values of x for which the graph of y_1 lies below the x-axis.

For Exercise 25, graph $y_1 = \dfrac{x-1}{x-2}$ and $y_2 = 3$. Then determine the values of x for which the graph of y_1 lies above the graph of y_2.

49. $\quad x^2 - 2x \leq 2$

$\quad x^2 - 2x - 2 \leq 0$

The solutions of $x^2 - 2x - 2 = 0$ are found using the quadratic formula. They are $1 \pm \sqrt{3}$, or about 2.7 and -0.7. These numbers divide the number line into three intervals as shown:

We try test numbers in each interval.

A: Test -1, $y = (-1)^2 - 2(-1) - 2 = 1 > 0$

B: Test 0, $y = 0^2 - 2 \cdot 0 - 2 = -2 < 0$

C: Test 3, $y = 3^2 - 2 \cdot 3 - 2 = 1 > 0$

The expression is negative for all values of x in interval B. The inequality symbol is $\leq$, so we must also include the intercepts. The solution set is $\{x | 1 - \sqrt{3} \leq x \leq 1 + \sqrt{3}\}$, or $[1 - \sqrt{3}, 1 + \sqrt{3}]$.

51. $\quad x^4 + 2x^2 > 0$

$\quad x^2(x^2 + 2) > 0$

$x^2 > 0$ for all $x \neq 0$, and $x^2 + 2 > 0$ for all values of x. Then $x^2(x^2 + 2) > 0$ for all $x \neq 0$. The solution set is $\{x | x \neq 0\}$, or the set of all real numbers except 0, or $(-\infty, 0) \cup (0, \infty)$.

53. $\left| \dfrac{x+2}{x-1} \right| < 3$

$\quad -3 < \dfrac{x+2}{x-1} < 3$

We rewrite the inequality using "and."

$-3 < \dfrac{x+2}{x-1} \ and \ \dfrac{x+2}{x-1} < 3$

We will solve each inequality and then find the intersection of their solution sets.

Solve: $-3 < \dfrac{x+2}{x-1}$

Solve the related equation.

$$-3 = \dfrac{x+2}{x-1}$$

$$-3x + 3 = x + 2$$

$$1 = 4x$$

$$\dfrac{1}{4} = x$$

Find the numbers for which the rational expression is undefined.

$$x - 1 = 0$$

$$x = 1$$

Use the numbers $\dfrac{1}{4}$ and 1 to divide the number line into intervals as shown:

Try test numbers in each interval.

A: Test 0,

$$-3 < \dfrac{x+2}{x-1}$$

$$-3 \ ? \ \dfrac{0+2}{0-1}$$

$$\Big| \ -2 \quad \text{TRUE}$$

Interval A is part of the solution set.

B: Test $\dfrac{1}{2}$,

$$-3 < \dfrac{x+2}{x-1}$$

$$-3 \; ? \; \dfrac{\dfrac{1}{2}+2}{\dfrac{1}{2}-1}$$

$$\dfrac{\dfrac{5}{2}}{-\dfrac{1}{2}}$$

$$-5 \quad \text{FALSE}$$

Interval B is not part of the solution set.

C: Test 2,

$$-3 < \dfrac{x+2}{x-1}$$

$$-3 \; ? \; \dfrac{2+2}{2-1}$$

$$4 \quad \text{TRUE}$$

Interval C is part of the solution set.

The solution set of $-3 < \dfrac{x+2}{x-1}$ is $\left\{x \middle| x < \dfrac{1}{4} \; or \; x > 1\right\}$, or $\left(-\infty, \dfrac{1}{4}\right) \cup (1, \infty)$.

Solve: $\dfrac{x+2}{x-1} > 3$

Solve the related equation.

$$\dfrac{x+2}{x-1} = 3$$

$$x + 2 = 3x - 3$$

$$5 = 2x$$

$$\dfrac{5}{2} = x$$

From our work above we know that the rational expression is undefined for 1.

Use the numbers $\dfrac{5}{2}$ and 1 to divide the number line into intervals as shown:

Try test numbers in each interval.

A: Test 0,

$$\dfrac{x+2}{x-1} < 3$$

$$\dfrac{0+2}{0-1} \; ? \; 3$$

$$-2 \quad | \quad \text{TRUE}$$

Interval A is part of the solution set.

B: Test 2,

$$\dfrac{x+2}{x-1} < 3$$

$$\dfrac{2+2}{2-1} \; ? \; 3$$

$$4 \quad | \quad \text{FALSE}$$

Interval B is not part of the solution set.

C: Test 3,

$$\dfrac{x+2}{x-1} < 3$$

$$\dfrac{3+2}{3-1} \; ? \; 3$$

$$\dfrac{5}{2} \quad | \quad \text{TRUE}$$

Interval C is part of the solution set.

The solution set of $\dfrac{x+2}{x-1} < 3$ is $\left\{x \middle| x < 1 \; or \; x > \dfrac{5}{2}\right\}$, or $(-\infty, 1) \cup \left(\dfrac{5}{2}, \infty\right)$.

The solution set of the original inequality is

$\left\{x \middle| x < \dfrac{1}{4} \; or \; x > 1\right\} \cap \left\{x \middle| x < 1 \; or \; x > \dfrac{5}{2}\right\}$, or

$\left\{x \middle| x < \dfrac{1}{4} \; or \; x > \dfrac{5}{2}\right\}$, or $\left(-\infty, \dfrac{1}{4}\right) \cup \left(\dfrac{5}{2}, \infty\right)$.

55. a) Solve: $-16t^2 + 32t + 1920 > 1920$

$$-16t^2 + 32t > 0$$

$$t^2 - 2t < 0$$

$$t(t - 2) < 0$$

The solutions of $t(t-2) = 0$ are 0 and 2. They divide the number line into three intervals as shown:

Try test numbers in each interval.

A: Test -1, $y = -1(-1-2) = 3 > 0$

B: Test 1, $y = 1(1-2) = -1 < 0$

C: Test 3, $y = 3(3-2) = 3 > 0$

The expression is negative for all values of t in interval B. The solution set is $\{t | 0 < t < 2\}$, or $(0, 2)$.

b) Solve: $-16t^2 + 32t + 1920 < 640$

$$-16t^2 + 32t + 1280 < 0$$

$$t^2 - 2t - 80 > 0$$

$$(t - 10)(t + 8) > 0$$

The solutions of $(t-10)(t+8) = 0$ are 10 and -8. They divide the number line into three intervals as shown:

Try test numbers in each interval.

A: Test -10, $y = (-10-10)(-10+8) = 40 > 0$

B: Test 0, $y = (0-10)(0+8) = -80 < 0$

C: Test 20, $y = (20-10)(20+8) = 280 > 0$

The expression is positive for all values of t in intervals A and C. However, since negative values of t have no meaning in this problem, we disregard interval A. Thus, the solution set is $\{t | t > 10\}$, or $(10, \infty)$.

Chapter 7 Vocabulary Reinforcement

1. The equation $x^2 = 2x - 8$ is an example of a <u>quadratic</u> equation.

2. The inequality $\dfrac{1}{x} < 7$ is an example of a <u>rational</u> inequality.

3. We can <u>complete</u> the square for $y^2 - 8y$ by adding 16.

4. The expression $b^2 - 4ac$ in the quadratic formula is called the <u>discriminant</u>.

5. The equation $m^6 - m^3 - 12 = 0$ is <u>quadratic</u> in form.

6. The graph of a quadratic function is a <u>parabola</u>.

7. The vertical line $x = 0$ is the line of <u>symmetry</u> for the graph of $y = x^2$.

8. The maximum or minimum value of a quadratic function is the y-coordinate of the <u>vertex</u>.

Chapter 7 Concept Reinforcement

1. False; for $f(x) = -(-x^2 - 8x - 3) = x^2 + 8x + 3$, we have $a > 0$, so the graph opens up.

2. True; see page 628 in the text.

3. False; for $f(x) = -3(x+2)^2 - 5 = -3[x - (-2)]^2 - 5$, we have $h < 0$, so the graph is a translation of the graph of $f(x) = -3x^2 - 5$ to the *left*.

Chapter 7 Study Guide

1. $(x-2)^2 = -9$
$$x - 2 = \sqrt{-9} \quad or \quad x - 2 = -\sqrt{-9}$$
$$x - 2 = 3i \quad or \quad x - 2 = -3i$$
$$x = 2 + 3i \quad or \quad x = 2 - 3i$$
The solutions are $2 \pm 3i$.

2. $x^2 - 12x + 31 = 0$
$$x^2 - 12x = -31$$
$$x^2 - 12x + 36 = -31 + 36 \quad \left(\frac{1}{2}(-12) = -6; (-6)^2 = 36\right)$$
$$(x-6)^2 = 5$$
$$x - 6 = \sqrt{5} \quad or \quad x - 6 = -\sqrt{5}$$
$$x = 6 + \sqrt{5} \quad or \quad x = 6 - \sqrt{5}$$
The solutions are $6 \pm \sqrt{5}$.

3. $\quad x^2 - 10x = -23$
$$x^2 - 10x + 23 = 0$$
$$a = 1,\ b = -10,\ c = 23$$
$$x = \frac{-b \pm \sqrt{b^2 - 4ac}}{2a} = \frac{-(-10) \pm \sqrt{(-10)^2 - 4 \cdot 1 \cdot 23}}{2 \cdot 1}$$
$$x = \frac{10 \pm \sqrt{100 - 92}}{2} = \frac{10 \pm \sqrt{8}}{2} = \frac{10 \pm \sqrt{4 \cdot 2}}{2}$$
$$x = \frac{10 \pm 2\sqrt{2}}{2} = \frac{2(5 \pm \sqrt{2})}{2} = 5 \pm \sqrt{2}$$
The exact solutions are $5 \pm \sqrt{2}$.
We now find the approximate solutions.
$$5 + \sqrt{2} \approx 6.414$$
$$5 - \sqrt{2} \approx 3.586$$

4. a) $\quad x^2 - 3x = 7$
$$x^2 - 3x - 7 = 0$$
$$a = 1,\ b = -3,\ c = -7$$
$$b^2 - 4ac = (-3)^2 - 4 \cdot 1 \cdot (-7) = 9 + 28 = 37$$
Since $b^2 - 4ac > 0$, there are two real solutions.

   b) $2x^2 - 5x + 5 = 0$
$$a = 2,\ b = -5,\ c = 5$$
$$b^2 - 4ac = (-5)^2 - 4 \cdot 2 \cdot 5 = 25 - 40 = -15$$
Since $b^2 - 4ac < 0$, there are two nonreal solutions.

5. $\quad x = -\dfrac{2}{5} \quad or \quad x = 3$
$$5x = -2 \quad or \quad x - 3 = 0$$
$$5x + 2 = 0 \quad or \quad x - 3 = 0$$
$$(5x + 2)(x - 3) = 0$$
$$5x^2 - 13x - 6 = 0$$

6. $(x^2 - 3)^2 - 5(x^2 - 3) - 6 = 0$
Let $u = x^2 - 3$.
$$u^2 - 5u - 6 = 0$$
$$(u - 6)(u + 1) = 0$$
$$u - 6 = 0 \quad or \quad u + 1 = 0$$
$$u = 6 \quad or \quad u = -1$$
$$x^2 - 3 = 6 \quad or \quad x^2 - 3 = -1$$
$$x^2 = 9 \quad or \quad x^2 = 2$$
$$x = \sqrt{9} \ or \ x = -\sqrt{9} \ or \ x = \sqrt{2} \ or \ x = -\sqrt{2}$$
$$x = 3 \quad or \quad x = -3 \quad or \quad x = \sqrt{2} \ or \ x = -\sqrt{2}$$
The solutions are ± 3 and $\pm\sqrt{2}$.

7. $f(x) = -x^2 - 2x - 3$
$$f(x) = -(x^2 + 2x + 3)$$
$$f(x) = -(x^2 + 2x) - 3$$
$$f(x) = -(x^2 + 2x + 1 - 1) - 3$$
$$f(x) = -(x^2 + 2x + 1) - (-1) - 3$$
$$f(x) = -(x + 1)^2 - 2$$
$$f(x) = -[x - (-1)]^2 + (-2)$$

The vertex is $(-1, -2)$, and the line of symmetry is $x = -1$. The coefficient of x^2 is negative, so the graph opens down. Thus, -2 is the maximum value of the function.

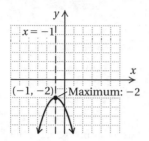

8. $f(x) = x^2 - 6x + 4$

$f(0) = 0^2 - 6 \cdot 0 + 4 = 4$, so the y-intercept is $(0, 4)$. To find the x-intercepts, we solve $f(x) = 0$.

$$x^2 - 6x + 4 = 0$$
$$a = 1,\ b = -6,\ c = 4$$
$$x = \frac{-b \pm \sqrt{b^2 - 4ac}}{2a} = \frac{-(-6) \pm \sqrt{(-6)^2 - 4 \cdot 1 \cdot 4}}{2 \cdot 1}$$
$$x = \frac{6 \pm \sqrt{36 - 16}}{2} = \frac{6 \pm \sqrt{20}}{2} = \frac{6 \pm \sqrt{4 \cdot 5}}{2}$$
$$x = \frac{6 \pm 2\sqrt{5}}{2} = \frac{2(3 \pm \sqrt{5})}{2} = 3 \pm \sqrt{5}$$

The x-intercepts are $(3 - \sqrt{5}, 0)$ and $(3 + \sqrt{5}, 0)$.

9. $\qquad x^2 + 40 > 14x$

$$x^2 - 14x + 40 > 0$$
$$(x - 4)(x - 10) > 0$$

The solutions of $(x - 4)(x - 10) = 0$ are 4 and 10.

These numbers divide the real-number line into three intervals as shown.

We try a test number in each interval.

A: Test 0, $0^2 - 14 \cdot 0 + 40 = 40 > 0$

B: Test 5, $5^2 - 14 \cdot 5 + 40 = -5 < 0$

C: Test 11, $11^2 - 14 \cdot 11 + 40 = 7 > 0$

The expression is positive for all numbers in intervals A and C. The solution set is $\{x | x < 4\ or\ x > 10\}$, or $(-\infty, 4) \cup (10, \infty)$.

10. $\dfrac{x + 7}{x - 5} \geq 3$

Solve the related equation.

$$\frac{x + 7}{x - 5} = 3$$
$$x + 7 = 3(x - 5)$$
$$x + 7 = 3x - 15$$
$$22 = 2x$$
$$11 = x$$

Now find the numbers for which the rational expression is not defined.

$$x - 5 = 0$$
$$x = 5$$

Use the numbers 11 and 5 to divide the number line into intervals as shown.

Try a test number in each interval.

A: Test 0, $\qquad \dfrac{x + 7}{x - 5} \geq 3$

$$\begin{array}{c|c} \dfrac{0 + 7}{0 - 5}\ ?\ 3 & \\ -\dfrac{7}{5} & \text{FALSE} \end{array}$$

The number 0 is not a solution of the inequality, so interval A is not part of the solution set.

B: Test 6, $\qquad \dfrac{x + 7}{x - 5} \geq 3$

$$\begin{array}{c|c} \dfrac{6 + 7}{6 - 5}\ ?\ 3 & \\ 13 & \text{TRUE} \end{array}$$

The number 6 is a solution of the inequality, so interval B is in the solution set.

C: Test 12, $\qquad \dfrac{x + 7}{x - 5} \geq 3$

$$\begin{array}{c|c} \dfrac{12 + 7}{12 - 5}\ ?\ 3 & \\ \dfrac{19}{7} & \\ 2\dfrac{5}{7} & \text{FALSE} \end{array}$$

The number 12 is not a solution of the inequality, so interval C is not in the solution set.

The solution set includes interval B. The number 11 is also included since the inequality symbol is $\geq$ and 11 is the solution of the related equation. The number 5 is not included because the rational expression is not defined for 5. The solution set is $\{x | 5 < x \leq 11\}$, or $(5, 11]$.

Chapter 7 Review Exercises

1. a) $\quad 2x^2 - 7 = 0$

$$2x^2 = 7$$
$$x^2 = \frac{7}{2}$$
$$x = \sqrt{\frac{7}{2}} \quad or \quad x = -\sqrt{\frac{7}{2}}$$
$$x = \sqrt{\frac{7}{2} \cdot \frac{2}{2}} \quad or \quad x = -\sqrt{\frac{7}{2} \cdot \frac{2}{2}}$$
$$x = \frac{\sqrt{14}}{2} \quad or \quad x = -\frac{\sqrt{14}}{2}$$

The solutions are $\pm\dfrac{\sqrt{14}}{2}$.

b) The real-number solutions of the equation $2x^2 - 7 = 0$ are the first coordinates of the x-intercepts of the graph of $f(x) = 2x^2 - 7$. Thus, the x-intercepts are $\left(-\dfrac{\sqrt{14}}{2}, 0\right)$ and $\left(\dfrac{\sqrt{14}}{2}, 0\right)$.

2. $14x^2 + 5x = 0$

$x(14x + 5) = 0$

$x = 0 \ \ or \ \ 14x + 5 = 0$

$x = 0 \ \ or \ \ \ \ \ \ 14x = -5$

$x = 0 \ \ or \ \ \ \ \ \ \ \ \ x = -\dfrac{5}{14}$

The solutions are 0 and $-\dfrac{5}{14}$.

3. $x^2 - 12x + 27 = 0$

$(x - 3)(x - 9) = 0$

$x - 3 = 0 \ \ or \ \ x - 9 = 0$

$x = 3 \ \ or \ \ \ \ \ \ x = 9$

The solutions are 3 and 9.

4. $x^2 - 7x + 13 = 0$

$a = 1, b = -7, c = 13$

$x = \dfrac{-b \pm \sqrt{b^2 - 4ac}}{2a}$

$x = \dfrac{-(-7) \pm \sqrt{(-7)^2 - 4 \cdot 1 \cdot 13}}{2 \cdot 1} = \dfrac{7 \pm \sqrt{49 - 52}}{2}$

$x = \dfrac{7 \pm \sqrt{-3}}{2} = \dfrac{7 \pm i\sqrt{3}}{2} = \dfrac{7}{2} \pm i\dfrac{\sqrt{3}}{2}$

The solutions are $\dfrac{7}{2} \pm \dfrac{\sqrt{3}}{2}i$.

5. $4x^2 + 6x = 1$

$4x^2 + 6x - 1 = 0$

$a = 4, b = 6, c = -1$

$x = \dfrac{-b \pm \sqrt{b^2 - 4ac}}{2a}$

$x = \dfrac{-6 \pm \sqrt{6^2 - 4 \cdot 4(-1)}}{2 \cdot 4} = \dfrac{-6 \pm \sqrt{36 + 16}}{8}$

$x = \dfrac{-6 \pm \sqrt{52}}{8} = \dfrac{-6 \pm \sqrt{4 \cdot 13}}{8} = \dfrac{-6 \pm 2\sqrt{13}}{8}$

$x = \dfrac{2(-3 \pm \sqrt{13})}{2 \cdot 4} = \dfrac{-3 \pm \sqrt{13}}{4}$

The solutions are $\dfrac{-3 \pm \sqrt{13}}{4}$.

6. $4x(x - 1) + 15 = x(3x + 4)$

$4x^2 - 4x + 15 = 3x^2 + 4x$

$x^2 - 8x + 15 = 0$

$(x - 3)(x - 5) = 0$

$x - 3 = 0 \ \ or \ \ x - 5 = 0$

$x = 3 \ \ or \ \ \ \ \ \ x = 5$

The solutions are 3 and 5.

7. $x^2 + 4x + 1 = 0$

$a = 1, b = 4, c = 1$

$x = \dfrac{-b \pm \sqrt{b^2 - 4ac}}{2a}$

$x = \dfrac{-4 \pm \sqrt{4^2 - 4 \cdot 1 \cdot 1}}{2 \cdot 1} = \dfrac{-4 \pm \sqrt{16 - 4}}{2}$

$x = \dfrac{-4 \pm \sqrt{12}}{2} = \dfrac{-4 \pm 2\sqrt{3}}{2}$

$x = \dfrac{2(-2 \pm \sqrt{3})}{2} = -2 \pm \sqrt{3}$

We can use a calculator to approximate the solutions:

$-2 + \sqrt{3} \approx 0.268; \ -2 - \sqrt{3} \approx -3.732$

The solutions are $-2 \pm \sqrt{3}$, or approximately 0.268 and -3.732.

8. $\dfrac{x}{x-2} + \dfrac{4}{x-6} = 0$, LCM is $(x-2)(x-6)$

$(x-2)(x-6)\left(\dfrac{x}{x-2} + \dfrac{4}{x-6}\right) = (x - 2)(x - 6) \cdot 0$

$x(x - 6) + 4(x - 2) = 0$

$x^2 - 6x + 4x - 8 = 0$

$x^2 - 2x - 8 = 0$

$(x - 4)(x + 2) = 0$

$x - 4 = 0 \ \ or \ \ x + 2 = 0$

$x = 4 \ \ or \ \ \ \ \ \ x = -2$

Both numbers check. The solutions are 4 and -2.

9. $\dfrac{x}{4} - \dfrac{4}{x} = 2$, LCM is $4x$

$4x\left(\dfrac{x}{4} - \dfrac{4}{x}\right) = 4x \cdot 2$

$x^2 - 16 = 8x$

$x^2 - 8x - 16 = 0$

$a = 1, b = -8, c = -16$

$x = \dfrac{-(-8) \pm \sqrt{(-8)^2 - 4 \cdot 1 \cdot (-16)}}{2 \cdot 1} = \dfrac{8 \pm \sqrt{64 + 64}}{2}$

$x = \dfrac{8 \pm \sqrt{128}}{2} = \dfrac{8 \pm 8\sqrt{2}}{2}$

$x = \dfrac{2(4 \pm 4\sqrt{2})}{2} = 4 \pm 4\sqrt{2}$

Both numbers check. The solutions are $4 \pm 4\sqrt{2}$.

10. $15 = \dfrac{8}{x+2} - \dfrac{6}{x-2}$, LCM is $(x+2)(x-2)$

$(x+2)(x-2) \cdot 15 = (x+2)(x-2)\left(\dfrac{8}{x+2} - \dfrac{6}{x-2}\right)$

$15(x^2 - 4) = 8(x - 2) - 6(x + 2)$

$15x^2 - 60 = 8x - 16 - 6x - 12$

$15x^2 - 60 = 2x - 28$

$15x^2 - 2x - 32 = 0$

$a = 15, b = -2, c = -32$

$$x = \frac{-(-2) \pm \sqrt{(-2)^2 - 4 \cdot 15 \cdot (-32)}}{2 \cdot 15} = \frac{2 \pm \sqrt{4 + 1920}}{30}$$

$$x = \frac{2 \pm \sqrt{1924}}{30} = \frac{2 \pm 2\sqrt{481}}{30}$$

$$x = \frac{2(1 \pm \sqrt{481})}{2 \cdot 15} = \frac{1 \pm \sqrt{481}}{15}$$

Both numbers check. The solutions are $\frac{1 \pm \sqrt{481}}{15}$.

11. $x^2 + 6x + 2 = 0$

$$x^2 + 6x = -2$$

$$x^2 + 6x + 9 = -2 + 9 \qquad \left(\frac{1}{2} \cdot 6 = 3; 3^2 = 9\right)$$

$$(x + 3)^2 = 7$$

$$x + 3 = \sqrt{7} \qquad or \quad x + 3 = -\sqrt{7}$$

$$x = -3 + \sqrt{7} \quad or \qquad x = -3 - \sqrt{7}$$

The solutions are $-3 \pm \sqrt{7}$.

12. $\qquad V(T) = 48T^2$

$$39 = 48T^2 \qquad \text{Substituting 39 for } V(T)$$

$$0.8125 = T^2$$

$$\sqrt{0.8125} = T$$

$$0.901 \approx T$$

The hang time is 0.901 sec.

13. Familiarize. Let $l =$ the length of the screen, in cm. Then $l - 5 =$ the width.

Translate. We will use the formula for the area of a rectangle, $A = lw$.

$$l(l - 5) = 126$$

Solve. We solve the equation.

$$l(l - 5) = 126$$

$$l^2 - 5l = 126$$

$$l^2 - 5l - 126 = 0$$

$$(l - 14)(l + 9) = 0$$

$$l - 14 = 0 \quad or \quad l + 9 = 0$$

$$l = 14 \quad or \qquad l = -9$$

Check. Since the length cannot be negative, we check only 14. If $l = 14$, then $l - 5 = 14 - 5 = 9$. If the length is 14 cm and the width is 9 cm, the width is 5 cm less than the length and the area is $14 \cdot 9$, or 126 cm². The answer checks.

State. The length is 14 cm, and the width is 9 cm.

14. Familiarize. Using the labels on the drawing in the text, we let $x =$ the width of the mat, in inches. Then the dimensions of the picture are $16 - x - x$ by $12 - x - x$, or $16 - 2x$ by $12 - 2x$.

Translate. We use the formula for the area of a rectangle, $A = lw$.

$$(16 - 2x)(12 - 2x) = 140$$

Solve. We solve the equation.

$$(16 - 2x)(12 - 2x) = 140$$

$$192 - 56x + 4x^2 = 140$$

$$4x^2 - 56x + 52 = 0$$

$$x^2 - 14x + 13 = 0 \qquad \text{Dividing by 4}$$

$$(x - 1)(x - 13) = 0$$

$$x - 1 = 0 \quad or \quad x - 13 = 0$$

$$x = 1 \quad or \qquad x = 13$$

Check. Since the matted picture measures 12 in. by 16 in., the width of the mat cannot be 13. If the width of the mat is 1 in., then the dimensions of the picture are $12 - 2 \cdot 1$, or 10, by $16 - 2 \cdot 1$, or 14. Thus, the area of the picture is $14 \cdot 10$, or 140 cm². The answer checks.

State. The mat is 1 in. wide.

15. Familiarize. We first make a drawing, labeling it with the known and unknown information. We can also organize the information in a table. We let r represent the speed and t the time for the first part of the trip.

$$\underset{\text{50 mi}}{r \text{ mph} \qquad t \text{ hr}} \bullet \underset{\text{80 mi}}{r - 10 \text{ mph} \quad 3 - t \text{ hr}}$$

Trip	Distance	Speed	Time
1st part	50	r	t
2nd part	80	$r - 10$	$3 - t$

Translate. Using $r = \dfrac{d}{t}$, we get two equations from the table, $r = \dfrac{50}{t}$ and $r - 10 = \dfrac{80}{3 - t}$.

Solve. We substitute $\dfrac{50}{t}$ for r in the second equation and solve for t.

$$\frac{50}{t} - 10 = \frac{80}{3 - t}, \quad \text{LCD is } t(3 - t)$$

$$t(3 - t)\left(\frac{50}{t} - 10\right) = t(3 - t) \cdot \frac{80}{3 - t}$$

$$50(3 - t) - 10t(3 - t) = 80t$$

$$150 - 50t - 30t + 10t^2 = 80t$$

$$150 - 80t + 10t^2 = 80t$$

$$10t^2 - 160t + 150 = 0$$

$$t^2 - 16t + 15 = 0 \qquad \text{Dividing by 10}$$

$$(t - 1)(t - 15) = 0$$

$$t = 1 \quad or \quad t = 15$$

Check. Since the time cannot be negative (If $t = 15$, $3 - t = -12$.), we check only 1 hr. If $t = 1$, then $3 - t = 2$. The speed of the first part is $\dfrac{50}{1}$, or 50 mph. The speed of the second part is $\dfrac{80}{2}$, or 40 mph. The speed of the second part is 10 mph slower than the first part. The value checks.

State. The speed of the first part was 50 mph, and the speed of the second part was 40 mph.

16. $x^2 + 3x - 6 = 0$

$a = 1, b = 3, c = -6$

We compute the discriminant.

$b^2 - 4ac = 3^2 - 4 \cdot 1 \cdot (-6) = 9 + 24 = 33$

Since $b^2 - 4ac > 0$, there are two real solutions.

17. $x^2 + 2x + 5 = 0$

$a = 1, b = 2, c = 5$

We compute the discriminant.

$b^2 - 4ac = 2^2 - 4 \cdot 1 \cdot 5 = 4 - 20 = -16$

Since $b^2 - 4ac < 0$, there are two nonreal solutions.

18.
$$x = \frac{1}{5} \quad or \quad x = -\frac{3}{5}$$

$$x - \frac{1}{5} = 0 \quad or \quad x + \frac{3}{5} = 0$$

$5x - 1 = 0 \quad or \quad 5x + 3 = 0 \quad$ Clearing fractions

$(5x - 1)(5x + 3) = 0$

$25x^2 + 10x - 3 = 0$

19. Since -4 is the only solution, it must be a double solution.

$$x = -4 \quad or \quad x = -4$$

$x + 4 = 0 \quad or \quad x + 4 = 0$

$(x + 4)(x + 4) = 0$

$x^2 + 8x + 16 = 0$

20.
$$N = 3\pi\sqrt{\frac{1}{p}}$$

$$\frac{N}{3\pi} = \sqrt{\frac{1}{p}}$$

$$\left(\frac{N}{3\pi}\right)^2 = \left(\sqrt{\frac{1}{p}}\right)^2$$

$$\frac{N^2}{9\pi^2} = \frac{1}{p}$$

$$pN^2 = 9\pi^2 \qquad \text{Multiplying by } 9\pi^2 p$$

$$p = \frac{9\pi^2}{N^2}$$

21.
$$2A = \frac{3B}{T^2}$$

$$2AT^2 = 3B \qquad \text{Multiplying by } T^2$$

$$T^2 = \frac{3B}{2A}$$

$$T = \sqrt{\frac{3B}{2A}}$$

22. $x^4 - 13x^2 + 36 = 0$

Let $u = x^2$.

$u^2 - 13u + 36 = 0$

$(u - 4)(u - 9) = 0$

$u = 4 \quad or \quad u = 9$

$x^2 = 4 \quad or \quad x^2 = 9$

$x = \pm 2 \quad or \quad x = \pm 3$

All four numbers check. The solutions are 2, -2, 3, and -3.

23. $15x^{-2} - 2x^{-1} - 1 = 0$

Let $u = x^{-1}$.

$15u^2 - 2u - 1 = 0$

$(5u + 1)(3u - 1) = 0$

$$u = -\frac{1}{5} \quad or \quad u = \frac{1}{3}$$

$$x^{-1} = -\frac{1}{5} \quad or \quad x^{-1} = \frac{1}{3}$$

$$\frac{1}{x} = -\frac{1}{5} \quad or \quad \frac{1}{x} = \frac{1}{3}$$

$$-5x \cdot \frac{1}{x} = -5x\left(-\frac{1}{5}\right) \quad or \quad 3x \cdot \frac{1}{x} = 3x \cdot \frac{1}{3}$$

$$-5 = x \quad or \quad 3 = x$$

Both numbers check. The solutions are -5 and 3.

24. $(x^2 - 4)^2 - (x^2 - 4) - 6 = 0$

Let $u = x^2 - 4$.

$u^2 - u - 6 = 0$

$(u - 3)(u + 2) = 0$

$u = 3 \quad or \quad u = -2$

$x^2 - 4 = 3 \quad or \quad x^2 - 4 = -2$

$x^2 = 7 \quad or \quad x^2 = 2$

$x = \pm\sqrt{7} \quad or \quad x = \pm\sqrt{2}$

All four numbers check. The solutions are $\pm\sqrt{7}$ and $\pm\sqrt{2}$.

25. $x - 13\sqrt{x} + 36 = 0$

Let $u = \sqrt{x}$.

$u^2 - 13u + 36 = 0$

$(u - 4)(u - 9) = 0$

$u = 4 \quad or \quad u = 9$

$\sqrt{x} = 4 \quad or \quad \sqrt{x} = 9$

$(\sqrt{x})^2 = 4^2 \quad or \quad (\sqrt{x})^2 = 9^2$

$x = 16 \quad or \quad x = 81$

Both numbers check. The solutions are 16 and 81.

26. $f(x) = -\frac{1}{2}(x - 1)^2 + 3$

a) Vertex: $(1, 3)$

b) Line of symmetry: $x = 1$

c) Since $-\frac{1}{2} < 0$, we know that 3 is a maximum value.

d) We plot a few points and draw the curve.

x	$f(x)$
-3	-5
-1	1
1	3
3	1
5	-5

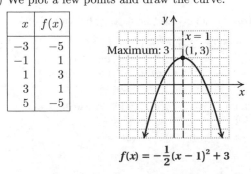

$$f(x) = -\frac{1}{2}(x - 1)^2 + 3$$

27. $f(x) = x^2 - x + 6 = (x^2 - x) + 6$

We complete the square inside the parentheses. We take half the x-coefficient and square it.

$$\frac{1}{2}(-1) = -\frac{1}{2} \text{ and } \left(-\frac{1}{2}\right)^2 = \frac{1}{4}$$

We add $\frac{1}{4} - \frac{1}{4}$ inside the parentheses.

$$\begin{aligned} f(x) &= \left(x^2 - x + \frac{1}{4} - \frac{1}{4}\right) + 6 \\ &= \left(x^2 - x + \frac{1}{4}\right) - \frac{1}{4} + 6 \\ &= \left(x - \frac{1}{2}\right)^2 + \frac{23}{4} \end{aligned}$$

a) Vertex: $\left(\dfrac{1}{2}, \dfrac{23}{4}\right)$

b) Line of symmetry: $x = \dfrac{1}{2}$

c) Since the coefficient of the x^2-term, 1, is positive, we know that $\dfrac{23}{4}$ is a minimum.

d) We plot a few points and draw the curve.

x	$f(x)$
-2	10
-1	8
0	6
1	6
2	8

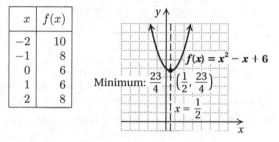

28. $f(x) = -3x^2 - 12x - 8 = -3(x^2 + 4x) - 8$

We complete the square inside the parentheses. We take half the x-coefficient and square it.

$$\frac{1}{2} \cdot 4 = 2 \text{ and } 2^2 = 4$$

We add $4 - 4$ inside the parentheses.

$$\begin{aligned} f(x) &= -3(x^2 + 4x + 4 - 4) - 8 \\ &= -3(x^2 + 4x + 4) + (-3)(-4) - 8 \\ &= -3(x + 2)^2 + 12 - 8 \\ &= -3(x + 2)^2 + 4 \\ &= -3[x - (-2)]^2 + 4 \end{aligned}$$

a) Vertex: $(-2, 4)$

b) Line of symmetry: $x = -2$

c) Since $-3 < 0$, we know that 4 is a maximum.

d) We plot a few points and draw the curve.

x	$f(x)$
-4	-8
-3	1
-2	4
-1	1
0	-8

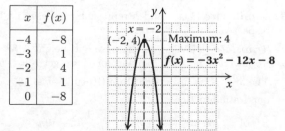

29. $f(x) = x^2 - 9x + 14$

$f(0) = 0^2 - 9 \cdot 0 + 14 = 14$, so the y-intercept is $(0, 14)$. To find the x-intercepts we solve $x^2 - 9x + 14 = 0$. Factoring and using the principle of zero products gives us $x = 2 \text{ or } x = 7$. Thus, the x-intercepts are $(2, 0)$ and $(7, 0)$.

30. $g(x) = x^2 - 4x - 3$

$g(0) = 0^2 - 4 \cdot 0 - 3 = -3$, so the y-intercept is $(0, -3)$. To find the x-intercepts we solve $x^2 - 4x - 3 = 0$. Using the quadratic formula we get $x = 2 \pm \sqrt{7}$. Thus, the x-intercepts are $(2 - \sqrt{7}, 0)$ and $(2 + \sqrt{7}, 0)$.

31. *Familiarize.* Let x and y represent the numbers.

Translate. The difference of the numbers is 22, so we have $x - y = 22$. Solve for x, we get $x = y + 22$. The product of the numbers is xy. Substituting $y + 22$ for x in the product, we get a quadratic function:

$$P = xy = (y + 22)y = y^2 + 22y$$

Carry out. The coefficient of y^2 is positive, so the graph of the function opens up and a minimum exists. We complete the square in order to find the vertex of the quadratic function.

$$\begin{aligned} P &= y^2 + 22y \\ &= y^2 + 22y + 121 - 121 \\ &= (y + 11)^2 - 121 \\ &= [y - (-11)]^2 + (-121) \end{aligned}$$

The vertex is $(-11, -121)$. This tells us that the minimum product is -121. The minimum occurs when $y = -11$. Note that when $y = -11$, $x = y + 22 = -11 + 22 = 11$, so the numbers that yield the minimum product are 11 and -11.

Check. We could find the value of the function for some values of y less than -11 and for some greater than -11, determining that the minimum value we found is smaller than these function values. We could also use the graph of the function to check the minimum value. Our answer checks.

State. The minimum product is -121. The numbers 11 and -11 yield this product.

32. We look for a function of the form $ax^2 + bx + c = 0$. Substituting the data points, we get

$$\begin{aligned} -2 &= a \cdot 0^2 + b \cdot 0 + c, \\ 3 &= a \cdot 1^2 + b \cdot 1 + c, \\ 7 &= a \cdot 3^2 + b \cdot 3 + c, \end{aligned}$$

or

$$\begin{aligned} -2 &= c, \\ 3 &= a + b + c, \\ 7 &= 9a + 3b + c. \end{aligned}$$

Solving this system, we get $(-1, 6, -2)$.

Thus, the desired function is $f(x) = -x^2 + 6x - 2$.

33. $(x + 2)(x - 1)(x - 2) > 0$

The solutions of $(x + 2)(x - 1)(x - 2) = 0$ are -2, 1, and 2. They divide the real-number line into four intervals as shown:

We try test numbers in each interval.

A: Test -3: $(-3+2)(-3-1)(-3-2) = -20 < 0$

B: Test 0: $(0+2)(0-1)(0-2) = 4 > 0$

C: Test 1.5: $(1.5+2)(1.5-1)(1.5-2) = -0.875 < 0$

D: Test 3: $(3+2)(3-1)(3-2) = 10 > 0$

The expression is positive for all values of x in intervals B and D. Thus, the solution set is $\{x \mid -2 < x < 1 \ or \ x > 2\}$, or $(-2, 1) \cup (2, \infty)$.

34. $\dfrac{(x+4)(x-1)}{x+2} < 0$

Solve the related equation.

$$\frac{(x+4)(x-1)}{x+2} = 0$$
$$(x+4)(x-1) = 0$$
$$x = -4 \ or \ x = 1$$

Find the numbers for which the rational expression is undefined.

$$x + 2 = 0$$
$$x = -2$$

Use the numbers -4, 1, and -2 to divide the number line into intervals as shown:

Try test numbers in each interval.

A: Test -5,

$$\frac{(x+4)(x-1)}{x+2} < 0$$

$$\frac{(-5+4)(-5-1)}{-5+2} \ ? \ 0$$

$$\frac{-1(-6)}{-3}$$

$$-2 \ \Big| \ \text{TRUE}$$

Interval A is part of the solution set.

B: Test -3,

$$\frac{(x+4)(x-1)}{x+2} < 0$$

$$\frac{(-3+4)(-3-1)}{-3+2} \ ? \ 0$$

$$\frac{1(-4)}{-1}$$

$$4 \ \Big| \ \text{FALSE}$$

Interval B is not part of the solution set.

C: Test 0,

$$\frac{(x+4)(x-1)}{x+2} < 0$$

$$\frac{(0+4)(0-1)}{0+2} \ ? \ 0$$

$$\frac{4(-1)}{2}$$

$$-2 \ \Big| \ \text{TRUE}$$

Interval C is part of the solution set.

D: Test 2,

$$\frac{(x+4)(x-1)}{x+2} < 0$$

$$\frac{(2+4)(2-1)}{2+2} \ ? \ 0$$

$$\frac{6 \cdot 1}{4}$$

$$\frac{3}{2} \ \Big| \ \text{FALSE}$$

Interval D is not part of the solution set.

The solution set is $\{x \mid x < -4 \ or \ -2 < x < 1\}$, or $(-\infty, -4) \cup (-2, 1)$.

35. a) We look for a function of the form $N(x) = ax^2 + bx + c$ where $N(x)$ represents the number of live births per 1000 women and x represents the age of the woman. We substitute the data points.

$$34 = a \cdot 16^2 + b \cdot 16 + c,$$
$$113.9 = a \cdot 27^2 + b \cdot 27 + c,$$
$$35.4 = a \cdot 37^2 + b \cdot 37 + c,$$

or

$$34 = 256a + 16b + c,$$
$$113.9 = 729a + 27b + c,$$
$$35.4 = 1369a + 37b + c$$

Solving the system of equations, we get $a \approx -0.720$, $b \approx 38.211$, and $c \approx -393.127$. Thus, the desired function is $N(x) = -0.720x^2 + 38.211x - 393.127$.

b) $N(30) = -0.720(30)^2 + 38.211(30) - 393.127$

≈ 105

36. $x^2 - 10x + 25 = 0$

$a = 1$, $b = -10$, $c = 25$

$b^2 - 4ac = (-10)^2 - 4 \cdot 1 \cdot 25 = 100 - 100 = 0$

Since $b^2 - 4ac = 0$, there is only one solution. It is a real number. Answer B is correct.

37. $2x^2 - 6x + 5 = 0$

$a = 2$, $b = -6$, $c = 5$

$$x = \frac{-b \pm \sqrt{b^2 - 4ac}}{2a} = \frac{-(-6) \pm \sqrt{(-6)^2 - 4 \cdot 2 \cdot 5}}{2 \cdot 2}$$

$$x = \frac{6 \pm \sqrt{36 - 40}}{4} = \frac{6 \pm \sqrt{-4}}{4} = \frac{6 \pm 2i}{4}$$

$$x = \frac{2(3 \pm i)}{2 \cdot 2} = \frac{3 \pm i}{2} = \frac{3}{2} \pm \frac{1}{2}i, \ or \ \frac{3}{2} \pm \frac{i}{2}$$

Answer D is correct.

38. *Familiarize.* Let b and h represent the base and height, respectively, in centimeters.

Translate. The sum of the base and height is 38 cm, so we have $b + h = 38$. Solving for b, we get $b = -h + 38$. The area is given by $\dfrac{1}{2}bh$. Substituting $-h + 38$, we get a quadratic function:

$$A = \frac{1}{2}bh = \frac{1}{2}(-h + 38)h = -\frac{1}{2}h^2 + 19h$$

Carry out. The coefficient of h^2 is negative, so the graph of the function opens down and a maximum exists. We find the coordinates of the vertex.

$$h = -\frac{b}{2a} = -\frac{19}{2\left(-\frac{1}{2}\right)} = -\frac{19}{-1} = 19$$

When $h = 19$, $A = -\frac{1}{2}(19)^2 + 19 \cdot 19 = 180.5$. The vertex is $(19, 180.5)$. Thus the maximum area is 180.5 cm^2. This occurs when $h = 19$ cm and $b = -h + 38 = -19 + 38 = 19$ cm.

Check. We could complete the square to find the vertex. We could also use the graph of the function to check the maximum value. The answer checks.

State. The maximum area of 180.5 cm^2 occurs when the base is 19 cm and the height is 19 cm.

39. ***Familiarize***. Let x represent one of the numbers. Then $\sqrt{x}$ represents the other number.

Translate. The average of the two numbers is 171, so we have

$$\frac{x + \sqrt{x}}{2} = 171.$$

Solve. We solve the equation.

$$\frac{x + \sqrt{x}}{2} = 171$$
$$x + \sqrt{x} = 342$$
$$x + \sqrt{x} - 342 = 0$$

Let $u = \sqrt{x}$.

$$u^2 + u - 342 = 0$$
$$(u + 19)(u - 18) = 0$$
$$u = -19 \quad or \quad u = 18$$
$$\sqrt{x} = -19 \quad or \quad \sqrt{x} = 18$$

No solution or $x = 324$

Check. Since $\sqrt{x}$ denotes the positive square root of x, the equation $\sqrt{x} = -19$ has no solution. If $x = 324$, then $\sqrt{x} = \sqrt{324} = 18$. Since $\frac{324 + 18}{2} = \frac{342}{2} = 171$, the answer checks.

State. The numbers are 324 and 18.

Chapter 7 Discussion and Writing Exercises

1. Yes; for any quadratic function $f(x) = ax^2 + bx + c$, $f(0) = c$ so the graph of every quadratic function has a y-intercept, $(0, c)$.

2. If the leading coefficient is positive, the graph of the function opens up and hence has a minimum value. If the leading coefficient is negative, the graph of the function opens down and hence has a maximum value.

3. When an input of $f(x) = (x+3)^2$ is 3 less than (or 3 units to the left of) an input of $f(x) = x^2$, then the outputs are the same. In addition, for any input, the output of $f(x) = (x+3)^2 - 4$ is 4 less than (or 4 units down from) the output of $f(x) = (x-3)^2$. Thus the graph of $f(x) = (x+3)^2 - 4$ looks like the graph of $f(x) = x^2$ translated 3 units to the left and 4 units down.

4. Find a quadratic function $f(x)$ whose graph lies entirely above the x-axis or a quadratic function $g(x)$ whose graph lies entirely below the x-axis. Then write $f(x) < 0$, $f(x) \le 0$, $g(x) > 0$, or $g(x) \ge 0$. For example, the quadratic inequalities $x^2 + 1 < 0$ and $-x^2 - 5 \ge 0$ have no solution.

5. No; if the vertex is off the x-axis, then due to symmetry the graph has either no x-intercept or two x-intercepts.

6. The x-coordinate of the vertex lies halfway between the x-coordinates of the x-intercepts. The function must be evaluated for this value of x in order to determine the maximum or minimum value.

Chapter 7 Test

1. a) $3x^2 - 4 = 0$

$$3x^2 = 4$$
$$x^2 = \frac{4}{3}$$
$$x = -\sqrt{\frac{4}{3}} \quad or \quad x = \sqrt{\frac{4}{3}}$$
$$x = -\sqrt{\frac{4}{3} \cdot \frac{3}{3}} \quad or \quad x = \sqrt{\frac{4}{3} \cdot \frac{3}{3}}$$
$$x = -\frac{2\sqrt{3}}{3} \quad or \quad x = \frac{2\sqrt{3}}{3}$$

The solutions are $-\frac{2\sqrt{3}}{3}$ and $\frac{2\sqrt{3}}{3}$, or $\pm\frac{2\sqrt{3}}{3}$.

b) The real-number solutions of the equation $3x^2 - 4 = 0$ are the first coordinates of the x-intercepts of the graph of $f(x) = 3x^2 - 4$. Thus, the x-intercepts are $\left(-\frac{2\sqrt{3}}{3}, 0\right)$ and $\left(\frac{2\sqrt{3}}{3}, 0\right)$.

2. $x^2 + x + 1 = 0$

$a = 1$, $b = 1$, $c = 1$

$$x = \frac{-b \pm \sqrt{b^2 - 4ac}}{2a}$$
$$x = \frac{-1 \pm \sqrt{1^2 - 4 \cdot 1 \cdot 1}}{2 \cdot 1} = \frac{-1 \pm \sqrt{1 - 4}}{2}$$
$$x = \frac{-1 \pm \sqrt{-3}}{2} = \frac{-1 \pm i\sqrt{3}}{2} = -\frac{1}{2} \pm i\frac{\sqrt{3}}{2}$$

The solutions are $-\frac{1}{2} \pm i\frac{\sqrt{3}}{2}$.

3. $x - 8\sqrt{x} + 7 = 0$

Let $u = \sqrt{x}$ and think of x as $(\sqrt{x})^2$, or u^2.

$u^2 - 8u + 7 = 0$ Substituting

$(u - 1)(u - 7) = 0$

$u - 1 = 0$ *or* $u - 7 = 0$

$u = 1$ *or* $u = 7$

Now we substitute $\sqrt{x}$ for u and solve these equations.

$\sqrt{x} = 1$ *or* $\sqrt{x} = 7$

$x = 1$ *or* $x = 49$

Both numbers check. The solutions are 1 and 49.

4. $4x(x - 2) - 3x(x + 1) = -18$

$4x^2 - 8x - 3x^2 - 3x = -18$

$x^2 - 11x = -18$

$x^2 - 11x + 18 = 0$

$(x - 2)(x - 9) = 0$

$x - 2 = 0$ *or* $x - 9 = 0$

$x = 2$ *or* $x = 9$

Both numbers check. The solutions are 2 and 9.

5. $4x^4 - 17x^2 + 15 = 0$

Let $u = x^2$ and think of x^4 as $(x^2)^2$, or u^2.

$4u^2 - 17u + 15 = 0$ Substituting

$(4u - 5)(u - 3) = 0$

$4u - 5 = 0$ *or* $u - 3 = 0$

$4u = 5$ *or* $u = 3$

$u = \dfrac{5}{4}$ *or* $u = 3$

Now substitute x^2 for u and solve these equations.

$x^2 = \dfrac{5}{4}$ *or* $x^2 = 3$

$x = \pm\dfrac{\sqrt{5}}{2}$ *or* $x = \pm\sqrt{3}$

All four numbers check. They are the solutions.

6. $x^4 + 4x = 2$

$x^2 + 4x - 2 = 0$

$x = \dfrac{-4 \pm \sqrt{4^2 - 4 \cdot 1 \cdot (-2)}}{2 \cdot 1} = \dfrac{-4 \pm \sqrt{16 + 8}}{2}$

$x = \dfrac{-4 \pm \sqrt{24}}{2} = \dfrac{-4 \pm \sqrt{4 \cdot 6}}{2}$

$x = \dfrac{-4 \pm 2\sqrt{6}}{2} = \dfrac{2(-2 \pm \sqrt{6})}{2 \cdot 1} = -2 \pm \sqrt{6}$

We can use a calculator to approximate the solutions:

$-2 - \sqrt{6} \approx -4.449; \ -2 + \sqrt{6} \approx 0.449$

The solutions are $-2 \pm \sqrt{6}$, or approximately -4.449 and 0.449.

7. $\dfrac{1}{4 - x} + \dfrac{1}{2 + x} = \dfrac{3}{4}$ LCM is $(4-x)(2+x)$

$(4-x)(2+x)\left(\dfrac{1}{4-x} + \dfrac{1}{2+x}\right) = (4 - x)(2 + x) \cdot \dfrac{3}{4}$

$2 + x + 4 - x = \dfrac{3}{4}(8 + 2x - x^2)$

$6 = \dfrac{3}{4}(8 + 2x - x^2)$

$\dfrac{4}{3} \cdot 6 = \dfrac{4}{3} \cdot \dfrac{3}{4}(8 + 2x - x^2)$

$8 = 8 + 2x - x^2$

$x^2 - 2x = 0$

$x(x - 2) = 0$

$x = 0$ *or* $x - 2 = 0$

$x = 0$ *or* $x = 2$

Both numbers check. The solutions are 0 and 2.

8. $x^2 - 4x + 1 = 0$

$x^2 - 4x = -1$

$x^2 - 4x + 4 = -1 + 4$ Adding 4: $\left(\dfrac{-4}{2}\right)^2 = (-2)^2 = 4$

$(x - 2)^2 = 3$

$x - 2 = -\sqrt{3}$ *or* $x - 2 = \sqrt{3}$

$x = 2 - \sqrt{3}$ *or* $x = 2 + \sqrt{3}$

The solutions are $2 \pm \sqrt{3}$.

9. $s(t) = 16t^2$

$723 = 16t^2$

$\dfrac{723}{16} = t^2$

$45.1875 = t^2$

$\sqrt{45.1875} = t$

$6.7 \approx t$

It will take an object about 6.7 sec to fall from the top.

10. ***Familiarize.*** Let r = the speed of the boat in still water and let t = the time of the trip upriver. Then $4 - t$ = the time of the return trip downriver. We organize the information in a table.

	Distance	Speed	Time
Upriver	3	$r - 2$	t
Downriver	3	$r + 2$	$4 - t$

Translate. Using $t = d/r$ in each row of the table, we have two equations:

$t = \dfrac{3}{r - 2}$ and $4 - t = \dfrac{3}{r + 2}$.

Solve. We substitute $\dfrac{3}{r - 2}$ for t in the second equation and solve for r.

$$4 - \frac{3}{r-2} = \frac{3}{r+2}$$
$$\text{LCM} = (r-2)(r+2)$$
$$(r-2)(r+2)\left(4 - \frac{3}{r-2}\right) = (r-2)(r+2) \cdot \frac{3}{r+2}$$
$$4(r-2)(r+2) - 3(r+2) = 3(r-2)$$
$$4(r^2-4) - 3r - 6 = 3r - 6$$
$$4r^2 - 16 - 3r - 6 = 3r - 6$$
$$4r^2 - 3r - 22 = 3r - 6$$
$$4r^2 - 6r - 16 = 0$$
$$2r^2 - 3r - 8 = 0 \quad \text{Dividing by 2}$$
$$r = \frac{-(-3) \pm \sqrt{(-3)^2 - 4 \cdot 2 \cdot (-8)}}{2 \cdot 2}$$
$$r = \frac{3 \pm \sqrt{9 + 64}}{4}$$
$$r = \frac{3 \pm \sqrt{73}}{4}$$
$$r \approx 2.89 \ or \ r \approx -1.39$$

Check. Since negative speed has no meaning in this application, we check only 2.89. If $r \approx 2.89$, then the speed upriver is about $2.89 - 2$, or 0.89 mph, and the time it takes to travel 3 mi is about $3/0.89 \approx 3.37$ hr. The speed downriver is about $2.89 + 2$, or 4.89 mph, and the time it takes to travel 3 mi is about $3/4.89 \approx 0.61$ hr. The total time is about $3.37 + 0.61$, or $3.98 \approx 4$ hr, so the answer checks.

State. The speed of the boat in still water must be about 2.89 mph.

11. **Familiarize**. Let $l =$ the length of the board and $w =$ the width, in cm. Then the perimeter of the board is $2l + 2w$ and the area is $l \cdot w$.

Translate. Using the formula for perimeter, we have:
$$2l + 2w = 28$$
$$2l = 28 - 2w$$
$$l = \frac{28 - 2w}{2}$$
$$l = 14 - w$$

Substituting $14 - w$ for l in the formula for area, we get a quadratic function.
$$A = l \cdot w = (14 - w)w = 14w - w^2, \text{ or } -w^2 + 14w$$

Carry out. We complete the square in order to find the vertex of the quadratic function.
$$A = -w^2 + 14w$$
$$= -(w^2 - 14w)$$
$$= -(w^2 - 14w + 49 - 49)$$
$$= -(w^2 - 14w + 49) + (-1)(-49)$$
$$= -(w - 7)^2 + 49$$

The vertex is $(7, 49)$. The coefficient of w^2 is negative, so the graph of the function is a parabola that opens down. This tells us that the function has a maximum value and that value occurs when $w = 7$. When $w = 7$, $l = 14 - w = 14 - 7 = 7$.

Check. We could find the value of the function for some values of w less than 7 and for some values greater than 7, determining that the maximum value we found, 49, is larger than these function values. We could also use the graph of the function to check the maximum value. Our answer checks.

State. Dimensions of 7 cm by 7 cm will provide the maximum area.

12. $V(T) = 48T^2$
$$43 = 48T^2$$
$$\frac{43}{48} = T^2$$
$$\sqrt{\frac{43}{48}} = T$$
$$0.946 \approx T$$

Robinson's hang time is about 0.946 sec.

13. $x^2 + 5x + 17 = 0$

$a = 1$, $b = 5$, $c = 17$

We compute the discriminant.

$b^2 - 4ac = 5^2 - 4 \cdot 1 \cdot 17 = 25 - 68 = -43$

Since $b^2 - 4ac < 0$, there are two nonreal solutions.

14. The solutions are $\sqrt{3}$ and $3\sqrt{3}$.
$$x = \sqrt{3} \ \ or \ \ \ \ \ \ \ x = 3\sqrt{3}$$
$$x - \sqrt{3} = 0 \ \ or \ \ x - 3\sqrt{3} = 0$$
$$(x - \sqrt{3})(x - 3\sqrt{3}) = 0 \ \ \text{Principle of zero}$$
$$\text{products}$$
$$x^2 - 3\sqrt{3}x - \sqrt{3}x + 3 \cdot 3 = 0$$
$$x^2 - 4\sqrt{3}x + 9 = 0$$

15. $V = 48T^2$
$$\frac{V}{48} = T^2$$
$$\sqrt{\frac{V}{48}} = T$$

We can rationalize the denominator.
$$\sqrt{\frac{V}{48}} = \sqrt{\frac{V}{48} \cdot \frac{3}{3}} = \sqrt{\frac{3V}{144}} = \frac{\sqrt{3V}}{12}$$

Thus, we also have $T = \frac{\sqrt{3V}}{12}$.

16. $f(x) = -x^2 - 2x$

We complete the square.
$$f(x) = -(x^2 + 2x)$$
$$= -(x^2 + 2x + 1 - 1) \quad \left(\frac{2}{2}\right)^2 = 1^2 = 1;$$
$$\text{add } 1 - 1$$
$$= -(x^2 + 2x + 1) + (-1)(-1)$$
$$= -(x + 1)^2 + 1$$
$$= -[x - (-1)]^2 + 1$$

a) The vertex is $(-1, 1)$.

b) The line of symmetry is $x = -1$.

c) The coefficient of x^2 is negative, so the graph opens down. This tells us that 1 is a maximum value.

d) We plot some points and draw the curve.

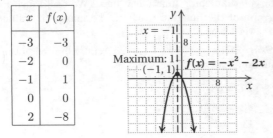

x	$f(x)$
-3	-3
-2	0
-1	1
0	0
2	-8

17. $f(x) = 4x^2 - 24x + 41$

$= 4(x^2 - 6x) + 41$

We complete the square inside the parentheses.

$f(x) = 4(x^2 - 6x + 9 - 9) + 41 \qquad \left(\dfrac{-6}{2}\right)^2 = (-3)^2 = 9;$

$\qquad\qquad\qquad\qquad\qquad\qquad\qquad \text{add } 9 - 9$

$= 4(x^2 - 6x + 9) + 4(-9) + 41$

$= 4(x - 3)^2 - 36 + 41$

$= 4(x - 3)^2 + 5$

a) The vertex is $(3, 5)$.

b) The line of symmetry is $x = 3$.

c) The coefficient of x^2 is positive, so the graph opens up. This tells us that 5 is a minimum value.

d) We plot some points and draw the curve.

x	$f(x)$
2	9
3	5
4	9
5	21

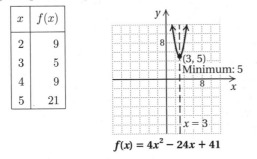

$f(x) = 4x^2 - 24x + 41$

18. $f(x) = -x^2 + 4x - 1$

The y-intercept is $(0, f(0))$. Since $f(0) = -0^2 + 4 \cdot 0 - 1 = -1$, the y-intercept is $(0, -1)$.

To find the x-intercepts, we solve $-x^2 + 4x - 1 = 0$. Using the quadratic formula we get $x = 2 \pm \sqrt{3}$. Thus, the x-intercepts are $(2 - \sqrt{3}, 0)$ and $(2 + \sqrt{3}, 0)$.

19. *Familiarize.* Let x and y represent the numbers.

Translate. The difference of the numbers is 8, so we have $x - y = 8$. Solve for x, we get $x = y + 8$. The product of the numbers is xy. Substituting $y + 8$ for x in the product, we get a quadratic function:

$P = xy = (y + 8)y = y^2 + 8y$

Carry out. The coefficient of y^2 is positive, so the graph of the function opens up and a minimum exists. We complete the square in order to find the vertex of the quadratic function.

$P = y^2 + 8y$

$= y^2 + 8y + 16 - 16$

$= (y + 4)^2 - 16$

$= [y - (-4)]^2 + (-16)$

The vertex is $(-4, -16)$. This tells us that the minimum product is -16. The minimum occurs when $y = -4$. Note that when $y = -4$, $x = y + 8 = -4 + 8 = 4$, so the numbers that yield the minimum product are 4 and -4.

Check. We could find the value of the function for some values of y less than -4 and for some values greater than -4, determining that the minimum value we found is smaller than these function values. We could also use the graph of the function to check the minimum value. Our answer checks.

State. The minimum product is -16. The numbers 4 and -4 yield this product.

20. We look for a function of the form $f(x) = ax^2 + bx + c$. Substituting the data points, we get

$0 = a \cdot 0^2 + b \cdot 0 + c,$

$0 = a \cdot 3^2 + b \cdot 3 + c,$

$2 = a \cdot 5^2 + b \cdot 5 + c,$

or

$0 = c,$

$0 = 9a + 3b + c,$

$2 = 25a + 5b + c.$

Solving this system, we get

$a = \dfrac{1}{5},\ b = -\dfrac{3}{5},\ \text{and } c = 0.$

Then the function we are looking for is

$f(x) = \dfrac{1}{5}x^2 - \dfrac{3}{5}x.$

21. a) Substituting the data points, we get

$18.5 = a \cdot 0^2 + b \cdot 0 + c,$

$20.7 = a \cdot 6^2 + b \cdot 6 + c,$

$8.7 = a \cdot 12^2 + b \cdot 12 + c,$

or

$18.5 = c,$

$20.7 = 36a + 6b + c,$

$8.7 = 144a + 12b + c.$

Solving this system, we get

$a = -\dfrac{71}{360},\ b = \dfrac{31}{20},\ \text{and } c = 18.5.$

Then the function we are looking for is

$A(x) = -\dfrac{71}{360}x^2 + \dfrac{31}{20}x + 18.5,$

where x is the number of years after 2000 and $A(x)$ is in thousands.

b) In 2011, $x = 2011 - 2000 = 11$.

$A(11) = -\dfrac{71}{360} \cdot 11^2 + \dfrac{31}{20} \cdot 11 + 18.5$

$\approx 11.7 \text{ thousand adoptions}$

22.
$$x^2 < 6x + 7$$
$$x^2 - 6x - 7 < 0$$
$$(x+1)(x-7) < 0$$

The solutions of $(x+1)(x-7) = 0$ are -1 and 7. They divide the real-number line into three intervals as shown:

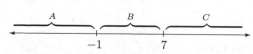

We try test numbers in each interval.

A: Test -2, $y = (-2+1)(-2-7) = 9 > 0$

B: Test 0, $y = (0+1)(0-7) = -7 < 0$

C: Test 8, $y = (8+1)(8-7) = 9 > 0$

The expression is negative for all real numbers in interval B. The solution set is $\{x| -1 < x < 7\}$, or $(-1, 7)$.

23. $\dfrac{x-5}{x+3} < 0$

Solve the related equation.
$$\frac{x-5}{x+3} = 0$$
$$x - 5 = 0$$
$$x = 5$$

Find the numbers for which the rational expression is undefined.
$$x + 3 = 0$$
$$x = -3$$

Use the numbers 5 and -3 to divide the number line into intervals as shown:

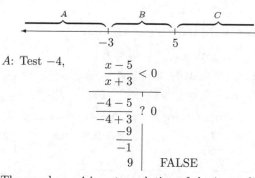

A: Test -4,
$$\frac{x-5}{x+3} < 0$$
$$\frac{-4-5}{-4+3} \; ? \; 0$$
$$\frac{-9}{-1}$$
$$9 \;\Big|\; \text{FALSE}$$

The number -4 is not a solution of the inequality, so the interval A is not part of the solution set.

B: Test 0,
$$\frac{x-5}{x+3} < 0$$
$$\frac{0-5}{0+3} \; ? \; 0$$
$$\frac{-5}{3}$$
$$-\frac{5}{3} \;\Big|\; \text{TRUE}$$

The number 0 is a solution of the inequality, so the interval B is part of the solution set.

C: Test 6,
$$\frac{x-5}{x+3} < 0$$
$$\frac{6-5}{6+3} \; ? \; 0$$
$$\frac{1}{9} \;\Big|\; \text{FALSE}$$

The number 6 is not a solution of the inequality, so the interval C is not part of the solution set. The solution set is $\{x| -3 < x < 5\}$, or $(-3, 5)$.

24. $\dfrac{x-2}{(x+3)(x-1)} \geq 0$

Solve the related equation.
$$\frac{x-2}{(x+3)(x-1)} = 0$$
$$x - 2 = 0$$
$$x = 2$$

Find the numbers for which the rational expression is undefined.
$$(x+3)(x-1) = 0$$
$$x = -3 \; or \; x = 1$$

Use the numbers 2, -3, and 1 to divide the number line into intervals as shown:

We try test numbers in each interval.

A: Test -4,
$$\frac{x-2}{(x+3)(x-1)} \geq 0$$
$$\frac{-4-2}{(-4+3)(-4-1)} \; ? \; 0$$
$$\frac{-6}{-1(-5)}$$
$$-\frac{6}{5} \;\Big|\; \text{FALSE}$$

Interval A is not part of the solution set.

B: Test 0,
$$\frac{x-2}{(x+3)(x-1)} \geq 0$$
$$\frac{0-2}{(0+3)(0-1)} \; ? \; 0$$
$$\frac{-2}{3(-1)}$$
$$\frac{2}{3} \;\Big|\; \text{TRUE}$$

Interval B is part of the solution set.

C: Test 1.5,
$$\frac{x-2}{(x+3)(x-1)} \geq 0$$
$$\frac{1.5-2}{(1.5+3)(1.5-1)} \; ? \; 0$$
$$\frac{-0.5}{4.5(0.5)}$$
$$-0.\overline{2} \;\Big|\; \text{FALSE}$$

Interval C is not part of the solution set.

D: Test 3,

$$\dfrac{x-2}{(x+3)(x-1)} \geq 0$$

$$\dfrac{3-2}{(3+3)(3-1)} \;?\; 0$$

$$\dfrac{1}{6 \cdot 2}$$

$$\dfrac{1}{12} \quad \text{TRUE}$$

Interval D is part of the solution set.

The solution set includes intervals B and D. The number 2 is also included because the inequality symbol is $\geq$ and 2 is the solution of the related equation. The numbers -3 and 1 are not included because the rational expression is undefined for these numbers. The solution set is $\{x | -3 < x < 1 \text{ or } x \geq 2\}$, or $(-3, 1) \cup [2, \infty)$.

25.
$$x = \dfrac{i}{2} \quad \text{or} \quad x = -\dfrac{i}{2}$$

$$x - \dfrac{i}{2} = 0 \quad \text{or} \quad x + \dfrac{i}{2} = 0$$

Then we have

$$\left(x - \dfrac{i}{2}\right)\left(x + \dfrac{i}{2}\right) = 0$$

$$x^2 - \dfrac{i^2}{4} = 0$$

$$x^2 - \left(\dfrac{-1}{4}\right) = 0$$

$$x^2 + \dfrac{1}{4} = 0$$

$$4x^2 + 1 = 0 \qquad \text{Multiplying by 4}$$

Answer C is correct.

26. We look for a function $f(x) = ax^2 + bx + c$. We substitute the known points.

$$0 = a(-2)^2 + b(-2) + c,$$
$$0 = a \cdot 7^2 + b \cdot 7 + c,$$
$$8 = a \cdot 0^2 + b \cdot 0 + c,$$
or
$$0 = 4a - 2b + c,$$
$$0 = 49a + 7b + c,$$
$$8 = c.$$

Solving this system, we get $\left(-\dfrac{4}{7}, \dfrac{20}{7}, 8\right)$. Thus, the function is $f(x) = -\dfrac{4}{7}x^2 + \dfrac{20}{7}x + 8$.

The coefficient of x^2 is negative, so the function has a maximum value. To find it we find the vertex of the function.

$$-\dfrac{b}{2a} = -\dfrac{20/7}{2(-4/7)} = -\dfrac{20/7}{-8/7} = \dfrac{20}{7} \cdot \dfrac{7}{8} = \dfrac{5}{2}$$

$$f\left(\dfrac{5}{2}\right) = -\dfrac{4}{7}\left(\dfrac{5}{2}\right)^2 + \dfrac{20}{7}\left(\dfrac{5}{2}\right) + 8 = \dfrac{81}{7}$$

The maximum value is $\dfrac{81}{7}$.

27. $kx^2 + 3x - k = 0$

First we substitute -2 for x and find k.

$$k(-2)^2 + 3(-2) - k = 0$$
$$4k - 6 - k = 0$$
$$3k - 6 = 0$$
$$3k = 6$$
$$k = 2$$

Now we have:

$$2x^2 + 3x - 2 = 0$$
$$(2x - 1)(x + 2) = 0$$
$$2x - 1 = 0 \quad \text{or} \quad x + 2 = 0$$
$$2x = 1 \quad \text{or} \qquad x = -2$$
$$x = \dfrac{1}{2} \quad \text{or} \qquad x = -2$$

The other solution is $\dfrac{1}{2}$.

Cumulative Review Chapters 1 - 7

1. Let $d =$ the distance that a hole in one shot would travel, in yards. We use the Pythagorean theorem.

$$a^2 + b^2 = d^2$$
$$177^2 + 383^2 = d^2$$
$$31,329 + 146,689 = d^2$$
$$178,018 = d^2$$
$$\sqrt{178,018} = d$$
$$422 \approx d$$

The shot would travel about 422 yd.

2. $(4 + 8x^2 - 5x) - (-2x^2 + 3x - 2) =$
$4 + 8x^2 - 5x + 2x^2 - 3x + 2 =$
$10x^2 - 8x + 6$

3. $(2x^2 - x + 3)(x - 4) = (2x^2 - x + 3)x + (2x^2 - x + 3)(-4) =$
$2x^3 - x^2 + 3x - 8x^2 + 4x - 12 = 2x^3 - 9x^2 + 7x - 12$

4.
$$\dfrac{a^2 - 16}{5a - 15} \cdot \dfrac{2a - 6}{a + 4} = \dfrac{(a^2 - 16)(2a - 6)}{(5a - 15)(a + 4)}$$
$$= \dfrac{(a + 4)(a - 4)(2)(a - 3)}{5(a - 3)(a + 4)}$$
$$= \dfrac{(a + 4)(a - 3)}{(a + 4)(a - 3)} \cdot \dfrac{2(a - 4)}{5}$$
$$= \dfrac{2(a - 4)}{5}$$

5.
$$\frac{y}{y^2 - y - 42} \div \frac{y^2}{y - 7} = \frac{y}{y^2 - y - 42} \cdot \frac{y - 7}{y^2}$$

$$= \frac{y(y - 7)}{(y^2 - y - 42)(y^2)}$$

$$= \frac{y(y - 7) \cdot 1}{(y - 7)(y + 6) \cdot y \cdot y}$$

$$= \frac{y(y - 7)}{y(y - 7)} \cdot \frac{1}{y(y + 6)}$$

$$= \frac{1}{y(y + 6)}$$

6.
$$\frac{2}{m + 1} + \frac{3}{m - 5} - \frac{m^2 - 1}{m^2 - 4m - 5}$$

$$= \frac{2}{m + 1} + \frac{3}{m - 5} - \frac{(m + 1)(m - 1)}{(m - 5)(m + 1)}$$

$$= \frac{2}{m + 1} + \frac{3}{m - 5} - \frac{m - 1}{m - 5}, \text{ LCD is } (m + 1)(m - 5)$$

$$= \frac{2}{m + 1} \cdot \frac{m - 5}{m - 5} + \frac{3}{m - 5} \cdot \frac{m + 1}{m + 1} - \frac{m - 1}{m - 5} \cdot \frac{m + 1}{m + 1}$$

$$= \frac{2(m - 5) + 3(m + 1) - (m - 1)(m + 1)}{(m + 1)(m - 5)}$$

$$= \frac{2m - 10 + 3m + 3 - (m^2 - 1)}{(m + 1)(m - 5)}$$

$$= \frac{2m - 10 + 3m + 3 - m^2 + 1}{(m + 1)(m - 5)}$$

$$= \frac{-m^2 + 5m - 6}{(m + 1)(m - 5)}$$

$$= \frac{-(m^2 - 5m + 6)}{(m + 1)(m - 5)}$$

$$= \frac{-(m - 3)(m - 2)}{(m + 1)(m - 5)}$$

7. We will use synthetic division.

$$(9x^3 + 5x^2 + 2) \div (x + 2) = (9x^3 + 5x^2 + 0x + 2) \div [x - (-2)]$$

$$\begin{array}{r|rrrr} -2 & 9 & 5 & 0 & 2 \\ & & -18 & 26 & -52 \\ \hline & 9 & -13 & 26 & -50 \end{array}$$

The answer is $9x^2 - 13x + 26$, R -50, or $9x^2 - 13x + 26 + \frac{-50}{x + 2}$.

8.
$$\frac{\frac{1}{x} - \frac{1}{y}}{x + y} = \frac{\frac{1}{x} - \frac{1}{y}}{x + y} \cdot \frac{xy}{xy}$$

$$= \frac{\frac{1}{x} \cdot xy - \frac{1}{y} \cdot xy}{xy(x + y)}$$

$$= \frac{y - x}{xy(x + y)}$$

9. $\sqrt{0.36} = 0.6$

(Remember $\sqrt{}$ indicates the nonnegative square root.)

10. $\sqrt{9x^2 - 36x + 36} = \sqrt{9(x^2 - 4x + 4)} = \sqrt{9(x - 2)^2} = 3(x - 2)$

11. $6\sqrt{45} - 3\sqrt{20} = 6\sqrt{9 \cdot 5} - 3\sqrt{4 \cdot 5}$

$$= 6 \cdot 3\sqrt{5} - 3 \cdot 2\sqrt{5}$$

$$= 18\sqrt{5} - 6\sqrt{5}$$

$$= 12\sqrt{5}$$

12.
$$\frac{2\sqrt{3} - 4\sqrt{2}}{\sqrt{2} - 3\sqrt{6}} = \frac{2\sqrt{3} - 4\sqrt{2}}{\sqrt{2} - 3\sqrt{6}} \cdot \frac{\sqrt{2} + 3\sqrt{6}}{\sqrt{2} + 3\sqrt{6}}$$

$$= \frac{2\sqrt{6} + 6\sqrt{18} - 4 \cdot 2 - 12\sqrt{12}}{2 - 9 \cdot 6}$$

$$= \frac{2\sqrt{6} + 6\sqrt{9 \cdot 2} - 8 - 12\sqrt{4 \cdot 3}}{2 - 54}$$

$$= \frac{2\sqrt{6} + 6 \cdot 3\sqrt{2} - 8 - 12 \cdot 2\sqrt{3}}{-52}$$

$$= \frac{2\sqrt{6} + 18\sqrt{2} - 8 - 24\sqrt{3}}{-52}$$

$$= \frac{2(\sqrt{6} + 9\sqrt{2} - 4 - 12\sqrt{3})}{2(-26)}$$

$$= \frac{\sqrt{6} + 9\sqrt{2} - 4 - 12\sqrt{3}}{-26}$$

13. $(8^{2/3})^4 = 8^{(2/3) \cdot 4} = 8^{8/3} = (\sqrt[3]{8})^8 = 2^8 = 256$

14. $(3 + 2i)(5 - i) = 15 - 3i + 10i - 2i^2 = 15 + 7i - 2(-1) = 15 + 7i + 2 = 17 + 7i$

15. $\frac{6 - 2i}{3i} = \frac{6 - 2i}{3i} \cdot \frac{-3i}{-3i} = \frac{-18i + 6i^2}{-9i^2} = \frac{-18i + 6(-1)}{-9(-1)} = \frac{-18i - 6}{9} = -2i - \frac{2}{3}$, or $-\frac{2}{3} - 2i$

16. Using trial and error or the ac-method, we have $2t^2 - 7t - 30 = (2t + 5)(t - 6)$.

17. $a^2 + 3a - 54$

We look for two numbers whose product is -54 and whose sum is 3. The numbers we want are 9 and -6.

$$a^2 + 3a - 54 = (a + 9)(a - 6)$$

18. $-3a^3 + 12a^2 = -3a^2 \cdot a - 3a^2(-4) = -3a^2(a - 4)$

19. $64a^2 - 9b^2 = (8a)^2 - (3b)^2 = (8a + 3b)(8a - 3b)$

20. $3a^2 - 36a + 108 = 3(a^2 - 12a + 36) = 3(a - 6)^2$

21. $\frac{1}{27}a^3 - 1 = \left(\frac{1}{3}a\right)^3 - 1^3 = \left(\frac{1}{3}a - 1\right)\left(\frac{1}{9}a^2 + \frac{1}{3}a + 1\right)$

22.
$$24a^3 + 18a^2 - 20a - 15$$

$$= 6a^2(4a + 3) - 5(4a + 3)$$

$$= (4a + 3)(6a^2 - 5)$$

23.
$$(x + 1)(x - 1) + (x + 1)(x + 2)$$

$$= (x + 1)(x - 1 + x + 2)$$

$$= (x + 1)(2x + 1)$$

24. $3(4x - 5) + 6 = 3 - (x + 1)$

$\quad\quad 12x - 15 + 6 = 3 - x - 1$

$\quad\quad\quad\quad 12x - 9 = -x + 2$

$\quad\quad\quad\quad 13x - 9 = 2$

$\quad\quad\quad\quad\quad 13x = 11$

$\quad\quad\quad\quad\quad\quad x = \dfrac{11}{13}$

The solution is $\dfrac{11}{13}$.

25. $F = \dfrac{mv^2}{r}$

$\quad rF = mv^2$

$\quad\quad r = \dfrac{mv^2}{F}$

26. $5 - 3(2x + 1) \le 8x - 3$

$\quad\quad 5 - 6x - 3 \le 8x - 3$

$\quad\quad\quad -6x + 2 \le 8x - 3$

$\quad\quad\quad\quad\quad 2 \le 14x - 3$

$\quad\quad\quad\quad\quad 5 \le 14x$

$\quad\quad\quad\quad \dfrac{5}{14} \le x$

The solution set is $\left\{ x \middle| x \ge \dfrac{5}{14} \right\}$, or $\left[\dfrac{5}{14}, \infty \right)$.

27. $3x - 2 < -6 \quad or \quad x + 3 > 9$

$\quad\quad 3x < -4 \quad or \quad\quad x > 6$

$\quad\quad\quad x < -\dfrac{4}{3} \quad or \quad\quad x > 6$

The solution set is $\left\{ x \middle| x < -\dfrac{4}{3} \text{ or } x > 6 \right\}$, or

$\left(-\infty, -\dfrac{4}{3} \right) \cup (6, \infty)$.

28. $|4x - 1| \le 14$

$\quad -14 \le 4x - 1 \le 14$

$\quad -13 \le 4x \le 15$

$\quad -\dfrac{13}{4} \le x \le \dfrac{15}{4}$

The solution set is $\left\{ x \middle| -\dfrac{13}{4} \le x \le \dfrac{15}{4} \right\}$, or $\left[-\dfrac{13}{4}, \dfrac{15}{4} \right]$.

29. $5x + 10y = -10, \quad (1)$

$\quad -2x - 3y = 5 \quad\quad (2)$

Multiply Equation (1) by 2 and Equation (2) by 5 and then add.

$\quad\quad 10x + 20y = -20$

$\quad\quad \underline{-10x - 15y = 25}$

$\quad\quad\quad\quad\quad 5y = 5$

$\quad\quad\quad\quad\quad y = 1$

Now substitute 1 for y in one of the original equations and solve for x. We will use Equation (1).

$\quad\quad 5x + 10 \cdot 1 = -10$

$\quad\quad\quad 5x + 10 = -10$

$\quad\quad\quad\quad 5x = -20$

$\quad\quad\quad\quad x = -4$

The solution is $(-4, 1)$.

30. $2x + \ y \ - \ z = 9, \quad (1)$

$\quad 4x - 2y + \ z = -9, \quad (2)$

$\quad 2x - \ y + 2z = -12 \quad (3)$

First we will add Equations (1) and (2) to eliminate z.

$\quad\quad 2x + \ y - z = 9$

$\quad\quad \underline{4x - 2y + z = -9}$

$\quad\quad 6x - \ y \quad\quad = 0 \quad (4)$

Now multiply Equation (1) by 2 and add it to Equation (3) to eliminate z from another pair of equations.

$\quad\quad 4x + 2y - 2z = 18$

$\quad\quad \underline{2x - \ y + 2z = -12}$

$\quad\quad 6x + \ y \quad\quad = 6 \quad (5)$

Now solve the system of Equations (4) and (5).

$\quad\quad 6x - y = 0 \quad (4)$

$\quad\quad \underline{6x + y = 6} \quad (5)$

$\quad\quad 12x \quad\quad = 6 \quad$ Adding

$\quad\quad\quad\quad x = \dfrac{1}{2}$

Substitute $\dfrac{1}{2}$ for x in Equation (4) and solve for y.

$\quad\quad 6 \cdot \dfrac{1}{2} - y = 0$

$\quad\quad\quad 3 - y = 0$

$\quad\quad\quad\quad 3 = y$

Finally, substitute $\dfrac{1}{2}$ for x and 3 for y in Equation (2) and solve for z.

$\quad\quad 4 \cdot \dfrac{1}{2} - 2 \cdot 3 + z = -9$

$\quad\quad\quad\quad 2 - 6 + z = -9$

$\quad\quad\quad\quad\quad -4 + z = -9$

$\quad\quad\quad\quad\quad z = -5$

The solution is $\left(\dfrac{1}{2}, 3, -5 \right)$.

31. $10x^2 + 28x - 6 = 0$

$\quad 2(5x^2 + 14x - 3) = 0$

$\quad 2(5x - 1)(x + 3) = 0$

$\quad 5x - 1 = 0 \quad or \quad x + 3 = 0$

$\quad 5x = 1 \quad or \quad\quad x = -3$

$\quad x = \dfrac{1}{5} \quad or \quad\quad x = -3$

The solutions are $\dfrac{1}{5}$ and -3.

32. $\dfrac{2}{n} - \dfrac{7}{n} = 3$

$$-\dfrac{5}{n} = 3$$

$$-5 = 3n$$

$$-\dfrac{5}{3} = n$$

The solutions is $-\dfrac{5}{3}$.

33.
$$\dfrac{1}{2x-1} = \dfrac{3}{5x}, \text{ LCM is } 5x(2x-1)$$

$$5x(2x-1) \cdot \dfrac{1}{2x-1} = 5x(2x-1) \cdot \dfrac{3}{5x}$$

$$5x = 3(2x-1)$$

$$5x = 6x - 3$$

$$-x = -3$$

$$x = 3$$

The number 3 checks. It is the solution.

34.
$$A = \dfrac{mh}{m+a}$$

$$A(m+a) = mh$$

$$Am + Aa = mh$$

$$Aa = mh - Am$$

$$Aa = m(h-A)$$

$$\dfrac{Aa}{h-A} = m$$

35. $\sqrt{2x-1} = 6$

$$(\sqrt{2x-1})^2 = 6^2$$

$$2x - 1 = 36$$

$$2x = 37$$

$$x = \dfrac{37}{2}$$

The number $\dfrac{37}{2}$ checks. It is the solution.

36. $\sqrt{x-2} + 1 = \sqrt{2x-6}$

$$(\sqrt{x-2}+1)^2 = (\sqrt{2x-6})^2$$

$$x - 2 + 2\sqrt{x-2} + 1 = 2x - 6$$

$$x - 1 + 2\sqrt{x-2} = 2x - 6$$

$$2\sqrt{x-2} = x - 5$$

$$(2\sqrt{x-2})^2 = (x-5)^2$$

$$4(x-2) = x^2 - 10x + 25$$

$$4x - 8 = x^2 - 10x + 25$$

$$0 = x^2 - 14x + 33$$

$$0 = (x-3)(x-11)$$

$$x - 3 = 0 \ \ or \ \ x - 11 = 0$$

$$x = 3 \ \ or \qquad x = 11$$

The number 3 does not check, but 11 does. It is the solution.

37. $16(t-1) = t(t+8)$

$$16t - 16 = t^2 + 8t$$

$$0 = t^2 - 8t + 16$$

$$0 = (t-4)(t-4)$$

$$t - 4 = 0 \ \ or \ \ t - 4 = 0$$

$$t = 4 \ \ or \qquad t = 4$$

The solution is 4.

38. $x^2 - 3x + 16 = 0$

$a = 1, \ b = -3, \ c = 16$

$$x = \dfrac{-b \pm \sqrt{b^2 - 4ac}}{2a} = \dfrac{-(-3) \pm \sqrt{(-3)^2 - 4 \cdot 1 \cdot 16}}{2 \cdot 1}$$

$$x = \dfrac{3 \pm \sqrt{9 - 64}}{2} = \dfrac{3 \pm \sqrt{-55}}{2}$$

$$x = \dfrac{3 \pm \sqrt{55}i}{2} = \dfrac{3}{2} \pm \dfrac{\sqrt{55}}{2}i$$

The solutions are $\dfrac{3}{2} \pm \dfrac{\sqrt{55}}{2}i$.

39.
$$\dfrac{18}{x+1} - \dfrac{12}{x} = \dfrac{1}{3}, \text{ LCM is } 3x(x+1)$$

$$3x(x+1)\left(\dfrac{18}{x+1} - \dfrac{12}{x}\right) = 3x(x+1) \cdot \dfrac{1}{3}$$

$$3x \cdot 18 - 3 \cdot 12(x+1) = x(x+1)$$

$$54x - 36x - 36 = x^2 + x$$

$$18x - 36 = x^2 + x$$

$$0 = x^2 - 17x + 36$$

$a = 1, \ b = -17, \ c = 36$

$$x = \dfrac{-b \pm \sqrt{b^2 - 4ac}}{2a} = \dfrac{-(17) \pm \sqrt{(-17)^2 - 4 \cdot 1 \cdot 36}}{2 \cdot 1}$$

$$x = \dfrac{17 \pm \sqrt{189 - 144}}{2} = \dfrac{17 \pm \sqrt{45}}{2}$$

The solutions are $\dfrac{17 \pm \sqrt{45}}{2}$.

40.
$$P = \sqrt{a^2 - b^2}$$

$$P^2 = (\sqrt{a^2 - b^2})^2$$

$$P^2 = a^2 - b^2$$

$$P^2 + b^2 = a^2$$

$$\sqrt{P^2 + b^2} = a \quad \text{Taking the positive square root}$$

41. $\dfrac{(x+3)(x+2)}{(x-1)(x+1)} < 0$

First solve the related equation.

$$\dfrac{(x+3)(x+2)}{(x-1)(x+1)} = 0$$

$$(x+3)(x+2) = 0$$

$$x + 3 = 0 \quad or \ \ x + 2 = 0$$

$$x = -3 \ \ or \qquad x = -2$$

Now find the numbers for which the rational expression is not defined.

$$(x-1)(x+1) = 0$$

$$x - 1 = 0 \quad or \quad x + 1 = 0$$
$$x = 1 \quad or \quad x = -1$$

Use the numbers -3, -2, 1, and -1 to divide the number line into intervals as shown.

Test a number in each interval.

A: Test -4,
$$\frac{(x+3)(x+2)}{(x-1)(x+1)} < 0$$

$$\frac{(-4+3)(-4+2)}{(-4-1)(-4+1)} \; ? \; 0$$
$$\frac{-1(-2)}{(-5)(-3)} \; \Big|$$
$$\frac{2}{15} \; \Big| \quad \text{FALSE}$$

Interval A is not part of the solution set.

B: Test -2.5,
$$\frac{(x+3)(x+2)}{(x-1)(x+1)} < 0$$

$$\frac{(-2.5+3)(-2.5+2)}{(-2.5-1)(-2.5+1)} \; ? \; 0$$
$$\frac{0.5(-0.5)}{-3.5(-1.5)} \; \Big|$$
$$\frac{-0.25}{5.25} \; \Big| \quad \text{TRUE}$$

Interval B is part of the solution set.

C: Test -1.5,
$$\frac{(x+3)(x+2)}{(x-1)(x+1)} < 0$$

$$\frac{(-1.5+3)(-1.5+2)}{(-1.5-1)(-1.5+1)} \; ? \; 0$$
$$\frac{1.5(0.5)}{-2.5(-0.5)} \; \Big|$$
$$\frac{0.75}{1.25} \; \Big| \quad \text{FALSE}$$

Interval C is not part of the solution set.

D: Test 0,
$$\frac{(x+3)(x+2)}{(x-1)(x+1)} < 0$$

$$\frac{(0+3)(0+2)}{(0-1)(0+1)} \; ? \; 0$$
$$\frac{3 \cdot 2}{-1 \cdot 1} \; \Big|$$
$$-6 \; \Big| \quad \text{TRUE}$$

Interval D is part of the solution set.

E: Test 2,
$$\frac{(x+3)(x+2)}{(x-1)(x+1)} < 0$$

$$\frac{(2+3)(2+2)}{(2-1)(2+1)} \; ? \; 0$$
$$\frac{5 \cdot 4}{1 \cdot 3} \; \Big|$$
$$\frac{20}{3} \; \Big| \quad \text{FALSE}$$

Interval E is not part of the solution set.

The solution set is $\{x | -3 < x < -2 \; or \; -1 < x < 1\}$, or $(-3, -2) \cup (-1, 1)$.

42.
$$4x^2 - 25 > 0$$
$$(2x+5)(2x-5) > 0$$

The solutions of $(2x+5)(2x-5) = 0$ are $-\frac{5}{2}$ and $\frac{5}{2}$. Use these numbers to divide the number line into intervals as shown.

Test a point in each interval.

A: Test -3, $4(-3)^2 - 25 = 11 > 0$

B: Test 0, $4 \cdot 0^2 - 25 = -25 < 0$

C: Test 3, $4 \cdot 3^2 - 25 = 11 > 0$

The expression is positive for numbers in intervals A and C. The solution set is $\left\{ x \middle| x < -\frac{5}{2} \; or \; x > \frac{5}{2} \right\}$, or $\left(-\infty, -\frac{5}{2} \right) \cup \left(\frac{5}{2}, \infty \right)$.

43. Graph $x + y = 2$.

We find some ordered pairs that are solutions of the equation, plot points, and draw the graph.

x	y
-3	5
0	2
2	0

44. Graph $y \geq 6x - 5$.

First graph the equation $y = 6x - 5$ using a solid line since the inequality symbol is $\geq$. Then test a point not on the line. We will test $(0, 0)$.

$$\frac{y \geq 6x - 5}{0 \ ? \ 6 \cdot 0 - 5}$$
$$\begin{vmatrix} \\ \end{vmatrix} \quad -5 \qquad \text{TRUE}$$

We shade the half-plane that contains $(0, 0)$.

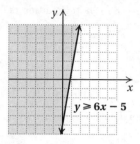

45. Graph $x < -3$.

First graph the equation $x = -3$ using a dashed line since the inequality symbol is $<$. Then test a point not on the line. We will test $(0, 0)$.

$$\frac{y < -3}{0 \ ? \ -3 \ \text{FALSE}}$$
$$\begin{vmatrix} \\ \end{vmatrix}$$

We shade the half-plane that does not contain $(0, 0)$.

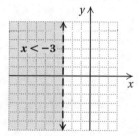

46. Graph $3x - y > 6$,

$\qquad 4x + y \leq 3$.

We graph $3x - y = 6$ using a dashed line and $4x + y = 3$ using a solid line. Indicate the region for each inequality using arrows. Shade the region where the solution sets overlap.

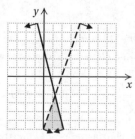

47. Graph $f(x) = x^2 - 1 = (x - 0)^2 - 1$.

The graph is a parabola with vertex $(0, -1)$. We find some points on both sides of the vertex and draw the graph.

x	$f(x)$
0	-1
-2	3
-1	0
1	0
2	3

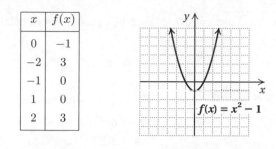

$f(x) = x^2 - 1$

48. Graph $f(x) = -2x^2 + 3 = -2(x - 0)^2 - (-3)$.

The graph is a parabola with vertex $(0, -3)$. We find some points on both sides of the vertex and draw the graph.

x	$f(x)$
0	3
-2	-5
-1	1
1	1
2	-5

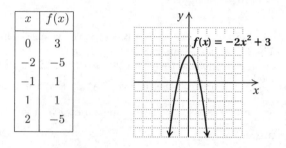

$f(x) = -2x^2 + 3$

49. We will use the point-slope equation.

$$y - y_1 = m(x - x_1)$$
$$y - 2 = \frac{1}{2}[x - (-4)]$$
$$y - 2 = \frac{1}{2}(x + 4)$$
$$y - 2 = \frac{1}{2}x + 2$$
$$y = \frac{1}{2}x + 4$$

50. First we find the slope of the given line.

$$3x + y = 4$$
$$y = -3x + 4$$

The slope is -3. The slope of a line parallel to this line will be -3 also. We are given the y-intercept, $(0, 1)$, so we substitute -3 for m and 1 for b in the slope-intercept equation $y = mx + b$ to get $y = -3x + 1$.

51. *Familiarize.* Let $r =$ the speed of the boat in still water, in km/h. We organize the information in a table.

	Distance	Speed	Time
Upriver	60	$r - 4$	t_1
Downriver	60	$r + 4$	t_2

Translate. Using $t = d/r$ in both rows of the table, we have two equations:

$$t_1 = \frac{60}{r - 4} \text{ and } t_2 = \frac{60}{r + 4}.$$

Since the total time of the trip is 8 hr, we have

$t_1 + t_2 = 8$, or

$$\frac{60}{r - 4} + \frac{60}{r + 4} = 8.$$

Solve. We multiply both sides of the equation by the LCM of the denominators, $(r-4)(r+4)$.

$$(r-4)(r+4)\left(\frac{60}{r-4}+\frac{60}{r+4}\right)=(r-4)(r+4)(8)$$
$$60(r+4)+60(r-4)=8(r^2-16)$$
$$60r+240+60r-240=8r^2-128$$
$$120r=8r^2-128$$
$$0=8r^2-120r-128$$
$$0=8(r^2-15r-16)$$
$$0=8(r-16)(r+1)$$
$$r-16=0 \quad or \quad r+1=0$$
$$r=16 \quad or \qquad r=-1$$

Check. Since the speed cannot be negative, we check only 16. The boat's speed upriver is $16-4$, or 12 km/h, and the speed downriver is $16+4$, or 20 km/h. At 12 km/h it takes $\frac{60}{12}$, or 5 hr, to travel upriver. At 20 km/h it takes $\frac{60}{20}$, or 3 hr, to travel downriver. Since 5 hr $+$ 3 hr $=$ 8 hr, the answer checks.

State. The speed of the boat in still water in 16 km/h.

52. **Familiarize**. Let l and w represent the length and width of the room, respectively, in feet.

    Translate. We use the equations for the perimeter and the area of a rectangle.
    $$2l+2w=56,$$
    $$A=l\cdot w$$
    Now we express A as a function of a single variable.
    $$2l+2w=56$$
    $$2l=56-2w$$
    $$l=\frac{56-2w}{2}$$
    $$l=28-w$$
    Then $A=l\cdot w=(28-w)w=28w-w^2=-w^2+28w$.

    Carry out. The graph of $A=-w^2+28w$ is a parabola that opens down, so there is a maximum value. The first coordinate of the vertex is
    $$-\frac{b}{2a}=-\frac{28}{2(-1)}=14.$$
    The second coordinate of the vertex is
    $$-w^2+28w=-14^2+28\cdot 14=196,$$
    so the maximum area of 196 ft^2 occurs when $w=14$ ft and $l=28-14=14$ ft.

    Check. We could try a value of w less than 14 and another greater than 14 in $A=-w^2+28w$. We could also use the graph to check the solution. The answer checks.

    State. The dimensions 14 ft by 14 ft yield the maximum area of 196 ft^2.

53. **Familiarize**. Let $h=$ the length of a side of the hexagon and $s=$ the length of a side of the square. Then the perimeter of the hexagon is $6h$ and the perimeter of the square is $4s$.

Translate. The perimeters are the same, so we have
$$6h=4s.$$
A side of the hexagon is 3 less than a side of the square, so we have
$$h=s-3.$$
We have a system of equations
$$6h=4s, \quad (1)$$
$$h=s-3. \quad (2)$$
Solve. Substitute $s-3$ for h in Equation (1) and solve for s.
$$6(s-3)=4s$$
$$6s-18=4s$$
$$-18=-2s$$
$$9=s$$
Now substitute 9 for s in Equation 2 and find h.
$$h=9-3=6$$
The perimeters are $6\cdot 6$, or 36, and $4\cdot 9$, or 36.

Check. We see above that the perimeters are the same. Also, 6 is 3 less than 9, so the answer checks.

State. The perimeter of each polygon is 36.

54. **Familiarize**. Let $n=$ the number of hours it takes the faster pipe to fill the tank working alone. Then $n+4=$ the slower pipe's time.

Translate. We use the work principle.
$$\frac{1\frac{1}{2}}{n}+\frac{1\frac{1}{2}}{n+4}=1, \text{ or}$$
$$\frac{1.5}{n}+\frac{1.5}{n+4}=1$$
Solve. We solve the equation.
$$\frac{1.5}{n}+\frac{1.5}{n+4}=1, \text{ LCM is } n(n+4)$$
$$n(n+4)\left(\frac{1.5}{n}+\frac{1.5}{n+4}\right)=n(n+4)\cdot 1$$
$$1.5(n+4)+1.5n=n^2+4n$$
$$1.5n+6+1.5n=n^2+4n$$
$$3n+6=n^2+4n$$
$$0=n^2+n-6$$
$$0=(n+3)(n-2)$$
$$n+3=0 \quad or \quad n-2=0$$
$$n=-3 \quad or \qquad n=2$$
Check. Since the time cannot be negative, we check only 2. We verify the work principle.
$$\frac{1.5}{2}+\frac{1.5}{2+4}=\frac{1.5}{2}+\frac{1.5}{6}=0.75+0.25=1$$
The answer checks.

State. It will take the faster pipe 2 hr to fill the tank working alone.

55. $f(x) = 5x^2 - 20x + 15$

$f(x) = 5(x^2 - 4x) + 15$

$f(x) = 5(x^2 - 4x + 4 - 4) + 15$

$f(x) = 5(x^2 - 4x + 4) + 5(-4) + 15$

$f(x) = 5(x - 2)^2 - 20 + 15$

$f(x) = 5(x - 2)^2 - 5$

Answer A is correct.

56. $f(x) = x^4 - 6x^2 - 16$

We use the discriminant to determine the nature of the solutions of $x^4 - 6x^2 - 16 = 0$.

$b^2 - 4ac = (-6)^2 - 4 \cdot 1 \cdot (-16) = 36 + 64 = 100$

Since $b^2 - 4ac > 0$, the equation has two different real solutions and the graph of $f(x) = x^4 - 6x^2 - 16$ has two x-intercepts. Answer B is correct.

57.
$$\frac{2x + 1}{x} = 3 + 7\sqrt{\frac{2x + 1}{x}}$$

$$\frac{2x + 1}{x} - 7\sqrt{\frac{2x + 1}{x}} - 3 = 0$$

Let $u = \sqrt{\dfrac{2x + 1}{x}}$.

$u^2 - 7u - 3 = 0$

$a = 1,\ b = -7,\ c = -3$

$$u = \frac{-b \pm \sqrt{b^2 - 4ac}}{2a} = \frac{-(-7) \pm \sqrt{(-7)^2 - 4 \cdot 1 \cdot (-3)}}{2 \cdot 1}$$

$$u = \frac{7 \pm \sqrt{49 + 12}}{2} = \frac{7 \pm \sqrt{61}}{2}$$

$$\sqrt{\frac{2x + 1}{x}} = \frac{7 - \sqrt{61}}{2} \ or \ \sqrt{\frac{2x + 1}{x}} = \frac{7 + \sqrt{61}}{2}$$

First we solve the equation on the left.

$$\left(\sqrt{\frac{2x + 1}{x}}\right)^2 = \left(\frac{7 - \sqrt{61}}{2}\right)^2$$

$$\frac{2x + 1}{x} = \frac{49 - 14\sqrt{61} + 61}{4}$$

$$\frac{2x + 1}{x} = \frac{110 - 14\sqrt{61}}{4}$$

$4(2x + 1) = x(110 - 14\sqrt{61})$ Multiplying by $4x$

$8x + 4 = 110x - 14x\sqrt{61}$

$4 = 102x - 14x\sqrt{61}$

$4 = 2x(51 - 7\sqrt{61})$

$$\frac{2}{51 - 7\sqrt{61}} = x \qquad \text{Dividing by } 2(51 - 7\sqrt{61})$$

Now we solve the other equation.

$$\left(\sqrt{\frac{2x + 1}{x}}\right)^2 = \left(\frac{7 + \sqrt{61}}{2}\right)^2$$

$$\frac{2x + 1}{x} = \frac{49 + 14\sqrt{61} + 61}{4}$$

$$\frac{2x + 1}{x} = \frac{110 + 14\sqrt{61}}{4}$$

$4(2x + 1) = x(110 + 14\sqrt{61})$ Multiplying by $4x$

$8x + 4 = 110x + 14x\sqrt{61}$

$4 = 102x + 14x\sqrt{61}$

$4 = 2x(51 + 7\sqrt{61})$

$$\frac{2}{51 + 7\sqrt{61}} = x \qquad \text{Dividing by } 2(51 + 7\sqrt{61})$$

Only $\dfrac{2}{51 + 7\sqrt{61}}$ checks. If we rationalize the denominator, we get $\dfrac{-51 + 7\sqrt{61}}{194}$.

58.
$$\frac{a^3}{8} + \frac{8b^3}{729} = \left(\frac{a}{2}\right)^3 + \left(\frac{2b}{9}\right)^3$$

$$= \left(\frac{a}{2} + \frac{2b}{9}\right)\left(\frac{a^2}{9} - \frac{ab}{9} + \frac{4b^2}{81}\right)$$

Chapter 8

Exponential and Logarithmic Functions

Exercise Set 8.1

RC1. True; see page 665 in the text.

RC3. False; see page 669 in the text.

1. Graph: $f(x) = 2^x$

We compute some function values and keep the results in a table.

$f(0) = 2^0 = 1$

$f(1) = 2^1 = 2$

$f(2) = 2^2 = 4$

$f(-1) = 2^{-1} = \dfrac{1}{2^1} = \dfrac{1}{2}$

$f(-2) = 2^{-2} = \dfrac{1}{2^2} = \dfrac{1}{4}$

x	$f(x)$
0	1
1	2
2	4
3	8
-1	$\dfrac{1}{2}$
-2	$\dfrac{1}{4}$
-3	$\dfrac{1}{8}$

Next we plot these points and connect them with a smooth curve.

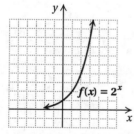

3. Graph: $f(x) = 5^x$

We compute some function values and keep the results in a table.

$f(0) = 5^0 = 1$

$f(1) = 5^1 = 5$

$f(2) = 5^2 = 25$

$f(-1) = 5^{-1} = \dfrac{1}{5^1} = \dfrac{1}{5}$

$f(-2) = 5^{-2} = \dfrac{1}{5^2} = \dfrac{1}{25}$

x	$f(x)$
0	1
1	5
2	25
-1	$\dfrac{1}{5}$
-2	$\dfrac{1}{25}$

Next we plot these points and connect them with a smooth curve.

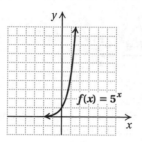

5. Graph: $f(x) = 2^{x+1}$

We compute some function values and keep the results in a table.

$f(0) = 2^{0+1} = 2^1 = 2$

$f(-1) = 2^{-1+1} = 2^0 = 1$

$f(-2) = 2^{-2+1} = 2^{-1} = \dfrac{1}{2^1} = \dfrac{1}{2}$

$f(-3) = 2^{-3+1} = 2^{-2} = \dfrac{1}{2^2} = \dfrac{1}{4}$

$f(1) = 2^{1+1} = 2^2 = 4$

$f(2) = 2^{2+1} = 2^3 = 8$

x	$f(x)$
0	2
-1	1
-2	$\dfrac{1}{2}$
-3	$\dfrac{1}{4}$
1	4
2	8

Next we plot these points and connect them with a smooth curve.

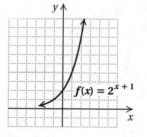

7. Graph: $f(x) = 3^{x-2}$

We compute some function values and keep the results in a table.

$f(0) = 3^{0-2} = 3^{-2} = \dfrac{1}{3^2} = \dfrac{1}{9}$

$f(1) = 3^{1-2} = 3^{-1} = \dfrac{1}{3^1} = \dfrac{1}{3}$

$f(2) = 3^{2-2} = 3^0 = 1$

$f(3) = 3^{3-2} = 3^1 = 3$

$f(4) = 3^{4-2} = 3^2 = 9$

$f(-1) = 3^{-1-2} = 3^{-3} = \dfrac{1}{3^3} = \dfrac{1}{27}$

$f(-2) = 3^{-2-2} = 3^{-4} = \dfrac{1}{3^4} = \dfrac{1}{81}$

x	$f(x)$
0	$\dfrac{1}{9}$
1	$\dfrac{1}{3}$
2	1
3	3
4	9
-1	$\dfrac{1}{27}$
-2	$\dfrac{1}{81}$

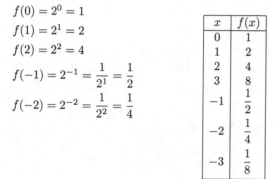

Next we plot these points and connect them with a smooth curve.

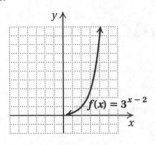

9. Graph: $f(x) = 2^x - 3$

We construct a table of values. Then we plot the points and connect them with a smooth curve.

$f(0) = 2^0 - 3 = 1 - 3 = -2$

$f(1) = 2^1 - 3 = 2 - 3 = -1$

$f(2) = 2^2 - 3 = 4 - 3 = 1$

$f(3) = 2^3 - 3 = 8 - 3 = 5$

$f(-1) = 2^{-1} - 3 = \dfrac{1}{2} - 3 = -\dfrac{5}{2}$

$f(-2) = 2^{-2} - 3 = \dfrac{1}{4} - 3 = -\dfrac{11}{4}$

x	$f(x)$
0	-2
1	-1
2	1
3	5
-1	$-\dfrac{5}{2}$
-2	$-\dfrac{11}{4}$

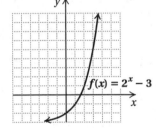

11. Graph: $f(x) = 5^{x+3}$

We construct a table of values. Then we plot the points and connect them with a smooth curve.

$f(0) = 5^{0+3} = 5^3 = 125$

$f(-1) = 5^{-1+3} = 5^2 = 25$

$f(-2) = 5^{-2+3} = 5^1 = 5$

$f(-3) = 5^{-3+3} = 5^0 = 1$

$f(-4) = 5^{-4+3} = 5^{-1} = \dfrac{1}{5}$

$f(-5) = 5^{-5+3} = 5^{-2} = \dfrac{1}{25}$

x	$f(x)$
0	125
-1	25
-2	5
-3	1
-4	$\dfrac{1}{5}$
-5	$\dfrac{1}{25}$

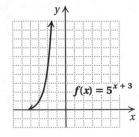

13. Graph: $f(x) = \left(\dfrac{1}{2}\right)^x$

We construct a table of values. Then we plot the points and connect them with a smooth curve.

$f(0) = \left(\dfrac{1}{2}\right)^0 = 1$

$f(1) = \left(\dfrac{1}{2}\right)^1 = \dfrac{1}{2}$

$f(2) = \left(\dfrac{1}{2}\right)^2 = \dfrac{1}{4}$

$f(3) = \left(\dfrac{1}{2}\right)^3 = \dfrac{1}{8}$

$f(-1) = \left(\dfrac{1}{2}\right)^{-1} = \dfrac{1}{\left(\dfrac{1}{2}\right)^1} = \dfrac{1}{\dfrac{1}{2}} = 2$

$f(-2) = \left(\dfrac{1}{2}\right)^{-2} = \dfrac{1}{\left(\dfrac{1}{2}\right)^2} = \dfrac{1}{\dfrac{1}{4}} = 4$

$f(-3) = \left(\dfrac{1}{2}\right)^{-3} = \dfrac{1}{\left(\dfrac{1}{2}\right)^3} = \dfrac{1}{\dfrac{1}{8}} = 8$

x	$f(x)$
0	1
1	$\dfrac{1}{2}$
2	$\dfrac{1}{4}$
3	$\dfrac{1}{8}$
-1	2
-2	4
-3	8

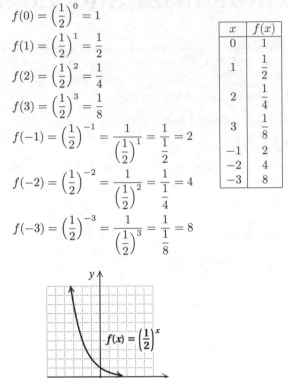

15. Graph: $f(x) = \left(\dfrac{1}{5}\right)^x$

We construct a table of values. Then we plot the points and connect them with a smooth curve.

$f(0) = \left(\dfrac{1}{5}\right)^0 = 1$

$f(1) = \left(\dfrac{1}{5}\right)^1 = \dfrac{1}{5}$

$f(2) = \left(\dfrac{1}{5}\right)^2 = \dfrac{1}{25}$

$f(-1) = \left(\dfrac{1}{5}\right)^{-1} = \dfrac{1}{\dfrac{1}{5}} = 5$

$f(-2) = \left(\dfrac{1}{5}\right)^{-2} = \dfrac{1}{\dfrac{1}{25}} = 25$

x	$f(x)$
0	1
1	$\dfrac{1}{5}$
2	$\dfrac{1}{25}$
-1	5
-2	25

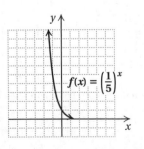

17. Graph: $f(x) = 2^{2x-1}$

We construct a table of values. Then we plot the points and connect them with a smooth curve.

$f(0) = 2^{2 \cdot 0 - 1} = 2^{-1} = \dfrac{1}{2}$

$f(1) = 2^{2 \cdot 1 - 1} = 2^1 = 2$

$f(2) = 2^{2 \cdot 2 - 1} = 2^3 = 8$

$f(-1) = 2^{2(-1)-1} = 2^{-3} = \dfrac{1}{8}$

$f(-2) = 2^{2(-2)-1} = 2^{-5} = \dfrac{1}{32}$

x	$f(x)$
0	$\dfrac{1}{2}$
1	2
2	8
-1	$\dfrac{1}{8}$
-2	$\dfrac{1}{32}$

19. Graph: $x = 2^y$

We can find ordered pairs by choosing values for y and then computing values for x.

For $y = 0$, $x = 2^0 = 1$.

For $y = 1$, $x = 2^1 = 2$.

For $y = 2$, $x = 2^2 = 4$.

For $y = 3$, $x = 2^3 = 8$.

For $y = -1$, $x = 2^{-1} = \dfrac{1}{2^1} = \dfrac{1}{2}$.

For $y = -2$, $x = 2^{-2} = \dfrac{1}{2^2} = \dfrac{1}{4}$.

For $y = -3$, $x = 2^{-3} = \dfrac{1}{2^3} = \dfrac{1}{8}$.

x	y
1	0
2	1
4	2
8	3
$\dfrac{1}{2}$	-1
$\dfrac{1}{4}$	-2
$\dfrac{1}{8}$	-3

    (1) Choose values for y.

    (2) Compute values for x.

We plot these points and connect them with a smooth curve.

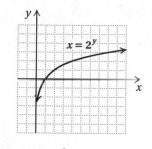

21. Graph: $x = \left(\dfrac{1}{2}\right)^y$

We can find ordered pairs by choosing values for y and then computing values for x. Then we plot these points and connect them with a smooth curve.

For $y = 0$, $x = \left(\dfrac{1}{2}\right)^0 = 1$.

For $y = 1$, $x = \left(\dfrac{1}{2}\right)^1 = \dfrac{1}{2}$.

For $y = 2$, $x = \left(\dfrac{1}{2}\right)^2 = \dfrac{1}{4}$.

For $y = 3$, $x = \left(\dfrac{1}{2}\right)^3 = \dfrac{1}{8}$.

For $y = -1$, $x = \left(\dfrac{1}{2}\right)^{-1} = \dfrac{1}{\frac{1}{2}} = 2$.

For $y = -2$, $x = \left(\dfrac{1}{2}\right)^{-2} = \dfrac{1}{\frac{1}{4}} = 4$.

For $y = -3$, $x = \left(\dfrac{1}{2}\right)^3 = \dfrac{1}{\frac{1}{8}} = 8$.

x	y
1	0
$\dfrac{1}{2}$	1
$\dfrac{1}{4}$	2
$\dfrac{1}{8}$	3
2	-1
4	-2
8	-3

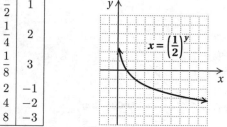

23. Graph: $x = 5^y$

We can find ordered pairs by choosing values for y and then computing values for x. Then we plot these points and connect them with a smooth curve.

For $y = 0$, $x = 5^0 = 1$.

For $y = 1$, $x = 5^1 = 5$.

For $y = 2$, $x = 5^2 = 25$.

For $y = -1$, $x = 5^{-1} = \dfrac{1}{5}$.

For $y = -2$, $x = 5^{-2} = \dfrac{1}{25}$.

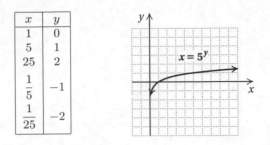

x	y
1	0
5	1
25	2
$\frac{1}{5}$	-1
$\frac{1}{25}$	-2

25. Graph $y = 2^x$ (see Exercise 1) and $x = 2^y$ (see Exercise 19) using the same set of axes.

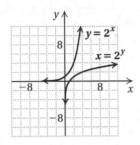

27. a) We substitute \$50,000 for P and 2%, or 0.02, for r in the formula $A = P(1 + r)^t$:

$$A(t) = \$50,000(1 + 0.02)^t = \$50,000(1.02)^t$$

b) $A(0) = \$50,000(1.02)^0 = \$50,000$

$A(1) = \$50,000(1.02)^1 = \$51,000$

$A(2) = \$50,000(1.02)^2 = \$52,020$

$A(4) = \$50,000(1.02)^4 \approx \$54,121.61$

$A(8) = \$50,000(1.02)^8 \approx \$58,582.97$

$A(10) = \$50,000(1.02)^{10} \approx \$60,949.72$

$A(20) = \$50,000(1.02)^{20} \approx \$74,297.37$

c)

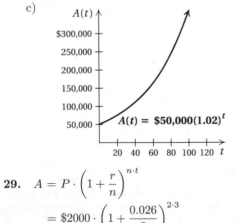

29. $A = P \cdot \left(1 + \dfrac{r}{n}\right)^{n \cdot t}$

$= \$2000 \cdot \left(1 + \dfrac{0.026}{2}\right)^{2 \cdot 3}$

$= \$2000(1.013)^6$

$\approx \$2161.16$

31. $A = P \cdot \left(1 + \dfrac{r}{n}\right)^{n \cdot t}$

$= \$4500 \cdot \left(1 + \dfrac{0.036}{4}\right)^{4(4.5)}$

$= \$4500 \cdot (1.009)^{18}$

$\approx \$5287.54$

33. a) In 2009, $t = 2009 - 2008 = 1$.

$A(1) = 234(2.43)^1 \approx 569$ outlets

In 2011, $t = 2011 - 2008 = 3$.

$A(3) = 234(2.43)^3 \approx 3358$ outlets

In 2012, $t = 2012 - 2008 = 4$.

$A(4) = 234(2.43)^4 \approx 8159$ outlets

b)

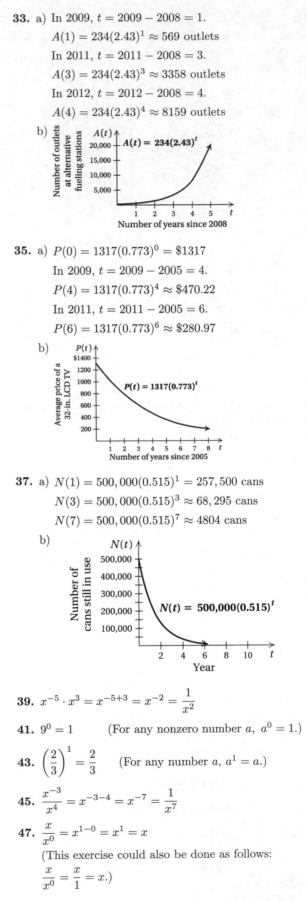

35. a) $P(0) = 1317(0.773)^0 = \$1317$

In 2009, $t = 2009 - 2005 = 4$.

$P(4) = 1317(0.773)^4 \approx \$470.22$

In 2011, $t = 2011 - 2005 = 6$.

$P(6) = 1317(0.773)^6 \approx \$280.97$

b)

37. a) $N(1) = 500,000(0.515)^1 = 257,500$ cans

$N(3) = 500,000(0.515)^3 \approx 68,295$ cans

$N(7) = 500,000(0.515)^7 \approx 4804$ cans

b)

39. $x^{-5} \cdot x^3 = x^{-5+3} = x^{-2} = \dfrac{1}{x^2}$

41. $9^0 = 1$ (For any nonzero number a, $a^0 = 1$.)

43. $\left(\dfrac{2}{3}\right)^1 = \dfrac{2}{3}$ (For any number a, $a^1 = a$.)

45. $\dfrac{x^{-3}}{x^4} = x^{-3-4} = x^{-7} = \dfrac{1}{x^7}$

47. $\dfrac{x}{x^0} = x^{1-0} = x^1 = x$

(This exercise could also be done as follows:

$\dfrac{x}{x^0} = \dfrac{x}{1} = x$.)

49. $(5^{\sqrt{2}})^{2\sqrt{2}} = 5^{\sqrt{2}\cdot 2\sqrt{2}} = 5^4$, or 625

51. Graph: $y = 2^x + 2^{-x}$

Construct a table of values, thinking of y as $f(x)$. Then plot these points and connect them with a curve.

$f(0) = 2^0 + 2^{-0} = 1 + 1 = 2$

$f(1) = 2^1 + 2^{-1} = 2 + \dfrac{1}{2} = 2\dfrac{1}{2}$

$f(2) = 2^2 + 2^{-2} = 4 + \dfrac{1}{4} = 4\dfrac{1}{4}$

$f(3) = 2^3 + 2^{-3} = 8 + \dfrac{1}{8} = 8\dfrac{1}{8}$

$f(-1) = 2^{-1} + 2^{-(-1)} = \dfrac{1}{2} + 2 = 2\dfrac{1}{2}$

$f(-2) = 2^{-2} + 2^{-(-2)} = \dfrac{1}{4} + 4 = 4\dfrac{1}{4}$

$f(-3) = 2^{-3} + 2^{-(-3)} = \dfrac{1}{8} + 8 = 8\dfrac{1}{8}$

x	y, or $f(x)$
0	2
1	$2\dfrac{1}{2}$
2	$4\dfrac{1}{4}$
3	$8\dfrac{1}{8}$
-1	$2\dfrac{1}{2}$
-2	$4\dfrac{1}{4}$
-3	$8\dfrac{1}{8}$

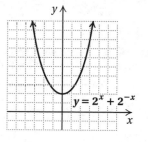

$y = 2^x + 2^{-x}$

53. $y = \left|\left(\dfrac{1}{2}\right)^x - 1\right|$

Construct a table of values, thinking of y as $f(x)$. Then plot these points and connect them with a curve.

$f(-4) = \left|\left(\dfrac{1}{2}\right)^{-4} - 1\right| = |16 - 1| = |15| = 15$

$f(-2) = \left|\left(\dfrac{1}{2}\right)^{-2} - 1\right| = |4 - 1| = |3| = 3$

$f(-1) = \left|\left(\dfrac{1}{2}\right)^{-1} - 1\right| = |2 - 1| = |1| = 1$

$f(0) = \left|\left(\dfrac{1}{2}\right)^{0} - 1\right| = |1 - 1| = |0| = 0$

$f(1) = \left|\left(\dfrac{1}{2}\right)^{1} - 1\right| = \left|\dfrac{1}{2} - 1\right| = \left|-\dfrac{1}{2}\right| = \dfrac{1}{2}$

$f(2) = \left|\left(\dfrac{1}{2}\right)^{2} - 1\right| = \left|\dfrac{1}{4} - 1\right| = \left|-\dfrac{3}{4}\right| = \dfrac{3}{4}$

$f(3) = \left|\left(\dfrac{1}{2}\right)^{3} - 1\right| = \left|\dfrac{1}{8} - 1\right| = \left|-\dfrac{7}{8}\right| = \dfrac{7}{8}$

x	y, or $f(x)$
-4	15
-2	3
-1	1
0	0
1	$\dfrac{1}{2}$
2	$\dfrac{3}{4}$
3	$\dfrac{7}{8}$

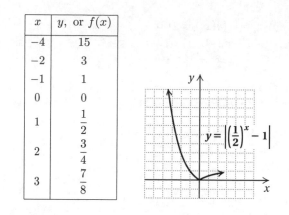

$y = \left|\left(\dfrac{1}{2}\right)^x - 1\right|$

55. Construct a table of values for each equation and then draw the graphs on the same set of axes.

For $y = 3^{-(x-1)}$:

x	y
-3	81
-2	27
-1	9
0	3
1	1
2	$\dfrac{1}{3}$
3	$\dfrac{1}{9}$
4	$\dfrac{1}{27}$

For $x = 3^{-(y-1)}$:

x	y
81	-3
27	-2
9	-1
3	0
1	1
$\dfrac{1}{3}$	2
$\dfrac{1}{9}$	3
$\dfrac{1}{27}$	4

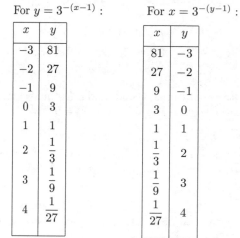

$y = 3^{-(x-1)}$

$y = x$

$x = 3^{-(y-1)}$

57. Left to the student

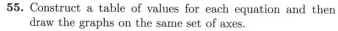

Exercise Set 8.2

RC1. Any set of ordered pairs is a <u>relation</u>.

RC3. If the graph of a function passes the <u>horizontal</u>-line test, the inverse of the function is also a function.

RC5. The function $g(x) = x - 10$ is the <u>inverse</u> of $f(x) = x + 10$.

1. $(f \circ g)(x) = f(g(x)) = f(6 - 4x) = 2(6 - 4x) - 3 =$
$$12 - 8x - 3 = -8x + 9$$

$(g \circ f)(x) = g(f(x)) = g(2x - 3) = 6 - 4(2x - 3) =$
$$6 - 8x + 12 = -8x + 18$$

3. $(f \circ g)(x) = f(g(x)) = f(2x - 1) = 3(2x - 1)^2 + 2 =$
$$3(4x^2 - 4x + 1) + 2 = 12x^2 - 12x + 3 + 2 = 12x^2 - 12x + 5$$

$(g \circ f)(x) = g(f(x)) = g(3x^2 + 2) = 2(3x^2 + 2) - 1 =$
$$6x^2 + 4 - 1 = 6x^2 + 3$$

5. $(f \circ g)(x) = f(g(x)) = f\left(\dfrac{2}{x}\right) = 4\left(\dfrac{2}{x}\right)^2 - 1 =$
$$4\left(\dfrac{4}{x^2}\right) - 1 = \dfrac{16}{x^2} - 1$$

$(g \circ f)(x) = g(f(x)) = g(4x^2 - 1) = \dfrac{2}{4x^2 - 1}$

7. $(f \circ g)(x) = f(g(x)) = f(x^2 - 5) = (x^2 - 5)^2 + 5 =$
$$x^4 - 10x^2 + 25 + 5 = x^4 - 10x^2 + 30$$

$(g \circ f)(x) = g(f(x)) = g(x^2 + 5) = (x^2 + 5)^2 - 5 =$
$$x^4 + 10x^2 + 25 - 5 = x^4 + 10x^2 + 20$$

9. $h(x) = (5 - 3x)^2$

This is $5 - 3x$ raised to the second power, so the two most obvious functions are $f(x) = x^2$ and $g(x) = 5 - 3x$.

11. $h(x) = \sqrt{5x + 2}$

This is the square root of $5x + 2$, so the two most obvious functions are $f(x) = \sqrt{x}$ and $g(x) = 5x + 2$.

13. $h(x) = \dfrac{1}{x - 1}$

This is the reciprocal of $x - 1$, so the two most obvious functions are $f(x) = \dfrac{1}{x}$ and $g(x) = x - 1$.

15. $h(x) = \dfrac{1}{\sqrt{7x + 2}}$

This is the reciprocal of the square root of $7x + 2$. Two functions that can be used are $f(x) = \dfrac{1}{\sqrt{x}}$ and $g(x) = 7x + 2$.

17. $h(x) = (\sqrt{x} + 5)^4$

This is $\sqrt{x} + 5$ raised to the fourth power, so the two most obvious functions are $f(x) = x^4$ and $g(x) = \sqrt{x} + 5$.

19. To find the inverse of the given relation we interchange the first and second coordinates of each ordered pair. The inverse of the relation is $\{(2, 1), (-3, 6), (-5, -3)\}$.

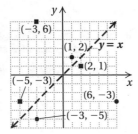

21. We interchange x and y to obtain an equation of the inverse of the relation. It is $x = 2y + 6$. The x-values in the first table become the y-values in the second table. We have

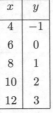

x	y
4	-1
6	0
8	1
10	2
12	3

We graph the original relation and its inverse. Since there is no horizontal line that crosses the graph more than once, the function is one-to-one.

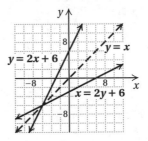

23. The graph of $f(x) = x - 5$ is shown below. Since no horizontal line crosses the graph more than once, the function is one-to-one.

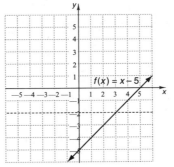

25 The graph of $f(x) = x^2 - 2$ is shown below. There are many horizontal lines that cross the graph more than once, so the function is not one-to-one.

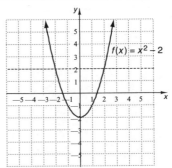

27. The graph of $g(x) = |x| - 3$ is shown below. There are many horizontal lines that cross the graph more than once, so the function is not one-to-one.

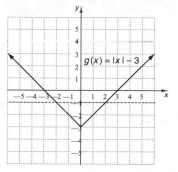

29. The graph of $g(x) = 3^x$ is shown below. Since no horizontal line crosses the graph more than once, the function is one-to-one.

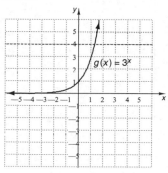

31. The graph of $f(x) = 5x - 2$ is shown below. It passes the horizontal-line test, so it is one-to-one.

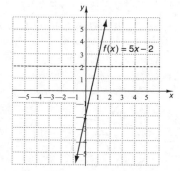

We find a formula for the inverse.

1. Replace $f(x)$ by y: $y = 5x - 2$
2. Interchange x and y: $x = 5y - 2$
3. Solve for y: $x + 2 = 5y$
$$\frac{x+2}{5} = y$$
4. Replace y by $f^{-1}(x)$: $f^{-1}(x) = \dfrac{x+2}{5}$

33. The graph of $f(x) = \dfrac{-2}{x}$ is shown below. It passes the horizontal-line test, so it is one-to-one.

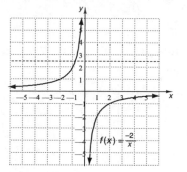

We find a formula for the inverse.

1. Replace $f(x)$ by y: $y = \dfrac{-2}{x}$
2. Interchange x and y: $x = \dfrac{-2}{y}$
3. Solve for y: $y = \dfrac{-2}{x}$
4. Replace y by $f^{-1}(x)$: $f^{-1}(x) = \dfrac{-2}{x}$

35. The graph of $f(x) = \dfrac{4}{3}x + 7$ is shown below. It passes the horizontal line test, so it is one-to-one.

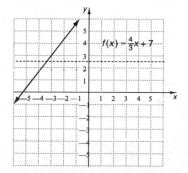

We find a formula for the inverse.

1. Replace $f(x)$ by y: $y = \dfrac{4}{3}x + 7$
2. Interchange x and y: $x = \dfrac{4}{3}y + 7$
3. Solve for y: $x - 7 = \dfrac{4}{3}y$
$$\frac{3}{4}(x - 7) = y$$
4. Replace y by $f^{(-1)}(x)$: $f^{-1}(x) = \dfrac{3}{4}(x - 7)$

37. The graph of $f(x) = \dfrac{2}{x+5}$ is shown below. It passes the horizontal line test, so it is one-to-one.

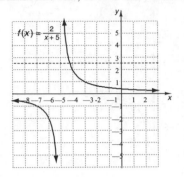

We find a formula for the inverse.

1. Replace $f(x)$ by y: $y = \dfrac{2}{x+5}$

2. Interchange x and y: $x = \dfrac{2}{y+5}$

3. Solve for y: $x(y+5) = 2$

$$y + 5 = \frac{2}{x}$$

$$y = \frac{2}{x} - 5$$

4. Replace y by $f^{-1}(x)$: $f^{-1}(x) = \dfrac{2}{x} - 5$

39. The graph of $f(x) = 5$ is shown below. The horizontal line $y = 5$ crosses the graph more than once, so the function is not one-to-one.

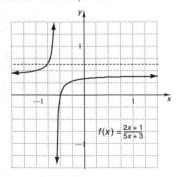

41. The graph of $f(x) = \dfrac{2x+1}{5x+3}$ is shown below. It passes the horizontal line test, so it is one-to-one.

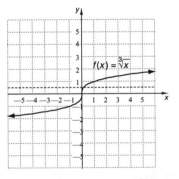

We find a formula for the inverse.

1. Replace $f(x)$ by y: $y = \dfrac{2x+1}{5x+3}$

2. Interchange x and y: $x = \dfrac{2y+1}{5y+3}$

3. Solve for y: $5xy + 3x = 2y + 1$

$$5xy - 2y = 1 - 3x$$

$$y(5x - 2) = 1 - 3x$$

$$y = \frac{1-3x}{5x-2}$$

4. Replace y by $f^{-1}(x)$: $f^{-1}(x) = \dfrac{1-3x}{5x-2}$

43. The graph of $f(x) = x^3 - 1$ is shown below. It passes the horizontal line test, so it is one-to-one.

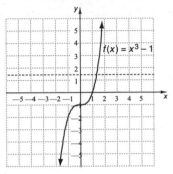

1. Replace $f(x)$ by y: $y = x^3 - 1$

2. Interchange x and y: $x = y^3 - 1$

3. Solve for y: $x + 1 = y^3$

$$\sqrt[3]{x+1} = y$$

4. Replace y by $f^{-1}(x)$: $f^{-1}(x) = \sqrt[3]{x+1}$

45. The graph of $f(x) = \sqrt[3]{x}$ is shown below. It passes the horizontal line test, so it is one-to-one.

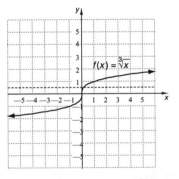

1. Replace $f(x)$ by y: $y = \sqrt[3]{x}$

2. Interchange x and y: $x = \sqrt[3]{y}$

3. Solve for y: $x^3 = y$

4. Replace y by $f^{-1}(x)$: $f^{-1}(x) = x^3$

47. $f(x) = \dfrac{1}{2}x - 3$

1. Replace $f(x)$ by y: $y = \dfrac{1}{2}x - 3$

2. Interchange x and y: $x = \dfrac{1}{2}y - 3$

3. Solve for y: $x = \dfrac{1}{2}y - 3$

$$x + 3 = \dfrac{1}{2}y$$

$$2x + 6 = y$$

4. Replace y by $f^{-1}(x)$: $f^{-1}(x) = 2x + 6$

Now we find the values of $f(x)$ for the table on the left.

$$f(-4) = \dfrac{1}{2}(-4) - 3 = -2 - 3 = -5$$

$$f(0) = \dfrac{1}{2} \cdot 0 - 3 = -3$$

$$f(2) = \dfrac{1}{2} \cdot 2 - 3 = 1 - 3 = -2$$

$$f(4) = \dfrac{1}{2} \cdot 4 - 3 = 2 - 3 = -1$$

The x-values in the table on the left become the values of $f^{-1}(x)$ in the table on the right.

x	$f(x)$
-4	-5
0	-3
2	-2
4	-1

x	$f^{-1}(x)$
-5	-4
-3	0
-2	2
-1	4

We now graph $f(x) = \dfrac{1}{2}x - 3$. The graph of f^{-1} can be obtained by reflecting the graph of f across the line $y = x$ or by plotting the ordered pairs in the table on the right.

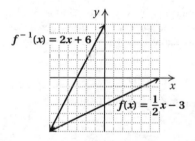

49. $f(x) = x^3$

1. Replace $f(x)$ by y: $y = x^3$

2. Interchange x and y: $x = y^3$

3. Solve for y: $x = y^3$

$$\sqrt[3]{x} = y$$

4. Replace y by $f^{-1}(x)$: $f^{-1}(x) = \sqrt[3]{x}$

We fill in the table on the left by finding the value of $f(x)$ for each given x-value. We can fill in the table on the right by interchanging the values of x and $f(x)$.

x	$f(x)$
0	0
1	1
2	8
3	27
-1	-1
-2	-8
-3	-27

x	$f^{-1}(x)$
0	0
1	1
8	2
27	3
-1	-1
-8	-2
-27	-3

We now graph $f(x) = x^3$. The graph of f^{-1} can be obtained by reflecting the graph of f across the line $y = x$ or by plotting the ordered pairs in the table on the right.

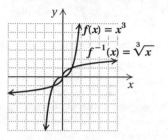

51. We check to see that $(f^{-1} \circ f)(x) = x$ and $(f \circ f^{-1})(x) = x$.

$$(f^{-1} \circ f)(x) = f^{-1}(f(x)) = f^{-1}\left(\dfrac{4}{5}x\right) =$$

$$\dfrac{5}{4} \cdot \dfrac{4}{5}x = x$$

$$(f \circ f^{-1})(x) = f(f^{-1}(x)) = f\left(\dfrac{5}{4}x\right) =$$

$$\dfrac{4}{5} \cdot \dfrac{5}{4}x = x$$

53. We check to see that $(f^{-1} \circ f)(x) = x$ and $(f \circ f^{-1})(x) = x$.

$$(f^{-1} \circ f)(x) = f^{-1}(f(x)) = 2\left(\dfrac{x+7}{2}\right) - 7 = x + 7 - 7 = x$$

$$(f \circ f^{-1})(x) = f(f^{-1}(x)) = \dfrac{(2x-7)+7}{2} = \dfrac{2x}{2} = x$$

55. We check to see that $(f^{-1} \circ f)(x) = x$ and $(f \circ f^{-1})(x) = x$.

$$(f^{-1} \circ f)(x) = f^{-1}(f(x)) = f^{-1}\left(\dfrac{1-x}{x}\right) =$$

$$\dfrac{1}{\dfrac{1-x}{x}+1} = \dfrac{1}{\dfrac{1-x}{x}+1} \cdot \dfrac{x}{x} = \dfrac{x}{1-x+x} =$$

$$\dfrac{x}{1} = x$$

$$(f \circ f^{-1})(x) = f(f^{-1}(x)) = f\left(\dfrac{1}{x+1}\right) =$$

$$\dfrac{1 - \dfrac{1}{x+1}}{\dfrac{1}{x+1}} = \dfrac{1 - \dfrac{1}{x+1}}{\dfrac{1}{x+1}} \cdot \dfrac{x+1}{x+1} =$$

$$\dfrac{x+1-1}{1} = \dfrac{x}{1} = x$$

57. The function $f(x) = 3x$ multiplies an input by 3, so the inverse would divide an input by 3. We have $f^{-1}(x) = \dfrac{x}{3}$.

Now we check to see that $(f^{-1} \circ f)(x) = x$ and $(f \circ f^{-1})(x) = x$.

$$(f^{-1} \circ f)(x) = f^{-1}(f(x)) = f^{-1}(3x) = \dfrac{3x}{3} = x$$

$$(f \circ f^{-1})(x) = f(f^{-1}(x)) = f\left(\dfrac{x}{3}\right) = 3 \cdot \dfrac{x}{3} = x$$

The inverse is correct.

59. The function $f(x) = -x$ takes the opposite of an input so the inverse would also take the opposite of an input. We have $f^{-1}(x) = -x$.

Now we check to see that $(f^{-1} \circ f)(x) = x$ and $(f \circ f^{-1})(x) = x$.

$(f^{-1} \circ f)(x) = f^{-1}(f(x)) = f^{-1}(-x) = -(-x) = x$

$(f \circ f^{-1})(x) = f(f^{-1}(x)) = f(-x) = -(-x) = x$

The inverse is correct.

61. The function $f(x) = \sqrt[3]{x-5}$ subtracts 5 from an input and then takes the cube root of the difference, so the inverse would cube an input and then add 5. We have $f^{-1}(x) = x^3 + 5$.

Now we check to see that $(f^{-1} \circ f)(x) = x$ and $(f \circ f^{-1})(x) = x$.

$(f^{-1} \circ f)(x) = f^{-1}(f(x)) = f^{-1}(\sqrt[3]{x-5}) = (\sqrt[3]{x-5})^3 + 5 = x - 5 + 5 = x$

$(f \circ f^{-1})(x) = f(f^{-1}(x)) = f(x^3 + 5) = \sqrt[3]{x^3 + 5 - 5} = \sqrt[3]{x^3} = x$

The inverse is correct.

63. a) $f(8) = 8 + 32 = 40$

Size 40 in France corresponds to size 8 in the U.S.

$f(10) = 10 + 32 = 42$

Size 42 in France corresponds to size 10 in the U.S.

$f(14) = 14 + 32 = 46$

Size 46 in France corresponds to size 14 in the U.S.

$f(18) = 18 + 32 = 50$

Size 50 in France corresponds to size 18 in the U.S.

b) The graph of $f(x) = x + 32$ is shown below. It passes the horizontal-line test, so the function is one-to-one and, hence, has an inverse that is a function.

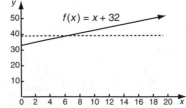

We now find a formula for the inverse.

1. Replace $f(x)$ by y: $y = x + 32$

2. Interchange x and y: $x = y + 32$

3. Solve for y: $x - 32 = y$

4. Replace y by $f^{-1}(x)$: $f^{-1}(x) = x - 32$

c) $f^{-1}(40) = 40 - 32 = 8$

Size 8 in the U.S. corresponds to size 40 in France.

$f^{-1}(42) = 42 - 32 = 10$

Size 10 in the U.S. corresponds to size 42 in France.

$f^{-1}(46) = 46 - 32 = 14$

Size 14 in the U.S. corresponds to size 46 in France.

$f^{-1}(50) = 50 - 32 = 18$

Size 18 in the U.S. corresponds to size 50 in France.

65. $\sqrt[6]{a^2} = a^{2/6} = a^{1/3} = \sqrt[3]{a}$

67. $\sqrt[12]{64x^6y^6} = (2^6 x^6 y^6)^{1/12} = 2^{1/2} x^{1/2} y^{1/2} = (2xy)^{1/2} = \sqrt{2xy}$

69. $i^{79} = i^{78} \cdot i = (i^2)^{39} \cdot i = (-1)^{39} \cdot i = -1 \cdot i = -i$

71. $\sqrt{2400} = \sqrt{400 \cdot 6} = \sqrt{400}\sqrt{6} = 20\sqrt{6}$

73. $(2y - 5)^2 = (2y)^2 - 2 \cdot 2y \cdot 5 + 5^2 = 4y^2 - 20y + 25$

75. Graph the functions in a square window and determine whether one is a reflection of the other across the line $y = x$. The graphs show that these functions are not inverses of each other.

77. Graph the functions in a square window and determine whether one is a reflection of the other across the line $y = x$. The graphs show that these functions are inverses of each other.

79. (1) C; (2) A; (3) B; (4) D

81. Reflect the graph of f across the line $y = x$.

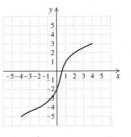

83. $f(x) = \dfrac{1}{2}x + 3$, $g(x) = 2x - 6$

Since $(f \circ g)(x) = x$ and $(g \circ f)(x) = x$, the functions are inverse.

Exercise Set 8.3

RC1. Let $\log_{10} 100 = x$. Then

$$10^x = 100$$
$$10^x = 10^2$$
$$x = 2.$$

The correct answer is $\underline{2}$.

RC3. Let $\log_2 8 = x$. Then
$$2^x = 8$$
$$2^x = 2^3$$
$$x = 3.$$
The correct answer is $\underline{3}$.

1. Graph: $f(x) = \log_2 x$

The equation $f(x) = y = \log_2 x$ is equivalent to $2^y = x$. We can find ordered pairs by choosing values for y and computing the corresponding x-values.

For $y = 0$, $x = 2^0 = 1$.

For $y = 1$, $x = 2^1 = 2$.

For $y = 2$, $x = 2^2 = 4$.

For $y = 3$, $x = 2^3 = 8$.

For $y = -1$, $x = 2^{-1} = \dfrac{1}{2}$.

For $y = -2$, $x = 2^{-2} = \dfrac{1}{4}$.

x, or 2^y	y
1	0
2	1
4	2
8	3
$\dfrac{1}{2}$	-1
$\dfrac{1}{4}$	-2
$\dfrac{1}{8}$	-3

We plot the set of ordered pairs and connect the points with a smooth curve.

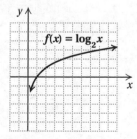

3. Graph: $f(x) = \log_{1/3} x$

The equation $f(x) = y = \log_{1/3} x$ is equivalent to $\left(\dfrac{1}{3}\right)^y = x$. We can find ordered pairs by choosing values for y and computing the corresponding x-values.

For $y = 0$, $x = \left(\dfrac{1}{3}\right)^0 = 1$.

For $y = 1$, $x = \left(\dfrac{1}{3}\right)^1 = \dfrac{1}{3}$.

For $y = 2$, $x = \left(\dfrac{1}{3}\right)^2 = \dfrac{1}{9}$.

For $y = -1$, $x = \left(\dfrac{1}{3}\right)^{-1} = 3$.

For $y = -2$, $x = \left(\dfrac{1}{3}\right)^{-2} = 9$.

x, or $\left(\dfrac{1}{3}\right)^y$	y
1	0
$\dfrac{1}{3}$	1
$\dfrac{1}{9}$	2
3	-1
9	-2

We plot the set of ordered pairs and connect the points with a smooth curve.

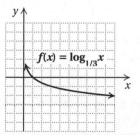

5. Graph $f(x) = 3^x$ (see Exercise Set 8.1, Exercise 2) and $f^{-1}(x) = \log_3 x$ on the same set of axes. We can obtain the graph of f^{-1} by reflecting the graph of f across the line $y = x$.

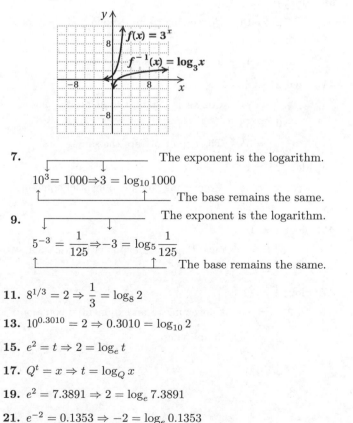

7. The exponent is the logarithm.
$$10^3 = 1000 \Rightarrow 3 = \log_{10} 1000$$
The base remains the same.

9. The exponent is the logarithm.
$$5^{-3} = \dfrac{1}{125} \Rightarrow -3 = \log_5 \dfrac{1}{125}$$
The base remains the same.

11. $8^{1/3} = 2 \Rightarrow \dfrac{1}{3} = \log_8 2$

13. $10^{0.3010} = 2 \Rightarrow 0.3010 = \log_{10} 2$

15. $e^2 = t \Rightarrow 2 = \log_e t$

17. $Q^t = x \Rightarrow t = \log_Q x$

19. $e^2 = 7.3891 \Rightarrow 2 = \log_e 7.3891$

21. $e^{-2} = 0.1353 \Rightarrow -2 = \log_e 0.1353$

23.

The logarithm is the exponent.

$$w = \log_4 10 \Rightarrow 4^w = 10$$

The base remains the same.

25.

The logarithm is the exponent.

$$\log_6 36 = 2 \Rightarrow 6^2 = 36$$

The base remains the same.

27. $\log_{10} 0.01 = -2 \Rightarrow 10^{-2} = 0.01$

29. $\log_{10} 8 = 0.9031 \Rightarrow 10^{0.9031} = 8$

31. $\log_e 100 = 4.6052 \Rightarrow e^{4.6052} = 100$

33. $\log_t Q = k \Rightarrow t^k = Q$

35. $\log_3 x = 2$

$\quad 3^2 = x \quad$ Converting to an exponential equation

$\quad 9 = x \quad$ Computing 3^2

37. $\log_x 16 = 2$

$\quad x^2 = 16 \quad$ Converting to an exponential equation

$\quad x = 4$ or $x = -4 \quad$ Principle of square roots

$\log_4 16 = 2$ because $4^2 = 16$. Thus, 4 is a solution. Since all logarithm bases must be positive, $\log_{-4} 16$ is not defined and -4 is not a solution.

39. $\log_2 16 = x$

$\quad 2^x = 16 \quad$ Converting to an exponential equation

$\quad 2^x = 2^4$

$\quad x = 4 \quad$ The exponents are the same.

41. $\log_3 27 = x$

$\quad 3^x = 27 \quad$ Converting to an exponential equation

$\quad 3^x = 3^3$

$\quad x = 3 \quad$ The exponents are the same.

43. $\log_x 25 = 1$

$\quad x^1 = 25 \quad$ Converting to an exponential equation

$\quad x = 25$

45. $\log_3 x = 0$

$\quad 3^0 = x \quad$ Converting to an exponential equation

$\quad 1 = x$

47. $\log_2 x = -1$

$\quad 2^{-1} = x \quad$ Converting to an exponential equation

$\quad \dfrac{1}{2} = x \quad$ Simplifying

49. $\log_8 x = \dfrac{1}{3}$

$\quad 8^{1/3} = x$

$\quad 2 = x$

51. Let $\log_{10} 100 = x$. Then

$\quad 10^x = 100$

$\quad 10^x = 10^2$

$\quad x = 2$

Thus, $\log_{10} 100 = 2$.

53. Let $\log_{10} 0.1 = x$. Then

$\quad 10^x = 0.1 = \dfrac{1}{10}$

$\quad 10^x = 10^{-1}$

$\quad x = -1$

Thus, $\log_{10} 0.1 = -1$.

55. Let $\log_{10} 1 = x$. Then

$\quad 10^x = 1$

$\quad 10^x = 10^0 \quad (10^0 = 1)$

$\quad x = 0$

Thus, $\log_{10} 1 = 0$.

57. Let $\log_5 625 = x$. Then

$\quad 5^x = 625$

$\quad 5^x = 5^4$

$\quad x = 4$

Thus, $\log_5 625 = 4$.

59. Think of the meaning of $\log_7 49$. It is the exponent to which you raise 7 to get 49. That exponent is 2. Therefore, $\log_7 49 = 2$.

61. Think of the meaning of $\log_2 8$. It is the exponent to which you raise 2 to get 8. That exponent is 3. Therefore, $\log_2 8 = 3$.

63. Let $\log_9 \dfrac{1}{81} = x$. Then

$\quad 9^x = \dfrac{1}{81}$

$\quad 9^x = 9^{-2}$

$\quad x = -2$

Thus, $\log_9 81 = -2$.

65. Let $\log_8 1 = x$. Then

$\quad 8^x = 1$

$\quad 8^x = 8^0 \quad (8^0 = 1)$

$\quad x = 0$

Thus, $\log_8 1 = 0$.

67. Let $\log_e e = x$. Then

$\quad e^x = e$

$\quad e^x = e^1$

$\quad x = 1$

Thus, $\log_e e = 1$.

69. Let $\log_{27} 9 = x$. Then

$$27^x = 9$$
$$(3^3)^x = 3^2$$
$$3^{3x} = 3^2$$
$$3x = 2$$
$$x = \frac{2}{3}$$

Thus, $\log_{27} 9 = \frac{2}{3}$.

71. 4.8970

73. -0.1739

75. Does not exist as a real number

77. 0.9464

79. $6 = 10^{0.7782}$; $84 = 10^{1.9243}$; $987,606 = 10^{5.9946}$; $0.00987606 = 10^{-2.0054}$; $98,760.6 = 10^{4.9946}$; $70,000,000 = 10^{7.8451}$; $7000 = 10^{3.8451}$

81.
$$\frac{t^2 - 9}{t^2 - 4t + 4} \cdot \frac{3t - 6}{3t + 9} = \frac{(t+3)(t-3)}{(t-2)(t-2)} \cdot \frac{3(t-2)}{3(t+3)}$$
$$= \frac{(t+3)(t-3)(3)(t-2)}{(t-2)(t-2)(3)(t+3)}$$
$$= \frac{(t+3)(3)(t-2)}{(t+3)(3)(t-2)} \cdot \frac{t-3}{t-2}$$
$$= \frac{t-3}{t-2}$$

83.
$$\frac{3}{x^2 - x - 2} \quad \frac{1}{2x^2 + 3x + 1}$$
$$= \frac{3}{(x-2)(x+1)} - \frac{1}{(2x+1)(x+1)},$$

LCD is $(x-2)(x+1)(2x+1)$

$$= \frac{3}{(x-2)(x+1)} \cdot \frac{2x+1}{2x+1} - \frac{1}{(2x+1)(x+1)} \cdot \frac{x-2}{x-2}$$
$$= \frac{3(2x+1)}{(x-2)(x+1)(2x+1)} - \frac{x-2}{(2x+1)(x+1)(x-2)}$$
$$= \frac{3(2x+1) - (x-2)}{(x-2)(x+1)(2x+1)}$$
$$= \frac{6x+3 - x + 2}{(x-2)(x+1)(2x+1)}$$
$$= \frac{5x+5}{(x-2)(x+1)(2x+1)}$$
$$= \frac{5(x+1)}{(x-2)(x+1)(2x+1)}$$
$$= \frac{x+1}{x+1} \cdot \frac{5}{(x-2)(2x+1)}$$
$$= \frac{5}{(x-2)(2x+1)}$$

85. $3x^3 - 24x^2 + 48x = 3x(x^2 - 8x + 16)$
$$= 3x(x^2 - 2 \cdot x \cdot 4 + 4^2)$$
$$= 3x(x-4)^2$$

87. $2x^3 - 3x^2 + 2x - 3 = x^2(2x - 3) + 1 \cdot (2x - 3)$
$$= (2x - 3)(x^2 + 1)$$

89. Graph: $f(x) = \log_3 |x + 1|$

x	$f(x)$
0	0
2	1
8	2
-2	0
-4	1
-9	2

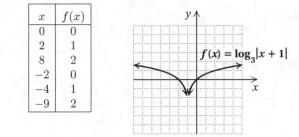

91. $\log_{125} x = \frac{2}{3}$

$$125^{2/3} = x$$
$$(5^3)^{2/3} = x$$
$$5^2 = x$$
$$25 = x$$

93. $\log_{128} x = \frac{5}{7}$

$$128^{5/7} = x$$
$$(2^7)^{5/7} = x$$
$$2^5 = x$$
$$32 = x$$

95. $\log_8(2x + 1) = -1$

$$8^{-1} = 2x + 1$$
$$\frac{1}{8} = 2x + 1$$
$$1 = 16x + 8 \quad \text{Multiplying by 8}$$
$$-7 = 16x$$
$$-\frac{7}{16} = x$$

97. Let $\log_{1/4} \frac{1}{64} = x$. Then

$$\left(\frac{1}{4}\right)^x = \frac{1}{64}$$
$$\left(\frac{1}{4}\right)^x = \left(\frac{1}{4}\right)^3$$
$$x = 3$$

Thus, $\log_{1/4} \frac{1}{64} = 3$.

99.
$$\log_{10}(\log_4(\log_3 81))$$
$$= \log_{10}(\log_4 4) \qquad (\log_3 81 = 4)$$
$$= \log_{10} 1 \qquad\qquad (\log_4 4 = 1)$$
$$= 0$$

101. Let $\log_{1/5} 25 = x$. Then

$$\left(\frac{1}{5}\right)^x = 25$$
$$(5^{-1})^x = 25$$
$$5^{-x} = 5^2$$
$$-x = 2$$
$$x = -2$$

Thus, $\log_{1/5} 25 = -2$.

Exercise Set 8.4

RC1. $\log_2(8 \cdot 4) = \log_2 8 + \log_2 4$ Property 1

The correct answer is (b).

RC3. $\log_2 4^8 = 8 \log_2 4$ Property 2

The correct answer is (d).

RC5. $\log_2 4^3 = 3 \log_2 4$ Property 2

The correct answer is (e).

1. $\log_2 (32 \cdot 8) = \log_2 32 + \log_2 8$ Property 1

3. $\log_4 (64 \cdot 16) = \log_4 64 + \log_4 16$ Property 1

5. $\log_a Qx = \log_a Q + \log_a x$ Property 1

7. $\log_b 3 + \log_b 84 = \log_b (3 \cdot 84)$ Property 1
$$= \log_b 252$$

9. $\log_c K + \log_c y = \log_c K \cdot y$ Property 1
$$= \log_c Ky$$

11. $\log_c y^4 = 4 \log_c y$ Property 2

13. $\log_b t^6 = 6 \log_b t$ Property 2

15. $\log_b C^{-3} = -3 \log_b C$ Property 2

17. $\log_a \dfrac{67}{5} = \log_a 67 - \log_a 5$ Property 3

19. $\log_b \dfrac{2}{5} = \log_b 2 - \log_b 5$ Property 3

21. $\log_c 22 - \log_c 3 = \log_c \dfrac{22}{3}$ Property 3

23. $\log_a x^2 y^3 z$
$$= \log_a x^2 + \log_a y^3 + \log_a z \quad \text{Property 1}$$
$$= 2 \log_a x + 3 \log_a y + \log_a z \quad \text{Property 2}$$

25. $\log_b \dfrac{xy^2}{z^3}$
$$= \log_b xy^2 - \log_b z^3 \quad\quad\quad \text{Property 3}$$
$$= \log_b x + \log_b y^2 - \log_b z^3 \quad \text{Property 1}$$
$$= \log_b x + 2 \log_b y - 3 \log_b z \quad \text{Property 2}$$

27. $\log_c \sqrt[3]{\dfrac{x^4}{y^3 z^2}}$
$$= \log_c \left(\frac{x^4}{y^3 z^2}\right)^{1/3}$$
$$= \frac{1}{3} \log_c \frac{x^4}{y^3 z^2} \quad\quad\quad\quad \text{Property 2}$$
$$= \frac{1}{3}(\log_c x^4 - \log_c y^3 z^2) \quad\quad \text{Property 3}$$
$$= \frac{1}{3}[\log_c x^4 - (\log_c y^3 + \log_c z^2)] \quad \text{Property 1}$$
$$= \frac{1}{3}(\log_c x^4 - \log_c y^3 - \log_c z^2) \quad \begin{array}{l}\text{Removing}\\\text{parentheses}\end{array}$$
$$= \frac{1}{3}(4 \log_c x - 3 \log_c y - 2 \log_c z) \quad \text{Property 2}$$
$$= \frac{4}{3} \log_c x - \log_c y - \frac{2}{3} \log_c z$$

29. $\log_a \sqrt[4]{\dfrac{m^8 n^{12}}{a^3 b^5}}$
$$= \log_a \left(\frac{m^8 n^{12}}{a^3 b^5}\right)^{1/4}$$
$$= \frac{1}{4} \log_a \frac{m^8 n^{12}}{a^3 b^5} \quad\quad\quad \text{Property 2}$$
$$= \frac{1}{4}(\log_a m^8 n^{12} - \log_a a^3 b^5) \quad\quad \text{Property 3}$$
$$= \frac{1}{4}[\log_a m^8 + \log_a n^{12} - (\log_a a^3 + \log_a b^5)] \text{ Property 1}$$
$$= \frac{1}{4}(\log_a m^8 + \log_a n^{12} - \log_a a^3 - \log_a b^5)$$
$$\quad\quad\quad\quad\quad\quad\quad\quad \text{Removing parentheses}$$
$$= \frac{1}{4}(\log_a m^8 + \log_a n^{12} - 3 - \log_a b^5) \quad \text{Property 4}$$
$$= \frac{1}{4}(8 \log_a m + 12 \log_a n - 3 - 5 \log_a b) \quad \text{Property 2}$$
$$= 2 \log_a m + 3 \log_a n - \frac{3}{4} - \frac{5}{4} \log_a b$$

31. $\dfrac{2}{3} \log_a x - \dfrac{1}{2} \log_a y$
$$= \log_a x^{2/3} - \log_a y^{1/2} \quad \text{Property 2}$$
$$= \log_a \frac{x^{2/3}}{y^{1/2}}, \text{ or} \quad\quad\quad \text{Property 3}$$
$$\log_a \frac{\sqrt[3]{x^2}}{\sqrt{y}}$$

33. $\log_a 2x + 3(\log_a x - \log_a y)$
$$= \log_a 2x + 3 \log_a x - 3 \log_a y$$
$$= \log_a 2x + \log_a x^3 - \log_a y^3 \quad \text{Property 2}$$
$$= \log_a 2x^4 - \log_a y^3 \quad\quad\quad \text{Property 1}$$
$$= \log_a \frac{2x^4}{y^3} \quad\quad\quad\quad\quad \text{Property 3}$$

35. $\log_a \dfrac{a}{\sqrt{x}} - \log_a \sqrt{ax}$

$= \log_a ax^{-1/2} - \log_a a^{1/2}x^{1/2}$

$= \log_a \dfrac{ax^{-1/2}}{a^{1/2}x^{1/2}}$ Property 3

$= \log_a \dfrac{a^{1/2}}{x}$, or

$\log_a \dfrac{\sqrt{a}}{x}$

37. $\log_b 15 = \log_b (3 \cdot 5)$

$= \log_b 3 + \log_b 5$ Property 1

$= 1.099 + 1.609$

$= 2.708$

39. $\log_b \dfrac{5}{3} = \log_b 5 - \log_b 3$ Property 3

$= 1.609 - 1.099$

$= 0.51$

41. $\log_b \dfrac{1}{5} = \log_b 1 - \log_b 5$ Property 3

$= 0 - 1.609$ $(\log_b 1 = 0)$

$= -1.609$

43. $\log_b \sqrt{b} = \log_b b^{1/2}$ $(\sqrt{b} = b^{1/2})$

$= \dfrac{1}{2}$ Property 4

45. $\log_b 5b = \log_b 5 + \log_b b$ Property 1

$= 1.609 + 1$ $(\log_b b = 1)$

$= 2.609$

47. $\log_b 2$ cannot be found using the properties of logarithms. $(\log_b 2 = \log_b(5 - 3) \neq \log_b 5 - \log_b 3.)$

49. $\log_e e^t = t$ Property 4

51. $\log_p p^5 = 5$ Property 4

53. $\log_2 2^7 = x$

$7 = x$ Property 4

55. $\log_e e^x = -7$

$x = -7$ Property 4

57. $i^{29} = i^{28} \cdot i = (i^2)^{14} \cdot i = (-1)^{14} \cdot i = 1 \cdot i = i$

59. $(2 + i)(2 - i) = 4 - i^2 = 4 - (-1) = 4 + 1 = 5$

61. $(7 - 8i) - (-16 + 10i) = 7 - 8i + 16 - 10i = 23 - 18i$

63. $(8 + 3i)(-5 - 2i) = -40 - 16i - 15i - 6i^2 =$
$-40 - 16i - 15i + 6 = -34 - 31i$

65. Enter $y_1 = \log\ x^2$ and $y_2 = (\log\ x)(\log\ x)$ and show that the graphs are different and that the y-values in a table of values are not the same.

67. $\log_a (x^8 - y^8) - \log_a (x^2 + y^2)$

$= \log_a \dfrac{x^8 - y^8}{x^2 + y^2}$ Property 3

$= \log_a \dfrac{(x^4 + y^4)(x^2 + y^2)(x + y)(x - y)}{x^2 + y^2}$ Factoring

$= \log_a [(x^4 + y^4)(x + y)(x - y)]$ Simplifying

$= \log_a (x^6 - x^4y^2 + x^2y^4 - y^6)$ Multiplying

69. $\log_a \sqrt{1 - s^2}$

$= \log_a (1 - s^2)^{1/2}$

$= \dfrac{1}{2}\ \log_a (1 - s^2)$

$= \dfrac{1}{2}\ \log_a [(1 - s)(1 + s)]$

$= \dfrac{1}{2}\ \log_a (1 - s) + \dfrac{1}{2}\ \log_a (1 + s)$

71. False. For example, let $a = 10$, $P = 100$, and $Q = 10$.

$\dfrac{\log 100}{\log 10} = \dfrac{2}{1} = 2$, but

$\log \dfrac{100}{10} = \log 10 = 1$.

73. True, by Property 1

75. False. For example, let $a = 2$, $P = 1$, and $Q = 1$.

$\log_2(1 + 1) = \log_2 2 = 1$, but

$\log_2 1 + \log_2 1 = 0 + 0 = 0$.

Chapter 8 Mid-Chapter Review

1. False; see the graph in Example 4 on page 668 in the text.

2. True; see page 683 in the text.

3. False; if $\log_a 0 = 1$, then $a^1 = 0$. This is true only when $a = 0$ but a logarithmic base a must be a positive number.

4. True by property 3 of logarithms

5. $\log_5 x = 3$

$5^3 = x$

$125 = x$

6. $\log_a 2 = 0.648$ and $\log_a 9 = 2.046$

a) $\log_a 18 = \log_a(2 \cdot 9)$

$= \log_a 2 + \log_a 9$

$= 0.648 + 2.046$

$= 2.694$

b) $\log_a \dfrac{1}{2} = \log_a 1 - \log_a 2$

$= 0 - 0.648$

$= -0.648$

7. Graph $f(x) = 3^{x-1}$.

We construct a table of values. Then we plot the points and connect them with a smooth curve.

$f(0) = 3^{0-1} = 3^{-1} = \dfrac{1}{3}$

$f(1) = 3^{1-1} = 3^0 = 1$

$f(2) = 3^{2-1} = 3^1 = 3$

$f(-1) = 3^{-1-1} = 3^{-2} = \dfrac{1}{3^2} = \dfrac{1}{9}$

$f(-2) = 3^{-2-1} = 3^{-3} = \dfrac{1}{3^3} = \dfrac{1}{27}$

x	$f(x)$
0	$\dfrac{1}{3}$
1	1
2	3
-1	$\dfrac{1}{9}$
-2	$\dfrac{1}{27}$

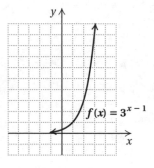

$f(x) = 3^{x-1}$

8. Graph $f(x) = \left(\dfrac{3}{4}\right)^x$.

We construct a table of values. Then we plot the points and connect them with a smooth curve.

$f(0) = \left(\dfrac{3}{4}\right)^0 = 1$

$f(1) = \left(\dfrac{3}{4}\right)^1 = \dfrac{3}{4}$

$f(2) = \left(\dfrac{3}{4}\right)^2 = \dfrac{9}{16}$

$f(-1) = \left(\dfrac{3}{4}\right)^{-1} = \dfrac{4}{3}$

$f(-2) = \left(\dfrac{3}{4}\right)^{-2} = \left(\dfrac{4}{3}\right)^2 = \dfrac{16}{9}$

x	$f(x)$
0	1
1	$\dfrac{3}{4}$
2	$\dfrac{9}{16}$
-1	$\dfrac{4}{3}$
-2	$\dfrac{16}{9}$

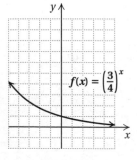

$f(x) = \left(\dfrac{3}{4}\right)^x$

9. Graph $f(x) = \log_4 x$.

The equation $f(x) = y = \log_4 x$ is equivalent to $4^y = x$. We can find ordered pairs by choosing values for y and computing the corresponding x-values.

For $y = 0, x = 4^0 = 1$.

For $y = 1, x = 4^1 = 4$.

For $y = 2, x = 4^2 = 16$.

For $y = -1, x = 4^{-1} = \dfrac{1}{4}$.

For $y = -2, x = 4^{-2} = \dfrac{1}{4^2} = \dfrac{1}{16}$.

x, or 4^y	y, or $f(x)$
1	0
4	1
16	2
$\dfrac{1}{4}$	-1
$\dfrac{1}{16}$	-2

We plot these ordered pairs and connect the points with a smooth curve.

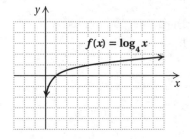

$f(x) = \log_4 x$

10. Graph $f(x) = \log_{1/4} x$.

The equation $f(x) = y = \log_{1/4} x$ is equivalent to $\left(\dfrac{1}{4}\right)^y = x$. We can find ordered pairs by choosing values for y and computing the corresponding x-values.

For $y = 0, x = \left(\dfrac{1}{4}\right)^0 = 1$.

For $y = 1, x = \left(\dfrac{1}{4}\right)^1 = \dfrac{1}{4}$.

For $y = 2, x = \left(\dfrac{1}{4}\right)^2 = \dfrac{1}{16}$.

For $y = -1, x = \left(\dfrac{1}{4}\right)^{-1} = \left(\dfrac{4}{1}\right)^1 = 4$.

For $y = -2, x = \left(\dfrac{1}{4}\right)^{-2} = \left(\dfrac{4}{1}\right)^2 = 16$.

x, or $\left(\dfrac{1}{4}\right)^y$	y, or $f(x)$
1	0
$\dfrac{1}{4}$	1
$\dfrac{1}{16}$	2
4	-1
16	-2

We plot these ordered pairs and connect the points with a smooth curve.

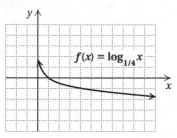

$$f(x) = \log_{1/4} x$$

11. a) $A(t) = P(1 + r)^t$

$A(t) = \$500(1 + 0.04)^t = \$500(1.04)^t$

b) $A(0) = \$500(1.04)^0 = \$500$

$A(4) = \$500(1.04)^4 \approx \$584.93$

$A(10) = \$500(1.04)^{10} \approx \$740.12$

12. $A = P \cdot \left(1 + \dfrac{r}{n}\right)^{n \cdot t}$

$= \$1500\left(1 + \dfrac{0.035}{4}\right)^{4(1.5)}$

$= \$1500(1.00875)^6$

$\approx \$1580.49$

13. The graph of $f(x) = 3x + 1$ is shown below. It passes the horizontal line test and is one-to-one.

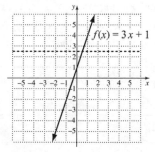

$f(x) = 3x + 1$

1. Replace $f(x)$ with y: $y = 3x + 1$
2. Interchange x and y: $x = 3y + 1$
3. Solve for y: $x = 3y + 1$

    $x - 1 = 3y$

    $\dfrac{x - 1}{3} = y$

4. Replace y with $f^{-1}(x)$: $f^{-1}(x) = \dfrac{x - 1}{3}$

14. The graph of $f(x) = x^3 + 2$ is shown below. It passes the horizontal line test and is one-to-one.

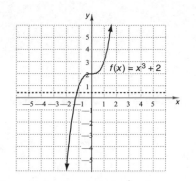

$$f(x) = x^3 + 2$$

1. Replace $f(x)$ with y: $y = x^3 + 2$
2. Interchange x and y: $x = y^3 + 2$
3. Solve for y: $x = y^3 + 2$

    $x - 2 = y^3$

    $\sqrt[3]{x - 2} = y$

4. Replace y with $f^{-1}(x)$: $f^{-1}(x) = \sqrt[3]{x - 2}$

15. $(f \circ g)(x) = f(g(x)) = f(3 - x) = 2(3 - x) - 5 = 6 - 2x - 5 = 1 - 2x$

$(g \circ f)(x) = g(f(x)) = g(2x - 5) = 3 - (2x - 5) = 3 - 2x + 5 = 8 - 2x$

16. $(f \circ g)(x) = f(g(x)) = f(3x - 1) = (3x - 1)^2 + 1 = 9x^2 - 6x + 1 + 1 = 9x^2 - 6x + 2$

$(g \circ f)(x) = g(f(x)) = g(x^2 + 1) = 3(x^2 + 1) - 1 = 3x^2 + 3 - 1 = 3x^2 + 2$

17. $h(x) = \dfrac{3}{x + 4}$

This is 3 divided by $x + 4$, so the two functions could be $f(x) = \dfrac{3}{x}$ and $g(x) = x + 4$.

18. $h(x) = \sqrt{6x - 7}$

This is the square root of $6x - 7$, so the two functions could be $f(x) = \sqrt{x}$ and $g(x) = 6x - 7$.

19. We check to see that $(f^{-1} \circ f)(x) = x$ and $(f \circ f^{-1})(x) = x$.

$(f^{-1} \circ f)(x) = f^{-1}(f(x)) = f^{-1}\left(\dfrac{x}{3}\right) = 3 \cdot \dfrac{x}{3} = x$

$(f \circ f^{-1})(x) = f(f^{-1}(x)) = f(3x) = \dfrac{3x}{3} = x$

The inverse is correct.

20. We check to see that $(f^{-1} \circ f)(x) = x$ and $(f \circ f^{-1})(x) = x$.

$(f^{-1} \circ f)(x) = f^{-1}(f(x)) = f^{-1}(\sqrt[3]{x + 4}) = (\sqrt[3]{x + 4})^3 - 4 = x + 4 - 4 = x$

$(f \circ f^{-1})(x) = f(f^{-1}(x)) = f(x^3 - 4) = \sqrt[3]{(x^3 - 4) + 4} = \sqrt[3]{x^3} = x$

The inverse is correct.

21. $7^3 = 343 \Rightarrow 3 = \log_7 343$

22. $3^{-4} = \dfrac{1}{81} \Rightarrow -4 = \log_3 \dfrac{1}{81}$

23. $\log_6 12 = t \Rightarrow 6^t = 12$

24. $\log_n T = m \Rightarrow n^m = T$

25. $\log_4 64 = x$

$\qquad 4^x = 64$

$\qquad 4^x = 4^3$

$\qquad\quad x = 3$

26. $\log_x \dfrac{1}{4} = -2$

$\qquad x^{-2} = \dfrac{1}{4}$

$\qquad x^{-2} = \dfrac{1}{2^2}$

$\qquad x^{-2} = 2^{-2}$

$\qquad\quad x = 2$

27. Let $\log_7 49 = x$.

$\qquad 7^x = 49$

$\qquad 7^x = 7^2$

$\qquad\ x = 2$

28. Let $\log_2 32 = x$.

$\qquad 2^x = 32$

$\qquad 2^x = 2^5$

$\qquad\ x = 5$

29. $\log 243.7 \approx 2.3869$

30. $\log 0.23 \approx -0.6383$

31. $\log_b \dfrac{2xy^2}{z^3}$

$= \log_b(2xy^2) - \log_b z^3$

$= \log_b 2 + \log_b x + \log_b y^2 - \log_b z^3$

$= \log_b 2 + \log_b x + 2\log_b y - 3\log_b z$

32. $\log_a \sqrt[3]{\dfrac{x^2 y^5}{z^4}}$

$= \log_a \left(\dfrac{x^2 y^5}{z^4}\right)^{1/3}$

$= \dfrac{1}{3} \log_a \left(\dfrac{x^2 y^5}{z^4}\right)$

$= \dfrac{1}{3}[\log_a(x^2 y^5) - \log_a z^4]$

$= \dfrac{1}{3}(\log_a x^2 + \log_a y^5 - \log_a z^4)$

$= \dfrac{1}{3}(2\log_a x + 5\log_a y - 4\log_a z)$

$= \dfrac{2}{3}\log_a x + \dfrac{5}{3}\log_a y - \dfrac{4}{3}\log_a z$

33. $\log_a x - 2\log_a y + \dfrac{1}{2}\log_a z = \log_a x - \log_a y^2 + \log_a z^{1/2}$

$= \log_a \dfrac{xz^{1/2}}{y^2}, \text{ or}$

$\log_a \dfrac{x\sqrt{z}}{y^2}$

34. $\log_m(b^2 - 16) - \log_m(b + 4) = \log_m \dfrac{b^2 - 16}{b + 4}$

$= \log_m \dfrac{(b+4)(b-4)}{b+4}$

$= \log_m(b - 4)$

35. $\log_8 1 = 0 \qquad (\log_a 1 = 0)$

36. $\log_3 3 = 1 \qquad (\log_a a = 1)$

37. $\log_a a^{-3} = -3 \qquad (\log_a a^k = k)$

38. $\log_c c^5 = 5 \qquad (\log_a a^k = k)$

39. $V^{-1}(t)$ could be used to predict when the value of the stamp will be t, where $V^{-1}(t)$ is the number of years after 1999.

40. $\log_a b$ is the number to which a is raised to get b. Since $\log_a b = c$, then $a^c = b$.

41. Express $\dfrac{x}{5}$ as $x \cdot 5^{-1}$ and then use the product rule and the power rule to get $\log_a\left(\dfrac{x}{5}\right) = \log_a(x \cdot 5^{-1}) =$ $\log_a x + \log_a 5^{-1} = \log_a x + (-1)\log_a 5 = \log_a x - \log_a 5$.

42. The student didn't subtract the logarithm of the entire denominator after using the quotient rule. The correct procedure is as follows:

$\log_b \dfrac{1}{x} = \log_b \dfrac{x}{xx}$

$\qquad\quad = \log_b x - \log_b xx$

$\qquad\quad = \log_b x - (\log_b x + \log_b x)$

$\qquad\quad = \log_b x - \log_b x - \log_b x$

$\qquad\quad = -\log_b x$

(Note that $-\log_b x$ is equivalent to $\log_b 1 - \log_b x$.)

Exercise Set 8.5

RC1. The correct answer is (b) by the change-of-base formula.

RC3. The correct answer is (b) by the change-of-base formula.

1. 0.6931

3. 4.1271

5. 8.3814

7. −5.0832

9. −1.6094

11. Does not exist

13. −1.7455

15. 1

17. 15.0293

19. 0.0305

21. 109.9472

23. 5

25. We will use common logarithms for the conversion. Let $a = 10$, $b = 6$, and $M = 100$ and substitute in the change-of-base formula.

$$\log_b M = \frac{\log_a M}{\log_a b}$$

$$\log_6 100 = \frac{\log_{10} 100}{\log_{10} 6}$$

$$\approx \frac{2}{0.7782}$$

$$\approx 2.5702$$

27. We will use common logarithms for the conversion. Let $a = 10$, $b = 2$, and $M = 100$ and substitute in the change-of-base formula.

$$\log_2 100 = \frac{\log_{10} 100}{\log_{10} 2}$$

$$\approx \frac{2}{0.3010}$$

$$\approx 6.6439$$

29. We will use natural logarithms for the conversion. Let $a = e$, $b = 7$, and $M = 65$ and substitute in the change-of-base formula.

$$\log_7 65 = \frac{\ln 65}{\ln 7}$$

$$\approx \frac{4.1744}{1.9459}$$

$$\approx 2.1452$$

31. We will use natural logarithms for the conversion. Let $a = e$, $b = 0.5$, and $M = 5$ and substitute in the change-of-base formula.

$$\log_{0.5} 5 = \frac{\ln 5}{\ln 0.5}$$

$$\approx \frac{1.6094}{-0.6931}$$

$$\approx -2.3219$$

33. We will use common logarithms for the conversion. Let $a = 10$, $b = 2$, and $M = 0.2$ and substitute in the change-of-base formula.

$$\log_2 0.2 = \frac{\log_{10} 0.2}{\log_{10} 2}$$

$$\approx \frac{-0.6990}{0.3010}$$

$$\approx -2.3219$$

35. We will use natural logarithms for the conversion. Let $a = e$, $b = \pi$, and $M = 200$.

$$\log_\pi 200 = \frac{\ln 200}{\ln \pi}$$

$$\approx \frac{5.2983}{1.1447}$$

$$\approx 4.6285$$

If $\ln 200$ and $\ln \pi$ are not rounded before the division is performed, the result is 4.6284.

37. Graph: $f(x) = e^x$

We find some function values with a calculator. We use these values to plot points and draw the graph.

x	$f(x)$
0	1
1	2.7
2	7.4
3	20.1
-1	0.4
-2	0.1
-3	0.05

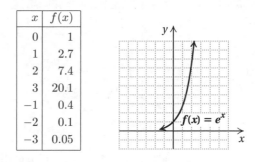

39. Graph: $f(x) = e^{-0.5x}$

We find some function values, plot points, and draw the graph.

x	$f(x)$
0	1
1	0.61
2	0.37
-1	1.65
-2	2.72
-3	4.48
-4	7.39

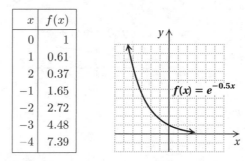

41. Graph: $f(x) = e^{x-1}$

We find some function values, plot points, and draw the graph.

x	$f(x)$
0	0.4
1	1
2	2.7
3	7.4
4	20.1
-1	0.1
-2	0.05

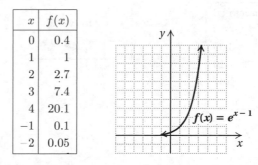

43. Graph: $f(x) = e^{x+2}$

We find some function values, plot points, and draw the graph.

x	$f(x)$
1	20.1
0	7.4
-2	1
-3	0.4
-4	0.1

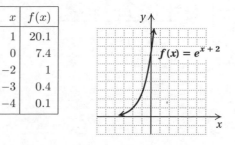

45. Graph: $f(x) = e^x - 1$

We find some function values, plot points, and draw the graph.

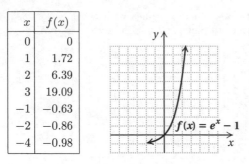

x	$f(x)$
0	0
1	1.72
2	6.39
3	19.09
-1	-0.63
-2	-0.86
-4	-0.98

47. Graph: $f(x) = \ln(x + 2)$

We find some function values, plot points, and draw the graph.

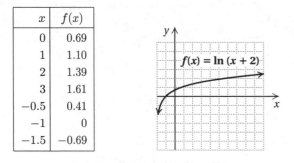

x	$f(x)$
0	0.69
1	1.10
2	1.39
3	1.61
-0.5	0.41
-1	0
-1.5	-0.69

49. Graph: $f(x) = \ln(x - 3)$

We find some function values, plot points, and draw the graph.

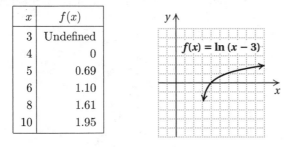

x	$f(x)$
3	Undefined
4	0
5	0.69
6	1.10
8	1.61
10	1.95

51. Graph: $f(x) = 2 \ln x$

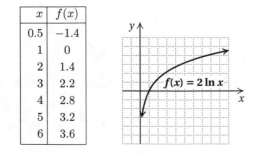

x	$f(x)$
0.5	-1.4
1	0
2	1.4
3	2.2
4	2.8
5	3.2
6	3.6

53. Graph: $f(x) = \dfrac{1}{2} \ln x + 1$

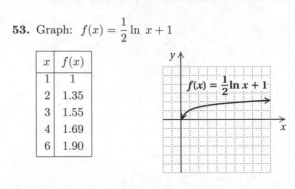

x	$f(x)$
1	1
2	1.35
3	1.55
4	1.69
6	1.90

55. Graph: $f(x) = |\ln x|$

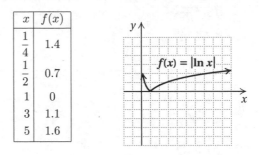

x	$f(x)$
$\dfrac{1}{4}$	1.4
$\dfrac{1}{2}$	0.7
1	0
3	1.1
5	1.6

57. $x^{1/2} - 6x^{1/4} + 8 = 0$

Let $u = x^{1/4}$.

$$u^2 - 6u + 8 = 0 \qquad \text{Substituting}$$

$$(u - 4)(u - 2) = 0$$

$$u = 4 \quad \text{or} \quad u = 2$$

$$x^{1/4} = 4 \quad \text{or} \quad x^{1/4} = 2$$

$$x = 256 \text{ or} \qquad x = 16 \qquad \text{Raising both sides}$$
$$\text{to the fourth power}$$

Both numbers check. The solutions are 256 and 16.

59. $x - 18\sqrt{x} + 77 = 0$

Let $u = \sqrt{x}$.

$$u^2 - 18u + 77 = 0 \qquad \text{Substituting}$$

$$(u - 7)(u - 11) = 0$$

$$u = 7 \quad \text{or} \quad u = 11$$

$$\sqrt{x} = 7 \quad \text{or} \quad \sqrt{x} = 11$$

$$x = 49 \text{ or} \qquad x = 121 \qquad \text{Squaring both sides}$$

Both numbers check. The solutions are 49 and 121.

61. Domain: $(-\infty, \infty)$, range: $[0, \infty)$

63. Domain: $(-\infty, \infty)$, range: $(-\infty, 100)$

65. $f(x)$ can be calculated for positive values of $2x - 5$. We have:

$$2x - 5 > 0$$

$$2x > 5$$

$$x > \frac{5}{2}$$

The domain is $\left\{ x \middle| x > \dfrac{5}{2} \right\}$, or $\left(\dfrac{5}{2}, \infty \right)$.

Exercise Set 8.6

RC1. $\quad 2^x = 6$

$\qquad \log 2^x = \log 6$

$\qquad x \log 2 = \log 6$

$\qquad x = \dfrac{\log 6}{\log 2} \approx 2.5850$

The given statement is false.

RC3. $\log_2 4 = x$

$\qquad 2^x = 4$

$\qquad 2^x = 2^2$

$\qquad x = 2$

The given statement is false.

RC5. $\log_8 x = 1$

$\qquad 8^1 = x$

$\qquad 8 = x$

The given statement is true.

1. $2^x = 8$

$\quad 2^x = 2^3$

$\quad x = 3 \qquad$ The exponents are the same.

3. $4^x = 256$

$\quad 4^x = 4^4$

$\quad x = 4 \qquad$ The exponents are the same.

5. $2^{2x} = 32$

$\quad 2^{2x} = 2^5$

$\quad 2x = 5$

$\quad x = \dfrac{5}{2}$

7. $3^{5x} = 27$

$\quad 3^{5x} = 3^3$

$\quad 5x = 3$

$\quad x = \dfrac{3}{5}$

9. $\qquad 2^x = 11$

$\quad \log 2^x = \log 11 \qquad$ Taking the common logarithm on both sides

$\quad x \log 2 = \log 11 \qquad$ Property 2

$\quad x = \dfrac{\log 11}{\log 2}$

$\quad x \approx 3.4594$

11. $\qquad 2^x = 43$

$\quad \log 2^x = \log 43 \qquad$ Taking the common logarithm on both sides

$\quad x \log 2 = \log 43 \qquad$ Property 2

$\quad x = \dfrac{\log 43}{\log 2}$

$\quad x \approx 5.4263$

13. $\qquad 5^{4x-7} = 125$

$\qquad 5^{4x-7} = 5^3$

$\quad 4x - 7 = 3 \qquad$ The exponents are the same.

$\qquad 4x = 10$

$\qquad x = \dfrac{10}{4}, \text{ or } \dfrac{5}{2}$

15. $\qquad 3^{x^2} \cdot 3^{4x} = \dfrac{1}{27}$

$\qquad 3^{x^2+4x} = 3^{-3}$

$\qquad x^2 + 4x = -3$

$\quad x^2 + 4x + 3 = 0$

$\quad (x+3)(x+1) = 0$

$\quad x = -3 \; or \; x = -1$

17. $\qquad 4^x = 8$

$\quad (2^2)^x = 2^3$

$\qquad 2^{2x} = 2^3$

$\qquad 2x = 3 \qquad$ The exponents are the same.

$\qquad x = \dfrac{3}{2}$

19. $\qquad e^t = 100$

$\quad \ln e^t = \ln 100 \qquad$ Taking ln on both sides

$\qquad t = \ln 100 \qquad$ Property 4

$\qquad t \approx 4.6052 \qquad$ Using a calculator

21. $\qquad e^{-t} = 0.1$

$\quad \ln e^{-t} = \ln 0.1 \qquad$ Taking ln on both sides

$\qquad -t = \ln 0.1 \qquad$ Property 4

$\qquad -t \approx -2.3026$

$\qquad t \approx 2.3026$

23. $\qquad e^{-0.02t} = 0.06$

$\quad \ln e^{-0.02t} = \ln 0.06 \qquad$ Taking ln on both sides

$\quad -0.02t = \ln 0.06 \qquad$ Property 4

$\qquad t = \dfrac{\ln 0.06}{-0.02}$

$\qquad t \approx \dfrac{-2.8134}{-0.02}$

$\qquad t \approx 140.6705$

25. $\qquad 2^x = 3^{x-1}$

$\quad \log 2^x = \log 3^{x-1}$

$\quad x \log 2 = (x-1) \log 3$

$\quad x \log 2 = x \log 3 - \log 3$

$\quad \log 3 = x \log 3 - x \log 2$

$\quad \log 3 = x(\log 3 - \log 2)$

$\qquad \dfrac{\log 3}{\log 3 - \log 2} = x$

$\quad \dfrac{0.4771}{0.4771 - 0.3010} \approx x$

$\qquad 2.7095 \approx x$

27. $(3.6)^x = 62$

$\log (3.6)^x = \log 62$

$x \log 3.6 = \log 62$

$x = \dfrac{\log 62}{\log 3.6}$

$x \approx 3.2220$

29. $\log_4 x = 4$

$x = 4^4$ Writing an equivalent exponential equation

$x = 256$

31. $\log_2 x = -5$

$x = 2^{-5}$ Writing an equivalent exponential equation

$x = \dfrac{1}{32}$

33. $\log x = 1$ The base is 10.

$x = 10^1$

$x = 10$

35. $\log x = -2$ The base is 10.

$x = 10^{-2}$

$x = \dfrac{1}{100}$

37. $\ln x = 2$

$x = e^2 \approx 7.3891$

39. $\ln x = -1$

$x = e^{-1}$

$x = \dfrac{1}{e} \approx 0.3679$

41. $\log_3 (2x + 1) = 5$

$2x + 1 = 3^5$ Writing an equivalent exponential equation

$2x + 1 = 243$

$2x = 242$

$x = 121$

43. $\log x + \log (x - 9) = 1$ The base is 10.

$\log_{10} [x(x - 9)] = 1$ Property 1

$x(x - 9) = 10^1$

$x^2 - 9x = 10$

$x^2 - 9x - 10 = 0$

$(x - 10)(x + 1) = 0$

$x = 10 \ or \ x = -1$

Check: For 10:

$\dfrac{\log x + \log (x - 9) = 1}{}$

$\log 10 + \log (10 - 9) \ ? \ 1$

$\log 10 + \log 1 \ \big|$

$1 + 0 \ \big|$

$1 \ \big| $ TRUE

For -1:

$\dfrac{\log x + \log (x - 9) = 1}{}$

$\log(-1) + \log (-1 - 9) \ ? \ 1$ FALSE

The number -1 does not check, because negative numbers do not have logarithms. The solution is 10.

45. $\log x - \log (x + 3) = -1$ The base is 10.

$\log_{10} \dfrac{x}{x + 3} = -1$ Property 3

$\dfrac{x}{x + 3} = 10^{-1}$

$\dfrac{x}{x + 3} = \dfrac{1}{10}$

$10x = x + 3$

$9x = 3$

$x = \dfrac{1}{3}$

The answer checks. The solution is $\dfrac{1}{3}$.

47. $\log_2 (x + 1) + \log_2 (x - 1) = 3$

$\log_2 [(x + 1)(x - 1)] = 3$ Property 1

$(x + 1)(x - 1) = 2^3$

$x^2 - 1 = 8$

$x^2 = 9$

$x = \pm 3$

The number 3 checks, but -3 does not. The solution is 3.

49. $\log_4 (x + 6) - \log_4 x = 2$

$\log_4 \dfrac{x + 6}{x} = 2$ Property 3

$\dfrac{x + 6}{x} = 4^2$

$\dfrac{x + 6}{x} = 16$

$x + 6 = 16x$

$6 = 15x$

$\dfrac{2}{5} = x$

The answer checks. The solution is $\dfrac{2}{5}$.

51. $\log_4 (x + 3) + \log_4 (x - 3) = 2$

$\log_4 [(x + 3)(x - 3)] = 2$ Property 1

$(x + 3)(x - 3) = 4^2$

$x^2 - 9 = 16$

$x^2 = 25$

$x = \pm 5$

The number 5 checks, but -5 does not. The solution is 5.

53. $\log_3 (2x - 6) - \log_3 (x + 4) = 2$

$$\log_3 \frac{2x - 6}{x + 4} = 2 \qquad \text{Property 3}$$

$$\frac{2x - 6}{x + 4} = 3^2$$

$$\frac{2x - 6}{x + 4} = 9$$

$$2x - 6 = 9x + 36$$

$$\qquad \text{Multiplying by } (x + 4)$$

$$-42 = 7x$$

$$-6 = x$$

Check:

$$\frac{\log_3 (2x - 6) - \log_3 (x + 4) = 2}{\log_3 [2(-6) - 6] - \log_3 (-6 + 4) \; ? \; 2}$$

$$\log_3 (-18) - \log_3 (-2) \; \Big| \qquad \text{FALSE}$$

The number -6 does not check, because negative numbers do not have logarithms. There is no solution.

55. $-3 \le x - 12 < 4$

$9 \le x \le 16 \qquad$ Adding 12

The solution set is $\{x | 9 \le x \le 16\}$, or $[9, 16]$.

57. $x^2 - x = 12$

$x^2 - x - 12 = 0$

$(x - 4)(x + 3) = 0$

$x - 4 = 0 \;$ or $\; x + 3 = 0$

$x = 4 \;$ or $\qquad x = -3$

The solutions are 4 and -3.

59. $\sqrt{n - 1} = 8$

$(\sqrt{n - 1})^2 = 8^2$

$n - 1 = 64$

$n = 65$

The number 65 checks. It is the solution.

61. Find the first coordinate of the point of intersection of $y_1 = \ln x$ and $y_2 = \log x$. The value of x for which the natural logarithm of x is the same as the common logarithm of x is 1.

63. a) 0.3770

b) -1.9617

c) 0.9036

d) -1.5318

65. $2^{2x} + 128 = 24 \cdot 2^x$

$2^{2x} - 24 \cdot 2^x + 128 = 0$

Let $u = 2^x$.

$u^2 - 24u + 128 = 0$

$(u - 8)(u - 16) = 0$

$u = 8 \;$ or $\; u = 16$

$2^x = 8 \;$ or $\; 2^x = 16 \quad$ Replacing u with 2^x

$2^x = 2^3 \;$ or $\; 2^x = 2^4$

$x = 3 \;$ or $\; x = 4$

The solutions are 3 and 4.

67. $8^x = 16^{3x + 9}$

$(2^3)^x = (2^4)^{3x + 9}$

$2^{3x} = 2^{12x + 36}$

$3x = 12x + 36$

$-36 = 9x$

$-4 = x$

69. $\log_6 (\log_2 x) = 0$

$\log_2 x = 6^0$

$\log_2 x = 1$

$x = 2^1$

$x = 2$

71. $\log_5 \sqrt{x^2 - 9} = 1$

$\sqrt{x^2 - 9} = 5^1$

$x^2 - 9 = 25 \qquad$ Squaring both sides

$x^2 = 34$

$x = \pm\sqrt{34}$

Both numbers check. The solutions are $\pm\sqrt{34}$.

73. $\log (\log x) = 5 \qquad$ The base is 10.

$\log x = 10^5$

$\log x = 100,000$

$x = 10^{100,000}$

The number checks. The solution is $10^{100,000}$.

75. $\log x^2 = (\log x)^2$

$2 \log x = (\log x)^2$

$0 = (\log x)^2 - 2 \log x$

Let $u = \log x$.

$0 = u^2 - 2u$

$0 = u(u - 2)$

$u = 0 \;$ or $\; u = 2$

$\log x = 0 \;$ or $\; \log x = 2$

$x = 10^0 \;$ or $\; x = 10^2$

$x = 1 \;$ or $\; x = 100$

Both numbers check. The solutions are 1 and 100.

77. $\log_a a^{x^2 + 4x} = 21$

$x^2 + 4x = 21 \qquad$ Property 4

$x^2 + 4x - 21 = 0$

$(x + 7)(x - 3) = 0$

$x = -7 \;$ or $\; x = 3$

Both numbers check. The solutions are $= -7$ and 3.

79. $3^{2x} - 8 \cdot 3^x + 15 = 0$

Let $u = 3^x$ and substitute.

$u^2 - 8u + 15 = 0$

$(u - 5)(u - 3) = 0$

$\quad u = 5 \qquad or \quad u = 3$

$\quad 3^x = 5 \qquad or \quad 3^x = 3 \qquad$ Substituting 3^x for u

$\quad \log 3^x = \log 5 \quad or \quad 3^x = 3^1$

$\quad x\,\log 3 = \log 5 \quad or \quad x = 1$

$\quad x = \dfrac{\log 5}{\log 3} \quad or \quad x = 1, \text{ or}$

$\quad x \approx 1.4650 \quad or \quad x = 1$

Both numbers check. Note that we can also express $\dfrac{\log 5}{\log 3}$ as $\log_3 5$ using the change-of-base formula.

Exercise Set 8.7

RC1. (b); see page 733 in the text.

RC3. (c); see page 733 in the text.

1. $L = 10 \cdot \log \dfrac{I}{I_0}$

$\quad = 10 \cdot \log \dfrac{10^{6.8}}{10^{-12}} \qquad$ Substituting

$\quad = 10 \cdot \log(10^{18.8})$

$\quad = 10(18.8)$

$\quad = 188$

The sound level is 188 dB.

3. First we find the intensity of the sound of the quieter dishwasher.

$L = 10 \cdot \log \dfrac{I}{I_0}$

$45 = 10 \cdot \log \dfrac{I}{10^{-12}}$

$4.5 = \log \dfrac{I}{10^{-12}}$

$4.5 = \log I - \log 10^{-12}$

$4.5 = \log I - (-12)$

$4.5 = \log I + 12$

$-7.5 = \log I$

$10^{-7.5} = I \qquad$ Converting to an exponential equation

$3.2 \times 10^{-8} \approx I$

The intensity of the sound of the quieter dishwasher is $10^{-7.5}$ W/m^2, or about 3.2×10^{-8} W/m^2.

Now we find the intensity of the sound of the noisier dishwasher.

$L = 10 \cdot \log \dfrac{I}{I_0}$

$60 = 10 \cdot \log \dfrac{I}{10^{-12}}$

$6 = \log \dfrac{I}{10^{-12}}$

$6 = \log I - \log 10^{-12}$

$6 = \log I - (-12)$

$6 = \log I + 12$

$-6 = \log I$

$10^{-6} = I \qquad$ Converting to an exponential equation

The intensity of the sound of the noisier dishwasher is 10^{-6} W/m^2.

5. $\text{pH} = -\log[\text{H}^+]$

$\quad = -\log[1.6 \times 10^{-7}]$

$\quad \approx -(-6.795880)$

$\quad \approx 6.8$

The pH of milk is about 6.8.

7. $\quad\quad\quad \text{pH} = -\log[\text{H}^+]$

$\quad\quad\quad 7.8 = -\log[\text{H}^+]$

$\quad\quad -7.8 = \log[\text{H}^+]$

$\quad 10^{-7.8} = [\text{H}^+]$

$1.58 \times 10^{-8} \approx [\text{H}^+]$

The hydrogen ion concentration is about 1.58×10^{-8} moles per liter.

9. $\quad 616,500 = 616.5$ thousands

$\quad\quad w(P) = 0.37 \ln P + 0.05$

$\quad w(616.5) = 0.37 \ln 616.5 + 0.05$

$\quad\quad\quad\quad \approx 2.43$ ft/sec

11. $\quad 2,707,120 = 2707.12$ thousands

$\quad\quad\quad w(P) = 0.37 \ln P + 0.05$

$\quad w(2707.12) = 0.37 \ln 2707.12 + 0.05$

$\quad\quad\quad\quad\quad \approx 2.97$ ft/sec

13. $R(t) = 294(1.074)^t$

a) In 2010, $t = 2010 - 1985 = 25$.

$\quad R(25) = 294(1.074)^{25} \approx 1752$ riverbanks

b)
$$1000 = 294(1.074)t$$

$$\frac{1000}{294} = 1.074^t$$

$$\ln\left(\frac{1000}{294}\right) = \ln 1.074^t$$

$$\ln\left(\frac{1000}{294}\right) = t \ln 1.074$$

$$\frac{\ln\left(\dfrac{1000}{294}\right)}{\ln 1.074} = t$$

$$17 \approx t$$

There were otters on 1000 riverbanks 17 years after 1985, or in 2002.

c)
$$2(294) = 294(1.074)^t$$

$$2 = (1.074)^t$$

$$\ln 2 = \ln (1.074)^t$$

$$\ln 2 = t \ln 1.074$$

$$\frac{\ln 2}{\ln 1.074} = t$$

$$9.7 \approx t$$

The doubling time is about 9.7 years.

15. $H(t) = 56(1.67)^t$

a) In 2012, $t = 2012 - 2000 = 12$.
$$H(12) = 56(1.67)^{12} \approx 26,350 \text{ veterans}$$

b)
$$15,000 = 56(1.67)^t$$

$$\frac{15,000}{56} = (1.67)^t$$

$$\ln\left(\frac{15,000}{56}\right) = \ln (1.67)^t$$

$$\ln\left(\frac{15,000}{56}\right) = t \ln 1.67$$

$$\frac{\ln\left(\dfrac{15,000}{56}\right)}{\ln 1.67} = t$$

$$11 \approx t$$

There were 15,000 veterans who were homeless or at risk of becoming homeless about 11 years after 2000, or in 2011.

c)
$$2(56) = 56(1.67)^t$$

$$2 = (1.67)^t$$

$$\ln 2 = \ln (1.67)^t$$

$$\ln 2 = t \ln 1.67$$

$$\frac{\ln 2}{\ln 1.67} = t$$

$$1.4 \approx t$$

The doubling time is about 1.4 yr.

17. a) $P(t) = P_0\, e^{0.03t}$

b) To find the balance after one year, replace P_0 with 5000 and t with 1. We find $P(1)$:
$$P(1) = 5000\, e^{0.03(1)} \approx \$5152.27$$

To find the balance after 2 years, replace P_0 with 5000 and t with 2. We find $P(2)$:
$$P(2) = 5000\, e^{0.03(2)} \approx \$5309.18$$

To find the balance after 10 years, replace P_0 with 5000 and t with 10. We find $P(10)$:
$$P(10) = 5000 e^{0.03(10)} \approx \$6749.29$$

c) To find the doubling time, replace P_0 with 5000 and $P(t)$ with 10,000 and solve for t.
$$10,000 = 5000\, e^{0.03t}$$

$$2 = e^{0.03t}$$

$$\ln 2 = \ln e^{0.03t} \quad \text{Taking the natural loga-}$$
$$\qquad\qquad\qquad \text{rithm on both sides}$$

$$\ln 2 = 0.03t \quad \text{Finding the logarithm of}$$
$$\qquad\qquad\qquad \text{the base to a power}$$

$$\frac{\ln 2}{0.03} = t$$

$$23.1 \approx t$$

The investment will double in about 23.1 years.

19. a) $P(t) = 7.1e^{0.011t}$, where $P(t)$ is in billions and t is the number of years after 2013.

b) In 2016, $t = 2016 - 2013 = 3$.
$$P(3) = 7.1e^{0.011(3)} \approx 7.3 \text{ billion}$$

c)
$$15 = 7.1e^{0.011t}$$

$$\frac{15}{7.1} = e^{0.011t}$$

$$\ln\left(\frac{15}{7.1}\right) = \ln e^{0.011t}$$

$$\ln\left(\frac{15}{7.1}\right) = 0.011t$$

$$\frac{\ln\left(\dfrac{15}{7.1}\right)}{0.011} = t$$

$$68 \approx t$$

The world population will reach 15 billion about 68 years after 2013, or in 2081.

d)
$$2(7.1) = 7.1e^{0.011t}$$

$$2 = e^{0.011t}$$

$$\ln 2 = \ln e^{0.011t}$$

$$\ln 2 = 0.011t$$

$$\frac{\ln 2}{0.011} = t$$

$$63.0 \approx t$$

The doubling time is about 63.0 yr.

21. a)
$$C(t) = C_0 e^{kt}$$
$$189 = 2e^{k \cdot 76}$$
$$94.5 = e^{76k}$$
$$\ln 94.5 = \ln e^{76k}$$
$$\ln 94.5 = 76k$$
$$\frac{\ln 94.5}{76} = k$$
$$0.060 \approx k$$

The exponential growth rate is about 0.060 or 6.0%.

The exponential growth function is $C(t) = 2e^{0.060t}$, where t is the number of years after 1935.

b) In 2013, $t = 2013 - 1935 = 78$.
$$C(78) = 2e^{0.060(78)} \approx 216 \text{ pages}$$

c)
$$250 = 2e^{0.060t}$$
$$125 = e^{0.060t}$$
$$\ln 125 = \ln e^{0.060t}$$
$$\ln 125 = 0.060t$$
$$\frac{\ln 125}{0.060} = t$$
$$80 \approx t$$

There will be 250 pages in the instruction book about 80 years after 1935, or in 2015.

23. If the scrolls had lost 22.3% of their carbon-14 from an initial amount P_0, then 77.7%(P_0) is the amount present. To find the age t of the scrolls, we substitute 77.7%(P_0), or $0.777P_0$, for $P(t)$ in the carbon-14 decay function and solve for t.
$$P(t) = P_0 e^{-0.00012t}$$
$$0.777P_0 = P_0 e^{-0.00012t}$$
$$0.777 = e^{-0.00012t}$$
$$\ln 0.777 = \ln e^{-0.00012t}$$
$$-0.2523 \approx -0.00012t$$
$$t \approx \frac{-0.2523}{-0.00012} \approx 2103$$

The scrolls are about 2103 years old.

25. The function $P(t) = P_0 e^{-kt}$, $k > 0$, can be used to model decay. For iodine-131, $k = 9.6\%$, or 0.096. To find the half-life we substitute 0.096 for k and $\frac{1}{2}P_0$ for $P(t)$, and solve for t.
$$\frac{1}{2}P_0 = P_0 e^{-0.096t}, \text{ or } \frac{1}{2} = e^{-0.096t}$$
$$\ln \frac{1}{2} = \ln e^{-0.096t} = -0.096t$$
$$t = \frac{\ln 0.5}{-0.096} \approx \frac{-0.6931}{-0.096} \approx 7.2 \text{ days}$$

27. The function $P(t) = P_0 e^{-kt}$, $k > 0$, can be used to model decay. We substitute $\frac{1}{2}P_0$ for $P(t)$ and 1 for t and solve for the decay rate k.
$$\frac{1}{2}P_0 = P_0 e^{-k \cdot 1}$$
$$\frac{1}{2} = e^{-k}$$
$$\ln \frac{1}{2} = \ln e^{-k}$$
$$-0.693 \approx -k$$
$$0.693 \approx k$$

The decay rate is 0.693, or 69.3% per year.

29. a) $2012 - 2007 = 5$, so we have the data points $(0, 451)$ and $(5, 198)$.
$$A(t) = A_0 e^{-kt}$$
$$198 = 451 e^{-k \cdot 5}$$
$$\frac{198}{451} = e^{-5k}$$
$$\ln \frac{198}{451} = \ln e^{-5k}$$
$$\ln \frac{198}{451} = -5k$$
$$\frac{\ln \frac{198}{451}}{-5} = k$$
$$0.165 \approx k$$

The exponential decay rate is about 0.165, or 16.5%. The exponential decay function is $A(t) = 451 e^{-0.165t}$, where $A(t)$ is in millions and t is the number of years after 2007.

b) In 2014, $t = 2014 - 2007 = 7$.
$$A(7) = 451 e^{-0.165(7)} \approx 142.1 \text{ million albums}$$

c)
$$300 = 451 e^{-0.165t}$$
$$\frac{300}{451} = e^{-0.165t}$$
$$\ln \frac{300}{451} = \ln e^{-0.165t}$$
$$\ln \frac{300}{451} = -0.165t$$
$$\frac{\ln \frac{300}{451}}{-0.165} = t$$
$$2 \approx t$$

There were 300 million albums sold in a physical format about 2 years after 2007, or in 2009.

31. a) We have the data points $(0, 2.431)$ and $(10, 2.356)$.

$$P(t) = P_0 e^{-kt}$$

$$2.356 = 2.431 e^{-k \cdot 10}$$

$$\frac{2.356}{2.431} = e^{-10k}$$

$$\ln \frac{2.356}{2.431} = \ln e^{-10k}$$

$$\ln \frac{2.356}{2.431} = -10k$$

$$\frac{\ln \frac{2.356}{2.431}}{-10} = k$$

$$0.003 \approx k$$

The exponential decay rate is about 0.003, or 0.3%. The exponential decay function is $P(t) = 2.431 e^{-0.003t}$, where $P(t)$ is in millions and t is the number of years after 2000.

b) In 2020, $t = 2020 - 2000 = 20$.

$$P(20) = 2.431 e^{-0.003(20)} \approx 2.289 \text{ million}$$

c)

$$2.2 = 2.431 e^{-0.003t}$$

$$\frac{2.2}{2.431} = e^{-0.003t}$$

$$\ln \frac{2.2}{2.431} = \ln e^{-0.003t}$$

$$\ln \frac{2.2}{2.431} = -0.003t$$

$$\frac{\ln \frac{2.2}{2.431}}{-0.003} = t$$

$$33 \approx t$$

The population will reach 2.2 million about 33 years after 2000, or in 2033.

33. a) $2004 - 1990 = 14$, so we have the data points $(0, 15.1)$ and $(14, 36.7)$.

$$V(t) = V_0 e^{kt}$$

$$36.7 = 15.1 e^{k \cdot 14}$$

$$\frac{36.7}{15.1} = e^{14k}$$

$$\ln \frac{36.7}{15.1} = \ln e^{14k}$$

$$\ln \frac{36.7}{15.1} = 14k$$

$$\frac{\ln \frac{36.7}{15.1}}{14} = k$$

$$0.063 \approx k$$

The exponential growth rate is about 0.063, or 6.3%. The exponential growth function is $V(t) = 15.1 e^{0.063t}$, where $V(t)$ is in millions of dollars and t is the number of years after 1990.

b) In 2015, $t = 2015 - 1990 = 25$.

$$V(25) = 15.1 e^{0.063(25)} \approx \$72.9 \text{ million}$$

c)

$$2(15.1) = 15.1 e^{0.063t}$$

$$2 = e^{0.063t}$$

$$\ln 2 = \ln e^{0.063t}$$

$$\ln 2 = 0.063t$$

$$\frac{\ln 2}{0.063} = t$$

$$11.0 \approx t$$

The doubling time is about 11.0 years.

d)

$$25 = 15.1 e^{0.063t}$$

$$\frac{25}{15.1} = e^{0.063t}$$

$$\ln \frac{25}{15.1} = \ln e^{0.063t}$$

$$\ln \frac{25}{15.1} = 0.063t$$

$$\frac{\ln \frac{25}{15.1}}{0.063} = t$$

$$8 \approx t$$

The value of the cabinet first exceeded \$25 million about 8 years after 1990, or in 1998.

35. $5x + 6y = -2$,   (1)
   $3x + 10y = 2$ (2)

Multiply Equation (1) by 5 and Equation (2) by -3 and then add.

$$\begin{array}{rcl} 25x + 30y &=& -10 \\ -9x - 30y &=& -6 \\ \hline 16x &=& -16 \\ x &=& -1 \end{array}$$

Substitute -1 for x in one of the original equations and solve for y. We will use Equation (1).

$$5x + 6y = -2$$

$$5(-1) + 6y = -2$$

$$-5 + 6y = -2$$

$$6y = 3$$

$$y = \frac{1}{2}$$

The solution is $\left(-1, \frac{1}{2}\right)$.

37. $x^2 + 2x + 3 = 0$

$a = 1, b = 2, c = 3$

$$x = \frac{-b \pm \sqrt{b^2 - 4ac}}{2a}$$

$$x = \frac{-2 \pm \sqrt{2^2 - 4 \cdot 1 \cdot 3}}{2 \cdot 1} = \frac{-2 \pm \sqrt{4 - 12}}{2}$$

$$x = \frac{-2 \pm \sqrt{-8}}{2} = \frac{-2 \pm 2i\sqrt{2}}{2}$$

$$x = \frac{2(-1 \pm i\sqrt{2})}{2 \cdot 1} = -1 \pm \sqrt{2}i$$

The solutions are $-1 \pm \sqrt{2}i$.

39.
$$\frac{7}{x^2 - 5x} - \frac{2}{x - 5} = \frac{4}{x}$$

$$\frac{7}{x(x - 5)} - \frac{2}{x - 5} = \frac{4}{x}, \text{ LCD is } x(x - 5)$$

$$x(x - 5)\left(\frac{7}{x(x - 5)} - \frac{2}{x - 5}\right) = x(x - 5) \cdot \frac{4}{x}$$

$$7 - 2x = 4(x - 5)$$
$$7 - 2x = 4x - 20$$
$$27 = 6x$$
$$\frac{9}{2} = x$$

The number $\frac{9}{2}$ checks. It is the solution.

41. $-0.937, 1.078, 58.770$

43. $-0.767, 2, 4$

45. First we find k. When $t = 24,360$, $P(t) = 0.5P_0$.

$$0.5P_0 = P_0 e^{-k \cdot 24,360}$$
$$0.5 = e^{-24,360k}$$
$$\ln 0.5 = \ln e^{-24,360k}$$
$$\ln 0.5 = -24,360k$$
$$\frac{\ln 0.5}{-24,360} = k$$
$$0.0000285 \approx k$$

Now we have a function for the decay of plutonium-239.

$$P(t) = P_0 e^{-0.0000285t}$$

If a fuel rod has lost 90% of its plutonium, then 10% of the initial amount is still present. We substitute and solve for t.

$$0.1P_0 = P_0 e^{-0.0000285t}$$
$$0.1 = e^{-0.0000285t}$$
$$\ln 0.1 = \ln e^{-0.0000285t}$$
$$\ln 0.1 = -0.0000285t$$
$$\frac{\ln 0.1}{-0.0000285} = t$$
$$80,792 \approx t$$

It will take about 80,792 yr for the fuel rod of plutonium -239 to lose 90% of its radioactivity.

If the value of k is not rounded but rather is stored in the calculator and used for the computation of t, the result will be about 80,922 yr.

Chapter 8 Vocabulary Reinforcement

1. The function given by $f(x) = 6^x$ is an example of an exponential function.

2. The inverse of a function given by a set of ordered pairs is found by interchanging the first and second coordinates in each ordered pair.

3. When interest is paid on interest previously earned, it is called compound interest.

4. Base-10 logarithms are called common logarithms.

5. The logarithm of a number is an exponent.

6. A quantity's half-life is the amount of time necessary for half of the quantity to decay.

Chapter 8 Concept Reinforcement

1. True; $f(0) = a^0 = 1$, so the y-intercept is $(0, 1)$.

2. False; see page 683 in the text.

3. True; see page 699 in the text.

4. True; $\log 78 \approx 1.8921 \Rightarrow 10^{1.8921} \approx 78$ by the definition of a logarithm.

5. False; $\ln 35 = \ln(7 \cdot 5) = \ln 7 + \ln 5$

6. True; see page 717 in the text.

Chapter 8 Study Guide

1. Graph: $f(x) = 2^x$

We compute some function values and keep the results in a table.

$$f(0) = 2^0 = 1$$
$$f(1) = 2^1 = 2$$
$$f(2) = 2^2 = 4$$
$$f(-1) = 2^{-1} = \frac{1}{2^1} = \frac{1}{2}$$
$$f(-2) = 2^{-2} = \frac{1}{2^2} = \frac{1}{4}$$

x	$f(x)$
0	1
1	2
2	4
3	8
-1	$\frac{1}{2}$
-2	$\frac{1}{4}$
-3	$\frac{1}{8}$

Next we plot these points and connect them with a smooth curve.

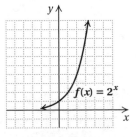

2. $(f \circ g)(x) = f(g(x)) = f(4x + 1) = 2(4x + 1) = 8x + 2$

$(g \circ f)(x) = g(f(x)) = g(2x) = 4 \cdot 2x + 1 = 8x + 1$

3. $h(x) = \dfrac{1}{3x + 2}$

Two functions that can be used are $f(x) = \dfrac{1}{x}$ and $g(x) = 3x + 2$. Answers may vary.

4. The graph of $f(x) = 3^x$ is shown below.

No horizontal line crosses the graph more than once, so the function is one-to-one.

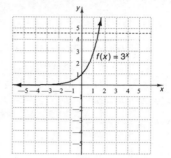

5. The graph of $g(x) = 4 - x$ is shown below.

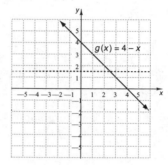

No horizontal line crosses the graph more than once, so the function is one-to-one. We find a formula for the inverse.

1. Replace $g(x)$ with y: $y = 4 - x$

2. Interchange x and y: $x = 4 - y$

3. Solve for y: $x = 4 - y$
$$y + x = 4$$
$$y = 4 - x$$

4. Replace y with $g^{-1}(x)$: $g^{-1}(x) = 4 - x$

6. We graph $f(x) = 2x + 1$ and then draw its reflection across the line $y = x$.

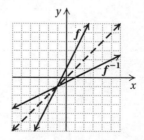

7. Graph $y = \log_5 x$.

The equation $y = \log_5 x$ is equivalent to $5^y = x$. We choose some values for y and find the corresponding x-values. Then we plot points and connect them with a smooth curve.

For $y = -2, x = 5^{-2} = \dfrac{1}{5^2} = \dfrac{1}{25}$.

For $y = -1, x = 5^{-1} = \dfrac{1}{5}$.

For $y = 0, x = 5^0 = 1$.

For $y = 1, x = 5^1 = 5$.

For $y = 2, x = 5^2 = 25$.

x	y
$\dfrac{1}{25}$	-2
$\dfrac{1}{5}$	-1
1	0
5	1
25	2

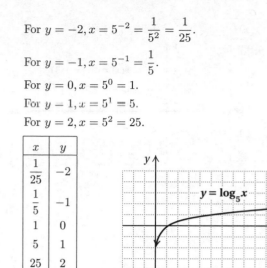

8.
$$\log_a \sqrt[5]{\frac{x^3}{y^2}} = \log_a \left(\frac{x^3}{y^2}\right)^{1/5}$$
$$= \frac{1}{5} \log_a \frac{x^3}{y^2}$$
$$= \frac{1}{5}(\log_a x^3 - \log_a y^2)$$
$$= \frac{1}{5}(3 \log_a x - 2 \log_a y)$$
$$= \frac{3}{5} \log_a x - \frac{2}{5} \log_a y$$

9. $\dfrac{1}{2} \log_a x - 3 \log_a y = \log_a x^{1/2} - \log_a y^3$
$$= \log_a \frac{x^{1/2}}{y^3}, \text{ or } \log_a \frac{\sqrt{x}}{y^3}$$

10. Graph: $f(x) = e^x - 1$

We find some function values, plot points, and draw the graph.

x	$f(x)$
0	0
1	1.72
2	6.39
3	19.09
-1	-0.63
-2	-0.86
-4	-0.98

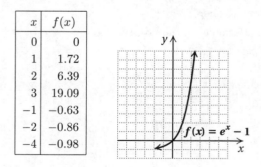

11. Graph: $f(x) = \ln(x + 3)$

We find some function values, plot points, and draw the graph.

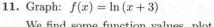

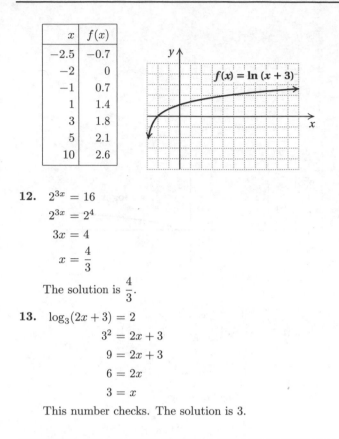

x	$f(x)$
-2.5	-0.7
-2	0
-1	0.7
1	1.4
3	1.8
5	2.1
10	2.6

12. $2^{3x} = 16$

$2^{3x} = 2^4$

$3x = 4$

$x = \dfrac{4}{3}$

The solution is $\dfrac{4}{3}$.

13. $\log_3(2x + 3) = 2$

$3^2 = 2x + 3$

$9 = 2x + 3$

$6 = 2x$

$3 = x$

This number checks. The solution is 3.

Chapter 8 Review Exercises

1. We interchange the coordinates of the ordered pairs. The inverse of the relation is

$\{(2, -4), (-7, 5), (-2, -1), (11, 10)\}$.

2. The graph of $f(x) = 4 - x^2$ fails the horizontal-line test, so it is not one-to-one.

3. The graph of $g(x) = \dfrac{2x - 3}{7}$ passes the horizontal-line test, so it is one-to-one.

We find a formula for the inverse.

1. Replace $g(x)$ by y: $y = \dfrac{2x - 3}{7}$

2. Interchange x and y: $x = \dfrac{2y - 3}{7}$

3. Solve for y: $7x = 2y - 3$

$7x + 3 = 2y$

$\dfrac{7x + 3}{2} = y$

4. Replace y by $g^{-1}(x)$: $g^{-1}(x) = \dfrac{7x + 3}{2}$

4. The graph of $f(x) = 8x^3$ passes the horizontal-line test, so it is one-to-one.

We find a formula for the inverse.

1. Replace $f(x)$ by y: $y = 8x^3$

2. Interchange x and y: $x = 8y^3$

3. Solve for y: $\dfrac{x}{8} = y^3$

$\dfrac{\sqrt[3]{x}}{2} = y$

4. Replace y by $f^{-1}(x)$: $f^{-1}(x) = \dfrac{\sqrt[3]{x}}{2}$, or $\dfrac{1}{2}\sqrt[3]{x}$

5. The graph of $f(x) = \dfrac{4}{3 - 2x}$ passes the horizontal-line test, so it is one-to-one.

We find a formula for the inverse.

1. Replace $f(x)$ by y: $y = \dfrac{4}{3 - 2x}$

2. Interchange x and y: $x = \dfrac{4}{3 - 2y}$

3. Solve for y: $x(3 - 2y) = 4$

$3x - 2xy = 4$

$-2xy = 4 - 3x$

$y = \dfrac{4 - 3x}{-2x}$, or $\dfrac{3x - 4}{2x}$

4. Replace y by $f^{-1}(x)$: $f^{-1}(x) = \dfrac{3x - 4}{2x}$

6. First graph $f(x) = x^3 + 1$. Then, to graph the inverse, reflect the graph of the function across the line $y = x$.

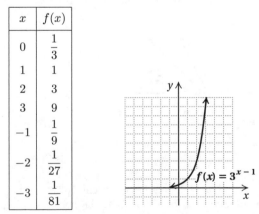

7. Graph: $f(x) = 3^{x-1}$

We make a table of values. Then we plot the points and connect them with a smooth curve.

x	$f(x)$
0	$\dfrac{1}{3}$
1	1
2	3
3	9
-1	$\dfrac{1}{9}$
-2	$\dfrac{1}{27}$
-3	$\dfrac{1}{81}$

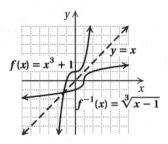

8. Graph $f(x) = \log_3 x$, or $y = \log_3 x$

The equation $f(x) = y = \log_3 x$ is equivalent to $3^y = x$. We find ordered pairs by choosing values for y and computing the corresponding x-values. Then we plot the set of ordered pairs and connect the points with a smooth curve.

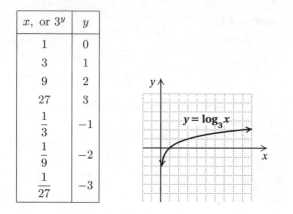

x, or 3^y	y
1	0
3	1
9	2
27	3
$\frac{1}{3}$	-1
$\frac{1}{9}$	-2
$\frac{1}{27}$	-3

$y = \log_3 x$

9. Graph: $f(x) = e^{x+1}$

We find some function values using a calculator. Then we use these values to plot points and draw the graph.

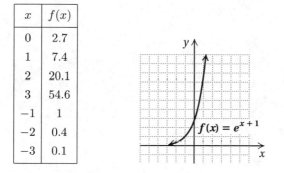

x	$f(x)$
0	2.7
1	7.4
2	20.1
3	54.6
-1	1
-2	0.4
-3	0.1

$f(x) = e^{x+1}$

10. Graph: $f(x) = \ln (x - 1)$

We find some function values using a calculator. Then we use these values to plot points and draw the graph.

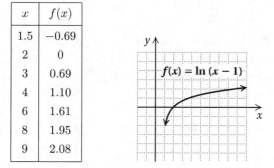

x	$f(x)$
1.5	-0.69
2	0
3	0.69
4	1.10
6	1.61
8	1.95
9	2.08

$f(x) = \ln (x - 1)$

11. $f \circ g(x) = f(g(x)) = f(3x-5) = (3x-5)^2 = 9x^2 - 30x + 25$

$g \circ f(x) = g(f(x)) = g(x^2) = 3x^2 - 5$

12. $h(x) = \sqrt{4 - 7x}$

This is the square root of $4 - 7x$, so the two most obvious functions are $f(x) = \sqrt{x}$ and $g(x) = 4 - 7x$.

13. $10^4 = 10,000 \Rightarrow 4 = \log 10,000$

14. $25^{1/2} = 5 \Rightarrow \frac{1}{2} = \log_{25} 5$

15. $\log_4 16 = x \Rightarrow 4^x = 16$

16. $\log_{1/2} 8 = -3 \Rightarrow \left(\frac{1}{2}\right)^{-3} = 8$

17. $\log_3 9 = x$

$3^x = 9$

$3^x = 3^2$

$x = 2$

18. $\log_{10} \frac{1}{10} = x$

$10^x = \frac{1}{10}$

$10^x = 10^{-1}$

$x = -1$

19. $\log_m m = 1$ since $m^1 = m$.

20. $\log_m 1 = 0$ since $m^0 = 1$.

21. $\log \left(\frac{78}{43,112}\right) \approx -2.7425$

22. $\log (-4)$ does not exist as a real number.

23. $\log_a x^4 y^2 z^3$

$= \log_a x^4 + \log_a y^2 + \log_a z^3$

$= 4 \log_a x + 2 \log_a y + 3 \log_a z$

24. $\log \sqrt[4]{\frac{z^2}{x^3 y}}$

$= \log \left(\frac{z^2}{x^3 y}\right)^{1/4}$

$= \frac{1}{4} \log \frac{z^2}{x^3 y}$

$= \frac{1}{4}[\log z^2 - \log (x^3 y)]$

$= \frac{1}{4}[2 \log z - (\log x^3 + \log y)]$

$= \frac{1}{4}(2 \log z - \log x^3 - \log y)$

$= \frac{1}{4}(2 \log z - 3 \log x - \log y)$

$= \frac{1}{2} \log z - \frac{3}{4} \log x - \frac{1}{4} \log y$

25. $\log_a 8 + \log_a 15 = \log_a (8 \cdot 15) = \log_a 120$

26. $\frac{1}{2} \log a - \log b - 2 \log c$

$= \log a^{1/2} - \log b - \log c^2$

$= \log a^{1/2} - (\log b + \log c^2)$

$= \log a^{1/2} - \log (bc^2)$

$= \log \frac{a^{1/2}}{bc^2}, \text{ or } \log \frac{\sqrt{a}}{bc^2}$

27. $\log_m m^{17} = 17$ by Property 4

28. $\log_m m^{-7} = -7$ by Property 4

29. $\log_a 28 = \log_a (2^2 \cdot 7)$

$\qquad = \log_a 2^2 + \log_a 7$

$\qquad = 2 \log_a 2 + \log_a 7$

$\qquad = 2(1.8301) + 5.0999$

$\qquad = 8.7601$

30. $\log_a 3.5 = \log_a \left(\dfrac{7}{2} \right)$

$\qquad = \log_a 7 - \log_a 2$

$\qquad = 5.0999 - 1.8301$

$\qquad = 3.2698$

31. $\log_a \sqrt{7} = \log_a 7^{1/2}$

$\qquad = \dfrac{1}{2} \log_a 7$

$\qquad = \dfrac{1}{2}(5.0999)$

$\qquad = 2.54995$

32. $\log_a \dfrac{1}{4} = \log_a 4^{-1}$

$\qquad = \log_a (2^2)^{-1}$

$\qquad = \log_a 2^{-2}$

$\qquad = -2 \log_a 2$

$\qquad = -2(1.8301)$

$\qquad = -3.6602$

33. $\ln 0.06774 \approx -2.6921$

34. $e^{-0.98} \approx 0.3753$

35. $e^{2.91} \approx 18.3568$

36. $\ln 1 = 0$

37. $\ln 0$ does not exist.

38. $\ln e = 1$

39. $\log_5 2 = \dfrac{\log 2}{\log 5} \approx 0.4307$

40. $\log_{12} 70 = \dfrac{\ln 70}{\ln 12} \approx 1.7097$

41. $\log_3 x = -2$

$\qquad 3^{-2} = x$

$\qquad \dfrac{1}{3^2} = x$

$\qquad \dfrac{1}{9} = x$

42. $\log_x 32 = 5$

$\qquad x^5 = 32$

$\qquad x^5 = 2^5$

$\qquad x = 2$

43. $\log x = -4$

$\qquad 10^{-4} = x$

$\qquad \dfrac{1}{10^4} = x$

$\qquad \dfrac{1}{10,000} = x$

44. $3 \ln x = -6$

$\qquad \ln x = -2$

$\qquad e^{-2} = x$

$\qquad 0.1353 \approx x$

45. $4^{2x-5} = 16$

$\qquad 4^{2x-5} = 4^2$

$\qquad 2x - 5 = 2$

$\qquad 2x = 7$

$\qquad x = \dfrac{7}{2}$

46. $2^{x^2} \cdot 2^{4x} = 32$

$\qquad 2^{x^2+4x} = 2^5$

$\qquad x^2 + 4x = 5$

$\qquad x^2 + 4x - 5 = 0$

$\qquad (x-1)(x+5) = 0$

$\qquad x = 1 \;\; or \;\; x = -5$

47. $4^x = 8.3$

$\qquad \log 4^x = \log 8.3$

$\qquad x \log 4 = \log 8.3$

$\qquad x = \dfrac{\log 8.3}{\log 4}$

$\qquad x \approx 1.5266$

48. $e^{-0.1t} = 0.03$

$\qquad \ln e^{-0.1t} = \ln 0.03$

$\qquad -0.1t = \ln 0.03$

$\qquad t = \dfrac{\ln 0.03}{-0.1}$

$\qquad t \approx 35.0656$

49. $\log_4 16 = x$

$\qquad 4^x = 16$

$\qquad 4^x = 4^2$

$\qquad x = 2$

50. $\log_4 x + \log_4 (x-6) = 2$

$\qquad \log_4 [x(x-6)] = 2$

$\qquad 4^2 = x(x-6)$

$\qquad 16 = x^2 - 6x$

$\qquad 0 = x^2 - 6x - 16$

$\qquad 0 = (x-8)(x+2)$

$x = 8 \;\; or \;\; x = -2$

The number 8 checks, but -2 does not. The solution is 8.

51. $\log_2(x+3) - \log_2(x-3) = 4$

$$\log_2 \frac{x+3}{x-3} = 4$$

$$2^4 = \frac{x+3}{x-3}$$

$$16 = \frac{x+3}{x-3}$$

$$(x-3) \cdot 16 = (x-3) \cdot \frac{x+3}{x-3}$$

$$16x - 48 = x + 3$$

$$15x = 51$$

$$x = \frac{51}{15} = \frac{17}{5}$$

The number $\frac{17}{5}$ checks. It is the solution.

52.
$$\log_3(x-4) = 3 - \log_3(x+4)$$

$$\log_3(x-4) + \log_3(x+4) = 3$$

$$\log_3[(x-4)(x+4)] = 3$$

$$3^3 = (x-4)(x+4)$$

$$27 = x^2 - 16$$

$$43 = x^2$$

$$x = -\sqrt{43} \quad or \quad x = \sqrt{43}$$

The number $-\sqrt{43}$ does not check, but $\sqrt{43}$ does. The solution is $\sqrt{43}$.

53. $L = 10 \cdot \log \dfrac{I}{I_0}$

$$L = 10 \cdot \log \frac{10^{1.7}}{10^{-12}}$$

$$= 10 \cdot \log 10^{13.7}$$

$$= 10 \cdot 13.7$$

$$= 137 \text{ dB}$$

54. $S(t) = 159(1.44)t$

a) $S(0) = 159(1.44)^0 = \$159$ billion

In 2014, $t = 2014 - 2012 = 2$.

$S(2) = 159(1.44)^2 \approx \$330$ billion

In 2016, $t = 2016 - 2012 = 4$.

$S(4) = 159(1.44)^4 \approx \$684$ billion

b)
$$1000 = 159(1.44)^t$$

$$\frac{1000}{159} = 1.44^t$$

$$\ln \frac{1000}{159} = \ln 1.44^t$$

$$\ln \frac{1000}{159} = t \ln 1.44$$

$$\frac{\ln \dfrac{1000}{159}}{\ln 1.44} = t$$

$$5 \approx t$$

$1000 billion in sales will be influenced by smartphones about 5 years after 2012, or in 2017.

c)
$$2(159) = 159(1.44)^t$$

$$2 = 1.44^t$$

$$\ln 2 = \ln 1.44^t$$

$$\ln 2 = t \ln 1.44$$

$$\frac{\ln 2}{\ln 1.44} = t$$

$$1.9 \approx t$$

The doubling time is about 1.9 years.

d)

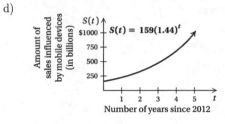

55. a) We start with the exponential growth function $V(t) - V_0 e^{kt}$, where t is the number of years after 2011.

Substituting 40,000 for V_0, we have

$$V(t) = 40,000 e^{kt}.$$

To find k observe that the value of the investment was \$53,000 in 2014, or 3 yr after 2011. We substitute and solve for k.

$$53,000 = 40,000 e^{k \cdot 3}$$

$$1.325 = e^{3k}$$

$$\ln 1.325 = \ln e^{3k}$$

$$\ln 1.325 = 3k$$

$$\frac{\ln 1.325}{3} = k$$

$$0.094 \approx k$$

Thus the exponential growth function is $V(t) = 40,000 e^{0.094t}$.

b) In 2021, $t = 2021 - 2011 = 10$.

$$V(10) = 40,000 e^{0.094(10)} \approx \$102,399$$

c)
$$85,000 = 40,000 e^{0.094t}$$

$$2.125 = e^{0.094t}$$

$$\ln 2.125 = \ln e^{0.094t}$$

$$\ln 2.125 = 0.094t$$

$$\frac{\ln 2.125}{0.094} = t$$

$$8 \approx t$$

The value of the investment will first reach \$85,000 about 8 yr after 2011, or in 2019.

56.
$$2\,P_0 = P_0\,e^{k\cdot 3}$$
$$2 = e^{3k}$$
$$\ln 2 = \ln e^{3k}$$
$$\ln 2 = 3k$$
$$\frac{\ln 2}{3} = k$$
$$0.231 \approx k$$

The exponential growth rate was about 0.231, or 23.1%.

57.
$$2(7600) = 7600\,e^{0.034t}$$
$$2 = e^{0.034t}$$
$$\ln 2 = \ln e^{0.034t}$$
$$\ln 2 = 0.034t$$
$$\frac{\ln 2}{0.034} = t$$
$$20.4 \text{ yr} \approx t$$

58. If the skeleton had lost 34% of its carbon-14 from an initial amount of P_0, then $66\%\,(P_0)$ is the amount present. We use the carbon-14 decay function to find the age of the skeleton.
$$P(t) = P_0\,e^{-0.00012t}$$
$$0.66\,P_0 = P_0\,e^{-0.00012t}$$
$$0.66 = e^{-0.00012t}$$
$$\ln 0.66 = \ln e^{-0.00012t}$$
$$\ln 0.66 = -0.00012t$$
$$\frac{\ln 0.66}{-0.00012} = t$$
$$3463 \text{ yr} \approx t$$

59. The inverse of $f(x) = a^x$ is given by $f^{-1}(x) = \log_a x$, so the inverse of $f(x) = 5^x$ is $f^{-1}(x) = \log_5 x$. Answer C is correct.

60.
$$\log\,(x^2 - 9) - \log(x + 3) = 1$$
$$\log \frac{x^2 - 9}{x + 3} = 1$$
$$10^1 = \frac{x^2 - 9}{x + 3}$$
$$10(x + 3) = x^2 - 9$$
$$10x + 30 = x^2 - 9$$
$$0 = x^2 - 10x - 39$$
$$0 = (x - 13)(x + 3)$$
$$x - 13 = 0 \quad or \quad x + 3 = 0$$
$$x = 13 \quad or \qquad x = -3$$

The number 13 checks, but -3 does not. The solution is 13. Answer D is correct.

61.
$$\ln\,(\ln\,x) = 3$$
$$e^3 = \ln\,x$$
$$x = e^{e^3}$$

62.
$$5^{x+y} = 25 \qquad\qquad 2^{2x-y} = 64$$
$$5^{x+y} = 5^2 \qquad\qquad 2^{2x-y} = 2^6$$
$$x + y = 2 \qquad\qquad 2x - y = 6$$

We have a system of equations. We solve using the elimination method.
$$
\begin{array}{rl}
x + y = 2, & (1) \\
2x - y = 6 & (2) \\
\hline
3x \quad = 8 & \text{Adding} \\
x = \dfrac{8}{3} &
\end{array}
$$

Now substitute $\dfrac{8}{3}$ for x in Equation (1) and solve for y.
$$\frac{8}{3} + y = 2$$
$$y = -\frac{2}{3}$$

The solution is $\left(\dfrac{8}{3}, -\dfrac{2}{3}\right)$.

Chapter 8 Discussion and Writing Exercises

1. Reflect the graph of $f(x) = e^x$ across the line $y = x$ and then translate it up one unit.

2. Christina mistakenly thinks that, because negative numbers do not have logarithms, negative numbers cannot be solutions of logarithmic equations.

3.
$$C(x) = \frac{100 + 5x}{x}$$
$$y = \frac{100 + 5x}{x} \qquad \text{Replace } C(x).$$
$$x = \frac{100 + 5y}{y} \qquad \text{Interchange variables.}$$
$$y = \frac{100}{x - 5} \qquad \text{Solve for } y.$$
$$C^{-1}(x) = \frac{100}{x - 5} \qquad \text{Replace } y.$$

$C^{-1}(x)$ gives the number of people in the group, where x is the cost per person, in dollars.

4. To solve $\ln\,x = 3$, graph $f(x) = \ln\,x$ and $g(x) = 3$ on the same set of axes. The solution is the first coordinate of the point of intersection of the two graphs.

5. You cannot take the logarithm of a negative number because logarithm bases are positive and there is no real-number power to which a positive number can be raised to yield a negative number.

6. Answers will vary.

Chapter 8 Test

1. Graph: $f(x) = 2^{x+1}$

We compute some function values and keep the results in a table.
$$f(0) = 2^{0+1} = 2^1 = 2$$

$f(-1) = 2^{-1+1} = 2^0 = 1$

$f(-2) = 2^{-2+1} = 2^{-1} = \dfrac{1}{2^1} = \dfrac{1}{2}$

$f(-3) = 2^{-3+1} = 2^{-2} = \dfrac{1}{2^2} = \dfrac{1}{4}$

$f(1) = 2^{1+1} = 2^2 = 4$

$f(2) = 2^{2+1} = 2^3 = 8$

x	$f(x)$
0	2
-1	1
-2	$\dfrac{1}{2}$
-3	$\dfrac{1}{4}$
1	4
2	8

Next we plot these points and connect them with a smooth curve.

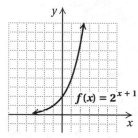

2. Graph: $y = \log_2 x$

The equation $y = \log_2 x$ is equivalent to $2^y = x$. We can find ordered pairs by choosing values for y and computing the corresponding x-values.

For $y = 0$, $x = 2^0 = 1$.

For $y = 1$, $x = 2^1 = 2$.

For $y = 2$, $x = 2^2 = 4$.

For $y = 3$, $x = 2^3 = 8$.

For $y = -1$, $x = 2^{-1} = \dfrac{1}{2}$.

For $y = -2$, $x = 2^{-2} = \dfrac{1}{4}$.

x, or 2^y	y
1	0
2	1
4	2
8	3
$\dfrac{1}{2}$	-1
$\dfrac{1}{4}$	-2
$\dfrac{1}{8}$	-3

 (1) Select y.

 (2) Compute x.

We plot the set of ordered pairs and connect the points with a smooth curve.

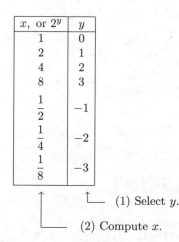

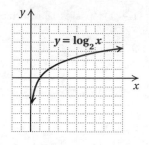

3. Graph: $f(x) = e^{x-2}$

We find some function values with a calculator. Use these values to plot points and draw the graph.

x	$f(x)$
-1	0.05
0	0.14
1	0.47
2	1
3	2.72
4	7.39

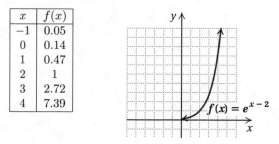

4. Graph: $f(x) = \ln(x - 4)$

We find some function values, plot points, and draw the graph.

x	$f(x)$
4.1	-2.30
4.5	-0.69
5	0
6	0.69
7	1.10

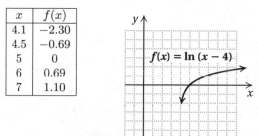

5. $f(x) = x + x^2$, $g(x) = 5x - 2$

$(f \circ g)(x) = f(g(x)) = 5x - 2 + (5x - 2)^2 =$
$5x - 2 + 25x^2 - 20x + 4 = 25x^2 - 15x + 2$

$(g \circ f)(x) = g(f(x)) = 5(x + x^2) - 2 = 5x + 5x^2 - 2$, or
$5x^2 + 5x - 2$

6. Interchange the first and second coordinates of each ordered pair.

$\{(3, -4), (-8, 5), (-3, -1), (12, 10)\}$

7. The graph of $f(x) = 4x - 3$ passes the horizontal-line test, so it is one-to-one.

We find a formula for the inverse.

1. Replace $f(x)$ by y: $\quad y = 4x - 3$

2. Interchange x and y: $\quad x = 4y - 3$

3. Solve for y: $\quad x + 3 = 4y$

$$\dfrac{x + 3}{4} = y$$

4. Replace y by $f^{-1}(x)$: $\quad f^{-1}(x) = \dfrac{x + 3}{4}$

8. The graph of $f(x) = (x+1)^3$ passes the horizontal-line test, so it is one-to-one.

We find a formula for the inverse.

1. Replace $f(x)$ by y: $y = (x+1)^3$

2. Interchange x and y: $x = (y+1)^3$

3. Solve for y: $\sqrt[3]{x} = y + 1$

$\qquad\qquad\qquad \sqrt[3]{x} - 1 = y$

4. Replace y by $f^{-1}(x)$: $f^{-1}(x) = \sqrt[3]{x} - 1$

9. The graph of $f(x) = 2 - |x|$ is shown below. It fails the horizontal-line test, so it is not one-to-one.

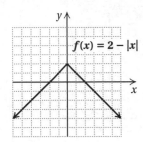

$f(x) = 2 - |x|$

10. $256^{1/2} = 16 \Rightarrow \dfrac{1}{2} = \log_{256} 16$

11. $m = \log_7 49 \Rightarrow 7^m = 49$

12. Let $\log_5 125 = x$. Then

$\qquad 5^x = 125$

$\qquad 5^x = 5^3$

$\qquad\ \ x = 3$

Thus, $\log_5 125 = 3$.

13. Since $\log_a a^k = k$, we have $\log_t t^{23} = 23$.

14. Since $\log_a 1 = 0$, we have $\log_p 1 = 0$.

15. $\log 0.0123 \approx -1.9101$

16. $\log(-5)$ does not exist as a real number.

17. $\qquad \log \dfrac{a^3 b^{1/2}}{c^2}$

$= \log (a^3 b^{1/2}) - \log c^2$

$= \log a^3 + \log b^{1/2} - \log c^2$

$= 3 \log a + \dfrac{1}{2} \log b - 2 \log c$

18. $\qquad \dfrac{1}{3} \log_a x - 3 \log_a y + 2 \log_a z$

$= \log_a x^{1/3} - \log_a y^3 + \log_a z^2$

$= \log_a x^{1/3} + \log_a z^2 - \log_a y^3$

$= \log_a (x^{1/3} z^2) - \log_a y^3$

$= \log_a \dfrac{x^{1/3} z^2}{y^3}$

19. $\log_a \dfrac{2}{7} = \log_a 2 - \log_a 7$

$\qquad\quad = 0.301 - 0.845$

$\qquad\quad = -0.544$

20. $\log_a 12 = \log_a (2 \cdot 6)$

$\qquad\quad = \log_a 2 + \log_a 6$

$\qquad\quad = 0.301 + 0.778$

$\qquad\quad = 1.079$

21. $\ln 807.39 \approx 6.6938$

22. $e^{4.68} \approx 107.7701$

23. $\ln 1 = 0$

24. $\log_{18} 31 = \dfrac{\log 31}{\log 18} \approx \dfrac{1.4914}{1.2553} \approx 1.1881$

25. $\log_x 25 = 2$

$\qquad\quad 25 = x^2$

$\qquad\quad\ \ 5 = x$ Taking the positive square root

The solution is 5.

26. $\log_4 x = \dfrac{1}{2}$

$\qquad\quad x = 4^{1/2} = \sqrt{4}$

$\qquad\quad x = 2$

The solution is 2.

27. $\log x = 4$

$\qquad\ x = 10^4$

$\qquad\ x = 10{,}000$

The solution is 10,000.

28. $\ln x = \dfrac{1}{4}$

$\qquad x = e^{1/4}$

$\qquad x \approx 1.2840$

The solution is $e^{1/4}$, or about 1.2840.

29. $\qquad 7^x = 1.2$

$\qquad \ln 7^x = \ln 1.2$

$\qquad x \ln 7 = \ln 1.2$

$\qquad\qquad x = \dfrac{\ln 1.2}{\ln 7}$

$\qquad\qquad x \approx \dfrac{0.1823}{1.9459}$

$\qquad\qquad x \approx 0.0937$

The solution is $\dfrac{\ln 1.2}{\ln 7}$, or about 0.0937.

30. $\log(x^2 - 1) - \log(x - 1) = 1$

$\qquad\qquad \log \dfrac{x^2 - 1}{x - 1} = 1$

$\qquad \log \dfrac{(x+1)(x-1)}{x-1} = 1$

$\qquad\qquad\ \log(x + 1) = 1$

$\qquad\qquad\qquad\ x + 1 = 10^1$

$\qquad\qquad\qquad\qquad\ x = 9$

The number 9 checks. It is the solution.

31. $\log_5 x + \log_5 (x + 4) = 1$

$\log_5 [x(x + 4)] = 1$

$x(x + 4) = 5^1$

$x^2 + 4x = 5$

$x^2 + 4x - 5 = 0$

$(x + 5)(x - 1) = 0$

$x = -5 \ or \ x = 1$

The number -5 does not check, but 1 does. The solution is 1.

32. pH $= -\log [\text{H}^+]$

$= -\log (6.3 \times 10^{-5})$

$\approx -(-4.2007)$

≈ 4.2

The pH is approximately 4.2.

33. $H(t) = 2.37(1.076)^t$

a) In 2012, $t = 2012 - 2008 = 4$.

$H(4) = 2.37(1.076)^4 \approx \$3.18$ trillion

b) $5 = 2.37(1.076)^t$

$\dfrac{5}{2.37} = 1.076^t$

$\log \dfrac{5}{2.37} = \log 1.076^t$

$\log \dfrac{5}{2.37} = t \log 1.076$

$\dfrac{\log \dfrac{5}{2.37}}{\log 1.076} = t$

$10 \approx t$

Spending on health care will be \$5 trillion about 10 yr after 2008, or in 2018.

c) $2(2.37) = 4.74$

Then we have:

$4.74 = 2.37(1.076)^t$

$2 = 1.076^t$

$\log 2 = \log 1.076^t$

$\log 2 = t \log 1.076$

$\dfrac{\log 2}{\log 1.076} = t$

$9.5 \approx t$

The doubling time is about 9.5 yr.

34. a) $P(t) = P_0 e^{kt}$

$1150.27 = 1000e^{k \cdot 5}$

$1.15027 = e^{5k}$

$\ln 1.15027 = \ln e^{5k}$

$\ln 1.15027 = 5k$

$\dfrac{\ln 1.15027}{5} = k$

$0.028 \approx k$

The interest rate is about 0.028, or 2.8%. The exponential growth function is $P(t) = 1000e^{0.028t}$.

b) $P(8) = 1000e^{0.028(8)} \approx \$1251.07$

c) $1439 = 1000e^{0.028t}$

$1.439 = e^{0.028t}$

$\ln 1.439 = \ln e^{0.028t}$

$\ln 1.439 = 0.028t$

$\dfrac{\ln 1.439}{0.028} = t$

$13 \approx t$

The balance will be \$1439 after 13 yr.

d) $2000 = 1000e^{0.028t}$

$2 = e^{0.028t}$

$\ln 2 = \ln e^{0.028t}$

$\ln 2 = 0.028t$

$\dfrac{\ln 2}{0.028} = t$

$24.8 \approx t$

The doubling time is about 24.8 yr.

35. $2 P_0 = P_0 e^{k \cdot 23}$

$2 = e^{23k}$

$\ln 2 = \ln e^{23k}$

$\ln 2 = 23k$

$\dfrac{\ln 2}{23} = k$

$0.03 \approx k$

The exponential growth rate was about 0.03, or about 3%.

36. If the bone has lost 43% of its carbon-14 from an initial amount of P_0, then $57\%(P_0)$, or $0.57P_0$, is the amount present. We use the carbon-14 decay function and solve for t.

$0.57P_0 = P_0 \ e^{-0.00012t}$

$0.57 = e^{-0.00012t}$

$\ln 0.57 = \ln e^{-0.00012t}$

$\ln 0.57 = -0.00012t$

$\dfrac{\ln 0.57}{-0.00012} = t$

$4684 \approx t$

The bone is about 4684 years old.

37. $\log(3x - 1) + \log x = 1$

$\log[(3x - 1)x] = 1$

$\log(3x^2 - x) = 1$

$10^1 = 3x^2 - x$

$10 = 3x^2 - x$

$0 = 3x^2 - x - 10$

$0 = (3x + 5)(x - 2)$

$3x + 5 = 0 \quad or \quad x - 2 = 0$

$3x = -5 \quad or \qquad x = 2$

$x = -\dfrac{5}{3} \quad or \qquad x = 2$

Check: For $-\dfrac{5}{3}$:

$$\dfrac{\log(3x-1)+\log x = 1}{\log\left[3\left(-\dfrac{5}{3}\right)-1\right]+\log\left(-\dfrac{5}{3}\right) \ ? \ 1 \ \text{UNDEFINED}}$$

For 2:

$$\dfrac{\log(3x-1)+\log x = 1}{\begin{array}{c|c}\log(3\cdot2-1)+\log 2 \ ? \ 1 & \\ \log(6-1)+\log 2 & \\ \log 5+\log 2 & \\ \log(5\cdot2) & \\ \log 10 & \\ 1 & \text{TRUE}\end{array}}$$

The number $-\dfrac{5}{3}$ does not check, but 2 does. The solution is 2. We see that there is exactly one solution, and it is positive. Thus, the correct answer is B.

38. $\log_3 |2x-7| = 4$

$|2x-7| = 3^4$

$|2x-7| = 81$

$2x-7 = -81 \ \ or \ \ 2x-7 = 81$

$2x = -74 \ \ or \ \ \ \ \ \ 2x = 88$

$x = -37 \ \ or \ \ \ \ \ \ \ \ x = 44$

Both numbers check. The solutions are -37 and 44.

39. $\log_a \dfrac{\sqrt[3]{x^2 z}}{\sqrt[3]{y^2 z^{-1}}}$

$= \log_a \sqrt[3]{\dfrac{x^2 z}{y^2 z^{-1}}}$

$= \log_a \left(\dfrac{x^2 z^2}{y^2}\right)^{1/3}$

$= \dfrac{1}{3}\log_a\left(\dfrac{x^2 z^2}{y^2}\right)$

$= \dfrac{1}{3}[\log_a(x^2 z^2) - \log_a y^2]$

$= \dfrac{1}{3}(\log_a x^2 + \log_a z^2 - \log_a y^2)$

$= \dfrac{1}{3}(2\log_a x + 2\log_a z - 2\log_a y)$

$= \dfrac{1}{3}(2\cdot2 + 2\cdot4 - 2\cdot3)$

$= \dfrac{1}{3}(4+8-6)$

$= \dfrac{1}{3}(6)$

$= 2$

Cumulative Review Chapters 1 - 8

1. $8(2x-3) = 6-4(2-3x)$

$16x-24 = 6-8+12x$

$16x-24 = -2+12x$

$4x = 22$

$x = \dfrac{11}{2}$

The solution is $\dfrac{11}{2}$.

2. $x(x-3) = 10$

$x^2-3x = 10$

$x^2-3x-10 = 0$

$(x-5)(x+2) = 0$

$x-5 = 0 \ \ or \ \ x+2 = 0$

$x = 5 \ \ or \ \ \ \ \ \ \ x = -2$

The solutions are 5 and -2.

3. $4x-3y = 15, \quad (1)$

$3x+5y = 4 \quad \ \ (2)$

We multiply Equation (1) by 5 and Equation (2) by 3 and then add to eliminate y.

$$\begin{array}{r}20x-15y = 75 \\ \underline{9x+15y = 12} \\ 29x \ \ \ \ \ \ \ \ = 87\end{array}$$

$x = 3$

Now substitute 3 for x in one of the original equations and solve for y. We use Equation (2).

$3\cdot3+5y = 4$

$9+5y = 4$

$5y = -5$

$y = -1$

The solution is $(3,-1)$.

4. $x+y-3z = -1, \quad (1)$

$2x-y+ \ \ z = 4, \quad \ \ (2)$

$-x-y+ \ \ z = 1 \quad \ \ (3)$

$$\begin{array}{r}x+y-3z = -1 \quad (1) \\ \underline{2x-y+ \ \ z = 4} \quad (2) \\ 3x \ \ \ \ \ \ - 2z = 3 \quad (4) \ \text{Adding}\end{array}$$

$$\begin{array}{r}x+y-3z = -1 \quad (1) \\ \underline{-x-y+ \ \ z = 1} \quad (3) \\ -2z = 0 \quad \text{Adding}\end{array}$$

$z = 0$

$3x-2\cdot0 = 3 \quad \text{Substituting in (4)}$

$3x = 3$

$x = 1$

$1 + y - 3 \cdot 0 = -1$ Substituting in (1)

$1 + y = -1$

$y = -2$

The solution is $(1, -2, 0)$.

5.

$$\frac{7}{x^2 - 5x} - \frac{2}{x - 5} = \frac{4}{x}$$

$$\frac{7}{x(x - 5)} - \frac{2}{x - 5} = \frac{4}{x}, \text{ LCD is } x(x - 5)$$

$$x(x - 5)\left(\frac{7}{x(x - 5)} - \frac{2}{x - 5}\right) = x(x - 5) \cdot \frac{4}{x}$$

$$7 - 2x = 4(x - 5)$$

$$7 - 2x = 4x - 20$$

$$-6x = -27$$

$$x = \frac{9}{2}$$

The number $\frac{9}{2}$ checks. It is the solution.

6.

$$\sqrt{x - 1} = \sqrt{x + 4} - 1$$

$$(\sqrt{x - 1})^2 = (\sqrt{x + 4} - 1)^2$$

$$x - 1 = x + 4 - 2\sqrt{x + 4} + 1$$

$$x - 1 = x + 5 - 2\sqrt{x + 4}$$

$$-6 = -2\sqrt{x + 4}$$

$$3 = \sqrt{x + 4} \qquad \text{Dividing by } -2$$

$$3^2 = (\sqrt{x + 4})^2$$

$$9 = x + 4$$

$$5 = x$$

The number 5 checks. It is the solution.

7. $x - 8\sqrt{x} + 15 = 0$

Let $u = \sqrt{x}$.

$$u^2 - 8u + 15 = 0$$

$$(u - 3)(u - 5) = 0$$

$$u = 3 \quad or \quad u = 5$$

$$\sqrt{x} = 3 \quad or \quad \sqrt{x} = 5$$

$$x = 9 \quad or \quad x = 25$$

Both numbers check. The solutions are 9 and 25.

8. $x^4 - 13x^2 + 36 = 0$

Let $u = x^2$.

$$u^2 - 13u + 36 = 0$$

$$(u - 4)(u - 9) = 0$$

$$u = 4 \quad or \quad u = 9$$

$$x^2 = 4 \quad or \quad x^2 = 9$$

$$x = \pm 2 \quad or \quad x = \pm 3$$

All four numbers check. The solutions are ± 2 and ± 3.

9. $\log_8 x = 1$

$$8^1 = x$$

$$8 = x$$

10.

$$3^{5x} = 7$$

$$\log 3^{5x} = \log 7$$

$$5x \cdot \log 3 = \log 7$$

$$x = \frac{\log 7}{5 \log 3}$$

$$x \approx 0.3542$$

11. $\log x - \log(x - 8) = 1$

$$\log \frac{x}{x - 8} = 1$$

$$10^1 = \frac{x}{x - 8}$$

$$10(x - 8) = x \qquad \text{Multiplying by } x - 8$$

$$10x - 80 = x$$

$$-80 = -9x$$

$$\frac{80}{9} = x$$

The number $\frac{80}{9}$ checks. It is the solution.

12.

$$x^2 + 4x > 5$$

$$x^2 + 4x - 5 > 0$$

$$(x + 5)(x - 1) > 0$$

The solutions of $(x + 5)(x - 1) = 0$ are -5 and 1. They divide the real-number line with three intervals as shown:

We try test numbers in each interval.

A: Test -6, $(-6 + 5)(-6 - 1) = 7 > 0$

B: Test 0, $(0 + 5)(0 - 1) = -5 < 0$

C: Test 2, $(2 + 5)(2 - 1) = 7 > 0$

The expression is positive for all numbers in intervals A and C. The solution set is $\{x | x < -5 \text{ or } x > 1\}$, or $(-\infty, -5) \cup (1, \infty)$.

13. $|2x - 3| \geq 9$

$$2x - 3 \leq -9 \quad or \quad 2x - 3 \geq 9$$

$$2x \leq -6 \quad or \qquad 2x \geq 12$$

$$x \leq -3 \quad or \qquad x \geq 6$$

The solution set is $\{x | x \leq -3 \text{ or } x \geq 6\}$, or $(-\infty, -3] \cup [6, \infty)$.

14.

$$x^2 + 6x = 11$$

$$x^2 + 6x - 11 = 0$$

$$a = 1, \ b = 6, \ c = -11$$

$$x = \frac{-b \pm \sqrt{b^2 - 4ac}}{2a}$$

$$x = \frac{-6 \pm \sqrt{6^2 - 4 \cdot 1 \cdot (-11)}}{2 \cdot 1} = \frac{-6 \pm \sqrt{36 + 44}}{2}$$

$$x = \frac{-6 \pm \sqrt{80}}{2} = \frac{-6 \pm 4\sqrt{5}}{2}$$

$$x = \frac{2(-3 \pm 2\sqrt{5})}{2} = -3 \pm 2\sqrt{5}$$

The solutions are $-3 \pm 2\sqrt{5}$.

15.
$$D = \frac{ab}{b+a}$$

$D(b+a) = ab$　　　Multiplying by $b+a$

$Db + Da = ab$

$Db = ab - Da$

$Db = a(b - D)$

$$\frac{Db}{b-D} = a$$

16.
$$\frac{1}{p} + \frac{1}{q} = \frac{1}{f}$$

$$pqf\left(\frac{1}{p} + \frac{1}{q}\right) = pqf \cdot \frac{1}{f}$$

$$qf + pf = pq$$

$$pf = pq - qf$$

$$pf = q(p - f)$$

$$\frac{pf}{p-f} = q$$

17. $f(x) = \dfrac{-4}{3x^2 - 5x - 2}$

The numbers excluded from the domain are those for which the denominator is 0.

$$3x^2 - 5x - 2 = 0$$

$$(3x+1)(x-2) = 0$$

$3x + 1 = 0$　　or　$x - 2 = 0$

$3x = -1$　or　　　$x = 2$

$x = -\dfrac{1}{3}$　or　　　$x = 2$

The domain is $\left(-\infty, -\dfrac{1}{3}\right) \cup \left(-\dfrac{1}{3}, 2\right) \cup (2, \infty)$.

18. *Familiarize*. Let $t =$ the number of minutes it will take to do the job, working together.

***Translate*.** We use the work principle.

$$\frac{t}{10} + \frac{t}{12} = 1$$

***Solve*.** We solve the equation.

$$60\left(\frac{t}{10} + \frac{t}{12}\right) = 60 \cdot 1$$

$$6t + 5t = 60$$

$$11t = 60$$

$$t = \frac{60}{11}, \text{ or } 5\frac{5}{11}$$

***Check*.** We verify the work principle.

$$\frac{60/11}{10} + \frac{60/11}{12} = \frac{60}{11} \cdot \frac{1}{10} + \frac{60}{11} \cdot \frac{1}{12} = \frac{6}{11} + \frac{5}{11} = 1$$

***State*.** It would take Anne and Clay $5\dfrac{5}{11}$ min to do the job, working together.

19. $S(t) = 78 - 15\log(t+1)$

a)　$S(0) = 78 - 15\log(0+1)$

　　　$= 78 - 15\log 1$

　　　$= 78 - 15 \cdot 0$

　　　$= 78$

b)　$S(4) = 78 - 15\log(4+1)$

　　　$= 78 - 15\log 5$

　　　≈ 67.5

20. *Familiarize*. Let $x =$ the amount of Swim Clean and $y =$ the amount of Pure Swim that should be used, in liters. We organize the information in a table.

	Swim Clean	Pure Swim	Mixture
Amount	x	y	100 L
Percent of acid	30%	80%	50%
Amount of acid	$0.3x$	$0.8y$	0.5(100), or 50 L

***Translate*.** The first and last rows of the table yield two equations.

$$x + y = 100,$$

$$0.3x + 0.8y = 50$$

After clearing decimals we have the following system of equations.

$$x + y = 100, \quad (1)$$

$$3x + 8y = 500 \quad (2)$$

***Solve*.** We first multiply Equation (1) by -3 and then add.

$$-3x - 3y = -300$$

$$\underline{3x + 8y = 500}$$

$$5y = 200$$

$$y = 40$$

Now we substitute in Equation (1) and solve for x.

$$x + 40 = 100$$

$$x = 60$$

***Check*.** 60 L + 40 L = 100 L. The amount of acid in the mixture is $0.3(60) + 0.8(40) = 18 + 32 = 50$ L. The answer checks.

***State*.** 60 L of Swim Clean and 40 L of Pure Swim should be used.

21. *Familiarize*. Let $r =$ the speed of the stream, in km/h. Then the boat's speed downstream is $5 + r$, and the speed upstream is $5 - r$. We organize the information in a table.

	Distance	Speed	Time
Downstream	42	$5 + r$	t
Upstream	12	$5 - r$	t

***Translate*.** Using $t = d/r$ in each row of the table, we can equate the two expressions for time.

$$\frac{42}{5+r} = \frac{12}{5-r}$$

Solve. We solve the equation.

$$\frac{42}{5+r} = \frac{12}{5-r}$$

$$(5+r)(5-r) \cdot \frac{42}{5+r} = (5+r)(5-r) \cdot \frac{12}{5-r}$$

$$42(5-r) = 12(5+r)$$

$$210 - 42r = 60 + 12r$$

$$-54r = -150$$

$$r = \frac{25}{9}, \text{ or } 2\frac{7}{9}$$

Check. If the speed of the stream is $\frac{25}{9}$ km/h, then the boat's speed downstream is $5 + \frac{25}{9}$, or $\frac{70}{9}$ km/h and it travels 42 km downstream in $\frac{42}{70/9} = 42 \cdot \frac{9}{70} = \frac{27}{5}$, or $5\frac{2}{5}$ hr. The boat's speed upstream is $5 - \frac{25}{9}$, or $\frac{20}{9}$ km/h and it travels 12 km upstream in $\frac{12}{20/9} = 12 \cdot \frac{9}{20} = \frac{27}{5}$, or $5\frac{2}{5}$ hr. Since the times are the same, the answer checks.

State. The speed of the stream is $2\frac{7}{9}$ km/h.

22. a) $P(t) = 201\,e^{0.0086t}$, where $P(t)$ is in millions and t is the number of years after 2013.

    b) In 2015, $t = 2015 - 2013 = 2$.

      $P(2) = 201\,e^{0.0086(2)} \approx 204.5$ million

      In 2020, $t = 2020 - 2013 = 7$.

      $P(7) = 201\,e^{0.0086(7)} \approx 213.5$ million

    c) $\quad 2(201) = 201\,e^{0.0086t}$

$$2 = e^{0.0086t}$$

$$\ln 2 = \ln e^{0.0086t}$$

$$\ln 2 = 0.0086t$$

$$\frac{\ln 2}{0.0086} = t$$

$$80.6 \approx t$$

    The doubling time is about 80.6 years.

23. *Familiarize*. Let $x =$ the width of the sidewalk, in feet. We make a drawing.

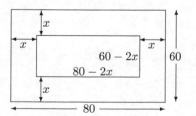

The length and width of the new lawn are represented by $80 - 2x$ and $60 - 2x$, respectively.

Translate. We use the formula for the area of a rectangle, $A = lw$.

$$2400 = (80 - 2x)(60 - 2x)$$

Solve. We solve the equation.

$$2400 = (80 - 2x)(60 - 2x)$$

$$2400 = 4800 - 280x + 4x^2$$

$$0 = 4x^2 - 280x + 2400$$

$$0 = x^2 - 70x + 600 \qquad \text{Dividing by 4}$$

$$0 = (x - 60)(x - 10)$$

$$x - 60 = 0 \quad or \quad x - 10 = 0$$

$$x = 60 \quad or \qquad x = 10$$

Check. 60 cannot be a solution because $80 - 2 \cdot 60 = -40$ and $60 - 2 \cdot 60 = -60$ and the dimensions of the new lawn cannot be negative. If $x = 10$, then the dimensions of the new lawn are $80 - 2 \cdot 10$, or 60, by $60 - 2 \cdot 10$, or 40, and the area is $60 \cdot 40$, or 2400 ft^2. The answer checks.

State. The sidewalk is 10 ft wide.

24.
$$y = \frac{kx^2}{z}$$

$$2 = \frac{k \cdot 5^2}{100}$$

$$2 = \frac{k \cdot 25}{100}$$

$$\frac{100}{25} \cdot 2 = k$$

$$8 = k \qquad \text{Constant of variation}$$

$$y = \frac{8x^2}{z} \qquad \text{Equation of variation}$$

$$y = \frac{8 \cdot 3^2}{4} = \frac{8 \cdot 9}{4} = 18$$

25. Graph $5x = 15 + 3y$.

We find the intercepts. To find the x-intercept we let $y = 0$ and solve for x.

$$5x = 15 + 3 \cdot 0$$

$$5x = 15$$

$$x = 3$$

The x-intercept is $(3,0)$.

To find the y-intercept we let $x = 0$ and solve for y.

$$5 \cdot 0 = 15 + 3y$$

$$0 = 15 + 3y$$

$$-15 = 3y$$

$$-5 = y$$

The y-intercept is $(0, -5)$.

We plot these points and draw the line. A third point could be found as a check.

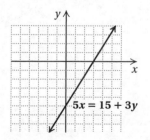

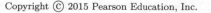

26. Graph $-2x - 3y \leq 6$.

We first graph $-2x - 3y = 6$. We draw the line solid since the inequality symbol is $\leq$. Then test the point $(0,0)$ to determine if it is a solution.

$$\frac{-2x - 3y \leq 6}{-2 \cdot 0 - 3 \cdot 0 \;?\; 6}$$
$$0 \;\Big|\; \text{TRUE}$$

Since $0 \leq 6$ is true, we shade the half-plane that contains $(0,0)$.

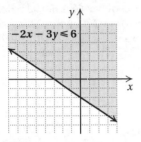

27. Graph $f(x) = 2x^2 - 4x - 1 = 2(x^2 - 2x) - 1$

We complete the square inside the parentheses. We take half the x-coefficient and square it.

$$\frac{1}{2}(-2) = -1 \text{ and } (-1)^2 = 1$$

Then we add $1 - 1$ inside the parentheses.

$$\begin{aligned} f(x) &= 2(x^2 - 2x + 1 - 1) - 1 \\ &= 2(x^2 - 2x + 1) + 2(-1) - 1 \\ &= 2(x - 1)^2 - 3 \\ &= 2(x - 1)^2 + (-3) \end{aligned}$$

The vertex is $(1, -3)$, and the line of symmetry is $x = 1$. We plot some points and draw the curve.

x	$f(x)$
1	-3
-1	5
0	-1
2	-1
3	5

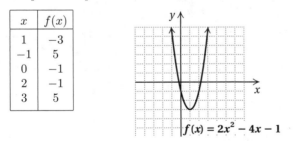

28. Graph $f(x) = 3^x$.

We find some function values, plot points, and then connect them with a smooth curve.

x	$f(x)$
0	1
1	3
2	9
3	27
-1	$\dfrac{1}{3}$
-2	$\dfrac{1}{9}$
-3	$\dfrac{1}{27}$

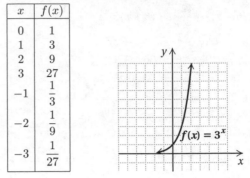

29. Graph $f(x) = \log_3 x$, or $y = \log_3 x$

The equation $f(x) = y = \log_3 x$ is equivalent to $3^y = x$. We find ordered pairs by choosing values for y and computing the corresponding x-values. Then we plot the set of ordered pairs and connect the points with a smooth curve.

x, or 3^y	y
1	0
3	1
9	2
27	3
$\dfrac{1}{3}$	-1
$\dfrac{1}{9}$	-2
$\dfrac{1}{27}$	-3

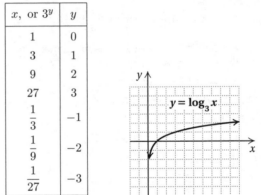

30. $(11x^2 - 6x - 3) - (3x^2 + 5x - 2)$
$$= 11x^2 - 6x - 3 + (-3x^2 - 5x + 2)$$
$$= 8x^2 - 11x - 1$$

31. $(3x^2 - 2y)^2 = (3x^2)^2 - 2 \cdot 3x^2 \cdot 2y + (2y)^2$
$$= 9x^4 - 12x^2y + 4y^2$$

32. $(5a + 3b)(2a - 3b)$
$$= 10a^2 - 15ab + 6ab - 9b^2 \quad \text{FOIL}$$
$$= 10a^2 - 9ab - 9b^2$$

33. $\dfrac{x^2 + 8x + 16}{2x + 6} \div \dfrac{x^2 + 3x - 4}{x^2 - 9}$

$$= \dfrac{x^2 + 8x + 16}{2x + 6} \cdot \dfrac{x^2 - 9}{x^2 + 3x - 4}$$

$$= \dfrac{(x + 4)(x + 4)(x + 3)(x - 3)}{2(x + 3)(x + 4)(x - 1)}$$

$$= \dfrac{(x + 4)(x + 4)(x + 3)(x - 3)}{2(x + 3)(x + 4)(x - 1)}$$

$$= \dfrac{(x + 4)(x - 3)}{2(x - 1)}$$

34. $\dfrac{1+\dfrac{3}{x}}{x-1-\dfrac{12}{x}} = \dfrac{1+\dfrac{3}{x}}{x-1-\dfrac{12}{x}} \cdot \dfrac{x}{x}$

$= \dfrac{\left(1+\dfrac{3}{x}\right)x}{\left(x-1-\dfrac{12}{x}\right)x}$

$= \dfrac{x+3}{x^2-x-12}$

$= \dfrac{x+3}{(x+3)(x-4)}$

$= \dfrac{(x+3)\cdot 1}{(x+3)(x-4)}$

$= \dfrac{1}{x-4}$

35. $\dfrac{3}{x+6} - \dfrac{2}{x^2-36} + \dfrac{4}{x-6}$

$= \dfrac{3}{x+6} - \dfrac{2}{(x+6)(x-6)} + \dfrac{4}{x-6},$ LCM is $(x+6)(x-6)$

$= \dfrac{3}{x+6} \cdot \dfrac{x-6}{x-6} - \dfrac{2}{(x+6)(x-6)} + \dfrac{4}{x-6} \cdot \dfrac{x+6}{x+6}$

$= \dfrac{3(x-6) - 2 + 4(x+6)}{(x+6)(x-6)}$

$= \dfrac{3x-18-2+4x+24}{(x+6)(x-6)}$

$= \dfrac{7x+4}{(x+6)(x-6)}$

36. $1 - 125x^3$

$= 1^3 - (5x)^3$ Difference of cubes

$= (1-5x)(1^2 + 1\cdot 5x + (5x)^2)$

$= (1-5x)(1+5x+25x^2)$

37. $6x^2 + 8xy - 8y^2$

$= 2(3x^2 + 4xy - 4y^2)$

$= 2(3x-2y)(x+2y)$ Using the FOIL method or the ac-method

38. $x^4 - 4x^3 + 7x - 28$

$= x^3(x-4) + 7(x-4)$

$= (x-4)(x^3+7)$

39. $2m^2 + 12mn + 18n^2$

$= 2(m^2 + 6mn + 9n^2)$

$= 2(m+3n)^2$

40. $x^4 - 16y^4$

$= (x^2)^2 - (4y^2)^2$ Difference of squares

$= (x^2 + 4y^2)(x^2 - 4y^2)$ Difference of squares

$= (x^2 + 4y^2)(x+2y)(x-2y)$

41. $h(x) = -3x^2 + 4x + 8$

$h(-2) = -3(-2)^2 + 4(-2) + 8$

$= -3\cdot 4 - 8 + 8$

$= -12 - 8 + 8$

$= -12$

42.

$$
\begin{array}{r|rrrrr}
3 & 1 & -5 & 2 & 0 & -6 \\
  &   & 3 & -6 & -12 & -36 \\
\hline
  & 1 & -2 & -4 & -12 & -42 \\
\end{array}
$$

The answer is $x^3 - 2x^2 - 4x - 12 + \dfrac{-42}{x-3}$.

43. $\sqrt{7xy^3} \cdot \sqrt{28x^2y} = \sqrt{196x^3y^4} = \sqrt{196x^2y^4 \cdot x} =$
$\sqrt{196x^2y^4} \cdot \sqrt{x} = 14xy^2\sqrt{x}$

44. $\dfrac{\sqrt[3]{40xy^8}}{\sqrt[3]{5xy}} = \sqrt[3]{\dfrac{40xy^8}{5xy}} = \sqrt[3]{8y^7} = \sqrt[3]{8y^6 \cdot y} =$
$\sqrt[3]{8y^6}\ \sqrt[3]{y} = 2y^2\ \sqrt[3]{y}$

45. $\dfrac{3-\sqrt{y}}{2-\sqrt{y}} = \dfrac{3-\sqrt{y}}{2-\sqrt{y}} \cdot \dfrac{2+\sqrt{y}}{2+\sqrt{y}}$

$= \dfrac{6 + 3\sqrt{y} - 2\sqrt{y} - y}{4 - y}$

$= \dfrac{6 + \sqrt{y} - y}{4 - y}$

46. $(1 + i\sqrt{3})(6 - 2i\sqrt{3})$

$= 6 - 2i\sqrt{3} + 6i\sqrt{3} - 2i^2 \cdot 3$

$= 6 - 2i\sqrt{3} + 6i\sqrt{3} - 2(-1)\cdot 3$

$= 6 - 2i\sqrt{3} + 6i\sqrt{3} + 6$

$= 12 + 4i\sqrt{3}$

47. The function $f(x) = 7 - 2x$ passes the horizontal-line test, so it is one-to-one and thus has an inverse. We find a formula for the inverse.

1. Replace $f(x)$ by y: $y = 7 - 2x$

2. Interchange x and y: $x = 7 - 2y$

3. Solve for y: $x - 7 = -2y$

$\dfrac{x-7}{-2} = y$

4. Replace y by $f^{-1}(x)$: $f^{-1}(x) = \dfrac{x-7}{-2},$ or $\dfrac{7-x}{2}$

48. First solve the equation for y to determine the slope of the given line.

$2x + y = 6$

$y = -2x + 6$

The slope of the given line is -2. The slope of the perpendicular line is the opposite of the reciprocal of -2, or $\dfrac{1}{2}$. We will use the slope-intercept equation to find the desired equation. Substitute $\dfrac{1}{2}$ for m, -3 for x, and 5 for y.

$y = mx + b$

$5 = \dfrac{1}{2}(-3) + b$

$5 = -\dfrac{3}{2} + b$

$\dfrac{13}{2} = b$

Thus we have $y = \dfrac{1}{2}x + \dfrac{13}{2}$.

49.
$$3 \log x - \frac{1}{2} \log y - 2 \log z$$
$$= \log x^3 - \log y^{1/2} - \log z^2$$
$$= \log x^3 - (\log y^{1/2} + \log z^2)$$
$$= \log x^3 - \log (y^{1/2} z^2)$$
$$= \log \frac{x^3}{y^{1/2} z^2}, \text{ or } \log \frac{x^3}{z^2 \sqrt{y}}$$

50. $\log_a 5 = x \Rightarrow a^x = 5$

51. $\log 0.05566 \approx -1.2545$

52. $10^{2.89} \approx 776.2471$

53. $\ln 12.78 \approx 2.5479$

54. $e^{-1.4} \approx 0.2466$

55.
$$f(x) = -2x^2 + 28x - 9$$
$$= -2(x^2 - 14x) - 9$$
$$= -2(x^2 - 14x + 49 - 49) - 9 \quad \frac{1}{2}(-14) = -7$$
$$\text{and } (-7)^2 = 49$$
$$= -2(x^2 - 14x + 49) + (-2)(-49) - 9$$
$$= -2(x - 7)^2 + 98 - 9$$
$$= -2(x - 7)^2 + 89$$

Answer D is correct.

56.
$$B = 2a(b^2 - c^2)$$
$$B = 2ab^2 - 2ac^2$$
$$2ac^2 = 2ab^2 - B$$
$$c^2 = \frac{2ab^2 - B}{2a}$$
$$c = \sqrt{\frac{2ab^2 - B}{2a}}$$

Answer D is correct.

57.
$$\frac{5}{3x - 3} + \frac{10}{3x + 6} = \frac{5x}{x^2 + x - 2}$$
$$\frac{5}{3(x - 1)} + \frac{10}{3(x + 2)} = \frac{5x}{(x + 2)(x - 1)}$$
$$\text{LCD is } 3(x - 1)(x + 2)$$
$$3(x-1)(x+2)\left(\frac{5}{3(x-1)} + \frac{10}{3(x+2)}\right) =$$
$$3(x-1)(x+2) \cdot \frac{5x}{(x+2)(x-1)}$$
$$5(x + 2) + 10(x - 1) = 3 \cdot 5x$$
$$5x + 10 + 10x - 10 = 15x$$
$$15x = 15x$$

We get an equation that is true for all values of x. Thus the solutions of the original equation are all real numbers except those for which a denominator is 0. A denominator is 0 when $x = 1$ or when $x = -2$, so all real numbers except 1 and -2 are solutions.

58.
$$\log \sqrt{3x} = \sqrt{\log 3x}$$
$$\log (3x)^{1/2} = \sqrt{\log 3x}$$
$$\frac{1}{2} \log 3x = \sqrt{\log 3x}$$
$$\left(\frac{1}{2} \log 3x\right)^2 = (\sqrt{\log 3x})^2$$
$$\frac{1}{4}(\log 3x)^2 = \log 3x$$
$$\frac{1}{4}(\log 3x)^2 - \log 3x = 0$$

Let $u = \log 3x$.
$$\frac{1}{4}u^2 - u = 0$$
$$u\left(\frac{1}{4}u - 1\right) = 0$$
$$u = 0 \quad or \quad \frac{1}{4}u - 1 = 0$$
$$u = 0 \quad or \quad \frac{1}{4}u = 1$$
$$u = 0 \quad or \quad u = 4$$
$$\log 3x = 0 \quad or \quad \log 3x = 4$$
$$10^0 = 3x \quad or \quad 10^4 = 3x$$
$$1 = 3x \quad or \quad 10,000 = 3x$$
$$\frac{1}{3} = x \quad or \quad \frac{10,000}{3} = x$$

Both numbers check. The solutions are $\frac{1}{3}$ and $\frac{10,000}{3}$.

59. *Familiarize.* Let $r =$ the speed of the train and $t =$ the time of the trip. We organize the information in a table.

	Distance	Speed	Time
Actual trip	280	r	t
Faster trip	280	$r + 5$	$t - 1$

Translate. Using Time = Distance/Speed in each row of the table, we have two equations.
$$\frac{280}{r} = t, \quad \frac{280}{r + 5} = t - 1$$

Solve. We substitute $\frac{280}{r}$ for t in the second equation and solve for t.
$$\frac{280}{r + 5} = \frac{280}{r} - 1, \text{ LCD is } r(r + 5)$$
$$r(r + 5) \cdot \frac{280}{r + 5} = r(r + 5)\left(\frac{280}{r} - 1\right)$$
$$280r = 280(r + 5) - r(r + 5)$$
$$280r = 280r + 1400 - r^2 - 5r$$
$$r^2 + 5r - 1400 = 0$$
$$(r + 40)(r - 35) = 0$$
$$r = -40 \quad or \quad r = 35$$

Check. The speed cannot be negative, so we check only 35. At 35 mph, the train would travel 280 mi in 280/35, or 8 hr. At $35 + 5$, or 40 mph, the train would travel 280 mi in 280/40, or 7 hr. Since 7 hr is 1 hr less than 8 hr, the answer checks.

State. The actual speed of the train is 35 mph.

Chapter 9

Conic Sections

Exercise Set 9.1

RC1. True; see page 758 in the text.

RC3. True; see page 762 in the text.

RC5. True; see page 764 in the text.

1. Graph: $y = x^2$

The graph is a parabola. The vertex is $(0,0)$; the line of symmetry is $x = 0$. The curve opens upward. We choose some x-values on both sides of the vertex and compute the corresponding y-values. Then we plot the points and graph the parabola.

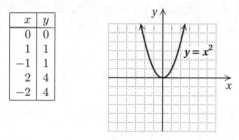

x	y
0	0
1	1
-1	1
2	4
-2	4

3. Graph: $x = y^2 + 4y + 1$

We complete the square to find the vertex.
$$x = (y^2 + 4y + 4 - 4) + 1$$
$$= (y^2 + 4y + 4) - 4 + 1$$
$$= (y + 2)^2 - 3, \text{ or}$$
$$= [y - (-2)]^2 + (-3)$$

The graph is a parabola. The vertex is $(-3, -2)$; the line of symmetry is $y = -2$. The curve opens to the right.

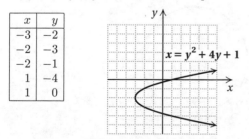

x	y
-3	-2
-2	-3
-2	-1
1	-4
1	0

5. Graph: $y = -x^2 + 4x - 5$

We use the formula to find the first coordinate of the vertex:
$$x = -\frac{b}{2a} = -\frac{4}{2(-1)} = 2$$

Then $y = -x^2 + 4x - 5 = -(2)^2 + 4(2) - 5 = -1$.

The vertex is $(2, -1)$; the line of symmetry is $x = 2$. The curve opens downward.

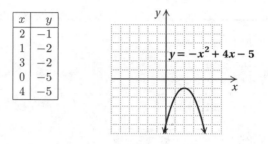

x	y
2	-1
1	-2
3	-2
0	-5
4	-5

7. Graph: $x = -3y^2 - 6y - 1$

We complete the square to find the vertex.
$$x = -3(y^2 + 2y) - 1$$
$$= -3(y^2 + 2y + 1 - 1) - 1$$
$$= -3(y^2 + 2y + 1) + 3 - 1$$
$$= -3(y + 1)^2 + 2$$
$$= -3[y - (-1)]^2 + 2$$

The graph is a parabola. The vertex is $(2, -1)$; the line of symmetry is $y = -1$. The curve opens to the left.

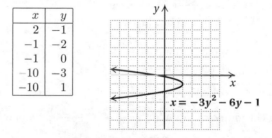

x	y
2	-1
-1	-2
-1	0
-10	-3
-10	1

9. $d = \sqrt{(x_2 - x_1)^2 + (y_2 - y_1)^2}$

$d = \sqrt{(2 - 6)^2 + [-7 - (-4)]^2}$ Substituting
$$= \sqrt{(-4)^2 + (-3)^2}$$
$$= \sqrt{25} = 5$$

11. $d = \sqrt{(x_2 - x_1)^2 + (y_2 - y_1)^2}$

$d = \sqrt{(5 - 0)^2 + [-6 - (-4)]^2}$
$$= \sqrt{5^2 + (-2)^2}$$
$$= \sqrt{29} \approx 5.385$$

13. $d = \sqrt{(x_2 - x_1)^2 + (y_2 - y_1)^2}$

$d = \sqrt{(-9 - 9)^2 + (-9 - 9)^2}$
$$= \sqrt{(-18)^2 + (-18)^2}$$
$$= \sqrt{648} \approx 25.456$$

15. $d = \sqrt{(x_2 - x_1)^2 + (y_2 - y_1)^2}$

$d = \sqrt{(-4.3 - 2.8)^2 + [-3.5 - (-3.5)]^2}$
$$= \sqrt{(-7.1)^2 + 0^2} = \sqrt{(-7.1)^2}$$
$$= 7.1$$

17. $d = \sqrt{(x_2 - x_1)^2 + (y_2 - y_1)^2}$

$d = \sqrt{\left(\dfrac{5}{7} - \dfrac{1}{7}\right)^2 + \left(\dfrac{1}{14} - \dfrac{11}{14}\right)^2}$

$= \sqrt{\left(\dfrac{4}{7}\right)^2 + \left(-\dfrac{5}{7}\right)^2}$

$= \sqrt{\dfrac{16}{49} + \dfrac{25}{49}}$

$= \sqrt{\dfrac{41}{49}}$

$= \dfrac{\sqrt{41}}{7} \approx 0.915$

19. $d = \sqrt{[56 - (-23)]^2 + (-17 - 10)^2}$

$= \sqrt{79^2 + (-27)^2} = \sqrt{6970} \approx 83.487$

21. $d = \sqrt{(a - 0)^2 + (b - 0)^2}$

$= \sqrt{a^2 + b^2}$

23. $d = \sqrt{(-\sqrt{7} - \sqrt{2})^2 + [\sqrt{5} - (-\sqrt{3})]^2}$

$= \sqrt{7 + 2\sqrt{14} + 2 + 5 + 2\sqrt{15} + 3}$

$= \sqrt{17 + 2\sqrt{14} + 2\sqrt{15}} \approx 5.677$

25. $d = \sqrt{[1000 - (-2000)]^2 + (-240 - 580)^2}$

$= \sqrt{3000^2 + (-820)^2}$

$= \sqrt{9,672,400} \approx 3110.048$

27. Using the midpoint formula $\left(\dfrac{x_1 + x_2}{2}, \dfrac{y_1 + y_2}{2}\right)$, we obtain

$\left(\dfrac{-1 + 4}{2}, \dfrac{9 + (-2)}{2}\right)$, or $\left(\dfrac{3}{2}, \dfrac{7}{2}\right)$.

29. Using the midpoint formula $\left(\dfrac{x_1 + x_2}{2}, \dfrac{y_1 + y_2}{2}\right)$, we obtain

$\left(\dfrac{3 + (-3)}{2}, \dfrac{5 + 6}{2}\right)$, or $\left(\dfrac{0}{2}, \dfrac{11}{2}\right)$, or $\left(0, \dfrac{11}{2}\right)$.

31. Using the midpoint formula $\left(\dfrac{x_1 + x_2}{2}, \dfrac{y_1 + y_2}{2}\right)$, we obtain

$\left(\dfrac{-10 + 8}{2}, \dfrac{-13 + (-4)}{2}\right)$, or $\left(\dfrac{-2}{2}, \dfrac{-17}{2}\right)$, or $\left(-1, -\dfrac{17}{2}\right)$.

33. Using the midpoint formula $\left(\dfrac{x_1 + x_2}{2}, \dfrac{y_1 + y_2}{2}\right)$, we obtain

$\left(\dfrac{-3.4 + 2.9}{2}, \dfrac{8.1 + (-8.7)}{2}\right)$, or $\left(\dfrac{-0.5}{2}, \dfrac{-0.6}{2}\right)$, or

$(-0.25, -0.3)$.

35. Using the midpoint formula $\left(\dfrac{x_1 + x_2}{2}, \dfrac{y_1 + y_2}{2}\right)$, we obtain

$\left(\dfrac{\dfrac{1}{6} + \left(-\dfrac{1}{3}\right)}{2}, \dfrac{-\dfrac{3}{4} + \dfrac{5}{6}}{2}\right)$, or $\left(\dfrac{-\dfrac{1}{6}}{2}, \dfrac{\dfrac{1}{12}}{2}\right)$, or $\left(-\dfrac{1}{12}, \dfrac{1}{24}\right)$.

37. Using the midpoint formula $\left(\dfrac{x_1 + x_2}{2}, \dfrac{y_1 + y_2}{2}\right)$, we obtain

$\left(\dfrac{\sqrt{2} + \sqrt{3}}{2}, \dfrac{-1 + 4}{2}\right)$, or $\left(\dfrac{\sqrt{2} + \sqrt{3}}{2}, \dfrac{3}{2}\right)$.

39. $\qquad (x + 1)^2 + (y + 3)^2 = 4$

$[x - (-1)]^2 + [y - (-3)]^2 = 2^2$ Standard form

The center is $(-1, -3)$, and the radius is 2.

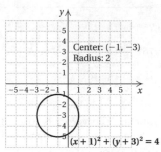

41. $\qquad (x - 3)^2 + y^2 = 2$

$(x - 3)^2 + (y - 0)^2 = (\sqrt{2})^2$ Standard form

The center is $(3, 0)$, and the radius is $\sqrt{2}$.

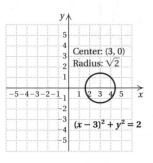

43. $\qquad x^2 + y^2 = 25$

$(x - 0)^2 + (y - 0)^2 = 5^2$ Standard form

The center is $(0, 0)$, and the radius is 5.

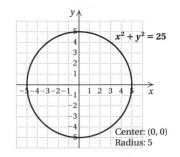

45. $(x - h)^2 + (y - k)^2 = r^2$ Standard form

$\quad (x - 0)^2 + (y - 0)^2 = 7^2$ Substituting

$\qquad\qquad\quad x^2 + y^2 = 49$

47. $\quad (x - h)^2 + (y - k)^2 = r^2$ Standard form

$[x - (-5)]^2 + (y - 3)^2 = (\sqrt{7})^2$ Substituting

$\qquad (x + 5)^2 + (y - 3)^2 = 7$

49.
$$x^2 + y^2 + 8x - 6y - 15 = 0$$
$$(x^2 + 8x) + (y^2 - 6y) - 15 = 0$$
Regrouping

$$(x^2 + 8x + 16 - 16) + (y^2 - 6y + 9 - 9) - 15 = 0$$
Completing the square twice

$$(x^2 + 8x + 16) + (y^2 - 6y + 9) - 16 - 9 - 15 = 0$$
$$(x + 4)^2 + (y - 3)^2 = 40$$
$$[x - (-4)]^2 + (y - 3)^2 = (\sqrt{40})^2$$
$$[x - (-4)]^2 + (y - 3)^2 = (2\sqrt{10})^2$$

The center is $(-4, 3)$, and the radius is $2\sqrt{10}$.

51.
$$x^2 + y^2 - 8x + 2y + 13 = 0$$
$$(x^2 - 8x) + (y^2 + 2y) + 13 = 0$$
Regrouping

$$(x^2 - 8x + 16 - 16) + (y^2 + 2y + 1 - 1) + 13 = 0$$
Completing the square twice

$$(x^2 - 8x + 16) + (y^2 + 2y + 1) - 16 - 1 + 13 = 0$$
$$(x - 4)^2 + (y + 1)^2 = 4$$
$$(x - 4)^2 + [y - (-1)]^2 = 2^2$$

The center is $(4, -1)$, and the radius is 2.

53.
$$x^2 + y^2 - 4x = 0$$
$$(x^2 - 4x) + y^2 = 0$$
$$(x^2 - 4x + 4 - 4) + y^2 = 0$$
$$(x^2 - 4x + 4) + y^2 - 4 = 0$$
$$(x - 2)^2 + y^2 = 4$$
$$(x - 2)^2 + (y - 0)^2 = 2^2$$

The center is $(2, 0)$, and the radius is 2.

55. We use the elimination method.
$$
\begin{array}{rl}
x - y = 7 & (1) \\
x + y = 11 & (2) \\
\hline
2x \quad\;\; = 18 & \text{Adding} \\
x = 9 &
\end{array}
$$

Substitute 9 for x in one of the original equations and solve for y.
$$x + y = 11 \quad (2)$$
$$9 + y = 11 \quad \text{Substituting}$$
$$y = 2$$
The solution is $(9, 2)$.

57.
$$
\begin{array}{rl}
y = 3x - 2, & (1) \\
2x - 4y = 50 & (2)
\end{array}
$$

Substitute $3x - 2$ for y in Equation (2) and solve for y.
$$2x - 4(3x - 2) = 50$$
$$2x - 12x + 8 = 50$$
$$-10x = 42$$
$$x = -\frac{21}{5}$$

Substitute $-\dfrac{21}{5}$ for x in Equation (1) and compute y.
$$y = 3\left(-\frac{21}{5}\right) - 2 = -\frac{63}{5} - 2 = -\frac{73}{5}$$
The solution is $\left(-\dfrac{21}{5}, -\dfrac{73}{5}\right)$.

59.
$$
\begin{array}{rl}
-4x + 12y = -9, & (1) \\
x - 3y = 2 & (2)
\end{array}
$$
We use the elimination method.
$$
\begin{array}{rl}
-4x + 12y = -9 & (1) \\
4x - 12y = 8 & \text{Multiplying (2) by 4} \\
\hline
0 = -1 & \text{Adding}
\end{array}
$$
We get a false equation. The system of equations has no solution.

61. $x^2 - 16 = x^2 - 4^2 = (x + 4)(x - 4)$

63. $64p^2 - 81q^2 = (8p)^2 - (9q)^2 = (8p + 9q)(8p - 9q)$

65.

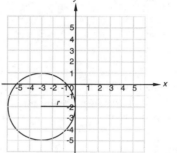

The center is $(-3, -2)$ and the radius is 3.
$$
\begin{array}{ll}
(x - h)^2 + (y - k)^2 = r^2 & \text{Standard form} \\
[x - (-3)]^2 + [y - (-2)]^2 = 3^2 & \text{Substituting} \\
(x + 3)^2 + (y + 2)^2 = 9 &
\end{array}
$$

67.
$$
\begin{aligned}
d &= \sqrt{(x_2 - x_1)^2 + (y_2 - y_1)^2} \\
d &= \sqrt{[6m - (-2m)]^2 + (-7n - n)^2} \quad \text{Substituting} \\
&= \sqrt{(8m)^2 + (-8n)^2} \\
&= \sqrt{64m^2 + 64n^2} \\
&= \sqrt{64(m^2 + n^2)} \\
&= 8\sqrt{m^2 + n^2}
\end{aligned}
$$

69. The distance between $(-8, -5)$ and $(6, 1)$ is
$$\sqrt{(-8 - 6)^2 + (-5 - 1)^2} = \sqrt{196 + 36} = \sqrt{232}.$$
The distance between $(6, 1)$ and $(-4, 5)$ is
$$\sqrt{[6 - (-4)]^2 + (1 - 5)^2} = \sqrt{100 + 16} = \sqrt{116}.$$
The distance between $(-4, 5)$ and $(-8, -5)$ is
$$\sqrt{[-4 - (-8)]^2 + [5 - (-5)]^2} = \sqrt{16 + 100} = \sqrt{116}.$$
Since $(\sqrt{116})^2 + (\sqrt{116})^2 = (\sqrt{232})^2$, the points are vertices of a right triangle.

71. a) Use the fact that the center of the circle $(0, k)$ is equidistant from the points $(-575, 0)$ and $(0, 19.5)$.

$$\sqrt{(-575-0)^2+(0-k)^2} = \sqrt{(0-0)^2+(19.5-k)^2}$$
$$\sqrt{330,625+k^2} = \sqrt{380.25-39k+k^2}$$
$$330,625+k^2 = 380.25-39k+k^2$$
Squaring both sides
$$330,244.75 = -39k$$
$$-8467.8 \approx k$$

Then the center of the circle is about $(0, -8467.8)$.

b) To find the radius we find the distance from the center, $(0, -8467.8)$ to any one of the points $(-575, 0)$, $(0, 19.5)$, or $(575, 0)$. We use $(0, 19.5)$.
$$r = \sqrt{(0-0)^2 + [19.5-(-8467.8)]^2} \approx$$
8487.3 mm

Exercise Set 9.2

RC1. A; see page 771 in the text.

RC3. B; see page 774 in the text.

1. $\dfrac{x^2}{9} + \dfrac{y^2}{36} = 1$

$\dfrac{x^2}{3^2} + \dfrac{y^2}{6^2} = 1$

The x-intercepts are $(-3, 0)$ and $(3, 0)$, and the y-intercepts are $(0, -6)$ and $(0, 6)$. We plot these points and connect them with an oval-shaped curve.

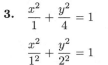

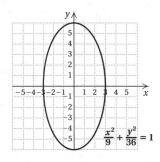

3. $\dfrac{x^2}{1} + \dfrac{y^2}{4} = 1$

$\dfrac{x^2}{1^2} + \dfrac{y^2}{2^2} = 1$

The x-intercepts are $(-1, 0)$ and $(1, 0)$, and the y-intercepts are $(0, -2)$ and $(0, 2)$. We plot these points and connect them with an oval-shaped curve.

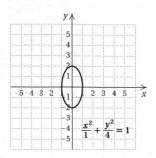

5. $4x^2 + 9y^2 = 36$

$\dfrac{x^2}{9} + \dfrac{y^2}{4} = 1$ Dividing by 36

$\dfrac{x^2}{3^2} + \dfrac{y^2}{2^2} = 1$

The x-intercepts are $(-3, 0)$ and $(3, 0)$, and the y-intercepts are $(0, -2)$ and $(0, 2)$. We plot these points and connect them with an oval-shaped curve.

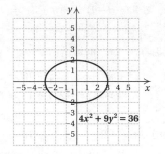

7. $x^2 + 4y^2 = 4$

$\dfrac{x^2}{4} + \dfrac{y^2}{1} = 1$ Dividing by 4

$\dfrac{x^2}{2^2} + \dfrac{y^2}{1^2} = 1$

The x-intercepts are $(-2, 0)$ and $(2, 0)$, and the y-intercepts are $(0, -1)$ and $(0, 1)$. We plot these points and connect them with an oval-shaped curve.

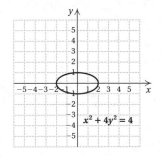

9. $2x^2 + 3y^2 = 6$

$\dfrac{x^2}{3} + \dfrac{y^2}{2} = 1$ Multiplying by $\dfrac{1}{6}$

$\dfrac{x^2}{(\sqrt{3})^2} + \dfrac{y^2}{(\sqrt{2})^2} = 1$

The x-intercepts are $(\sqrt{3}, 0)$ and $(-\sqrt{3}, 0)$, and the y-intercepts are $(0, \sqrt{2})$ and $(0, -\sqrt{2})$. We plot these points and connect them with an oval-shaped curve.

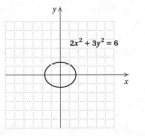

11. $12x^2 + 5y^2 - 120 = 0$

$$12x^2 + 5y^2 = 120$$

$$\frac{x^2}{10} + \frac{y^2}{24} = 1 \quad \text{Multiplying by } \frac{1}{120}$$

$$\frac{x^2}{(\sqrt{10})^2} + \frac{y^2}{(\sqrt{24})^2} = 1$$

The x-intercepts are $(\sqrt{10}, 0)$ and $(-\sqrt{10}, 0)$, or about $(3.162, 0)$ and $(-3.162, 0)$. The y-intercepts are $(0, \sqrt{24})$ and $(0, -\sqrt{24})$, or about $(0, 4.899)$ and $(0, -4.899)$. We plot these points and connect them with an oval-shaped curve.

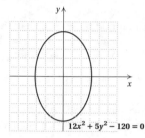

$12x^2 + 5y^2 - 120 = 0$

13. $\dfrac{(x-2)^2}{9} + \dfrac{(y-1)^2}{25} = 1$

$$\frac{(x-2)^2}{3^2} + \frac{(y-1)^2}{5^2} = 1$$

The center of the ellipse is $(2, 1)$. Note that $a = 3$ and $b = 5$. We locate the center and then plot the points $(2+3, 1)$ $(2-3, 1)$, $(2, 1+5)$, and $(2, 1-5)$, or $(5, 1)$, $(-1, 1)$, $(2, 6)$, and $(2, -4)$. Connect these points with an oval-shaped curve.

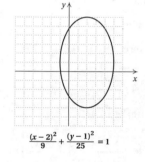

$\frac{(x-2)^2}{9} + \frac{(y-1)^2}{25} = 1$

15. $\dfrac{(x+1)^2}{16} + \dfrac{(y+2)^2}{25} = 1$

$$\frac{[x-(-1)]^2}{4^2} + \frac{[y-(-2)]^2}{5^2} = 1$$

The center of the ellipse is $(-1, -2)$. Note that $a = 4$ and $b = 5$. We locate the center and then plot the points $(-1+4, -2)$, $(-1-4, -2)$, $(-1, -2+5)$, and $(-1, -2-5)$, or $(3, -2)$, $(-5, -2)$, $(-1, 3)$, and $(-1, -7)$. Connect these points with an oval-shaped curve.

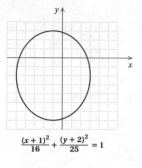

$\frac{(x+1)^2}{16} + \frac{(y+2)^2}{25} = 1$

17. $12(x-1)^2 + 3(y+2)^2 = 48$

$$\frac{(x-1)^2}{4} + \frac{(y+2)^2}{16} = 1$$

$$\frac{(x-1)^2}{2^2} + \frac{(y-(-2))^2}{4^2} = 1$$

The center of the ellipse is $(1, -2)$. Note that $a = 2$ and $b = 4$. We locate the center and then plot the points $(1+2, -2)$, $(1-2, -2)$, $(1, -2+4)$, and $(1, -2-4)$, or $(3, -2)$, $(-1, -2)$, $(1, 2)$, and $(1, -6)$. Connect these points with an oval-shaped curve.

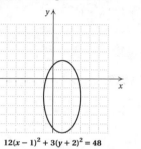

$12(x-1)^2 + 3(y+2)^2 = 48$

19. $(x+3)^2 + 4(y+1)^2 - 10 = 6$

$$(x+3)^2 + 4(y+1)^2 = 16$$

$$\frac{(x+3)^2}{16} + \frac{(y+1)^2}{4} = 1$$

$$\frac{[x-(-3)]^2}{4^2} + \frac{[y-(-1)]^2}{2^2} = 1$$

The center of the ellipse is $(-3, -1)$. Note that $a = 4$ and $b = 2$. We locate the center and then plot the points $(-3+4, -1)$, $(-3-4, -1)$, $(-3, -1+2)$, and $(-3, -1-2)$, or $(1, -1)$, $(-7, -1)$, $(-3, 1)$, and $(-3, -3)$. Connect these points with an oval-shaped curve.

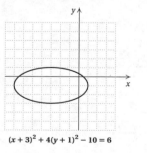

$(x+3)^2 + 4(y+1)^2 - 10 = 6$

21. $3x^2 - 2x + 7 = 0$

$a = 3,\ b = -2,\ c = 7$

$x = \dfrac{-b \pm \sqrt{b^2 - 4ac}}{2a}$

$x = \dfrac{-(-2) \pm \sqrt{(-2)^2 - 4 \cdot 3 \cdot 7}}{2 \cdot 3}$

$x = \dfrac{2 \pm \sqrt{4 - 84}}{6} = \dfrac{2 \pm \sqrt{-80}}{6}$

$x = \dfrac{2 \pm 4i\sqrt{5}}{6} = \dfrac{1 \pm 2i\sqrt{5}}{3}$

The solutions are $\dfrac{1 + 2i\sqrt{5}}{3}$ and $\dfrac{1 - 2i\sqrt{5}}{3}$.

23. $x^2 + x + 2 = 0$

$a = 1,\ b = 1,\ c = 2$

$x = \dfrac{-b \pm \sqrt{b^2 - 4ac}}{2a}$

$x = \dfrac{-1 \pm \sqrt{1^2 - 4 \cdot 1 \cdot 2}}{2 \cdot 1}$

$x = \dfrac{-1 \pm \sqrt{1 - 8}}{2} = \dfrac{-1 \pm \sqrt{-7}}{2}$

$x = \dfrac{-1 \pm i\sqrt{7}}{2}$

The solutions are $\dfrac{-1 + i\sqrt{7}}{2}$ and $\dfrac{-1 - i\sqrt{7}}{2}$.

25. $x^2 + 2x - 17 = 0$

$a = 1,\ b = 2,\ c = -17$

$x = \dfrac{-b \pm \sqrt{b^2 - 4ac}}{2a}$

$x = \dfrac{-2 \pm \sqrt{2^2 - 4 \cdot 1 \cdot (-17)}}{2 \cdot 1}$

$x = \dfrac{-2 \pm \sqrt{72}}{2} = \dfrac{-2 \pm 6\sqrt{2}}{2}$

$x = -1 \pm 3\sqrt{2}$

The solutions are $-1 + 3\sqrt{2} \approx 3.2$ and $-1 - 3\sqrt{2} \approx -5.2$.

27. $3x^2 - 12x + 7 = 10 - x^2 + 5x$

$4x^2 - 17x - 3 = 0$

$a = 4,\ b = -17,\ c = -3$

$x = \dfrac{-b \pm \sqrt{b^2 - 4ac}}{2a}$

$x = \dfrac{-(-17) \pm \sqrt{(-17)^2 - 4 \cdot 4 \cdot (-3)}}{2 \cdot 4}$

$x = \dfrac{17 \pm \sqrt{337}}{8}$

The solutions are $\dfrac{17 + \sqrt{337}}{8} \approx 4.4$ and $\dfrac{17 - \sqrt{337}}{8} \approx -0.2$.

29. $a^{-t} = b \Rightarrow -t = \log_a b$

31. $\ln 24 = 3.1781 \Rightarrow e^{3.1781} = 24$

33. Plot the given points.

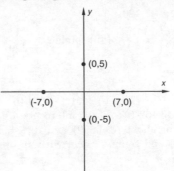

From the location of these points, we see that the ellipse that contains them is centered at the origin with $a = 7$ and $b = 5$. We write the equation of the ellipse:

$$\frac{x^2}{49} + \frac{y^2}{25} = 1$$

35. We have a vertical ellipse centered at the origin with $a = 6/2$, or 3, and $b = 10/2$, or 5. Then the equation is $\dfrac{x^2}{3^2} + \dfrac{y^2}{5^2} = 1$, or $\dfrac{x^2}{9} + \dfrac{y^2}{25} = 1$.

Chapter 9 Mid-Chapter Review

1. True; $y - x^2 = 5$, or $y = x^2 + 5$, is of the form $y = ax^2 + bx + c$, $a > 0$, so the graph is a parabola that opens up.

2. True; see page 771 in the text.

3. True; the center of the ellipse is $(1, 4)$.

4. False; the center of the circle is $(1, 4)$.

5. a)　$d = \sqrt{(x_2 - x_1)^2 + (y_2 - y_1)^2}$

$= \sqrt{[4 - (-6)]^2 + (-1 - 2)^2} = \sqrt{10^2 + (-3)^2}$

$= \sqrt{100 + 9} = \sqrt{109} \approx 10.440$

　　b)　$\left(\dfrac{x_1 + x_2}{2}, \dfrac{y_1 + y_2}{2} \right) = \left(\dfrac{-6 + 4}{2}, \dfrac{2 + (-1)}{2} \right) =$

$\left(\dfrac{-2}{2}, \dfrac{1}{2} \right) = \left(-1, \dfrac{1}{2} \right)$

6.　　　　$x^2 + y^2 - 20x + 4y + 79 = 0$

　　　　　$x^2 - 20x + y^2 + 4y = -79$

$x^2 - 20x + 100 + y^2 + 4y + 4 = -79 + 100 + 4$

　　　　　　$(x - 10)^2 + (y + 2)^2 = 25$

　　　　　$(x - 10)^2 + (y - (-2))^2 = 5^2$

Center: $(10, -2)$; radius: 5

7.　$d = \sqrt{(2 - 5)^2 + [-9 - (-6)]^2}$

$= \sqrt{(-3)^2 + (-3)^2}$

$= \sqrt{9 + 9} = \sqrt{18}$

$= \sqrt{9 \cdot 2} = 3\sqrt{2} \approx 4.243$

8. $d = \sqrt{(-8 - 2.3)^2 + (4.2 - 8)^2}$

$= \sqrt{(-10.3)^2 + (3.8)^2}$

$= \sqrt{106.09 + 14.44}$

$= \sqrt{120.53} \approx 10.979$

9. $d = \sqrt{(-\sqrt{5} - 0)^2 + (0 - \sqrt{6})^2}$

$= \sqrt{(-\sqrt{5})^2 + (-\sqrt{6})^2}$

$= \sqrt{5 + 6}$

$= \sqrt{11} \approx 3.317$

10. $\left(\dfrac{-11 + (-8)}{2}, \dfrac{3 + 12}{2}\right) = \left(-\dfrac{19}{2}, \dfrac{15}{2}\right)$

11. $\left(\dfrac{-\dfrac{5}{6} + \dfrac{1}{2}}{2}, \dfrac{\dfrac{1}{3} + \dfrac{5}{12}}{2}\right) = \left(\dfrac{-\dfrac{5}{6} + \dfrac{3}{6}}{2}, \dfrac{\dfrac{4}{12} + \dfrac{5}{12}}{2}\right) =$

$\left(\dfrac{-\dfrac{2}{6}}{2}, \dfrac{\dfrac{9}{12}}{2}\right) = \left(-\dfrac{2}{6} \cdot \dfrac{1}{2}, \dfrac{9}{12} \cdot \dfrac{1}{2}\right) = \left(-\dfrac{1}{6}, \dfrac{3}{8}\right)$

12. $\left(\dfrac{7.2 + (-10.2)}{2}, \dfrac{-4.6 + (-3.2)}{2}\right) = \left(\dfrac{-3}{2}, \dfrac{-7.8}{2}\right) =$

$(-1.5, -3.9)$

13. $x^2 + y^2 = 121$

$(x - 0)^2 + (y - 0)^2 = 11^2$

Center: $(0, 0)$; radius: 11

14. $(x - 13)^2 + (y + 9)^2 = 109$

$(x - 13)^2 + (y - (-9))^2 = (\sqrt{109})^2$

Center: $(13, -9)$; radius: $\sqrt{109}$

15. $x^2 + (y - 5)^2 = 14$

$(x - 0)^2 + (y - 5)^2 = (\sqrt{14})^2$

Center: $(0, 5)$; radius: $\sqrt{14}$

16. $x^2 + y^2 + 6x - 14y + 42 = 0$

$x^2 + 6x + y^2 - 14y = -42$

$x^2 + 6x + 9 + y^2 - 14y + 49 = -42 + 9 + 49$

$(x + 3)^2 + (y - 7)^2 = 16$

$(x - (-3))^2 + (y - 7)^2 = 4^2$

Center: $(-3, 7)$; radius: 4

17. $(x - h)^2 + (y - k)^2 = r^2$

$(x - 0)^2 + (y - 0)^2 = 1^2$

$x^2 + y^2 = 1$

18. $(x - h)^2 + (y - k)^2 = r^2$

$\left(x - \left(-\dfrac{1}{2}\right)\right)^2 + \left(y - \dfrac{3}{4}\right)^2 = \left(\dfrac{9}{2}\right)^2$

$\left(x + \dfrac{1}{2}\right)^2 + \left(y - \dfrac{3}{4}\right)^2 = \dfrac{81}{4}$

19. $(x - h)^2 + (y - k)^2 = r^2$

$(x - (-8))^2 + (y - 6)^2 = (\sqrt{17})^2$

$(x + 8)^2 + (y - 6)^2 = 17$

20. $(x - h)^2 + (y - k)^2 = r^2$

$(x - 3)^2 + (y - (-5))^2 = (2\sqrt{5})^2$

$(x - 3)^2 + (y + 5)^2 = 20$

21. $\dfrac{x^2}{4} + \dfrac{y^2}{36} = 1$

$\dfrac{x^2}{2^2} + \dfrac{y^2}{6^2} = 1$

This is an equation of an ellipse. The x-intercepts are $(-2, 0)$ and $(2, 0)$, and the y-intercepts are $(0, -6)$ and $(0, 6)$. We plot these points and connect them with an oval-shaped curve.

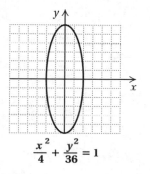

$$\dfrac{x^2}{4} + \dfrac{y^2}{36} = 1$$

22. $y = x^2 + 2x - 1$

This is an equation of a parabola. We use the formula to find the first coordinate of the vertex.

$$x = -\dfrac{b}{2a} = -\dfrac{2}{2 \cdot 1} = -1$$

When $x = -1$, $y = (-1)^2 + 2(-1) - 1 = 1 - 2 - 1 = -2$.

The vertex is $(-1, 2)$; the line of symmetry is $x = -1$. The curve opens up because the coefficient of x^2 is positive.

x	y
-1	2
-2	-1
0	-1
-3	2
1	2

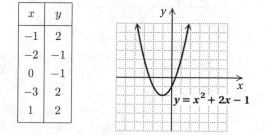

$$y = x^2 + 2x - 1$$

23. $(x-1)^2 + (y+2)^2 = 9$

$(x-1)^2 + (y-(-2))^2 = 3^2$

This is an equation of a circle. The graph is a circle with center $(1, -2)$ and radius 3.

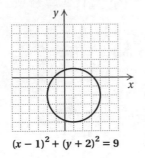

$$(x-1)^2 + (y+2)^2 = 9$$

24. $x = y^2 - 2$

$x = (y-0)^2 - 2$

This is an equation of a parabola. The vertex is $(-2, 0)$; the line of symmetry is $y = 0$. The parabola opens to the right.

x	y
-2	0
-1	1
-1	-1
2	2
2	-2

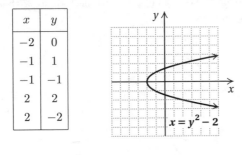

$$x = y^2 - 2$$

25. $x^2 + y^2 = \dfrac{9}{4}$

$x^2 + y^2 = \left(\dfrac{3}{2}\right)^2$

This is an equation of a circle with center $(0, 0)$ and radius $\dfrac{3}{2}$.

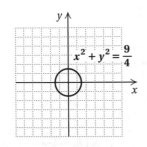

$$x^2 + y^2 = \frac{9}{4}$$

26. $\dfrac{(x-1)^2}{4} + \dfrac{(y+3)^2}{9} = 1$

$\dfrac{(x-1)^2}{2^2} + \dfrac{(y-(-3))^2}{3^2} = 1$

This is an equation of an ellipse with center $(1, -3)$. Note that $a = 2$ and $b = 3$. We locate the center and then plot the points $(1+2, -3)$, $(1-2, -3)$, $(1, -3+3)$, and $(1, -3-3)$, or $(3, -3)$, $(-1, -3)$, $(1, 0)$, and $(1, -6)$. Connect these points with an oval-shaped curve.

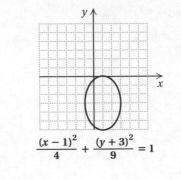

$$\frac{(x-1)^2}{4} + \frac{(y+3)^2}{9} = 1$$

27. $\dfrac{x^2}{16} + \dfrac{y^2}{1} = 1$

$\dfrac{x^2}{4^2} + \dfrac{y^2}{1^2} = 1$

This is an equation of an ellipse with center $(0, 0)$. The x-intercepts are $(-4, 0)$ and $(4, 0)$, and the y-intercepts are $(0, -1)$ and $(0, 1)$. We connect these points with an oval-shaped curve.

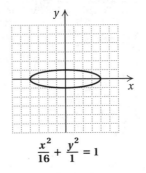

$$\frac{x^2}{16} + \frac{y^2}{1} = 1$$

28. $y = 6 - x^2$

$y = -1 \cdot (x-0)^2 + 6$

This is an equation of a parabola with vertex $(0, 6)$. The parabola opens down because the coefficient of x^2 is negative.

x	y
0	6
-1	5
1	5
-3	-3
3	-3

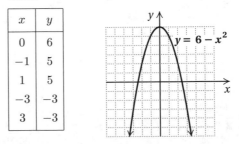

29. One method is to graph $y = ax^2 + bx + c$ and then use the DrawInv feature to graph the inverse relation, $x = ay^2 + by + c$. Another method is to use the quadratic formula to solve $x = ay^2 + by + c$, or $ay^2 + by + c - x = 0$. The solutions are $\dfrac{-b \pm \sqrt{b^2 - 4a(c-x)}}{2a}$. Then graph

$$y_1 = \frac{-b + \sqrt{b^2 - 4a(c-x)}}{2a} \text{ and}$$

$$y_2 = \frac{-b - \sqrt{b^2 - 4a(c-x)}}{2a} \text{ on the same screen.}$$

30. No; a circle is defined to be the set of points in a plane that are a fixed distance from the center. Thus, unless $r = 0$ and the "circle" is one point, the center is not part of the circle.

31. Bank shots originating at one focus (the tiny dot) are deflected to the other focus (the hole).

32. a)

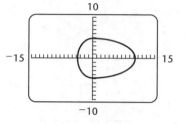

b) Some other factors are the wind speed, amount of rainfall in the preceding months, and the composition of the forest.

Exercise Set 9.3

RC1. B

RC3. A

RC5. F

1. $\dfrac{y^2}{9} - \dfrac{x^2}{9} = 1$

$\dfrac{y^2}{3^2} - \dfrac{x^2}{3^2} = 1$

$a = 3$ and $b - 3$, so the asymptotes are $y - \dfrac{3}{3}x$ and $y = -\dfrac{3}{3}x$, or $y = x$ and $y = -x$.

Replacing x with 0 and solving for y, we get $y = \pm 3$, so the intercepts are $(0, 3)$ and $(0, -3)$.

We plot the intercepts and draw smooth curves through them that approach the asymptotes.

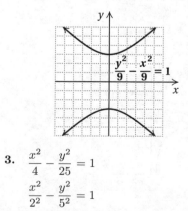

3. $\dfrac{x^2}{4} - \dfrac{y^2}{25} = 1$

$\dfrac{x^2}{2^2} - \dfrac{y^2}{5^2} = 1$

$a = 2$ and $b = 5$, so the asymptotes are $y = \dfrac{5}{2}x$ and $y = -\dfrac{5}{2}x$.

Replacing y with 0 and solving for x, we get $x = \pm 2$, so the intercepts are $(2, 0)$ and $(-2, 0)$.

We plot the intercepts and draw smooth curves through them that approach the asymptotes.

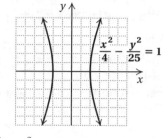

5. $\dfrac{y^2}{36} - \dfrac{x^2}{9} = 1$

$\dfrac{y^2}{6^2} - \dfrac{x^2}{3^2} = 1$

$a = 3$ and $b = 6$, so the asymptotes are $y = \dfrac{6}{3}x$ and $y = -\dfrac{6}{3}x$, or $y = 2x$ and $y = -2x$.

Replacing x with 0 and solving for y, we get $y = \pm 6$, so the intercepts are $(0, 6)$ and $(0, -6)$.

We plot the intercepts and draw smooth curves through them that approach the asymptotes.

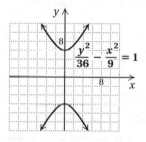

7. $y^2 - x^2 = 25$

$\dfrac{y^2}{25} - \dfrac{x^2}{25} = 1$

$\dfrac{y^2}{5^2} - \dfrac{x^2}{5^2} = 1$

$a = 5$ and $b = 5$, so the asymptotes are $y = \dfrac{5}{5}x$ and $y = -\dfrac{5}{5}x$, or $y = x$ and $y = -x$.

Replacing x with 0 and solving for y, we get $y = \pm 5$, so the intercepts are $(0, 5)$ and $(0, -5)$.

We plot the intercepts and draw smooth curves through them that approach the asymptotes.

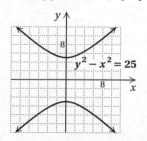

9. $x^2 = 1 + y^2$

$x^2 - y^2 = 1$

$\dfrac{x^2}{1^2} - \dfrac{y^2}{1^2} = 1$

$a = 1$ and $b = 1$, so the asymptotes are $y = \dfrac{1}{1}x$ and $y = -\dfrac{1}{1}x$, or $y = x$ and $y = -x$.

Replacing y with 0 and solving for x, we get $x = \pm 1$, so the intercepts are $(1, 0)$ and $(-1, 0)$.

We plot the intercepts and draw smooth curves through them that approach the asymptotes.

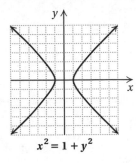

$x^2 = 1 + y^2$

11. $25x^2 - 16y^2 = 400$

$\dfrac{x^2}{16} - \dfrac{y^2}{25} = 1$ Multiplying by $\dfrac{1}{400}$

$\dfrac{x^2}{4^2} - \dfrac{y^2}{5^2} = 1$

$a = 4$ and $b = 5$, so the asymptotes are $y = \dfrac{5}{4}x$ and $y = -\dfrac{5}{4}x$.

Replacing y with 0 and solving for x, we get $x = \pm 4$, so the intercepts are $(4, 0)$ and $(-4, 0)$.

We plot the intercepts and draw smooth curves through them that approach the asymptotes.

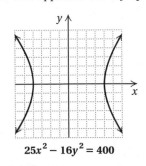

$25x^2 - 16y^2 = 400$

13. $xy = -4$

$y = -\dfrac{4}{x}$ Solving for y

We find some solutions, keeping the results in a table.

x	y
$\dfrac{1}{2}$	-8
1	-4
2	-2
4	-1
8	$-\dfrac{1}{2}$
$-\dfrac{1}{2}$	8
-1	4
-2	2
-8	$\dfrac{1}{2}$

Note that we cannot use 0 for x. The x-axis and the y-axis are the asymptotes.

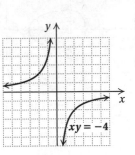

$xy = -4$

15. $xy = 3$

$y = \dfrac{3}{x}$ Solving for y

We find some solutions, keeping the results in a table.

x	y
$\dfrac{1}{3}$	9
$\dfrac{1}{2}$	6
1	3
3	1
6	$\dfrac{1}{2}$
9	$\dfrac{1}{3}$
$-\dfrac{1}{3}$	-9
$-\dfrac{1}{2}$	-6
-1	-3
-3	-1
-6	$-\dfrac{1}{2}$
-9	$-\dfrac{1}{3}$

Note that we cannot use 0 for x. The x-axis and the y-axis are the asymptotes.

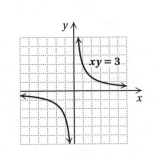

$xy = 3$

17. $xy = -2$

$y = -\dfrac{2}{x}$ Solving for y

x	y
$\dfrac{1}{2}$	-4
1	-2
2	-1
4	$-\dfrac{1}{2}$
$-\dfrac{1}{2}$	4
-1	2
-2	1
-4	$\dfrac{1}{2}$

Note that we cannot use 0 for x. The x-axis and the y-axis are the asymptotes.

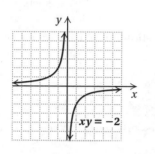

19. $xy = \dfrac{1}{2}$

$y = \dfrac{1}{2x}$ Solving for y

x	y
$\dfrac{1}{8}$	4
$\dfrac{1}{4}$	2
$\dfrac{1}{2}$	1
1	$\dfrac{1}{2}$
2	$\dfrac{1}{4}$
$-\dfrac{1}{8}$	-4
$-\dfrac{1}{4}$	-2
$-\dfrac{1}{2}$	1
1	$\dfrac{1}{2}$
2	$\dfrac{1}{4}$

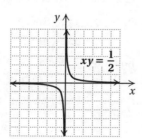

21.
$$3(x-1) - 4(x-2) \le 5 + x$$
$$3x - 3 - 4x + 8 \le 5 + x$$
$$-x + 5 \le 5 + x$$
$$5 \le 5 + 2x$$
$$0 \le 2x$$
$$0 \le x$$

The solution set is $\{x | x \ge 0\}$, or $[0, \infty)$.

23.
$$-4 < 5 - y < 2$$
$$-9 < -y < -3$$
$$9 > y > 3 \quad \text{Reversing the inequality symbol}$$

The solution set is $\{y | 3 < y < 9\}$, or $(3, 9)$.

25.
$$2x^2 + 5x = 3$$
$$2x^2 + 5x - 3 = 0$$
$$(2x - 1)(x + 3) = 0$$
$$2x - 1 = 0 \quad or \quad x + 3 = 0$$
$$2x = 1 \quad or \quad\quad x = -3$$
$$x = \frac{1}{2} \quad or \quad\quad x = -3$$

The solutions are $\dfrac{1}{2}$ and -3.

27.
$$\sqrt{x + 5} = x - 1$$
$$(\sqrt{x + 5})^2 = (x - 1)^2$$
$$x + 5 = x^2 - 2x + 1$$
$$0 = x^2 - 3x - 4$$
$$0 = (x - 4)(x + 1)$$
$$x - 4 = 0 \quad or \quad x + 1 = 0$$
$$x = 4 \quad or \quad\quad x = -1$$

The number 4 checks but -1 does not. The solution is 4.

29. Left to the student

31.
$$x^2 + y^2 - 10x + 8y - 40 = 0$$
$$x^2 - 10x + y^2 + 8y = 40$$
$$(x^2 - 10x + 25) + (y^2 + 8y + 16) = 40 + 25 + 16$$
$$(x - 5)^2 + (y + 4)^2 = 81$$

The graph is a circle.

33.
$$1 - 3y = 2y^2 - x$$
$$x = 2y^2 + 3y - 1$$

The graph is a parabola.

35.
$$4x^2 + 25y^2 - 8x - 100y + 4 = 0$$
$$4x^2 - 8x + 25y^2 - 100y = -4$$
$$4(x^2 - 2x + 1) + 25(y^2 - 4y + 4) = -4 + 4 + 100$$
$$4(x - 1)^2 + 25(y - 2)^2 = 100$$
$$\frac{(x - 1)^2}{25} + \frac{(y - 2)^2}{4} = 1$$

The graph is an ellipse.

37. $x^2 + y^2 = 8$

The graph is a circle.

39. $x - \dfrac{3}{y} = 0$

$xy - 3 = 0$

$xy = 3$

The graph is a hyperbola.

41. $3x^2 + 5y^2 + x^2 = y^2 + 49$

$4x^2 + 4y^2 = 49$

$x^2 + y^2 = \dfrac{49}{4}$

The graph is a circle.

Exercise Set 9.4

RC1. True; see page 791 in the text.

RC3. False; see Example 4 on page 792 in the text.

1. $x^2 + y^2 = 100,$ (1)

   $y - x = 2$ (2)

First solve Equation (2) for y.

   $y = x + 2$ (3)

Then substitute $x + 2$ for y in Equation (1) and solve for x.

$$x^2 + y^2 = 100$$
$$x^2 + (x + 2)^2 = 100$$
$$x^2 + x^2 + 4x + 4 = 100$$
$$2x^2 + 4x - 96 = 0$$
$$x^2 + 2x - 48 = 0 \qquad \text{Multiplying by } \dfrac{1}{2}$$
$$(x + 8)(x - 6) = 0$$

$x + 8 = 0$ *or* $x - 6 = 0$ Principle of zero products

     $x = -8$ *or* $x = 6$

Now substitute these numbers into Equation (3) and solve for y.

$y = -8 + 2 = -6$

$y = 6 + 2 = 8$

The pairs $(-8, -6)$ and $(6, 8)$ check, so they are the solutions.

3. $9x^2 + 4y^2 = 36,$ (1)

  $3x + 2y = 6$ ·   (2)

First solve Equation (2) for x.

  $3x = 6 - 2y$

   $x = 2 - \dfrac{2}{3}y$ (3)

Then substitute $2 - \dfrac{2}{3}y$ for x in Equation (1) and solve for y.

$$9x^2 + 4y^2 = 36$$
$$9\left(2 - \dfrac{2}{3}y\right)^2 + 4y^2 = 36$$
$$9\left(4 - \dfrac{8}{3}y + \dfrac{4}{9}y^2\right) + 4y^2 = 36$$
$$36 - 24y + 4y^2 + 4y^2 = 36$$
$$8y^2 - 24y = 0$$
$$y^2 - 3y = 0$$
$$y(y - 3) = 0$$

$y = 0$ *or* $y - 3 = 0$ Principle of zero products

$y = 0$ *or* $y = 3$

Now substitute these numbers into Equation (3) and solve for x.

$x = 2 - \dfrac{2}{3} \cdot 0 = 2$

$x = 2 - \dfrac{2}{3} \cdot 3 = 2 - 2 = 0$

The pairs $(2, 0)$ and $(0, 3)$ check, so they are the solutions.

5. $y^2 = x + 3,$ (1)

  $2y = x + 4$ (2)

First solve Equation (2) for x.

  $2y - 4 = x$ (3)

Then substitute $2y - 4$ for x in Equation (1) and solve for y.

$$y^2 = x + 3$$
$$y^2 = (2y - 4) + 3$$
$$y^2 = 2y - 1$$
$$y^2 - 2y + 1 = 0$$
$$(y - 1)(y - 1) = 0$$

$y - 1 = 0$ *or* $y - 1 = 0$

  $y = 1$ *or* $y = 1$

Now substitute 1 for y in Equation (3) and solve for x.

 $2 \cdot 1 - 4 = x$

   $-2 = x$

The pair $(-2, 1)$ checks. It is the solution.

7. $x^2 - xy + 3y^2 = 27,$ (1)

  $x - y = 2$ (2)

First solve Equation (2) for y.

  $x - 2 = y$ (3)

Then substitute $x - 2$ for y in Equation (1) and solve for x.

$$x^2 - xy + 3y^2 = 27$$
$$x^2 - x(x - 2) + 3(x - 2)^2 = 27$$
$$x^2 - x^2 + 2x + 3x^2 - 12x + 12 = 27$$
$$3x^2 - 10x - 15 = 0$$
$$x = \dfrac{-(-10) \pm \sqrt{(-10)^2 - 4(3)(-15)}}{2 \cdot 3}$$

$$x = \frac{10 \pm \sqrt{100 + 180}}{6} = \frac{10 \pm \sqrt{280}}{6}$$

$$x = \frac{10 \pm 2\sqrt{70}}{6} = \frac{5 \pm \sqrt{70}}{3}$$

Now substitute these numbers in Equation (3) and solve for y.

$$y = \frac{5 + \sqrt{70}}{3} - 2 = \frac{-1 + \sqrt{70}}{3}$$

$$y = \frac{5 - \sqrt{70}}{3} - 2 = \frac{-1 - \sqrt{70}}{3}$$

The pairs $\left(\frac{5 + \sqrt{70}}{3}, \frac{-1 + \sqrt{70}}{3}\right)$ and

$\left(\frac{5 - \sqrt{70}}{3}, \frac{-1 - \sqrt{70}}{3}\right)$ check, so they are the solutions.

9. $x^2 - xy + 3y^2 = 5,$ (1)

$\quad x - y = 2$ (2)

First solve Equation (2) for y.

$$x - 2 = y \qquad (3)$$

Then substitute $x - 2$ for y in Equation (1) and solve for x.

$$x^2 - xy + 3y^2 = 5$$
$$x^2 - x(x - 2) + 3(x - 2)^2 = 5$$
$$x^2 - x^2 + 2x + 3x^2 - 12x + 12 = 5$$
$$3x^2 - 10x + 7 = 0$$
$$(3x - 7)(x - 1) = 0$$

$3x - 7 = 0 \quad or \quad x - 1 = 0$

$\quad x = \dfrac{7}{3} \quad or \qquad x = 1$

Now substitute these numbers in Equation (3) and solve for y.

$$y = \frac{7}{3} - 2 = \frac{1}{3}$$
$$y = 1 - 2 = -1$$

The pairs $\left(\frac{7}{3}, \frac{1}{3}\right)$ and $(1, -1)$ check, so they are the solutions.

11. $a + b = -6,$ (1)

$\quad ab = -7$ (2)

First solve Equation (1) for a.

$$a = -b - 6 \qquad (3)$$

Then substitute $-b - 6$ for a in Equation (2) and solve for b.

$$(-b - 6)b = -7$$
$$-b^2 - 6b = -7$$
$$0 = b^2 + 6b - 7$$
$$0 = (b + 7)(b - 1)$$

$b + 7 = 0 \quad or \quad b - 1 = 0$

$\quad b = -7 \quad or \qquad b = 1$

Now substitute these numbers in Equation (3) and solve for a.

$$a = -(-7) - 6 = 1$$
$$a = -1 - 6 = -7$$

The pairs $(-7, 1)$ and $(1, -7)$ check, so they are the solutions.

13. $2a + b = 1,$ (1)

$\quad b = 4 - a^2$ (2)

Equation (2) is already solved for b. Substitute $4 - a^2$ for b in Equation (1) and solve for a.

$$2a + 4 - a^2 = 1$$
$$0 = a^2 - 2a - 3$$
$$0 = (a - 3)(a + 1)$$

$a - 3 = 0 \quad or \quad a + 1 = 0$

$\quad a = 3 \quad or \qquad a = -1$

Substitute these numbers in Equation (2) and solve for b.

$$b = 4 - 3^2 = -5$$
$$b = 4 - (-1)^2 = 3$$

The pairs $(3, -5)$ and $(-1, 3)$ check.

15. $x^2 + y^2 = 5,$ (1)

$\quad x - y = 8$ (2)

First solve Equation (2) for x.

$$x = y + 8 \qquad (3)$$

Then substitute $y + 8$ for x in Equation (1) and solve for y.

$$(y + 8)^2 + y^2 = 5$$
$$y^2 + 16y + 64 + y^2 = 5$$
$$2y^2 + 16y + 59 = 0$$

$$y = \frac{-16 \pm \sqrt{(16)^2 - 4(2)(59)}}{2 \cdot 2}$$

$$y = \frac{-16 \pm \sqrt{-216}}{4} = \frac{-16 \pm 6i\sqrt{6}}{4}$$

$$y = \frac{-8 \pm 3i\sqrt{6}}{2}, \text{ or } -4 \pm \frac{3}{2}i\sqrt{6}$$

Now substitute these numbers in Equation (3) and solve for x.

$$x = -4 + \frac{3}{2}i\sqrt{6} + 8 = 4 + \frac{3}{2}i\sqrt{6}, \text{ or } \frac{8 + 3i\sqrt{6}}{2}$$

$$x = -4 - \frac{3}{2}i\sqrt{6} + 8 = 4 - \frac{3}{2}i\sqrt{6}, \text{ or } \frac{8 - 3i\sqrt{6}}{2}$$

The pairs $\left(4 + \frac{3}{2}i\sqrt{6}, -4 + \frac{3}{2}i\sqrt{6}\right)$ and

$\left(4 - \frac{3}{2}i\sqrt{6}, -4 - \frac{3}{2}i\sqrt{6}\right)$, or $\left(\frac{8 + 3i\sqrt{6}}{2}, \frac{-8 + 3i\sqrt{6}}{2}\right)$ and

$\left(\frac{8 - 3i\sqrt{6}}{2}, \frac{-8 - 3i\sqrt{6}}{2}\right)$ check.

17. $x^2 + y^2 = 25,$ (1)

$\quad y^2 = x + 5$ (2)

We substitute $x + 5$ for y^2 in Equation (1) and solve for x.

$$x^2 + y^2 = 25$$
$$x^2 + (x + 5) = 25$$
$$x^2 + x - 20 = 0$$
$$(x + 5)(x - 4) = 0$$
$$x + 5 = 0 \quad or \quad x - 4 = 0$$
$$x = -5 \quad or \quad x = 4$$

We substitute these numbers for x in either Equation (1) or Equation (2) and solve for y. Here we use Equation (2).

$y^2 = -5 + 5 = 0$ and $y = 0$.

$y^2 = 4 + 5 = 9$ and $y = \pm 3$.

The pairs $(-5, 0)$, $(4, 3)$ and $(4, -3)$ check.

19. $x^2 + y^2 = 9,$ (1)

       $x^2 - y^2 = 9$ (2)

Here we use the elimination method.

$$\begin{array}{ll} x^2 + y^2 = \;\; 9 & (1) \\ \underline{x^2 - y^2 = \;\; 9} & (2) \\ 2x^2 \qquad = 18 & \text{Adding} \\ x^2 = \;\; 9 & \\ x = \pm 3 & \end{array}$$

If $x = 3$, $x^2 = 9$, and if $x = -3$, $x^2 = 9$, so substituting 3 or -3 in Equation (1) gives us

$$x^2 + y^2 = 9$$
$$9 + y^2 = 9$$
$$y^2 = 0$$
$$y = 0.$$

The pairs $(3, 0)$ and $(-3, 0)$ check.

21. $x^2 + y^2 = 20,$ (1)

       $xy = 8$ (2)

First we solve Equation (2) for y.

$$y = \frac{8}{x}$$

Then substitute $\dfrac{8}{x}$ for y in Equation (1) and solve for x.

$$x^2 + \left(\frac{8}{x}\right)^2 = 20$$
$$x^2 + \frac{64}{x^2} = 20$$
$$x^4 + 64 = 20x^2 \quad \text{Multiplying by } x^2$$
$$x^4 - 20x^2 + 64 = 0$$
$$u^2 - 20u + 64 = 0 \quad \text{Letting } u = x^2$$
$$(u - 4)(u - 16) = 0$$
$$u - 4 = 0 \quad or \quad u - 16 = 0$$
$$u = 4 \quad or \qquad u = 16$$

We now substitute x^2 for u and solve for x.

$$x^2 = 4 \quad or \quad x^2 = 16$$
$$x = \pm 2 \quad or \quad x = \pm 4$$

Since $y = \dfrac{8}{x}$, if $x = 2$, $y = 4$; if $x = -2$, $y = -4$; if $x = 4$, $y = 2$; if $x = -4$, $y = -2$. The pairs $(2, 4)$, $(-2, -4)$, $(4, 2)$, and $(-4, -2)$ check. They are the solutions.

23. $x^2 + y^2 = 13,$ (1)

       $xy = 6$ (2)

First we solve Equation (2) for y.

$$y = \frac{6}{x}$$

Then substitute $\dfrac{6}{x}$ for y in Equation (1) and solve for x.

$$x^2 + \left(\frac{6}{x}\right)^2 = 13$$
$$x^2 + \frac{36}{x^2} = 13$$
$$x^4 + 36 = 13x^2 \quad \text{Multiplying by } x^2$$
$$x^4 - 13x^2 + 36 = 0$$
$$u^2 - 13u + 36 = 0 \quad \text{Letting } u = x^2$$
$$(u - 9)(u - 4) = 0$$
$$u - 9 = 0 \quad or \quad u - 4 = 0$$
$$u = 9 \quad or \qquad u = 4$$

We now substitute x^2 for u and solve for x.

$$x^2 = 9 \quad or \quad x^2 = 4$$
$$x = \pm 3 \quad or \quad x = \pm 2$$

Since $y = \dfrac{6}{x}$, if $x = 3$, $y = 2$; if $x = -3$, $y = -2$; if $x = 2$, $y = 3$; if $x = -2$, $y = -3$. The pairs $(3, 2)$, $(-3, -2)$, $(2, 3)$, and $(-2, -3)$ check. They are the solutions.

25. $2xy + 3y^2 = 7,$ (1)

       $3xy - 2y^2 = 4$ (2)

$$\begin{array}{ll} 6xy + \;\; 9y^2 = 21 & \text{Multiplying (1) by 3} \\ \underline{-6xy + \;\; 4y^2 = -8} & \text{Multiplying (2) by } -2 \\ 13y^2 = 13 & \\ y^2 = 1 & \\ y = \pm 1 & \end{array}$$

Substitute for y in Equation (1) and solve for x.

When $y = 1$: $2 \cdot x \cdot 1 + 3 \cdot 1^2 = 7$

                      $2x = 4$

                       $x = 2$

When $y = -1$: $2 \cdot x \cdot (-1) + 3 \cdot (-1)^2 = 7$

                      $-2x = 4$

                       $x = -2$

The pairs $(2, 1)$ and $(-2, -1)$ check. They are the solutions.

27. $4a^2 - 25b^2 = 0,$ (1)

    $2a^2 - 10b^2 = 3b + 4$ (2)

$$4a^2 - 25b^2 = 0$$
$$\underline{-4a^2 + 20b^2 = -6b - 8} \quad \text{Multiplying (2) by } -2$$
$$-5b^2 = -6b - 8$$
$$0 = 5b^2 - 6b - 8$$
$$0 = (5b + 4)(b - 2)$$
$$5b + 4 = 0 \quad or \quad b - 2 = 0$$
$$b = -\frac{4}{5} \quad or \quad b = 2$$

Substitute for b in Equation (1) and solve for a.

When $b = -\frac{4}{5}$: $4a^2 - 25\left(-\frac{4}{5}\right)^2 = 0$

$$4a^2 = 16$$
$$a^2 = 4$$
$$a = \pm 2$$

When $b = 2$: $4a^2 - 25(2)^2 = 0$

$$4a^2 = 100$$
$$a^2 = 25$$
$$a = \pm 5$$

The pairs $\left(2, -\frac{4}{5}\right)$, $\left(-2, -\frac{4}{5}\right)$, $(5, 2)$ and $(-5, 2)$ check. They are the solutions.

29. $ab - b^2 = -4,$ (1)

    $ab - 2b^2 = -6$ (2)

$$ab - b^2 = -4$$
$$\underline{-ab + 2b^2 = \quad 6} \quad \text{Multiplying (2) by } -1$$
$$b^2 = \quad 2$$
$$b = \pm\sqrt{2}$$

Substitute for b in Equation (1) and solve for a.

When $b = \sqrt{2}$: $a(\sqrt{2}) - (\sqrt{2})^2 = -4$

$$a\sqrt{2} = -2$$
$$a = -\frac{2}{\sqrt{2}} = -\sqrt{2}$$

When $b = -\sqrt{2}$: $a(-\sqrt{2}) - (-\sqrt{2})^2 = -4$

$$-a\sqrt{2} = -2$$
$$a = \frac{2}{\sqrt{2}} = \sqrt{2}$$

The pairs $(-\sqrt{2}, \sqrt{2})$ and $(\sqrt{2}, -\sqrt{2})$ check. They are the solutions.

31. $x^2 + y^2 = 25,$ (1)

    $9x^2 + 4y^2 = 36$ (2)

$$-4x^2 - 4y^2 = -100 \quad \text{Multiplying (1) by } -4$$
$$\underline{9x^2 + 4y^2 = \quad 36}$$
$$5x^2 = -64$$

$$x^2 = -\frac{64}{5}$$
$$x = \pm\sqrt{-\frac{64}{5}} = \pm\frac{8i}{\sqrt{5}}$$
$$x = \pm\frac{8i\sqrt{5}}{5} \quad \text{Rationalizing the denominator}$$

Substituting $\frac{8i\sqrt{5}}{5}$ or $-\frac{8i\sqrt{5}}{5}$ for x in Equation (1) and solving for y gives us

$$-\frac{64}{5} + y^2 = 25$$
$$y^2 = \frac{189}{5}$$
$$y = \pm\sqrt{\frac{189}{5}} = \pm 3\sqrt{\frac{21}{5}}$$
$$y = \pm\frac{3\sqrt{105}}{5}. \quad \text{Rationalizing the denominator}$$

The pairs $\left(\frac{8i\sqrt{5}}{5}, \frac{3\sqrt{105}}{5}\right)$, $\left(-\frac{8i\sqrt{5}}{5}, \frac{3\sqrt{105}}{5}\right)$, $\left(\frac{8i\sqrt{5}}{5}, -\frac{3\sqrt{105}}{5}\right)$, and $\left(-\frac{8i\sqrt{5}}{5}, -\frac{3\sqrt{105}}{5}\right)$, check. They are the solutions.

33. *Familiarize.* Let l and w represent the length and the width of the glass, in centimeters, respectively.

Translate. The perimeter is 28 cm, so we have $2l + 2w = 28$. The length of a diagonal is 10 cm, so we use the Pythagorean theorem to write another equation: $l^2 + w^2 = 10^2$, or $l^2 + w^2 = 100$.

Solve. We solve the system of equations.

$$2l + 2w = 28$$
$$l^2 + w^2 = 100 \quad (2)$$

First we solve Equation (1) for l.

$$2l + 2w = 28$$
$$2l = 28 - 2w$$
$$l = 14 - w$$

Now substitute $14 - w$ for l in Equation (2) and solve for w.

$$(14 - w)^2 + w^2 = 100$$
$$196 - 28w + w^2 + w^2 = 100$$
$$2w^2 - 28w + 196 = 100$$
$$2w^2 - 28w + 96 = 0$$
$$w^2 - 14w + 48 = 0 \quad \text{Dividing by 2}$$
$$(w - 8)(w - 6) = 0$$
$$w - 8 = 0 \quad or \quad w - 6 = 0$$
$$w = 8 \quad or \quad w = 6$$

If $w = 8$, then $l = 14 - 8 = 6$. If $w = 6$, then $l = 14 - 6 = 8$. Since length is usually considered to be longer than width, we have the solution $l = 8$ and $w = 6$, or $(8, 6)$.

Check. If $l = 8$ and $w = 6$, the perimeter is $2 \cdot 8 + 2 \cdot 6$, or 28. The length of a diagonal is $8^2 + 6^2 = 64 + 36 = 100 = 10^2$. The answer checks.

State. The length is 8 cm, and the width is 6 cm.

35. *Familiarize.* We first make a drawing. We let l = the length and w = the width of the rectangle.

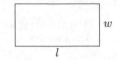

Translate.

   Area: $lw = 14$

   Perimeter: $2l + 2w = 18$, or $l + w = 9$

Solve. We solve the system.

Solve the second equation for l: $l = 9 - w$

Substitute $9 - w$ for l in the first equation and solve for w.

$$(9 - w)w = 14$$
$$9w - w^2 = 14$$
$$0 = w^2 - 9w + 14$$
$$0 = (w - 7)(w - 2)$$

$w - 7 = 0$ *or* $w - 2 = 0$

$\ \ \ \ w = 7$ *or* $\ \ \ \ \ w = 2$

If $w = 7$, then $l = 9 - 7$, or 2. If $w = 2$, then $l = 9 - 2$, or 7. Since length is usually considered to be longer than the width, we have the solution $l = 7$ and $w = 2$, or $(7, 2)$.

Check. If $l = 7$ and $w = 2$, the area is $7 \cdot 2$, or 14. The perimeter is $2 \cdot 7 + 2 \cdot 2$, or 18. The numbers check.

State. The length is 7 in., and the width is 2 in.

37. *Familiarize.* Let l = the length and w = the width of the rectangle. Then the length of a diagonal is $\sqrt{l^2 + w^2}$.

Translate.

The diagonal   is   1 ft   longer than   the length.

$\ \ \ \ \downarrow\ \ \ \ \ \ \ \ \ \downarrow\ \ \ \downarrow\ \ \ \ \ \ \ \downarrow\ \ \ \ \ \ \ \ \ \ \ \downarrow$

$\sqrt{l^2 + w^2}\ \ \ =\ \ \ 1\ \ \ \ \ +\ \ \ \ \ \ \ \ l$

The diagonal   is   3 ft   longer than   twice the width.

$\ \ \ \ \downarrow\ \ \ \ \ \ \ \ \ \downarrow\ \ \ \downarrow\ \ \ \ \ \ \ \downarrow\ \ \ \ \ \ \ \downarrow$

$\sqrt{l^2 + w^2}\ \ \ =\ \ \ 3\ \ \ \ \ +\ \ \ \ \ 2\cdot\ \ \ \ w$

Solve. We solve the system of equations.

$$\sqrt{l^2 + w^2} = 1 + l, \qquad (1)$$
$$\sqrt{l^2 + w^2} = 3 + 2w \qquad (2)$$

We substitute $3 + 2w$ for $\sqrt{l^2 + w^2}$ in Equation (1) and solve for l.

$$3 + 2w = 1 + l$$
$$2 + 2w = l \qquad (3)$$

Now we substitute $2 + 2w$ for l in Equation (2).

$$\sqrt{(2 + 2w)^2 + w^2} = 3 + 2w$$
$$\sqrt{4 + 8w + 4w^2 + w^2} = 3 + 2w$$
$$\sqrt{5w^2 + 8w + 4} = 3 + 2w$$
$$(\sqrt{5w^2 + 8w + 4})^2 = (3 + 2w)^2$$
$$5w^2 + 8w + 4 = 9 + 12w + 4w^2$$
$$w^2 - 4w - 5 = 0$$
$$(w - 5)(w + 1) = 0$$

$w - 5 = 0$ *or* $w + 1 = 0$

$\ \ \ \ w = 5$ *or* $\ \ \ \ \ \ \ \ w = -1$

Since the width cannot be negative, we consider only 5. We substitute 5 for w in Equation (3) and solve for l.

$$l = 2 + 2 \cdot 5 = 12$$

Check. The length of a diagonal of a 12 ft by 5 ft rectangle is $\sqrt{12^2 + 5^2} = \sqrt{169} = 13$. This is 1 ft longer than the length, 12, and 3 ft longer than twice the width, $2 \cdot 5$, or 10. The numbers check.

State. The length is 12 ft, and the width is 5 ft.

39. *Familiarize.* We first make a drawing. We let l = the length and w = the width.

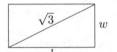

Translate.

Area: $\sqrt{2} = lw$ \qquad (1)

Using the Pythagorean theorem: $l^2 + w^2 = (\sqrt{3})^2$, or

$$l^2 + w^2 = 3 \qquad (2)$$

Solve. We solve the system of equations. First we solve Equation (1) for w.

$$\frac{\sqrt{2}}{l} = w$$

Then we substitute $\dfrac{\sqrt{2}}{l}$ for w in Equation (2) and solve for l.

$$l^2 + \left(\frac{\sqrt{2}}{l}\right)^2 = 3$$
$$l^2 + \frac{2}{l^2} = 3$$
$$l^4 + 2 = 3l^2 \qquad \text{Multiplying by } l^2$$
$$l^4 - 3l^2 + 2 = 0$$
$$u^2 - 3u + 2 = 0 \qquad \text{Letting } u = l^2$$
$$(u - 2)(u - 1) = 0$$

$u - 2 = 0$ \qquad *or* \quad $u - 1 = 0$

$\ \ \ \ u = 2$ \qquad *or* \qquad $u = 1$

$\ \ \ l^2 = 2$ \qquad *or* \qquad $l^2 - 1$ \quad Substituting l^2 for u

$\ \ l = \pm\sqrt{2}$ \quad *or* \qquad $l = \pm 1$

Length cannot be negative, we only need to consider $l = \sqrt{2}$ or $l = 1$. Since $w = \sqrt{2}/l$, if $l = \sqrt{2}$, $w = 1$ and if $l = 1$, $w = \sqrt{2}$. Length is usually considered to be longer than width, so we have the solution $l = \sqrt{2}$ and $w = 1$, or $(\sqrt{2}, 1)$.

Check. If $l = \sqrt{2}$ and $w = 1$, the area is $\sqrt{2} \cdot 1$, or $\sqrt{2}$. Also $(\sqrt{2})^2 + 1^2 = 2 + 1 = 3 = (\sqrt{3})^2$. The numbers check.

State. The length is $\sqrt{2}$ m, and the width is 1 m.

41. Familiarize. We let $x =$ the length of a side of one peanut bed and $y =$ the length of a side of the other peanut bed. Make a drawing.

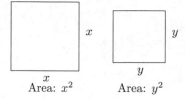

Area: x^2 Area: y^2

Translate.

The sum of the areas is 832 ft^2.

$$x^2 + y^2 = 832$$

The difference of the areas is 320 ft^2.

$$x^2 - y^2 = 320$$

Solve. We solve the system of equations.

$$\begin{array}{rl} x^2 + y^2 = & 832 \\ x^2 - y^2 = & 320 \\ \hline 2x^2 \quad\quad = & 1152 \quad \text{Adding} \\ x^2 = & 576 \\ x = & \pm 24 \end{array}$$

Since length cannot be negative, we consider only $x = 24$. Substitute 24 for x in the first equation and solve for y.

$$24^2 + y^2 = 832$$
$$576 + y^2 = 832$$
$$y^2 = 256$$
$$y = \pm 16$$

Again, we consider only the positive value, 16. The possible solution is $(24, 16)$.

Check. The areas of the beds are 24^2, or 576, and 16^2, or 256. The sum of the areas is $576 + 256$, or 832. The difference of the areas is $576 - 256$, or 320. The values check.

State. The lengths of the beds are 24 ft and 16 ft.

43. Familiarize. Let $l =$ the length and $h =$ the height, in cm.

Translate. Since the ratio of the length to the height is 4 to 3, we have one equation:

$$\frac{l}{h} = \frac{4}{3}$$

The Pythagorean theorem gives us a second equation:

$$l^2 + h^2 = 31^2, \text{ or } l^2 + h^2 = 961$$

We have a system of equations.

$$\frac{l}{h} = \frac{4}{3}, \quad\quad (1)$$
$$l^2 + h^2 = 961 \quad (2)$$

Solve. We solve the system of equations. First we solve Eq. (1) for l:

$$\frac{l}{h} = \frac{4}{3}$$
$$l = \frac{4}{3}h \quad (3)$$

Now substitute $\frac{4}{3}h$ for l in Eq. (2) and solve for h.

$$l^2 + h^2 = 961$$
$$\left(\frac{4}{3}h\right)^2 + h^2 = 961$$
$$\frac{16}{9}h^2 + h^2 = 961$$
$$\frac{25}{9}h^2 = 961$$
$$h^2 = \frac{961 \cdot 9}{25} = \frac{8649}{25}$$
$$h = \frac{93}{5} \quad \text{or} \quad h = -\frac{93}{5}$$

Since the height cannot be negative, we consider only $\frac{93}{5}$. Substitute $\frac{93}{5}$ for h in Eq. (3) and find l.

$$l = \frac{4}{3} \cdot \frac{93}{5} = \frac{124}{5}$$

Check. The ratio of $124/5$ to $93/5$ is $\dfrac{124/5}{93/5} = \dfrac{124}{93} = \dfrac{4}{3}$. Also $\left(\dfrac{124}{5}\right)^2 + \left(\dfrac{93}{5}\right)^2 = 961 = 31^2$. The answer checks.

State. The length is $\dfrac{124}{5}$ cm, or 24.8 cm, and the height is $\dfrac{93}{5}$ cm, or 18.6 cm.

45. $f(x) = 10^x$

The function passes the horizontal-line test, so it has an inverse.

1. Replace $f(x)$ by y: $y = 10^x$

2. Interchange x and y: $x = 10^y$

3. Solve for y: $\log x = \log 10^y$

$$\log x = y \log 10$$
$$\log x = y$$

4. Replace y by $f^{-1}(x)$: $f^{-1}(x) = \log x$

47. $f(x) = |x|$

Since a horizontal line can cross the graph more than once, the function is not one-to-one so it does not have an inverse.

49. $f(x) = \dfrac{x-2}{x+3}$

The function passes the horizontal-line test, so it has an inverse.

1. Replace $f(x)$ by y: $y = \dfrac{x-2}{x+3}$

2. Interchange x and y: $x = \dfrac{y-2}{y+3}$

3. Solve for y: $xy + 3x = y - 2$
$$3x + 2 = y - xy$$
$$3x + 2 = y(1 - x)$$
$$\frac{3x+2}{1-x} = y$$

4. Replace y by $f^{-1}(x)$: $f^{-1}(x) = \dfrac{3x+2}{1-x}$

51. $a + b = \dfrac{5}{6}$,   (1)

$\dfrac{a}{b} + \dfrac{b}{a} = \dfrac{13}{6}$ (2)

$b = \dfrac{5}{6} - a = \dfrac{5-6a}{6}$ Solving Eq. (1) for b

$\dfrac{a}{\dfrac{5-6a}{6}} + \dfrac{\dfrac{5-6a}{6}}{a} = \dfrac{13}{6}$ Substituting for b
in Eq. (2)

$\dfrac{6a}{5-6a} + \dfrac{5-6a}{6a} = \dfrac{13}{6}$

$36a^2 + 25 - 60a + 36a^2 = 65a - 78a^2$

$150a^2 - 125a + 25 = 0$

$6a^2 - 5a + 1 = 0$

$(3a - 1)(2a - 1) = 0$

$a = \dfrac{1}{3} \ or \ a = \dfrac{1}{2}$

Substitute for a and solve for b.

When $a = \dfrac{1}{3}$, $b = \dfrac{5 - 6\left(\dfrac{1}{3}\right)}{6} = \dfrac{1}{2}$.

When $a = \dfrac{1}{2}$, $b = \dfrac{5 - 6\left(\dfrac{1}{2}\right)}{6} = \dfrac{1}{3}$.

The pairs $\left(\dfrac{1}{3}, \dfrac{1}{2}\right)$ and $\left(\dfrac{1}{2}, \dfrac{1}{3}\right)$ check. They are the solutions.

53. ***Familiarize***. Let $x =$ the length of the longer piece of wire and $y =$ the length of the shorter piece. Then the lengths of the sides of the squares are $\dfrac{x}{4}$ and $\dfrac{y}{4}$.

Translate. The length of the wire is 100 cm, so we have $x + y = 100$. The area of one square is 144 cm^2 greater than the area of the other, so we have a second equation:
$$\left(\frac{x}{4}\right)^2 - \left(\frac{y}{4}\right)^2 = 144.$$

Solve. The solution of the system of equations is $(61.52, 38.48)$.

Check. The sum of the lengths is $61.52 + 38.48$, or 100 cm. The length of a side of the larger square is $61.52/4$, or 15.38, and the area is $(15.38)^2$, or 236.5444 cm^2. The length of a side of the larger square is $38.48/4$, or 9.62, and the area is $(9.62)^2$, or 92.5444 cm^2. The larger area is 144 cm^2 greater than the smaller area. The answer checks.

State. The wire should be cut so that the length of one piece is 61.52, or $61\dfrac{13}{25}$ cm, and the other is 38.48, or $38\dfrac{12}{25}$ cm.

55. $R = C$
$$100x + x^2 = 80x + 1500$$
$$x^2 + 20x - 1500 = 0$$
$$(x - 30)(x + 50) = 0$$
$$x = 30 \ or \ x = -50$$

Since the number of units cannot be negative, the solution of the problem is 30. Thus, 30 units must be sold in order to break even.

Chapter 9 Vocabulary Reinforcement

1. A circle is the set of all points in a plane that are a fixed distance from a point in that plane.

2. The graph of $xy = 9$ is a hyperbola.

3. A horizontal parabola opens to the left or right.

4. In the equation of a parabola, the point (h, k) represents the vertex of the parabola.

5. In the equation of a circle, the point (h, k) represents the center of the circle.

6. The point halfway between the vertices of a hyperbola is the center of the hyperbola.

Chapter 9 Concept Reinforcement

1. False; the graph of $x - 2y^2 = 3$, or $x = 2y^2 + 3$ is a parabola that opens to the right. See page 760 in the text.

2. True; consider the system of equations consisting of the parabola $y = x^2 - 5$ and the circle $x^2 + y^2 = 11$. The four pairs $(\sqrt{2}, -3)$, $(-\sqrt{2}, -3)$, $(\sqrt{3}, -2)$, and $(-\sqrt{3}, -2)$ are solutions. We could also sketch many pairs of circles and parabolas that intersect at four points.

3. False; see page 781 in the text.

4. True; see page 764 in the text.

Chapter 9 Study Guide

1. $y = -x^2 - 4x - 1$

This is an equation of a parabola. We use the formula to find the first coordinate of the vertex.

$$x = -\frac{b}{2a} = -\frac{-4}{2(-1)} = -2$$

When $x = -2$, $y = -(-2)^3 - 4(-2) - 1 = -4 + 8 - 1 = 3$.

The vertex is $(-2, 3)$; the line of symmetry is $x = -2$. The curve opens down since the coefficient of x^2 is negative.

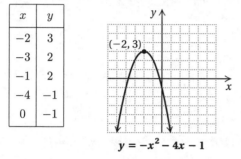

x	y
-2	3
-3	2
-1	2
-4	-1
0	-1

$$y = -x^2 - 4x - 1$$

2. $d = \sqrt{[-1 - (-2)]^2 + (7 - 10)^2}$

$= \sqrt{1^2 + (-3)^2}$

$= \sqrt{1 + 9}$

$= \sqrt{10} \approx 3.162$

3. $\left(\dfrac{17 + (-9)}{2}, \dfrac{-14 + (-2)}{2}\right) = \left(\dfrac{8}{2}, \dfrac{-16}{2}\right) = (4, -8)$

4. $(x - 2)^2 + (y + 1)^2 = 16$

$(x - 2)^2 + (y - (-1))^2 = 4^2$

Center: $(2, -1)$; radius, 4

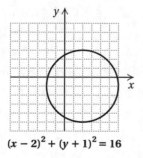

$$(x - 2)^2 + (y + 1)^2 = 16$$

5. $(x - h)^2 + (y - k)^2 = r^2$

$(x - 0)^2 + (y - 3)^2 = 6^2$

$x^2 + (y - 3)^2 = 36$

6. $25x^2 + 4y^2 = 100$

$\dfrac{1}{100}(25x^2 + 4y^2) = \dfrac{1}{100} \cdot 100$

$\dfrac{x^2}{4} + \dfrac{y^2}{25} = 1$

$\dfrac{x^2}{2^2} + \dfrac{y^2}{5^2} = 1$

This is an equation of an ellipse with center $(0, 0)$. The x-intercepts are $(-2, 0)$ and $(2, 0)$, and the y-intercepts are $(0, -5)$ and $(0, 5)$.

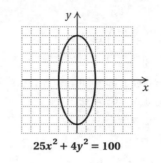

$$25x^2 + 4y^2 = 100$$

7. $\dfrac{x^2}{9} - \dfrac{y^2}{25} = 1$

$\dfrac{x^2}{3^2} - \dfrac{y^2}{5^2} = 1$, $a = 3$, $b = 5$

This is an equation of a hyperbola. The asymptotes are $y = -\dfrac{b}{a}x$ and $y = \dfrac{b}{a}x$, or $y = -\dfrac{5}{3}x$ and $y = \dfrac{5}{3}x$. We sketch them.

The hyperbola has center $(0, 0)$ and a horizontal axis. The vertices are the x-intercepts. We find them.

$$\frac{x^2}{9} - \frac{0^2}{25} = 1$$

$$\frac{x^2}{9} = 1$$

$$x^2 = 9$$

$$x = \pm 3$$

The x-intercepts are $(-3, 0)$ and $(3, 0)$. There are no y-intercepts. We plot the intercepts and sketch the graph.

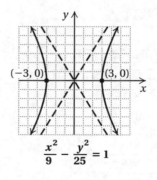

$$\frac{x^2}{9} - \frac{y^2}{25} = 1$$

8. $\dfrac{x^2}{36} + \dfrac{y^2}{4} = 1$, $\quad$ (1)

$3y - x = 6 \quad$ (2)

First we solve Equation (2) for x.

$3y - 6 = x \quad$ (3)

Then we substitute $3y - 6$ for x in Equation (1) and solve for y.

$$\frac{(3y-6)^2}{36} + \frac{y^2}{4} = 1$$

$$\frac{9y^2 - 36y + 36}{36} + \frac{y^2}{4} = 1$$

$$36\left(\frac{9y^2 - 36y + 36}{36} + \frac{y^2}{4}\right) = 36 \cdot 1$$

$$9y^2 - 36y + 36 + 9y^2 = 36$$

$$18y^2 - 36y + 36 = 36$$

$$18y^2 - 36y = 0$$

$$18y(y-2) = 0$$

$$18y = 0 \quad or \quad y - 2 = 0$$

$$y = 0 \quad or \quad\quad y = 2$$

Now substitute in Equation (3) to find the x-value that corresponds to each y-value.

For $y = 0$, $x = 3 \cdot 0 - 6 = -6$.

For $y = 2$, $x = 3 \cdot 2 - 6 = 6 - 6 = 0$.

The solutions are $(-6, 0)$ and $(0, 2)$.

Chapter 9 Review Exercises

1. $d = \sqrt{(x_2 - x_1)^2 + (y_2 - y_1)^2}$
$d = \sqrt{(6-2)^2 + (6-6)^2}$
$= \sqrt{4^2 + 0^2}$
$= \sqrt{4^2} = 4$

2. $d = \sqrt{(x_2 - x_1)^2 + (y_2 - y_1)^2}$
$d = \sqrt{[-5 - (-1)]^2 + (4-1)^2}$
$= \sqrt{(-4)^2 + 3^2}$
$= \sqrt{25} = 5$

3. $d = \sqrt{(x_2 - x_1)^2 + (y_2 - y_1)^2}$
$d = \sqrt{(4.7 - 1.4)^2 + (-5.3 - 3.6)^2}$
$= \sqrt{(3.3)^2 + (-8.9)^2}$
$= \sqrt{90.1} \approx 9.492$

4. $d = \sqrt{(x_2 - x_1)^2 + (y_2 - y_1)^2}$
$d = \sqrt{(-1 - 2)^2 + (a - 3a)^2}$
$= \sqrt{(-3)^2 + (-2a)^2}$
$= \sqrt{9 + 4a^2}$

5. $\left(\frac{x_1 + x_2}{2}, \frac{y_1 + y_2}{2}\right) = \left(\frac{1+7}{2}, \frac{6+6}{2}\right) =$
$\left(\frac{8}{2}, \frac{12}{2}\right) = (4, 6)$

6. $\left(\frac{x_1 + x_2}{2}, \frac{y_1 + y_2}{2}\right) = \left(\frac{-1 + (-5)}{2}, \frac{1+4}{2}\right) =$
$\left(\frac{-6}{2}, \frac{5}{2}\right) = \left(-3, \frac{5}{2}\right)$

7. $\left(\frac{x_1 + x_2}{2}, \frac{y_1 + y_2}{2}\right) = \left(\frac{1 + \frac{1}{2}}{2}, \frac{\sqrt{3} + (-\sqrt{2})}{2}\right) =$
$\left(\frac{\frac{3}{2}}{2}, \frac{\sqrt{3} - \sqrt{2}}{2}\right) = \left(\frac{3}{4}, \frac{\sqrt{3} - \sqrt{2}}{2}\right)$

8. $\left(\frac{x_1 + x_2}{2}, \frac{y_1 + y_2}{2}\right) = \left(\frac{2 + (-1)}{2}, \frac{3a + a}{2}\right) =$
$\left(\frac{1}{2}, \frac{4a}{2}\right) = \left(\frac{1}{2}, 2a\right)$

9. $(x + 2)^2 + (y - 3)^2 = 2$
$[x - (-2)]^2 + (y - 3)^2 = (\sqrt{2})^2$ Standard form
Center: $(-2, 3)$
Radius: $\sqrt{2}$

10. $(x - 5)^2 + y^2 = 49$
$(x - 5)^2 + (y - 0)^2 = 7^2$ Standard form
Center: $(5, 0)$
Radius: 7

11. $x^2 + y^2 - 6x - 2y + 1 = 0$
$(x^2 - 6x) + (y^2 - 2y) + 1 = 0$
    Regrouping
$(x^2 - 6x + 9 - 9) + (y^2 - 2y + 1 - 1) + 1 = 0$
    Completing the square twice
$(x^2 - 6x + 9) + (y^2 - 2y + 1) - 9 - 1 + 1 = 0$
$(x - 3)^2 + (y - 1)^2 = 9$
$(x - 3)^2 + (y - 1)^2 = 3^2$
Center: $(3, 1)$
Radius: 3

12. $x^2 + y^2 + 8x - 6y - 10 = 0$
$(x^2 + 8x) + (y^2 - 6y) - 10 = 0$
    Regrouping
$(x^2 + 8x + 16 - 16) + (y^2 - 6y + 9 - 9) - 10 = 0$
    Completing the square twice
$(x^2 + 8x + 16) + (y^2 - 6y + 9) - 16 - 9 - 10 = 0$
$(x + 4)^2 + (y - 3)^2 = 35$
$[x - (-4)]^2 + (y - 3)^2 = (\sqrt{35})^2$
Center: $(-4, 3)$
Radius: $\sqrt{35}$

13. $(x - h)^2 + (y - k)^2 = r^2$ Standard form
$[x - (-4)]^2 + (y - 3)^2 = (4\sqrt{3})^2$ Substituting
$(x + 4)^2 + (y - 3)^2 = 48$

14. $(x - h)^2 + (y - k)^2 = r^2$
$(x - 7)^2 + [y - (-2)]^2 = (2\sqrt{5})^2$
$(x - 7)^2 + (y + 2)^2 = 20$

15. $\dfrac{x^2}{16} + \dfrac{y^2}{4} = 1$

$$\dfrac{x^2}{4^2} + \dfrac{y^2}{2^2} = 1$$

The graph is an ellipse. The x-intercepts are $(-4, 0)$ and $(4, 0)$, and the y-intercepts are $(0, -2)$ and $(0, 2)$. We plot these points and connect them with an oval-shaped curve.

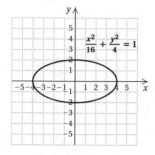

16. $\dfrac{y^2}{9} - \dfrac{x^2}{4} = 1$

$$\dfrac{y^2}{3^2} - \dfrac{x^2}{2^2} = 1$$

The graph is a hyperbola. $a = 2$ and $b = 3$, so the asymptotes are $y = \dfrac{3}{2}x$ and $y = -\dfrac{3}{2}x$. Replacing x with 0 and solving for y, we get $y = \pm 3$, so the intercepts are $(0, 3)$ and $(0, -3)$.

We plot the intercepts and draw smooth curves through them that approach the asymptotes.

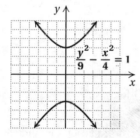

17. $x^2 + y^2 = 16$

$$(x - 0)^2 + (y - 0)^2 = 4^2$$

The graph is a circle. The center is $(0, 0)$, and the radius is 4.

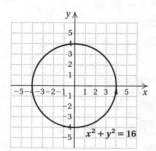

18. $x = y^2 + 2y - 2$

The graph is a parabola. We complete the square.

$$\begin{aligned}
x &= (y^2 + 2y + 1 - 1) - 2 \\
&= (y^2 + 2y + 1) - 1 - 2 \\
&= (y + 1)^2 - 3 \\
&= [y - (-1)]^2 + (-3)
\end{aligned}$$

The vertex is $(-3, -1)$; the line of symmetry is $y = -1$. The curve opens to the right.

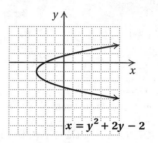

19. $y = -2x^2 - 2x + 3$

The graph is a parabola. We complete the square.

$$\begin{aligned}
y &= -2(x^2 + x) + 3 \\
&= -2\left(x^2 + x + \dfrac{1}{4} - \dfrac{1}{4}\right) + 3 \\
&= -2\left(x^2 + x + \dfrac{1}{4}\right) + (-2)\left(-\dfrac{1}{4}\right) + 3 \\
&= -2\left(x + \dfrac{1}{2}\right)^2 + \dfrac{7}{2} \\
&= -2\left[x - \left(-\dfrac{1}{2}\right)\right]^2 + \dfrac{7}{2}
\end{aligned}$$

The vertex is $\left(-\dfrac{1}{2}, \dfrac{7}{2}\right)$; the line of symmetry is $x = -\dfrac{1}{2}$. The curve opens down.

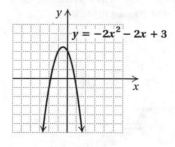

20. $x^2 + y^2 + 2x - 4y - 4 = 0$

The graph is a circle. We regroup and complete the square twice.

$$\begin{aligned}
(x^2 + 2x + 1 - 1) + (y^2 - 4y + 4 - 4) - 4 &= 0 \\
(x^2 + 2x + 1) + (y^2 - 4y + 4) - 1 - 4 - 4 &= 0 \\
(x + 1)^2 + (y - 2)^2 &= 9 \\
[x - (-1)]^2 + (y - 2)^2 &= 3^2
\end{aligned}$$

The center is $(-1, 2)$, and the radius is 3.

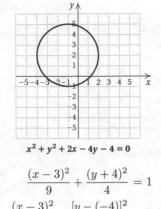

$$x^2 + y^2 + 2x - 4y - 4 = 0$$

21. $\dfrac{(x-3)^2}{9} + \dfrac{(y+4)^2}{4} = 1$

$\dfrac{(x-3)^2}{3^2} + \dfrac{[y-(-4)]^2}{2^2} = 1$

The graph is an ellipse with center $(3, -4)$. Note that $a = 3$ and $b = 2$. We locate the center and then plot the points $(3 - 3, 2)$, $(3 + 3, 2)$, $(3, -4 - 2)$, and $(3, -4 + 2)$, or $(0, 2)$, $(6, 2)$, $(3, -6)$, and $(3, -2)$. Connect these points with an oval-shaped curve.

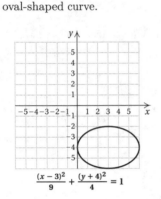

$$\dfrac{(x-3)^2}{9} + \dfrac{(y+4)^2}{4} = 1$$

22. $xy = 9$

$y = \dfrac{9}{x}$

We find some solutions. Note that x cannot be 0. The x-axis and the y-axis are the asymptotes.

x	y
-6	$-\dfrac{3}{2}$
-3	-3
-1	-9
1	9
3	3
6	$\dfrac{3}{2}$

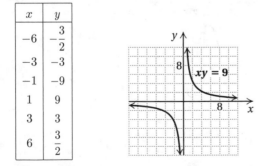

23. $x + y^2 = 2y + 1$

$x = -y^2 + 2y + 1$

The graph is a parabola. We complete the square.

$x = -(y^2 - 2y + 1 - 1) + 1$

$ = -(y^2 - 2y + 1) + (-1)(-1) + 1$

$ = -(y - 1)^2 + 2$

The vertex is $(2, 1)$; the line of symmetry is $y = 1$. The graph opens to the left.

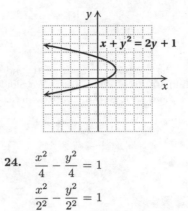

24. $\dfrac{x^2}{4} - \dfrac{y^2}{4} = 1$

$\dfrac{x^2}{2^2} - \dfrac{y^2}{2^2} = 1$

The graph is a hyperbola. We have $a = 2$ and $b = 2$, so the asymptotes are $y = \dfrac{2}{2}x$ and $y = -\dfrac{2}{2}x$, or $y = x$ and $y = -x$. Replacing y with 0 and solving for x, we get $x = \pm 2$, so the intercepts are $(2, 0)$ and $(-2, 0)$.

We plot the intercepts and draw smooth curves through them that approach the asymptotes.

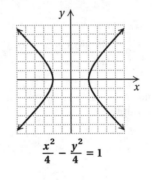

$$\dfrac{x^2}{4} - \dfrac{y^2}{4} = 1$$

25. $x^2 - y^2 = 33,$ (1)

$x + y = 11$ (2)

First we solve Equation (2) for y.

$y = 11 - x$ (3)

Then substitute $11 - x$ for y in Equation (1) and solve for x.

$$x^2 - y^2 = 33$$
$$x^2 - (11 - x)^2 = 33$$
$$x^2 - (121 - 22x + x^2) = 33$$
$$x^2 - 121 + 22x - x^2 = 33$$
$$-121 + 22x = 33$$
$$22x = 154$$
$$x = 7$$

Now substitute 7 for x in Equation (3) and find y.

$y = 11 - 7 = 4$

The pair $(7, 4)$ checks, so it is the solution.

26. $x^2 - 2x + 2y^2 = 8,$ (1)

$2x + y = 6$ (2)

First we solve Equation (2) for y.

$y = 6 - 2x$ (3)

Then substitute $6 - 2x$ for y in Equation (1) and solve for x.

$$x^2 - 2x + 2(6 - 2x)^2 = 8$$
$$x^2 - 2x + 2(36 - 24x + 4x^2) = 8$$
$$x^2 - 2x + 72 - 48x + 8x^2 = 8$$
$$9x^2 - 50x + 72 = 8$$
$$9x^2 - 50x + 64 = 0$$
$$(9x - 32)(x - 2) = 0$$
$$9x - 32 = 0 \quad or \quad x - 2 = 0$$
$$x = \frac{32}{9} \quad or \qquad x = 2$$

Now substitute these numbers in Equation (3) and find y.

$$y = 6 - 2 \cdot \frac{32}{9} = 6 - \frac{64}{9} = -\frac{10}{9}$$
$$y = 6 - 2 \cdot 2 = 6 - 4 = 2$$

The pairs $\left(\frac{32}{9}, -\frac{10}{9} \right)$ and $(2, 2)$ check, so they are the solutions.

27. $x^2 - y = 3,$ (1)

$\quad\ 2x - y = 3$ (2)

We multiply Equation (2) by -1 and then add.

$$x^2 - \ y = 3$$
$$\underline{-2x + \ y = -3}$$
$$x^2 - 2x = 0$$

Then we have:

$$x^2 - 2x = 0$$
$$x(x - 2) = 0$$
$$x = 0 \ \ or \ \ x = 2$$

We substitute these numbers in one of the original equations and solve for y. We use Equation (2).

$$2 \cdot 0 - y = 3 \qquad\qquad 2 \cdot 2 - y = 3$$
$$-y = 3 \qquad\qquad\ 4 - y = 3$$
$$y = -3 \qquad\qquad\quad -y = -1$$
$$y = -3 \qquad\qquad\qquad\ y = 1$$

The pairs $(0, -3)$ and $(2, 1)$ check, so they are the solutions.

28. $x^2 + y^2 = 25$ (1)

$\underline{\quad x^2 - y^2 = 7 \quad\ (2)}$

$2x^2 \qquad\ = 32$ Adding

$$x^2 = 16$$
$$x = \pm 4$$

If $x = 4$, $x^2 = 16$, and if $x = -4$, $x^2 = 16$, so substituting 4 or -4 in Equation (1) gives us

$$16 + y^2 = 25$$
$$y^2 = 9$$
$$y = \pm 3.$$

The pairs $(4, 3)$, $(4, -3)$, $(-4, 3)$, and $(-4, -3)$ check.

29. $x^2 - y^2 = 3,$ (1)

$\quad\ y = x^2 - 3$ (2)

Substitute $x^2 - 3$ for y in Equation (1) and solve for x.

$$x^2 - (x^2 - 3)^2 = 3$$
$$x^2 - (x^4 - 6x^2 + 9) = 3$$
$$x^2 - x^4 + 6x^2 - 9 = 3$$
$$0 = x^4 - 7x^2 + 12$$

Let $u = x^2$.

$$0 = u^2 - 7u + 12$$
$$0 = (u - 3)(u - 4)$$
$$u = 3 \quad or \quad u = 4$$
$$x^2 = 3 \quad or \ x^2 = 4$$
$$x = \pm\sqrt{3} \ or \quad x = \pm 2$$

When $x = \pm\sqrt{3}$, $x^2 = 3$ and, substituting in Equation (2), we get $y = 3 - 3 = 0$.

When $x = \pm 2$, $x^2 = 4$ and, substituting in Equation (2), we get $y = 4 - 3 = 1$.

The pairs $(\sqrt{3}, 0)$, $(-\sqrt{3}, 0)$, $(2, 1)$, and $(-2, 1)$ check.

30. $x^2 + y^2 = 18,$ (1)

$\quad\ 2x + y = 3$ (2)

First we solve Equation (2) for y.

$$y = 3 - 2x \quad (3)$$

Now substitute $3 - 2x$ for y in Equation (1) and solve for x.

$$x^2 + (3 - 2x)^2 = 18$$
$$x^2 + 9 - 12x + 4x^2 = 18$$
$$5x^2 - 12x - 9 = 0$$
$$(5x + 3)(x - 3) = 0$$
$$5x + 3 = 0 \quad or \quad x - 3 = 0$$
$$x = -\frac{3}{5} \ or \qquad x = 3$$

Substitute these numbers in Equation (3) and find y.

$$y = 3 - 2\left(-\frac{3}{5} \right) = 3 + \frac{6}{5} = \frac{21}{5}$$
$$y = 3 - 2 \cdot 3 = 3 - 6 = -3$$

The pairs $\left(-\frac{3}{5}, \frac{21}{5} \right)$ and $(3, -3)$ check.

31. $x^2 + y^2 = 100,$ (1)

$\quad\ 2x^2 - 3y^2 = -120$ (2)

Multiply Equation (1) by 3 and then add.

$3x^2 + 3y^2 = 300$

$\underline{2x^2 - 3y^2 = -120}$

$5x^2 \qquad\ = 180$

$$x^2 = 36$$
$$x = \pm 6$$

When $x = 6$, $x^2 = 36$, and when $x = -6$, $x^2 = 36$, so substituting 6 or -6 in Equation (1) give us

$$36 + y^2 = 100$$
$$y^2 = 64$$
$$y = \pm 8.$$

The pairs $(6, 8)$, $(6, -8)$, $(-6, 8)$, and $(-6, -8)$ check.

32. $x^2 + 2y^2 = 12$,   (1)

$xy = 4$ (2)

First we solve Equation (2) for y.

$$y = \frac{4}{x} \quad (3)$$

Then substitute $\frac{4}{x}$ for y in Equation (1) and solve for x.

$$x^2 + 2\left(\frac{4}{x}\right)^2 = 12$$

$$x^2 + 2\left(\frac{16}{x^2}\right) = 12$$

$$x^2 + \frac{32}{x^2} = 12$$

$$x^4 + 32 = 12x^2 \quad \text{Multiplying by } x^2$$

$$x^4 - 12x^2 + 32 = 0$$

$$u^2 - 12u + 32 = 0 \qquad \text{Letting } u = x^2$$

$$(u - 4)(u - 8) = 0$$

$$u = 4 \quad or \quad u = 8$$

$$x^2 = 4 \quad or \quad x^2 = 8$$

$$x = \pm 2 \quad or \quad x = \pm\sqrt{8} = \pm 2\sqrt{2}$$

Now we use Equation (3) to find the y-value that corresponds to each value of x.

When $x = 2$, $y = \frac{4}{2} = 2$.

When $x = -2$, $y = \frac{4}{-2} = -2$.

When $x = 2\sqrt{2}$, $y = \frac{4}{2\sqrt{2}} = \frac{2}{\sqrt{2}} = \sqrt{2}$.

When $x = -2\sqrt{2}$, $y = \frac{4}{-2\sqrt{2}} = -\frac{2}{\sqrt{2}} = -\sqrt{2}$.

The pairs $(2, 2)$, $(-2, -2)$, $(2\sqrt{2}, \sqrt{2})$, and $(-2\sqrt{2}, -\sqrt{2})$ check.

33. **Familiarize**. Let l and w represent the length and width, respectively, of the carton, in inches.

Translate. The area is 12 in^2, so we have $lw = 12$. The diagonal is 5 in., so the Pythagorean theorem gives us a second equation, $l^2 + w^2 = 5^2$, or $l^2 + w^2 = 25$.

Solve. We solve the system of equations.

$$lw = 12, \qquad (1)$$

$$l^2 + w^2 = 25 \quad (2)$$

First we solve Equation (1) for l.

$$l = \frac{12}{w}$$

Then substitute $\frac{12}{w}$ for l in Equation (2) and solve for w.

$$\left(\frac{12}{w}\right)^2 + w^2 = 25$$

$$\frac{144}{w^2} + w^2 = 25$$

$$144 + w^4 = 25w^2 \quad \text{Multiplying by } w^2$$

$$w^4 - 25w^2 + 144 = 0$$

$$u^2 - 25u + 144 = 0 \quad \text{Letting } u = w^2$$

$$(u - 9)(u - 16) = 0$$

$$u - 9 = 0 \quad or \quad u - 16 = 0$$

$$u = 9 \quad or \qquad u = 16$$

$$w^2 = 9 \quad or \qquad w^2 = 16$$

$$w = \pm 3 \quad or \qquad w = \pm 4$$

Since the width cannot be negative, -3 and -4 cannot be solutions. If $w = 3$, $l = \frac{12}{3} = 4$; if $w = 4$, $l = \frac{12}{4} = 3$. Since length is usually considered to be longer than width, we let $l = 4$ and $w = 3$.

Check. If $l = 4$ and $w = 3$, the area is $4 \cdot 3$, or 12 in^2; also $4^3 + 3^2 = 25 = 5^2$, so the answer checks.

State. The length of the carton is 4 in., and the width is 3 in.

34. **Familiarize**. Using the labels on the drawing in the text, let r_1 = the radius of the larger flower bed, in feet, and let r_2 = the radius of the smaller bed.

Translate. The sum of the areas is 130π ft^2, so we have $\pi r_1{}^2 + \pi r_2{}^2 = 130\pi$, or $r_1{}^2 + r_2{}^2 = 130$. The difference of the circumferences is 16π ft, so we also have $2\pi r_1 - 2\pi r_2 = 16\pi$, or $r_1 - r_2 = 8$.

Solve. We solve the system of equations.

$$r_1{}^2 + r_2{}^2 = 130, \quad (1)$$

$$r_1 - r_2 = 8 \qquad (2)$$

First solve Equation (2) for r_1.

$$r_1 = 8 + r_2$$

Substitute $8 + r_2$ for r_1 and Equation (1) and solve for r_2.

$$(8 + r_2)^2 + r_2{}^2 = 130$$

$$64 + 16r_2 + r_2{}^2 + r_2{}^2 = 130$$

$$2r_2{}^2 + 16r_2 + 64 = 130$$

$$2r_2{}^2 + 16r_2 - 66 = 0$$

$$r_2{}^2 + 8r_2 - 33 = 0 \quad \text{Dividing by 2}$$

$$(r_2 + 11)(r_2 - 3) = 0$$

$$r_2 = -11 \quad or \quad r_2 = 3$$

Since the radius cannot be negative, -11 cannot be a solution. If $r_2 = 3$, then $r_1 = 8 + 3 = 11$.

Check. The sum of the area is $\pi \cdot 11^2 + \pi \cdot 3^2 = 121\pi + 9\pi = 130\pi$, and the difference of the circumferences is $2\pi \cdot 11 - 2\pi \cdot 3 = 22\pi - 6\pi = 16\pi$. The answer checks.

State. The radii are 11 ft and 3 ft.

35. **Familiarize**. Let x and y represent the positive integers.

Translate. The sum of the numbers is 12, so we have $x + y = 12$. The sum of the reciprocals is $\frac{3}{8}$, so we also have $\frac{1}{x} + \frac{1}{y} = \frac{3}{8}$.

Solve. We solve the system of equations.

$$x + y = 12, \quad (1)$$

$$\frac{1}{x} + \frac{1}{y} = \frac{3}{8} \quad (2)$$

First we solve Equation (1) for x.

$$x = 12 - y$$

Now substitute $12 - y$ for x in Equation (2) and solve for y.

$$\frac{1}{12 - y} + \frac{1}{y} = \frac{3}{8}, \text{ LCD is } 8y(12 - y)$$

$$8y(12 - y)\left(\frac{1}{12 - y} + \frac{1}{y}\right) = 8y(12 - y) \cdot \frac{3}{8}$$

$$8y + 8(12 - y) = 3y(12 - y)$$

$$8y + 96 - 8y = 36y - 3y^2$$

$$3y^2 - 36y + 96 = 0$$

$$y^2 - 12y + 32 = 0 \quad \text{Dividing by 3}$$

$$(y - 4)(y - 8) = 0$$

$y = 4 \ \ or \ \ y = 8$

If $y = 4$, then $x = 12 - 4 = 8$; if $y = 8$, then $x = 12 - 8 = 4$. In either case, the numbers are 4 and 8.

Check. $4 + 8 = 12$ and $\dfrac{1}{4} + \dfrac{1}{8} = \dfrac{2}{8} + \dfrac{1}{8} = \dfrac{3}{8}$. The answer checks.

State. The numbers are 4 and 8.

36. Familiarize. Let $l =$ the length of the garden and $w =$ the width, in meters.

Translate. The perimeter is 38 m, so we have $2l + 2w = 38$, or $l + w = 19$. The area is 84 m^2, so we have $lw = 84$.

Solve. We solve the system of equations.

$$l + w = 19, \quad (1)$$
$$lw = 84 \qquad (2)$$

First we solve Equation (1) for l.

$$l = 19 - w \quad (3)$$

Then substitute $19 - w$ for l in Equation (2) and solve for w.

$$(19 - w)w = 84$$
$$19w - w^2 = 84$$
$$0 = w^2 - 19w + 84$$
$$0 = (w - 7)(w - 12)$$

$w = 7 \ \ or \ \ w = 12$

If $w = 7$, then $l = 19 - 7 = 12$. If $w = 12$, then $l = 19 - 12 = 7$. Since length is usually considered to be longer than width, we have $l = 12$ and $w = 7$.

Check. If $l = 12$ and $w = 7$, the perimeter is $2 \cdot 12 + 2 \cdot 7$, or 38 m, and the area is $12 \cdot 7$, or 84 m^2. The answer checks.

State. The length of the garden is 12 m, and the width is 7 m.

37. The graph of $y = \dfrac{1}{2}x^2 + 1$ is a parabola that opens up. We solve the second equation for y.

$$2x - 3y = -6$$
$$-3y = -2x - 6$$
$$-\frac{1}{3}(-3y) = -\frac{1}{3}(-2x - 6)$$
$$y = \frac{2}{3}x + 2$$

The graph of this equation is a line that slants up from left to right. The only graph that represents both of these equations is B.

38.
$$x^2 + y^2 + 6x - 16y + 66 = 0$$
$$x^2 + 6x + y^2 - 16y = -66$$
$$x^2 + 6x + 9 + y^2 - 16y + 64 = -66 + 9 + 64$$
$$(x + 3)^2 + (y - 8)^2 = 7$$
$$(x - (-3))^2 + (y - 8)^2 = (\sqrt{7})^2$$

Center: $(-3, 8)$; radius: $\sqrt{7}$

Answer D is correct.

39. $4x^2 - x - 3y^2 = 9, \quad (1)$

$-x^2 + x + y^2 = 2 \quad (2)$

First solve Equation (2) for y^2.

$$y^2 = x^2 - x + 2 \quad (3)$$

Substitute $x^2 - x + 2$ for y^2 in Equation (1) and solve for x.

$$4x^2 - x - 3(x^2 - x + 2) = 9$$
$$4x^2 - x - 3x^2 + 3x - 6 = 9$$
$$x^2 + 2x - 6 = 9$$
$$x^2 + 2x - 15 = 0$$
$$(x + 5)(x - 3) = 0$$

$x = -5 \ \ or \ \ x = 3$

Now use Equation (3) to find y.

When $x = -5$:
$$y^2 = (-5)^2 - (-5) + 2$$
$$y^2 = 25 + 5 + 2$$
$$y^2 = 32$$
$$y = \pm\sqrt{32} = \pm 4\sqrt{2}$$

When $x = 3$:
$$y^2 = 3^2 - 3 + 2$$
$$y^2 = 9 - 3 + 2$$
$$y^2 = 8$$
$$y = \pm\sqrt{8} = \pm 2\sqrt{2}$$

The solutions are $(-5, 4\sqrt{2})$, $(-5, -4\sqrt{2})$, $(3, 2\sqrt{2})$, and $(3, -2\sqrt{2})$.

40. We substitute the given points in the standard form of the equation of a circle, $(x - h)^2 + (y - k)^2 = r^2$.

For $(-2, -4)$:
$$(-2 - h)^2 + (-4 - k)^2 = r^2$$
$$4 + 4h + h^2 + 16 + 8k + k^2 = r^2$$
$$h^2 + 4h + k^2 + 8k + 20 = r^2$$
$$h^2 + 4h + k^2 + 8k = r^2 - 20 \quad (1)$$

For $(5, -5)$:
$$(5 - h)^2 + (-5 - k)^2 = r^2$$
$$25 - 10h + h^2 + 25 + 10k + k^2 = r^2$$
$$h^2 - 10h + k^2 + 10k + 50 = r^2$$
$$h^2 - 10h + k^2 + 10k = r^2 - 50 \quad (2)$$

For $(6, 2)$:
$$(6 - h)^2 + (2 - k)^2 = r^2$$
$$36 - 12h + h^2 + 4 - 4k + k^2 = r^2$$
$$h^2 - 12h + k^2 - 4k + 40 = r^2$$
$$h^2 - 12h + k^2 - 4k = r^2 - 40 \quad (3)$$

We solve the system of equations (1), (2), and (3). First we will eliminate the squared terms from two pairs of equations. Multiply Equation (1) by -1 and add it to Equation (2) and to Equation (3).

$$\begin{aligned} -h^2 - 4h - k^2 - 8k &= -r^2 + 20 \\ h^2 - 10h + k^2 + 10k &= r^2 - 50 \quad (2) \\ \hline -14h \qquad\quad + 2k &= -30 \quad (4) \end{aligned}$$

$$\begin{aligned} -h^2 - 4h - k^2 - 8k &= -r^2 + 20 \\ h^2 - 12h + k^2 - 4k &= r^2 - 40 \\ \hline -16h \qquad\quad - 12k &= -20 \quad (5) \end{aligned}$$

Now solve the system of equations (4) and (5). Multiply Equation (4) by 6 and then add.

$$\begin{aligned} -84h + 12k &= -180 \\ -16h - 12k &= -20 \\ \hline -100h \qquad\quad &= -200 \\ h &= 2 \end{aligned}$$

Substitute 2 for h in Equation (4) and solve for k.
$$-14 \cdot 2 + 2k = -30$$
$$-28 + 2k = -30$$
$$2k = -2$$
$$k = -1$$

Now substitute 2 for h and -1 for k in one of the original equations and solve for r^2. We will use Equation (1).
$$2^2 + 4 \cdot 2 + (-1)^2 + 8(-1) = r^2 - 20$$
$$4 + 8 + 1 - 8 = r^2 - 20$$
$$5 = r^2 - 20$$
$$25 = r^2$$

Then the equation of the circle is $(x-2)^2 + [y-(-1)]^2 = 25$, or $(x - 2)^2 + (y + 1)^2 = 25$.

41. The center of the ellipse is the midpoint of the segment joining the x-intercepts or the y-intercepts. We will use the x-intercepts to find the center.
$$\left(\frac{-7 + 7}{2}, \frac{0 + 0}{2}\right) = \left(\frac{0}{2}, \frac{0}{2}\right) = (0, 0).$$

We also see that $a = 7$ and $b = 3$, so we have:
$$\frac{(x - 0)^2}{7^2} + \frac{(y - 0)^2}{3^2} = 1, \text{ or}$$
$$\frac{x^2}{49} + \frac{y^2}{9} = 1$$

42. Let $(x, 0)$ represent the coordinates of the point we want to find. We use the distance formula.
$$\sqrt{(-3 - x)^2 + (4 - 0)^2} = \sqrt{(5 - x)^2 + (6 - 0)^2}$$
$$\sqrt{9 + 6x + x^2 + 16} = \sqrt{25 - 10x + x^2 + 36}$$
$$\sqrt{x^2 + 6x + 25} = \sqrt{x^2 - 10x + 61}$$
$$(\sqrt{x^2 + 6x + 25})^2 = (\sqrt{x^2 - 10x + 61})^2$$
$$x^2 + 6x + 25 = x^2 - 10x + 61$$
$$16x = 36$$
$$x = \frac{9}{4}$$

The point is $\left(\frac{9}{4}, 0\right)$.

43.
$$-y + 4x^2 = 5 - 2x$$
$$-y = -4x^2 - 2x + 5$$
$$y = 4x^2 + 2x - 5$$
The graph is a parabola.

44. $xy = -6$

The graph is a hyperbola.

45.
$$\frac{x^2}{23} + \frac{y^2}{23} = 1$$
$$x^2 + y^2 = 23 \quad \text{Multiplying by 23}$$
The graph is a circle.

46.
$$43 - 12x^2 + y^2 = 21x^2 + 2y^2$$
$$43 = 33x^2 + y^2$$
$$1 = \frac{x^2}{43/33} + \frac{y^2}{43}$$
The graph is an ellipse.

47.
$$3x^2 + 3y^2 = 170$$
$$x^2 + y^2 = \frac{170}{3}$$
The graph is a circle.

48.
$$\frac{x^2}{8} - \frac{y^2}{2} = 1$$
The graph is a hyperbola.

Chapter 9 Discussion and Writing Exercises

1. Earlier we studied systems of linear equations. In this chapter, we studied systems of two equations in which at least one equation is of second degree.

2. Parabolas of the form $y = ax^2 + bx + c$ and hyperbolas of the form $xy = c$ pass the vertical-line test, so they are functions. Circles, ellipses, parabolas of the form $x = ay^2 + by + c$, and hyperbolas of the form $\frac{x^2}{a^2} - \frac{y^2}{b^2} = 1$ or $\frac{y^2}{b^2} - \frac{x^2}{a^2} = 1$ fail the vertical-line test, and hence are not functions.

3. The graph of a parabola has one branch whereas the graph of a hyperbola has two branches. A hyperbola has asymptotes but a parabola does not.

4. The asymptotes are $y = x$ and $y = -x$, because for $a = b$, $\pm\dfrac{b}{a} = \pm 1$.

Chapter 9 Test

1. $d = \sqrt{(x_2 - x_1)^2 + (y_2 - y_1)^2}$
$d = \sqrt{[6 - (-6)]^2 + (8 - 2)^2}$
$\quad = \sqrt{12^2 + 6^2}$
$\quad = \sqrt{180} \approx 13.416$

2. $d = \sqrt{(x_2 - x_1)^2 + (y_2 - y_1)^2}$
$d = \sqrt{(-3 - 3)^2 + [a - (-a)]^2}$
$\quad = \sqrt{(-6)^2 + (2a)^2}$
$\quad = \sqrt{36 + 4a^2}$
$\quad = \sqrt{4(9 + a^2)}$
$\quad = 2\sqrt{9 + a^2}$

3. We use the formula $\left(\dfrac{x_1 + x_2}{2}, \dfrac{y_1 + y_2}{2} \right)$.
$\left(\dfrac{-6 + 6}{2}, \dfrac{2 + 8}{2} \right) = \left(\dfrac{0}{2}, \dfrac{10}{2} \right) = (0, 5)$

4. We use the formula $\left(\dfrac{x_1 + x_2}{2}, \dfrac{y_1 + y_2}{2} \right)$.
$\left(\dfrac{3 + (-3)}{2}, \dfrac{-a + a}{2} \right) = \left(\dfrac{0}{2}, \dfrac{0}{2} \right) = (0, 0)$

5. $\quad (x + 2)^2 + (y - 3)^2 = 64$
$[x - (-2)]^2 + (y - 3)^2 = 8^2$ Standard form
The center is $(-2, 3)$; the radius is 8.

6. $\quad\quad\quad x^2 + y^2 + 4x - 6y + 4 = 0$
$\quad\quad (x^2 + 4x) + (y^2 - 6y) + 4 = 0$
$(x^2 + 4x + 4 - 4) + (y^2 - 6y + 9 - 9) + 4 = 0$
$\quad\quad\quad\quad\quad\quad\quad$ Completing the square
$(x^2 + 4x + 4) + (y^2 - 6y + 9) - 4 - 9 + 4 = 0$
$\quad\quad\quad\quad\quad (x + 2)^2 + (y - 3)^2 = 9$
$\quad\quad\quad\quad [x - (-2)]^2 + (y - 3)^2 = 3^2$
$\quad\quad\quad\quad\quad\quad\quad\quad$ Standard form
The center is $(-2, 3)$; the radius is 3.

7. $\quad\quad (x - h)^2 + (y - k)^2 = r^2$
$[x - (-2)]^2 + [y - (-5)]^2 = (3\sqrt{2})^2$
$\quad\quad (x + 2)^2 + (y + 5)^2 = 18$

8. Graph: $y = x^2 - 4x - 1$
The graph is a parabola. We find the first coordinate of the vertex.
$-\dfrac{b}{2a} = -\dfrac{-4}{2 \cdot 1} = -\dfrac{-4}{2} = 2$

Then $y = 2^2 - 4 \cdot 2 - 1 = 4 - 8 - 1 = -5$.

The vertex is $(2, -5)$; the line of symmetry is $x = 2$. The curve opens up.

x	y
-1	4
0	-1
1	-4
2	-5
4	-1

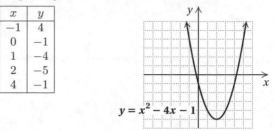

$y = x^2 - 4x - 1$

9. $\quad\quad\quad\quad x^2 + y^2 = 36$
$(x - 0)^2 + (y - 0)^2 = 6^2$ Standard form

The graph is a circle. The center is $(0, 0)$, and the radius is 6.

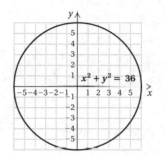

$x^2 + y^2 = 36$

10. $\dfrac{x^2}{9} - \dfrac{y^2}{4} = 1$
$\dfrac{x^2}{3^2} - \dfrac{y^2}{2^2} = 1$

The graph is a hyperbola. We have $a = 3$ and $b = 2$, so the asymptotes are $y = \dfrac{2}{3}x$ and $y = -\dfrac{2}{3}x$. Replacing y with 0 and solving for x, we get $x = \pm 3$, so the intercepts are $(3, 0)$ and $(-3, 0)$.

We plot the intercepts and draw smooth curves through them that approach the asymptotes.

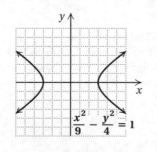

$\dfrac{x^2}{9} - \dfrac{y^2}{4} = 1$

11. $\dfrac{(x + 2)^2}{16} + \dfrac{(y - 3)^2}{9} = 1$
$\dfrac{[x - (-2)]^2}{4^2} + \dfrac{(y - 3)^2}{3^2} = 1$

The graph is an ellipse. The center is $(-2, 3)$. Note that $a = 4$ and $b = 3$. We locate the center and then plot the points $(-2 + 4, 3)$, $(-2 - 4, 3)$, $(-2, 3 - 3)$, and $(-2, 3 + 3)$, or

$(2, 3)$, $(-6, 3)$, $(-2, 0)$, and $(-2, 6)$. Connect these points with an oval-shaped curve.

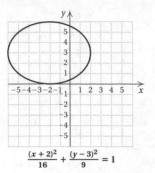

$$\frac{(x+2)^2}{16} + \frac{(y-3)^2}{9} = 1$$

12.
$$x^2 + y^2 - 4x + 6y + 4 = 0$$
$$(x^2 - 4x) + (y^2 + 6y) + 4 = 0$$
$$(x^2 - 4x + 4 - 4) + (y^2 + 6y + 9 - 9) + 4 = 0$$
$$(x^2 - 4x + 4) + (y^2 + 6y + 9) - 4 - 9 + 4 = 0$$
$$(x - 2)^2 + (y + 3)^2 = 9$$
$$(x - 2)^2 + [y - (-3)]^2 = 3^2$$

This is the equation of a circle with center $(2, -3)$ and radius 3.

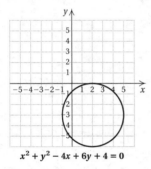

$$x^2 + y^2 - 4x + 6y + 4 = 0$$

13. $9x^2 + y^2 = 36$

$$\frac{x^2}{4} + \frac{y^2}{36} = 1 \quad \text{Dividing by 36}$$

$$\frac{x^2}{2^2} + \frac{y^2}{6^2} = 1$$

The graph is an ellipse. The x-intercepts are $(-2, 0)$ and $(2, 0)$ and the y-intercepts are $(0, -6)$ and $(0, 6)$. We plot these points and connect them with an oval-shaped curve.

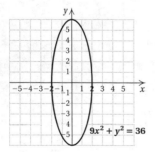

$$9x^2 + y^2 = 36$$

14. $xy = 4$

$$y = \frac{4}{x} \quad \text{Solving for } y$$

The graph is a hyperbola. We find some solutions, keeping the results in a table.

x	y
$\frac{1}{2}$	8
1	4
2	2
4	1
8	$\frac{1}{2}$
$-\frac{1}{2}$	-8
-1	-4
-2	-2
-8	$-\frac{1}{2}$

Note that we cannot use 0 for x. The x-axis and the y-axis are the asymptotes.

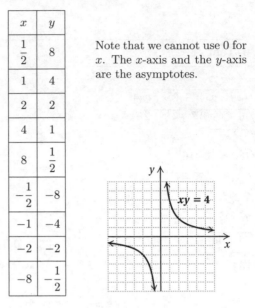

$xy = 4$

15. Graph: $x = -y^2 + 4y$

We complete the square.
$$x = -(y^2 - 4y)$$
$$= -(y^2 - 4y + 4 - 4)$$
$$= -(y^2 - 4y + 4) + (-1)(-4)$$
$$= -(y - 2)^2 + 4$$

The graph is a parabola. The vertex is $(4, 2)$; the line of symmetry is $y = 2$. The curve opens to the left.

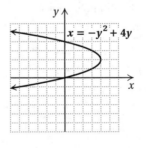

$x = -y^2 + 4y$

16. $\frac{x^2}{16} + \frac{y^2}{9} = 1, \quad (1)$

$3x + 4y = 12 \quad (2)$

First we multiply Equation (1) by 144 to clear the fractions.

$$144\left(\frac{x^2}{16} + \frac{y^2}{9}\right) = 144 \cdot 1$$

$$9x^2 + 16y^2 = 144 \quad (3)$$

Now we solve Equation (2) for x.

$$3x + 4y = 12$$
$$3x = -4y + 12$$
$$x = \frac{-4y + 12}{3}$$

Substitute $\frac{-4y + 12}{3}$ for x in Equation (3) and solve for y.

$$9\left(\frac{-4y + 12}{3}\right)^2 + 16y^2 = 144$$
$$9\left(\frac{16y^2 - 96y + 144}{9}\right) + 16y^2 = 144$$
$$16y^2 - 96y + 144 + 16y^2 = 144$$
$$32y^2 - 96y = 0$$
$$32y(y - 3) = 0$$
$$y = 0 \ or \ y = 3$$

When $y = 0$, $x = \frac{-4 \cdot 0 + 12}{3} = \frac{12}{3} = 4$.

When $y = 3$, $x = \frac{-4 \cdot 3 + 12}{3} = \frac{-12 + 12}{3} = \frac{0}{3} = 0$.

Both pairs check. The solutions are $(4, 0)$ and $(0, 3)$.

17.
$$x^2 + y^2 = 16, \quad (1)$$
$$\frac{x^2}{16} - \frac{y^2}{9} = 1 \quad (2)$$

First we multiply Equation (2) by 144 to clear the fractions.

$$144\left(\frac{x^2}{16} - \frac{y^2}{9}\right) = 144 \cdot 1$$
$$9x^2 - 16y^2 = 144 \quad (3)$$

Now we multiply Equation (1) by 16 and add it to Equation (3).

$$16x^2 + 16y^2 = 256$$
$$\underline{9x^2 - 16y^2 = 144}$$
$$25x^2 \qquad = 400$$
$$x^2 = 16$$
$$x = -4 \ or \ x = 4$$

When $x = -4$, we have:

$$(-4)^2 + y^2 = 16$$
$$16 + y^2 = 16$$
$$y^2 = 0$$
$$y = 0$$

When $x = 4$, we have:

$$4^2 + y^2 = 16$$
$$16 + y^2 = 16$$
$$y^2 = 0$$
$$y = 0$$

Both pairs check. The solutions are $(-4, 0)$ and $(4, 0)$.

18. Familiarize. Let $l =$ the length and $w =$ the width, in feet.

Translate. The diagonal is 20 ft, so we have $l^2 + w^2 = 20^2$ or $l^2 + w^2 = 400$. The perimeter is 56 ft, so we have $2l + 2w = 56$.

Solve. We solve the system of equations.

$$l^2 + w^2 = 400, \quad (1)$$
$$2l + 2w = 56 \quad (2)$$

First solve Equation (2) for l.

$$2l + 2w = 56$$
$$2l = 56 - 2w$$
$$l = 28 - w$$

Now substitute $28 - w$ for l in Equation (1) and solve for w.

$$l^2 + w^2 = 400$$
$$(28 - w)^2 + w^2 = 400$$
$$784 - 56w + w^2 + w^2 = 400$$
$$2w^2 - 56w + 784 = 400$$
$$2w^2 - 56w + 384 = 0$$
$$w^2 - 28w + 192 = 0 \qquad \text{Dividing by 2}$$
$$(w - 16)(w - 12) = 0$$
$$w = 16 \ or \ w = 12$$

If $w = 16$, then $l = 28 - 16 = 12$.

If $w = 12$, then $l = 28 - 12 = 16$.

Check. Since we usually consider length to be longer than width, we check $l = 16$ and $w = 12$. Since $16^2 + 12^2 = 256 + 144 = 400 = 20^2$ and $2 \cdot 16 + 2 \cdot 12 = 32 + 24 = 56$, the answer checks.

State. The length is 16 ft, and the width is 12 ft.

19. Familiarize. Let $x =$ the amount Peggyann invested and let $r =$ the interest rate. Then $x + 240 =$ the amount Sally Jean invested and $\frac{5}{6}r =$ the interest rate. The amount of interest Peggyann earned was $x \cdot r \cdot 1$, or xr, and Sally Jean earned $(x + 240) \cdot \frac{5}{6}r \cdot 1$, or $(x + 240) \cdot \frac{5}{6}r$.

Translate. Each investment earned \$72 interest, so we have two equations.

$$xr = 72, \qquad (1)$$
$$(x + 240) \cdot \frac{5}{6}r = 72 \quad (2)$$

Solve. First solve Equation (1) for x.

$$xr = 72$$
$$x = \frac{72}{r} \quad (3)$$

Now substitute $\frac{72}{r}$ for x in Equation (2) and solve for r.

$$\left(\frac{72}{r} + 240\right) \cdot \frac{5}{6}r = 72$$
$$60 + 200r = 72$$
$$200r = 12$$
$$r = 0.06, \text{ or } 6\%$$

Substitute 0.06 for r in Equation (3) and find x.

$$x = \frac{72}{0.06} = 1200$$

Check. If Peggyann invested \$1200 at 6% interest for 1 yr, the interest earned was \$1200 · 0.06 · 1 = \$72. The answer checks.

State. Peggyann invested \$1200 at 6% interest.

20. **Familiarize**. Let l = the length and w = the width, in yards.

Translate. The diagonal is $5\sqrt{5}$ yd, so we have $l^2 + w^2 = (5\sqrt{5})^2$, or $l^2 + w^2 = 125$. The area is 22 yd^2, so we have $lw = 22$.

Solve. We solve the system of equations.

$$l^2 + w^2 = 125, \quad (1)$$
$$lw = 22 \qquad (2)$$

First solve Equation (2) for l.

$$l = \frac{22}{w}$$

Then substitute $\frac{22}{w}$ for l in Equation (1) and solve for w.

$$\left(\frac{22}{w}\right)^2 + w^2 = 125$$
$$\frac{484}{w^2} + w^2 = 125$$
$$484 + w^4 = 125w^2 \quad \text{Multiplying by } w^2$$
$$w^4 - 125w^2 + 484 = 0$$
$$u^2 - 125u + 484 = 0 \qquad \text{Letting } u = w^2$$
$$(u - 121)(u - 4) = 0$$
$$u = 121 \quad or \quad u = 4$$
$$w^2 = 121 \quad or \quad w^2 = 4$$
$$w = \pm 11 \quad or \quad w = \pm 2$$

Check. The dimensions cannot be negative, so -11 and -2 cannot be solutions. If $w = 11$, $l = \frac{22}{11} = 2$. If $w = 2$, $l = \frac{22}{2} = 11$. Since we usually consider length to be longer than width, we check $l = 11$ and $w = 2$. We have $11^2 + 2^2 = 121 + 4 = 125 = (5\sqrt{5})^2$ and the area is 11 · 2, or 22 yd^2. The answer checks.

State. The length is 11 yd, and the width is 2 yd.

21. **Familiarize**. Let x and y represent the lengths of the sides of the squares, in meters. Then the areas are x^2 and y^2.

Translate. The sum of the areas is 8 m^2, so we have $x^2 + y^2 = 8$. The difference of the areas is 2 m^2, so we have $x^2 - y^2 = 2$.

Solve. We solve the system of equations.

$$\begin{array}{rl} x^2 + y^2 = 8, & (1) \\ x^2 - y^2 = 2 & (2) \\ \hline 2x^2 \phantom{{}+y} = 10 & \\ x^2 = 5 & \\ x = \pm\sqrt{5} & \end{array}$$

Since the length of a side cannot be negative we consider only $\sqrt{5}$. Substitute $\sqrt{5}$ for x in Equation (1) and solve for y.

$$(\sqrt{5})^2 + y^2 = 8$$
$$5 + y^2 = 8$$
$$y^2 = 3$$
$$y = \pm\sqrt{3}$$

Again, we consider only the positive value. The possible solution is $(\sqrt{5}, \sqrt{3})$.

Check. The areas of the squares are $(\sqrt{5})^2$, or 5 m^2, and $(\sqrt{3})^2$, or 3 m^2. The sum of the areas is $5 + 3$, or 8 m^2, and the difference is $5 - 3$, or 2 m^2. The answer checks.

State. The lengths of the sides of the squares are $\sqrt{5}$ m and $\sqrt{3}$ m.

22. The graph of $\dfrac{x^2}{9} - \dfrac{y^2}{4} = 1$ is a hyperbola with intercepts $(-3, 0)$ and $(3, 0)$. The graph of $x^2 + y^2 = 16$ is a circle with center $(0, 0)$ and radius 4. Graph B represents the solution set of the system of equations.

23. The center of the ellipse is the midpoint of the segment joining the vertices. We find it.

$$\left(\frac{1 + 11}{2}, \frac{3 + 3}{2}\right) = \left(\frac{12}{2}, \frac{6}{2}\right) = (6, 3)$$

Thus, we know that $h = 6$ and $k = 3$.

Since the vertex on the left is $(1, 3)$, we know that $h - a = 6 - a = 1$, so $a = 5$. Since the points $(6, 0)$ and $(6, 6)$ lie directly below and above the center, we know that $k - b = 3 - b = 0$, or $b = 3$. We write the equation of the ellipse.

$$\frac{(x - h)^2}{a^2} + \frac{(y - k)^2}{b^2} = 1$$
$$\frac{(x - 6)^2}{5^2} + \frac{(y - 3)^2}{3^2} = 1, \text{ or } \frac{(x - 6)^2}{25} + \frac{(y - 3)^2}{9} = 1$$

24. Let (x, y) be a point whose distance from $(8, 0)$ is 10. Then we have:

$$\sqrt{(x - 8)^2 + (y - 0)^2} = 10$$
$$\sqrt{(x - 8)^2 + y^2} = 10$$
$$(x - 8)^2 + y^2 = 100 \quad \text{Squaring both sides}$$

Then the set of all points whose distance from $(8, 0)$ is 10 is $\{(x, y) | (x - 8)^2 + y^2 = 100\}$.

We could also observe that the set of points whose distance from $(8, 0)$ is 10 are the points on the circle with center $(8, 0)$ and radius 10, or $(x - 8)^2 + (y - 0)^2 = 10^2$, or $(x - 8)^2 + y^2 = 100$. This gives us the same result as above.

25. **Familiarize**. Let x and y represent the numbers.

Translate. The sum of the numbers is 36, so we have $x + y = 36$. The product of the numbers is 4, so we have $xy = 4$.

Solve. We solve the system of equations.

$$x + y = 36, \quad (1)$$
$$xy = 4 \qquad (2)$$

First we solve Equation (1) for y.

$$y = 36 - x$$

Substitute $36 - x$ for y in Equation (2) and solve for x.

$$x(36 - x) = 4$$
$$36x - x^2 = 4$$
$$0 = x^2 - 36x + 4$$
$$x = \frac{-(-36) \pm \sqrt{(-36)^2 - 4 \cdot 1 \cdot 4}}{2 \cdot 1}$$
$$= \frac{36 \pm \sqrt{1296 - 16}}{2} = \frac{36 \pm \sqrt{1280}}{2}$$
$$= \frac{36 \pm \sqrt{256 \cdot 5}}{2} = \frac{36 \pm 16\sqrt{5}}{2}$$
$$= \frac{2(18 \pm 8\sqrt{5})}{2} = 18 \pm 8\sqrt{5}$$

If $x = 18 - 8\sqrt{5}$, then $y = 36 - (18 - 8\sqrt{5}) = 36 - 18 + 8\sqrt{5} = 18 + 8\sqrt{5}$.

If $x = 18 + 8\sqrt{5}$, then $y = 36 - (18 + 8\sqrt{5}) = 36 - 18 - 8\sqrt{5} = 18 - 8\sqrt{5}$.

In either case we see that the numbers are $18 + 8\sqrt{5}$ and $18 - 8\sqrt{5}$.

Check. The sum of the numbers is $18 + 8\sqrt{5} + 18 - 8\sqrt{5} = 36$. The product is $(18 + 8\sqrt{5})(18 - 8\sqrt{5}) = 324 - 320 = 4$. The answer checks.

Now we find the sum of the reciprocals.

$$\frac{1}{18 + 8\sqrt{5}} + \frac{1}{18 - 8\sqrt{5}} = \frac{18 - 8\sqrt{5} + 18 + 8\sqrt{5}}{(18 + 8\sqrt{5})(18 - 8\sqrt{5})}$$
$$= \frac{36}{4} = 9$$

State. The sum of the reciprocals of the numbers is 9.

26. Let $(0, y)$ represent the point on the y-axis that is equidistant from $(-3, -5)$ and $(4, -7)$. Then we have:

$$\sqrt{(-3 - 0)^2 + (-5 - y)^2} = \sqrt{(4 - 0)^2 + (-7 - y)^2}$$
$$\sqrt{9 + 25 + 10y + y^2} = \sqrt{16 + 49 + 14y + y^2}$$
$$\sqrt{34 + 10y + y^2} = \sqrt{65 + 14y + y^2}$$
$$34 + 10y + y^2 = 65 + 14y + y^2$$
$$34 + 10y = 65 + 14y$$
$$-4y = 31$$
$$y = -\frac{31}{4}$$

The point is $\left(0, -\frac{31}{4}\right)$.

Cumulative Review Chapters 1 - 9

1.
$$\frac{1}{3}x - \frac{1}{5} \geq \frac{1}{5}x - \frac{1}{3}$$
$$15\left(\frac{1}{3}x - \frac{1}{5}\right) \geq 15\left(\frac{1}{5}x - \frac{1}{3}\right)$$
$$5x - 3 \geq 3x - 5$$
$$2x \geq -2$$
$$x \geq -1$$

The solution set is $\{x | x \geq -1\}$, or $[-1, \infty)$.

2. $|x| > 6.4$

$x < -6.4 \ \ or \ \ x > 6.4$

The solution set is $\{x | x < -6.4 \ or \ x > 6.4\}$, or $(-\infty, -6.4) \cup (6.4, \infty)$.

3. $3 \leq 4x + 7 < 31$

$$-4 \leq 4x < 24$$
$$-1 \leq x < 6$$

The solution set is $\{x | -1 \leq x < 6\}$, or $[-1, 6)$.

4.
$$3x + y = 4, \quad (1)$$
$$\underline{-6x - y = -3} \quad (2)$$
$$-3x = 1 \quad \text{Adding}$$
$$x = -\frac{1}{3}$$

Substitute $-\frac{1}{3}$ for x in Equation (1) and solve for y.

$$3\left(-\frac{1}{3}\right) + y = 4$$
$$-1 + y = 4$$
$$y = 5$$

The solution is $\left(-\frac{1}{3}, 5\right)$.

5. $x^4 - 13x^2 + 36 = 0$

Let $u = x^2$.

$$u^2 - 13u + 36 = 0$$
$$(u - 4)(u - 9) = 0$$
$$u - 4 = 0 \quad or \quad u - 9 = 0$$
$$u = 4 \quad or \quad u = 9$$
$$x^2 = 4 \quad or \quad x^2 = 9 \quad \text{Substituting } x^2 \text{ for } u$$
$$x = \pm 2 \quad or \quad x = \pm 3$$

The solutions are 2, -2, 3, and -3.

6.
$$2x^2 = x + 3$$
$$2x^2 - x - 3 = 0$$
$$(2x - 3)(x + 1) = 0$$
$$2x - 3 = 0 \quad or \quad x + 1 = 0$$
$$2x = 3 \quad or \quad x = -1$$
$$x = \frac{3}{2} \quad or \quad x = -1$$

The solutions are $\frac{3}{2}$ and -1.

7.
$$3x - \frac{6}{x} = 7$$
$$x\left(3x - \frac{6}{x}\right) = x \cdot 7$$
$$3x^2 - 6 = 7x$$
$$3x^2 - 7x - 6 = 0$$
$$(3x + 2)(x - 3) = 0$$
$$3x + 2 = 0 \quad or \quad x - 3 = 0$$
$$3x = -2 \quad or \quad x = 3$$
$$x = -\frac{2}{3} \quad or \quad x = 3$$

Both numbers check. The solutions are $-\frac{2}{3}$ and 3.

8.
$$\sqrt{x + 5} = x - 1$$
$$(\sqrt{x + 5})^2 = (x - 1)^2$$
$$x + 5 = x^2 - 2x + 1$$
$$0 = x^2 - 3x - 4$$
$$0 = (x - 4)(x + 1)$$
$$x - 4 = 0 \quad or \quad x + 1 = 0$$
$$x = 4 \quad or \quad x = -1$$

The number 4 checks but -1 does not, so the solution is 4.

9.
$$x(x + 10) = -21$$
$$x^2 + 10x = -21$$
$$x^2 + 10x + 21 = 0$$
$$(x + 3)(x + 7) = 0$$
$$x + 3 = 0 \quad or \quad x + 7 = 0$$
$$x = -3 \quad or \quad x = -7$$

The solutions are -3 and -7.

10. $2x^2 + x + 1 = 0$
$$a = 2, \ b = 1, \ c = 1$$
$$x = \frac{-b \pm \sqrt{b^2 - 4ac}}{2a}$$
$$x = \frac{-1 \pm \sqrt{1^2 - 4 \cdot 2 \cdot 1}}{2 \cdot 2} = \frac{-1 \pm \sqrt{1 - 8}}{4}$$
$$= \frac{-1 \pm \sqrt{-7}}{4} = \frac{-1 \pm i\sqrt{7}}{4}$$
$$= -\frac{1}{4} \pm i\frac{\sqrt{7}}{4}$$

The solutions are $-\frac{1}{4} \pm i\frac{\sqrt{7}}{4}$.

11.
$$7^x = 30$$
$$\log 7^x = \log 30$$
$$x \log 7 = \log 30$$
$$x = \frac{\log 30}{\log 7} \approx 1.748$$

12. $\frac{x + 1}{x - 2} > 0$
Solve the related equation.

$$\frac{x + 1}{x - 2} = 0$$
$$x + 1 = 0$$
$$x = -1$$

Find the numbers for which the rational expression is undefined.
$$x - 2 = 0$$
$$x = 2$$

Use the numbers -1 and 2 to divide the number line into intervals as shown:

Try test numbers in each interval.

A: Test -2,
$$\frac{x + 1}{x - 2} > 0$$
$$\frac{-2 + 1}{-2 - 2} \ ? \ 0$$
$$\frac{-1}{-4}$$
$$\frac{1}{4} \quad \bigg| \quad \text{TRUE}$$

The number -2 is a solution of the inequality, so the interval A is part of the solution set.

B: Test 0,
$$\frac{x + 1}{x - 2} > 0$$
$$\frac{0 + 1}{0 - 2} \ ? \ 0$$
$$-\frac{1}{2} \quad \bigg| \quad \text{FALSE}$$

The number 0 is not a solution of the inequality, so the interval B is not part of the solution set.

C: Test 4,
$$\frac{x + 1}{x - 2} > 0$$
$$\frac{4 + 1}{4 - 2} \ ? \ 0$$
$$\frac{5}{2} \quad \bigg| \quad \text{TRUE}$$

The number 4 is a solution of the inequality, so the interval C is part of the solution set. The solution set is
$\{x | x < -1 \ or \ x > 2\}$, or $(-\infty, -1) \cup (2, \infty)$.

13.
$$\log_3 x = 2$$
$$3^2 = x$$
$$9 = x$$

14.
$$x^2 - 1 \geq 0$$
$$(x + 1)(x - 1) \geq 0$$

The solutions of $(x + 1)(x - 1) = 0$ are -1 and 1. They divide the real-number line into three intervals as shown:

We try a test number in each interval.

A: Test -2, $(-2+1)(-2-1) = 3 > 0$

B: Test 0, $(0+1)(0-1) = -1 < 0$

C: Test 2, $(2+1)(2-1) = 3 > 0$

The expression is positive for all numbers in intervals A and C. The inequality symbol is $\geq$, so we need to include the x-intercepts. The solution set is $\{x | x \leq -1 \ or \ x \geq 1\}$, or $(-\infty, -1] \cup [1, \infty)$.

15. $\log_2 x + \log_2 (x+7) = 3$

$$\log_2 x(x+7) = 3$$
$$2^3 = x(x+7)$$
$$8 = x^2 + 7x$$
$$0 = x^2 + 7x - 8$$
$$0 = (x+8)(x-1)$$

$x + 8 = 0 \quad or \quad x - 1 = 0$

$x = -8 \quad or \quad x = 1$

The number -8 does not check but 1 does, so the solution is 1.

16.
$$\frac{1}{p} + \frac{1}{q} = \frac{1}{f}$$

$$pqf\left(\frac{1}{p} + \frac{1}{q}\right) = pqf \cdot \frac{1}{f} \quad \text{Multiplying by } pqf$$

$$pqf \cdot \frac{1}{p} + pqf \cdot \frac{1}{q} = pq$$

$$qf + pf = pq$$

$$qf = pq - pf \quad \text{Subtracting } pf$$

$$qf = p(q - f) \quad \text{Factoring}$$

$$\frac{qf}{q - f} = p \quad \text{Dividing by } q - f$$

17. $\quad x - y + 2z = 3, \quad (1)$

$\quad -x \quad\ \ + z = 4, \quad (2)$

$\quad 2x + y - z = -3 \quad (3)$

First we add Equations (1) and (3).

$$x - y + 2z = 3$$
$$\underline{2x + y - z = -3}$$
$$3x \quad\ \ + z = 0 \quad (4)$$

Now we solve the system of Equations (2) and (4).

$-x + z = 4, \quad (2)$

$3x + z = 0 \quad (4)$

Multiply Equation (2) by -1 and then add.

$$x - z = -4$$
$$\underline{3x + z = 0}$$
$$4x \quad\ \ = -4$$
$$x = -1$$

Now substitute -1 for x in Equation (2) and solve for z.

$$-(-1) + z = 4$$
$$1 + z = 4$$
$$z = 3$$

Finally substitute -1 for x and 3 for z in Equation (3) and solve for y.

$$2(-1) + y - 3 = -3$$
$$-2 + y - 3 = -3$$
$$y - 5 = -3$$
$$y = 2$$

The solution is $(-1, 2, 3)$.

18. $\quad -x^2 + 2y^2 = 7, \quad (1)$

$\quad \underline{x^2 + y^2 = 5} \quad (2)$

$\qquad\qquad 3y^2 = 12 \quad \text{Adding}$

$\qquad\qquad\ \ y^2 = 4$

$\qquad\qquad\ \ \ y = \pm 2$

When $y = -2$, $y^2 = 4$, and when $y = 2$, $y^2 = 4$, so we can substitute 4 for y^2 in Equation (2) and solve for x.

$$x^2 + 4 = 5$$
$$x^2 = 1$$
$$x = \pm 1$$

The solutions are $(1, 2)$, $(1, -2)$, $(-1, 2)$, and $(-1, -2)$.

19.
$$\frac{3}{x-3} - \frac{x+2}{x^2 + 2x - 15} = \frac{1}{x+5}$$

$$\frac{3}{x-3} - \frac{x+2}{(x+5)(x-3)} = \frac{1}{x+5}, \quad \text{LCD is } (x-3)(x+5)$$

$$(x-3)(x+5)\left(\frac{3}{x-3} - \frac{x+2}{(x+5)(x-3)}\right) =$$
$$(x-3)(x+5) \cdot \frac{1}{x+5}$$

$$3(x+5) - (x+2) = x - 3$$
$$3x + 15 - x - 2 = x - 3$$
$$2x + 13 = x - 3$$
$$x = -16$$

The number -16 checks. It is the solution.

20.
$$P = \frac{3}{4}(M + 2N)$$

$$4P = 3(M + 2N) \quad \text{Multiplying by } 4$$

$$4P = 3M + 6N$$

$$4P - 3M = 6N$$

$$\frac{4P - 3M}{6} = N$$

21. $N(t) = 77(1.019)^t$

a) In 2016, $t = 2016 - 2000 = 16$.

$\quad N(16) = 77(1.019)^{16} \approx 104$ million barrels per day

$\quad$ In 2025, $t = 2025 - 2000 = 25$.

$\quad N(25) = 77(1.019)^{25} \approx 123$ million barrels per day

b) $N_0 = 77$, so $2N_0 = 2 \cdot 77 = 154$.

$$154 = 77(1.019)^t$$
$$2 = 1.019^t$$
$$\log 2 = \log 1.019^t$$
$$\log 2 = t \log 1.019$$
$$\frac{\log 2}{\log 1.019} = t$$
$$37 \text{ yr} \approx t$$

c) Use the points found in parts (a) and (b) and any additional points as needed to graph the function.

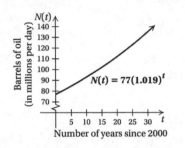

22. a) $A(t) = \$50,000(1 + 0.04)^t = \$50,000(1.04)^t$

b) $A(0) = \$50,000(1.04)^0 = \$50,000$

$A(4) = \$50,000(1.04)^4 \approx \$58,492.93$

$A(8) = \$50,000(1.04)^8 \approx \$68,428.45$

$A(10) = \$50,000(1.04)^{10} \approx \$74,012.21$

c) Use the points found in part (b) and any additional points as needed to graph the function.

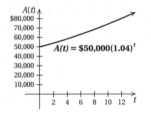

23.
$$(2x+3)(x^2 - 2x - 1)$$
$$= (2x+3)x^2 + (2x+3)(-2x) + (2x+3)(-1)$$
$$= 2x^3 + 3x^2 - 4x^2 - 6x - 2x - 3$$
$$= 2x^3 - x^2 - 8x - 3$$

24.
$$(3x^2 + x^3 - 1) - (2x^3 + x + 5)$$
$$= (3x^2 + x^3 - 1) + (-2x^3 - x - 5)$$
$$= -x^3 + 3x^2 - x - 6$$

25.
$$\frac{2m^2 + 11m - 6}{m^3 + 1} \cdot \frac{m^2 - m + 1}{m + 6}$$
$$= \frac{(2m^2 + 11m - 6)(m^2 - m + 1)}{(m^3 + 1)(m + 6)}$$
$$= \frac{(2m - 1)(m + 6)(m^2 - m + 1)}{(m + 1)(m^2 - m + 1)(m + 6)}$$
$$= \frac{(m + 6)(m^2 - m + 1)}{(m + 6)(m^2 - m + 1)} \cdot \frac{2m - 1}{m + 1}$$
$$= \frac{2m - 1}{m + 1}$$

26.
$$\frac{x}{x - 1} + \frac{2}{x + 1} - \frac{2x}{x^2 - 1}$$
$$= \frac{x}{x-1} + \frac{2}{x+1} - \frac{2x}{(x+1)(x-1)}, \text{ LCD is } (x-1)(x+1)$$
$$= \frac{x}{x - 1} \cdot \frac{x + 1}{x + 1} + \frac{2}{x + 1} \cdot \frac{x - 1}{x - 1} - \frac{2x}{(x + 1)(x - 1)}$$
$$= \frac{x(x + 1) + 2(x - 1) - 2x}{(x - 1)(x + 1)}$$
$$= \frac{x^2 + x + 2x - 2 - 2x}{(x - 1)(x + 1)}$$
$$= \frac{x^2 + x - 2}{(x - 1)(x + 1)}$$
$$= \frac{(x + 2)(x - 1)}{(x - 1)(x + 1)}$$
$$= \frac{(x + 2)(x - 1)}{(x - 1)(x + 1)}$$
$$= \frac{x + 2}{x + 1}$$

27.
$$\frac{1 - \dfrac{5}{x}}{x - 4 - \dfrac{5}{x}} = \frac{1 - \dfrac{5}{x}}{x - 4 - \dfrac{5}{x}} \cdot \frac{x}{x}$$
$$= \frac{\left(1 - \dfrac{5}{x}\right)x}{\left(x - 4 - \dfrac{5}{x}\right)x}$$
$$= \frac{x - 5}{x^2 - 4x - 5}$$
$$= \frac{x - 5}{(x + 1)(x - 5)}$$
$$= \frac{(x - 5) \cdot 1}{(x + 1)(x - 5)}$$
$$= \frac{1}{x + 1}$$

28.

$$\begin{array}{r}
x^3 + 2x^2 - 2x + 1 \\
x + 1 \overline{\smash{\big)}\ x^4 + 3x^3 + 0x^2 - x + 4} \\
\underline{x^4 + x^3} \\
2x^3 + 0x^2 \\
\underline{2x^3 + 2x^2} \\
-2x^2 - x \\
\underline{-2x^2 - 2x} \\
x + 4 \\
\underline{x + 1} \\
3
\end{array}$$

The answer is $x^3 + 2x^2 - 2x + 1 + \dfrac{3}{x + 1}$.

29. $\dfrac{\sqrt{75x^5 y^2}}{\sqrt{3xy}} = \sqrt{\dfrac{75x^5 y^2}{3xy}} = \sqrt{25x^4 y} = \sqrt{25x^4} \cdot \sqrt{y} = 5x^2\sqrt{y}$

30. $4\sqrt{50} - 3\sqrt{18} = 4\sqrt{25 \cdot 2} - 3\sqrt{9 \cdot 2} = 4\sqrt{25}\sqrt{2} - 3\sqrt{9}\sqrt{2} = 4 \cdot 5\sqrt{2} - 3 \cdot 3\sqrt{2} = 20\sqrt{2} - 9\sqrt{2} = 11\sqrt{2}$

31. $(16^{3/2})^{1/2} = 16^{3/4} = (2^4)^{3/4} = 2^3 = 8$

32. $(2 - i\sqrt{2})(5 + 3i\sqrt{2}) = 10 + 6i\sqrt{2} - 5i\sqrt{2} - 3i^2 \cdot 2$
$$= 10 + i\sqrt{2} + 6 \quad (i^2 = -1)$$
$$= 16 + i\sqrt{2}$$

33. $\dfrac{5 + i}{2 - 4i} = \dfrac{5 + i}{2 - 4i} \cdot \dfrac{2 + 4i}{2 + 4i}$
$$= \dfrac{10 + 20i + 2i + 4i^2}{4 - 16i^2}$$
$$= \dfrac{10 + 22i - 4}{4 + 16}$$
$$= \dfrac{6 + 22i}{20}$$
$$= \dfrac{3}{10} + \dfrac{11}{10}i$$

34. $g(t) = -700t + 53{,}000$

a) In 2006, $t = 2006 - 2000 = 6$.

$g(6) = -700 \cdot 6 + 53{,}000 = 48{,}800 \text{ ft}^2$

In 2010, $t = 2010 - 2000 = 10$.

$g(10) = -700 \cdot 10 + 53{,}000 = 46{,}000 \text{ ft}^2$

In 2012, $t = 2012 - 2000 = 12$.

$g(12) = -700 \cdot 12 + 53{,}000 = 44{,}600 \text{ ft}^2$

b) Use the points found in part (a) to graph the function.

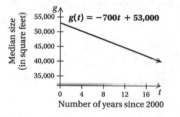

c) The equation is in slope-intercept form, so we see that the y-intercept is $(0, 53{,}000)$.

d) The equation is in slope-intercept form, so we see that the slope is -700.

e) The slope is -700, so the rate of change is a decrease of 700 ft^2 per year.

35. First we will find the slope of the line.
$$m = \dfrac{0 - 4}{-1 - 1} = \dfrac{-4}{-2} = 2$$

We will use the point-slope formula and use the point $(-1, 0)$.

$$y - y_1 = m(x - x_1)$$
$$y - 0 = 2(x - (-1))$$
$$y = 2(x + 1)$$
$$y = 2x + 2$$

36. First solve the given equation for y to determine its slope.
$$2x - y = 3$$
$$-y = -2x + 3$$
$$y = 2x - 3$$

The slope of the given line is 2. The slope of the perpendicular line is the opposite of the reciprocal of 2, or $-\dfrac{1}{2}$. We will use the slope-intercept equation.

$$y = mx + b$$
$$2 = -\dfrac{1}{2} \cdot 1 + b$$
$$2 = -\dfrac{1}{2} + b$$
$$\dfrac{5}{2} = b$$

Finally we substitute $-\dfrac{1}{2}$ for m and $\dfrac{5}{2}$ for b in the slope-intercept equation to find the desired equation.

$$y = -\dfrac{1}{2}x + \dfrac{5}{2}$$

37. $4y - 3x = 12$

We will find the intercepts. To find the x-intercept we let $y = 0$ and solve for x.

$$4 \cdot 0 - 3x = 12$$
$$-3x = 12$$
$$x = -4$$

The x-intercept is $(-4, 0)$.

To find the y-intercept, let $x = 0$ and solve for y.

$$4y - 3 \cdot 0 = 12$$
$$4y = 12$$
$$y = 3$$

The y-intercept is $(0, 3)$.

We plot these points and draw the line. A third point can be plotted as a check.

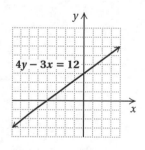

38. $y < -2$

First we graph $y = 2$. We draw the line dashed since the inequality symbol is $<$. Test the point $(0, 0)$ to determine if it is a solution.

$$\dfrac{y < -2}{0 \ ? \ -2 \quad \text{FALSE}}$$

Since $0 < -2$ is false, we shade the half-plane that does not contain $(0, 0)$.

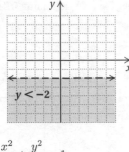

39. $\dfrac{x^2}{9} + \dfrac{y^2}{25} = 1$

$\dfrac{x^2}{3^2} + \dfrac{y^2}{5^2} = 1$

This is an equation of an ellipse with center at $(0,0)$. The x-intercepts are $(-3,0)$ and $(3,0)$. The y-intercepts are $(0,-5)$ and $(0,5)$. We plot the intercepts and connect them with an oval-shaped curve.

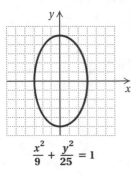

$$\dfrac{x^2}{9} + \dfrac{y^2}{25} = 1$$

40. $x^2 + y^2 = 2.25$

$x^2 + y^2 = 1.5^2$

This is an equation of a circle with center $(0,0)$ and radius 1.5.

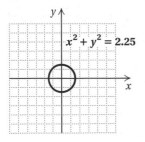

41. $x + y \leq 0,$

$\qquad x \geq -4,$

$\qquad y \geq -1$

Shade the intersection of the graphs of the three inequalities.

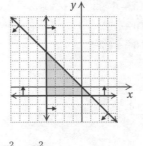

42. $\dfrac{x^2}{25} - \dfrac{y^2}{16} = 1$

$\dfrac{x^2}{5^2} - \dfrac{y^2}{4^2} = 1, \ a = 5, \ b = 4$

This is an equation of a hyperbola. The asymptotes are $y = -\dfrac{b}{a}x$ and $y = \dfrac{b}{a}x$, or $y = -\dfrac{5}{4}x$ and $y = \dfrac{5}{4}x$. The hyperbola has center $(0,0)$ and a horizontal axis. The vertices are the x-intercepts. We find them.

$$\dfrac{x^2}{25} - \dfrac{0^2}{16} = 1$$

$$\dfrac{x^2}{25} = 1$$

$$x^2 = 25$$

$$x = \pm 5$$

The x-intercepts are $(-5,0)$ and $(5,0)$. There are no y-intercepts. We plot the intercepts and draw the graph.

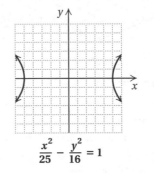

$$\dfrac{x^2}{25} - \dfrac{y^2}{16} = 1$$

43. $(x-1)^2 + (y+1)^2 = 9$

$(x-1)^2 + [y-(-1)]^2 = 3^2$ Standard form

The graph is a circle with center $(1,-1)$ and radius 3.

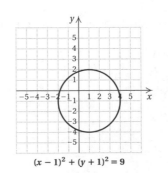

$$(x-1)^2 + (y+1)^2 = 9$$

44. $f(x) = 2x^2 - 8x + 9 = 2(x^2 - 4x) + 9$

We complete the square inside the parentheses. We take half the x-coefficient and square it.

$$\frac{1}{2}(-4) = -2 \text{ and } (-2)^2 = 4$$

Then we add $4 - 4$ inside the parentheses.

$$\begin{aligned} f(x) &= 2(x^2 - 4x + 4 - 4) + 9 \\ &= 2(x^2 - 4x + 4) + 2(-4) + 9 \\ &= 2(x - 2)^2 + 1 \end{aligned}$$

Vertex: $(2, 1)$

Line of symmetry: $x = 2$

We plot a few points and draw the graph.

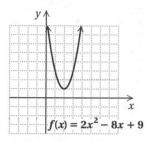

45. $x = 3.5$

This is an equation of a vertical line with x-intercept $(3.5, 0)$.

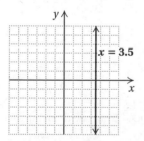

46. $x = y^2 + 1 = (y - 0)^2 + 1$

The graph is a parabola with vertex $(1, 0)$ and line of symmetry $y = 0$. The curve opens to the right. We plot a few points and draw the graph.

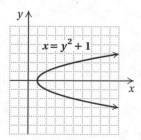

47. Graph: $f(x) = e^{-x}$

We find some function values with a calculator. We use these values to plot points and draw the graph.

x	e^{-x}
0	1
-1	2.7
-2	7.4
-3	20.1
1	0.4
2	0.1
3	0.05

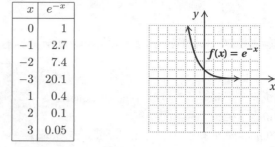

48. $f(x) = \log_2 x$

The equation $f(x) = y = \log_2 x$ is equivalent to $2^y = x$. We can find ordered pairs by choosing values for y and computing the corresponding x-values.

For $y = 0$, $x = 2^0 = 1$.

For $y = 1$, $x = 2^1 = 2$.

For $y = 2$, $x = 2^2 = 4$.

For $y = 3$, $x = 2^3 = 8$.

For $y = -1$, $x = 2^{-1} = \dfrac{1}{2}$.

For $y = -2$, $x = 2^{-2} = \dfrac{1}{4}$.

x, or 2^y	y
1	0
2	1
4	2
8	3
$\dfrac{1}{2}$	-1
$\dfrac{1}{4}$	-2

           (1) Select y.

           (2) Compute x.

We plot the set of ordered pairs and connect the points with a smooth curve.

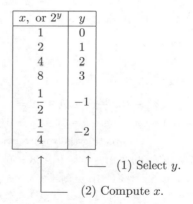

49. $\qquad 2x^4 - 12x^3 + x - 6$

$\qquad = 2x^3(x - 6) + (x - 6) \quad$ Factoring by grouping

$\qquad = (x - 6)(2x^3 + 1)$

50. $3a^2 - 12ab - 135b^2$

$= 3(a^2 - 4ab - 45b^2)$

$= 3(a - 9b)(a + 5b)$ Factoring by trial and error

51. $x^2 - 17x + 72$

$= (x - 8)(x - 9)$ Factoring by trial and error

52. $81m^4 - n^4$

$= (9m^2)^2 - (n^2)^2$ Difference of squares

$= (9m^2 + n^2)(9m^2 - n^2)$

$= (9m^2 + n^2)[(3m)^2 - n^2]$ Difference of squares

$= (9m^2 + n^2)(3m + n)(3m - n)$

53. $16x^2 - 16x + 4$

$= 4(4x^2 - 4x + 1)$ Square of a binomial

$= 4(2x - 1)^2$

54. $81a^3 - 24$

$= 3(27a^3 - 8)$

$= 3[(3a)^3 - 2^3]$ Difference of cubes

$= 3(3a - 2)[(3a)^2 + 3a \cdot 2 + 2^2]$

$= 3(3a - 2)(9a^2 + 6a + 4)$

55. $10x^2 + 66x - 28$

$= 2(5x^2 + 33x - 14)$

$= 2(5x - 2)(x + 7)$ FOIL or ac-method

56. $6x^3 + 27x^2 - 15x$

$= 3x(2x^2 + 9x - 5)$

$= 3x(2x - 1)(x + 5)$ FOIL or ac-method

57.
$$x^2 - 16x + y^2 + 6y + 68 = 0$$
$$(x^2 - 16x + 64 - 64) + (y^2 + 6y + 9 - 9) + 68 = 0$$

Completing the square twice

$$(x^2 - 16x + 64) + (y^2 + 6y + 9) - 64 - 9 + 68 = 0$$
$$(x - 8)^2 + (y + 3)^2 = 5$$
$$(x - 8)^2 + [y - (-3)]^2 = (\sqrt{5})^2$$

The center is $(8, -3)$ and the radius is $\sqrt{5}$.

58. $f(x) = 2x - 3$ passes the horizontal-line test so it is one-to-one and has an inverse that is a function. We find a formula for the inverse.

1. Replace $f(x)$ by y: $y = 2x - 3$

2. Interchange x and y: $x = 2y - 3$

3. Solve for y: $x + 3 = 2y$

$$\frac{x + 3}{2} = y$$

4. Replace y by $f^{-1}(x)$: $f^{-1}(x) = \dfrac{x + 3}{2}$, or $\dfrac{1}{2}(x + 3)$

59. $z = \dfrac{kx}{y^3}$

$5 = \dfrac{k \cdot 4}{2^3}$

$5 = \dfrac{4k}{8}$

$10 = k$ Variation constant

$z = \dfrac{10x}{y^3}$ Equation of variation

$z = \dfrac{10 \cdot 10}{5^3}$

$z = \dfrac{100}{125} = \dfrac{4}{5}$

60. $f(x) = x^3 - 2$

$f(-2) = (-2)^3 - 2 = -8 - 2 = -10$

61. $d = \sqrt{(x_2 - x_1)^2 + (y_2 - y_1)^2}$

$d = \sqrt{(8 - 2)^2 + (9 - 1)^2}$

$= \sqrt{6^2 + 8^2} = \sqrt{36 + 64}$

$= \sqrt{100} = 10$

62. $\left(\dfrac{x_1 + x_2}{2}, \dfrac{y_1 + y_2}{2}\right) = \left(\dfrac{-1 + 3}{2}, \dfrac{-3 + 0}{2}\right) =$

$\left(\dfrac{2}{2}, \dfrac{-3}{2}\right) = \left(1, -\dfrac{3}{2}\right)$

63. $\dfrac{5 + \sqrt{a}}{3 - \sqrt{a}} = \dfrac{5 + \sqrt{a}}{3 - \sqrt{a}} \cdot \dfrac{3 + \sqrt{a}}{3 + \sqrt{a}}$

$= \dfrac{15 + 5\sqrt{a} + 3\sqrt{a} + a}{9 - a}$

$= \dfrac{15 + 8\sqrt{a} + a}{9 - a}$

64. $f(x) = \dfrac{4x - 3}{3x^2 + x}$

To find the numbers excluded from the domain, we set the denominator equal to 0 and solve.

$3x^2 + x = 0$

$x(3x + 1) = 0$

$x = 0$ $\ or$ $3x + 1 = 0$

$x = 0$ $\ or$ $\qquad 3x = -1$

$x = 0$ $\ or$ $\qquad x = -\dfrac{1}{3}$

Then the domain is $\left(-\infty, -\dfrac{1}{3}\right) \cup \left(-\dfrac{1}{3}, 0\right) \cup (0, \infty)$.

65. $3a^2 + a = 2$

$3a^2 + a - 2 = 0$

$(3a - 2)(a + 1) = 0$

$3a - 2 = 0$ $\ or$ $a + 1 = 0$

$3a = 2$ $\ or$ $\qquad a = -1$

$a = \dfrac{2}{3}$ $\ or$ $\qquad a = -1$

The values of a for which $f(a) = 2$ are $\dfrac{2}{3}$ and -1.

66. *Familiarize*. Let c = the number of books purchased annually. Then limited members would pay $20 + 20c$ and preferred members would pay $40 + 15c$.

Translate.

$$\underbrace{\text{Cost for preferred members}}_{40 + 15c} \quad \underbrace{\text{is less than}}_{<} \quad \underbrace{\text{cost for limited members.}}_{20 + 20c}$$

Solve. We solve the inequality.

$$40 + 15c < 20 + 20c$$
$$20 < 5c$$
$$4 < c$$

Check. When 4 books are purchased the cost for preferred members is $40 + 15 \cdot 4$, or \$100, and the cost for limited members is $20 + 20 \cdot 4$, or \$100, so the costs are the same. When more than 4 books are purchased, say 5 books, preferred members pay $40 + 15 \cdot 5$, or \$115, and limited members pay $20 + 20 \cdot 5$, or \$120, so preferred members pay less. This partial check shows that the answer is probably correct.

State. When more than 4 books are purchased annually, it is less expensive to be a preferred member.

67. *Familiarize*. Let d = the distance the trains travel before the passenger train overtakes the freight train, in miles, and let t = the number of hours the freight train travels before being overtaken. Since it is 9 hr from 2 A.M. to 11 A.M., $t - 9$ = the number of hours the passenger train travels. The speed of the passenger train is $2 \cdot 34$, or 68 mph. We organize the information in a table.

Train	Distance	Speed	Time
Freight	d	34	t
Passenger	d	68	$t - 9$

Translate. Using $d = rt$ in each row of the table we have two equations.

$$d = 34t, \qquad (1)$$
$$d = 68(t - 9) \quad (2)$$

Solve. We substitute $34t$ for d in Equation (2) and solve for t.

$$34t = 68(t - 9)$$
$$34t = 68t - 612$$
$$-34t = -612$$
$$t = 18$$

When $t = 18$, $d = 34 \cdot 18 = 612$.

Check. At 34 mph, in 18 hr the freight train travels $34 \cdot 18 = 612$ mph. At 68 mph, in $18 - 9$, or 9 hr, the passenger train travels $68 \cdot 9$, or 612 mi. Since the distances are the same, the answer checks.

State. The passenger train will overtake the freight train 612 mi from the station.

68. *Familiarize*. Let s = the length of a side of the octagon. Then $3s - 2$ = the length of a side of the pentagon. The perimeter of the pentagon is $5(3s - 2)$ and the perimeter of the octagon is $8s$.

Translate.

$$\underbrace{\text{Perimeter of pentagon}}_{5(3s - 2)} \quad \underbrace{\text{is the same as}}_{=} \quad \underbrace{\text{perimeter of octagon.}}_{8s}$$

Solve. We solve the equation.

$$5(3s - 2) = 8s$$
$$15s - 10 = 8s$$
$$-10 = -7s$$
$$\frac{10}{7} = s$$

Check. If $s = \dfrac{10}{7}$, then the perimeter of the octagon is $8 \cdot \dfrac{10}{7}$, or $\dfrac{80}{7}$, or $11\dfrac{3}{7}$. The length of a side of the pentagon is $3 \cdot \dfrac{10}{7} - 2$, or $\dfrac{30}{7} - \dfrac{14}{7}$, or $\dfrac{16}{7}$, and the perimeter is $5 \cdot \dfrac{16}{7}$, or $\dfrac{80}{7}$, or $11\dfrac{3}{7}$. Since the perimeters are the same, the answer is correct.

State. The perimeter of each figure is $11\dfrac{3}{7}$.

69. *Familiarize*. Let a and b represent the number of liters of solutions A and B to be used, respectively. We organize the information in a table.

	A	B	Mixture
Amount	a	b	80
Percent of ammonia	6%	2%	3.2%
Amount of ammonia	0.06a	0.02b	0.032(80), or 2.56

The first and third rows of the table give us two equations.

$$a + b = 80,$$
$$0.06a + 0.02b = 2.56$$

After clearing decimals we have the following system of equations.

$$a + b = 80, \quad (1)$$
$$6a + 2b = 256 \quad (2)$$

Solve. First multiply Equation (1) by -2 and then add.

$$\begin{aligned}
-2a - 2b &= -160 \\
\underline{6a + 2b} &= \underline{256} \\
4a &= 96 \\
a &= 24
\end{aligned}$$

Now substitute 24 for a in Equation (1) and solve for b.

$$24 + b = 80$$
$$b = 56$$

Check. $24 + 56 = 80$ L. The amount of ammonia in the mixture is $0.06(24) + 0.02(56) = 1.44 + 1.12 = 2.56$ L. The answer checks.

State. 24 L of solution A and 56 L of solution B should be used.

70. Familiarize. Let $r =$ the speed of the plane in still air. Then $r + 30 =$ the speed with the wind and $r - 30 =$ the speed against the wind. Let $t =$ the time it takes to make the trip in each direction. We organize the information in a table.

	Distance	Speed	Time
With wind	190	$r + 30$	t
Against wind	160	$r - 30$	t

Translate. Using $t = d/r$ in each row of the table, we can equate the two expressions for time.
$$\frac{190}{r + 30} = \frac{160}{r - 30}$$

Solve. We solve the equation.
$$(r + 30)(r - 30) \cdot \frac{190}{r + 30} = (r + 30)(r - 30) \cdot \frac{160}{r - 30}$$
$$190(r - 30) = 160(r + 30)$$
$$190r - 5700 = 160r + 4800$$
$$30r = 10,500$$
$$r = 350$$

Check. If the speed of the plane in still air is 350 mph, then the speed with the wind is $350 + 30$, or 380 mph, and it takes $190/380$, or $1/2$ hr, to fly 190 mi. The speed against the wind is $350 - 30$, or 320 mph, and it takes $160/320$, or $1/2$ hr, to fly 160 mi. The times are the same, so the answer checks.

State. The speed of the plane in still air is 350 mph.

71. Familiarize. Let $t =$ the time, in minutes, it would take to do the job, working together.

Translate. We will use the work principle.
$$\frac{t}{21} + \frac{t}{14} = 1$$

Solve. We solve the equation.
$$42\left(\frac{t}{21} + \frac{t}{14}\right) = 42 \cdot 1$$
$$2t + 3t = 42$$
$$5t = 42$$
$$t = \frac{42}{5}, \text{ or } 8\frac{2}{5}$$

Check. We verify the work principle.
$$\frac{42/5}{21} + \frac{42/5}{14} = \frac{42}{5} \cdot \frac{1}{21} + \frac{42}{5} \cdot \frac{1}{14} = \frac{2}{5} + \frac{3}{5} = 1$$

State. It would take $8\frac{2}{5}$ min to do the job, working together.

72.
$$F = \frac{kv^2}{r}$$
$$8 = \frac{k \cdot 1^2}{10}$$
$$8 = \frac{k}{10}$$
$$80 = k$$
$$F = \frac{80v^2}{r} \qquad \text{Equation of variation}$$
$$F = \frac{80 \cdot 2^2}{16}$$
$$F = \frac{80 \cdot 4}{16}$$
$$F = 20$$

73. Familiarize. Let l and w represent the length and width of the rectangle, respectively, in feet.

Translate. The perimeter is 34 ft, so we have $2l + 2w = 34$, or $l + w = 17$. We use the Pythagorean theorem to get a second equation:
$$l^2 + w^2 = 13^2, \text{ or } l^2 + w^2 = 169.$$

Solve. We solve the system of equations.
$$l + w = 17, \quad (1)$$
$$l^2 + w^2 = 169 \quad (2)$$

First we solve Equation (1) for l: $l = 17 - w$. Then substitute $17 - w$ for l in Equation (2) and solve for w.
$$(17 - w)^2 + w^2 = 169$$
$$289 - 34w + w^2 + w^2 = 169$$
$$2w^2 - 34w + 120 = 0$$
$$w^2 - 17w + 60 = 0 \qquad \text{Dividing by 2}$$
$$(w - 5)(w - 12) = 0$$
$$w = 5 \ \text{ or } \ w = 12$$

If $w = 5$, then $l = 17 - 5 = 12$.

If $w = 12$, then $l = 17 - 12 = 5$.

Since we usually consider length to be longer than width, we have $l = 12$ and $w = 5$.

Check. The perimeter is $2 \cdot 12 + 2 \cdot 5 = 34$ ft. Also, $12^2 + 5^2 = 144 + 25 = 169 = 13^2$, so the answer checks.

State. The dimensions are 12 ft by 5 ft.

74. Familiarize. Let l and w represent the length and width of the rug, respectively, in feet.

Translate. The diagonal is 25 ft so, using the Pythagorean theorem, we have $l^2 + w^2 = 25^2$, or $l^2 + w^2 = 625$. The area is 300 ft^2, so we have a second equation, $lw = 300$.

Solve. We solve the system of equations.
$$l^2 + w^2 = 625, \quad (1)$$
$$lw = 300 \qquad\quad (2)$$

First we solve Equation (2) for w: $w = \frac{300}{l}$. Then substitute $\frac{300}{l}$ for w in Equation (1) and solve for l.

$$l^2 + \left(\frac{300}{l}\right)^2 = 625$$

$$l^2 + \frac{90,000}{l^2} = 625$$

$$l^4 + 90,000 = 625l^2 \quad \text{Multiplying by } l^2$$

$$l^4 - 625l^2 + 90,000 = 0$$

Let $u = l^2$.

$$u^2 - 625u + 90,000 = 0$$

$$(u - 225)(u - 400) = 0$$

$$u - 225 = 0 \quad \text{or} \quad u - 400 = 0$$

$$u = 225 \quad \text{or} \quad u = 400$$

$$l^2 = 225 \quad \text{or} \quad l^2 = 400$$

$$l = \pm 15 \quad \text{or} \quad l = \pm 20$$

Since the length cannot be negative, -15 and -20 cannot be solutions. When $l = 15$, $w = 300/15 = 20$. When $w = 20$, $l = 300/20$, or 15. Since we usually consider length to be longer than width, we have $l = 20$ and $w = 15$.

Check. We have $20^2 + 15^2 = 400 + 225 = 625 = 25^2$. Also, the area is $20 \cdot 15$, or 300 ft^2. The answer checks.

State. The length is 20 ft, and the width is 15 ft.

75. Familiarize. Using the labels on the drawing in the text, we let $w = $ the width of the rectangular region, in feet, and $100 - 2w = $ the length.

Translate. We use the formula for the area of a rectangle.

$$A = (100 - 2w)w = 100w - 2w^2 = -2w^2 + 100w$$

Carry out. The graph of the function is a parabola that opens down (since $-2 < 0$), so it has a maximum value at the vertex. We complete the square to find the vertex.

$$A = -2(w^2 - 50w)$$
$$= -2(w^2 - 50w + 625 - 625)$$
$$= -2(w^2 - 50w + 625) + (-2)(-625)$$
$$= -2(w - 25)^2 + 1250$$

The vertex is $(25, 1250)$, so the maximum value is 1250.

Check. We can do a partial check by trying some values of w and determining that each yields a value of A that is less than 1250. We could also examine the graph of the function to check the maximum value.

State. The area of the largest region that can be fenced in is 1250 ft^2.

76. If a bone has lost 25% of its carbon-14 from an initial amount P_0, then 75% (P_0) is the amount present.

$$P(t) = P_0\, e^{-0.00012t}$$
$$0.75\, P_0 = P_0\, e^{-0.00012t}$$
$$0.75 = e^{-0.00012t}$$
$$\ln 0.75 = \ln e^{-0.00012t}$$
$$\ln 0.75 = -0.00012t$$
$$\frac{\ln 0.75}{-0.00012} = t$$
$$2397 \text{ yr} \approx t$$

77. $$W = \frac{k}{L}$$
$$1440 = \frac{k}{14}$$
$$20,160 = k$$
$$W = \frac{20,160}{L} \quad \text{Equation of variation}$$
$$W = \frac{20,160}{6}$$
$$W = 3360 \text{ kg}$$

78. First we find the slope of the line.
$$m = \frac{-3 - (-4)}{2 - 5} = \frac{1}{-3} = -\frac{1}{3}$$
We will use the point-slope equation and the point $(5, -4)$.
$$y - y_1 = m(x - x_1)$$
$$y - (-4) = -\frac{1}{3}(x - 5)$$
$$y + 4 = -\frac{1}{3}x + \frac{5}{3}$$
$$y = -\frac{1}{3}x - \frac{7}{3}$$

79. We find a function of the form $f(x) = ax^2 + bx + c$. Substituting the data points we get
$$4 = a(-2)^2 + b(-2) + c,$$
$$-6 = a(-5)^2 + b(-5) + c,$$
$$-3 = a \cdot 1^2 + b \cdot 1 + c,$$

or

$$4 = 4a - 2b + c,$$
$$-6 = 25a - 5b + c,$$
$$-3 = a + b + c.$$

Solving the system of equations, we get
$$\left(-\frac{17}{18}, -\frac{59}{18}, \frac{11}{9}\right).$$

Thus, the function is $f(x) = -\frac{17}{18}x^2 - \frac{59}{18}x + \frac{11}{9}$.

80. $10^6 = r \Rightarrow \log r = 6$

81. $\log_3 Q = x \Rightarrow 3^x = Q$

82. $$\frac{1}{5}(7 \log_b x - \log_b y - 8 \log_b z)$$
$$= \frac{1}{5}(\log_b x^7 - \log_b y - \log_b z^8)$$
$$= \frac{1}{5}[\log_b x^7 - (\log_b y + \log_b z^8)]$$
$$= \frac{1}{5}(\log_b x^7 - \log_b yz^8)$$
$$= \frac{1}{5}\log_b \frac{x^7}{yz^8}$$
$$= \log_b \left(\frac{x^7}{yz^8}\right)^{1/5}, \text{ or } \log_b \frac{x^{7/5}}{y^{1/5}z^{8/5}}$$

83.

$$\log_b \left(\frac{xy^5}{z}\right)^{-6}$$

$$= -6 \log_b \frac{xy^5}{z}$$

$$= -6(\log_b xy^5 - \log_b z)$$

$$= -6(\log_b x + \log_b y^5 - \log_b z)$$

$$= -6(\log_b x + 5 \log_b y - \log_b z)$$

$$= -6 \log_b x - 30 \log_b y + 6 \log_b z$$

84. Familiarize. Let x and y represent the numbers.

Translate. The sum of the numbers is 26, so we have $x + y = 26$. Solving for y, we get $y = 26 - x$. The product of the numbers is xy. Substituting $26 - x$ for y in the product, we get a quadratic function:

$$P = xy = x(26 - x) = 26x - x^2 = -x^2 + 26x$$

Carry out. The coefficient of x^2 is negative, so the graph of the function is a parabola that opens down and a maximum exists. We complete the square in order to find the vertex of the quadratic function.

$$P = -x^2 + 26x$$

$$= -(x^2 - 26x)$$

$$= -(x^2 - 26x + 169 - 169)$$

$$= -(x^2 - 26x + 169) + (-1)(-169)$$

$$= -(x - 13)^2 + 169$$

The vertex is $(13, 169)$. This tells us that the maximum product is 169.

Check. We could use the graph of the function to check the maximum value.

State. The maximum product is 169.

85. The graph of $f(x) = 4 - x^2$ is shown below. We use the horizontal-line test. Since it is possible for a horizontal line to intersect the graph more than once, the function is not one-to-one.

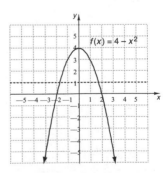

86. a) Locate 2 on the horizontal axis and the find the point on the graph for which 2 is the first coordinate. From that point, look to the vertical axis to find the corresponding y-coordinate. It is -5. Thus, $f(2) = -5$.

b) No endpoints are indicated, so we see that the graph extends indefinitely horizontally. Thus, the domain is the set of all real numbers, or $(-\infty, \infty)$.

c) To determine all x-values for which $f(x) = -5$, locate -5 on the vertical axis. From there look left and right to the graph to find any points for which -5 is the second coordinate. Four such points exist. Their x-coordinates are -2, -1, 1, and 2.

d) The smallest y-value is -7. No endpoints are indicated, so we see that the graph extends upward indefinitely from $(-7, 0)$. Thus, the range is $\{y | y \geq -7\}$, or $[-7, \infty)$.

87. a)

$$P(t) = P_0 e^{kt}$$

$$2,758,931 = 1,998,257 e^{k \cdot 12}$$

$$\frac{2,758,931}{1,998,257} = e^{12k}$$

$$\ln \left(\frac{2,758,931}{1,998,257}\right) = \ln e^{12k}$$

$$\ln \left(\frac{2,758,931}{1,998,257}\right) = 12k$$

$$\frac{\ln \left(\dfrac{2,758,931}{1,998,257}\right)}{12} = k$$

$$0.03 \approx k$$

Thus, the exponential growth rate is about 0.03 and the exponential growth function is $P(t) = 1,998,257 e^{0.03t}$, where t is the number of years since 2000.

b) In 2015, $t = 2015 - 2000 = 15$.

$$P(15) = 1,998,257 e^{0.03(15)} \approx 3,133,891$$

c)

$$3,500,000 = 1,998,257 e^{0.03t}$$

$$\frac{3,500,000}{1,998,257} = e^{0.03t}$$

$$\ln \left(\frac{3,500,000}{1,998,257}\right) = \ln e^{0.03t}$$

$$\ln \left(\frac{3,500,000}{1,998,257}\right) = 0.03t$$

$$\frac{\ln \left(\dfrac{3,500,000}{1,998,257}\right)}{0.03} = t$$

$$19 \approx t$$

The population will reach 3.5 million about 19 yr after 2000, or in 2019.

88.

$$\frac{9}{x} - \frac{9}{x+12} = \frac{108}{x^2 + 12x}$$

$$\frac{9}{x} - \frac{9}{x+12} = \frac{108}{x(x+12)}, \text{ LCM is } x(x+12)$$

$$x(x+12)\left(\frac{9}{x} - \frac{9}{x+12}\right) = x(x+12) \cdot \frac{108}{x(x+12)}$$

$$9(x+12) - 9x = 108$$

$$9x + 108 - 9x = 108$$

$$108 = 108$$

We get an equation that is true for all values of x. Thus all real numbers except those that make a denominator 0 are solutions of the equation. A denominator is 0 when $x = 0$ or $x = -12$, so all real numbers except 0 and -12 are solutions of the equation.

89. $\log_2 (\log_3 x) = 2$

$$2^2 = \log_3 x$$

$$4 = \log_3 x$$

$$x = 3^4$$

$$x = 81$$

The number 81 checks. It is the solution.

90. When $a^2 = b^2$, we have

$$\frac{x^2}{a^2} + \frac{y^2}{a^2} = 1, \text{ or } x^2 + y^2 = a^2.$$

Thus the graph is a circle with center $(0, 0)$ and radius $|a|$.

91. Let $d =$ the number of years Diaphantos lived. Then we have:

$$\frac{1}{6}d + \frac{1}{12}d + \frac{1}{7}d + 5 + 4 + \frac{1}{2}d = d$$

$$\frac{25}{28}d + 9 = d$$

$$9 = \frac{3}{28}d$$

$$\frac{28}{3} \cdot 9 = d$$

$$84 = d$$

Diaphantos lived 84 yr.